Atomic Numbers and Atomic Masses of the Elements

Based on $^{12}_{6}C$. Numbers in parentheses are the mass numbers of the most stable isotopes of radioactive elements.

Element	*Symbol*	*Atomic Number*	*Atomic Mass*
Actinium	Ac	89	(227)
Aluminum	Al	13	26.98
Americium	Am	95	(243)
Antimony	Sb	51	121.76
Argon	Ar	18	39.95
Arsenic	As	33	74.92
Astatine	At	85	(210)
Barium	Ba	56	137.33
Berkelium	Bk	97	(247)
Beryllium	Be	4	9.01
Bismuth	Bi	83	208.98
Bohrium	Bh	107	(264)
Boron	B	5	10.81
Bromine	Br	35	79.90
Cadmium	Cd	48	112.41
Calcium	Ca	20	40.08
Californium	Cf	98	(251)
Carbon	C	6	12.01
Cerium	Ce	58	140.12
Cesium	Cs	55	132.91
Chlorine	Cl	17	35.45
Chromium	Cr	24	52.00
Cobalt	Co	27	58.93
Copper	Cu	29	63.55
Curium	Cm	96	(248)
Dubnium	Db	105	(262)
Dysprosium	Dy	66	162.50
Einsteinium	Es	99	(252)
Erbium	Er	68	167.26
Europium	Eu	63	151.96
Fermium	Fm	100	(257)
Fluorine	F	9	19.00
Francium	Fr	87	(223)
Gadolinium	Gd	64	157.25
Gallium	Ga	31	69.72
Germanium	Ge	32	72.59
Gold	Au	79	196.97
Hafnium	Hf	72	178.49
Hassium	Hs	108	(269)
Helium	He	2	4.00
Holmium	Ho	67	164.93
Hydrogen	H	1	1.01
Indium	In	49	114.82
Iodine	I	53	126.90
Iridium	Ir	77	192.22
Iron	Fe	26	55.85
Krypton	Kr	36	83.80
Lanthanum	La	57	138.91
Lawrencium	Lr	103	(262)
Lead	Pb	82	207.19
Lithium	Li	3	6.94
Lutetium	Lu	71	174.97
Magnesium	Mg	12	24.30
Manganese	Mn	25	54.94
Meitnerium	Mt	109	(268)
Mendelevium	Md	101	(260)
Mercury	Hg	80	200.59
Molybdenum	Mo	42	95.94
Neodymium	Nd	60	144.24
Neon	Ne	10	20.18
Neptunium	Np	93	(237)
Nickel	Ni	28	58.69
Niobium	Nb	41	92.91
Nitrogen	N	7	14.01
Nobelium	No	102	(259)
Osmium	Os	76	190.23
Oxygen	O	8	16.00
Palladium	Pd	46	106.42
Phosphorus	P	15	30.97
Platinum	Pt	78	195.08
Plutonium	Pu	94	(242)
Polonium	Po	84	(209)
Potassium	K	1[illegible]	39.10
Praseodymium	Pr	59	140.91
Promethium	Pm	61	(145)
Protactinium	Pa	91	(231)
Radium	Ra	88	(226)
Radon	Rn	86	222)
Rhenium	Re	75	86.21
Rhodium	Rh	45	2.91
Rubidium	Rb	37	85.47
Ruthenium	Ru	44	101.07
Rutherfordium	Rf	104	(261)
Samarium	Sm	62	150.36
Scandium	Sc	21	44.96
Seaborgium	Sg	106	(266)
Selenium	Se	34	78.96
Silicon	Si	14	28.09
Silver	Ag	47	107.87
Sodium	Na	11	22.99
Strontium	Sr	38	87.62
Sulfur	S	16	32.07
Tantalum	Ta	73	180.95
Technetium	Tc	43	(98)
Tellurium	Te	52	127.60
Terbium	Tb	65	158.93
Thallium	Tl	81	204.38
Thorium	Th	90	(232)
Thulium	Tm	69	168.93
Tin	Sn	50	118.71
Titanium	Ti	22	47.87
Tungsten	W	74	183.84
Uranium	U	92	(238)
Vanadium	V	23	50.94
Xenon	Xe	54	131.29
Ytterbium	Yb	70	173.04
Yttrium	Y	39	88.91
Zinc	Zn	30	65.38
Zirconium	Zr	40	91.22

Essentials of General, Organic, and Biological Chemistry

H. Stephen Stoker
Weber State University

Australia • Brazil • Japan • Korea • Mexico • Singapore • Spain • United Kingdom • United States

Essentials of General, Organic, and Biological Chemistry

H. Stephen Stoker

Publisher: Charles Hartford
Executive Editor: Richard Stratton
Development Editor: Rita Lombard
Editorial Assistant: Rosemary Mack
Project Editor: Merrill Peterson
Senior Production/Design Coordinator: Jill Haber
Manufacturing Manager: Florence Cadran
Senior Marketing Manager: Katherine Greig
Marketing Associate: Alexandra Shaw

Library of Congress Control Number: 2001133353
ISBN-13: 978-0-618-19282-3
ISBN-10: 0-618-19282-4

Brooks/Cole
20 Davis Drive
Belmont, CA 94002-3098
USA

Cengage Learning is a leading provider of customized learning solutions with office locations around the globe, including Singapore, the United Kingdom, Australia, Mexico, Brazil, and Japan. Locate your local office at: **www.cengage.com/global.**

Cengage Learning products are represented in Canada by Nelson Education, Ltd.

To learn more about Brooks/Cole, visit **www.cengage.com/brookscole.**

Purchase any of our products at your local college store or at our preferred online store **www.cengagebrain.com.**

Printed in China by China Translation & Printing Services Limited
5 6 7 8 9 10 11 12 13 12 11 10

▶ Brief Contents

▶ CONTENTS

Preface

Essentials of General, Organic, and Biological Chemistry is a text written for students in the fields of nursing, allied health, biological sciences, agricultural sciences, food sciences, and public health, who are required to take a single course in the field of chemistry. As the title of the text implies, topic coverage in the text includes not only general chemistry but also an introduction to the fields of organic chemistry and biochemistry.

Chapters 1 through 9 of the text cover the general principles of chemistry including the nature of matter, the subatomic makeup of atoms, chemical bonding, and the nature of chemical reactions. Chapters 10 through 13 treat organic chemistry with a particular emphasis on those aspects of organic chemistry most helpful in understanding the biochemical concepts that are to follow. In the biochemistry portion of the text, Chapters 14 through 18, carbohydrates, lipids, proteins, and nucleic acids are considered. In addition, an introduction to metabolism is presented.

The students who will use this text most often have little or no background in chemistry and hence, approach the course with a good deal of trepidation. Thus, this text's development of chemical topics always starts at ground level. Though clearly some chemical principles cannot be divorced from mathematics, the amount and level of mathematics used is purposefully minimized.

Realizing that many instructors in a one-term course will cover only selected portions of the text, I have included Practice Questions and Problems at the end of each section. Each section thus, becomes a self-contained lesson or learning unit and can easily be assigned for student work. Also the visual emphasis in the book and the clear design help students by presenting the material in discrete manageable segments of information.

Focus on Biochemistry: Most students taking this course have a greater interest in the biochemistry portion of the course than the preceding two parts. But biochemistry, of course, cannot be understood without a knowledge of the fundamentals of organic chemistry, and understanding organic chemistry in turn depends on knowing the key concepts of general chemistry. Thus, in writing this text, I essentially started from the back and worked forward. I began by determining what topics would be considered in the biochemistry chapters and then tailored the organic and then general sections to support that presentation.

Emphasis on Visual Support: I believe strongly in visual reinforcement of key concepts in a textbook; thus, this book uses art and photos wherever possible to teach key concepts. The book uses artwork to make connections and highlight what's important for the student to know. Color is used in reaction equations to emphasize the portions of a molecule that undergo change. Colors are likewise assigned to things like valence shells and classes of compounds, to help students follow trends. Computer-generated, three-dimensional molecular models accompany many discussions in the organic and biochemistry sections of the text. Color photographs show applications of chemistry to help make concepts real and more readily remembered.

Visual summary features, called ***Chemistry at a Glance***, pull together material from several sections of a chapter to help students see the larger picture. See, for example, the Chemistry at a Glance on page 169, that summarizes intermolecular forces; the one on page 246 summarizing buffer solutions; the one on page 275 that summarizes the physical and chemical properties of alkanes, or the one on page 480 on DNA replication. The *Chemistry at a Glance* features serve both as overviews for the student reading the material for the first time and as review tools for the student preparing for exams.

Commitment to Student Learning. In addition to the study help *Chemistry at a Glance* offers, the text is built on a strong foundation of learning aids designed to help students master the course material:

- ***Learning Focus* features** are placed at the start of each section within a chapter to provide a framework for the section and to motivate student learning.
- **Problem-solving pedagogy:** Because problem solving is often difficult for students in this course to master, I have taken special care to provide support to help students build their skills:
  - Within the chapters, "Single Concept" ***Practice Questions and Problems*** at the end of each section refer to the section's contents and focus the student's attention on thoroughly learning one concept.
  - ***Chapter outlines*** give students a road map for where they are going.
  - ***In-text worked-out examples*** walk students through the thought process involved in problem solving.
  - ***End-of-chapter problems*** include Multiple Choice and True-False Practice Tests and Additional Problems (involving more than one concept).
- ***Chemical Portraits*** This feature "spotlights," in groups of three, selected chemical elements and compounds and explains their importance in students' everyday lives. Often, the focus is on how the elements or compounds relate to the functioning of the human body. This feature always ends with questions for the student to consider. The questions are linked to recommended resources and links on the student web site.
- ***Margin Notes*** liberally distributed throughout the text provide tips for remembering and distinguishing between concepts, highlight links across chapters, and describe interesting historical background information.
- ***Key Terms*** and their definitions are highlighted in the text when they are first presented, using boldface and italic type. Each definition appears in a complete sentence; students are never forced to deduce a definition from context. These key terms are also listed at the end of each chapter.
- ***End-of-chapter review aids*** feature multiple choice and true-false self-test questions, a set of additional multiple-choice problems, a chapter summary, and a glossary for "new terms" found in the chapter. These features provide students with reference materials and an opportunity to test their knowledge of key concepts learned in the chapter.

The Package

Study Help for Students Student Website (accessible through www.cengage.com/chemistry/stoker) Available free of charge, this dedicated website offers a wealth of resources to help students succeed, including:

- Chapter outlines
- Self-quizzing using the Online Testing Center
- Electronic flashcards of key terms, reactions, and concepts
- Additional real-life applications with relevant links
- Follow up to Chemical Portraits web questions
- Career-related information
- Chime Tutorial

Study Guide with Answers to Selected Problems (0-618-19283-2) by Danny V. White of American River College and Joanne A. White, includes, for each chapter, a brief overview, activities and practice problems to reinforce skills, and a practice test. The answers section includes answers for all odd-numbered end of chapter exercises.

Student CD-ROM (0-618-19287-5) This CD, included at no additional charge with all new texts, offers brief tutorials on basic mathematical concepts that appear in the text, including solving simple algebraic equations, scientific notation, conversions, reading a graph, and ratio and proportion.

Course Support For Instructors **Instructor Website (accessible through www.cengage.com/chemistry/stoker)** allows access to all student website resources (above) as well as additional instructor course/classroom resources such as downloadable PowerPoint slides, and useful links.

Instructor's Resource Manual with Test Bank (0-618-23697-X) by H. Stephen Stoker includes answers to all chapter exercises and a printed test bank of over 1,500 multiple choice and matching problems. The Test Bank is also available on CD-Rom (see the HMClass Prep with HMTesting Version 6.0).

HMClass Prep with HMTesting Version 6.0 CD-ROM Package (ISBN 0-618-19285-9) This package includes both HMClass Prep and HMTesting on one CD-ROM. It allows an instructor to access both lecture aids and testing software in one place. These components cannot be ordered separately.

- *HMClass Prep* includes everything an instructor will need to develop their lectures—PowerPoints slides with all text figures, tables, concept checks, and virtually all photos, as well as Instructor's Resource Guide and MS Word files of the Printed Test Bank.
- *HM Testing* Version 6.0 combines a flexible test-editing program with a comprehensive gradebook function for easy administration and tracking. It enables instructors to administer tests via network server or the web. The *HM Testing* database contains a wealth of algorithmically generated questions and can produce multiple-choice, true/false, fill-in-the-blank, and essay tests. Questions can be customized based on the chapter being covered, the question format, level of difficulty, and specific topics. *HM Testing* provides for the utmost security in accessing both test questions and grades.

Acknowledgments

I gratefully acknowledge the valuable comments of the following reviewers, whose thoughtful critiques guided my revision efforts: Kay Davis, Garden City Community College; David Hilton, Chipola Junior College; Tracy Whitehead, Henderson State University; Ronald T. Amel, Viterbo University; Kathleen Kiedrowicz, Carroll College; and James F. Nugent, Salve Regina University.

Special thanks go to Stephen Z. Goldberg, Adelphi University, for his help in ensuring this book's accuracy by reviewing manuscript, proofs, and artwork.

I also give special thanks to the people at the publishing company who guided the revision through various stages of development and production: Richard Stratton, Executive Editor, Chemistry, Rita Lombard, Development Editor, Charline Lake, Production Services Manager, Jill Haber, Senior Production/Design Coordinator, Katherine Greig, Senior Marketing Manager, Science, Alexandra Shaw, Marketing Associate, Science, who made especially significant suggestions for refining the features and text. I would also like to especially thank Rosemary Mack, Editorial Assistant, for her management of the ancillary program, Charlotte Miller, Art Editor, for her thoughtful and creative contributions to the illustration program, Naomi Kornhauser, Photo Editor, for helping to enrich the text with photographs and molecular models, Jean Hammond, Designer for her contemporary design suggestions, Merrill Peterson, Project Editor, and Michele Ostovar, Assistant Project Editor, for making the production process for this text a smooth one.

H. Stephen Stoker, Weber State University

Exciting Photo Program Careful selection of photos, like this one in Chapter 10, help students see the everyday application of the chemistry they are learning.

Chapter Outlines give students a road map for where they are going.

Learning Focus objectives, placed at the beginning of each section, break up the material into more digestible units for students, emphasizing the key idea or concept in that section.

CHAPTER OUTLINE

CHAPTER TEN

Saturated Hydrocarbons

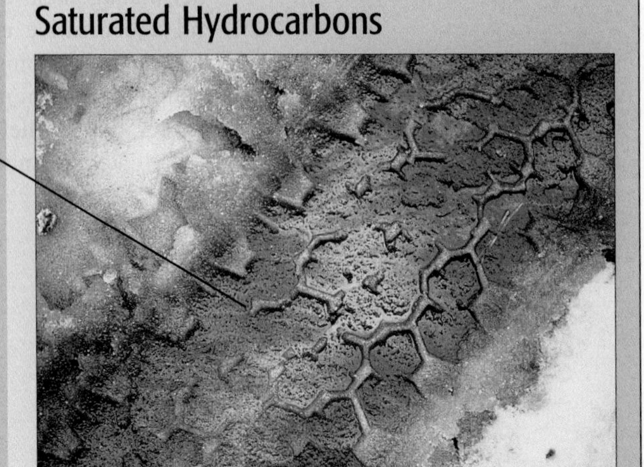

A tire tread in snow with oil (hydrocarbons) causing the iridescent pattern.

This chapter is the first of four that deal with the subject of organic chemistry and organic compounds. Organic compounds are the chemical basis for life itself, as well as an important component of the basis for our current high standard of living. Organic compounds are found in natural gas, petroleum, coal, gasoline, and many synthetic materials such as dyes, plastics, and clothing fibers. Proteins, carbohydrates, enzymes, and hormones are complex organic molecules.

Learning Focus

Understand the historical and modern definitions of the term *organic chemistry.*

▶ The historical origins of the terms *organic* and *inorganic* involve the following conceptual pairings:
organic—living *organisms*
*in*organic—*in*animate materials

10.1 Organic and Inorganic Compounds

During the latter part of the eighteenth century and the early part of the nineteenth century, chemists began to categorize compounds into two types: organic and inorganic. Compounds obtained from living organisms were called *organic* compounds, and compounds obtained from mineral constituents of the earth were called *inorganic* compounds.

During this early period, chemists believed that a special "vital force," supplied by a living organism, was necessary for the formation of an organic compound. This concept was proved incorrect in 1828 by the German chemist Friedrich Wöhler. Wöhler heated an aqueous solution of two inorganic compounds, ammonium chloride and silver cyanate, and obtained urea (a component of urine).

$$NH_4Cl + AgNCO \longrightarrow \underset{\text{urea}}{(NH_2)_2CO} + AgCl$$

▼ Naming of Phen

Besides being the name
name of the simplest me

Substituted phenols
gins with the hydroxyl g
to the next carbon atom

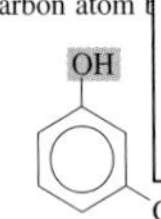

3-Chlorophenol (or *meta*-chlorophenol)　　4-Ethyl-2-methylphenol　　2,5-Dibromophenol

Chemically, the behavior of phenols differs considerably from that of alkyl alcohols. This is the reason for considering them a group separate from the other alcohols.

Chemical Portraits 19 **Some Commonly Encountered Alcohols**

Methanol (Methyl Alcohol) (CH_3—OH)

Profile: Methanol, a clear, colorless, flammable, poisonous liquid, has excellent solvent properties. It is the solvent of choice for paints, shellacs, and varnishes. For safety reasons, it is the fuel for race cars; methanol fires are easier to put out than gasoline fires. Methanol is sometimes called *wood alcohol,* terminology drawing attention to an early method for its preparation—heating wood in the absence of air.

Biochemical considerations: Drinking methanol is very dangerous; its metabolism in the human body produces formaldehyde and formic acid. Formaldehyde is toxic to the eye and as little as 1 oz can cause optic nerve damage and blindness. Formic acid causes acidosis.

What is the current method for the industrial production of methanol?

Ethanol (Ethyl Alcohol) (CH_3—CH_2—OH)

Profile: Ethanol is a clear, colorless, flammable liquid which—like methanol—is a good industrial solvent. Industrial ethanol is most often *denatured,* that is, it contains an added substance that makes it unfit to drink. It is also used as an octane-booster for gasoline.

Biochemical considerations: Drinking alcohol, the alcohol present in alcoholic beverages, is ethanol that has been produced through fermentation of grain products; hence its common name *grain alcohol.* Pure ethanol is tasteless; the taste of alcoholic beverages comes from other fermentation products present. A 70% ethanol solution is used as an antiseptic to cleanse an area of skin before giving an injection or taking a blood sample; in this function, ethanol destroys bacteria by coagulating their protein.

What are the "pros and cons" of the use of ethanol as a gasoline alternative?

Menthol (5-methyl-2-isopropylcyclohexanol)

Profile: Menthol, a substituted cyclohexanol, has a peppermint taste and odor. It occurs naturally in peppermint oil and can be made synthetically. In its pure state, menthol is a white crystalline solid. It is only slightly soluble in water.

Biochemical considerations: Topical application of menthol to the skin causes a refreshing, cooling sensation followed by a slight burning-and-prickling sensation. At the same time as cooling is perceived, it depresses the nerves for pain reception. The cooling sensation occurs independent of body temperature. Products that often contain menthol include throat sprays and lozenges, cough drops, chest-rub preparations, and aftershave lotions.

Why is menthol often added to toothpastes and mouthwashes?

See the text web site at **www.cengage.com/chemistry/stoker** for answers to the above questions and for further information.

Chemical Portraits This feature spotlights selected chemical elements and compounds and explains their importance in students' everyday lives. Often, the focus is on how the elements or compounds relate to the functioning of the human body. This feature always ends with questions for the student to consider that are linked to recommended resources on the student web site.

Chemistry at a Glance acts as a visual summary feature that pulls together material from a group of sections or a whole chapter to help students see the larger picture.

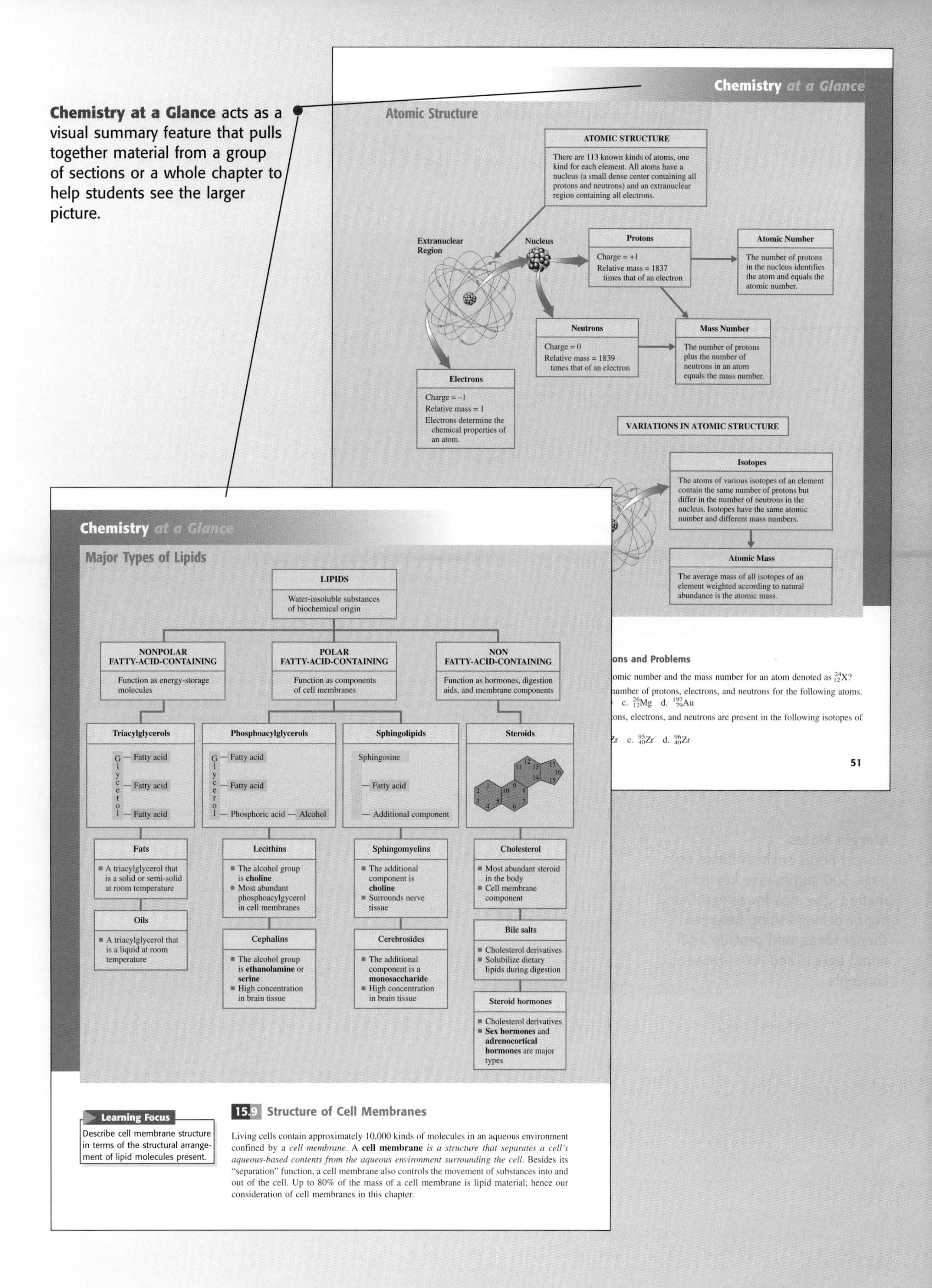

Chemistry at a Glance

Atomic Structure

ATOMIC STRUCTURE

There are 113 known kinds of atoms, one kind for each element. All atoms have a nucleus (a small dense center containing all protons and neutrons) and an extranuclear region containing all electrons.

Extranuclear Region

Nucleus

Protons

Charge = +1
Relative mass = 1837 times that of an electron

Atomic Number

The number of protons in the nucleus identifies the atom and equals the atomic number.

Neutrons

Charge = 0
Relative mass = 1839 times that of an electron

Mass Number

The number of protons plus the number of neutrons in an atom equals the mass number.

Electrons

Charge = −1
Relative mass = 1
Electrons determine the chemical properties of an atom.

VARIATIONS IN ATOMIC STRUCTURE

Isotopes

The atoms of various isotopes of an element contain the same number of protons but differ in the number of neutrons in the nucleus. Isotopes have the same atomic number and different mass numbers.

Atomic Mass

The average mass of all isotopes of an element weighted according to natural abundance is the atomic mass.

...ons and Problems

...omic number and the mass number for an atom denoted as $^{24}_{12}X$?

...umber of protons, electrons, and neutrons for the following atoms.

c. $^{26}_{12}Mg$ d. $^{197}_{79}Au$

...ons, electrons, and neutrons are present in the following isotopes of

...r c. $^{95}_{40}Zr$ d. $^{96}_{40}Zr$

51

Chemistry at a Glance

Major Types of Lipids

LIPIDS

Water-insoluble substances of biochemical origin

NONPOLAR FATTY-ACID-CONTAINING

Function as energy-storage molecules

POLAR FATTY-ACID-CONTAINING

Function as components of cell membranes

NON FATTY-ACID-CONTAINING

Function as hormones, digestion aids, and membrane components

Triacylglycerols

Glycerol — Fatty acid, Fatty acid, Fatty acid

Phosphoacylglycerols

Glycerol — Fatty acid, Fatty acid, Phosphoric acid — Alcohol

Sphingolipids

Sphingosine — Fatty acid — Additional component

Steroids

Fats

- A triacylglycerol that is a solid or semi-solid at room temperature

Oils

- A triacylglycerol that is a liquid at room temperature

Lecithins

- The alcohol group is **choline**
- Most abundant phosphoacylgycerol in cell membranes

Cephalins

- The alcohol group is **ethanolamine** or **serine**
- High concentration in brain tissue

Sphingomyelins

- The additional component is **choline**
- Surrounds nerve tissue

Cerebrosides

- The additional component is a **monosaccharide**
- High concentration in brain tissue

Cholesterol

- Most abundant steroid in the body
- Cell membrane component

Bile salts

- Cholesterol derivatives
- Solubilize dietary lipids during digestion

Steroid hormones

- Cholesterol derivatives
- **Sex hormones** and **adrenocortical hormones** are major types

Learning Focus

Describe cell membrane structure in terms of the structural arrangement of lipid molecules present.

15.9 Structure of Cell Membranes

Living cells contain approximately 10,000 kinds of molecules in an aqueous environment confined by a *cell membrane*. A **cell membrane** *is a structure that separates a cell's aqueous-based contents from the aqueous environment surrounding the cell*. Besides its "separation" function, a cell membrane also controls the movement of substances into and out of the cell. Up to 80% of the mass of a cell membrane is lipid material; hence our consideration of cell membranes in this chapter.

84 Chapter 4 Chemical Bonds

Example 4.1 **Determining the Number of Valence Electrons in an Atom**

Determine the number of valence electrons in atoms of each of the following elements.

a. ${}_{12}Mg$ **b.** ${}_{14}Si$ **c.** ${}_{33}As$

Solution

a. Atoms of the element magnesium have two valence electrons, as can be seen by examining magnesium's electron configuration.

$$1s^2 2s^2 2p^6 3s^2$$

Number of valence electrons

Highest value of the electron shell number

The highest value of the electron shell number is $n = 3$. Only two electrons are found in shell 3: the two electrons in the 3*s* subshell.

b. Atoms of the element silicon have four valence electrons.

$$1s^2 2s^2 2p^6 3s^2 3p^2$$

Number of valence electrons

Highest value of the electron shell number

Electrons in two different subshells can simultaneously be valence electrons. The highest shell number is 3, and both the 3*s* and the 3*p* subshells belong to this shell. Hence all of the electrons in both of these subshells are valence electrons.

c. Atoms of the element arsenic have five valence electrons.

$$1s^2 2s^2 2p^6 3s^2 3p^6 4s^2 3d^{10} 4p^3$$

Number of valence electrons

Highest value of the electron shell number

The 3*d* electrons are not counted as valence electrons because the 3*d* subshell is in shell 3, and this shell
and has maximum *n*

Scientists have devel
electrons present in atom
tures. A **Lewis structure**
electron placed around t
(all representative or nob
Figure 4.1. Lewis structu
(Figure 4.2) who first int

In positioning electr
sides of the elemental s
which side of the symbol

IA

H•

IIA

Li• •Be•

Na• •Mg•

K• •Ca•

Figure 4.1 Lewis structures for selected representative and noble-gas elements.

In-Text Worked-Out Examples

In-text worked-out examples walk students through the thought process involved in problem solving.

300 Chapter 11 Unsaturated Hydrocarbons

Monosubstituted benzene structures are often drawn with the substituent at the "12 o'clock" position, as in the previous structures. However, because all the hydrogen atoms in benzene are equivalent, it does not matter at which carbon of the ring the substituted group is located. Each of the following formulas represents toluene.

CH_3 CH_3 CH_3 CH_3

For monosubstituted benzene rings that have a group attached that is not easily named as a substituent, the benzene ring is often treated as a group attached to this substituent. In this reversed approach, the benzene ring attachment is called a *phenyl* group, and the compound is named according to the rules for naming alkanes, alkenes, and alkynes.

▸ The word *phenyl* comes from "phene," a European term used during the 1800s for benzene. The word is pronounced *fen*-nil.

$CH_2{=}CH{-}CH{-}CH_3$

3-Phenyl-1-butene

When two substituents, either the same or different, are attached to a benzene ring, three isomeric structures are possible.

X X X X X X

To distinguish among these three isomers, we must specify the positions of the substituents relative to one another. This can be done in either of two ways: by using numbers and by using nonnumerical prefixes.

When numbers are used, the three isomeric dimethylbenzenes have the first-listed set of names:

▸ *Cis–trans* isomerism is not possible for disubstituted benzenes. All 12 atoms of benzene are in the same plane—that is, benzene is a flat molecule. When a substituent group replaces an H atom, the atom that bonds the group to the ring is also in the plane of the ring.

CH_3 CH_3 CH_3 CH_3 CH_3 CH_3

1,2-Dimethylbenzene (*ortho*-dimethylbenzene) 1,3-Dimethylbenzene (*meta*-dimethylbenzene) 1,4-Dimethylbenzene (*para*-dimethylbenzene)

▸ Learn the meaning of the prefixes *ortho-*, *meta-*, and *para-*. These prefixes are extensively used in naming disubstituted benzenes.

X

← *ortho* to X

← *meta* to X

para to X

The prefix system uses the prefixes *ortho-*, *meta-*, and *para-* (abbreviated *o-*, *m-*, and *p-*).

Ortho- means 1,2 disubstitution; the substituents are on adjacent carbon atoms.
Meta- means 1,3 disubstitution; the substituents are one carbon removed from each other.
Para- means 1,4 disubstitution; the substituents are two carbons removed from each other (on opposite sides of the ring).

When prefixes are used, the three isomeric dimethylbenzenes have the second-listed set of names above.

Margin Notes

Margin notes such as those on page 300 summarize key information, give tips for remembering or distinguishing between similar ideas, and provide additional details and links between concepts.

ramifications in closed rooms that lack effective air circulation. The warmer and less dense air stays near the top of the room. This is desirable in the summer but not in the winter.

Practice Problems and Questions

6.16 According to Charles's law, indicate what will happen to the volume of a fixed amount of gas when
a. its temperature is increased at constant pressure
b. its temperature is decreased at constant pressure

6.17 What is the meaning of the phrase "two variables are directly proportional"?

6.18 On the basis of Charles's law, indicate what will happen to the volume of a gas when the Kelvin temperature of the gas is reduced to
a. one-half its original value
b. one-third its original value
c. one-fourth its original value
d. two-thirds its original value

6.19 Use Charles's law to calculate the value of the unknown temperature or volume.
a. $V_1 = 2.00$ L, $T_1 = 327°C$, $V_2 = 4.00$ L, $T_2 = ?$ °C
b. $V_1 = 2.00$ L, $T_1 = 327°C$, $V_2 = ?$ L, $T_2 = 27°C$
c. $V_1 = 2.00$ L, $T_1 = ?$ °C, $V_2 = 8.00$ L, $T_2 = 27°C$
d. $V_1 = ?$ L, $T_1 = 127°C$, $V_2 = 4.00$ L, $T_2 = -73°C$

6.20 At atmospheric pressure, a sample of H_2 gas has a volume of 2.73 L at 27°C. What volume, in liters, will the H_2 gas occupy if the temperature is increased to 127°C and the pressure is held constant?

6.21 A sample of N_2 gas occupies a volume of 375 mL at 25°C and a pressure of 2.0 atm. Determine the temperature, in degrees Celsius, at which the volume of the gas would be 525 mL at the same pressure.

Practice Problems and Questions
Within the chapters, "Single Concept" Practice Problems and Questions at the end of each section refer to the section's contents and focus the student's attention on thoroughly learning one concept.

CONCEPTS TO REMEMBER

Carbon atom bonding characteristics. Carbon atoms in organic compounds must have four bonds.

Types of hydrocarbons. Hydrocarbons, binary compounds of carbon and hydrogen, are of two types: saturated and unsaturated. In saturated hydrocarbons, all carbon–carbon bonds are single bonds. Unsaturated hydrocarbons have one or more carbon–carbon multiple bonds—double bonds, triple bonds, or both.

Alkanes. Alkanes are saturated hydrocarbons in which the carbon atom arrangement is that of an unbranched or branched chain. The formulas of all alkanes can be represented by the general formula C_nH_{2n+2}, where n is the number of carbon atoms present.

Structural formulas. Structural formulas are two-dimensional representations of the arrangement of the atoms in molecules. These formulas give complete information about the arrangement of the atoms in a molecule but not the spatial orientation of the atoms. Two types of structural formulas are commonly encountered: expanded and condensed.

Structural isomerism. Structural isomers are two or more compounds that have the same molecular formula but different structural formulas—that is, different arrangements of atoms within the molecule.

Conformations. Conformations are differing orientations of the same molecule made possible by free rotation about single bonds in the molecule.

Alkane nomenclature. The IUPAC name for an alkane is based on the longest continuous chain of carbon atoms in the molecule. A group of carbon atoms attached to the chain is an alkyl group. Both the position and the ider... name of the longest carb...

Cycloalkanes. Cycloal... least one cyclic arranger... las of all cycloalkanes can be represented by the general formula C_nH_{2n}, where n is the number of carbon atoms present.

Cycloalkane nomenclature. The IUPAC name for a cycloalkane is obtained by placing the prefix *cyclo-* before the alkane name that corresponds to the number of carbon atoms in the ring. Alkyl groups attached to the ring are located by using a ring-numbering system.

***Cis–trans* isomerism.** For certain disubstituted cycloalkanes, *cis–trans* isomers exist. *Cis–trans* isomers are compounds that have the same molecular and structural formulas but different arrangements of atoms in space because of restricted rotation about bonds.

Natural sources of saturated hydrocarbons. Natural gas and petroleum are the largest and most important natural sources of both alkanes and cycloalkanes.

Physical properties of saturated hydrocarbons. Saturated hydrocarbons are not soluble in water and have lower densities than water. Melting and boiling points increase with increasing carbon chain length or ring size.

Chemical properties of saturated hydrocarbons. Two important reactions that saturated h... halogenation. In combus... produce CO_2 and H_2O. ... which one or more hydr... by halogen atoms.

KEY REACTIONS AND EQUATIONS

1. Combustion (rapid reaction with O_2) of alkanes (Section 10.14)
$$\text{Alkane} + O_2 \longrightarrow CO_2 + H_2O$$

2. Halogenation of alk...
R—H + ...

KEY TERMS

Alkane (10.4)
Alkyl group (10.8)
Branched-chain alkane (10.6)
***Cis* isomer** (10.11)
***Cis–trans* isomers** (10.11)
Combustion reactions (10.14)
Condensed structural formula (10.5)
Conformations (10.7)
Continuous-chain alkane (10.6)
Cycloalkane (10.9)
Expanded structural formula (10.5)
Halogenation reaction (10.14)
Hydrocarbon (10.3)
Hydrocarbon derivative (10.3)
Line–angle drawing (10.9)
Organic chemistry (10.1)

ADDITIONAL PROBLEMS

18.78 Classify each of the following substances as a reactant in the citric acid cycle, a reactant in the electron transport chain, or a reactant in both the CAC and the ETC.
a. NAD^+ b. NADH c. O_2 d. Fumarate

18.79 Which of these substances—ATP, CoA, FAD, or NAD^+—contains the following subunits of structure? More than one choice may apply in a given situation.
a. Contains two ribose subunits
b. Contains two phosphate subunits
c. Contains one adenine subunit
d. Contains one ribitol subunit

18.80 At which of these locations—first fixed enzyme site, second fixed enzyme site, third fixed enzyme site, or mobile enzyme site—does each of the following ETC reactions occur?
a. $FADH_2$ + CoQ b. NADH + FMN
c. cyt $a_3 + O_2$ d. FeSP + CoQ

18.81 Classify each of the following substances as a reactant in the citric acid cycle, a reactant in glycolysis, or a reactant in both the citric acid cycle and glycolysis.
a. Oxaloacetate b. Dihydroxyacetone phosphate
c. 2-Phosphoglycerate d. Succinate

18.82 Classify each of the substances in Problem 18.80 as a C_6 molecule, a C_5 molecule, a C_4 molecule, or a C_3 molecule.

18.83 What is the difference between oxidative phosphorylation and substrate-level phosphorylation?

18.84 In the processing of 1 glucose molecule, what is the yield of ATP molecules from each of the following?
a. NADH produced during glycolysis
b. NADH produced during the oxidation of pyruvate to acetyl CoA
c. NADH produced during the citric acid cycle
d. $FADH_2$ produced during the citric acid cycle

18.85 Indicate how many reaction steps there are in
a. glycolysis
b. the citric acid cycle
c. the electron transport chain
d. the oxidation of pyruvate to acetyl CoA

PRACTICE TEST ▸ True/False

18.86 In catabolic reactions, large molecules are broken down into smaller molecules.

18.87 In mitochondria, both the matrix and the intermembrane space interface with the inner membrane.

18.88 ADP, NAD^+, and $FADH_2$ all have structures that contain two phosphate groups.

18.89 Both ATP and coenzyme A have structures that contain one of the B vitamins.

18.90 Glycolysis and the citric acid cycle are both components of the common metabolic pathway.

18.91 The intermediates in the citric acid cycle are all either C_6 or C_5 molecules.

18.92 CO_2 is produced in the citric acid cycle and also in the oxidation of pyruvate to acetyl CoA.

18.93 The first product in the citric acid cycle is citrate, and the first product in glycolysis is glucose 6-phosphate.

18.94 The "fuel" for the citric acid cycle is acetyl CoA.

18.95 In the electron transport chain, all of the enzymes are located at fixed sites.

18.96 Cytochromes are enzymes that contain iron atoms.

18.97 One of the end products of the electron transport chain is molecular O_2.

18.98 Most of the intermediates in glycolysis are C_3 molecules.

18.99 The amount of ATP produced during glycolysis exceeds that produced during operation of the electron transport chain and oxidative phosphorylation.

18.100 In the human body, under oxygen-deficient conditions, pyruvate is reduced to ethanol rather than oxidized to acetyl CoA.

PRACTICE TEST ▸ Multiple Choice

18.101 The correct notation for the reduced form of nicotinamide adenine dinucleotide is
a. NAD^+ b. NAD c. NADH d. $NADH_2$

18.102 Which of the following is a correct skeletal equation for a hydrolysis reaction involving adenosine phosphates?
a. ATP + $H_2O \longrightarrow$ ADP + 2 P_i
b. ADP + $H_2O \longrightarrow$ AMP + 2 P_i
c. ADP + $H_2O \longrightarrow$ ATP + P_i
d. ADP + $H_2O \longrightarrow$ AMP + P_i

18.103 Which of the following is a correct general description of the reaction that occurs in the first step of the citric acid c ycle?
a. $C_2 + C_4 \longrightarrow C_6$
b. $C_3 + C_3 \longrightarrow C_6$
c. $C_2 + C_2 + C_2 \longrightarrow C_6$
d. $C_4 + C_4 \longrightarrow C_6 + C_2$

18.104 The first two intermediates in the citric acid cycle are, respectively,
a. isocitrate and α-ketoglutarate
b. citrate and α-ketoglutarate
c. citrate and isocitrate
d. isocitrate and succinate

18.105 At which step in the electron transport chain does O_2 participate?
a. First step b. Second step
c. Next to last step d. Last step

18.106 How many fixed-enzyme sites are associated with the electron transport chain?
a. One b. Two c. Three d. Four

18.107 In which of the following listings of electron transport chain electron carriers are the electron carriers listed

End-of-Chapter Review Aids feature a chapter summary, a glossary for "new terms" found in the chapter, multiple choice and true-false self-test questions (that are paired with answers to the odd-numbered questions at the end of the text) that always involve only a single concept, and additional problems that involve more than one concept.

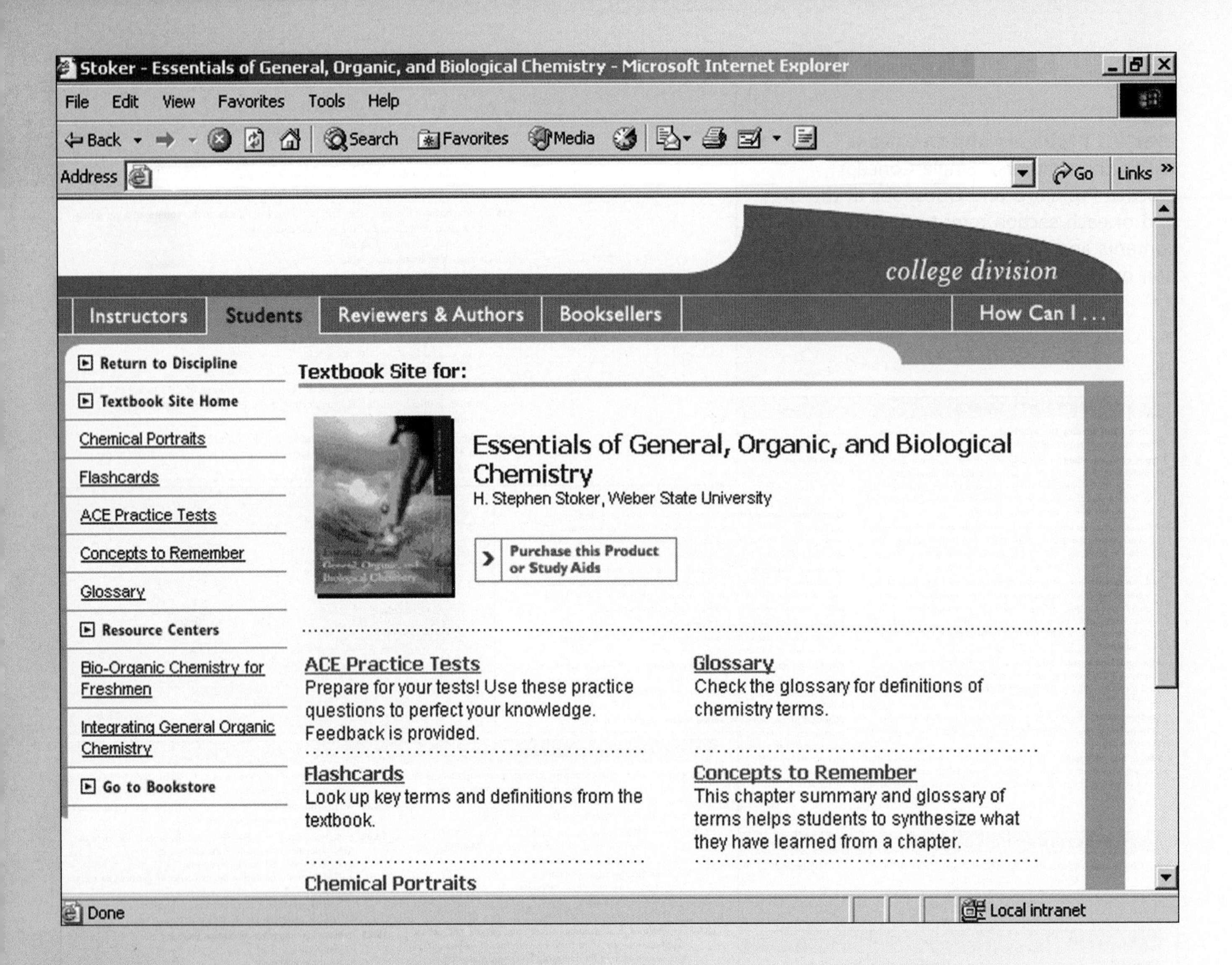

Student Website

A password protected Student Web site is accessible through www.cengage.com/chemistry/stoker. It includes a wealth of resources to help students in the course, including:

- Chapter outlines
- Self-quizzing using the Online Testing Center
- Electronic flashcards of key terms, reactions, and concepts
- Additional real-life applications with relevant links
- Follow up to the *Chemical Portraits* web questions
- Career-related information
- Chime Tutorial

CHAPTER OUTLINE

▶ CHAPTER ONE

Basic Concepts About Matter

Heat and light generation, as well as numerous changes in matter, occur as wood is burned in a bonfire.

In this chapter we address the question "What exactly is chemistry about?" In addition, we consider common terminology associated with the field of chemistry. Much of this terminology is introduced in the context of the ways in which matter is classified. Like all other sciences, chemistry has its own specific language. It is necessary to restrict the meanings of some words so that all chemists (and those who study chemistry) can understand a given description of a chemical phenomenon in the same way.

1.1 Chemistry: The Study of Matter

Learning Focus

Define the term *matter* and list the aspects of matter that are of particular concern to chemists.

Chemistry *is the field of study concerned with the characteristics, composition, and transformations of matter.* What is matter? **Matter** *is anything that has mass and occupies space.* The term *mass* refers to the amount of matter present in a sample.

Matter includes all things—both living and nonliving—that can be seen (such as plants, soil, and rocks) as well as things that cannot be seen (such as air and bacteria). Not considered to be matter are the various forms of energy, such as heat, light, and electricity. However, chemists must be concerned with energy as well as matter, because almost all changes that matter undergoes involve the release or absorption of energy.

▶ The universe is composed entirely of matter and energy.

The scope of chemistry is extremely broad, and it touches every aspect of our lives. An iron gate rusting, a chocolate cake baking, the diagnosis and treatment of a heart attack, the propulsion of a jet airliner, and the digesting of food all fall within the realm of chemistry (see Figure 1.1). The key to understanding such diverse processes is an understanding

Figure 1.1 Everyday activities such as making a batch of muffins involve chemistry.

of the fundamental nature of matter, which is what we are going to talk about for the rest of this chapter.

▶ The term *chemistry* is derived from the word *alchemy,* which denotes practices carried out during the Middle Ages in an attempt to transform something common into something precious (in particular, lead into gold). Alchemy originated in Alexandrian Egypt, and the term *alchemy* is derived from the Greek *al* ("the") and *khemia* (a native name for Egypt).

Practice Questions and Problems

1.1 Classify each of the following as matter or energy (nonmatter).
a. Air b. Pizza c. Sound d. Light e. Gold f. Virus

1.2 What three aspects of matter are chemists particularly interested in?

1.2 Physical States of Matter

Learning Focus

Characterize each of the three physical states of matter in terms of the definiteness or indefiniteness of its shape and volume.

Three physical states exist for matter: solid, liquid, and gas. The classification of a given matter sample in terms of physical state is based on whether its shape and volume are definite or indefinite.

A **solid** *is the physical state characterized by a definite shape and a definite volume.* A silver dollar has the same shape and volume whether it is placed in a large container or on a table top (Figure 1.2a). For solids in powdered or granulated forms, such as sugar or salt, a quantity of the solid takes the shape of the portion of the container it occupies, but each individual particle has a definite shape and volume. A **liquid** *is the physical state characterized by an indefinite shape and a definite volume.* A liquid always takes the shape of its container to the extent that it fills the container (Figure 1.2b). A **gas** *is the physical state characterized by an indefinite shape and an indefinite volume.* A gas always completely fills its container, adopting both its volume and its shape (Figure 1.2c).

▶ The *volume* of a sample of matter is a measure of the amount of space occupied by the sample.

Figure 1.2 (a) A solid has a definite shape and a definite volume. (b) A liquid has an indefinite shape—it takes the shape of its container—and a definite volume. (c) A gas has an indefinite shape and an indefinite volume—it assumes the shape and volume of its container.

Figure 1.3 Water can be found in the solid, liquid, and vapor (gaseous) forms simultaneously, as shown here at Yellowstone National Park.

The state of matter observed for a particular substance depends on its temperature, the surrounding pressure, and the strength of the forces holding its structural particles together. At the temperatures and pressures normally encountered on Earth, water is one of the few substances found in all three of its physical states: solid ice, liquid water, and gaseous steam (Figure 1.3). Under laboratory conditions, states other than those commonly observed can be attained for almost all substances. Oxygen, which is nearly always thought of as a gas, becomes a liquid at −183°C and a solid at −218°C. The metal iron is a gas at extremely high temperatures (above 3000°C).

Practice Questions and Problems

1.3 Give a characteristic that distinguishes
a. liquids from solids b. gases from liquids

1.4 Give a characteristic that is the same for
a. liquids and solids b. gases and liquids

1.5 Indicate whether each of the following would take the shape of its container and also have a definite volume.
a. Copper wire b. Oxygen gas c. Granulated sugar d. Liquid water

Learning Focus

Classify a given property of a substance as a physical property or a chemical property.

1.3 Properties of Matter

Various kinds of matter are distinguished from each other by their properties. **Properties** *are the distinguishing characteristics of a substance that are used in its identification and description.* Each substance has a unique set of properties that distinguishes it from all other substances. Properties of matter are of two general types: physical and chemical.

A **physical property** *is a characteristic of a substance that can be observed without changing the basic identity of the substance.* Common physical properties include color, odor, physical state (solid, liquid, or gas), melting point, boiling point, and hardness.

During the process of determining a physical property, the physical appearance of a substance may change, but the substance's identity does not. For example, it is impossible to measure the melting point of a solid without changing the solid into a liquid. Although the liquid's appearance is much different from that of the solid, the substance is still the same; its chemical identity has not changed. Hence melting point is a physical property.

▶ *Chemical* properties describe the ability of a substance to form new substances, either by reaction with other substances or by decomposition. *Physical* properties are properties associated with a substance's physical existence. They can be determined without reference to any other substance, and their determination causes no change in the identity of the substance.

A **chemical property** *is a characteristic of a substance that describes the way the substance undergoes or resists change to form a new substance.* For example, copper objects turn green when exposed to moist air for long periods of time (Figure 1.4); this is a chemical property of copper. The green coating formed on the copper is a new substance that results from the copper's reaction with oxygen, carbon dioxide, and water present in air. The properties of this new substance (the green coating) are very different from those of metallic copper. On the other hand, gold objects resist change when exposed to air for long periods of time. The lack of reactivity of gold with air is a chemical property of gold.

Most often the changes associated with chemical properties result from the interaction (reaction) of a substance with one or more other substances. However, the presence of a second substance is not an absolute requirement. Sometimes the presence of energy (usually heat or light) can trigger the change called decomposition. The fact that hydrogen peroxide, in the presence of either heat or light, decomposes into the substances water and oxygen is a chemical property of hydrogen peroxide.

When we specify chemical properties, we usually give conditions such as temperature and pressure because they influence the interactions between substances. For example, the gases oxygen and hydrogen are unreactive toward each other at room temperature, but they interact explosively at a temperature of several hundred degrees.

Figure 1.4 The green color of the Statue of Liberty (present before it was restored) results from the reaction of the copper skin of the statue with the components of air. The fact that copper will react with the components of air is a chemical property of copper.

Practice Questions and Problems

1.6 Classify each of the following properties of the substance magnesium as a physical property or a chemical property.
a. Is a solid at room temperature
b. Ignites upon heating in air
c. Melts at 651°C
d. Does not react with cold water

1.7 Indicate whether each of the following statements describes a physical property or a chemical property.
a. Aspirin tablets can be pulverized with a hammer.
b. Mercury is a liquid at room temperature.
c. Beryllium metal vapor is extremely toxic to humans.
d. Nitric acid discolors the skin by reacting with skin protein.

1.4 Changes in Matter

Learning Focus

Classify a given change that occurs in matter as a physical change or a chemical change.

Changes in matter are common and familiar occurrences. Changes take place when food is digested, paper is burned, and a pencil is sharpened. Like properties of matter, changes in matter are classified into two categories: physical and chemical.

A **physical change** *is a process in which a substance changes its physical appearance but not its chemical composition.* A new substance is never formed as a result of a physical change.

▶ Physical changes need not involve a change of state. Pulverizing an aspirin tablet into a powder and cutting a piece of adhesive tape into small pieces are physical changes that involve only the solid state.

A change in physical state is the most common type of physical change. Melting, freezing, evaporation, and condensation are all changes of state. In any of these processes, the composition of the substance undergoing change remains the same even though its physical state and appearance change. The melting of ice does not produce a new substance; the substance is water both before and after the change (see Figure 1.5). Similarly, the steam produced from boiling water is still water.

A **chemical change** *is a process in which a substance undergoes a change in chemical composition.* Chemical changes always involve conversion of the material or materials under consideration into one or more new substances, each of which has distinctly different properties and composition from the original materials. Consider, for example, the rusting of iron objects left exposed to moist air (Figure 1.6). The reddish brown sub-

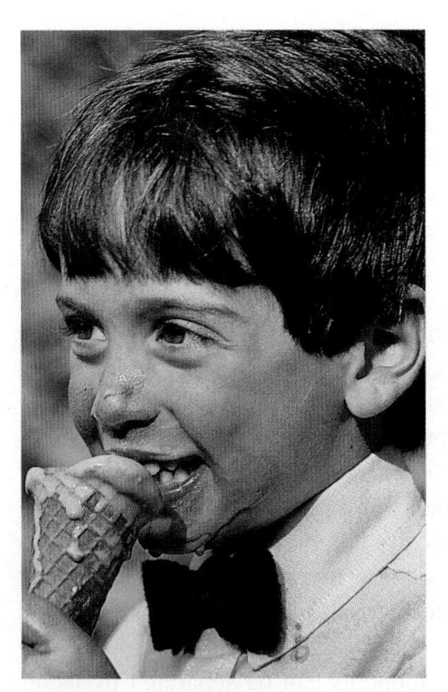

Figure 1.5 The melting of ice cream is a physical change.

Figure 1.6 As a result of chemical change, bright steel girders become rusty when exposed to moist air.

stance (the rust) that forms is a new substance with chemical properties that are obviously different from those of the original iron.

Chemists study the nature of changes in matter to learn how to bring about favorable changes and prevent undesirable ones. The control of chemical change has been a major factor in attainment of the modern standard of living now enjoyed by most people in the developed world. The many plastics, synthetic fibers, and prescription drugs now in common use are the result of controlled chemical change.

On the basis of the discussion in this section and the preceding one, some generalizations concerning the use of the terms *physical* and *chemical* are in order.

1. Whenever the term *physical* is used to modify another term, as in *physical property* or *physical change,* it always conveys the ideas that the composition (chemical identity) of the substance involved *did not change.*
2. Whenever the term *chemical* is used to modify another term, as in *chemical property* or *chemical change,* it always conveys the idea that the composition (chemical identity) of the substance(s) involved either *did change* or *successfully resisted change* as the result of an external challenge to its identity.

Example 1.1 Correct Use of the Terms *Physical* and *Chemical*

Correctly complete each of the following sentences by placing the word *physical* or *chemical* in the blank

a. The fact that pure aspirin melts at 143°C is a ____ property of aspirin.
b. The fact that potassium metal vigorously interacts with water to produce hydrogen gas is a ____ property of potassium.
c. Straightening a bent piece of iron with a hammer is an example of a ____ change.
d. The ignition of a match is an example of a ____ change.

Solution

a. Physical. Changing solid aspirin to liquid aspirin (melting) does not produce any new substances. We still have aspirin.
b. Chemical. A new substance, hydrogen, is produced.
c. Physical. The piece of iron is still a piece of iron.
d. Chemical. New gaseous substances, as well as heat and light, are produced as the match burns.

Practice Questions and Problems

1.8 Indicate whether each of the following statements describes a physical change or a chemical change.
a. An Alka-Seltzer tablet is dropped into water.
b. A table leg is fashioned from a piece of wood.
c. Water placed in the refrigerator is converted into ice cubes.
d. Leaves turn red in the autumn.

1.9 Correctly complete each of the following sentences by placing the word *physical* or *chemical* in the blank.
a. The destruction of a newspaper through burning involves a ____ change.
b. The grating of a piece of cheese is a ____ change.
c. The heating of a blue powdered material to produce a white glassy substance and a gas is a ____ change.
d. The crushing of some ice to make some ice chips is a ____ change.

Learning Focus

List the major differences between pure substances, heterogeneous mixtures, and homogeneous mixtures.

1.5 Pure Substances and Mixtures

In addition to its classification by physical state (Section 1.2), matter can also be classified in terms of its chemical composition as a pure substance or as a mixture. A **pure substance** *is a single kind of matter that cannot be separated into other kinds of matter by any physical means.* All samples of a pure substance contain only that substance and nothing else. Pure water is water and nothing else. Pure sucrose (table sugar) contains only that substance and nothing else.

▶ *Substance* is a general term used to denote any variety of matter. *Pure substance* is a specific term that applies only to matter that contains a single substance.

A pure substance always has a definite and constant composition. This invariant composition dictates that the properties of a pure substance are always the same under a given set of conditions. Collectively, these definite and constant physical and chemical properties constitute the means by which we identify the pure substance.

▶ All samples of a pure substance, no matter what their source, have the same properties under the same conditions.

A **mixture** *is a physical combination of two or more pure substances in which each substance retains its own chemical identity.* Components of a mixture retain their identity because they are physically mixed rather than chemically combined. Consider a mixture of small rock salt crystals and ordinary sand. Mixing these two substances changes neither the salt nor the sand in any way. The larger, colorless salt particles are easily distinguished from the smaller, light-gray sand granules.

One characteristic of any mixture is that its components can be separated by using physical means. In our salt–sand mixture, the larger salt crystals could be—though very tediously—"picked out" from the sand. A somewhat easier separation method would be to dissolve the salt in water, which would leave the undissolved sand behind. The salt could then be recovered by evaporation of the water. Figure 1.7 shows a heterogeneous mixture of potassium dichromate (orange crystals) and iron filings. A magnet can be used to separate the components of this mixture.

▶ Most naturally occurring samples of matter are mixtures. Gold and diamond are two of the few naturally occurring pure substances. Despite their scarcity in nature, numerous pure substances are known. They are obtained from natural mixtures by using various types of separation techniques or are synthesized in the laboratory from naturally occurring materials.

Another characteristic of a mixture is variable composition. Numerous different salt–sand mixtures, with compositions ranging from a slightly salty sand mixture to a slightly sandy salt mixture, could be made by varying the amounts of the two components.

Mixtures are subclassified as heterogeneous or homogeneous. This subclassification is based on visual recognition of the mixture's components. A **heterogeneous mixture** *contains visibly different phases (parts), each of which has different properties.* A nonuniform appearance is a characteristic of all heterogeneous mixtures. Examples include chocolate chip cookies and blueberry muffins. Naturally occurring heterogeneous mixtures include rocks, soils, and wood.

A **homogeneous mixture** *contains only one visibly distinct phase (part), which has uniform properties throughout.* The components present in a homogeneous mixture can-

(a)

(b)

Figure 1.7 (a) A magnet (on the left) and a mixture consisting of potassium dichromate (the orange crystals) and iron fillings. (b) The magnet can be used to separate the iron filings from the potassium dichromate.

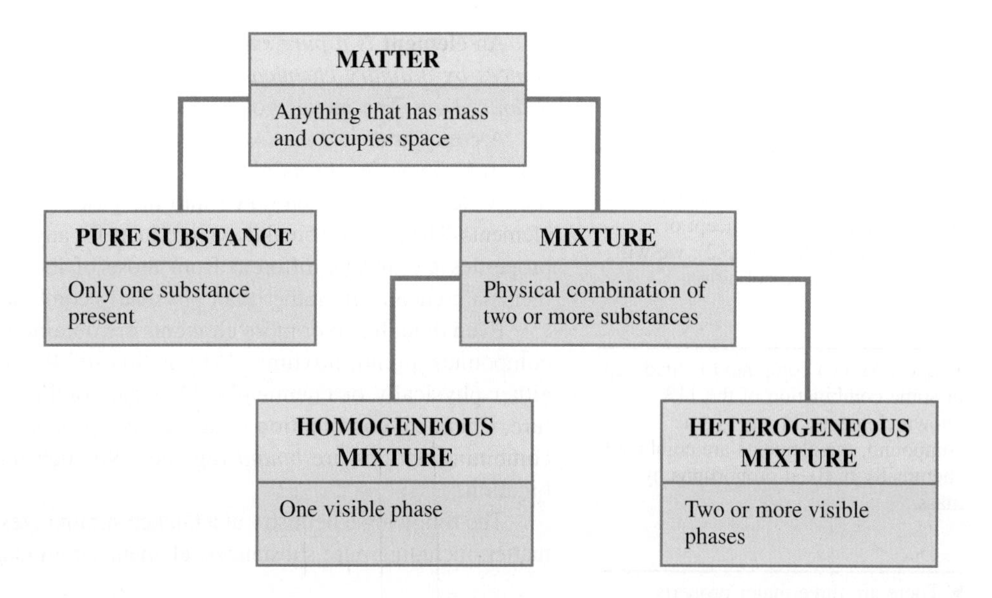

Figure 1.8 Matter falls into two basic classes: pure substances and mixtures. Mixtures, in turn, may be homogeneous or heterogeneous.

not be visually distinguished. A sugar–water mixture in which all of the sugar has dissolved has an appearance similar to that of pure water. Air is a homogeneous mixture of gases; motor oil and gasoline are multicomponent homogeneous mixtures of liquids; and metal alloys such as 14-karat gold (a mixture of copper and gold) are examples of homogeneous mixtures of solids.

Figure 1.8 summarizes what we have learned thus far about various classifications of matter.

Practice Questions and Problems

1.10 Classify each of the following statements as *true* or *false.*
a. All heterogeneous mixtures must contain three or more substances.
b. Pure substances cannot have a variable composition.
c. Substances maintain their identity in a heterogeneous mixture but not in a homogeneous mixture.
d. A homogeneous mixture contains only one visibly distinct phase.

1.11 Assign each of the following descriptions of matter to one of these categories: *heterogeneous mixture, homogeneous mixture,* or *pure substance.*
a. Two substances present, two phases present
b. Two substances present, one phase present
c. One substance present, one phase present
d. Three substances present, three phases present

1.12 Classify each of the following samples of matter as a *heterogeneous mixture, homogeneous mixture,* or *pure substance.*
a. Water and dissolved salt
b. Water and sand
c. Water, ice, and oil
d. Salt water and sugar water

Learning Focus
List the major differences between elements and compounds.

1.6 Elements and Compounds

Chemists have isolated and characterized an estimated 8.5 million pure substances. A very small number of these pure substances, 113 to be exact, are different from all of the others. They are elements. All of the rest, the remaining millions, are compounds. What distinguishes an element from a compound?

▶ Both elements and compounds are pure substances.

▶ The definition for the term *element* that is given here will do for now. After considering the concept of atomic number (Section 3.2), we will give a more precise definition.

▶ Every known compound is made up of some combination of the 113 known elements. In any given compound, the elements are combined chemically in fixed proportions by mass.

▶ There are three major property distinctions between compounds and mixtures.

1. Compounds have properties distinctly different from those of the substances that combined to form the compound. The components of mixtures retain their individual properties.
2. Compounds have a definite composition. Mixtures have a variable composition.
3. Physical methods are sufficient to separate the components of a mixture. The components of a compound cannot be separated by physical methods; chemical methods are required.

An **element** *is a pure substance that cannot be broken down into simpler pure substances by ordinary chemical means such as a reaction, an electric current, heat, or a beam of light.* The metals gold, silver, and copper are all elements.

A **compound** *is a pure substance that can be broken down into two or more simpler pure substances by chemical means.* Water is a compound. By means of an electric current, water can be broken down into the gases hydrogen and oxygen, both of which are elements. The ultimate breakdown products for any compound are elements. A compound's properties are always different from those of its component elements, because the elements are chemically rather than physically combined in the compound (Figure 1.9).

Even though two or more elements are obtained from decomposition of compounds, compounds are not mixtures. Why is this so? Remember, substances can be combined either physically or chemically. Physical combination of substances produces a mixture. Chemical combination of substances produces a compound, a substance in which combining entities are *bound* together. No such binding occurs during physical combination.

The following Chemistry at a Glance summarizes what we have learned thus far about matter, including pure substances, elements, compounds, and mixtures.

Practice Questions and Problems

1.13 Indicate whether each of the following statements is *true* or *false*.
a. Both elements and compounds are pure substances.
b. A compound results from the physical combination of two or more elements.
c. Compounds, but not elements, can have a variable composition.
d. A compound must contain at least two elements.

1.14 On the basis of the information given, classify each of the pure substances A through D as *elements* or *compounds,* or indicate that no such classification is possible because of insufficient information.
a. Substance A cannot be broken down into simpler substances by chemical means.
b. Substance B decomposes upon heating.
c. Heating substance C to 1000°C causes no change in it.
d. Substance D readily reacts with the element chlorine.

Figure 1.9 A pure substance can be either an element or a compound.

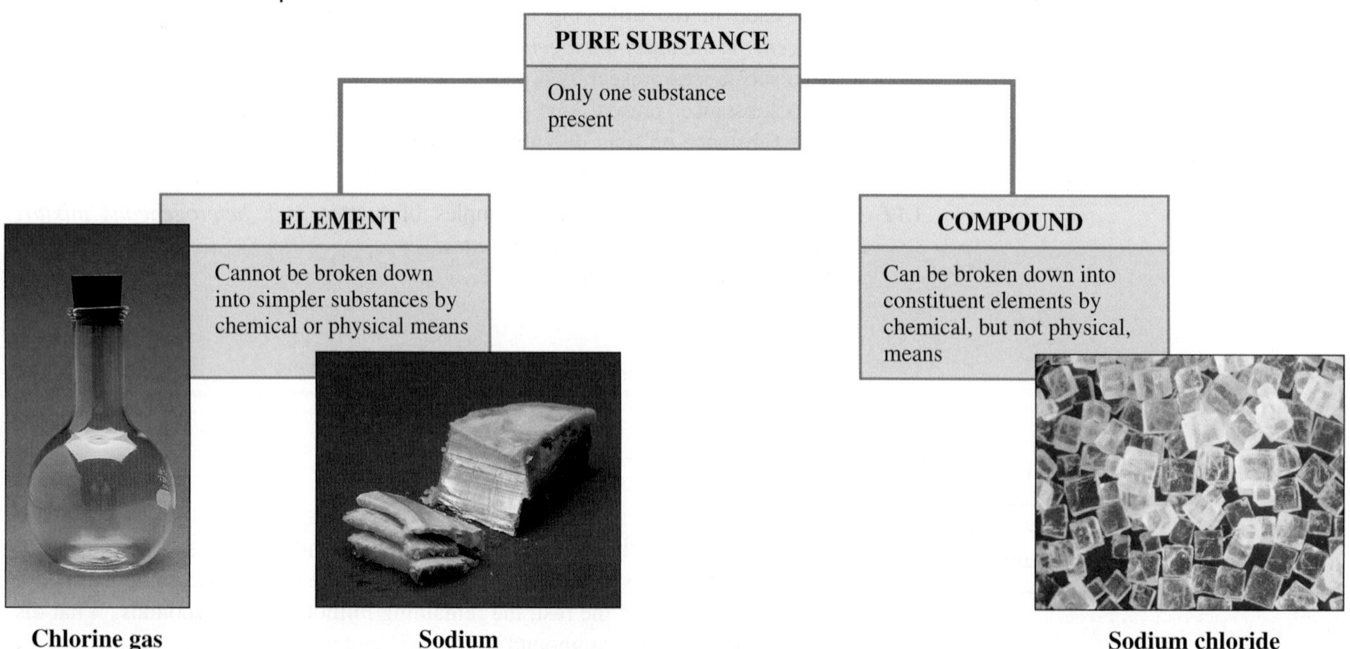
PURE SUBSTANCE
Only one substance present
ELEMENT
Cannot be broken down into simpler substances by chemical or physical means
COMPOUND
Can be broken down into constituent elements by chemical, but not physical, means
Chlorine gas
Sodium
Sodium chloride

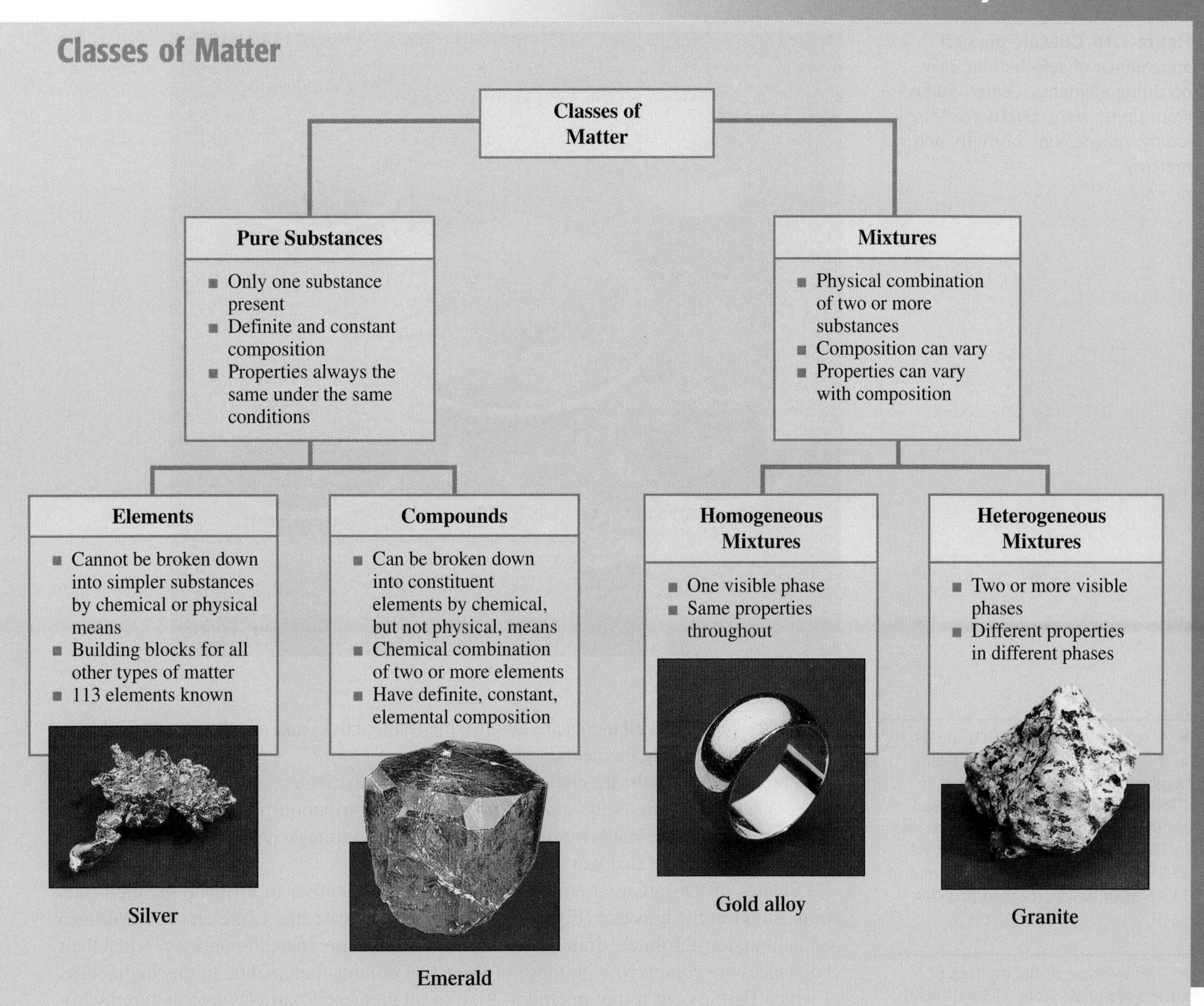

Classes of Matter
Classes of Matter
Pure Substances
Only one substance present
Definite and constant composition
Properties always the same under the same conditions
Mixtures
Physical combination of two or more substances
Composition can vary
Properties can vary with composition
Elements
Cannot be broken down into simpler substances by chemical or physical means
Building blocks for all other types of matter
113 elements known
Compounds
Can be broken down into constituent elements by chemical, but not physical, means
Chemical combination of two or more elements
Have definite, constant, elemental composition
Homogeneous Mixtures
One visible phase
Same properties throughout
Heterogeneous Mixtures
Two or more visible phases
Different properties in different phases
Silver
Emerald
Gold alloy
Granite

1.15 From the information given in the following equations, classify each of the substances A through F as *elements* or *compounds,* or indicate that no such classification is possible because of insufficient information.

a. $A + B \longrightarrow C$

b. $D \longrightarrow E + F$

Learning Focus

List general trends concerning the discovery and abundance of the elements.

1.7 Discovery and Abundance of the Elements

The discovery and isolation of the 113 known elements, the building blocks for all matter, have taken place over a period of several centuries. Most of the discoveries have occurred since 1700, the 1800s being the most active period.

Eighty-eight of the 113 elements occur naturally, and 25 have been synthesized in the laboratory by bombarding samples of naturally occurring elements with small particles. Figure 1.10 shows samples of selected naturally occurring elements. The synthetic

Figure 1.10 Outward physical appearance of selected naturally occurring elements. *Center:* Sulfur. *From upper right, clockwise:* Arsenic, iodine, magnesium, bismuth, and mercury.

(laboratory-produced) elements are all unstable (radioactive) and usually revert quickly to the naturally occurring elements.

The naturally occurring elements are not evenly distributed on Earth and in the universe. What is startling is the nonuniformity of the distribution. A small number of elements account for the majority of elemental particles (atoms). (An atom is the smallest particle of an element that can exist. See Section 1.9.)

Studies of the radiation emitted by stars enable scientists to estimate the elemental composition of the universe (Figure 1.11a). Results indicate that two elements, hydrogen and helium, are absolutely dominant. All other elements are mere "impurities" when their abundances are compared with those of these two dominant elements. In this big picture, in which Earth is but a tiny microdot, 91% of all elemental particles (atoms) are hydrogen, and almost all of the remaining 9% are helium.

▶ A student who attended a university in the year 1700 would have been taught that 13 elements existed. In 1750 he or she would have learned about 16 elements, in 1800 about 34, in 1850 about 59, in 1900 about 82, and in 1950 about 98. Today's total of 113 elements was reached in 1999.

▶ Any increase in the number of known elements from 113 will result from the production of additional synthetic elements. Current chemical theory strongly suggests that all naturally occurring elements have been identified. The isolation of the last of the known naturally occurring elements, rhenium, occurred in 1925.

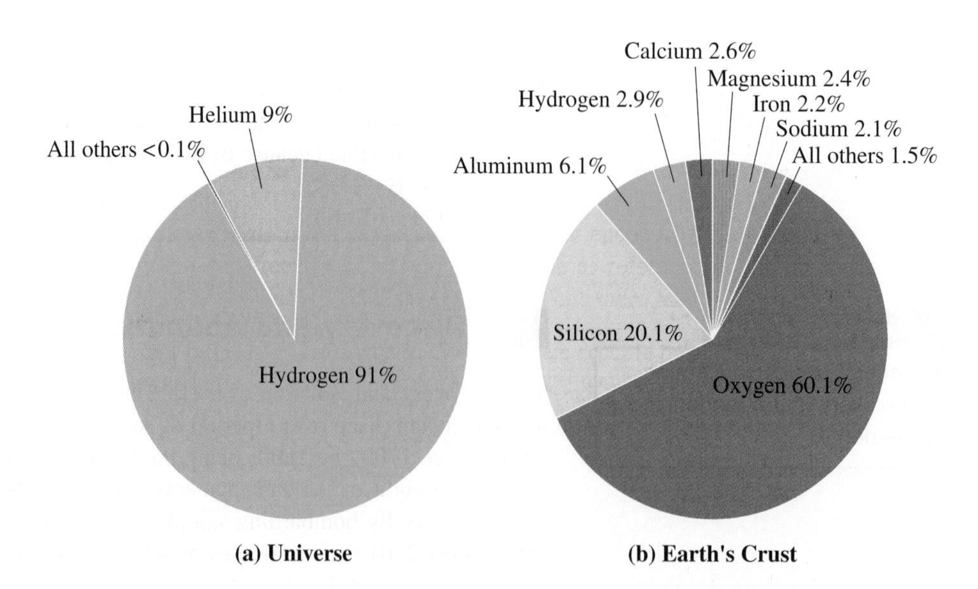

Figure 1.11 Abundance of elements in the universe and in Earth's crust (in atom percent).

If we narrow our view to the chemical world of humans—Earth's crust (its waters, atmosphere, and outer solid surface)—a different perspective emerges. Again, two elements dominate, but this time they are oxygen and silicon. Figure 1.11b provides information on elemental abundances for Earth's crust. The numbers given are atom percents—that is, the percentage of total atoms that are of a given type. Note that the eight elements listed (the only elements with atom percents greater than 1%) account for over 98% of total atoms in Earth's crust. Note also the dominance of oxygen and silicon; these two elements account for 80% of the atoms that make up the chemical world of humans.

The elements of Earth's crust are not distributed equally throughout it. Ore deposits of some elements are found only in localized regions of the crust. Thus not all countries have domestic access to all elements. Even the United States, with its wealth of natural resources, lacks domestic sources of certain elements. For example, the United States has no domestic production sources for the elements nickel, chromium, and cobalt. All new supplies of these elements must be imported from other countries.

To some observers, our nation's reliance on such imports constitutes a dangerous dependence, portending a "strategic materials crisis" equal in seriousness to the "energy crisis" associated with the importation of crude oil and petroleum products. Others see the U.S. position as manageable, though not without dangers and difficulties. All observers agree that if supplies of such imported elements were cut off, many readjustments would have to be made in lifestyles in the United States.

Chemical Portraits 1 profiles the "imported" elements nickel, chromium, and cobalt. The major use for all three of these elements is in metal alloys. (Alloys are solid-state homogeneous mixtures; see Section 1.5.) Import dependence for each of these elements is 60% or greater. (Import dependence is not 100% because significant supplies of these elements are now obtained from domestic recycling of scrap materials, such as spent metal alloys, that contain these elements.

Chemical Portraits 1 — Strategic Elements–Dependence on Imports

Nickel (Ni)

Profile: The United States imports 55–65% of its nickel, of which 39% comes from Canada, 15% comes from Norway, and 13% comes from Russia.

Uses: Nickel's primary use is in stainless steel, a type of steel that is corrosion resistant. Adding nickel to steel increases its workability and strength at high temperatures. Some nickel is used in making coins; a U.S. nickel (5 cents) contains 25% nickel and 75% copper. Although nickel is relatively scarce in the Earth's crust, it is believed that large deposits of this element exist within the Earth's core.

What is the importance of the domestic recycling of scrap metals containing nickel?

Chromium (Cr)

Profile: The United States imports 75–80% of its chromium, of which 46% comes from South Africa, 14% comes from Kazakhstan, and 10% comes from Russia.

Uses: Chromium's primary use is as an ingredient in stainless steel; it is the chromium present that gives such steel its corrosion resistance. Chromium atoms near the surface of the steel react with the oxygen of air to produce a chromium-oxygen compound. This compound forms a corrosion-resistant tenacious protective coating on the surface of the steel.

What properties of chromium are responsible for the practice of chrome-plating other metal objects?

Cobalt (Co)

Profile: The United States imports 73–76% of its cobalt, of which 23% comes from Norway, 20% comes from Finland, and 13% comes from Zambia.

Uses: Cobalt's primary use is as an ingredient in superalloys (alloys that can withstand higher temperatures than stainless steel can). Such superalloys are used primarily in aircraft gas turbine engines. Cobalt, like iron, has magnetic properties and it retains them at temperatures far higher than does iron. This leads to cobalt's use in small, powerful magnets for electrical equipment.

What role does the element cobalt play in human nutrition?

See the text web site at **www.cengage.com/chemistry/stoker** for answers to the above questions and for further information.

Table 1.1
Abundance of Elements in the Human Body (in atom percent)

Element	Percent of total number of atoms in the human body
hydrogen	63
oxygen	25.5
carbon	9.5
nitrogen	1.4
calcium	0.31
phosphorus	0.22
potassium	0.06
sulfur	0.05
chlorine	0.03
sodium	0.03
magnesium	0.01

The distribution of elements in the human body and other living systems is very different from that found in Earth's crust. This distribution is the result of living systems *selectively* taking up matter from their external environment, rather than simply accumulating matter representative of their surroundings.

Eleven elements are found in the human body in atom percent levels of 0.01 or greater, as shown in Table 1.1. The high abundances of hydrogen and oxygen in the body reflect its high water content. Hydrogen is over twice as abundant as oxygen, largely because water contains hydrogen and oxygen in a 2-to-1 atom ratio.

Practice Questions and Problems

1.16 Indicate whether each of the following statements about the known elements is *true* or *false*.
  a. The majority of the known elements have been discovered since 1900.
  b. At present, 108 elements are known.
  c. Elements that do not occur in nature can be produced in a laboratory setting.
  d. New elements have been identified within the last 10 years.

1.17 Indicate whether each of the following statements about elemental abundances is *true* or *false*.
  a. Silicon is the second most abundant element in Earth's crust.
  b. Oxygen is the most abundant element both in Earth's crust and in the human body.
  c. Hydrogen is the most abundant element in the universe but not in Earth's crust.
  d. Two elements account for over three-fourths of the atoms in Earth's crust.

1.8 Names and Chemical Symbols of the Elements

Learning Focus

For the common elements, given the name of the element write its chemical symbol, or given its chemical symbol write its name.

Each element has a unique name that, in most cases, was selected by its discoverer. Abbreviations called chemical symbols also exist for the names of the elements. A **chemical symbol** *is a one- or two-letter designation for an element derived from the name of the element.* These symbols are used more frequently than the elements' names. They can be written more quickly than the names, and they occupy less space. A complete list of the known elements and their chemical symbols is given in Table 1.2. The chemical symbols and names of the more frequently encountered elements are shown in color in this table.

▶ Learning the symbols of the more common elements is an important key to success in studying chemistry. Knowledge of chemical symbols is essential for writing chemical formulas (Section 1.10) and chemical equations (Section 5.6).

Note that the first letter of a chemical symbol is always capitalized and that the second is not. Two-letter symbols are often, but not always, the first two letters of the element's name.

Eleven elements have symbols that bear no relationship to the element's English-language name. In ten of these cases, the symbol is derived from the Latin name of the

Table 1.2
The Chemical Symbols for the Elements[a]
The symbols and names of the more frequently encountered elements are shown in red.

Ac	actinium	Ge	germanium	Pr	praseodymium
Ag	silver*	H	hydrogen	Pt	platinum
Al	aluminum	He	helium	Pu	plutonium
Am	americium	Hf	hafnium	Ra	radium
Ar	argon	Hg	mercury*	Rb	rubidium
As	arsenic	Ho	holmium	Re	rhenium
At	astatine	Hs	hassium	Rf	rutherfordium
Au	gold*	I	iodine	Rh	rhodium
B	boron	In	indium	Rn	radon
Ba	barium	Ir	iridium	Ru	ruthenium
Be	beryllium	K	potassium*	S	sulfur
Bh	bohrium	Kr	krypton	Sb	antimony*
Bi	bismuth	La	lanthanum	Sc	scandium
Bk	berkelium	Li	lithium	Se	selenium
Br	bromine	Lu	lutetium	Sg	seaborgium
C	carbon	Lr	lawrencium	Si	silicon
Ca	calcium	Md	mendelevium	Sm	samarium
Cd	cadmium	Mg	magnesium	Sn	tin*
Ce	cerium	Mn	manganese	Sr	strontium
Cf	californium	Mo	molybdenum	Ta	tantalum
Cl	chlorine	Mt	meitnerium	Tb	terbium
Cm	curium	N	nitrogen	Tc	technetium
Co	cobalt	Na	sodium*	Te	tellurium
Cr	chromium	Nb	niobium	Th	thorium
Cs	cesium	Nd	neodymium	Ti	titanium
Cu	copper*	Ne	neon	Tl	thallium
Db	dubnium	Ni	nickel	Tm	thulium
Dy	dysprosium	No	nobelium	U	uranium
Er	erbium	Np	neptunium	V	vanadium
Es	einsteinium	O	oxygen	W	tungsten*
Eu	europium	Os	osmium	Xe	xenon
F	fluorine	P	phosphorus	Y	yttrium
Fe	iron*	Pa	protactinium	Yb	ytterbium
Fm	fermium	Pb	lead*	Zn	zinc
Fr	francium	Pd	palladium	Zr	zirconium
Ga	gallium	Pm	promethium		
Gd	gadolinium	Po	polomium		

[a]Only 109 elements are listed in this table. Elements 110–112 and 114 discovered (synthesized) in the period 1994–1999, are yet to be named.

*These elements have symbols that were derived from non-English names.

element; in the case of the element tungsten, a German name is the symbol's source. Most of these elements have been known for hundreds of years and date back to the time when Latin was the language of scientists. Elements whose symbols are derived from non-English names are marked with an asterisk in Table 1.2.

Practice Questions and Problems

1.18 In which of the following sequences of elements do all of the elements have two-letter chemical symbols?

a. Magnesium, nitrogen, phosphorus
b. Bromine, iron, calcium
c. Aluminum, copper, chlorine
d. Boron, barium, beryllium

1.19 In which of the following sequences of elements do all of the elements have chemical symbols that start with a letter that is not the first letter of the element's English name?

a. Silver, gold, mercury
b. Copper, helium, neon
c. Cobalt, chromium, sodium
d. Potassium, iron, lead

1.9 Atoms and Molecules

Learning Focus

Define the term *atom* and be familiar with the various general types of *molecules* that exist.

Consider the process of subdividing a sample of the element gold (or any other element) into smaller and smaller pieces. It seems reasonable that eventually a "smallest possible piece" of gold would be reached that could not be divided further and still be the element gold. This smallest possible unit of gold is called a gold atom. An **atom** *is the smallest particle of an element that can exist and still have the properties of the element.*

A sample of any element is composed of atoms of a single type, those of that element. In contrast, a compound must have two or more types of atoms present, because by definition at least two elements must be present (Section 1.6).

No one ever has seen or ever will see an atom with the naked eye; they are simply too small for such observation. However, sophisticated electron microscopes, with magnification factors in the millions, have made it possible to photograph "images" of individual atoms (Figure 1.12).

Atoms are incredibly small particles. Atomic dimensions, although not directly measurable, can be calculated from measurements made on large-size samples of elements. The diameter of an atom is approximately four-billionths of an inch. If atoms of such diameter were arranged in a straight line, it would take 254 million of them to extend a distance of 1 inch (see Figure 1.13).

▶ Reasons for the tendency of atoms to aggregate into molecules and information on the binding forces involved are considered in Chapter 4.

Free atoms are rarely encountered in nature. Instead, under normal conditions of temperature and pressure, atoms are nearly always found together in aggregates or clusters ranging in size from two atoms to numbers too large to count. When the group or cluster of atoms is relatively small and bound together tightly, the resulting entity is called a molecule. A **molecule** *is a group of two or more atoms that functions as a unit because the atoms are tightly bound together.* This resultant "package" of atoms behaves in many ways as a single, distinct particle would.

Figure 1.12 A computer reconstruction of the surface of a sample of a solid as observed with a scanning tunneling microscope. The image reveals the regular pattern of individual atoms. The color was added to the image by computer.

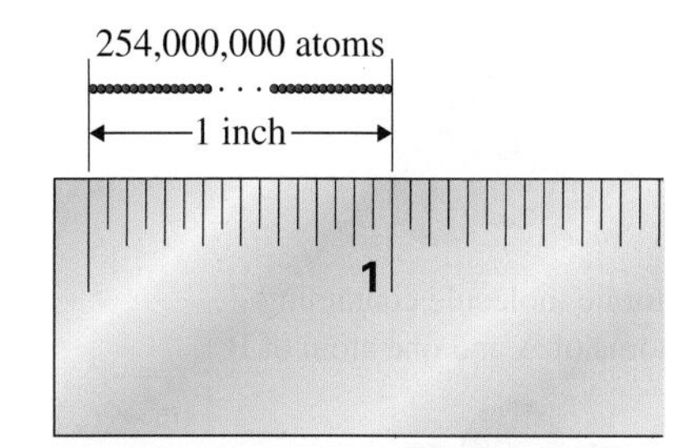

Figure 1.13 254 million atoms arranged in a straight line would extend a distance of approximately 1 inch.

▶ The Latin word *mole* means "a mass." The word *molecule* denotes a "little mass."

A **diatomic molecule** *is a molecule that contains two atoms.* It is the simplest type of molecule that can exist. Next is complexity are *triatomic* molecules. A **triatomic molecule** *is a molecule that contains three atoms.* The molecule present in water, the most common of all compounds, is triatomic; it contains two hydrogen atoms and one oxygen atom. Continuing on numerically, we have *tetratomic* molecules, *pentatomic* molecules, and so on.

The atoms contained in a molecule may all be of the same kind, or two or more kinds may be present. In accordance with this observation, molecules are classified into the two categories of *homoatomic* and *heteroatomic.* A **homoatomic molecule** *is a molecule in which all atoms present are of the same kind.* A substance containing homoatomic molecules must be an element. A **heteroatomic molecule** *is a molecule in which two or more kinds of atoms are present.* Substances that contain heteroatomic molecules must be compounds, because the presence of two or more kinds of atoms reflects the presence of two or more kinds of elements.

The fact that homoatomic molecules exist indicates that individual atoms are not always the preferred structural unit for an element. Nearly all of the oxygen present in air is in the form of diatomic molecules. Several other elements, including nitrogen (the other major constituent of air) and hydrogen, are also diatomic in the gaseous state. Chemical Portraits 2 profiles the "diatomic" elements oxygen, nitrogen, and hydrogen. The diatomic nature of these elements can be specified by using the notations O_2, N_2, and H_2 to represent molecules of these elements.

Figure 1.14 shows general models for four simple types of heteroatomic molecules. Comparison of parts (c) and (d) of this figure shows that molecules with the same number of atoms need not have the same arrangement of atoms.

Chemical Portraits 2 — Three Well-Known Gaseous Diatomic Elements

Oxygen (O_2)

Profile: Oxygen, the second most abundant gas in air, constitutes 21% by volume of clean, dry air. This colorless, odorless, tasteless gas is often referred to as the "life-giving" element because a person can live weeks without food, days without water, but only minutes without oxygen. To commercially obtain pure oxygen, air is liquefied at low temperatures, and its various components are separated according to their different boiling points.

Uses: The steel industry is the dominant outlet for industrial O_2. Here, impurities are removed from the steel by their reaction with O_2. Medical and life-support uses for oxygen consume less than 2% of commercial oxygen production.

What is the major use for liquid oxygen in the United States's space program?

Nitrogen (N_2)

Profile: Nitrogen, the most abundant gas in air, constitutes 78% by volume of clean, dry air. Like O_2, it is colorless, odorless, and tasteless, and it is obtained in the pure state through low-temperature extraction from liquid air. Unlike O_2, however, N_2 is a very unreactive gas.

Uses: The major industrial use for N_2 gas is as the nitrogen source for making nitrogen-containing fertilizers such as liquid ammonia and ammonium nitrate. Nitrogen is a nutrient required by all plants and plants cannot obtain it from air. Because of its unreactivity, another major use for N_2 is that of a "blanketing agent." Here it is used to protect substances that would react with O_2 or moisture in air.

When was nitrogen discovered and by whom?

Hydrogen (H_2)

Profile: Hydrogen, unlike N_2 and O_2, is not a constituent of air. Hydrogen molecules are the least massive of all molecules. Because of their small mass, gaseous H_2 molecules are able to acquire velocities sufficient for them to overcome the gravitational forces of the earth and escape to outer space; thus no H_2 is present in the atmosphere. Like O_2 and N_2, H_2 gas is colorless, odorless, and tasteless; like O_2, H_2 molecules are relatively reactive.

Uses: About 40% of commercial H_2 production is used to make ammonia, the starting material for nitrogen-fertilizer production. Decomposition of hydrogen-containing compounds such as water and methane is the source of such hydrogen.

What is the state of development for hydrogen-powered cars?

See the text web site at **www.cengage.com/chemistry/stoker** for answers to the above questions and for further information.

Figure 1.14 Depictions of various simple heteroatomic molecules using models. Spheres of different sizes and colors represent different kinds of atoms.

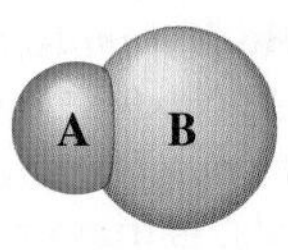

(a) A diatomic molecule containing one atom of A and one atom of B

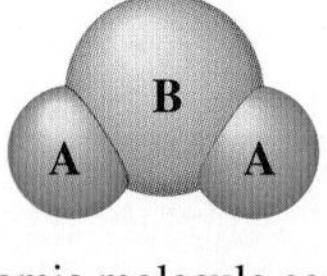

(b) A triatomic molecule containing two atoms of A and one atom of B

(c) A tetratomic molecule containing two atoms of A and two atoms of B

(d) A tetratomic molecule containing three atoms of A and one atom of B

Example 1.2 **Classifying Molecules Based on Numbers of and Types of Atoms**

Classify each of the following molecules as *diatomic, triatomic,* etc., and also as *homoatomic* or *heteroatomic.*

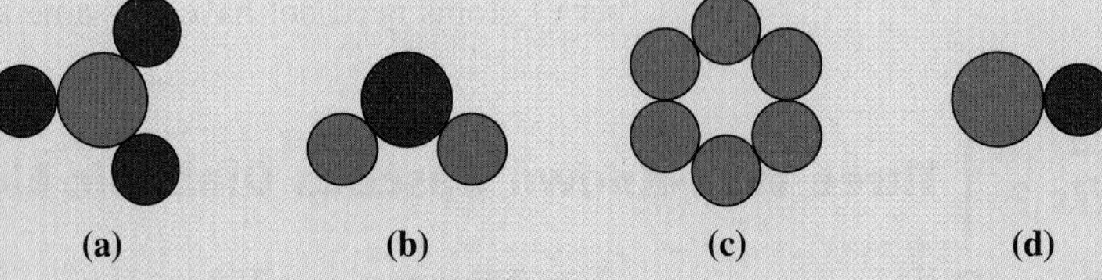

(a) (b) (c) (d)

Solution

a. Tetratomic and heteroatomic (four atoms; two kinds of atoms)
b. Triatomic and heteroatomic (three atoms; two kinds of atoms)
c. Hexatomic and homoatomic (six atoms; only one kind of atom)
d. Diatomic and heteroatomic (two atoms; two kinds of atoms)

► The concept that heteroatomic molecules are the building blocks for *all* compounds will have to be modified when certain solids, called ionic solids, are considered in Section 4.6.

A molecule is the smallest particle of a compound capable of a stable independent existence. Continued subdivision of a quantity of table sugar to yield smaller and smaller amounts would ultimately lead to the isolation of one single "unit" of table sugar: a molecule of table sugar. This table sugar molecule could not be broken down any further and still exhibit the physical and chemical properties of table sugar. The table sugar molecule could be broken down further by chemical (not physical) means to produce atoms, but if that occurred, we would no longer have table sugar. The *molecule* is the limit of *physical* subdivision. The *atom* is the limit of *chemical* subdivision.

Practice Questions and Problems

1.20 Indicate whether each of the following statements is *true* or *false.*
a. Triatomic molecules must contain at least two kinds of atoms.
b. A molecule of a compound must be heteroatomic.
c. Heteroatomic molecules do not maintain the properties of their constituent elements.
d. A molecule of an element may be homoatomic or heteroatomic, depending on which element is involved.

1.21 Which of the terms *heteroatomic, homoatomic, diatomic, triatomic, element,* and *compound* apply to each of the following molecules? (More than one term applies in each situation.)

a.  b. (Q—Z—X) c. (X—X) d. (X—Q—X)

Learning Focus

Interpret a chemical formula in terms of the number of elements and the number of atoms present.

1.10 Chemical Formulas

Information about compound composition can be presented in a concise way by using a chemical formula. A **chemical formula** *is a notation made up of the symbols of the elements present in a compound and numerical subscripts (located to the right of each symbol) that indicate the number of atoms of each element present in a molecule of the compound.*

The chemical formula for the compound aspirin is $C_9H_8O_4$. This chemical formula conveys the information that an aspirin molecule contains three different elements—carbon (C), hydrogen (H), and oxygen (O)—and 21 atoms—9 carbon atoms, 8 hydrogen atoms, and 4 oxygen atoms.

When only one atom of a particular element is present in a molecule of a compound, that element's symbol is written without a numerical subscript in the formula for the compound. The formula for rubbing alcohol, C_3H_6O, reflects this practice for the element oxygen.

In order to write formulas correctly, one must follow the capitalization rules for elemental symbols (Section 1.8). Making the error of capitalizing the second letter of an element's symbol can dramatically alter the meaning of a chemical formula. The formulas $CoCl_2$ and $COCl_2$ illustrate this point; the symbol Co stands for the element cobalt, whereas CO stands for one atom of carbon and one atom of oxygen.

Sometimes chemical formulas contain parentheses; an example is $Al_2(SO_4)_3$. The interpretation of this formula is straightforward; in a formula unit, there are present 2 aluminum (Al) atoms and 3 SO_4 groups. The subscript following the parentheses always indicates the number of units in the formula of the polyatomic entity inside the parentheses. In terms of atoms, the formula $Al_2(SO_4)_3$ denotes 2 aluminum (Al) atoms, $3 \times 1 = 3$ sulfur (S) atoms, and $3 \times 4 = 12$ oxygen (O) atoms. Example 1.3 contains further comments about chemical formulas that contain parentheses.

▶ Further information about the use of parentheses in formulas (when and why) will be presented in Section 4.19. The important concern now is being able to interpret formulas that contain parentheses in terms of total atoms present.

Example 1.3 Interpreting Chemical Formulas

For each of the following chemical formulas, state how many atoms of each element are present in one molecule of the substance.

a. HCN—hydrogen cyanide, a poisonous gas
b. $C_{18}H_{21}NO_3$—codeine, a pain-killing drug
c. $Ca_{10}(PO_4)_6(OH)_2$—hydroxyapatite, present in tooth enamel

Solution

a. One atom each of the elements hydrogen, carbon, and nitrogen is present. Remember that the subscript 1 is implied when no subscript is written.
b. This formula indicates that 18 carbon atoms, 21 hydrogen atoms, 1 nitrogen atom, and 3 oxygen atoms are present in one molecule of the compound.
c. There are 10 calcium atoms. The amounts of phosphorus, hydrogen, and oxygen are affected by the subscripts outside the parentheses. There are 6 phosphorus atoms and 2 hydrogen atoms present. Oxygen atoms are present in two locations in the formula. There are a total of 26 oxygen atoms: 24 from the PO_4 subunits (6×4) and 2 from the OH subunits (2×1).

Practice Questions and Problems

1.22 On the basis of its formula, classify each of the following substances as an element or a compound.

a. $LiCO_3$ b. CO c. Co d. O_3 e. $CoCl_2$ f. $COCl_2$

1.23 Determine the number of elements and the number of atoms present in molecules represented by the following formulas.

a. H_2CO_3 b. NH_4ClO_4 c. $CaSO_4$ d. C_4H_{10} e. $Be(CN)_2$ f. $Cu(NO_3)_2$

1.24 Write a chemical formula for each of the following substances using the information given for the substance.

a. A molecule of nicotine contains 10 atoms of carbon, 14 atoms of hydrogen, and 2 atoms of nitrogen.

b. A molecule of vitamin C contains 6 atoms of carbon, 8 atoms of hydrogen, and 6 atoms of oxygen.

1.25 What is the chemical formula for each of the following molecules?

a. 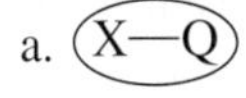

b. 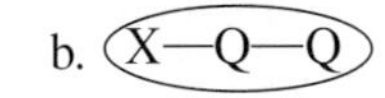

c. 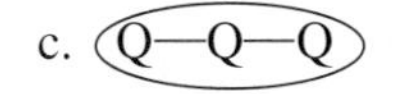

d. 

CONCEPTS TO REMEMBER

Chemistry. Chemistry is the field of study that is concerned with the characterization, composition, and transformations of matter.

Matter. Matter, the substances of the physical universe, is anything that has mass and occupies space. Matter exists in three physical states: solid, liquid, and gas.

Properties of matter. Properties, the distinguishing characteristics of a substance that are used in its identification and description, are of two types: physical and chemical. Physical properties are properties that we can observe without changing a substance into another substance. Chemical properties are properties that matter exhibits as it undergoes or resists changes in chemical composition. The failure of a substance to undergo change in the presence of another substance is considered a chemical property.

Changes in matter. Changes that can occur in matter are classified into two types: physical and chemical. A physical change is a process that does not alter the basic nature (chemical composition) of the substance under consideration. No new substances are ever formed as a result of a physical change. A chemical change is a process that involves a change in the basic nature (chemical composition) of the substance. Such changes always involve conversion of the material or materials under consideration into one or more new substances that have properties and composition distinctly different from those of the original materials.

Pure substances and mixtures. All specimens of matter are either pure substances or mixtures. A pure substance is a form of matter that always has a definite and constant composition. A mixture is a physical combination of two or more pure substances in which the pure substances retain their identity.

Types of mixtures. Mixtures can be classified as heterogeneous or homogeneous on the basis of the visual recognizability of the components present. A heterogeneous mixture contains visibly different parts or phases, each of which has different properties. A homogeneous mixture contains only one phase, which has uniform properties throughout it.

Types of pure substances. A pure substance can be classified as either an element or a compound on the basis of whether it can be broken down into two or more simpler substances by ordinary chemical means. Elements cannot be broken down into simpler substances. Compounds yield two or more simpler substances when broken down. There are 113 pure substances that qualify as elements. There are millions of compounds.

Atoms and molecules. An atom is the smallest particle of an element that can exist and still have the properties of the element. Free isolated atoms are rarely encountered in nature. Instead, atoms are almost always found together in aggregates or clusters. A molecule is a group of two or more atoms that functions as a unit because the atoms are tightly bound together.

Types of molecules. Molecules are of two types: homoatomic and heteroatomic. Homoatomic molecules are molecules in which all atoms present are of the same kind. A pure substance containing homoatomic molecules is an element. Heteroatomic molecules are molecules in which two or more different kinds of atoms are present. Pure substances that contain heteroatomic molecules must be compounds.

Chemical symbols. Chemical symbols are a shorthand notation for the names of the elements. Most consist of two letters; a few involve a single letter. The first letter of a chemical symbol is always capitalized, and the second letter is always lower-case.

Chemical formulas. Chemical formulas are used to specify compound composition in a concise manner. They consist of the symbols of the elements present in the compound and numerical subscripts (located to the right of each symbol) that indicate the number of atoms of each element present in a molecule of the compound.

KEY TERMS

Atom (1.9)
Chemical change (1.4)
Chemical formula (1.10)
Chemical property (1.3)
Chemical symbol (1.8)
Chemistry (1.1)
Compound (1.6)
Diatomic molecule (1.9)
Element (1.6)
Gas (1.2)
Heteroatomic molecule (1.9)
Heterogeneous mixture (1.5)
Homoatomic molecule (1.9)
Homogeneous mixture (1.5)
Liquid (1.2)
Matter (1.1)
Mixture (1.5)
Molecule (1.9)
Physical change (1.4)
Physical property (1.3)
Properties (1.3)
Pure substance (1.5)
Solid (1.2)
Triatomic molecule (1.9)

ADDITIONAL PROBLEMS

1.26 Assign each of the following descriptions of matter to one of the following categories: *element, compound,* or *mixture.*
a. One substance present, one phase present, substance cannot be decomposed by chemical means
b. One substance present, three elements present
c. Two substances present, two phases present
d. Two elements present, composition is definite and constant

1.27 Assign each of the following descriptions of matter to one of the following categories: *element, compound,* or *mixture.*
a. One substance present, one phase present, one kind of homoatomic molecule present
b. Two substances present, two phases present, all molecules are heteroatomic
c. One phase present, two kinds of homoatomic molecules present
d. One phase present, all molecules are triatomic, all molecules are heteroatomic, all molecules are identical

1.28 Indicate whether each of the following samples of matter is a *heterogeneous mixture,* a *homogeneous mixture,* a *compound,* or an *element.*
a. A colorless gas, only part of which reacts with hot iron
b. A "cloudy" liquid that separates into two layers upon standing for 2 hours
c. A green solid, all of which melts at the same temperature to produce a liquid that decomposes upon further heating
d. A colorless gas that cannot be separated into simpler substances using physical means and that reacts with copper to produce both a copper–nitrogen and a copper–oxygen compound

1.29 Classify each of the following pairs of substances as (1) two elements, (2) two compounds, (3) an element and a compound, or (4) a single pure substance.

a. (Q—X) and (Q—Q) b. (Q—X) and (X)

c. (Q) and (X) d. (Q—X) and (Q—X)

1.30 Write a formula for each of the following substances by using the information given about molecules of the substance.
a. Molecules are triatomic and contain the elements hydrogen, carbon, and nitrogen.
b. Molecules are heptatomic and contain 2 atoms of hydrogen, 1 atom of sulfur, and the element oxygen.
c. Molecules are triatomic and contain twice as many atoms of nitrogen as of oxygen.
d. Molecules are pentatomic and contain the elements hydrogen and nitrogen and 3 atoms of oxygen.

1.31 In each of the following pairs of formulas, indicate whether the first formula listed denotes *more total atoms, the same number of total atoms,* or *fewer total atoms* than the second formula listed.
a. HN_3 and NH_3
b. $CaSO_4$ and $Mg(OH)_2$
c. $NaClO_3$ and $Be(CN)_2$
d. $Be_3(PO_4)_2$ and $Mg(C_2H_3O_2)_2$

1.32 On the basis of the given information, determine the numerical value of the subscript x in each of the following chemical formulas.
a. BaS_2O_x; formula unit contains 6 atoms
b. $Al_2(SO_x)_3$; formula unit contains 17 atoms
c. SO_xCl_x; formula unit contains 5 atoms
d. $C_xH_{2x}Cl_x$; formula unit contains 8 atoms

1.33 A mixture contains the following five pure substances: N_2, N_2H_4, NH_3, CH_4, and CH_3Cl.
a. How many different kinds of molecules that contain four or fewer atoms are present in the mixture?
b. How many different kinds of atoms are present in the mixture?
c. How many total atoms are present in a mixture sample containing five molecules of each component?
d. How many total hydrogen atoms are present in a mixture sample containing four molecules of each component?

PRACTICE TEST ▸ True/False

1.34 An *indefinite* volume is a property of both gases and liquids.

1.35 When a substance undergoes a *chemical* change, it is always true that one or more new substances are produced.

1.36 The description "two substances present, two phases present" applies to all types of mixtures.

1.37 Scientists "suspect" that there are more *naturally occurring* elements yet to be discovered.

1.38 The chemical symbol for an element is always the first one or two letters in the element's name.

1.39 A compound results from the *physical* combination of two or more elements.

1.40 The most abundant elements in the universe and in Earth's crust are, respectively, hydrogen and oxygen.

1.41 Both homogeneous mixtures and heterogeneous mixtures can have a *variable* composition.

1.42 The descriptors *homoatomic* and *triatomic* both apply to molecules of the compound H_2O.

1.43 The properties "melts at 73°C" and "decomposes upon heating" are both examples of *chemical* properties of a substance.

1.44 A pure substance can never be decomposed into simpler pure substances by *chemical* means.

1.45 The elements gold, silver, and aluminum all have two-letter chemical symbols.

1.46 The chemical formulas NH_3 and HN_3 both denote the same compound.

1.47 The total number of atoms present in one formula unit of $(NH_4)_3PO_4$ is 18.

1.48 A *physical* change is a process in which a substance changes its physical appearance but not its chemical composition.

PRACTICE TEST ▶ Multiple Choice

1.49 Which of the following is a property of *both* liquids and solids?
a. Definite shape
b. Definite volume
c. Indefinite shape
d. Indefinite volume

1.50 In which of the following pairs of properties are both properties *chemical* properties?
a. Freezes at 5°C, flammable
b. Decomposes at 75°C, reacts with chlorine
c. Good reflector of light, blue in color
d. Has a high density, is very hard

1.51 When a substance undergoes a *chemical* change, it is always true that
a. it melts
b. it changes physical state
c. its chemical composition changes
d. heat is evolved

1.52 Which of the following statements is *correct?*
a. Elements, but not compounds, are pure substances.
b. Compounds, but not elements, are pure substances.
c. Both elements and compounds are pure substances.
d. Neither elements nor compounds are pure substances.

1.53 A pure substance A is found to change upon heating into two new pure substances B and C. Both B and C may be decomposed by chemical means. From this, we may conclude that
a. A is an element, and B and C are compounds
b. A is a compound, and B and C are elements
c. A, B, and C are all elements
d. A, B, and C are all compounds

1.54 In which of the following sequences of elements do each of the elements have a one-letter chemical symbol?
a. Lead, nitrogen, zinc
b. Potassium, fluorine, carbon
c. Tin, hydrogen, iodine
d. Oxygen, silicon, chlorine

1.55 Which one of the following statements is *incorrect*?
a. An atom is the smallest "piece" of an element that can exist and still have the properties of the element.
b. Free isolated atoms are rarely encountered in nature.
c. Atoms may be decomposed using chemical change.
d. Only one kind of atom may be present in a homoatomic molecule.

1.56 Which of the following pairings of terms is *incorrect*?
a. Element — homoatomic molecules
b. Pure substance — variable composition
c. Heterogeneous mixture — two or more regions with differing properties
d. Homogeneous mixture — two or more substances present

1.57 Which of the following classifications of matter could *not* contain heteroatomic molecules?
a. Heterogeneous mixture
b. Homogeneous mixture
c. Pure substance
d. Element

1.58 In which of the following pairs of formulas do the two members of the pair contain the same number of elements as well as the same number of atoms?
a. Hf and HF
b. $CoCl_2$ and $COCl_2$
c. SO_2 and SO_3
d. NH_4Cl and $BaSO_4$

CHAPTER OUTLINE

CHAPTER TWO

Measurements in Chemistry

Measurements can never be exact; there is always some degree of uncertainty.

It would be extremely difficult for a carpenter to build cabinets without being able to use hammers, saws, and drills. They are the tools of a carpenter's trade. Chemists also have "tools of the trade." The tool they use most is called *measurement*. Understanding measurement is indispensable in the study of chemistry. Questions such as "How much ... ?," "How long ... ?," and "How many ... ?" simply cannot be answered without resorting to measurements. This chapter will help you learn what you need to know to deal properly with measurement. Much of the material in the chapter is mathematical. This is necessary; measurements require the use of numbers.

2.1 Measurement Systems

Learning Focus

Understand why scientists prefer the metric system of units over the English system of units when making measurements.

We all make measurements on a routine basis. For example, measurements are involved in following a recipe for making brownies, in determining our height and weight, and in fueling a car with gasoline. **Measurement** *is the determination of the dimensions, capacity, quantity, or extent of something.* In chemical laboratories, the most common types of measurements are those of mass, volume, length, time, temperature, pressure, and concentration.

Two systems of measurement are in use in the United States at present: (1) the English system of units, and (2) the metric system of units. Common measurements of commerce, such as those used in a grocery store, are made in the *English system.* The units of this system include the inch, foot, pound, quart, and gallon. The *metric system* is used in scientific work. The units of this system include the gram, meter, and liter.

▶ The word *metric* is derived from the Greek word *metron,* which means "measure."

Figure 2.1 Metric system units are becoming increasingly evident on highway signs and consumer products.

The United States is in the process of voluntary conversion to the metric system for measurements of commerce. Metric system units now appear on numerous consumer products (Figure 2.1). Soft drinks now come in 1-, 2- and 3-liter containers. Road signs in some states display distances in both miles and kilometers. Canned and packaged goods such as cereals and mixes on grocery store shelves now have the masses of their contents listed in grams as well as in pounds and ounces.

Interrelationships between units of the same type, such as volume or length, are less complicated in the metric system than in the English system. Within the metric system, conversion from one unit size to another can be accomplished simply by multiplying or dividing by units of 10, because the metric system is a decimal unit system—that is, it is based on multiples of 10. The metric system is simply more convenient to use.

Practice Questions and Problems

2.1 List the most common types of measurements made in chemical laboratories.

2.2 What is the main reason why scientists prefer to use the metric measurement system instead of the English measurement system?

2.2 Metric System Units

Learning Focus

Recognize units of the metric system by name and abbreviation, and know the numerical meanings associated with various metric system prefixes.

In the metric system, there is one base unit for each type of measurement (length, mass, volume, and so on). The names of fractional parts of the base unit and multiples of the base unit are constructed by adding prefixes to the base unit. These prefixes indicate the size of the unit relative to the base unit. Table 2.1 lists common metric system prefixes, along with their symbols or abbreviations and mathematical meanings. The prefixes in color are the ones most frequently used.

The meaning of a metric system prefix is independent of the base unit it modifies and always remains constant. For example, the prefix *kilo-* always means 1000; a *kilo*second is 1000 seconds, a *kilo*watt is 1000 watts, and a *kilo*calorie is 1000 calories. Similarly, the prefix *nano-* always means one-billionth; a *nano*meter is one-billionth of a meter, a *nano*gram is one-billionth of a gram, and a *nano*liter is one-billionth of a liter.

▶ The use of numerical prefixes should not be new to you. Consider the use of the prefix *tri-* in the words *tri*angle, *tri*cycle, *tri*o, *tri*nity, and *tri*ple. Each of these words conveys the idea of three of something. The metric system prefixes are used in the same way.

Metric Length Units

The **meter** (m) *is the base unit of length in the metric system.* It is about the same size as the English yard; 1 meter equals 1.09 yards (Figure 2.2a). The prefixes listed in Table 2.1 enable us to derive other units of length from the meter. The kilometer (km) is 1000 times larger than the meter; the centimeter (cm) and millimeter (mm) are, respectively, one-hundredth and one-thousandth of a meter. Most laboratory length measurements are made in centimeters rather than meters because of the meter's relatively large size.

▶ *Length* is measured by determining the distance between two points.

Table 2.1
Common Metric System Prefixes with Their Symbols and Mathematical Meanings

	Prefix[a]	Symbol	Mathematical Meaning[b]
Multiples	giga-	G	1,000,000,000 (10^9, billion)
	mega-	M	1,000,000 (10^6, million)
	kilo-	k	1000 (10^3, thousand)
Fractional parts	deci-	d	0.1 (10^{-1}, one-tenth)
	centi-	c	0.01 (10^{-2}, one-hundredth)
	milli-	m	0.001 (10^{-3}, one-thousandth)
	micro-	μ (Greek mu)	0.000001 (10^{-6}, one-millionth)
	nano-	n	0.000000001 (10^{-9}, one-billionth)
	pico-	p	0.000000000001 (10^{-12}, one-trillionth)

[a]Other prefixes also are available but are less commonly used.
[b]The power-of-10 notation for denoting numbers is considered in Section 2.5.

Metric Mass Units

▶ *Mass* is measured by determining the amount of matter in an object.

The **gram** (g) *is the base unit of mass in the metric system.* It is a very small unit compared with the English ounce and pound (Figure 2.2b). It takes approximately 28 grams to equal 1 ounce and nearly 454 grams to equal 1 pound. Both grams and milligrams (mg) are commonly used in the laboratory, where the kilogram (kg) is generally too large.

The terms *mass* and *weight* are often used interchangeably in measurement discussions; technically, however, they have different meanings. **Mass** *is a measure of the total quantity of matter in an object.* **Weight** *is a measure of the force exerted on an object by the pull of gravity.*

The mass of a substance is a constant; the weight of an object varies with the object's geographical location. For example, matter at the equator weighs less than it would at the North Pole because the pull of gravity is less at the equator. Because Earth is not a perfect sphere, but bulges at the equator, the magnitude of gravitational attraction is less at the equator. An object would weigh less on the moon than on Earth because of the smaller size of the moon and the correspondingly lower gravitational attraction. Quantitatively, a 22.0-lb mass weighing 22.0 lb at Earth's North Pole would weigh 21.9 lb at Earth's equator and only 3.7 lb on the moon. In outer space, an astronaut may be weightless but never massless. In fact, he or she has the same mass in space as on Earth.

▶ Students often erroneously think that the terms *mass* and *weight* have the same meaning. *Mass* is a measure of the amount of material present in a sample. *Weight* is a measure of the force exerted on an object by the pull of gravity.

(a) Length	(b) Mass	(c) Volume
A meter is slightly larger than a yard.	A gram is a small unit compared to a pound.	A liter is slightly larger than a quart.
1 meter = 1.09 yards.	1 gram = 1/454 pound.	1 liter = 1.06 quarts.
A baseball bat is about 1 meter long.	Two pennies, five paperclips, and a marble have masses of about 5, 2, and 5 grams, respectively.	Most beverages are now sold by the liter rather than by the quart.

Figure 2.2 Comparisons of the base metric system units of length (meter), mass (gram), and volume (liter) with common objects.

Metric Volume Units

▶ Volume is measured by determining the amount of space occupied by a three-dimensional object.

The **liter** (L) *is the base unit of volume in the metric system.* The abbreviation for liter is a capital L rather than a lower-case l because a lower-case l is easily confused with the number 1. A liter is a volume equal to that occupied by a cube that is 10 centimeters on each side. Because the volume of a cube is calculated by multiplying length times width times height (which are all the same for a cube), we have

$$\begin{aligned} 1 \text{ liter} &= \text{volume of a cube with edge 10 cm} \\ &= 10 \text{ cm} \times 10 \text{ cm} \times 10 \text{ cm} \\ &= 1000 \text{ cm}^3 \end{aligned}$$

A liter is also equal to 1000 milliliters; the prefix *milli-* means one-thousandth. Therefore,

$$1000 \text{ mL} = 1000 \text{ cm}^3$$

Dividing both sides of this equation by 1000 reveals that

$$1 \text{ mL} = 1 \text{ cm}^3$$

▶ Another abbreviation for the unit cubic centimeter, used in medical situations, is cc.
$1 \text{ cm}^3 = 1 \text{ cc}$

Consequently, the units mL and cm^3 are the same. In practice, mL is used for volumes of liquids and gases, cm^3 for volumes of solids. Figure 2.3 shows the relationship between 1 mL (1 cm^3) and its parent unit, the liter, in terms of cubic measurements.

A liter and a quart have approximately the same volume; 1 liter equals 1.06 quarts (Figure 2.2c). The milliliter and deciliter (dL) are commonly used in the laboratory. Deciliter units are routinely encountered in clinical laboratory reports detailing the composition of body fluids. A deciliter is equal to 100 mL (0.100 L).

Figure 2.3 A cube 10 cm on a side has a volume of 1000 cm^3, which is equal to 1 L. A cube 1 cm on a side has a volume of 1 cm^3, which is equal to 1 mL.

Total volume of large cube = 1000 cm^3 = 1 L

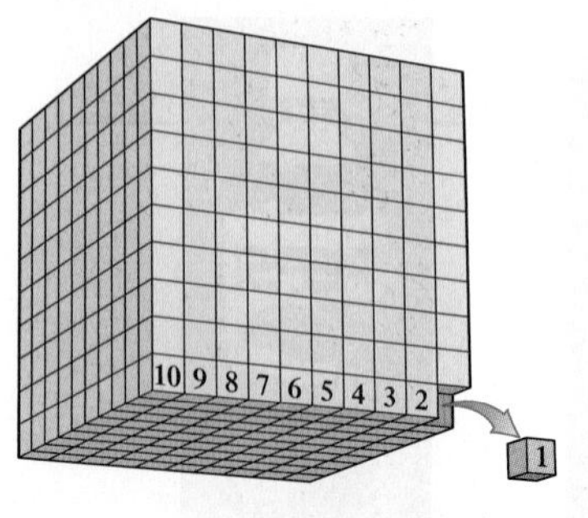

1 cm^3 = 1 mL

Example 2.1 Mass Units and Measurement

Complete the following table.

Type of measurement	Name of metric unit	Metric unit abbreviation
mass	________	kg
________	meter	________
________	cubic centimeter	________
________	________	mL

Solution

mass	kilogram	kg
length	meter	m
volume	cubic centimeter	cm^3
volume	milliliter	mL

Practice Questions and Problems

2.3 Write the name of the metric system prefix associated with each of the following mathematical meanings.
a. 10^3 b. 10^{-3} c. 10^{-6} d. 1/10

2.4 Provide the full unit name for each of the following abbreviations, and indicate what is being measured (time, mass, etc.)
a. cm b. kL c. mg d. ng

2.5 Arrange each of the following sets of units in order of increasing size (from smallest to largest).
a. Milligram, centigram, nanogram
b. Gigameter, megameter, kilometer
c. Microliter, deciliter, picoliter
d. Milligram, kilogram, microgram

Learning Focus

Classify a number as exact or inexact on the basis of the context of its use.

2.3 Exact and Inexact Numbers

In scientific work, numbers are grouped in two categories: *exact numbers* and *inexact numbers*. An **exact number** *has a value that has no uncertainty associated with it—that is, it is known exactly*. Exact numbers occur in definitions (for example, there are exactly 12 objects in a dozen, not 12.01 or 12.02); in counting (for example, there can be 7 people in a room, but never 6.99 or 7.03); and in simple fractions (for example, 1/3, 3/5, or 5/9).

An **inexact number** *has a value that has a degree of uncertainty associated with it*. Inexact numbers result anytime a measurement is made. It is impossible to make an *exact* measurement; some uncertainty will always be present. Flaws in construction of a measuring device, improper calibration of an instrument, and the skill (or lack of skill) of a person using a measuring device all contribute to error (uncertainty).

Practice Questions and Problems

2.6 Indicate whether the number in each of the following statements is an exact or an inexact number.
a. A classroom contains 32 chairs.
b. There are 60 seconds in a minute.
c. A bowl of cherries weighs 3.2 pounds.
d. A newspaper article contains 323 words.

2.7 Indicate whether each of the following quantities would involve an exact number or an inexact number.
a. The length of a swimming pool
b. The number of gummi bears in a bag
c. The number of inches in a foot
d. The surface area of a living room rug

2.8 A person is told that there are 12 inches in a foot and also that a piece of rope is 12 inches long. What is the fundamental difference between the value of 12 in these two pieces of information?

Learning Focus

Know the fundamental rule for recording measurements, and be able to determine the number of significant figures in a given measurement.

2.4 Uncertainty in Measurement and Significant Figures

As noted in the previous section, because of the limitations of the measuring device and the limited powers of observation of the individual making the measurement, every measurement carries a degree of uncertainty or error. Even when very elaborate measuring devices are used, some degree of uncertainty is always present.

Origin of Measurement Uncertainty

To illustrate how measurement uncertainty arises, let us consider how two different rulers, shown in Figure 2.4, are used to measure a given length. Using ruler A, we can say with certainty that the length of the rod is between 3 and 4 centimeters. We can further say that the actual length is closer to 4 centimeters and estimate it to be 3.7 centimeters. Ruler B has more subdivisions on its scale than ruler A. It is marked off in tenths of a centimeter instead of in centimeters. Using ruler B, we can definitely say that the length of the rod is between 3.7 and 3.8 centimeters and can estimate it to be 3.74 centimeters.

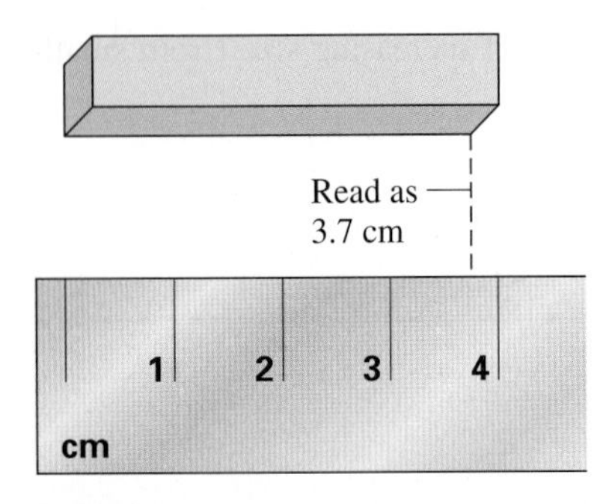

Ruler A

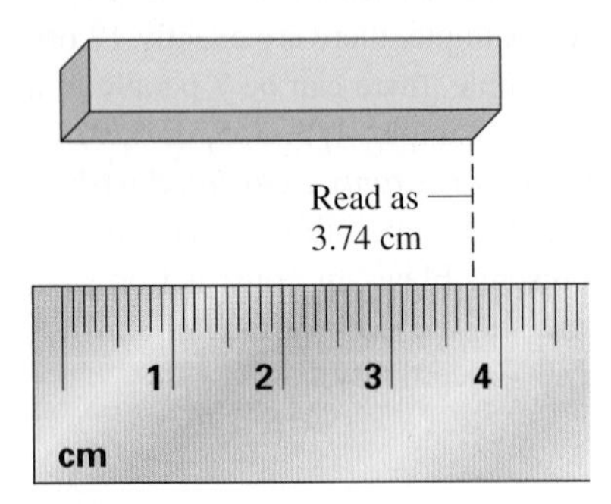

Ruler B

Figure 2.4 The scale on a measuring device determines the magnitude of the uncertainty for the recorded measurement. Measurements made with ruler A will have greater uncertainty than those made with ruler B.

Note how both length measurements (ruler A and ruler B) contain some digits (all those except the last one) that are exactly known and one digit (the last one) that is estimated. It is this last digit, the estimated one, that produces uncertainty in a measurement. Note also that the uncertainty in the second length measurement is less than that in the first one—an uncertainty in the hundredths place compared with an uncertainty in the tenths place. We say that the second measurement is more *precise* than the first one; that is, it has less uncertainty than the first measurement.

Only one estimated digit is ever recorded as part of a measurement. It would be incorrect for a scientist to report that the length of the metal rod in Figure 2.4 is 3.745 centimeters as read by using ruler B. The value 3.745 contains two estimated digits, the 4 and the 5, and indicates a measurement of precision greater than what is actually obtainable with that particular measuring device. Again, only one estimated digit is ever recorded as part of a measurement.

Because measurements are never exact, two types of information must be conveyed whenever a numerical value for a measurement is recorded: (1) the magnitude of the measurement and (2) the uncertainty of the measurement. The magnitude is indicated by the digit values. Uncertainty is indicated by the number of significant figures recorded. **Significant figures** *are the digits in any measurement that are known with certainty plus one digit that is uncertain.*

▶ The term *significant figures* is often verbalized in shortened form as "sig figs."

Guidelines for Determining Significant Figures

Recognizing the number of significant figures in a measured quantity is easy for measurements we make ourselves, because we know the type of instrument we are using and its limitations. However, when someone else makes the measurement, such information is often not available. In such cases, we follow a set of guidelines for determining the number of significant figures in a measured quantity.

1. In any measurement, all nonzero digits are significant.
2. *Zeros* may or may not be significant because zeros can be used in two ways: (1) to position a decimal point, and (2) to indicate a measured value. Zeros that perform the first function are not significant, and zeros that perform the second function are significant. When zeros are present in a measured number, we follow these rules:
   a. *Leading zeros,* those at the beginning of a number, are never significant.

      0.0141 has three significant figures.
      0.0000000048 has two significant figures.

   b. *Confined zeros,* those between nonzero digits, are always significant.

      3.063 has four significant figures.
      0.001004 has four significant figures.

   c. *Trailing zeros,* those at the end of a number, are significant if a decimal point is present in the number.

      56.00 has four significant figures.
      0.05050 has four significant figures.

   d. *Trailing zeros,* those at the end of a number, are not significant if the number lacks an explicitly shown decimal point.

      59,000,000 has two significant figures.
      6010 has three significant figures.

Practice Questions and Problems

2.9 Why are measured numbers restricted to a specific number of significant figures?

2.10 Indicate to what decimal position readings should be recorded (nearest 0.1, 0.01, etc.) for measurements made with the following devices.

a. A thermometer with a smallest scale marking of 1°C
b. A graduated cylinder with a smallest scale marking of 0.1 mL
c. A volumetric device with a smallest scale marking of 10 mL
d. A ruler with a smallest scale marking of 1 mm

2.11 Which is the estimated digit in each of the following numbers?
a. 2.31 cm b. 1.0 mL c. 25°C d. 1.23 g

2.12 Determine the number of significant figures in each of the following measured values.
a. 6.000 b. 0.0032 c. 0.01001 d. 65,400 e. 766.010 f. 0.03050

2.13 In which of the following pairs of numbers do both members of the pair contain the same number of significant figures?
a. 11.01 and 11.00 b. 2002 and 2020
c. 0.000066 and 660,000 d. 0.05700 and 0.05070

Learning Focus

Be able to adjust calculated answers obtained using measurements to the correct number of significant figures.

2.5 Significant Figures and Mathematical Operations

When measurements are added, subtracted, multiplied, or divided, consideration must be given to the number of significant figures in the computed result. Mathematical operations should not increase (or decrease) the precision of experimental measurements.

Hand-held electronic calculators generally "complicate" preciseness considerations because they are not programmed to take significant figures into account. Consequently, the digital readouts display more digits than are warranted (Figure 2.5). It is a mistake to record these extra digits, because they are not significant figures and hence are meaningless.

Figure 2.5 The digital readout on an electronic calculator usually shows more digits than are needed—and more than are acceptable. Calculators are not programmed to account for significant figures.

Rounding Off Numbers

When we obtain calculator answers that contain too many digits, it is necessary to drop the nonsignificant digits, a process that is called rounding off. **Rounding off** *is the process of deleting unwanted (nonsignificant) digits from calculated numbers.* There are two rules for rounding off numbers.

1. *If the first digit to be deleted is 4 or less, simply drop it and all the following digits.* For example, the number 3.724567 becomes 3.72 when rounded to three significant figures.
2. *If the first digit to be deleted is 5 or greater, that digit and all that follow are dropped and the last retained digit is increased by one.* The number 5.00673 becomes 5.01 when rounded to three significant figures.

Operational Rules

Significant-figure considerations in mathematical operations that involve measured numbers are governed by two rules, one for multiplication and division and one for addition and subtraction.

1. *In multiplication and division, the number of significant figures in the answer is the same as the number of significant figures in the measurement that contains the fewest significant figures.* For example,

Four significant figures (6.038) Three significant figures (2.57)

$$6.038 \times 2.57 = 15.51766 \quad \text{(calculator answer)}$$
$$= 15.5 \quad \text{(correct answer)}$$

Three significant figures (15.5)

The calculator answer is rounded to three significant figures because the measurement with the fewest significant figures (2.57) contains only three significant figures.

2. *In addition and subtraction, the answer has no more digits to the right of the decimal point than are found in the measurement with the fewest digits to the right of the decimal point.* For example,

$$\begin{array}{rl} 9.333 & \leftarrow \text{Uncertain digit (thousandths)} \\ +\ 1.4 & \leftarrow \text{Uncertain digit (tenths)} \\ \hline 10.733 & \text{(calculator answer)} \\ 10.7 & \text{(correct answer)} \\ \uparrow & \\ \text{Uncertain digit (tenths)} & \end{array}$$

The calculator answer is rounded to the tenths place because the uncertainty in the number 1.4 is in the tenths place.

▶ Concisely stated, the significant-figure operational rules are
× or ÷: Keep smallest number of significant figures in answer.
\+ or −: Keep smallest number of decimal places in answer.

Note the contrast between the rule for multiplication and division and the rule for addition and subtraction. In multiplication and division, significant figures are counted; in addition and subtraction, decimal places are counted. It is possible to gain or lose significant figures during addition or subtraction, but *never* during multiplication or division. In our previous sample addition problem, one of the input numbers (1.4) has two significant figures and the correct answer (10.7) has three significant figures. This is allowable in addition (and subtraction) because we are counting decimal places, not significant figures.

Example 2.2 Expressing Answers to the Proper Number of Significant Figures

Perform the following computations, expressing your answers to the proper number of significant figures.

a. 6.7321×0.0021 **b.** $\dfrac{16{,}340}{23.42}$ **c.** 6.000×4.000

d. $8.3 + 1.2 + 1.7$ **e.** $3.07 \times (17.6 - 13.73)$

Solution

a. The calculator answer to this problem is

$$6.7321 \times 0.0021 = 0.01413741$$

The input number with the least number of significant figures is 0.0021.

$$\underset{\text{Five significant figures}}{6.7321} \times \underset{\text{Two significant figures}}{0.0021}$$

Thus the calculator answer must be rounded to two significant figures.

$$\underset{\text{Calculator answer}}{0.01413741} \quad \text{becomes} \quad \underset{\text{Correct answer}}{0.014}$$

b. The calculator answer to this problem is

$$\frac{16{,}340}{23.42} = 697.69427$$

Both input numbers contain four significant figures. Thus the correct answer will also contain four significant figures.

$$\underset{\text{Calculator answer}}{697.69427} \quad \text{becomes} \quad \underset{\text{Correct answer}}{697.7}$$

c. The calculator answer to this problem is

$$6.000 \times 4.000 = 24$$

Both input numbers contain four significant figures. Thus the correct answer must also contain four significant figures:

24	becomes	24.00
Calculator answer		Correct answer

Note that here the calculator answer had too few significant figures. Most calculators cut off zeros after the decimal point even if these zeros are significant. Using too few significant figures in an answer is just as wrong as using too many.

d. The calculator answer to this problem is

$$8.3 + 1.2 + 1.7 = 11.2$$

All three input numbers have uncertainty in the tenths place. Thus the last retained digit in the correct answer will be that of tenths. (In this particular problem, the calculator answer and the correct answer are the same, a situation that does not occur very often.)

e. This problem involves the use of both multiplication and subtraction significant-figure rules. We do the subtraction first.

$$17.6 - 13.73 = 3.87 \quad \text{(calculator answer)}$$
$$= 3.9 \quad \text{(correct answer)}$$

This answer must be rounded to tenths because the input number 17.6 involves only tenths. We now do the multiplication.

$$3.07 \times 3.9 = 11.973 \quad \text{(calculator answer)}$$
$$= 12 \quad \text{(correct answer)}$$

The number 3.9 limits the answer to two significant figures.

Because exact numbers (Section 2.3) have no uncertainty associated with them, they possess an unlimited number of significant figures. Therefore, such numbers never limit the number of significant figures in a computational answer.

The Chemistry at a Glance on the following page summarizes the rules that govern which digits in a measurement are significant.

Practice Questions and Problems

2.14 Round off each of the following numbers to the number of significant figures indicated in parentheses.
a. 0.350763 (three) b. 653,899 (four)
c. 22.55555 (five) d. 0.277654 (four)

2.15 Without actually solving the problem, indicate the number of significant figures that should be present in the answers to the following multiplication and division problems.
a. $10.300 \times 0.30 \times 0.300$ b. $3300 \times 3330 \times 333.0$ c. $\frac{6.0}{33.0}$ d. $\frac{6.000}{33}$

2.16 Carry out the following multiplications and divisions, expressing your answer to the correct number of significant figures. Assume that all numbers are measured numbers.
a. $2.0000 \times 2.00 \times 0.0020$ b. 4.1567×0.00345
c. $\frac{533{,}000}{465{,}300}$ d. $\frac{4.670 \times 3.00}{2.450}$

2.17 Carry out the following additions and subtractions, expressing your answer to the correct number of significant figures.
a. $12 + 23 + 127$ b. $3.111 + 3.11 + 3.1$
c. $1237.6 + 23 + 0.12$ d. $43.65 - 23.7$

Chemistry *at a Glance*

Significant Figures

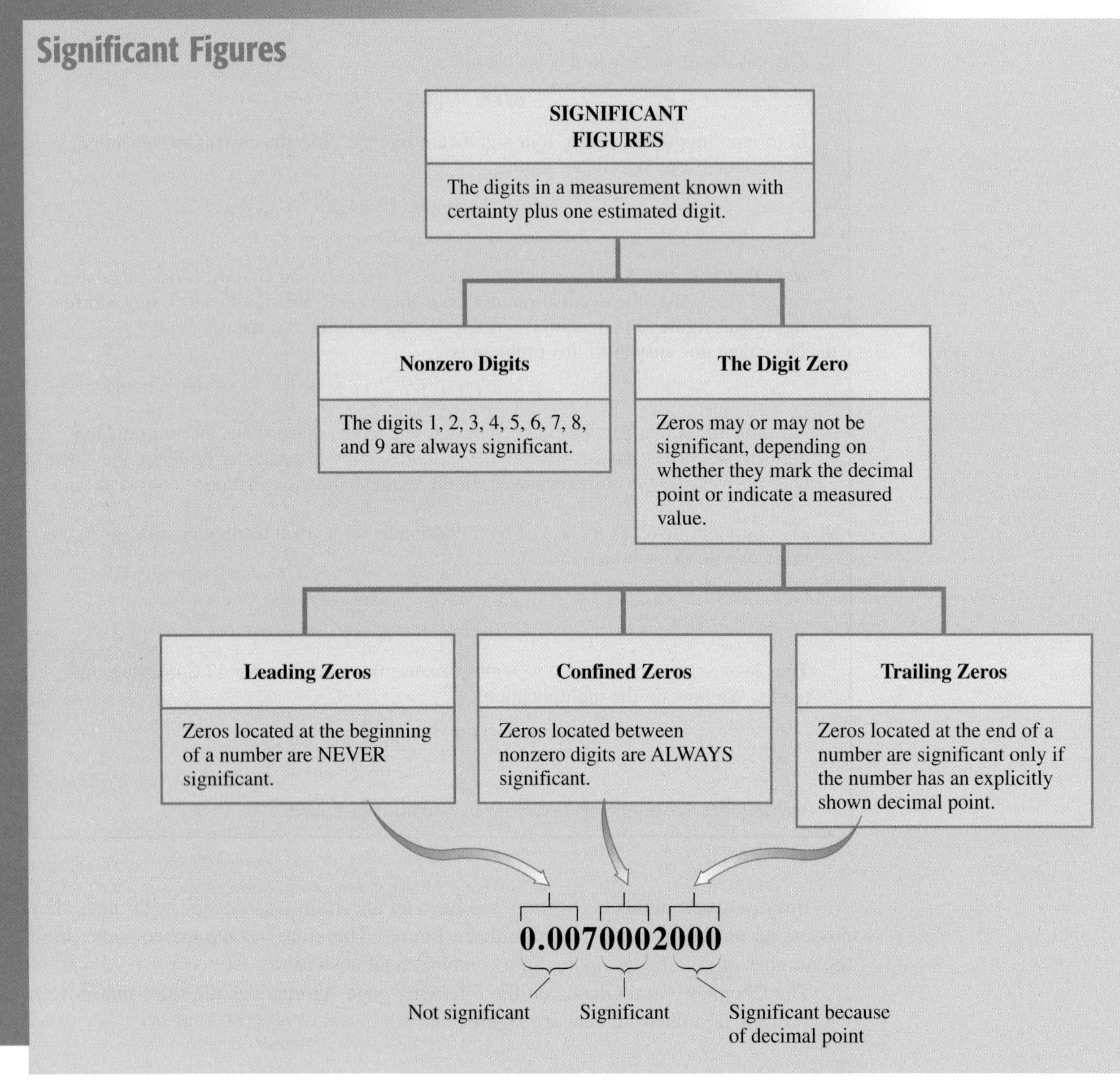

SIGNIFICANT FIGURES
The digits in a measurement known with certainty plus one estimated digit.
Nonzero Digits
The digits 1, 2, 3, 4, 5, 6, 7, 8, and 9 are always significant.
The Digit Zero
Zeros may or may not be significant, depending on whether they mark the decimal point or indicate a measured value.
Leading Zeros
Zeros located at the beginning of a number are NEVER significant.
Confined Zeros
Zeros located between nonzero digits are ALWAYS significant.
Trailing Zeros
Zeros located at the end of a number are significant only if the number has an explicitly shown decimal point.
0.0070002000
Not significant
Significant
Significant because of decimal point

Learning Focus

Be able to convert numbers from decimal notation to scientific notation.

2.6 Scientific Notation

Up to this point in the chapter, we have expressed all numbers in decimal notation, the everyday method for expressing numbers. Such notation becomes cumbersome for very large and very small numbers (which occur frequently in scientific work). For example, in one drop of blood, which is 92% water by mass, there are approximately

1,600,000,000,000,000,000,000 molecules

of water, each of which has a mass of

0.000000000000000000000030 gram

Recording such large and small numbers is not only time-consuming but also open to error; often, too many or too few zeros are recorded. Also, it is impossible to multiply or divide such numbers with most calculators because they can't accept that many digits. (Most calculators accept either 8 or 10 digits.)

▶ Scientific notation is also called exponential notation.

A method called *scientific notation* exists for expressing multidigit numbers involving many zeros in compact form. **Scientific notation** *is a system in which an ordinary decimal number is expressed as the product of a number between 1 and 10 and 10 raised to a power.* The ordinary decimal number is called a *coefficient* and is written first. The number 10 raised to a power is called an *exponential term.* The coefficient is always multiplied by the exponential term.

Coefficient — Exponent

$$1.07 \times 10^{4}$$

Multiplication sign — Exponential term

The two previously cited numbers that deal with molecules of water are expressed in scientific notation as

$$1.6 \times 10^{21} \text{ molecules}$$

and

$$3.0 \times 10^{-22} \text{ gram}$$

Obviously, scientific notation is a much more concise way of expressing numbers. Such scientific notation is compatible with most calculators.

Converting from Decimal to Scientific Notation

The procedure for converting a number from decimal notation to scientific notation has two parts.

1. *The decimal point in the decimal number is moved to the position behind the first nonzero digit.*
2. *The exponent for the exponential term is equal to the number of places the decimal point has been moved.* The exponent is positive if the original decimal number is 10 or greater and is negative if the original decimal number is less than 1. For numbers between 1 and 10, the exponent is zero.

The following two examples illustrate the use of these procedures:

$$93{,}000{,}000 = 9.3 \times 10^{7}$$

Decimal point is moved 7 places

$$0.0000037 = 3.7 \times 10^{-6}$$

Decimal point is moved 6 places

Significant Figures and Scientific Notation

How do significant-figure considerations affect scientific notation? The answer is simple. *Only significant figures become part of the coefficient.* The numbers 63, 63.0, and 63.00, which respectively have two, three, and four significant figures, when converted to scientific notation become, respectively,

▶ The decimal and scientific notation forms of a number *always* contain the same number of significant figures.

6.3×10^{1} (two significant figures)

6.30×10^{1} (three significant figures)

6.300×10^{1} (four significant figures)

Practice Questions and Problems

2.18 Express the following numbers in scientific notation.
a. 120.7 b. 0.0034 c. 231.00 d. 23,000

2.19 Which number in each pair of numbers is the larger of the two?
a. 1.0×10^{-3} or 1.0×10^{-6} b. 1.0×10^{3} or 1.0×10^{-2}
c. 6.3×10^{4} or 2.3×10^{4} d. 6.3×10^{-4} or 1.2×10^{-4}

2.20 How many significant figures are present in each of the following measured numbers?
a. 1.0×10^{2} b. 5.34×10^{5} c. 5.34×10^{-4} d. 6.000×10^{3}

Learning Focus

Understand what conversion factors are, and be able to use them and dimensional analysis to change from one unit to another.

2.7 Conversion Factors and Dimensional Analysis

With both the English unit and metric unit systems in common use in the United States, we often must change measurements from one system to their equivalent in the other system. The mathematical tool we use to accomplish this task is a general method of problem solving called *dimensional analysis*. Central to the use of dimensional analysis is the concept of conversion factors. A **conversion factor** *is a ratio that specifies how one unit of measurement is related to another.*

Conversion factors are derived from equations (equalities) that relate units. Consider the quantities "1 minute" and "60 seconds," both of which describe the same amount of time. We may write an equation describing this fact.

$$1 \text{ min} = 60 \text{ sec}$$

This fixed relationship is the basis for the construction of a pair of conversion factors that relate seconds and minutes.

$$\frac{1 \text{ min}}{60 \text{ sec}} \quad \text{and} \quad \frac{60 \text{ sec}}{1 \text{ min}}$$

These two quantities are the same.

Note that conversion factors always come in pairs, one member of the pair being the reciprocal of the other. Also note that the numerator and the denominator of a conversion factor always describe the same amount of whatever we are considering. One minute and 60 seconds denote the same amount of time.

Conversion Factors Within a System of Units

Most students are familiar with and have memorized numerous conversion factors within the English system of measurement (English-to-English conversion factors). Some of these factors, with only one member of a conversion factor pair being listed, are

$$\frac{12 \text{ in.}}{1 \text{ ft}} \quad \frac{3 \text{ ft}}{1 \text{ yd}} \quad \frac{4 \text{ qt}}{1 \text{ gal}} \quad \frac{16 \text{ oz}}{1 \text{ lb}}$$

▶ In order to avoid confusion with the word *in*, the abbreviation for inches, in., includes a period. This is the only unit abbreviation in which a period appears.

Such conversion factors contain an unlimited number of significant figures because the numbers within them arise from definitions.

Metric-to-metric conversion factors are similar to English-to-English conversion factors in that they arise from definitions. Individual conversion factors are derived from the meanings of the metric system prefixes (Section 2.2). For example, the set of conversion factors involving kilometer and meter come from the equality

$$1 \text{ kilometer} = 10^3 \text{ meters}$$

and those relating microgram and gram come from the equality

$$1 \text{ microgram} = 10^{-6} \text{ gram}$$

▶ In order to obtain metric-to-metric conversion factors, you need to know the meaning of the metric system prefixes in terms of powers of 10 (see Table 2.1).

The two pairs of conversion factors are

$$\frac{10^3 \text{ m}}{1 \text{ km}} \quad \text{and} \quad \frac{1 \text{ km}}{10^3 \text{ m}} \qquad \frac{1\ \mu\text{g}}{10^{-6} \text{ g}} \quad \text{and} \quad \frac{10^{-6} \text{ g}}{1\ \mu\text{g}}$$

Note that the numerical equivalent of the prefix is always associated with the base (unprefixed) unit in a metric-to-metric conversion factor.

The number 1 always goes with the *prefixed* unit.

$$\frac{1 \text{ mL}}{10^{-3} \text{ L}}$$

The power of 10 always goes with the *unprefixed* unit.

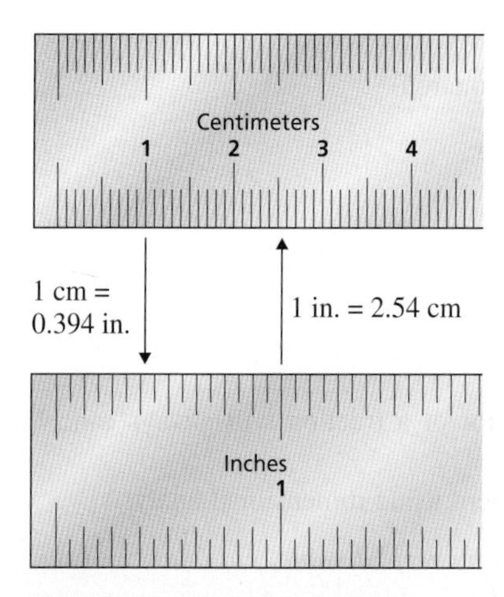

Figure 2.6 It is experimentally determined that 1 inch equals 2.54 centimeters or 1 centimeter equals 0.394 inch.

Conversion Factors Between Systems of Units

Conversion factors that relate metric units to English units and vice versa are not exact defined quantities because they involve two different systems of measurement. The numbers associated with these conversion factors must be determined experimentally (see Figure 2.6). Table 2.2 lists commonly encountered relationships between metric system and English system units. These few conversion factors are sufficient to solve most of the problems that we will encounter.

Metric-to-English conversion factors can be specified to differing numbers of significant figures. For example,

$$1.00 \text{ lb} = 454 \text{ g}$$

$$1.000 \text{ lb} = 453.6 \text{ g}$$

$$1.0000 \text{ lb} = 453.59 \text{ g}$$

In a problem-solving context, which "version" of a conversion factor is used depends on how many significant figures there are in the other numbers of the problem. Conversion factors should never limit the number of significant figures in the answer to a problem. The conversion factors in Table 2.2 are given to three significant figures, which is sufficient for the applications we will make of them.

Dimensional Analysis

Dimensional analysis *is a general problem-solving method in which the units associated with numbers are used as a guide in setting up calculations.* In this method, units are treated in the same way as numbers; that is, they can be multiplied, divided, or canceled. For example, just as

$$5 \times 5 = 5^2 \qquad \text{(5 squared)}$$

Table 2.2
Equalities and Conversion Factors That Relate the English and Metric Systems of Measurement

	Metric to English	English to Metric
Length		
1.00 inch = 2.54 centimeters	$\frac{1.00 \text{ in.}}{2.54 \text{ cm}}$	$\frac{2.54 \text{ cm}}{1.00 \text{ in.}}$
1.00 meter = 39.4 inches	$\frac{39.4 \text{ in.}}{1.00 \text{ m}}$	$\frac{1.00 \text{ m}}{39.4 \text{ in.}}$
1.00 kilometer = 0.621 mile	$\frac{0.621 \text{ mi}}{1.00 \text{ km}}$	$\frac{1.00 \text{ km}}{0.621 \text{ mi}}$
Mass		
1.00 pound = 454 grams	$\frac{1.00 \text{ lb}}{454 \text{ g}}$	$\frac{454 \text{ g}}{1.00 \text{ lb}}$
1.00 kilogram = 2.20 pounds	$\frac{2.20 \text{ lb}}{1.00 \text{ kg}}$	$\frac{1.00 \text{ kg}}{2.20 \text{ lb}}$
1.00 ounce = 28.3 grams	$\frac{1.00 \text{ oz}}{28.3 \text{ g}}$	$\frac{28.3 \text{ g}}{1.00 \text{ oz}}$
Volume		
1.00 quart = 0.946 liter	$\frac{1.00 \text{ qt}}{0.946 \text{ L}}$	$\frac{0.946 \text{ L}}{1.00 \text{ qt}}$
1.00 liter = 0.265 gallon	$\frac{0.265 \text{ gal}}{1.00 \text{ L}}$	$\frac{1.00 \text{ L}}{0.265 \text{ gal}}$
1.00 milliliter = 0.034 fluid ounce	$\frac{0.034 \text{ fl oz}}{1.00 \text{ mL}}$	$\frac{1.00 \text{ mL}}{0.034 \text{ fl oz}}$

we have

$$\text{cm} \times \text{cm} = \text{cm}^2 \quad \text{(cm squared)}$$

Also, just as the 3s cancel in the expression

$$\frac{\cancel{3} \times 5 \times 7}{\cancel{3} \times 2}$$

the centimeters cancel in the expression

$$\frac{\cancel{(\text{cm})} \times (\text{in.})}{\cancel{(\text{cm})}}$$

"Like units" found in the numerator and denominator of a fraction will always cancel, just as like numbers do.

The following steps show how to set up a problem using dimensional analysis.

Step 1: *Identify the known or given quantity (both numerical value and units) and the units of the new quantity to be determined.*

This information will always be found in the statement of the problem. Write an equation with the given quantity on the left and the units of the desired quantity on the right.

Step 2: *Multiply the given quantity by one or more conversion factors in such a manner that the unwanted (original) units are canceled, leaving only the desired units.*

The general format for the multiplication is

$$\text{(Information given)} \times \text{(conversion factors)} = \text{(information sought)}$$

The number of conversion factors depends on the individual problem.

Step 3: *Perform the mathematical operations indicated by the conversion factor setup.*

When performing the calculation, double-check to make sure that all units except the desired set have canceled.

Example 2.3 Unit Conversions Within the Metric System

A standard aspirin tablet contains 324 mg of aspirin. How many grams of aspirin are in a standard aspirin tablet?

Solution

Step 1: The given quantity is 324 mg, the mass of aspirin in the tablet. The unit of the desired quantity is grams.

$$324 \text{ mg} = ? \text{ g}$$

Step 2: Only one conversion factor will be needed to convert from milligrams to grams, one that relates milligrams to grams. The two forms of this conversion factor are

$$\frac{1 \text{ mg}}{10^{-3} \text{ g}} \quad \text{and} \quad \frac{10^{-3} \text{ g}}{1 \text{ mg}}$$

The second factor is used because it allows for cancellation of the milligram units, leaving us with grams as the new units.

$$324 \cancel{\text{mg}} \times \left(\frac{10^{-3} \text{ g}}{1 \cancel{\text{mg}}}\right) = ? \text{ g}$$

Step 3: Combining numerical terms as indicated generates the final answer.

$$\left(324 \times \frac{10^{-3}}{1}\right) \text{g} = 0.324 \text{ g}$$

Number from first factor — Numbers from second factor

Note that 10^{-3} is equal to 0.001 and that 324 times 0.001 is equal to 0.324.

The answer is given to three significant figures because the given quantity in the problem, 324 mg, has three significant figures. The conversion factor used arises from a definition and thus does not limit significant figures in any way.

Example 2.4 **Unit Conversions Between the Metric and English Systems**

Capillaries, the microscopic vessels that carry blood from small arteries to small veins, are on the average only 1 mm long. What is the average length of a capillary in inches?

Solution

Step 1: The given quantity is 1 mm, and the units of the desired quantity are inches.

$$1 \text{ mm} = ? \text{ in.}$$

Step 2: The conversion factor needed for a one-step solution, millimeters to inches, is not given in Table 2.2. However, a related conversion factor, meters to inches, is given. Therefore, we first convert millimeters to meters and then use the meters-to-inches conversion factor in Table 2.2.

$$\text{mm} \longrightarrow \text{m} \longrightarrow \text{in.}$$

The correct conversion factor setup is

$$1 \cancel{\text{mm}} \times \left(\frac{10^{-3} \cancel{\text{m}}}{1 \cancel{\text{mm}}}\right) \times \left(\frac{39.4 \text{ in.}}{1.00 \cancel{\text{m}}}\right) = ? \text{ in.}$$

All of the units except for inches cancel, which is what is needed. The information for the middle conversion factor was obtained from the meaning of the prefix *milli-*.

This setup illustrates the fact that sometimes the given units must be changed to intermediate units before common conversion factors, such as those found in Table 2.2, are applicable.

Step 3: Collecting the numerical factors and performing the indicated math gives

$$\left(\frac{1 \times 10^{-3} \times 39.4}{1 \times 1.00}\right) \text{in.} = 0.0394 \text{ in.} \quad \text{(calculator answer)}$$

$$= 0.04 \text{ in.} \quad \text{(correct answer)}$$

The calculator answer must be rounded to one significant figure because 1 mm, the given quantity, contains only one significant figure.

The next Chemistry at a Glance reviews what we have learned about conversion factors.

Practice Questions and Problems

2.21 Give both forms of the conversion factor that you would use to relate the following sets of units to each other.

a. Gram and kilogram
b. Meter and nanometer
c. Liter and milliliter
d. Centigram and gram

Chemistry *at a Glance*

Conversion Factors

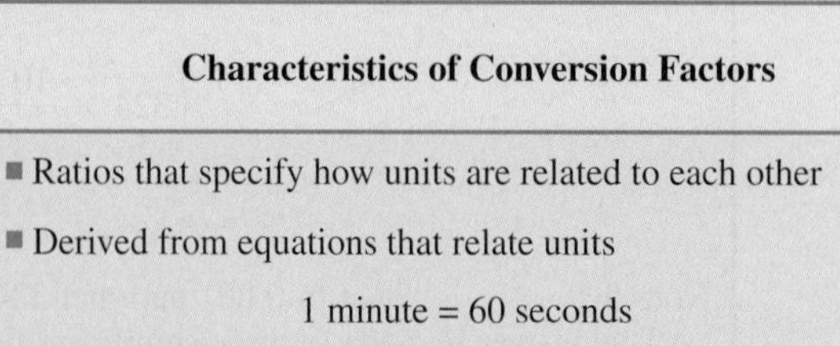

Characteristics of Conversion Factors

- Ratios that specify how units are related to each other
- Derived from equations that relate units

$$1 \text{ minute} = 60 \text{ seconds}$$

- Come in pairs, one member of the pair being the reciprocal of the other

$$\frac{1 \text{ min}}{60 \text{ sec}} \quad \text{and} \quad \frac{60 \text{ sec}}{1 \text{ min}}$$

- Conversion factors originate from two types of relationships: (1) defined relationships and (2) measured relationships

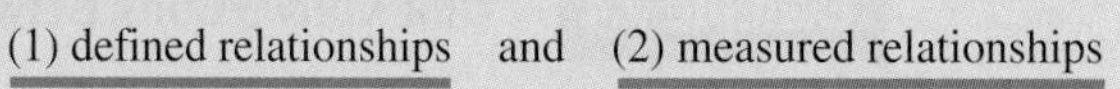

Conversion Factors from DEFINED Relationships

- All English-to-English and metric-to-metric conversion factors
- Such conversion factors have an unlimited number of significant figures

12 inches = 1 foot (exactly)
4 quarts = 1 gallon (exactly)
1 kilogram = 10^3 grams (exactly)

- Metric-to-metric conversion factors are derived using the meaning of the metric system prefixes

Prefixes that INCREASE Base Unit Size

kilo- 10^3
mega- 10^6
giga- 10^9

Prefixes that DECREASE Base Unit Size

deci- 10^{-1}
centi- 10^{-2}
milli- 10^{-3}
micro- 10^{-6}
nano- 10^{-9}

Conversion Factors from MEASURED Relationships

- All English-to-metric and metric-to-English conversion factors
- Such conversion factors have a specific number of significant figures, depending on the precision of the defining relationship

1.00 lb = 454 g (three sig figs)
1.000 lb = 453.6 g (four sig figs)
1.0000 lb = 453.59 g (five sig figs)

2.22 Give both forms of the conversion factor that you would use to relate the following sets of units to each other.
a. Inch and centimeter b. Gram and pound
c. Liter and quart d. Meter and inches

2.23 Write the equality from which each of the following conversion factors is obtained.
a. 60 sec/1 min b. 12 in./1 ft c. 2.54 cm/1.00 in. d. 454 g/1.00 lb

2.24 Using dimensional analysis and a single conversion factor, solve each of the following problems.
a. 160,000 centimeters = ? meters

b. 24 nanometers = ? meters
c. 0.0030 kilometer = ? meters
d. 3.00 millimeters = ? meters

2.25 Using dimensional analysis and a single conversion factor, solve each of the following problems.
a. 6.4 grams = ? pounds
b. 6.4 pounds = ? grams
c. 53 centimeters = ? inches
d. 3.5 quarts = ? liters

2.26 Using dimensional analysis and two conversion factors, solve each of the following problems.
a. 2.0 meters = ? feet
b. 3.2 yards = ? centimeters
c. 3.2 feet = ? meters
d. 4.3 centimeters = ? feet

2.27 The human stomach produces approximately 2500 mL of gastric juice per day. What is the volume, in liters, of gastric juice produced?

2.28 The mass of premature babies is often determined in grams. If a premature baby weighs 1550 g, what is its mass in pounds?

Learning Focus

Calculate the density of a substance, and be able to use density as a conversion factor to calculate the mass or volume of a substance.

2.8 Density

Density *is the ratio of the mass of an object to the volume occupied by that object.*

$$\text{Density} = \frac{\text{mass}}{\text{volume}}$$

People often speak of a substance as being heavier or lighter than another substance. What they actually mean is that the two substances have different densities; a specific volume of one substance is heavier or lighter than the same volume of the second substance (Figure 2.7). Equal masses of substances with different densities will occupy different volumes; the volume contrast is often very striking (see Figure 2.8).

(a)

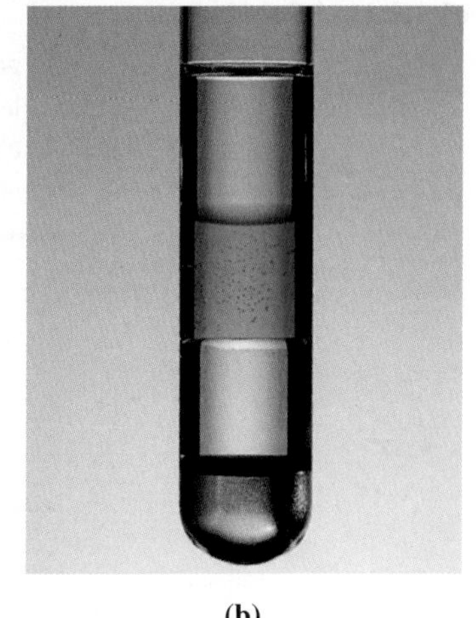

(b)

Figure 2.7 (a) The penny is less dense than the mercury it floats on. (b) Liquids that do not dissolve in one another and that have different densities float on one another, forming layers. The top layer is gasoline, with a density of about 0.8 g/mL. Next is water (plus food coloring), with a density of 1.0 g/mL. The next layer is carbon tetrachloride, with a density of 1.6 g/mL. The bottom layer is mercury, with a density of 13.6 g/mL.

Figure 2.8 Both of these items have a mass of 23 grams, but very different volumes. The volume differences result from the two items having different densities.

A correct density expression includes a number, a mass unit, and a volume unit. Although any mass and volume units can be used, densities are usually expressed in grams per cubic centimeter (g/cm^3) for solids, grams per milliliter (g/mL) for liquids, and grams per liter (g/L) for gases. Table 2.3 gives density values for a number of substances. Note that temperature must be specified with density values, because substances expand and contract with changes in temperature. For the same reason, the pressure of gases is also given with their density values.

Example 2.5 Calculating Density

A student determines that the mass of a 20.0-mL sample of olive oil is 18.4 g. What is the density of the olive oil in grams per milliliter?

Solution

To calculate density, we substitute the given mass and volume values into the defining formula for density.

$$\text{Density} = \frac{\text{mass}}{\text{volume}} = \frac{18.4\ \text{g}}{20.0\ \text{mL}} = \underset{\text{Calculator answer}}{0.92\ \frac{\text{g}}{\text{mL}}} = \underset{\text{Correct answer}}{0.920\ \frac{\text{g}}{\text{mL}}}$$

Because both input numbers contain three significant figures, the density is specified to three significant figures.

▶ Density may be used as a conversion factor to convert from mass to volume or vice versa.

Density can be used as a conversion factor that relates the volume of a substance to its mass. This use of density enables us to calculate the volume of a substance if we know its mass. Conversely, the mass can be calculated if the volume is known.

Density conversion factors, like all other conversion factors, have two reciprocal forms. For a density of 1.03 g/mL, the two conversion factor forms are

$$\frac{1.03\ \text{g}}{1\ \text{mL}} \quad \text{and} \quad \frac{1\ \text{mL}}{1.03\ \text{g}}$$

Table 2.3
Densities of Selected Substances

Solids (25°C)			
gold	19.3 g/cm^3	table salt	2.16 g/cm^3
lead	11.3 g/cm^3	bone	1.7–2.0 g/cm^3
copper	8.93 g/cm^3	table sugar	1.59 g/cm^3
aluminum	2.70 g/cm^3	wood, pine	0.30–0.50 g/cm^3
Liquids (25°C)			
mercury	13.55 g/mL	water	0.997 g/mL
milk	1.028–1.035 g/mL	olive oil	0.92 g/mL
blood plasma	1.027 g/mL	ethyl alcohol	0.79 g/mL
urine	1.003–1.030 g/mL	gasoline	0.56 g/mL
Gases (25°C and 1 atmosphere pressure)			
chlorine	3.17 g/L	nitrogen	1.25 g/L
carbon dioxide	1.96 g/L	methane	0.66 g/L
oxygen	1.42 g/L	hydrogen	0.08 g/L
air (dry)	1.29 g/L		

Example 2.6 **Converting from Mass to Volume by Using Density as a Conversion Factor**

Blood plasma has a density of 1.027 g/mL at 25°C. What volume, in milliliters, does 125 g of plasma occupy?

Solution

Step 1: The given quantity is 125 g of blood plasma. The units of the desired quantity are milliliters. Thus our starting point is

$$125 \text{ g} = ? \text{ mL}$$

Step 2: The conversion from grams to milliliters can be accomplished in one step because the given density, used as a conversion factor, directly relates grams to milliliters. Of the two conversion factor forms

$$\frac{1.027 \text{ g}}{1 \text{ mL}} \quad \text{and} \quad \frac{1 \text{ mL}}{1.027 \text{ g}}$$

we will use the latter, because it allows for cancellation of gram units, leaving milliliters.

$$125 \cancel{\text{g}} \times \left(\frac{1 \text{ mL}}{1.027 \cancel{\text{g}}}\right) = ? \text{ mL}$$

Step 3: Doing the necessary arithmetic gives us our answer:

$$\left(\frac{125 \times 1}{1.027}\right) \text{mL} = 121.71372 \text{ mL} \quad \text{(calculator answer)}$$

$$= 122 \text{ mL} \quad \text{(correct answer)}$$

Even though the given density contained four significant figures, the correct answer is limited to three significant figures. This is because the other given number, the mass of blood plasma, had only three significant figures.

At room temperature and pressure, most of the naturally occurring elements—over 85% of them—are solids. Only 2 elements are liquids: Br and Hg. There are 10 gaseous elements: H, He, N, O, F, Ne, Cl, Ar, Kr, and Xe. A consideration of element densities shows that osmium is the most dense of the "solid-state" elements, mercury the most dense of the "liquid-state" elements, and xenon the most dense of the "gaseous-state" elements. Chemical Portraits 3 profiles these three "most dense" elements.

Practice Questions and Problems

2.29 A sample of mercury is found to have a mass of 524.5 g and to have a volume of 38.72 cm^3. What is its density in grams per cubic centimeter?

2.30 Acetone, the solvent in nail polish remover, has a density of 0.791 g/mL. What is the volume, in milliliters, of 20.0 g of acetone?

2.31 Nickel metal has a density of 8.90 g/cm^3. How much, in grams, does 15 cm^3 of nickel metal weigh?

Chemical Portraits 3

Solid, Liquid, and Gaseous "Most Dense" Elements

Osmium (Os)

Profile: With a density of 22.6 g/cm^3, osmium is the densest natural element. This hard, brittle, bluish-white element is used in small amounts to harden metal alloys used for items such as phonograph needles, fountain pen tips, and electrical contacts.

Biochemical considerations: In bulk form, pure Os is unaffected by air or water. However, powdered Os slowly reacts with air to produce OsO_4, a strong smelling volatile compound capable of causing lung and skin problems. This "OsO_4 problem" makes the use of pure Os metal impractical. Its use as an alloying agent is, however, safe. The name *osmium* comes from the Greek word "osme" which means "smell."

The metal osmium is obtained as a "by-product" from the processing of what types of ores?

Mercury (Hg)

Profile: With a density of 13.6 g/cm^3, mercury is the densest "liquid-state" element. Its chief source is the ore *cinnabar* (HgS). A semi-solid mixture of silver, tin, and mercury, which hardens upon standing, has been used widely to make dental fillings. The hardening process produces the compounds Ag_5Hg_8 and Sn_7Hg_8, nontoxic forms of mercury.

Biochemical considerations: When Hg is "spilled" from a broken thermometer, it forms little balls that roll around without adhering to anything. Such spills represent a danger since Hg vapor originating from the "balls" is toxic when inhaled. Spill clean-up involves sprinkling the area with sufur, to form HgS, a nonvolatile Hg-compound that can be picked up with a vacuum system.

What is the antidote for mercury poisoning associated with ingestion of mercury-containing substances?

Xenon (Xe)

Profile: With a density of 5.89 g/cm^3—approximately 5 times that of air—Xe is the densest "gaseous-state" element. The least abundant of the natural elements, Xe occurs in trace amounts in the atmosphere, which is its sole source. Air liquefaction is required for its isolation. Chemically, Xe is an extremely *unreactive* element, reacting only with fluorine, the most reactive of all elements.

Uses: Halogen automobile headlights contain Xe and it is also used in high-intensity photographic flash tubes. The passing of an electric charge through Xe gas produces a brilliant white "lightning-like" flash. Xe in a vacuum tube produces a beautiful blue glow when excited by an electric discharge.

Why are Xe-lamps more economical to operate than standard light bulbs?

See the text web site at **www.cengage.com/chemistry/stoker** for answers to the above questions and for further information.

Learning Focus

List and use the relationships among the Fahrenheit, Celsius, and Kelvin temperature scales, and list the relationships among the heat energy units joules, calories, and Calories.

2.9 Temperature Scales and Heat Energy

Heat is a form of energy. Temperature is an indicator of the tendency of heat energy to be transferred. Heat energy flows from objects of higher temperature to objects of lower temperature.

Three different temperature scales are in common use: Celsius, Kelvin, and Fahrenheit (Figure 2.9). Both the Celsius and the Kelvin scales are part of the metric measurement system; the Fahrenheit scale belongs to the English measurement system. Degrees of different size and different reference points are what produce the various temperature scales.

The *Celsius scale* is the scale most commonly encountered in scientific work. The normal boiling and freezing points of water serve as reference points on this scale, the former having a value of 100° and the latter 0°. Thus there are 100 "degree intervals" between the two reference points.

The *Kelvin scale* is a close relative of the Celsius scale. Both have the same size of degree, and the number of degrees between the freezing and boiling points of water is the same. The two scales differ only in the numbers assigned to the reference points. On the Kelvin scale, the boiling point of water is 373 kelvins (K) and the freezing point of water is 273 K. The choice of these reference points makes all temperature readings on the Kelvin scale positive values. Note that the degree sign (°) is not used with the Kelvin scale. For example, we say that an object has a temperature of 350 K (*not* 350°K).

▶ Zero on the Kelvin scale is known as *absolute zero*. It corresponds to the lowest temperature allowed by nature. How fast particles (molecules) move depends on temperature. The colder it gets, the more slowly they move. At absolute zero, movement stops. Scientists in laboratories have been able to attain temperatures as low as 0.0001 K, but a temperature of 0 K is impossible.

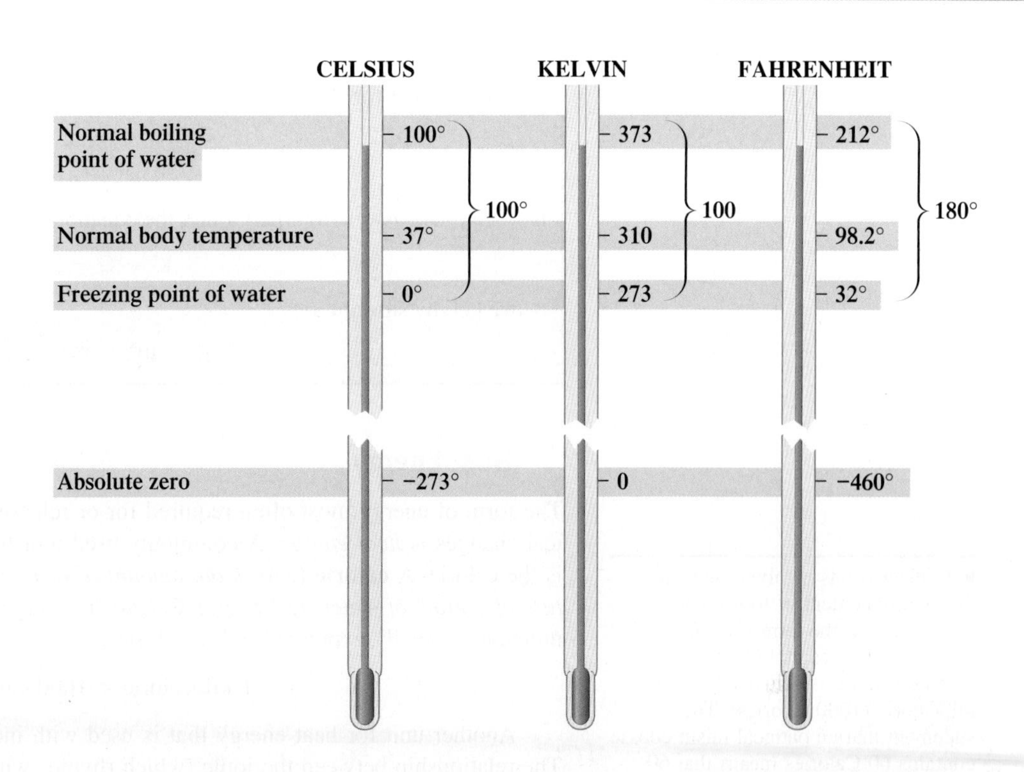

Figure 2.9 The relationships among the Celsius, Kelvin, and Fahrenheit temperature scales are determined by the degree sizes and the reference point values.

The *Fahrenheit scale* has a smaller degree size than the other two temperature scales. On this scale, there are 180 degrees between the freezing and boiling points of water as contrasted to 100 degrees on the other two scales. Thus the Celsius (and Kelvin) degree size is almost two times ($\frac{9}{5}$) larger than the Fahrenheit degree. Reference points on the Fahrenheit scale are 32° for the freezing point of water and 212° for the normal boiling point of water.

Conversions Between Temperature Scales

Because the size of the degree is the same, the relationship between the Kelvin and Celsius scales is very simple. No conversion factors are needed; all that is required is an adjustment for the differing numerical scale values. The adjustment factor is 273, the number of degrees by which the two scales are offset from one another.

$$\text{K} = {}^\circ\text{C} + 273$$
$$^\circ\text{C} = \text{K} - 273$$

The relationship between the Fahrenheit and Celsius scales can also be stated in an equation format.

$$^\circ\text{F} = \frac{9}{5}(^\circ\text{C}) + 32 \quad \text{or} \quad ^\circ\text{C} = \frac{5}{9}(^\circ\text{F} - 32)$$

Example 2.7 **Converting from One Temperature Scale to Another**

Body temperature for a person with a high fever is found to be 104°F. To what is this temperature equivalent on the following scales?

a. Celsius scale **b.** Kelvin scale

Solution

a. We substitute 104° for °F in the equation

$$^\circ\text{C} = \frac{5}{9}(^\circ\text{F} - 32)$$

Then solving for °C gives

$$°C = \frac{5}{9}(104 - 32) = \frac{5}{9}(72) = 40°$$

b. Using the answer from part **a** and the equation

$$K = °C + 273$$

we get, by substitution,

$$K = 40° + 273 = 313$$

Heat Energy

▶ In discussions involving nutrition, the energy content of foods, and dietary tables, the term *Calorie* (spelled with a capital C) is used. The dietetic Calorie is actually 1 kilocalorie (1000 calories). The statement that an oatmeal raisin cookie contains 60 Calories means that 60 kcal (60,000 cal) of energy is released when the cookie is metabolized (undergoes chemical change) within the body.

The form of energy most often required for or released by chemical reactions and physical changes is *heat energy*. A commonly used unit for the measurement of heat energy is the calorie. A **calorie** (cal) *is the amount of heat energy needed to raise the temperature of 1 gram of water by 1 degree Celsius*. For large amounts of heat energy, the measurement is usually expressed in kilocalories.

$$1 \text{ kilocalorie} = 1000 \text{ calories}$$

Another unit for heat energy that is used with increasing frequency is the joule (J). The relationship between the joule (which rhymes with *pool*) and the calorie is

$$1 \text{ calorie} = 4.184 \text{ joules}$$

Heat energy values in calories can be converted to joules by using the conversion factor

$$\frac{4.184\text{J}}{1 \text{ cal}}$$

Practice Questions and Problems

2.32 An oven for baking pizza operates at approximately 525°F. What is this temperature in degrees Celsius?

2.33 A comfortable temperature for bathtub water is 35°C. What is this temperature in degrees Fahrenheit?

2.34 The body temperature for a hypothermia victim is found to have dropped to 29°C. What is this temperature on the Kelvin scale?

2.35 Which quantity of heat energy in each of the following pairs of heat energy values is the larger?

a. 2.0 joules or 2.0 calories
b. 1.0 kilocalorie or 92 calories
c. 100 Calories or 100 calories
d. 2.3 Calories or 1000 kilocalories

CONCEPTS TO REMEMBER

The metric system. The metric system, the measurement system preferred by scientists, is a decimal system in which larger and smaller units of a quantity are related by factors of 10. Prefixes are used to designate relationships between the basic unit and larger or smaller units of a quantity. Units in the metric system include the gram (mass), liter (volume), and meter (length).

Exact and inexact numbers. Numbers are of two kinds: exact and inexact. An exact number has a value that has no uncertainty associated with it. Exact numbers occur in definitions, in counting, and in simple fractions. An inexact number has a value that has a degree of uncertainty associated with it. Inexact numbers are generated anytime a measurement is made.

Significant figures. Significant figures in a measurement are those digits that are certain, plus a last digit that has been estimated. The maximum number of significant figures possible in a measurement is determined by the design of the measuring device.

Calculations and significant figures. Calculations should never improve (or decrease) the precision of experimental measurements. In multiplication and division, the number of significant figures in the answer is the same as that in the measurement containing the fewest significant figures. In addition and subtraction, the answer has no more digits to the right of the decimal point than are found in the measurement with the fewest digits to the right of the decimal point.

Scientific notation. Scientific notation is a system for writing decimal numbers in a more compact form that greatly simplifies the mathematical operations of multiplication and division. In this system, numbers are expressed as the product of a number between 1 and 10 and 10 raised to a power.

Dimensional analysis. Dimensional analysis is a general problem-solving method in which the units associated with numbers are used as a guide in setting up calculations. A given quantity is multiplied by one or more conversion factors in such a manner that the unwanted (original) units are canceled, leaving only the desired units.

Density. Density is the ratio of the mass of an object to the volume occupied by that object. A correct density expression includes a number, a mass unit, and a volume unit.

Temperature scales. The three major temperature scales are the Celsius, Kelvin, and Fahrenheit scales. The size of the degree for the Celsius and Kelvin scale is the same. They differ only in the numerical values assigned to the reference points. The Fahrenheit scale has a smaller degree size than the other two temperature scales.

Heat energy. The most commonly used unit of measurement for heat energy is the calorie. A calorie is the amount of heat energy needed to raise the temperature of 1 gram of water by 1 degree Celsius.

KEY REACTIONS AND EQUATIONS

1. Density of a substance (Section 2.7)

$$\text{Density} = \frac{\text{mass}}{\text{volume}}$$

2. Conversion of temperature readings from one scale to another (Section 2.8)

$$\text{K} = {}^\circ\text{C} + 273 \qquad {}^\circ\text{C} = \text{K} - 273$$

$${}^\circ\text{F} = \frac{9}{5}({}^\circ\text{C}) + 32 \qquad {}^\circ\text{C} = \frac{5}{9}({}^\circ\text{F} - 32)$$

KEY TERMS

Calorie (2.9)
Conversion factor (2.7)
Density (2.8)
Dimensional analysis (2.7)
Exact number (2.3)
Gram (2.2)
Inexact number (2.3)
Liter (2.2)
Mass (2.2)
Measurement (2.1)
Meter (2.2)
Rounding off (2.5)
Scientific notation (2.6)
Significant figures (2.4)
Weight (2.2)

ADDITIONAL PROBLEMS

2.36 Round off the number 4.7205059 to the indicated number of significant figures.
a. Six b. Five c. Four d. Two

2.37 Write each of the following numbers in scientific notation to the number of significant figures indicated in parentheses.
a. 0.00300300 (three) b. 936,000 (two)
c. 23.5003 (three) d. 450,000,001 (six)

2.38 For each of the pairs of units listed, indicate whether the first unit is larger or smaller than the second unit, and then indicate how many times larger or smaller it is.
a. Milliliter, liter b. Kiloliter, microliter
c. Nanoliter, deciliter d. Centiliter, megaliter

2.39 Indicate how each of the following conversion factors should be interpreted in terms of significant figures present.
a. $\frac{2.540\text{ cm}}{1.000\text{ in.}}$ b. $\frac{453.6\text{ g}}{1.000\text{ lb}}$ c. $\frac{2.113\text{ pt}}{1.00\text{ L}}$ d. $\frac{10^{-9}\text{ m}}{1\text{ nm}}$

2.40 Without actually performing the following multiplications, specify the number of significant figures that the answer should contain. The first listed number is an exact number, and the rest of the numbers are measured numbers.
a. $2 \times 2.00 \times 3.00$ b. $32 \times 4.31 \times 52{,}000$
c. $3 \times 3.0 \times 3.00$ d. $323 \times 320 \times 3200$

2.41 How many significant figures must the number Q possess, in each case, to make the following mathematical equations valid from the standpoint of significant figures?
a. $6.000 \times Q = 4.0$ b. $6.000 \times 4.0 = Q$
c. $5.250 + Q = 7.03$ d. $0.7777 - Q = 0.011$

2.42 A 1-gram sample of a powdery white solid is found to have a volume of 2 cubic centimeters. Calculate the solid's density using the following uncertainty specifications, and express your answers in scientific notation.
a. 1.0 g and 2.0 cm^3 b. 1.000 g and 2.00 cm^3
c. 1.0000 g and 2.0000 cm^3 d. 1.000 g and 2.0000 cm^3

2.43 Which is the higher temperature, −10°C or 10°F?

2.44 An individual weighs 83.2 kg and is 1.92 m tall. What are the person's equivalent measurements in pounds and feet?

2.45 What is wrong with the statement "The number of objects is exactly 12.00"?

PRACTICE TEST ▶ True/False

2.46 The density of an object is the ratio of its mass to its height.

2.47 Two conversion factors, which have a reciprocal relationship, can be derived from the equality 24 hours = 1 day.

2.48 A micrometer is a smaller metric system unit than a picometer.

2.49 All of the zeros in the measured number 0.0040400 are significant.

2.50 The number 344,700, when rounded to three significant figures, becomes 345.

2.51 The meter is the base unit of mass in the metric system.

2.52 1 mL, 1 cm^3, and 1 cc are three representations for the same volume.

2.53 All numbers have a degree of uncertainty associated with them.

2.54 The answer for the problem (3.11 + 9.2) should have an uncertainty of tenths.

2.55 The number 3.21×10^2 is larger than the number 314.

2.56 The number 0.0030, when it is expressed in scientific notation, becomes 3×10^{-3}.

2.57 The metric system prefix *micro* has the mathematical meaning 10^{-6}.

2.58 Addition of the value 273 to a Fahrenheit scale temperature reading will convert it into a Kelvin scale temperature reading.

2.59 Readings from a ruler scale with a smallest scale marking of 1 mm should be estimated to the closest 0.1 mm.

2.60 Solving the problem "How many feet are there in 3.72 yards?" using dimensional analysis requires the use of the conversion factor (1 yd/3 ft).

PRACTICE TEST ▶ Multiple Choice

2.61 The "mathematical meanings" associated with the metric system prefixes *centi, milli,* and *kilo* are respectively,
a. 10^{-3}, 10^{-4}, 10^{-6}
b. 10^{-2}, 10^{-3}, 10^{6}
c. 10^{-2}, 10^{-3}, 10^{3}
d. 10^{2}, 10^{-3}, 10^{3}

2.62 To what decimal position should a volume measurement be recorded if the smallest markings on the measurement scale are tenths of a milliliter?
a. To the closest milliliter
b. To tenths of a milliliter
c. To hundredths of a milliliter
d. To thousandths of a milliliter

2.63 Which of the following statements about the "significance" of zeros in recorded measurements is *incorrect*?
a. Leading zeros are never significant.
b. Confined zeros are always significant.
c. Trailing zeros are not always significant.
d. Trailing zeros and confined zeros are not always significant.

2.64 The number 43250, when rounded off to three significant figures, becomes
a. 432
b. 433
c. 43200
d. 43300

2.65 The number 105.00, when expressed in scientific notation, becomes
a. 1.05×10^{-2}
b. 1.0500×10^{-2}
c. 1.05×10^{2}
d. 1.0500×10^{2}

2.66 The calculator answer obtained by multiplying the measurements 62.32 and 7.00 is 436.24. This answer
a. is correct as written
b. should be rounded to 436.2
c. should be rounded to 436
d. should be rounded to 440

2.67 The correct answer obtained by adding the measurements 8.1, 2.16, and 3.123 contains
a. two significant figures
b. three significant figures
c. four significant figures
d. five significant figures

2.68 According to dimensional analysis, which of the following is the correct setup for the problem "How many milligrams are there in 67 kilograms?"

a. $67 \text{ kg} \times \left(\frac{1 \text{ g}}{10^3 \text{ kg}}\right) \times \left(\frac{1 \text{ mg}}{10^{-3} \text{ g}}\right)$

b. $67 \text{ kg} \times \left(\frac{10^3 \text{ g}}{1 \text{ kg}}\right) \times \left(\frac{1 \text{ mg}}{10^{-3} \text{ g}}\right)$

c. $67 \text{ kg} \times \left(\frac{10^3 \text{ mg}}{1 \text{ kg}}\right) \times \left(\frac{10^{-3} \text{ g}}{1 \text{ mg}}\right)$

d. $67 \text{ kg} \times \left(\frac{1 \text{ g}}{10^3 \text{ kg}}\right) \times \left(\frac{10^{-3} \text{ kg}}{1 \text{ mg}}\right)$

2.69 If object A weighs 8 g and has a volume of 4 mL and object B weighs 12 g and has a volume of 3 mL, then
a. B is less dense than A
b. A and B have equal densities
c. B is twice as dense as A
d. B is four times as dense as A

2.70 Which of the following comparisons of the size of a degree on the major temperature scales is *correct*?
a. A kelvin is larger than a Celsius degree.
b. A Fahrenheit degree and a Celsius degree are equal in size.
c. A Fahrenheit degree is larger than a kelvin.
d. A Celsius degree and a kelvin are equal in size.

CHAPTER OUTLINE
3.1 Internal Structure of an Atom
3.2 Atomic Number and Mass Number
3.3 Isotopes and Atomic Masses
Chemistry at a Glance: Atomic Structure
3.4 The Periodic Law and the Periodic Table
3.5 Metals and Nonmetals
Chemical Portraits 4: Metals of Importance as Determined by Extent of Use
3.6 Electron Arrangements Within Atoms
3.7 Electron Configurations
3.8 The Electronic Basis for the Periodic Law and the Periodic Table
3.9 Classification of the Elements
Chemical Portraits 5: Noble Gases: The Most Unreactive of All Elements
Chemistry at a Glance: Element Classification Schemes and the Periodic Table
3.10 Nuclear Stability and Radioactivity
3.11 Half-Life
3.12 The Nature of Radioactive Emissions
3.13 Equations for Radioactive Decay
3.14 Biological Effects of Radiation
3.15 Nuclear Medicine
Chemistry at a Glance: Terminology Associated with Nuclear Reactions

CHAPTER THREE

Atomic Structure, the Periodic Table, and Radioactivity

Music consists of a series of tones that build octave after octave. Similarly, elements have properties that recur period after period.

In Chapter 1 we learned that all matter is made up of small particles called atoms and that 113 different types of atoms are known, each type of atom corresponding to a different element. Furthermore, we found that compounds result from the chemical combination of different types of atoms in various ratios and arrangements.

Until the last two decades of the nineteenth century, scientists believed that atoms were solid, indivisible spheres without an internal structure. Today, this model of the atom is known to be incorrect. Evidence from a variety of sources indicates that atoms are made up of even smaller particles called *subatomic particles*. In this chapter we consider the fundamental types of subatomic particles, how they arrange themselves within an atom, and the relationship between an atom's subatomic makeup and its chemical identity.

Learning Focus

Describe the internal structure of an atom in terms of subatomic particles present and give the fundamental properties for each of the types of subatomic particles.

3.1 Internal Structure of an Atom

Atoms possess internal structure; that is, they are made up of even smaller particles, which are called subatomic particles. **Subatomic particles** *are very small particles that are the building blocks from which atoms are made.* Three types of subatomic particles are found within atoms: electrons, protons, and neutrons. Key properties of these three types of particles are summarized in Table 3.1. An **electron** *is a subatomic particle that possesses a negative (−) electrical charge.* It is the smallest, in terms of mass, of the three types of subatomic particles. A **proton** *is a subatomic particle that possesses a positive (+) electrical charge.* Protons and electrons carry the *same amount*

Table 3.1 Charge and Mass Characteristics of Electrons, Protons, and Neutrons

	Electron	Proton	Neutron
Charge	−1	+1	0
Actual mass (g)	9.109×10^{-28}	1.673×10^{-24}	1.675×10^{-24}
Relative mass (based on the electron being 1 unit)	1	1837	1839

▶ Atoms of all 113 elements contain the same three types of subatomic particles. Different elements differ only in the numbers of the various subatomic particles they contain.

of charge; the charges, however, are opposite (positive versus negative). A **neutron** *is a subatomic particle that has no charge associated with it; that is, it is neutral.* Both protons and neutrons are massive particles compared to electrons; they are almost 2000 times heavier.

Arrangement of Subatomic Particles Within an Atom

The arrangement of subatomic particles within an atom is not haphazard. *All* protons and *all* neutrons present are found at the center of an atom in a very tiny volume called the *nucleus* (Figure 3.1). The **nucleus** *is the very small, dense, positively charged center of an atom.* A nucleus is always positively charged because it contains positively charged protons. Because the nucleus houses the heavy subatomic particles (protons and neutrons), almost all (over 99.9%) of the mass of an atom is concentrated in its nucleus. The small size of the nucleus, coupled with its large amount of mass, causes nuclear material to be extremely dense.

Closely resembling the term *nucleus* is the term *nucleon.* A **nucleon** *is any subatomic particle found in the nucleus of an atom.* Thus both protons and neutrons are nucleons, and the nucleus can be regarded as containing a collection of nucleons (protons and neutrons).

The outer (extranuclear) region of an atom contains all of the electrons. In this region, which accounts for most of the volume of an atom, the electrons move rapidly about the nucleus. The electrons are attracted to the positively charged protons of the nucleus by the forces that exist between particles of opposite charge. The motion of the electrons in the extranuclear region determines the volume (size) of the atom in the same way that the blade of a fan determines a volume by its circular motion. The volume occupied by the electrons is sometimes referred to as the *electron cloud.* Because electrons are negatively charged, the electron cloud is also negatively charged. Figure 3.1 illustrates the nuclear and extranuclear regions of an atom.

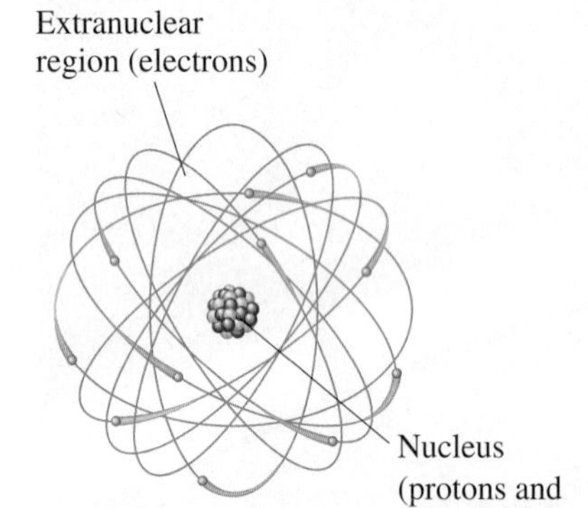

Figure 3.1 The protons and neutrons of an atom are found in the central nuclear region, or nucleus, and the electrons are found in an electron cloud outside the nucleus. Note that this figure is not drawn to scale; the correct scale would be comparable to a penny (the nucleus) in the center of a baseball field (the atom).

Charge Neutrality of an Atom

An atom as a whole is electrically neutral; that is, it has no *net* electrical charge. For this to be the case, the same amount of positive and negative charge must be present in the atom. Equal amounts of positive and negative charges cancel one another. Thus equal numbers of protons and electrons are present in an atom.

$$\text{Number of protons} = \text{number of electrons}$$

Size Relationships Within an Atom

To help you visualize the size relationships among the parts of an atom, imagine enlarging (magnifying) the nucleus until it is the size of a baseball (about 2.9 inches in diameter). If the nucleus were this large, the whole atom would have a diameter of approximately 2.5 miles. The electrons would still be smaller than the periods used to end sentences in this text, and they would move about at random within that 2.5-mile region.

The concentration of nearly all of the mass of an atom in the nucleus can also be illustrated by using our imagination. If a coin the same size as a copper penny contained copper nuclei (copper atoms stripped of their electrons) rather than copper atoms (which are mostly empty space), the coin would weigh 190,000,000 tons! Nuclei are indeed very dense matter.

Despite the existence of subatomic particles, we will continue to use the concept of atoms as the fundamental building blocks for all types of matter. Subatomic particles do not lead an independent existence for any appreciable length of time; they gain stability by joining together to form atoms.

Practice Questions and Problems

3.1 Indicate whether each of the following statements describes a *proton,* an *electron,* or a *neutron.*
a. Possesses a negative charge
b. Has no charge
c. Has a charge equal to, but opposite in sign from, that of an electron
d. Has a mass slightly less than that of a neutron

3.2 Describe the composition of the nucleus of an atom and its location relative to the atom as a whole.

3.3 Why does the nucleus of every atom have a positive charge?

3.4 What is the relationship between the number of protons and the number of electrons in an atom?

3.5 Indicate whether each of the following statements about the nucleus of an atom is *true* or *false.*
a. The nucleus of an atom contains all of the "heavy" subatomic particles.
b. The nucleus of an atom contains only neutrons.
c. The nucleus of an atom accounts for almost all of the volume of the atom.
d. The nucleus of an atom contains all of the subatomic particles present in an atom.

Learning Focus

Define the terms *atomic number* and *mass number* and, given these two numbers, know how to determine the number of protons, neutrons, and electrons present in an atom.

3.2 Atomic Number and Mass Number

The **atomic number** *of an atom is the number of protons in its nucleus.* Because an atom has the same number of electrons as protons (Section 3.1), the atomic number also specifies the number of electrons present. The symbol Z is used as a general designation for the atomic number.

Atomic number = number of protons = number of electrons = Z

▶ Atomic number and mass number are always *whole* numbers because they are obtained by counting whole objects (protons, neutrons, and electrons).

The **mass number** *of an atom is the sum of the numbers of protons and neutrons in its nucleus.* Thus the mass number gives the number of subatomic particles present in the nucleus. The mass of an atom is almost totally accounted for by the protons and neutrons present—hence the term *mass number.* The symbol A is used as a general designation for the mass number.

Mass number = number of protons + number of neutrons = A

The number and identity of subatomic particles present in an atom can be calculated from its atomic and mass numbers in the following manner.

Number of protons = atomic number = Z
Number of electrons = atomic number = Z
Number of neutrons = mass number − atomic number = A − Z

▶ The *sum* of the mass number and the atomic number for an atom corresponds to the total number of subatomic particles present in the atom (protons, neutrons, and electrons).

Note that neutron count is obtained by subtracting atomic number from mass number.

Example 3.1 Determining the Subatomic Particle Makeup of an Atom Given Its Atomic Number and Mass Number

An atom has an atomic number of 9 and a mass number of 19.

a. Determine the number of protons present.
b. Determine the number of neutrons present.
c. Determine the number of electrons present.

Solution

a. There are 9 protons because the atomic number is always equal to the number of protons present.
b. There are 10 neutrons because the number of neutrons is always obtained by subtracting the atomic number from the mass number.

$$\underbrace{(\text{Protons} + \text{neutrons})}_{\text{Mass number}} - \underbrace{\text{protons}}_{\text{Atomic number}} = \text{neutrons}$$

c. There are 9 electrons because the number of protons and the number of electrons are always the same in an atom.

Electrons and Chemical Properties

The chemical properties of an atom, which are the basis for its identification, are determined by the number and arrangement of the electrons about the nucleus. When two atoms interact, the outer part (electrons) of one interacts with the outer part (electrons) of the other. The small nuclear centers never come in contact with each other in a chemical reaction. The number of electrons about a nucleus may be considered to be determined by the number of protons in the nucleus; charge balance requires an equal number of the two (Section 3.1). Hence the number of protons (which is the atomic number) characterizes an atom. All atoms with the same atomic number have the same chemical properties and are atoms of the same element.

In Section 1.6, an element was defined as a pure substance that cannot be broken down into simpler substances by ordinary chemical means. Although this is a good historical definition for an element, we can now give a more rigorous definition by using the concept of atomic number. An **element** *is a pure substance in which all atoms present have the same atomic number.*

An alphabetical listing of the 113 known elements, with their atomic numbers as well as other information, is found on the inside front cover of this text. If you check the atomic number column in this tabulation, you will find an entry for each of the numbers in the sequence 1 to 112 plus 114. The highest-atomic-numbered element that occurs naturally is uranium (element 92); elements 93 to 112 and 114 have been made in the laboratory but are not found in nature (Section 1.7). The fact that there are no gaps in the numerical sequence 1 to 92 is interpreted by scientists to mean that there are no "missing elements" yet to be discovered in nature.

Practice Questions and Problems

3.6 Determine the atomic number and mass number for atoms with the following subatomic makeups.
a. 2 protons, 2 neutrons, and 2 electrons
b. 4 protons, 5 neutrons, and 4 electrons
c. 5 protons, 4 neutrons, and 5 electrons
d. 28 protons, 30 neutrons, and 28 electrons

3.7 Determine the number of protons, neutrons, and electrons present in atoms with the following characteristics.
a. Atomic number = 8 and mass number = 16
b. Mass number = 18 and $Z = 8$
c. Atomic number = 20 and $A = 44$
d. $A = 257$ and $Z = 100$

3.8 Indicate whether the *atomic number,* the *mass number,* or *both the atomic number and mass number* are needed to determine each of the following.
a. Number of protons in an atom
b. Number of neutrons in an atom

c. Number of nucleons in an atom
d. Total number of subatomic particles in an atom

3.9 What information about the subatomic particles present in an atom is obtained from each of the following?
a. Atomic number
b. Mass number
c. Mass number − atomic number
d. Mass number + atomic number

3.10 What is the definition for an element in terms of atomic number?

Learning Focus

Be able to define the term *isotope,* write the symbol for an isotope, and calculate the atomic mass for an element from isotopic masses and percent abundances.

3.3 Isotopes and Atomic Masses

Charge neutrality (Section 3.1) requires the presence in an atom of an equal number of protons and electrons. However, because neutrons have no electrical charge, their numbers in atoms do not have to be the same as the number of protons or electrons. Most atoms contain more neutrons than either protons or electrons.

Studies of atoms of various elements also show that the number of neutrons present in atoms of an element is not constant; it varies over a small range. This means that not all atoms of an element have to be identical. They must have the same number of protons and electrons, but they can differ in the number of neutrons.

Atoms of an element that differ in neutron count are called isotopes. **Isotopes** *are atoms of an element that have the same number of protons and electrons but different numbers of neutrons.* Different isotopes always have the same atomic number and different mass numbers.

▶ The word *isotope* comes from the Greek *iso,* meaning "equal," and *topos,* meaning "place." Isotopes occupy an equal place (location) in listings of elements because all isotopes of an element have the same atomic number.

Most elements found in nature exist in isotopic forms, with the number of naturally occurring isotopes ranging from two to ten. For example, all silicon atoms have 14 protons and 14 electrons. Most silicon atoms also contain 14 neutrons. However, some silicon atoms contain 15 neutrons and others contain 16 neutrons. Thus three different kinds of silicon atoms exist, that is, three silicon isotopes exist.

Isotopes of an element have the same chemical properties, but their physical properties are often slightly different. Isotopes of an element have the same chemical properties because they have the same number of electrons. They have slightly different physical properties because they have different numbers of neutrons and therefore different masses.

When it is necessary to distinguish between isotopes of an element, the following notation is used:

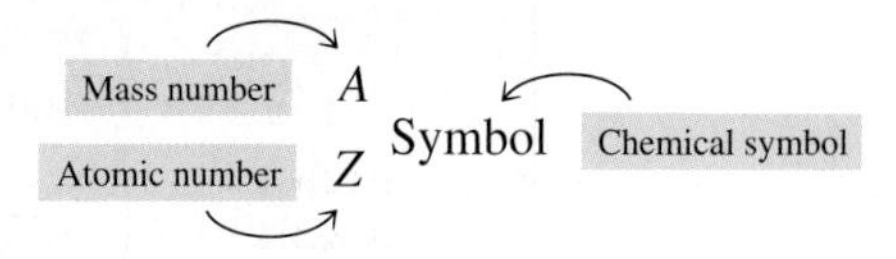

The atomic number is written as a *subscript* to the left of the elemental symbol for the atom. The mass number is written as a *superscript* to the left of the elemental symbol. Thus the three silicon isotopes are designated, respectively, as

$$^{28}_{14}\text{Si}, \quad ^{29}_{14}\text{Si}, \quad \text{and} \quad ^{30}_{14}\text{Si}$$

▶ There are a few elements for which all naturally occurring atoms have the same number of neutrons—that is, for which all atoms are identical. They include the elements Be, F, Na, Al, P, and Au.

Names for isotopes include the mass number. $^{28}_{14}$Si is called silicon-28, and $^{29}_{14}$Si is called silicon-29. The atomic number is not included in the name because it is the same for all isotopes of an element.

▶ A mass number, in contrast to an atomic number, lacks uniqueness. Atoms of different elements can have the same mass number. For example, carbon-14 and nitrogen-14 have the same mass numbers. Atoms of different elements, however, cannot have the same atomic number.

The various isotopes of a given element are of varying abundance; usually one isotope is predominant. Silicon is typical of this situation. The percentage abundances for its three isotopes are 92.21% ($^{28}_{14}$Si), 4.70% ($^{29}_{14}$Si), and 3.09% ($^{30}_{14}$Si). Percentage abundances are number percentages (numbers of atoms) rather than mass percentages. A sample of 10,000 silicon atoms contains 9221 $^{28}_{14}$Si atoms, 470 $^{29}_{14}$Si atoms, and 309 $^{30}_{14}$Si atoms.

There are 286 isotopes that occur naturally. In addition, over 2000 more have been synthesized in the laboratory via nuclear rather than chemical reactions. All these synthetic isotopes are unstable (radioactive). Despite their instability, many are used in chemical and biological research, as well as in medicine. The topic of radioactivity is considered in the last few sections of this chapter.

▶ An analogy involving isotopes and identical twins may be helpful: Identical twins need not weigh the same, even though they have identical "gene packages." Likewise, isotopes, even though they have different masses, have the same number of protons.

Atomic Masses

The existence of isotopes means that atoms of an element can have several different masses. For example, silicon atoms can have any one of three masses because there are three silicon isotopes. Which of these three silicon isotopic masses is used in situations in which the mass of the element silicon needs to be specified? The answer is none of them. Instead we use a *weighted-average mass* that takes into account the existence of isotopes and their relative abundances.

▶ The terms *atomic mass* and *atomic weight* are often used interchangeably. Atomic mass, however, is the correct term.

The *weighted-average mass* of the isotopes of an element is known as the element's atomic mass. An **atomic mass** *is the calculated average mass for the isotopes of an element, expressed on a scale using atoms of* $^{12}_{6}C$ *as the reference.* What we need to calculate an atomic mass are the masses of the various isotopes on the $^{12}_{6}C$ reference scale and the percentage abundance of each isotope.

The $^{12}_{6}C$ reference scale mentioned in the definition of *atomic mass* is a scale scientists have set up for comparing the masses of atoms. On this scale, the mass of a $^{12}_{6}C$ atom is defined to be exactly 12 atomic mass units (amu). The masses of all other atoms are then determined relative to that of $^{12}_{6}C$. For example, if an atom is twice as heavy as $^{12}_{6}C$, its mass is 24 amu, and if an atom weighs half as much as an atom of $^{12}_{6}C$, its mass is 6 amu.

Example 3.2 shows how an atomic mass is calculated by using the amu ($^{12}_{6}C$) scale, the percentage abundances of isotopes, and the number of isotopes of an element.

Example 3.2 Calculation of an Element's Atomic Mass

Naturally occurring chlorine exists in two isotopic forms, $^{35}_{17}Cl$ and $^{37}_{17}Cl$. The relative mass of $^{35}_{17}Cl$ is 34.97 amu, and its abundance is 75.53%; the relative mass of $^{37}_{17}Cl$ is 36.97 amu, and its abundance is 24.47%. What is the atomic mass of chlorine?

Solution

An element's atomic mass is calculated by multiplying the relative mass of each isotope by its fractional abundance and then totaling the products. The fractional abundance for an isotope is its percentage abundance converted to decimal form (divided by 100).

$$^{35}_{17}Cl: \quad \left(\frac{75.53}{100}\right) \times 34.97\text{ amu} = (0.7553) \times 34.97\text{ amu} = 26.41\text{ amu}$$

$$^{37}_{17}Cl: \quad \left(\frac{24.47}{100}\right) \times 36.97\text{ amu} = (0.2447) \times 36.97\text{ amu} = 9.047\text{ amu}$$

$$\text{Atomic mass of Cl} = (26.41 + 9.047)\text{ amu}$$
$$= 35.46\text{ amu}$$

This calculation involved an element containing just two isotopes. A similar calculation for an element having three isotopes would be carried out the same way, but it would have three terms in the final sum; an element possessing four isotopes would have four terms in the final sum.

The alphabetical listing of the known elements printed inside the front cover of this text gives the calculated atomic mass for each of the elements; it is the last column of numbers.

The accompanying Chemistry at a Glance summarizes all that we have said about atoms thus far.

Atomic Structure

ATOMIC STRUCTURE

There are 113 known kinds of atoms, one kind for each element. All atoms have a nucleus (a small dense center containing all protons and neutrons) and an extranuclear region containing all electrons.

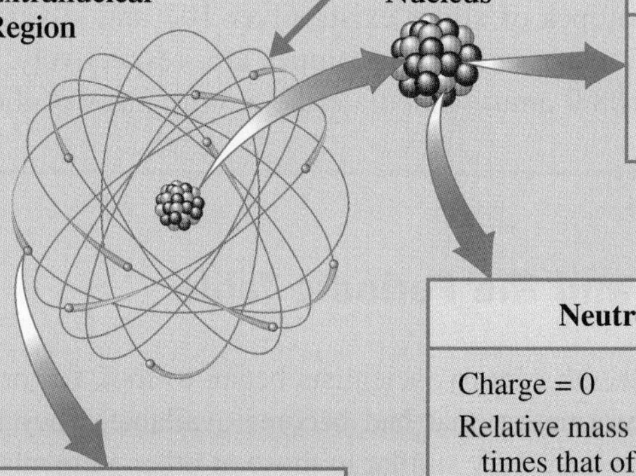

Extranuclear Region

Nucleus

Protons

Charge = +1
Relative mass = 1837 times that of an electron

Atomic Number

The number of protons in the nucleus identifies the atom and equals the atomic number.

Neutrons

Charge = 0
Relative mass = 1839 times that of an electron

Mass Number

The number of protons plus the number of neutrons in an atom equals the mass number.

Electrons

Charge = –1
Relative mass = 1
Electrons determine the chemical properties of an atom.

VARIATIONS IN ATOMIC STRUCTURE

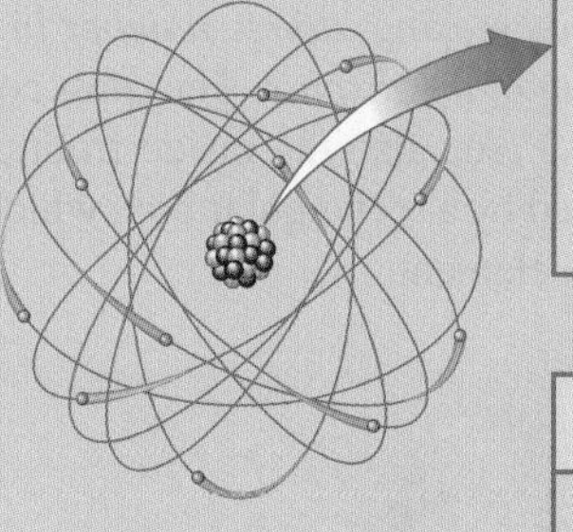

Isotopes

The atoms of various isotopes of an element contain the same number of protons but differ in the number of neutrons in the nucleus. Isotopes have the same atomic number and different mass numbers.

Atomic Mass

The average mass of all isotopes of an element weighted according to natural abundance is the atomic mass.

Practice Questions and Problems

3.11 What are the atomic number and the mass number for an atom denoted as $^{24}_{12}X$?

3.12 Determine the number of protons, electrons, and neutrons for the following atoms.
a. $^{12}_{6}C$ b. $^{16}_{8}O$ c. $^{26}_{12}Mg$ d. $^{197}_{79}Au$

3.13 How many protons, electrons, and neutrons are present in the following isotopes of zirconium?
a. $^{90}_{40}Zr$ b. $^{92}_{40}Zr$ c. $^{95}_{40}Zr$ d. $^{96}_{40}Zr$

3.14 Indicate whether the isotopes of an element differ or are the same in each of the following properties.
a. Atomic number b. Mass number
c. Number of neutrons d. Number of electrons

3.15 The atomic number of the element carbon (C) is 6. Write the complete symbol for each of the following carbon isotopes: carbon-12, carbon-13, and carbon-14.

3.16 What information about the isotopes of an element is needed before the atomic mass of the element can be calculated?

3.17 Most (92.58%) naturally occurring atoms of the element lithium have a mass of 7.02 amu. The remainder (7.42%) have a mass of 6.01 amu. Calculate the atomic mass of the element lithium.

3.18 Two naturally occurring isotopes of silver exist: silver-107 and silver-109. The percentage abundances and masses of these isotopes are, respectively, (51.82%, 106.9 amu) and (48.18%, 108.9 amu). Calculate the atomic mass of the element silver.

Learning Focus

Understand and state the periodic law. Understand the rationale behind the organization of the periodic table, relate the terms *group* and *period* to the periodic law, and indicate what general information about each element is given in the periodic table.

3.4 The Periodic Law and the Periodic Table

During the early part of the nineteenth century, scientists began to look for order in the increasing amount of chemical information that had become available. They knew that certain elements had properties that were very similar to those of other elements, and they sought reasons for these similarities in the hope that these similarities would suggest a method for arranging or classifying the elements.

In 1869, these efforts culminated in the discovery of what is now called the *periodic law,* proposed independently by the Russian chemist Dmitri Mendeleev (Figure 3.2) and the German chemist Julius Lothar Meyer. Given in its modern form, the **periodic law** *states that when elements are arranged in order of increasing atomic number, elements with similar properties occur at periodic (regularly recurring) intervals.*

A periodic table represents a compact graphical method for representing the behavior described by the periodic law. A **periodic table** *is a graphical display of the elements in order of increasing atomic number in which elements with similar properties fall in the same column of the display.* The most commonly used form of the periodic table is shown in Figure 3.3 (see also the inside front cover of the text). Within the table, each element is represented by a rectangular box, which contains the symbol, atomic number, and atomic mass of the element.

Figure 3.2 Dmitri Ivanovich Mendeleev (1834–1907). Mendeleev constructed a periodic table as part of his effort to systematize chemistry. He received many international honors for his work, but his reception at home in czarist Russia was mixed. Element 101 carries his name.

▶ Using the information on a periodic table, you can quickly determine the number of protons and electrons for atoms of an element. However, no information concerning neutrons is available from a periodic table; mass numbers are not part of the information given, because they are not unique to an element.

▶ The elements within a given periodic-table group show numerous similarities in properties, the degree of similarity varying from group to group. In no case are the group members "clones" of one another. Each element has some individual characteristics not found in other elements of the group. By analogy, the members of a human family often bear many resemblances to each other, but each member also has some (and often much) individuality.

Groups and Periods of Elements

The location of an element within the periodic table is specified by giving its group number and period number.

A **group** *in the periodic table is a vertical column of elements.* There are two notations in use for designating individual periodic-table groups. In the first notation, which has been in use for many years, groups are designated by using Roman numerals and the letters A and B. In the second notation, which has recently been recommended for use by an international scientific commission, the Arabic numbers 1 through 18 are used. Note that in Figure 3.3 both group notations are given at the top of each group. The elements with atomic numbers 8, 16, 34, 52, and 84 (O, S, Se, Te, and Po) constitute Group VIA (old notation) or Group 16 (new notation).

Several groups of elements have common (non-numerical) names that are used so frequently that they should be learned. The Group IA elements, except for hydrogen, are called the *alkali metals,* and the Group IIA elements the *alkaline earth metals.* On the opposite side of the periodic table from the IA and IIA elements are found the *halogens* (the Group VIIA elements) and the *noble gases* (Group VIIIA).

A **period** *in the periodic table is a horizontal row of elements.* For identification purposes, the periods are numbered sequentially with Arabic numbers, starting at the top of the periodic table. In Figure 3.3 period numbers are found on the left side of the table. The elements Na, Mg, Al, Si, P, S, Cl, and Ar are all members of Period 3, the third row of elements. Period 4 is the fourth row of elements, and so on. There are only two elements in Period 1, H and He.

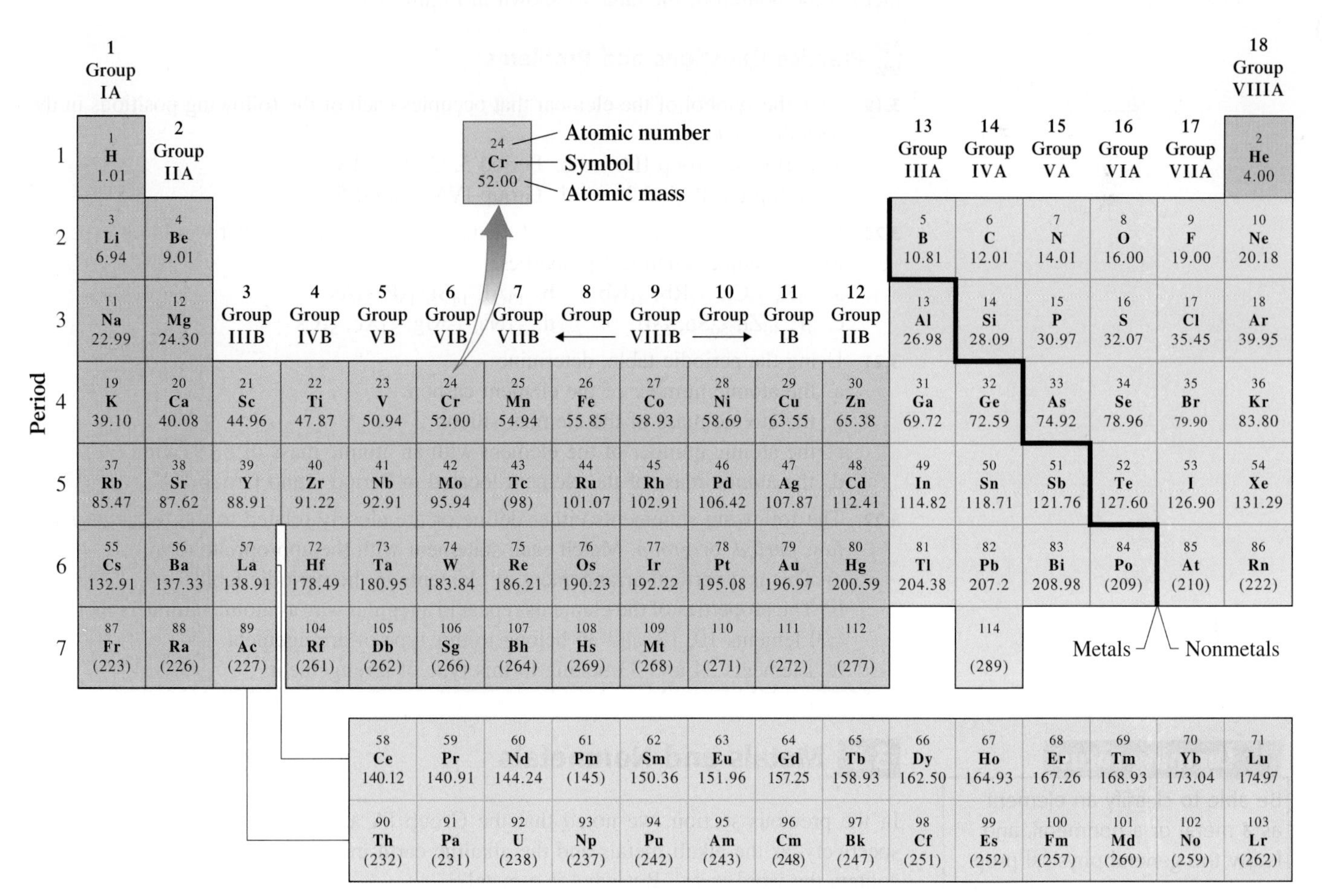

Figure 3.3 The periodic table of the elements is a graphical way to show relationships among the elements. Elements with similar chemical properties fall in the same vertical column.

1																															2
3	4																									5	6	7	8	9	10
11	12																									13	14	15	16	17	18
19	20	21															22	23	24	25	26	27	28	29	30	31	32	33	34	35	36
37	38	39															40	41	42	43	44	45	46	47	48	49	50	51	52	53	54
55	56	57	58	59	60	61	62	63	64	65	66	67	68	69	70	71	72	73	74	75	76	77	78	79	80	81	82	83	84	85	86
87	88	89	90	91	92	93	94	95	96	97	98	99	100	101	102	103	104	105	106	107	108	109	110	111	112		114				

Figure 3.4 In this periodic table, elements 58 through 71 and 90 through 103 (in color) are shown in their proper positions.

The location of any element in the periodic table is specified by giving its group number and its period number. The element gold, with an atomic number of 79, belongs to Group IB (or 11) and is in Period 6. The element nitrogen, with an atomic number of 7, belongs to Group VA (or 15) and is in Period 2.

The Shape of the Periodic Table

▶ When the statement "the first ten elements" is used, it means the first ten elements in the periodic table, the elements with atomic numbers 1 through 10.

Within the periodic table of Figure 3.3 the practice of arranging the elements according to increasing atomic number is violated in Groups IIIB and IVB. Element 72 follows element 57, and element 104 follows element 89. The missing elements, elements 58 through 71 and 90 through 103 are located in two rows at the bottom of the periodic table. Technically, the elements at the bottom of the table should be included in the body of the table, as shown in Figure 3.4. However, in order to have a more compact table, we place them at the bottom of the table as shown in Figure 3.3.

Practice Questions and Problems

3.19 Give the symbol of the element that occupies each of the following positions in the periodic table.

a. Period 4, Group IIA b. Period 5, Group VIB
c. Group IA, Period 2 d. Group IVA, Period 5

3.20 For each of the following sets of elements, choose the two that would be expected to have similar chemical properties.

a. ${}_{19}K$, ${}_{29}Cu$, ${}_{37}Rb$, ${}_{41}Nb$ b. ${}_{13}Al$, ${}_{14}Si$, ${}_{15}P$, ${}_{33}As$
c. ${}_{9}F$, ${}_{40}Zr$, ${}_{50}Sn$, ${}_{53}I$ d. ${}_{11}Na$, ${}_{12}Mg$, ${}_{54}Xe$, ${}_{55}Cs$

3.21 Using the periodic table, determine

a. the atomic number of the element carbon
b. the atomic mass of the element silicon
c. the atomic number of the element with an atomic mass of 88.91 amu
d. the atomic mass of the element located in Period 2 and Group IIA

3.22 The following statements either define or are closely related to the terms *periodic law, period,* or *group*. Match each statement with the appropriate term.

a. This is a vertical arrangement of elements in the periodic table.
b. The properties of the elements repeat in a regular way as atomic numbers increase.
c. Elements 10, 18, and 36 belong to this type of arrangement.
d. Elements 24 and 33 belong to this type of arrangement.

3.5 Metals and Nonmetals

Learning Focus

Be able to classify an element as a metal or a nonmetal, and know the general physical properties for metallic and nonmetallic elements.

In the previous section, we noted that the Group IA and IIA elements are known, respectively, as the alkali metals and the alkaline earth metals. Both of these designations contain the word *metal*. But what is a metal?

On the basis of selected physical properties, elements are classified into the categories metal and nonmetal. A **metal** *is an element that has the characteristic properties of luster,*

(a)

(b)

Figure 3.5 (a) Some familiar metals are aluminum, lead, tin, and zinc. (b) Some familiar nonmetals are sulfur (yellow), phosphorus (dark red), and bromine (reddish-brown).

thermal conductivity, electrical conductivity, and malleability. With the exception of mercury, all metals are solids at room temperature (25°C). Among the more familiar metals are the elements iron, aluminum, copper, silver, gold, lead, tin, and zinc (see Figure 3.5a). A **nonmetal** *is an element characterized by the absence of the properties of luster, thermal conductivity, electrical conductivity, and malleability*. Many of the nonmetals, such as hydrogen, oxygen, nitrogen, and the noble gases, are gases. The only nonmetal found as a liquid at room temperature is bromine. Solid nonmetals include carbon, iodine, sulfur, and phosphorus (Figure 3.5b).

Table 3.2 contrasts selected physical properties of metals and nonmetals. In many ways, the general properties of metals and nonmetals are opposites. Metals generally are lustrous (shine, reflect light), malleable (can be drawn into wires), ductile (can be rolled into sheets), and good thermal and electrical conductors. Nonmetals tend to lack these properties. Generally, the nonmetals have lower densities and lower melting points than metals.

Periodic Table Locations for Metals and Nonmetals

The majority of the elements are metals. Only 22 elements are nonmetals. It is not necessary to memorize which elements are nonmetals and which are metals; this information is obtainable from a periodic table (Figure 3.6). The steplike heavy line that runs through the right third of the periodic table separates the metals on the left from the nonmetals on the right. Note also that the element hydrogen is a nonmetal.

The fact that the vast majority of elements are metals in no way indicates that metals are more important than nonmetals. Most nonmetals are relatively common and are found in many important compounds. For example, water (H_2O) is a compound involving two nonmetals.

Table 3.2
Selected Physical Properties of Metals and Nonmetals

Metals	Nonmetals
1. High electrical conductivity that decreases with increasing temperature	1. Poor electrical conductivity (except carbon in the form of graphite)
2. High thermal conductivity	2. Good heat insulators (except carbon in the form of diamond)
3. Metallic gray or silver luster[a]	3. No metallic luster
4. Almost all are solids[b]	4. Solids, liquids, or gases
5. Malleable (can be hammered into sheets)	5. Brittle in solid state
6. Ductile (can be drawn into wires)	6. Nonductile

[a]Except copper and gold.
[b]Except mercury; cesium and gallium melt on a hot summer day (85°F) or when held in a person's hand.

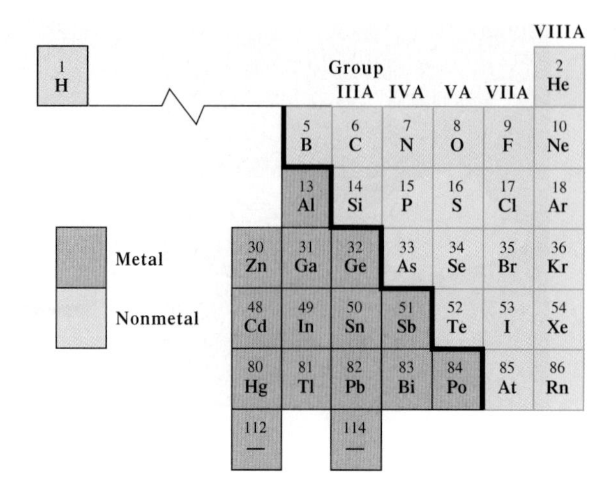

Figure 3.6 This portion of the periodic table shows the dividing line between metals and nonmetals. All elements that are not shown are metals.

An analysis of the abundance of the elements in Earth's crust (Figure 1.11) in terms of metals and nonmetals shows that the two most abundant elements, which account for 80.2% of all atoms, are nonmetals—oxygen and silicon. The four most abundant elements in the human body (Table 1.1), which comprise over 99% of all atoms in the body, are nonmetals—hydrogen, oxygen, carbon, and nitrogen.

The ranking of metals according to abundance does not reflect their importance in today's world. Abundance and importance (use) are two entirely different things. Some metals with low abundances in Earth's crust are more widely used than some of the more abundant metals. Why is this so? Obviously, the chemical properties of the various metals help determine their use. Another major factor is the availability of a metal's deposits

Chemical Portraits 4 — Metals of Importance as Determined by Extent of Use

Iron (Fe)

Profile: Iron is the most used of all metals. In fact, more Fe is consumed, both worldwide and in the United States, than all other metals combined. When pure, Fe is lustrous, silvery, and soft. However, pure Fe is seldom encountered; rather the end form for nearly all consumption of Fe is steel, an Fe alloy. When exposed to air and moisture, iron *rusts*; the reddish-brown compound formed is a hydrated form of Fe_2O_3.

Uses: Two types of steel exist: carbon steel and alloy steel. Both are metal alloys that contain iron and carbon. Alloy steels also contain small amounts of other metals added to improve specific characteristics of the steel.

What is the major function for iron present in the human body?

Aluminum (Al)

Profile: Aluminum, the most abundant metal on Earth, is the second-most-used metal (although its use is just 10% that of iron). Freshly cut aluminum appears silvery. However, its surface quickly changes to a dull gray as a thin film of aluminum oxide (Al_2O_3) forms, preventing further corrosion.

Uses: Since 1994, the transportation sector has been the largest U.S. market for aluminum. The second-largest market is containers and packaging—beverage cans, food containers, and household and institutional foil. In addition, 90% of all overhead electrical transmission lines in the United States are made of aluminum alloys.

What is the "driving force" for the recycling of aluminum-containing products?

Copper (Cu)

Profile: With a reddish-yellow color, copper is the only metal besides gold to have a natural color other than gray-silver. Usage for Cu, the third-most-used metal, is just 5% that of iron. Weathering coats Cu with a green film that protects it from further corrosion. This green film is a compound formed from the reaction of Cu with three components of the air: oxygen, carbon dioxide, and water.

Uses: Copper is one of the few elements that is mainly used in pure form rather than as an alloy or compound. Building wire (16%) and plumbing (14%) are Cu's top two markets. Important copper alloys include *bronze* (up to 33% zinc present) and *brass* (up to 18% tin present).

How much copper is present in the various types of United States coinage?

See the text web site at **www.cengage.com/chemistry/stoker** for answers to the above questions and for further information.

in Earth's crust. Some metals have been concentrated by natural processes into localized areas. Less abundant metals that are found in "concentrated form" are generally more used than more abundant metals that are not "concentrated." The difficulty and cost of isolating a metal from its natural sources also affect the extent of its use. Chemical Portraits 4 profiles the three "most used" metals: iron, aluminum, and copper.

Practice Questions and Problems

3.23 In which of the following pairs of elements are both members of the pair metals?
a. $_{17}$Cl and $_{35}$Br b. $_{13}$Al and $_{14}$Si c. $_{29}$Cu and $_{42}$Mo d. $_{30}$Zn and $_{83}$Bi

3.24 Identify the nonmetal in each of the following sets of elements.
a. S, Na, K b. Cu, Li, P c. Be, I, Ca d. Fe, Cl, Ga

3.25 Classify each of the following general physical properties as a property of metallic elements or of nonmetallic elements.
a. Malleable and ductile
b. Low electrical conductivity
c. High thermal conductivity
d. Good heat insulator

Learning Focus

Understand how the terms *electron shell, electron subshell,* and *electron orbital* are related and how they are used in describing electron arrangements within an atom.

3.6 Electron Arrangements Within Atoms

Current chemical theory indicates that as electrons move about an atom's nucleus, they are restricted to specific regions within the extranuclear portion of the atom. Such restrictions are determined by the amount of energy the electrons possess. Furthermore, chemical theory indicates that electron energies are limited to certain values and that a specific "behavior" is associated with each allowed energy value.

The space in which electrons move rapidly about a nucleus is divided into subspaces called *shells, subshells,* and *orbitals.*

Electron Shells

Electrons within an atom are grouped into main energy levels called electron shells. An **electron shell** *is a region of space about a nucleus that contains electrons that have approximately the same energy and that spend most of their time approximately the same distance from the nucleus.*

▶ Electrons that occupy the first electron shell are closer to the nucleus and have a lower energy than electrons in the second electron shell.

Electron shells are numbered 1, 2, 3, and so on, outward from the nucleus. Electron energy increases as the distance of the electron shell from the nucleus increases. An electron in shell 1 has the minimum amount of energy that an electron can have.

The maximum number of electrons that an electron shell can accommodate varies; the higher the shell number (n), the more electrons that can be present. In higher-energy shells the electrons are farther from the nucleus, and a greater volume of space is available for them; hence more electrons can be accommodated. (Conceptually, electron shells may be considered to be nested one inside another, somewhat like the layers of flavors inside a jawbreaker or similar type of candy.)

The lowest-energy shell ($n = 1$) accommodates a maximum of 2 electrons. In the second, third, and fourth shells, 8, 18, and 32 electrons, respectively, are allowed. The relationship among these numbers is given by the formula $2n^2$, where n is the shell number. For example, when $n = 4$, the quantity $2n^2 = 2(4)^2 = 32$.

Electron Subshells

Within each electron shell, electrons are further grouped into energy sublevels called electron subshells. An **electron subshell** *is a region of space within an electron shell that contains electrons that have the same energy.* We can draw an analogy between the relationship of shells and subshells and the physical layout of a high-rise apartment complex.

The shells are analogous to the floors of the apartment complex, and the subshells are the counterparts of the various apartments on each floor.

The number of subshells within a shell is the same as the shell number. Shell 1 contains one subshell, shell 2 contains two subshells, shell 3 contains three subshells, and so on.

Subshells within a shell differ in size (that is, the maximum number of electrons they can accommodate) and energy. The higher the energy of the contained electrons, the larger the subshell.

▶ The letters used to label the different types of subshells come from old spectroscopic terminology associated with the lines in the spectrum of the element hydrogen. These lines were denoted as *s*harp, *p*rincipal, *d*iffuse, and *f*undamental. Relationships exist between such lines and the arrangement of electrons in an atom.

Subshell size (type) is designated using the letters *s*, *p*, *d*, and *f*. Listed in this order, these letters denote subshells of increasing energy and size. The lowest-energy subshell within a shell is always the *s* subshell, the next highest is the *p* subshell, then the *d* subshell, and finally the *f* subshell. An *s* subshell can accommodate 2 electrons, a *p* subshell 6 electrons, a *d* subshell 10 electrons, and an *f* subshell 14 electrons.

Both a number and a letter are used in identifying subshells. The number gives the shell within which the subshell is located, and the letter gives the type of subshell. Shell 1 has only one subshell, the 1*s*. Shell 2 has two subshells, the 2*s* and 2*p*. Shell 3 has three subshells, the 3*s*, 3*p*, and 3*d*; and so on. Figure 3.7 summarizes the relationships between electron shells and electron subshells for the first four shells.

Electron Orbitals

Electron subshells have within them a certain, definite number of locations (regions of space), called electron orbitals, where electrons may be found. In our apartment complex analogy, if shells are the counterparts of floor levels and subshells are the apartments, then electron orbitals are the rooms of the apartments. An **electron orbital** *is a region of space within an electron subshell where an electron with a specific energy is most likely to be found.*

▶ An electron orbital is also often called an atomic orbital.

An electron orbital, independent of all other considerations, can accommodate a maximum of 2 electrons. Thus an *s* subshell (2 electrons) contains one orbital, a *p* subshell (6 electrons) contains three orbitals, a *d* subshell (10 electrons) contains five orbitals, and an *f* subshell (14 electrons) contains seven orbitals.

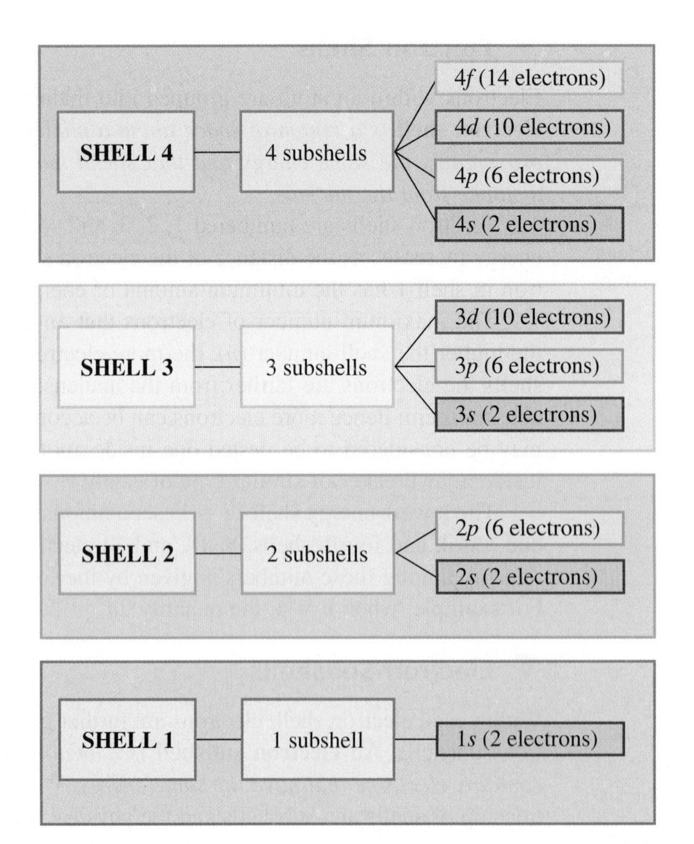

Figure 3.7 The number of subshells within a shell is equal to the shell number, as shown here for the first four shells. Each individual subshell is denoted with both a number (its shell) and a letter (the type of subshell it is in).

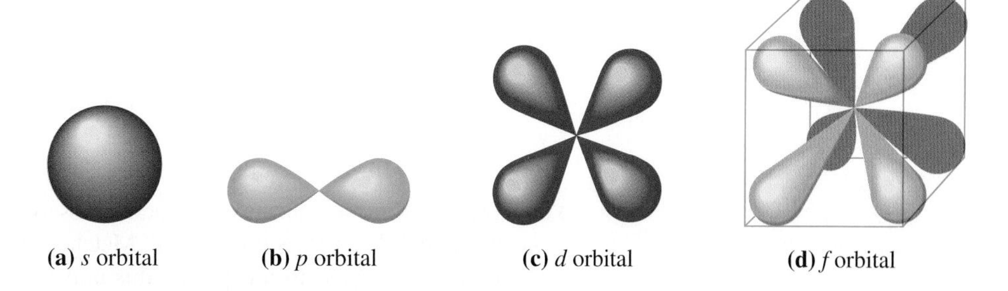

Figure 3.8 An *s* orbital has a spherical shape, a *p* orbital has two lobes, a *d* orbital has four lobes, and an *f* orbital has eight lobes. The *f* orbital is shown within a cube to illustrate that its lobes are directed toward the corners of a cube. Some *d* and *f* orbitals have shapes related to, but not identical to, those shown.

Orbitals have distinct shapes that are related to the type of subshell in which they are found. Note that we are talking not about the shape of an electron but, rather, about the shape of the region in which the electron is found. An orbital in an *s* subshell, which is called an *s* orbital, has a spherical shape (Figure 3.8a). Orbitals found in *p* subshells—*p* orbitals—have shapes similar to the "figure 8" of an ice skater (Figure 3.8b). More complex shapes involving four and eight lobes, respectively, are associated with *d* and *f* orbitals (Figures 3.8c and 3.8d). Some *d* and *f* orbitals have shapes related to, but not identical to, those shown in Figure 3.8.

Figure 3.9 which is an extension of Figure 3.7, summarizes the important relationships among electron shells, electron subshells, and electron orbitals.

Practice Questions and Problems

3.26 What is the maximum number of electrons that can occupy each of the following *electron shells*?
a. First shell b. Second shell c. Third shell d. Fifth shell

3.27 How many electron subshells are present in each of the following electron shells?
a. First shell b. Second shell c. Third shell d. Fourth shell

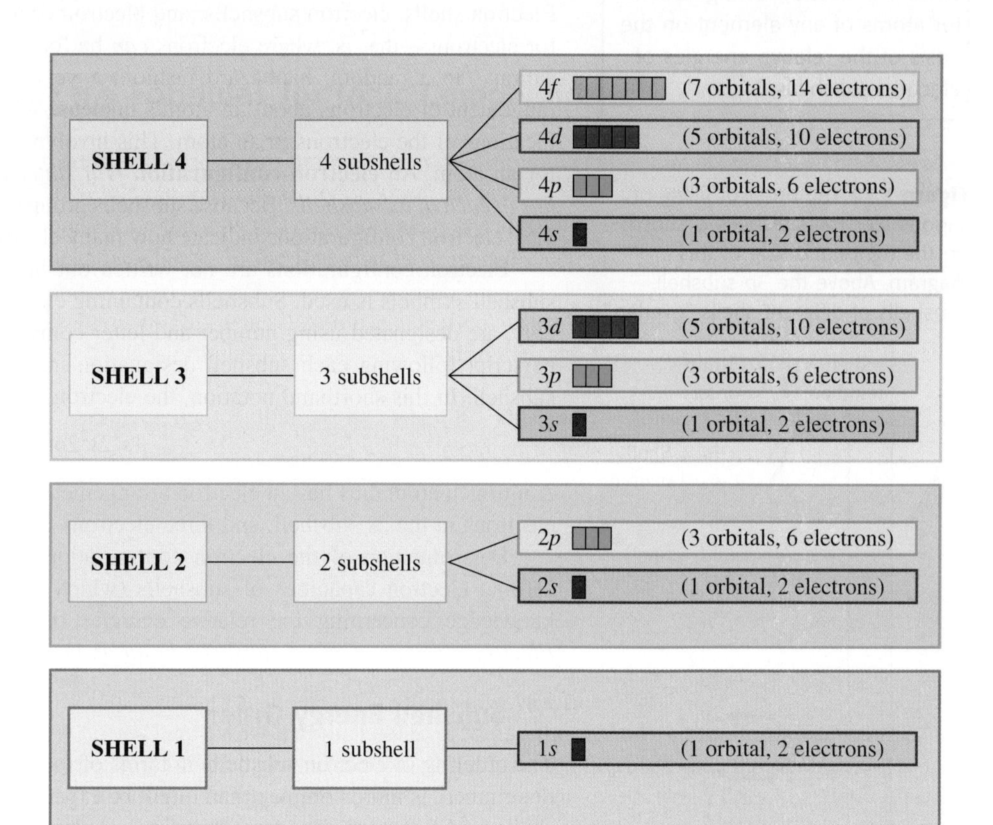

Figure 3.9 A summary of the interrelationships among electron shells, electron subshells, and electron orbitals for the first four shells. Similar relationship patterns exist for higher-numbered shells.

3.28 What is the maximum number of electrons that can occupy each of the following types of *electron subshells*?
a. *s* subshell b. *p* subshell c. *d* subshell d. *f* subshell

3.29 What is the maximum number of electrons that can occupy each of the following *electron orbitals*?
a. 2*s* b. 3*p* c. 3*d* d. 4*s*

3.30 The following statements define or are closely related to the terms *electron shell, electron subshell,* and *electron orbital.* Match each statement with the appropriate term.
a. In terms of electron capacity, this unit is the smallest of the three.
b. This unit can contain a maximum of 2 electrons.
c. This unit is designated using just a number.
d. The term *energy sublevel* is closely associated with this unit.

3.31 Indicate whether each of the following statements is *true* or *false*.
a. An orbital has a definite size and shape, which are related to the energy of the electrons it could contain.
b. All the orbitals in a subshell have the same energy.
c. All subshells accommodate the same number of electrons.
d. A 2*p* subshell and a 3*p* subshell can accommodate the same number of electrons.

3.32 Give the maximum number of electrons that can occupy each of the following electron-accommodating units.
a. One of the orbitals in the 2*p* subshell
b. One of the orbitals in the 3*d* subshell
c. The 4*p* subshell
d. The third shell

3.7 Electron Configurations

Learning Focus

Write the electron configuration for atoms of any element on the basis of the relative energies of electron subshells.

Electron shells, electron subshells, and electron orbitals describe "permissible" locations for electrons—that is, where electrons *can* be found. Electrons do not occupy these "locations" in a random, haphazard fashion; a very predictable pattern exists for the arrangement of electrons about an atom's nucleus. We are now ready to discuss the *actual* locations of the electrons in an atom. This involves specifying the *electron configuration* for an atom. An **electron configuration** *is a statement of how many electrons an atom has in each of its subshells.* Because subshells group electrons according to energy (Section 3.6), electron configurations indicate how many electrons an atom has of various energies.

Electron configurations are not written out in words; a shorthand system based on subshell symbols is used. Subshells containing electrons, listed in order of increasing energy, are designated using number-and-letter combinations (1*s*, 2*s*, 2*p*, and so on). A superscript following each subshell designation indicates the number of electrons in that subshell. In this shorthand notation, the electron configuration for nitrogen is

$$1s^2 2s^2 2p^3$$

A nitrogen atom thus has an electron arrangement of two electrons in the 1*s* subshell, two electrons in the 2*s* subshell, and three electrons in the 2*p* subshell.

Determination of the electron configuration for an atom requires knowledge concerning electron capacities of subshells (which we already have; see Section 3.6) and knowledge concerning the relative energies of subshells (which we now consider). *Electrons occupy electron subshells in an atom in order of increasing subshell energy.*

Figure 3.10 The order of filling of various electron subshells is shown on the right-hand side of this diagram. Above the 3*p* subshell, subshells of different shells "overlap."

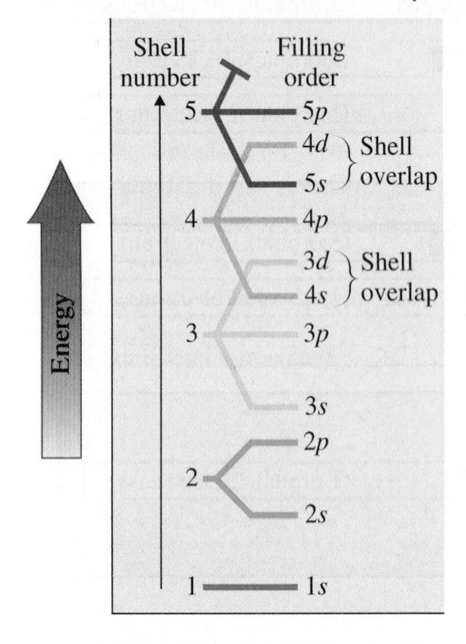

Subshell Energy Order

The ordering of electron subshells in terms of increasing energy, which is experimentally determined, is more complex than might be expected. This is because the energies of subshells in different shells often "overlap," as shown in Figure 3.10. This diagram shows, for example, that the 4*s* subshell has lower energy than the 3*d* subshell.

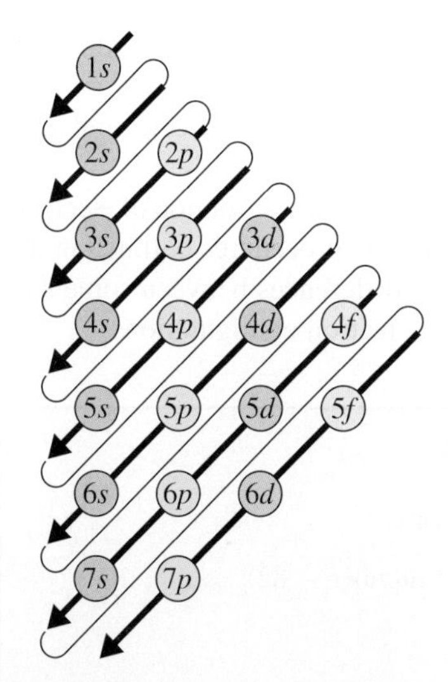

Figure 3.11 The order for filling electron subshells with electrons follows the order given by the arrows in this diagram. Start with the arrow at the top of the diagram and work toward the bottom of the diagram, moving from the bottom of one arrow to the top of the next-lower arrow.

A useful mnemonic (memory) device for remembering subshell filling order, which incorporates "overlap" situations such as those in Figure 3.10, is given in Figure 3.11. This diagram, which lists all subshells needed to specify the electron arrangements for all 113 elements, is constructed by locating all *s* subshells in column 1, all *p* subshells in column 2, and so on. Subshells that belong to the same shell are found in the same row. The order of subshell filling is given by following the diagonal arrows, starting at the top. The 1*s* subshell fills first. The second arrow points to (goes through) the 2*s* subshell, which fills next. The third arrow points to both the 2*p* and the 3*s* subshells. The 2*p* fills first, followed by the 3*s*. Any time a single arrow points to more than one subshell, we start at the tail of the arrow and work to its tip to determine the proper filling sequence.

Writing Electron Configurations

We are now ready to write electron configurations. Let us systematically consider electron configurations for the first few elements in the periodic table.

Hydrogen (atomic number = 1) has only one electron, which goes into the 1*s* subshell, which has the lowest energy of all subshells. Hydrogen's electron configuration is written as

$$1s^1$$

Helium (atomic number = 2) has two electrons, both of which occupy the 1*s* subshell. (Remember, an *s* subshell contains one orbital, and an orbital can accommodate two electrons.) Helium's electron configuration is

$$1s^2$$

Lithium (atomic number = 3) has three electrons, and the third electron cannot enter the 1*s* subshell because its maximum capacity is two electrons. (All *s* subshells are completely filled with two electrons.) The third electron is placed in the next-higher-energy subshell, the 2*s*. The electron configuration for lithium is

$$1s^22s^1$$

For *beryllium* (atomic number = 4), the additional electron is placed in the 2*s* subshell, which is now completely filled, giving beryllium the electron configuration

$$1s^22s^2$$

For *boron* (atomic number = 5), the 2*p* subshell, which is the subshell of next-higher energy (Figures 3.10 and 3.11) becomes occupied for the first time. Boron's electron configuration is

$$1s^22s^22p^1$$

A *p* subshell can accommodate six electrons because there are three orbitals within it (Section 3.6). The 2*p* subshell can thus accommodate the additional electrons found in the elements with atomic numbers 6 through 10: *carbon* (C), *nitrogen* (N), *oxygen* (O), *fluorine* (F), and *neon* (Ne). The electron configurations for these elements are

▶ The symbols $1s^2$, $2s^2$, and $2p^3$ are read as "one s two," "two s two," and "two p three," not as "one s squared," "two s squared," and "two p cubed."

$$\begin{aligned} \text{C: } & 1s^22s^22p^2 \\ \text{N: } & 1s^22s^22p^3 \\ \text{O: } & 1s^22s^22p^4 \\ \text{F: } & 1s^22s^22p^5 \\ \text{Ne: } & 1s^22s^22p^6 \end{aligned}$$

With *sodium* (atomic number = 11), the 3*s* subshell acquires an electron for the first time. Sodium's electron configuration is

$$1s^22s^22p^63s^1$$

Note the pattern that is developing in the electron configurations we have written so far. Each element has an electron configuration that is the same as the one just before it except for the addition of one electron.

▶ An *electron configuration* is a shorthand notation designating the subshells in an atom that are occupied by electrons. The sum of the superscripts in an electron configuration equals the total number of electrons present and hence must equal the atomic number of the element.

Electron configurations for other elements are obtained by simply extending the principles we have just illustrated. A subshell of lower energy is always filled before electrons are added to the next highest subshell; this continues until the correct number of electrons have been accommodated.

For a few elements in the middle of the periodic table, the actual distribution of electrons within subshells differs slightly from that obtained by using the procedures outlined in this section. These exceptions are caused by very small energy differences between some subshells and are not important in the uses we shall make of electron configurations.

Example 3.3 Writing an Electron Configuration

Write the electron configurations for the following elements.

a. Strontium (atomic number = 38) **b.** Lead (atomic number = 82)

Solution

a. The number of electrons in a strontium atom is 38. Remember that the atomic number gives the number of electrons (Section 3.2). We will need to fill subshells, in order of increasing energy, until 38 electrons have been accommodated.

The $1s$, $2s$, and $2p$ subshells fill first, accommodating a total of 10 electrons among them.

$$1s^22s^22p^6 \ldots$$

Next, according to Figures 3.10 and 3.11 the $3s$ subshell fills and then the $3p$ subshell.

$$1s^22s^22p^6\boxed{3s^23p^6}\ldots$$

We have accommodated 18 electrons at this point. We still need to add 20 more electrons to get our desired number of 38.

The $4s$ subshell fills next, followed by the $3d$ subshell, giving us 30 electrons at this point.

$$1s^22s^22p^63s^23p^6\boxed{4s^23d^{10}}\ldots$$

Note that the maximum electron population for d subshells is 10 electrons.

Eight more electrons are needed, which are added to the next two higher subshells, the $4p$ and the $5s$. The $4p$ subshell can accommodate 6 electrons, and the $5s$ can accommodate 2 electrons.

$$1s^22s^22p^63s^23p^64s^23d^{10}\boxed{4p^65s^2}$$

To double-check that we have the correct number of electrons, 38, we add the superscripts in our final electron configuration.

$$2 + 2 + 6 + 2 + 6 + 2 + 10 + 6 + 2 = 38$$

The sum of the superscripts in any electron configuration should add up to the atomic number if the configuration is for a neutral atom.

b. To write this configuration, we continue along the same lines as in part **a**, remembering that the maximum electron subshell populations are $s = 2$, $p = 6$, $d = 10$, and $f = 14$.

Lead, with an atomic number of 82, contains 82 electrons, which are added to subshells in the following order. (The line of numbers beneath the electron configuration is a running total of added electrons and is obtained by adding the superscripts up to that point. We stop when we have 82 electrons.)

$$1s^22s^22p^63s^23p^64s^23d^{10}4p^65s^24d^{10}5p^66s^24f^{14}5d^{10}6p^2$$

2 4 10 12 18 20 30 36 38 48 54 56 70 80 82

Running total of electrons added

Note in this electron configuration that the $6p$ subshell contains only 2 electrons, even though it can hold a maximum of 6. We put only 2 electrons in this subshell because that is sufficient to give 82 total electrons. If we had completely filled this subshell, we would have had 86 total electrons, which is too many.

Practice Questions and Problems

3.33 On the basis of the total number of electrons present, identify the elements whose electron configurations are
a. $1s^22s^22p^4$ b. $1s^22s^22p^6$ c. $1s^22s^22p^63s^23p^1$ d. $1s^22s^22p^63s^23p^64s^2$

3.34 Write the electron configurations for atoms of the following elements.
a. Carbon b. Scandium c. Arsenic d. Element with atomic number of 17

3.35 Indicate whether each of the following statements concerning the element whose electron configuration is $1s^22s^22p^63s^23p^5$ is *true* or *false*.
a. Atoms of the element contain 17 electrons.
b. Atoms of the element have electrons in five different subshells.
c. Atoms of the element have electrons in subshells in three different shells.
d. The identity of the element is sulfur.

Learning Focus

Be able to relate the periodic law and the shape of the periodic table to electron configurations for atoms of the elements.

3.8 The Electronic Basis for the Periodic Law and the Periodic Table

For many years, there was no explanation available for either the periodic law or why the periodic table has the shape that it has. We now know that the theoretical basis for both the periodic law and the periodic table is found in electronic theory. As we saw earlier in the chapter (Section 3.2), when two atoms interact, it is their electrons that interact. Thus the number and arrangement of electrons determine how an atom reacts with other atoms—that is, what its chemical properties are. The properties of the elements repeat themselves in a periodic manner because the arrangement of electrons about the nucleus of an atom follows a periodic pattern, as we saw in Section 3.7.

Electron Configurations and the Periodic Law

The periodic law (Section 3.4) points out that the properties of the elements repeat themselves in a regular manner when the elements are arranged in order of increasing atomic number. The elements that have similar chemical properties are placed under one another in vertical columns (groups) in the periodic table.

Groups of elements have similar chemical properties because of similarities in their electron configuration. *Chemical properties repeat themselves in a regular manner among the elements because electron configurations repeat themselves in a regular manner among the elements.*

To illustrate this correlation between similar chemical properties and similar electron configurations, let us look at the electron configurations of two groups of elements known to have similar chemical properties.

We begin with the elements lithium, sodium, potassium, and rubidium, all members of Group IA of the periodic table. The electron configurations for these elements are

▶ The electron arrangement in the outermost shell is the same for elements in the same group. This is why elements in the same group have similar chemical properties.

$$_{3}\text{Li}: \quad 1s^2\,2s^1$$
$$_{11}\text{Na}: \quad 1s^22s^22p^6\,3s^1$$
$$_{19}\text{K}: \quad 1s^22s^22p^63s^23p^6\,4s^1$$
$$_{37}\text{Rb}: \quad 1s^22s^22p^63s^23p^64s^23d^{10}4p^6\,5s^1$$

Note that each of these elements has one electron in its outermost shell. (The outermost shell is the shell with the highest number.) This similarity in outer-shell electron arrangements causes these elements to have similar chemical properties. In general, elements with similar outer-shell electron configurations have similar chemical properties.

Let us consider another group of elements known to have similar chemical properties: fluorine, chlorine, bromine, and iodine of Group VIIA of the periodic table. The electron configurations for these four elements are

$$_{9}F: \quad 1s^2 2s^2 2p^5$$
$$_{17}Cl: \quad 1s^2 2s^2 2p^6 3s^2 3p^5$$
$$_{35}Br: \quad 1s^2 2s^2 2p^6 3s^2 3p^6 4s^2 3d^{10} 4p^5$$
$$_{53}I: \quad 1s^2 2s^2 2p^6 3s^2 3p^6 4s^2 3d^{10} 4p^6 5s^2 4d^{10} 5p^5$$

Once again, similarities in electron configuration are readily apparent. This time, the repeating pattern involves an outermost *s* and *p* subshell containing seven electrons (shown in color). Remember that for Br and I, shell numbers 4 and 5 designate, respectively, electrons in the outermost shells.

Electron Configurations and the Periodic Table

One of the strongest pieces of supporting evidence for the assignment of electrons to shells, subshells, and orbitals is the periodic table itself. The basic shape and structure of this table, which was determined many years before electrons were even discovered, is consistent with and can be explained by electron configurations. Indeed, the specific location of an element in the periodic table can be used to obtain information about its electron configuration.

As the first step in linking electron configurations to the periodic table, let us analyze the general shape of the periodic table in terms of columns of elements. As shown in Figure 3.12, on the extreme left of the table, there are 2 columns of elements; in the center, there is a region containing 10 columns of elements; to the right there is a block of 6 columns of elements; and in the two rows at the bottom of the table, there are 14 columns of elements.

The number of columns of elements in the various regions of the periodic table—2, 6, 10, and 14—is the same as the maximum number of electrons that the various types of subshells can accommodate. We will see shortly that this is a very significant observation; the number matchup is no coincidence. The various columnar regions of the periodic table are called the *s* area (2 columns), the *p* area (6 columns), the *d* area (10 columns), and the *f* area (14 columns), as shown in Figure 3.12.

The concept of *distinguishing electrons* is the key to obtaining electron configuration information from the periodic table. The **distinguishing electron** *for an element is*

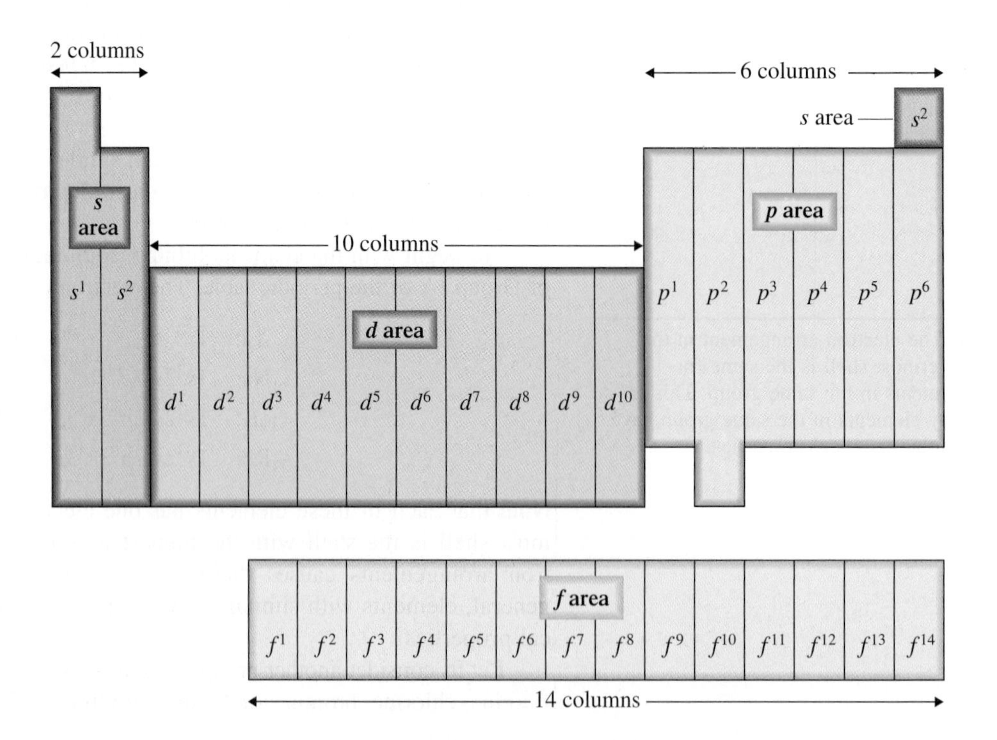

Figure 3.12 Electron configurations and the positions of elements in the periodic table. The periodic table can be divided into four areas that are 2, 6, 10, and 14 columns wide. The four areas contain elements whose distinguishing electron is located, respectively, in *s*, *p*, *d*, and *f* subshells. The extent of filling of the subshell that contains an element's distinguishing electron can be determined from the element's position in the periodic table.

the last electron that is added to its electron configuration when subshells are filled in order of increasing energy. This last electron is the one that causes an element's electron configuration to differ from that of the element immediately preceding it in the periodic table.

For all elements that are located in the *s* area of the periodic table, the distinguishing electron is always found in an *s* subshell. All *p* area elements have distinguishing electrons in *p* subshells. Similarly, elements in the *d* and *f* areas of the periodic table have distinguishing electrons located in *d* and *f* subshells, respectively. Thus the area location of an element in the periodic table can be used to determine the type of subshell that contains the distinguishing electron. Note that the element helium belongs to the *s* rather than the *p* area of the periodic table, even though its table position is on the right-hand side. (The reason for this placement of helium will be explained in Section 4.3).

The extent to which the subshell containing an element's distinguishing electron is filled can also be determined from the element's position in the periodic table. All elements in the first column of a specific area contain only one electron in the subshell; all elements in the second column contain two electrons in the subshell; and so on. Thus all elements in the first column of the *p* area (Group IIIA) have an electron configuration ending in p^1. Elements in the second column of the *p* area (Group IVA) have electron configurations ending in p^2; and so on. Similar relationships hold in other areas of the table, as shown in Figure 3.12.

Practice Questions and Problems

3.36 Indicate whether the elements represented by the given pairs of electron configurations have similar chemical properties.
a. $1s^22s^1$ and $1s^22s^2$
b. $1s^22s^22p^6$ and $1s^22s^22p^63s^23p^6$
c. $1s^22s^22p^3$ and $1s^22s^22p^63s^23p^64s^23d^3$
d. $1s^22s^22p^63s^23p^4$ and $1s^22s^22p^63s^23p^64s^23d^{10}4p^4$

3.37 Specify the location of each of the following elements in the periodic table in terms of *s* area, *p* area, *d* area, or *f* area.
a. Magnesium b. Copper c. Uranium d. Bromine

3.38 With the help of the periodic table, for each of the following elements, specify the extent to which the subshell containing the distinguishing electron is filled (s^2, p^3, d^5, etc.).
a. $_{13}Al$ b. $_{23}V$ c. $_{20}Ca$ d. $_{36}Kr$

3.9 Classification of the Elements

Learning Focus

Understand the classification system for elements that is based on electron configuration.

The elements can be classified in several ways. The two most common classification systems are

1. A system based on selected physical properties of the elements, in which they are described as metals or nonmetals. This classification scheme was discussed in Section 3.5.
2. A system based on the electron configurations of the elements, in which elements are described as *noble gas, representative, transition,* or *inner transition elements*.

The classification scheme based on electron configurations of the elements is depicted in Figure 3.13.

The **noble-gas elements** *are found in the far right column of the periodic table*. They are all gases at room temperature, and they have little tendency to form chemical compounds. With one exception, the distinguishing electron for a noble gas completes the *p* subshell; therefore, noble gases have electron configurations ending in p^6. The exception is helium, in which the distinguishing electron completes the first shell—a shell that has only two electrons. Helium's electron configuration is $1s^2$.

▶ The electron configurations of the noble gases will be an important focal point when we consider chemical bonding theory in Chapter 4.

Figure 3.13 A classification scheme for the elements based on their electron configurations. Representative elements occupy the *s* area and most of the *p* area shown in Figure 3.12. The noble-gas elements occupy the last column of the *p* area. The transition elements are found in the *d* area, and the inner transition elements are found in the *f* area.

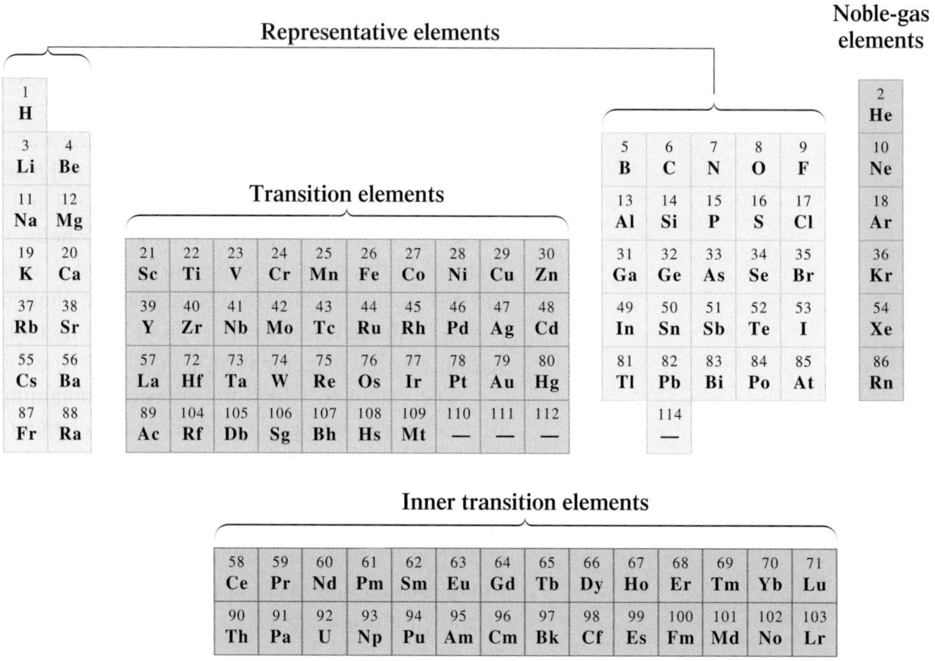

Representative elements | Transition elements | Noble-gas elements

Representative elements		Transition elements										Representative elements					Noble-gas elements
1 H																	2 He
3 Li	4 Be											5 B	6 C	7 N	8 O	9 F	10 Ne
11 Na	12 Mg											13 Al	14 Si	15 P	16 S	17 Cl	18 Ar
19 K	20 Ca	21 Sc	22 Ti	23 V	24 Cr	25 Mn	26 Fe	27 Co	28 Ni	29 Cu	30 Zn	31 Ga	32 Ge	33 As	34 Se	35 Br	36 Kr
37 Rb	38 Sr	39 Y	40 Zr	41 Nb	42 Mo	43 Tc	44 Ru	45 Rh	46 Pd	47 Ag	48 Cd	49 In	50 Sn	51 Sb	52 Te	53 I	54 Xe
55 Cs	56 Ba	57 La	72 Hf	73 Ta	74 W	75 Re	76 Os	77 Ir	78 Pt	79 Au	80 Hg	81 Tl	82 Pb	83 Bi	84 Po	85 At	86 Rn
87 Fr	88 Ra	89 Ac	104 Rf	105 Db	106 Sg	107 Bh	108 Hs	109 Mt	110 —	111 —	112 —		114 —				

Inner transition elements

58 Ce	59 Pr	60 Nd	61 Pm	62 Sm	63 Eu	64 Gd	65 Tb	66 Dy	67 Ho	68 Er	69 Tm	70 Yb	71 Lu
90 Th	91 Pa	92 U	93 Np	94 Pu	95 Am	96 Cm	97 Bk	98 Cf	99 Es	100 Fm	101 Md	102 No	103 Lr

The unreactive nature of the noble gases is reflected in their occurrence in nature in the form of individual uncombined atoms. There are no known compounds of the lighter noble gases—helium, neon, and argon—and only a very few compounds of the heavier noble gases—krypton, xenon, and radon. Chemical Portraits 5 profiles helium, argon, and krypton, the three noble gases that are totally unreactive (no compounds are known).

Chemical Portraits 5 Noble Gases: The Most Unreactive of All Elements

Helium (He)

Profile: Helium is the only element that was discovered extraterrestrially before being found on Earth. It was first detected in solar flares associated with the sun's outer surface. On Earth, almost all He is obtained from natural gas, which contains up to 0.3% He. Because of dwindling natural gas supplies, the U.S. government now stockpiles He to ensure its availability in the future. The amount of He present in the Earth's atmosphere is very low.

Uses: Although hydrogen is less dense than helium, He is preferred for use in balloons and lighter-than-air craft such as blimps because it is nonflammable.

Why is He substituted for N_2 in the breathing mixtures used by "deep-sea" divers?

Neon (Ne)

Profile: Neon, like all noble gases, is a colorless, odorless, nonflammable, nontoxic gas. It is three times more abundant than helium and is extracted from liquid air for commercial purposes. A helium-neon laser, emitting red light, was the first continuously operating gas laser.

Uses: Ne gas emits a bright red glow when an electric current is passed through it. This is the basis for its use in luminescent advertising signs and the designation of such signs as "neon signs." (The term *neon sign* has become a generic term that denotes all such advertising signs, even though many of them contain other noble gases than neon.).

What is the relationship between the element neon and bar-code scanners used in grocery stores?

Argon (Ar)

Profile: Argon, with a 1% by volume concentration, is the third-most-abundant atmospheric gas. This concentration is 400 times greater than that of all the other noble gases combined.

Uses: A 93% Ar-7% N_2 mixture is used to fill metal filament light bulbs. This mixture extends the life of the metal filament by slowing its vaporization, a process that causes the light bulb to "darken." Argon also finds use as a "blanketing agent" in high temperature metallurgical welding processes; the blanketing agent prevents the hot metal from reacting with oxygen. Ar is preferred over He for this purpose because its greater density provides better protection.

Why is 93% argon rather than 100% argon gas used in metal filament light bulbs?

See the text web site at **www.cengage.com/chemistry/stoker** for answers to the above questions and for further information.

Element Classification Schemes and the Periodic Table

CLASSIFICATION BY PHYSICAL PROPERTIES	
Nonmetals	■ No metallic luster ■ Poor electrical conductivity ■ Good heat insulators ■ Brittle and nonmalleable
Metals	■ Metallic gray or silver luster ■ High electrical and thermal conductivity ■ Malleable and ductile

CLASSIFICATION BY ELECTRONIC PROPERTIES	
Representative elements	■ Found in *s* area and first five columns of the *p* area ■ Some are metals, some nonmetals
Noble-gas elements	■ Found in last column of *p* area plus He (*s* area) ■ All are nonmetals
Transition elements	■ Found in *d* area ■ All are metals
Inner transition elements	■ Found in *f* area ■ All are metals

PERIODIC TABLE GROUPS WITH SPECIAL NAMES	
Alkali metals	■ Group IA elements (except for H, a nonmetal) ■ Electron configurations end in s^1
Alkaline earth metals	■ Group IIA elements ■ Electron configurations end in s^2
Halogens	■ Group VIIA ■ Electron configurations end in p^5
Noble gases	■ Group VIIIA elements ■ Electron configurations end in p^6, except for He, which ends in s^2

IA IIA VIIA VIIIA

The **representative elements** *are all the elements of the* s *and* p *areas of the periodic table, with the exception of the noble gases.* The distinguishing electron in these elements partially or completely fills an *s* subshell or partially fills a *p* subshell. The representative elements include most of the more common elements.

The **transition elements** *are all the elements of the* d *area of the periodic table.* Each has its distinguishing electron in a *d* subshell.

The **inner transition elements** *are all the elements of the* f *area of the periodic table.* Each has its distinguishing electron in an *f* subshell. There is very little variance in the properties of either the 4*f* or the 5*f* series of inner transition elements.

The Chemistry at a Glance above contrasts the three element classification schemes that have been considered so far in this chapter: by physical properties (Section 3.5), by electron configuration (Section 3.9), and by non-numerical periodic table group names (Section 3.4).

Practice Questions and Problems

3.39 Classify each of the following elements as a noble gas, a representative element, a transition element, or an inner transition element.
a. $_{15}P$ b. $_{18}Ar$ c. $_{79}Au$ d. $_{92}U$

3.40 Classify the element with each of the following electron configurations as a noble gas, a representative element, transition element, or inner transition element.
a. $1s^22s^22p^6$
b. $1s^22s^22p^63s^23p^4$
c. $1s^22s^22p^63s^23p^64s^23d^1$
d. $1s^22s^22p^63s^23p^64s^2$

Learning Focus

Understand what is meant by the term *radioactive atom.*

3.10 Nuclear Stability and Radioactivity

We now return to further discussion about the nucleus of an atom. In Section 3.1 we learned that an atomic nucleus, located at the center of an atom, is a small, dense structure that contains all of the protons and neutrons present in an atom. Numerous studies concerning atomic nuclei show that they may be divided into two categories based on nuclear stability. Some nuclei are stable and others are not. A **stable nucleus** *is a nucleus that does not easily undergo change.* Conversely, an **unstable nucleus** *is a nucleus that spontaneously undergoes change.* The spontaneous change that unstable nuclei undergo involves emission of radiation from the nucleus, a process by which an unstable nucleus can become more stable. The radiation emitted from unstable nuclei is called *radioactivity.* **Radioactivity** *is the radiation spontaneously emitted from an unstable nucleus.* Atoms that possess unstable nuclei are said to be *radioactive.* A **radioactive atom** *is an atom with an unstable nucleus from which radiation is spontaneously emitted.*

Of the 88 elements found in nature (Section 1.7), 29 have at least one naturally occurring isotope that has an unstable nucleus and is, therefore, a *radioactive* isotope. Radioactive isotopes are known for *all* 113 elements, even though they occur naturally for only the aforementioned 29 elements. This is because laboratory procedures have been developed by which scientists convert nonradioactive isotopes (stable nucleus) into radioactive isotopes (unstable nucleus).

No simple rule exists for predicting whether a particular nucleus is radioactive. However, considering some observations about those nuclei that are stable is helpful in understanding why some nuclei are stable and others are not. Two generalizations are readily apparent from a study of the properties of stable nuclei found in nature.

1. *There is a correlation between nuclear stability and the total number of nucleons found in a nucleus.* All nuclei with 84 or more protons are unstable. The largest stable nucleus known is that of $^{209}_{83}Bi$, a nucleus that contains 209 nucleons. It thus appears that there is a limit to the number of nucleons that can be packed into a stable nucleus.
2. *There is a correlation between nuclear stability and neutron-to-proton ratio in a nucleus.* The number of neutrons necessary to create a stable nucleus increases as the number of protons increases. For elements of low atomic number, neutron-to-proton ratios for stable nuclei are very close to 1. For heavier elements, stable nuclei have higher neutron-to-proton ratios, and the ratio reaches approximately 1.5 for the heaviest stable elements. These observations suggest that neutrons are at least partially responsible for the stability of a nucleus. It should be remembered that like charges repel each other and that most nuclei contain many protons (with identical positive charges) squeezed together into a very small volume. As the number of protons increases, the forces of repulsion between protons sharply increase.

Therefore, a greater number of neutrons is necessary to counteract the increased repulsions. Finally, at element 84, the repulsive forces become so great that nuclei are unstable regardless of the number of neutrons present.

Practice Questions and Problems

3.41 What physical manifestation indicates that an atom possesses an unstable nucleus?

3.42 What is the limit for nuclear stability in terms of number of nucleons present in a nucleus?

3.43 How do the neutron-to-proton ratios compare for stable nuclei of low atomic number and stable nuclei of high atomic number?

3.44 For which of the following elements would all isotopes be radioactive?
a. ${}_{76}Os$ b. ${}_{86}Rn$ c. ${}_{96}Cm$ d. ${}_{106}Sg$

Learning Focus

Understand the concept of half-life, and determine the amount of a radionuclide left after a given number of half-lives, or vice versa.

3.11 Half-Life

The term *radioactive isotope,* and its shortened form *radioisotope,* are alternative designations for the term *radioactive atom.* The process by which radioactive isotopes produce radiation is called *radioactive decay.* **Radioactive decay** *is the process whereby an unstable nucleus undergoes change as a result of the emission of radiation.* Not all radioactive isotopes decay at the same rate. Some decay very rapidly; others undergo change at extremely slow rates. This indicates that not all radioactive isotopes are equally unstable. The faster the decay rate, the lower the stability of a nucleus.

The concept of half-life is used to quantify the instability of an atomic nucleus. The **half-life** ($t_{1/2}$) *is the time required for 1/2 of any given quantity of a radioactive substance to undergo decay.* For example, if a radionuclide's half-life is 12 days and you have a 4.00-g sample of it, then after 1 half-life (12 days), only 2.00 g of the sample (1/2 of the original amount) will remain undecayed; the other half will have decayed into some other substance (Section 3.12). Similarly, during the next half-life, 1/2 of the 2.00 g remaining will decay, leaving 1/4 of the original atoms (1.00 g) unchanged. After 3 half-lives, 1/8 ($1/2 \times 1/2 \times 1/2$) of the original sample will remain undecayed. Figure 3.14 illustrates the radioactive decay curve for a radioisotope.

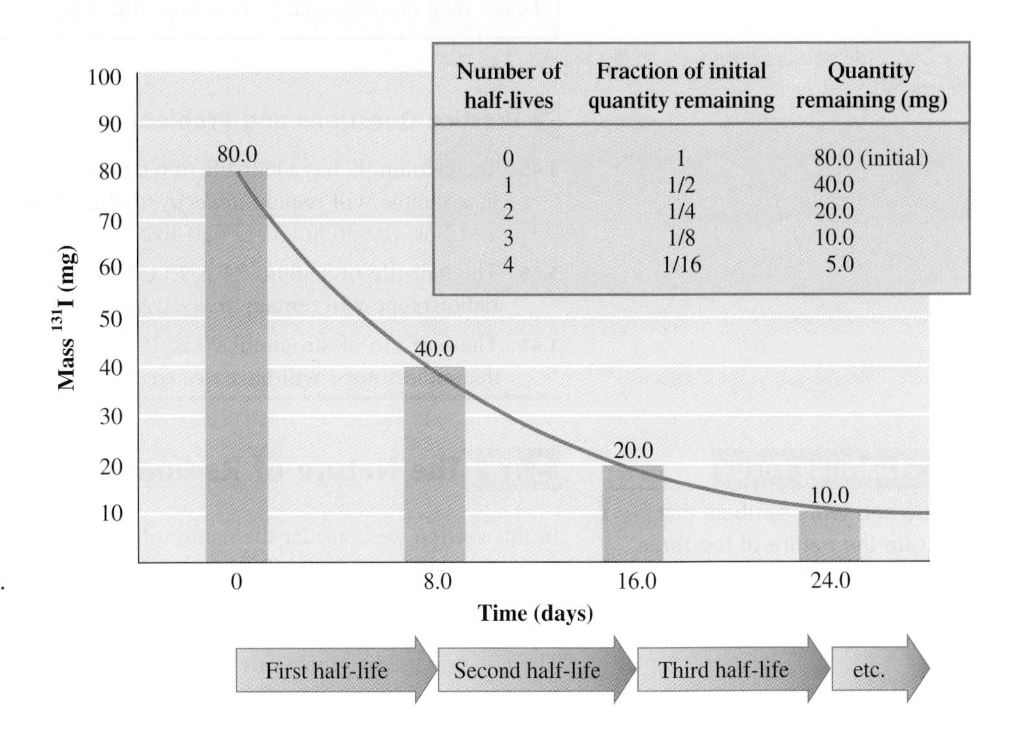

Number of half-lives	Fraction of initial quantity remaining	Quantity remaining (mg)
0	1	80.0 (initial)
1	1/2	40.0
2	1/4	20.0
3	1/8	10.0
4	1/16	5.0

Figure 3.14 Decay of 80.0 mg of ^{131}I, which has a half-life of 8.0 days. After each half-life period, the quantity of material present at the beginning of the period is reduced by half.

▶ Most radionuclides used in diagnostic medicine have short half-lives. This limits to a short time interval the exposure of the human body to radiation.

▶ The half-life for a radionuclide is independent of external conditions such as temperature, pressure, and state of chemical combination.

There is a wide range of half-lives for radionuclides. Half-lives as long as billions of years and as short as a fraction of a second have been determined (Table 3.3). Most naturally occurring radionuclides have long half-lives. However, some radionuclides with *short* half-lives are also found in nature. Naturally occurring mechanisms exist for the continual production of the short-lived species.

The decay rate (half-life) of a radionuclide is constant. It is independent of physical conditions such as temperature, pressure, and state of chemical combination. It depends only on the identity of the radionuclide. For example, radioactive sodium-24 decays at the same rate whether it is incorporated into NaCl, NaBr, Na_2SO_4, or $NaC_2H_3O_2$. If a nuclide is radioactive, nothing will stop it from decaying and nothing will increase or decrease its decay rate.

Table 3.3
Range of Half-lives Found for Naturally Occurring Radionuclides

Element	Half-life ($t_{1/2}$)
vanadium-50	6×10^{15} yr
platinum-190	6.9×10^{11} yr
uranium-238	4.5×10^{9} yr
uranium-235	7.1×10^{8} yr
thorium-230	7.5×10^{4} yr
lead-210	22 yr
bismuth-214	19.7 min
polonium-212	3.0×10^{-7} sec

Example 3.4 Using Half-life to Calculate the Amount of Radioisotope That Remains Undecayed After a Certain Time

Iodine-131 is a radionuclide that is frequently used in nuclear medicine. Among other things, it is used to detect fluid buildup in the brain. The half-life of iodine-131 is 8.0 days. How much of a 16.0-g sample of iodine-131 will remain undecayed after a period of 32 days?

Solution

First, we must determine the number of half-lives that have elapsed.

$$32 \text{ days} \times \left(\frac{1 \text{ half-life}}{8.0 \text{ days}} \right) = 4 \text{ half-lives}$$

Constructing a tabular summary of the amount of sample remaining after each of the elapsed half-lives yields the following data.

Number of half-lives	Fraction decayed	Amount undecayed
0	none	original amount (16.0 g)
1	1/2	$1/2 \times (16.0 \text{ g}) = 8.00$ g
2	$1/2 \times 1/2 = 1/4$	$1/4 \times (16.0 \text{ g}) = 4.00$ g
3	$1/2 \times 1/2 \times 1/2 = 1/8$	$1/8 \times (16.0 \text{ g}) = 2.00$ g
4	$1/2 \times 1/2 \times 1/2 \times 1/2 = 1/16$	$1/16 \times (16.0 \text{ g}) = 1.00$ g

Thus 1.00 g of undecayed ^{131}I remains after 4 half-lives (32 days) have elapsed.

Practice Questions and Problems

3.45 Technetium-99 has a half-life of 6.0 hours. What fraction of the technetium-99 atoms in a sample will remain *undecayed* after the following times?
a. 12 hr b. 36 hr c. 3 half-lives d. 6 half-lives

3.46 The half-life of sodium-24 is 15.0 hr. How many grams of a 4.00-g sample of this radioisotope will remain *undecayed* after 60.0 hr?

3.47 The half-life of strontium-90 is 28 years. How many grams of a 4.00-g sample of this radioisotope will have *decayed* after 112 years?

3.12 The Nature of Radioactive Emissions

Learning Focus

Name and write symbols that indicate the nature of the three most common types of radiation given off by unstable nuclei.

In this section we consider the nature of the three most common types of radiation emanating from the nuclei of radioactive atoms. These radiation types are alpha particles, beta particles, and gamma rays.

An **alpha particle** *is a particle in which two protons and two neutrons are present.* The notation used to represent an alpha particle is $^{4}_{2}\alpha$. The numerical subscript indicates that the charge on the particle is +2 (from the two protons). The numerical superscript

Table 3.4
Characteristics of the Three Most Common Types of Radiation Given Off by Radioactive Atoms

Type of radiation	Symbol for radiation	Charge	Relative mass
Alpha	${}^{4}_{2}\alpha$	+2	heavy
Beta	${}^{0}_{-1}\beta$	−1	light
Gamma	${}^{0}_{0}\gamma$	none	none

indicates a mass of 4 amu. Alpha particles are identical to the nuclei of helium-4 (${}^{4}_{2}He$) atoms; because of this, an alternative designation for an alpha particle is ${}^{4}_{2}He$.

A **beta particle** *is a particle whose charge and mass are identical to those of an electron.* However, beta particles are not extranuclear electrons; they are particles that have been produced inside the nucleus and then ejected. We will discuss this process in Section 3.13. The symbol used to represent a beta particle is ${}^{0}_{-1}\beta$. The numerical subscript indicates that the charge on the beta particle is −1; it is the same as that of an electron. The use of the superscript zero for the mass of a beta particle should be interpreted as meaning not that a beta particle has no mass but, rather, that the mass is very close to zero amu. The actual mass of a beta particle is 0.00055 amu.

A **gamma ray** *is a form of high energy radiation without mass or charge.* Gamma rays are very high-energy radiation, somewhat like X rays. The symbol for a gamma ray is ${}^{0}_{0}\gamma$.

Table 3.4 contrasts the physical characteristics of alpha, beta, and gamma types of radiation.

Practice Questions and Problems

3.48 Supply a complete symbol, with superscript and subscript, for each of the following types of radiation.
a. Alpha particle b. Beta particle c. Gamma ray

3.49 Give the charge and the mass (in amu) of each of the following types of radiation.
a. Alpha particle b. Beta particle c. Gamma ray

3.50 State the composition of an alpha particle in terms of protons and neutrons.

3.51 What is the relationship between the characteristics of a beta particle and those of an electron?

3.13 Equations for Radioactive Decay

Learning Focus

Write equations, balanced for mass number and atomic number, to represent various alpha, beta, and gamma decay processes. Understand what is meant by the terms *parent nucleus* and *daughter nucleus.*

Alpha, beta, and gamma emissions come from the nucleus of an atom. These spontaneous emissions alter nuclei; obviously, if a nucleus loses an alpha particle (two protons and two neutrons), it will not be the same as it was before the departure of the particle. In the case of alpha and beta emissions, the nuclear alteration causes the identity of atoms to change, forming a new element. The terms *parent nucleus* and *daughter nucleus* are often used when describing radioactive decay processes. A **parent nucleus** *is the nucleus that undergoes decay in a radioactive decay process.* A **daughter nucleus** *is the nucleus that is produced as a result of a radioactive decay process.*

Alpha Particle Decay

Alpha particle decay, which is the emission of an alpha particle from a nucleus, always results in the formation of a nucleus of a different element. The product nucleus has an atomic number that is 2 less than that of the original nucleus and a mass number that is 4 less than that of the original nucleus. We can represent alpha particle decay in general terms by the equation

$${}^{A}_{Z}X \longrightarrow {}^{4}_{2}\alpha + {}^{A-4}_{Z-2}Y$$

where X is the symbol for the nucleus of the original element undergoing decay and Y is the symbol of the element formed as a result of the decay.

▶ Loss of an alpha particle from an unstable nucleus results in (1) a decrease of 4 units in the mass number (A) and (2) a decrease of 2 units in the atomic number (Z).

▶ Note that the symbols in nuclear equations stand for *nuclei* rather than atoms. We do not worry about electrons when writing nuclear equations.

Let us write equations for two alpha particle decay processes. Both $^{211}_{83}Bi$ and $^{238}_{92}U$ are radioisotopes that undergo alpha particle decay. The nuclear equations for these two decay processes are

$$^{211}_{83}Bi \longrightarrow {}^{4}_{2}\alpha + {}^{207}_{81}Tl$$
$$^{238}_{92}U \longrightarrow {}^{4}_{2}\alpha + {}^{234}_{90}Th$$

where thallium-207 and thorium-234 are the daughter nuclei.

In a **balanced nuclear equation,** *the sums of the subscripts (atomic numbers or particle charges) on both sides of the equation are equal, and the sums of the superscripts (mass numbers) on both sides of the equation are equal.*

Both of our example equations are balanced. In the alpha decay of $^{211}_{83}Bi$, the subscripts on both sides total 83, and the superscripts total 211. For the alpha decay of $^{238}_{92}U$, the subscripts total 92 on both sides, and the superscripts total 238.

▶ The rules for balancing nuclear equations are
1. The sum of the subscripts must be the same on both sides of the equation.
2. The sum of the superscripts must be the same on both sides of the equation.

Beta Particle Decay

Beta particle decay always results in the formation of a nucleus of a different element. The mass number of the new nucleus is the same as that of the original atom. However, the atomic number has increased by 1 unit. The general equation for beta decay is

$$^{A}_{Z}X \longrightarrow {}^{0}_{-1}\beta + {}_{Z+1}^{A}Y$$

▶ Loss of a beta particle from an unstable nucleus results in (1) no change in the mass number (A) and (2) an increase of 1 unit in the atomic number (Z).

Specific examples of beta particle decay are

$$^{10}_{4}Be \longrightarrow {}^{0}_{-1}\beta + {}^{10}_{5}B$$
$$^{234}_{90}Th \longrightarrow {}^{0}_{-1}\beta + {}^{234}_{91}Pa$$

Both of these nuclear equations are balanced; superscripts and subscripts add to the same sums on both sides of the equation.

At this point in the discussion, you may be wondering how a nucleus, which is composed only of neutrons and protons, ejects a negative particle (beta particle) when no such particle is present in the nucleus. Explained simply, a neutron in the nucleus is transformed into a proton and a beta particle through a complex series of steps; that is,

$$\text{Neutron} \longrightarrow \text{proton} + \text{beta particle}$$
$$^{1}_{0}n \longrightarrow {}^{1}_{1}p + {}^{0}_{-1}\beta$$

Once it is formed within the nucleus, the beta particle is ejected with a high velocity. Note the symbols used to denote a neutron ($^{1}_{0}n$; no charge and a mass of 1 amu) and a proton ($^{1}_{1}p$; a +1 charge and a mass of 1 amu).

Gamma Ray Emission

For naturally occurring radioisotopes, gamma ray emission almost always takes place in conjunction with an alpha or a beta decay process; it never occurs independently. These gamma rays are often not included in the nuclear equation because they do not affect the balancing of the equation or the identity of the daughter nucleus. This can be seen from the following two nuclear equations.

▶ Gamma rays are to nuclear reactions what heat is to ordinary chemical reactions.

$$^{226}_{88}Ra \longrightarrow {}^{222}_{86}Rn + {}^{4}_{2}\alpha + {}^{0}_{0}\gamma$$

Balanced nuclear equation with gamma radiation included

$$^{226}_{88}Ra \longrightarrow {}^{222}_{86}Rn + {}^{4}_{2}\alpha$$

Balanced nuclear equation with gamma radiation omitted

▶ Among *synthetically* produced radioisotopes, pure "gamma emitters," radioisotopes that give off gamma rays but no alpha or beta particles, occur. These radioisotopes are important in diagnostic nuclear medicine. Pure "gamma emitters" are not found among naturally occurring radioisotopes.

The fact that gamma rays are often left out of balanced nuclear equations should not be interpreted to mean that such rays are not important in nuclear chemistry. On the contrary, gamma rays are more important than alpha and beta particles when the effects of external radiation exposure on living organisms are considered.

Example 3.5 **Writing Balanced Nuclear Equations, Given the Parent Nucleus and Its Mode of Decay**

Write a balanced nuclear equation for the decay of each of the following radioactive nuclei. The mode of decay is indicated in parentheses.

a. $^{70}_{31}Ga$ (beta emission) **b.** $^{144}_{60}Nd$ (alpha emission)
c. $^{248}_{100}Fm$ (alpha emission) **d.** $^{113}_{47}Ag$ (beta emission)

Solution

In each case, the atomic and mass numbers of the daughter nucleus are obtained by writing the symbols of the parent nucleus and the particle emitted by the nucleus (alpha or beta). Then the equation is balanced.

a. Let X represent the product of the radioactive decay, the daughter nucleus. Then

$$^{70}_{31}Ga \longrightarrow {}^{0}_{-1}\beta + X$$

The sums of the superscripts on both sides of the equation must be equal, so the superscript for X must be 70. In order for the sums of the subscripts on both sides of the equation to be equal, the subscript for X must be 32. Then $31 = (-1) + (32)$. As soon as we determine the subscript of X, we can obtain the identity of X by looking at a periodic table. The element with an atomic number of 32 is Ge (germanium). Therefore,

$$^{70}_{31}Ga \longrightarrow {}^{0}_{-1}\beta + {}^{70}_{32}Ge$$

b. Letting X represent the product of the radioactive decay, we have, for the alpha decay of $^{144}_{60}Nd$,

$$^{144}_{60}Nd \longrightarrow {}^{4}_{2}\alpha + X$$

We balance the equation by making the superscripts on each side of the equation total 144 and the subscripts total 60. We get

$$^{144}_{60}Nd \longrightarrow {}^{4}_{2}\alpha + {}^{140}_{58}Ce$$

c. Similarly, we write

$$^{248}_{100}Fm \longrightarrow {}^{4}_{2}\alpha + X$$

Balancing superscripts and subscripts, we get

$$^{248}_{100}Fm \longrightarrow {}^{4}_{2}\alpha + {}^{244}_{98}Cf$$

In alpha emission, the atomic number of the daughter nucleus always decreases by 2, and the mass number of the daughter nucleus always decreases by 4.

d. Finally, we write

$$^{113}_{47}Ag \longrightarrow {}^{0}_{-1}\beta + X$$

In beta emission, the atomic number of the daughter nucleus always increases by 1, and the mass number does not change from that of the parent. The balancing procedure gives us the result

$$^{113}_{47}Ag \longrightarrow {}^{0}_{-1}\beta + {}^{113}_{48}Cd$$

Practice Exercises and Questions

3.52 Write balanced nuclear equations for the alpha particle decay of the following radioisotopes.
a. $^{200}_{84}Po$ b. $^{244}_{96}Cm$ c. Curium-240 d. Uranium-238

3.53 Write balanced nuclear equations for the beta particle decay of the following radioisotopes.
a. $^{10}_{4}Be$ b. $^{77}_{32}Ge$ c. Iron-60 d. Sodium-25

3.54 What is the effect on the mass number and atomic number of the parent nucleus when alpha particle decay occurs?

3.55 What is the effect on the mass number and atomic number of the parent nucleus when beta particle decay occurs?

3.56 Supply the missing symbol in each of the following radioactive decay equations.
a. ${}^{34}_{14}Si \rightarrow {}^{34}_{15}P + ?$ b. $? \rightarrow {}^{28}_{13}Al + {}^{0}_{-1}\beta$
c. ${}^{252}_{99}Es \rightarrow {}^{248}_{97}Bk + ?$ d. ${}^{204}_{82}Pb \rightarrow ? + {}^{4}_{2}\alpha$

3.57 Identify the mode of decay for each of the following radioactive decays, where the parent and daughter nuclei are given.
a. Parent = platinum-190; daughter = osmium-186
b. Parent = oxygen-19; daughter = fluorine-19

3.14 Biological Effects of Radiation

Learning Focus

Contrast the biological effects of alpha, beta, and gamma radiation.

The alpha, beta, and gamma radiations produced from radioactive decay possess extremely high amounts of energy. It is this high-energy content that makes radiation dangerous to living organisms. As the radiations travel outward from their nuclear sources into the material surrounding the radioactive substance, this energy is dissipated through collisions with the atoms and molecules of the surrounding materials. In the great majority of radiation–atom and radiation–molecule interactions, electrons are knocked away from the atoms and molecules involved, producing very reactive species with chemistries much different from the species from which they were produced. It is the presence of these radiation-produced, very reactive species that leads to "radiation damage" in living cells.

The three types of naturally occurring radioactive emissions—alpha particles, beta particles, and gamma rays—differ in their ability to penetrate matter and cause cellular damage. Consequently, the extent of the biological effects of radiation depends on the type of radiation involved.

Alpha Particle Effects

Alpha particles are the most massive and also the slowest particles involved in natural radioactive decay processes. Maximum alpha particle velocities are on the order of one-tenth of the speed of light. For a given alpha-emitting radioisotope all alpha particles have the same energy; different alpha-emitting radioisotopes, however, produce alpha particles of differing energies.

▶ The speed of light, 3.0×10^8 m/sec (186,000 miles/sec), is the maximum limit of velocity. Objects cannot travel faster than the speed of light.

Because of their "slowness," alpha particles have low penetrating power and cannot penetrate the body's outer layers of skin. The major damage from alpha radiation occurs when alpha-emitting radioisotopes are ingested—for example, in contaminated food. There are no protective layers of skin within the body.

Beta Particle Effects

Unlike alpha particles, which are all emitted with the same discrete energy from a given radioisotope, beta particles emerge from a beta-emitting substance with a continuous range of energies up to a specific limit that is characteristic of the particular radioisotope. Maximum beta particle velocities are on the order of nine-tenths of the speed of light.

With their greater velocity, beta particles can penetrate much deeper than alpha particles and can cause severe skin burns if their source remains in contact with the skin for an appreciable time. Because of their much smaller size, they do not collide with as many molecules as do alpha particles. An alpha particle is approximately 8000 times heavier than a beta particle. A typical alpha particle travels about 6 cm in air and collides with 40,000 molecules and/or atoms and a typical beta particle travels 1000 cm in air and collides with about 2000 molecules and/or atoms. Internal exposure to beta radiation is as serious as internal alpha exposure.

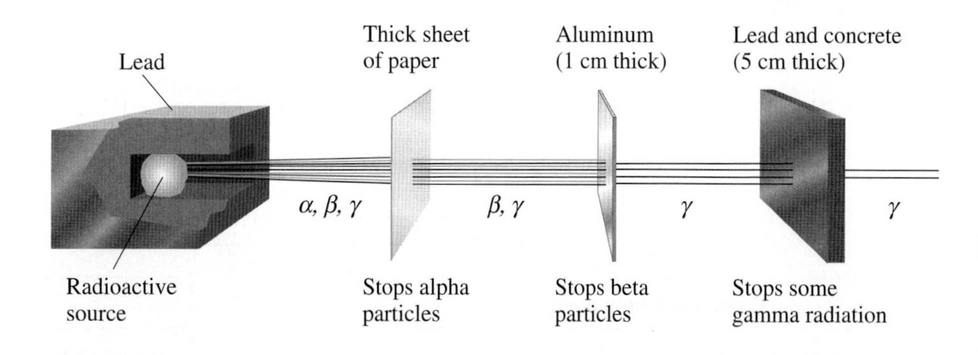

Figure 3.15 Alpha, beta, and gamma radiations differ in penetrating ability.

Gamma Radiation Effects

Gamma radiation is released at a velocity equal to that of the speed of light. Gamma rays readily penetrate deeply into organs, bone, and tissue.

Figure 3.15 contrasts the abilities of alpha, beta, and gamma radiations to penetrate paper, aluminum foil, and a thin layer of a lead–concrete mixture.

▶ X rays (which come from nonnuclear sources) and gamma rays are similar except that X rays are of lower energy. X rays used in diagnostic medicine have energies approximately 10% of that of gamma rays.

Practice Exercises and Questions

3.58 Contrast the abilities of alpha, beta, and gamma radiations to penetrate a thick sheet of paper.

3.59 Contrast the abilities of alpha, beta, and gamma radiations to penetrate human skin.

3.60 Contrast the velocities with which alpha, beta, and gamma radiations are emitted by unstable nuclei.

3.61 Contrast the distance that alpha and beta particles can travel in air, and the number of collisions they undergo in traveling that distance, before their excess energy is dissipated.

The next Chemistry at a Glance summarizes the terminology and concepts we have considered so far that are related to unstable nuclei and the radiations that they give off.

3.15 Nuclear Medicine

Learning Focus

Know the basic principles governing the use of radioisotopes in diagnostic and therapeutic nuclear medicine.

In medicine, radioisotopes are used both diagnostically and therapeutically. In diagnostic applications, technicians use small amounts of radioisotopes whose progress through the body or localization in specific organs can be followed. Larger quantities of radioisotopes are used in therapeutic applications.

▶ An additional use for radioisotopes in medicine, besides diagnostic and therapeutic uses, is as a source of power (Section 11.12). Cardiac pacemakers powered by plutonium-238 can remain in a patient for longer periods than those powered by chemical batteries without the additional surgery required to replace batteries.

Diagnostic Uses for Radioisotopes

The fundamental chemical principle behind the use of radioisotopes in diagnostic medical work is the fact that a radioactive nucleus of an element has the same chemical properties as a nonradioactive nucleus of the element. Thus body chemistry is not upset by the presence of a small amount of a radioactive substance whose nonradioactive form is already present in the body.

The criteria used in selecting radioisotopes for diagnostic procedures include the following:

1. At low concentrations (to minimize radiation damage), the radioisotope must be detectable by instrumentation placed outside the body. Almost all diagnostic radioisotopes are gamma emitters, because the penetrating power of alpha and beta particles is too low.

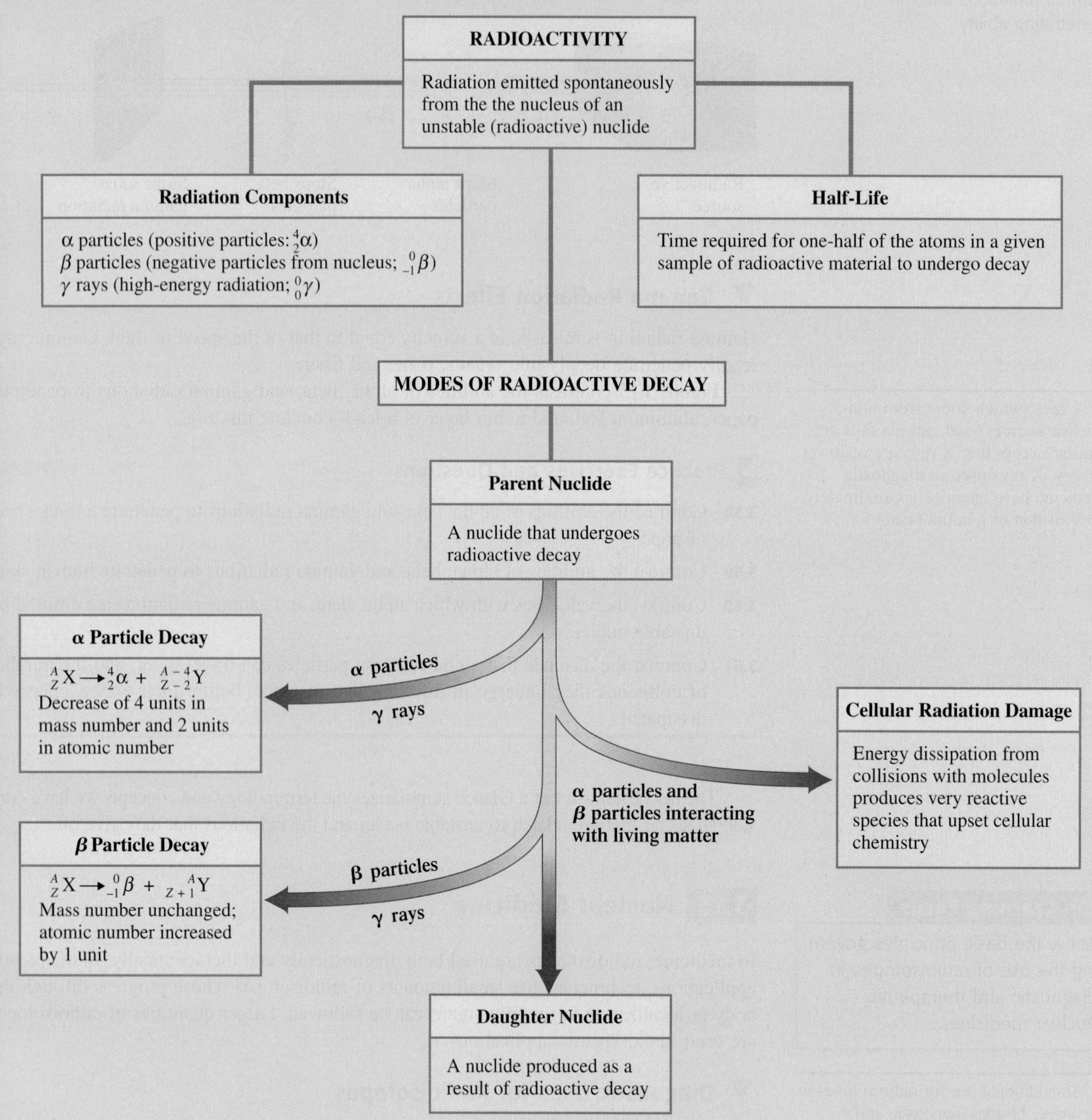

2. The radioisotope must have a short half-life so that the intensity of the radiation is sufficiently great to be detected. A short half-life also limits the time period of radiation exposure.
3. The radioisotope must have a known mechanism for elimination from the body so that the material does not remain in the body indefinitely.
4. The chemical properties of the radioisotope must be such that it is compatible with normal body chemistry. It must be able to be selectively transmitted to the part or system of the body that is under study.

The circulation of blood in the body can be followed by using radioactive sodium-24. A small amount of this isotope is injected into the bloodstream in the form of a sodium chloride solution. The movement of this radioisotope through the circulatory system can be followed easily with radiation detection equipment. If it takes longer than normal for the isotope to show up at a particular part of the body, this is an indication that the circulation is impaired at that spot.

Radiologists evaluate the functioning of the thyroid gland by administering iodine-131, usually in the form of a sodium iodide (NaI) solution. The radioactive iodine behaves in the same manner as ordinary iodine and is absorbed by the thyroid at a rate related to the activity of the gland. If a hypothyroid condition exists, then the amount accumulated is less than normal; and if a hyperthyroid condition exists, then a greater-than-average amount accumulates.

The size and shape of organs, as well as the presence of tumors, can be determined in some situations by scanning the organ in which a radioisotope tends to concentrate. Iodine-131 and technetium-99 are used to generate thyroid and brain scans, respectively. In the brain, technetium-99 concentrates in brain tumors more than in normal brain tissue; this helps radiologists determine the presence, size, and location of brain tumors. Figure 3.16 shows a brain scan obtained by using the radioisotope technetium-99. In this figure, the bright spot at the upper right indicates a tumor that has absorbed a greater amount of radioactive material than the normal brain tissue.

Therapeutic Uses for Radioisotopes

The objectives in therapeutic radioisotope use are entirely different from those for diagnostic procedures. The main objective in the therapeutic use of radioisotopes is to *selectively destroy* abnormal (usually cancerous) cells. The radioisotope is often, but not always, placed within the body. Therapeutic radioisotopes implanted in the body are usually alpha or beta emitters, because an intense dose of radiation in a small localized area is needed.

A commonly used implantation radioisotope that is effective in the localized treatment of tumors is yttrium-90, a beta emitter with a half-life of 64 hr. Yttrium-90 salts are implanted by inserting small, hollow needles into the tumor.

External, high-energy beams of gamma radiation are also extensively used in the treatment of certain cancers. Cobalt-60 is frequently used for this purpose; a beam of radiation is focused on the small area of the body where the tumor is located (see Figure 3.17).

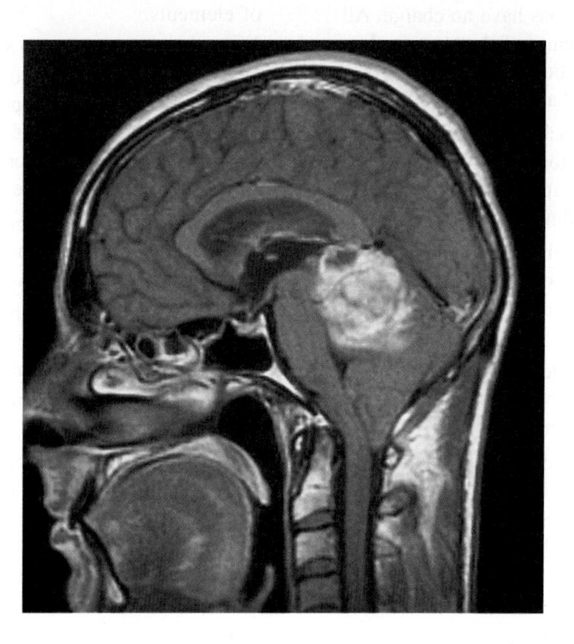

Figure 3.16 Brain scans, such as this one, are obtained using radioactive technetium-99, a laboratory-produced radioisotope.

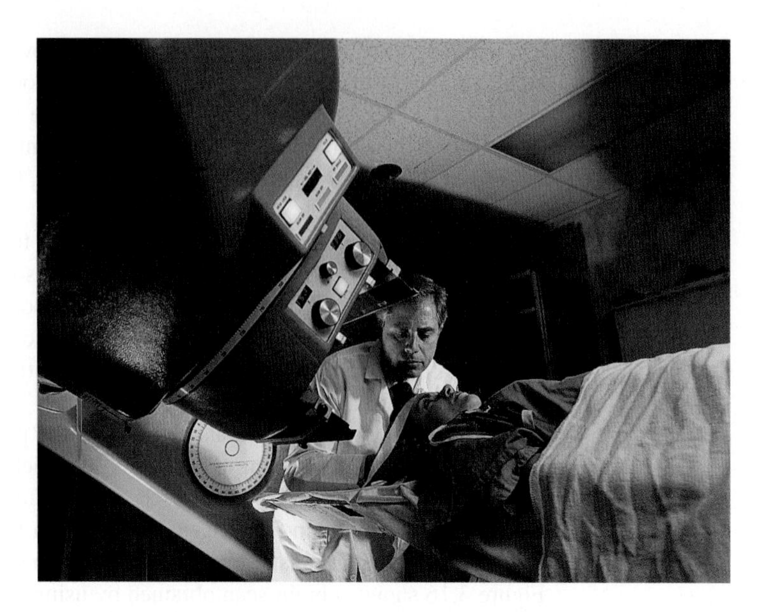

Figure 3.17 Cobalt-60 is used as a source of gamma radiation in radiation therapy.

Practice Exercises and Questions

3.62 Why are the radioisotopes used for diagnostic procedures usually gamma emitters?

3.63 Why do the radioisotopes used in diagnostic procedures nearly always have short half-lives?

3.64 How do the radioisotopes used for therapeutic purposes differ from the radioisotopes used for diagnostic purposes?

3.65 Contrast the different ways in which cobalt-60 and yttrium-90 are used in radiation therapy.

CONCEPTS TO REMEMBER

Subatomic particles. Subatomic particles, the very small building blocks from which atoms are made, are of three major types: electrons, protons, and neutrons. Electrons are negatively charged, protons are positively charged, and neutrons have no charge. All neutrons and protons are found at the center of the atom in the nucleus. The electrons occupy the region about the nucleus. Protons and neutrons have much larger masses than the electron.

Atomic number and mass number. Each atom has a characteristic atomic number and mass number. The atomic number is equal to the number of protons in the nucleus of the atom. The mass number is equal to the total number of protons and neutrons in the nucleus.

Isotopes. Isotopes are atoms that have the same number of protons and electrons but have different numbers of neutrons. The isotopes of an element always have the same atomic number and different mass numbers. Isotopes of an element have the same chemical properties.

Atomic mass. The atomic mass of an element is a calculated average mass. It depends on the percentage abundances and masses of the naturally occurring isotopes of the element.

Periodic law and periodic table. The periodic law states that when elements are arranged in order of increasing atomic number, elements with similar chemical properties occur at periodic (regularly recurring) intervals. The periodic table is a graphical representation of the behavior described by the periodic law. In a modern periodic table, vertical columns contain elements with similar chemical properties. A group in the periodic table is a vertical column of elements. A period in the periodic table is a horizontal row of elements.

Metals and nonmetals. Metals exhibit luster, thermal conductivity, electrical conductivity, and malleability. Nonmetals are characterized by the absence of the properties associated with metals. The majority of the elements are metals. The steplike heavy line that runs through the right third of the periodic table separates the metals on the left from the nonmetals on the right.

Electron shell. A shell contains electrons that have approximately the same energy and spend most of their time approximately the same distance from the nucleus.

Electron subshell. A subshell contains electrons that all have the same energy. The number of subshells in a particular shell is equal to the shell number. Each subshell can hold a specific maximum number of electrons. These values are 2, 6, 10, and 14 for *s*, *p*, *d*, and *f* subshells, respectively.

Electron orbital. An orbital is a region of space about a nucleus where an electron with a specific energy is most likely to be found. Each subshell consists of one or more orbitals. For *s*, *p*, *d*, and *f* subshells there are 1, 3, 5, and 7 orbitals, respectively. No more than two electrons may occupy any orbital.

Electron configuration. An electron configuration is a statement of how many electrons an atom has in each of its subshells. The principle that electrons normally occupy the lowest-energy subshell available is used to write electron configurations.

Electron configurations and the periodic law. Chemical properties repeat themselves in a regular manner among the elements because electron configurations repeat themselves in a regular manner among the elements.

Electron configurations and the periodic table. The groups of the periodic table consist of elements with similar electron configurations. Thus the location of an element in the periodic table can be used to obtain information about its electron configuration.

Classification system for the elements. On the basis of electron configuration, elements can be classified into four categories: noble gases (far right column of the periodic table); representative elements (*s* and *p* areas of the periodic table, with the exception of the noble gases); transition elements (*d* area of the periodic table); and inner transition elements (*f* area of the periodic table).

Nuclear stability and radioactivity. Some atoms possess nuclei that are unstable. To achieve stability, these unstable nuclei spontaneously emit energy (radiation). Such atoms are said to be radioactive.

Emissions from radioactive nuclei. The types of radiation emitted by naturally occurring radioactive nuclei are alpha, beta, and gamma. These radiations can be characterized by mass and charge values. Alpha particles carry a positive charge, beta particles a negative charge, and gamma radiation no charge.

Half-life. Every radioisotope decays at a characteristic rate given by its half-life. One half-life is the time required for half of any given quantity of a radioactive substance to undergo decay.

Balancing nuclear equations. Atomic numbers (subscripts) and mass numbers (superscripts) are always shown when nuclei are represented in a nuclear equation. The balancing of nuclear equations is based on both *charge* and *nucleon* conservation. Charge conversion means that the sum of the subscripts for the products must equal the sum of the subscripts for the reactants. Nucelon conservation means that the sum of the superscripts for the products equals the sum of the superscripts for the reactants.

Biological effects of radiation. The biological effects of radiation depend on the energy, size, and penetrating ability of the radiation. Alpha particles have the greatest size, and gamma rays have the greatest penetrating ability.

Nuclear medicine. Radioisotopes are used in medicine for both diagnosis and therapy. Diagnostic radioisotopes are generally gamma emitters, whereas therapeutic radioisotopes are often alpha and beta emitters. The choice of radioisotope is dictated by the purpose of its use as well as by the target organ.

KEY REACTIONS AND EQUATIONS

1. Relationships involving atomic number and mass number for a neutral atom (Section 3.2)
   Atomic number = number of protons = number of electrons
   Mass number = number of protons + number of neutrons
   Mass number = total number of subatomic particles in the nucleus
   Mass number − atomic number = number of neutrons
   Mass number + atomic number = total number of subatomic particles
2. Relationships involving electron shells, electron subshells, and electron orbitals (Section 3.6)
   Number of subshells in a shell = shell number
   Maximum number of electrons in an *s* subshell = 2
   Maximum number of electrons in a *p* subshell = 6
   Maximum number of electrons in a *d* subshell = 10
   Maximum number of electrons in an *f* subshell = 14
   Maximum number of electrons in an orbital = 2
3. Order of filling of subshells in terms of increasing energy (Section 3.7)
   1*s*, 2*s*, 2*p*, 3*s*, 3*p*, 4*s*, 3*d*, 4*p*, 5*s*, 4*d*, 5*p*, 6*s*, 4*f*, 5*d*, 6*p*, 7*s*, 5*f*, 6*d*, 7*p*
4. General equation for alpha decay (Section 3.13)
   $$^{A}_{Z}X \longrightarrow {}^{A-4}_{Z-2}Y + {}^{4}_{2}\alpha$$
5. General equation for beta decay (Section 3.13)
   $$^{A}_{Z}X \longrightarrow {}^{A}_{Z+1}Y + {}^{0}_{-1}\beta$$

KEY TERMS

Alpha particle (3.12)
Atomic mass (3.3)
Atomic number (3.2)
Balanced nuclear equation (3.13)
Beta particle (3.12)
Daughter nucleus (3.13)
Distinguishing electron (3.8)
Electron (3.1)
Electron configuration (3.7)
Electron orbital (3.6)
Electron shell (3.6)
Electron subshell (3.6)
Element (3.2)
Gamma ray (3.12)
Group (3.4)
Half-life (3.11)
Inner transition elements (3.9)
Isotopes (3.3)
Mass number (3.2)
Metal (3.5)
Neutron (3.1)
Noble-gas elements (3.9)
Nonmetal (3.5)
Nucleon (3.1)
Nucleus (3.1)
Parent nucleus (3.13)
Period (3.4)
Periodic law (3.4)
Periodic table (3.4)
Proton (3.1)
Radioactive atom (3.10)
Radioactive decay (3.11)
Radioactivity (3.10)
Representative elements (3.9)
Stable nucleus (3.10)
Subatomic particles (3.1)
Transition elements (3.9)
Unstable nucleus (3.10)

ADDITIONAL PROBLEMS

3.66 With the help of the periodic table, complete the following table.

Element	Symbol	Atomic number	Mass number	Number of protons	Number of neutrons
(a) ____	$^{3}_{2}He$	____	____	____	____
(b) nickel	____	____	60	____	____
(c) ____	____	18	37	____	____
(d) ____	____	____	90	____	52
(e) ____	$^{235}_{92}U$	____	____	____	____
(f) ____	____	____	____	17	20
(g) ____	____	____	232	94	____
(h) ____	$^{32}_{16}S$	____	____	____	____
(i) iron	____	____	56	____	____
(j) calcium	____	____	____	____	20

3.67 Write the complete symbol ($^{A}_{Z}E$), with the help of the periodic table, for atoms with the following characteristics.
a. Contains 20 electrons and 24 neutrons
b. Beryllium atom with one more neutron than proton
c. Silver atom that contains 157 subatomic particles
d. Beryllium atom that contains 9 nucleons

3.68 Characterize each of the following pairs of atoms as containing (1) the same number of neutrons, (2) the same number of electrons, or (3) the same total number of subatomic particles.
a. $^{13}_{6}C$ and $^{14}_{7}N$ b. $^{18}_{8}O$ and $^{19}_{9}F$
c. $^{37}_{17}Cl$ and $^{36}_{18}Ar$ d. $^{35}_{17}Cl$ and $^{37}_{17}Cl$

3.69 Using the information given in the following table, indicate whether each of the following pairs of atoms are isotopes.

	Atom A	Atom B	Atom C	Atom D
Number of protons	9	10	10	9
Number of neutrons	10	9	10	9
Number of electrons	9	10	10	9

a. A and B b. A and C c. A and D d. B and C

3.70 Three naturally occurring isotopes of magnesium exist: magnesium-24, magnesium-25, and magnesium-26. The percentage abundances and masses of these three isotopes are, respectively, (78.99%, 23.99 amu), (10.00%, 24.99 amu), and (11.01%, 25.98 amu). Calculate the atomic mass of magnesium.

3.71 The atomic mass of fluorine is 18.998 amu, and all fluorine atoms have this mass. The atomic mass of iron is 55.847 amu, and not a single iron atom has this mass. Explain.

3.72 Which of the six elements nitrogen, beryllium, argon, aluminum, silver, and gold belong(s) in each of the following classifications?
a. Period number and Roman-numeral group number are numerically equal
b. Readily conducts electricity and heat
c. Has an atomic mass greater than its atomic number
d. All atoms have a nuclear charge greater than +20

3.73 Write electron configurations for the following elements.
a. The Group IIIA element in the same period as $_{4}Be$
b. The Period 3 element in the same group as $_{5}B$
c. The lowest-atomic-numbered metal in Group IA
d. The highest-atomic-numbered metal that contains only *s* electrons

3.74 Write electron configurations for the following atoms.
a. An atom with 13 electrons
b. An atom with 13 protons
c. An atom with a +13 charge on its nucleus
d. An atom with an atomic number of 13

3.75 With the help of the periodic table, identify the element of lowest atomic number whose electron configuration
a. contains one or more *p* electrons
b. contains one or more *d* electrons
c. contains three or more *s* electrons
d. contains nine or more *p* electrons

3.76 Write nuclear equations for each of the following radioactive decay processes.
a. Thallium-206 is formed by beta emission.
b. Palladium-109 undergoes beta emission.
c. Plutonium-241 is formed by alpha emission.
d. Fermium-249 undergoes alpha emission.

3.77 Cobalt-55 has a half-life of 18 hours. How long will it take, in hours, for the following fractions of atoms in a cobalt-55 sample to undergo decay?
a. 7/8 b. 31/32 c. 63/64 d. 127/128

PRACTICE TEST ▶ True/False

3.78 All protons and neutrons present in an atom are found in its nucleus.

3.79 A nucleon is any subatomic particle that is uncharged (neutral).

3.80 The difference between the mass number and atomic number for an atom gives to the number of electrons present in the atom.

3.81 Isotopes of an element have the same chemical properties because their electron configurations are identical.

3.82 In the standard periodic table, elements with similar chemical properties are always found in the same period.

3.83 The majority of the elements are metals rather than nonmetals.

3.84 The maximum number of electrons that any electron subshell can accommodate is six.

3.85 Electron configurations give the number of electrons that are located in various electron orbitals.

3.86 The number of columns of elements in the *s* area of the periodic table is two.

3.87 The noble-gas elements are found in the far-right column of the periodic table.

3.88 Atoms that possess unstable nuclei are said to be radioactive.

3.89 In a radioactive sample, half of the radioactive atoms undergo decay during the first half-life, and the other half undergo decay during the second half-life.

3.90 Alpha particles are much heavier than beta particles and travel much faster.

3.91 Beta particle decay always results in the formation of an atom of a different element.

3.92 The composition of an alpha particle is two protons and two electrons.

PRACTICE TEST ▶ Multiple Choice

3.93 Which of the following collections of subatomic particles would have the greatest mass?
a. 4 electrons and 1 proton
b. 2 neutrons and 1 electron
c. 1 proton and 2 neutrons
d. 1 electron, 1 proton, and 1 neutron

3.94 The nucleus of an atom
a. contains all subatomic particles present in the atom
b. is negatively charged because of the presence of electrons
c. is neutral because it contains only neutrons
d. accounts for only a small amount of the total volume of an atom

3.95 An atom contains 26 protons, 30 neutrons, and 26 electrons. The atomic number and mass number for this atom are, respectively,
a. 30 and 26
b. 26 and 30
c. 26 and 56
d. 30 and 56

3.96 An atom of $^{27}_{14}Si$ contains
a. more protons than neutrons
b. more electrons than protons
c. the same equal number of protons and neutrons
d. more neutrons than electrons

3.97 Isotopes of a given element have
a. the same mass number but different numbers of protons
b. different mass numbers but the same number of protons
c. the same atomic number but different chemical properties
d. the same mass number but different chemical properties

3.98 Chlorine, which exists in nature in two isotopic forms, has an atomic mass of 35.5 amu. This means that
a. all chlorine atoms have masses of 35.5 amu
b. some, but not all, chlorine atoms have masses of 35.5 amu
c. 35.5 amu is the upper limit for the mass of a chlorine atom
d. no chlorine atoms have masses of 35.5 amu

3.99 Which of the following statements is consistent with the electron configuration $1s^22s^22p^63s^23p^6$?
a. There are 6 electrons present in a $3p$ orbital.
b. There are 6 electrons present in a $3p$ subshell.
c. There are 6 electrons present in a $3p$ shell.
d. There are 6 electrons present in the third shell.

3.100 The elements in group IVA of the periodic table all have electron configurations ending in
a. p^2
b. p^4
c. d^4
d. s^2

3.101 The daughter nucleus produced by the beta decay of $^{234}_{90}Th$ is
a. $^{230}_{89}Ac$ b. $^{234}_{91}Pa$ c. $^{234}_{92}U$ d. $^{230}_{88}Ra$

3.102 The loss of an alpha particle by a radioactive atom causes
a. no change in mass number
b. the atomic number to decrease by 4
c. the mass number to decrease by 2
d. the atomic number to decrease by 2

CHAPTER OUTLINE

CHAPTER FOUR

Chemical Bonds

Magnification of crystals of sodium chloride (table salt), one of the most commonly encountered ionic compounds.

As scientists study living organisms and the world in which we live, they rarely encounter free isolated atoms. Instead, under normal conditions of temperature and pressure, they nearly always find atoms associated in aggregates or clusters ranging in size from two atoms to numbers too large to count. In this chapter, we will explain why atoms tend to join together in larger units, and we will discuss the binding forces (chemical bonds) that hold them together.

As we examine the nature of attractive forces between atoms, we will discover that both the tendency and the capacity of an atom to be attracted to other atoms are dictated by its electron configuration.

4.1 Chemical Bonds

Learning Focus

Define the term *chemical bond* and describe the two major types of chemical bonds.

Chemical compounds are conveniently divided into two broad classes called *ionic compounds* and *molecular compounds.* Ionic and molecular compounds can be distinguished from each other on the basis of general physical properties. Ionic compounds tend to have high melting points (500°C – 2000°C) and are good conductors of electricity when they are in a molten (liquid) state or in solution. Molecular compounds, on the other hand, generally have much lower melting points and tend to be gases, liquids, or low-melting solids. They do not conduct electricity in the molten state. Ionic compounds, unlike molecular compounds, do not have molecules as their basic structural unit. Instead, an extended array of positively and negatively charged particles called *ions* is present (Section 4.8).

Some combinations of elements produce ionic compounds, whereas other combinations of elements form molecular compounds. What determines whether the interaction of two elements produces ions (an ionic compound) or molecules (a molecular compound)? To answer this question, we need to learn about chemical bonds. A **chemical bond** *is the attractive force that holds two atoms together in a more complex unit.* Chemical bonds form as a result of interactions between electrons found in the combining atoms. Thus the nature of chemical bonds is closely linked to electron configurations (Section 3.7).

▶ Another designation for *molecular compound* is *covalent compound.* The two designations are used interchangeably. The modifier *molecular* draws attention to the basic structural unit present (the molecule), and the modifier *covalent* focuses on the mode of bond formation (electron sharing).

Corresponding to the two broad categories of chemical compounds are two types of chemical attractive forces (chemical bonds): ionic bonds and covalent bonds. An **ionic bond** *is a chemical bond formed through the transfer of one or more electrons from one atom or group of atoms to another.* As its name suggests, the ionic bond model (electron transfer) is used in describing the attractive forces in ionic compounds. A **covalent bond** *is a chemical bond formed through the sharing of one or more pairs of electrons between two atoms.* The covalent bond model (electron sharing) is used in describing the attractions between atoms in molecular compounds.

▶ *Purely* ionic bonds involve a complete transfer of electrons from one atom to another. *Purely* covalent bonds involve equal sharing of electrons. Experimentally, it is found that most actual bonds have some degree of both ionic and covalent character. The exceptions are bonds between identical atoms; here, the bonding is purely covalent.

Even before we consider the details of these two bond models, it is important to emphasize that the concepts of ionic and covalent bonds are actually "convenience concepts." Most bonds are not 100% ionic or 100% covalent. Instead, most bonds have some degree of both ionic and covalent character—that is, some degree of both the transfer and the sharing of electrons. However, it is easiest to understand these intermediate bonds (the real bonds) by relating them to the pure or ideal bond types called ionic and covalent.

Two fundamental concepts are necessary for understanding both the ionic and the covalent bonding models.

1. Not all electrons in an atom are available for bonding. Those that are available are called *valence electrons.*
2. Certain arrangements of electrons are more stable than others, as is explained by the *octet rule.*

Section 4.2 addresses the concept of valence electrons, and Section 4.3 discusses the octet rule.

Practice Questions and Problems

4.1 What type of subatomic particle is involved in the interactions that lead to chemical bond formation?

4.2 Contrast the two general types of chemical bonds in terms of the mechanism by which they form.

4.3 Contrast the two general types of chemical compounds in terms of general physical properties.

4.2 Valence Electrons and Lewis Structures

Learning Focus

Determine the number of valence electrons a representative element has, given its location in the periodic table or its electron configuration, and write a Lewis structure for the element.

Certain electrons, called valence electrons, are particularly important in determining the bonding characteristics of a given atom. For representative and noble-gas elements, **valence electrons** *are the electrons in the outermost electron shell, which is the shell with the highest shell number (n).* Valence electrons are always found in either *s* or *p* subshells. Note the restriction on the use of this definition; it applies only to representative and noble-gas elements. Because most of the common elements are representative elements, this definition is quite useful. (We will not consider in this text the more complicated valence electron definitions for transition or inner transition elements; here, the presence of incompletely filled *inner d* or *f* subshells is a complicating factor.)

▶ The term *valence* is derived from the Latin word *valentia*, which means "capacity" (to form bonds).

The number of valence electrons in an atom of a representative element can be determined from the atom's electron configuration, as is illustrated in Example 4.1.

Example 4.1 Determining the Number of Valence Electrons in an Atom

Determine the number of valence electrons in atoms of each of the following elements.

a. $_{12}Mg$ **b.** $_{14}Si$ **c.** $_{33}As$

Solution

a. Atoms of the element magnesium have two valence electrons, as can be seen by examining magnesium's electron configuration.

$$1s^22s^22p^6\textcircled{3}s^{\textcircled{2}}$$

Number of valence electrons

Highest value of the electron shell number

The highest value of the electron shell number is $n = 3$. Only two electrons are found in shell 3: the two electrons in the $3s$ subshell.

b. Atoms of the element silicon have four valence electrons.

$$1s^22s^22p^6\textcircled{3}s^{\textcircled{2}}\textcircled{3}p^{\textcircled{2}}$$

Number of valence electrons

Highest value of the electron shell number

Electrons in two different subshells can simultaneously be valence electrons. The highest shell number is 3, and both the $3s$ and the $3p$ subshells belong to this shell. Hence all of the electrons in both of these subshells are valence electrons.

c. Atoms of the element arsenic have five valence electrons.

$$1s^22s^22p^63s^23p^6\textcircled{4}s^{\textcircled{2}}3d^{10}\textcircled{4}p^{\textcircled{3}}$$

Number of valence electrons

Highest value of the electron shell number

The $3d$ electrons are not counted as valence electrons because the $3d$ subshell is in shell 3, and this shell does not have maximum n value. Shell 4 is the outermost shell and has maximum n value.

Scientists have developed a shorthand system for designating the number of valence electrons present in atoms of an element. This system involves the use of Lewis structures. A **Lewis structure** *consists of an element's symbol with one dot for each valence electron placed around the elemental symbol.* Lewis structures for the first 20 elements (all representative or noble-gas elements), arranged as in the periodic table, are given in Figure 4.1. Lewis structures, named in honor of the American chemist Gilbert N. Lewis (Figure 4.2) who first introduced them, are also frequently called *electron-dot structures.*

In positioning electrons in a Lewis structure, always place one electron on all four sides of the elemental symbol before putting a second electron on any side. Note that which side of the symbol the dot is placed on is not critical. The following notations all

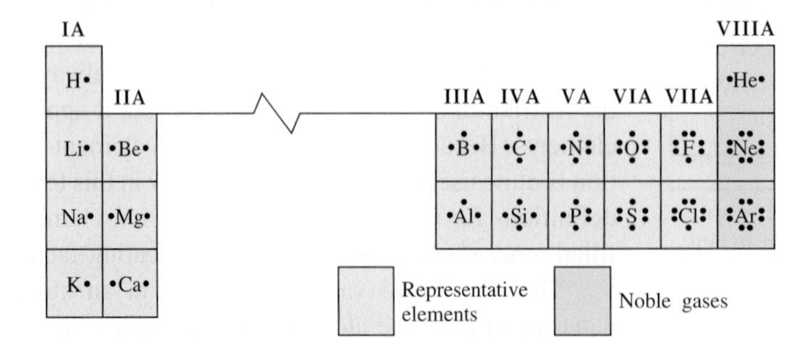

Figure 4.1 Lewis structures for selected representative and noble-gas elements.

Figure 4.2 Gilbert Newton Lewis (1875–1946), one of the foremost chemists of the twentieth century, made significant contributions in other areas of chemistry besides his pioneering work in describing chemical bonding. He formulated a generalized theory for describing acids and bases and was the first to isolate deuterium (heavy hydrogen).

have the same meaning:

$$\dot{\mathrm{Ca}}\cdot \qquad \underset{\cdot}{\mathrm{Ca}}\cdot \qquad \cdot\dot{\mathrm{Ca}} \qquad \cdot\mathrm{Ca}\cdot$$

Three important generalizations about valence electrons can be drawn from a study of the structures shown in Figure 4.1.

1. *Representative elements in the same group of the periodic table have the same number of valence electrons.* This should not be surprising. Elements in the same group in the periodic table have similar chemical properties as a result of their similar outer-shell electron configurations (Section 3.8). The electrons in the outermost shell are the valence electrons.
2. *The number of valence electrons for representative elements is the same as the Roman numeral periodic-table group number.* For example, the Lewis structures for oxygen and sulfur, which are both members of Group VIA, show six dots. Similarly, the Lewis structures of hydrogen, lithium, sodium, and potassium, which are all members of Group IA, show one dot.
3. *The maximum number of valence electrons for any element is eight.* Only the noble gases (Section 3.9), beginning with neon, have the maximum number of eight electrons. Helium, which has only two valence electrons, is the exception in the noble-gas family. Obviously, an element with a total of two electrons cannot have eight valence electrons. Although shells with n greater than 2 are capable of holding more than eight electrons, they do so only when they are no longer the outermost shell and are thus not the valence shell. For example, arsenic has 18 electrons in its third shell; however, shell 4 is the valence shell for arsenic.

Practice Questions and Problems

4.4 How many valence electrons do atoms with the following electron configurations have?
a. $1s^22s^2$ b. $1s^22s^22p^23s^2$ c. $1s^22s^22p^63s^23p^1$ d. $1s^22s^22p^63s^23p^64s^23d^{10}4p^2$

4.5 What are the group number, and the number of valence electrons present, for each of the following representative elements?
a. ${}_{3}\mathrm{Li}$ b. ${}_{10}\mathrm{Ne}$ c. ${}_{20}\mathrm{Ca}$ d. ${}_{53}\mathrm{I}$

4.6 What do the dots represent in a Lewis structure?

4.7 Write the group number and draw the Lewis structure for each of the following elements.
a. ${}_{12}\mathrm{Mg}$ b. ${}_{19}\mathrm{K}$ c. ${}_{15}\mathrm{P}$ d. ${}_{36}\mathrm{Kr}$

4.8 Each of the following Lewis structures represents a Period 2 element. With the help of the periodic table, determine the element's identity.
a. $\mathrm{X}\cdot$ b. $:\ddot{\mathrm{X}}:$ c. $\dot{\mathrm{X}}\cdot$ d. $\cdot\underset{\cdot}{\ddot{\mathrm{X}}}\cdot$

4.9 In each of the following sets of three elements, select the two elements that have the same number of valence electrons.
a. Be, C, Mg b. Li, N, P c. O, S, Cl d. Na, Ca, K

4.10 Write the Lewis structure for the Period 3 element with each of the following numbers of valence electrons.
a. two b. three c. five d. seven

Learning Focus

State the octet rule and understand the basis for this rule.

4.3 The Octet Rule

A key concept in modern elementary bonding theory is that certain arrangements of valence electrons are more stable than others. The term *stable* as used here refers to the idea that a system, which in this case is an arrangement of electrons, does not easily undergo spontaneous change.

The valence electron configurations of the noble gases (helium, neon, argon, krypton, xenon, and radon) are considered the *most stable of all valence electron configurations.* All of the noble gases except helium possess eight valence electrons, which is the maximum number possible. Helium's valence electron configuration is $1s^2$. All of the other noble gases possess ns^2np^6 valence electron configurations, where n has the maximum value found in the atom.

He: $\boxed{1s^2}$

Ne: $1s^2\boxed{2s^22p^6}$

Ar: $1s^22s^22p^6\boxed{3s^23p^6}$

Kr: $1s^22s^22p^63s^23p^6\boxed{4s^2}3d^{10}\boxed{4p^6}$

Xe: $1s^22s^22p^63s^23p^64s^23d^{10}4p^6\boxed{5s^2}4d^{10}\boxed{5p^6}$

Rn: $1s^22s^22p^63s^23p^64s^23d^{10}4p^65s^24d^{10}5p^6\boxed{6s^2}4f^{14}5d^{10}\boxed{6p^6}$

▶ The *outermost* electron shell of an atom is also called the *valence* electron shell.

Except for helium, all the noble-gas valence electron configurations have the outermost s and p subshells *completely filled.*

The conclusion that an ns^2np^6 configuration ($1s^2$ for helium) is the most stable of all valence electron configurations is based on the chemical properties of the noble gases. The noble gases are the *most unreactive* of all the elements. They are the only elemental gases found in nature in the form of individual uncombined atoms. There are no known compounds of helium, neon, and argon, and only a very few compounds of krypton, xenon, and radon are known. The noble gases have little or no tendency to form bonds to other atoms.

▶ Some compounds exist whose formulation is not consistent with the octet rule, but the vast majority of simple compounds have formulas that are consistent with its precepts.

Atoms of many elements that lack this very stable noble-gas valence electron configuration tend to acquire it through chemical reactions that result in compound formation. This observation is known as the **octet rule:** *In compound formation, atoms of elements lose, gain, or share electrons in such a way that their electron configurations become identical to that of the noble gas nearest them in the periodic table.*

Practice Problems and Questions

4.11 What are the properties of the noble gases that lead to the conclusion that they possess extremely stable electron arrangements?

4.12 What is the maximum number of valence electrons that a representative element/noble gas can possess, and which group of elements possesses this maximum number of valence electrons?

4.13 Which noble-gas element is an exception to the rule that noble-gas elements possess eight valence electrons?

4.14 What does the octet rule indicate happens to elements that lack a noble-gas electron configuration?

Learning Focus

Define the term *ion* and understand the notation used to denote ions.

4.4 The Ionic Bond Model

▶ The word *ion* is pronounced "eye-on."

Electron transfer between two or more atoms is central to the ionic bond model. This electron transfer process produces charged particles called ions. An **ion** *is an atom (or group of atoms) that is electrically charged as a result of the loss or gain of electrons.* An atom is neutral when the number of protons (positive charges) is equal to the number of electrons (negative charges). Loss or gain of electrons destroys this proton–electron balance and leaves a net charge on the atom.

If an atom *gains* one or more electrons, it becomes a *negatively* charged ion; excess negative charge is present because electrons outnumber protons. If an atom *loses* one or more electrons, it becomes a *positively* charged ion; more protons are present than electrons. There is excess positive charge (Figure 4.3). Note that the excess positive charge associated with a positive ion is never caused by proton gain but always by electron loss.

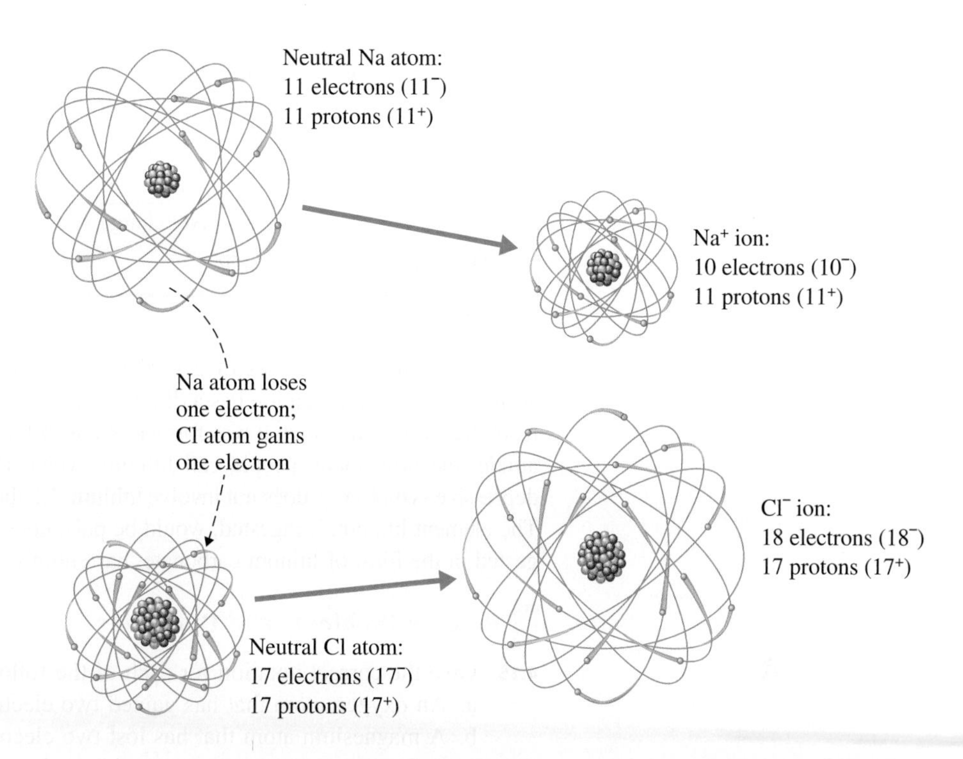

Figure 4.3 Loss of an electron from a sodium atom leaves it with one more proton than electrons, so it has a net electrical charge of +1. When chlorine gains an electron, it has one more electron than protons, so it has a net electrical charge of −1.

▶ An atom's nucleus *never* changes during the process of ion formation. The number of neutrons and protons remains constant.

▶ A loss of electrons by an atom always produces a positive ion. A gain of electrons by an atom always produces a negative ion.

If the number of protons remains constant and the number of electrons decreases, the result is net positive charge. The number of protons, which determines the identity of an element, never changes during ion formation.

The charge on an ion depends on the number of electrons that are lost or gained. Loss of one, two, or three electrons gives ions with +1, +2, and +3 charges, respectively. A gain of one, two, or three electrons gives ions with −1, −2, and −3 charges, respectively. (Ions that have lost or gained more than three electrons are very seldom encountered.)

The notation for charges on ions is a superscript placed to the right of the elemental symbol. Some examples of ion symbols are:

Positive ions: Na^+, K^+, Ca^{2+}, Mg^{2+}, Al^{3+}

Negative ions: Cl^-, Br^-, O^{2-}, S^{2-}, N^{3-}

Note that we use a single plus or minus sign to denote a charge of 1, instead of using the notation $^{1+}$ or $^{1-}$. Also note that in multicharged ions, the number precedes the charge sign; that is, the notation for a charge of plus two is $^{2+}$ rather than $^{+2}$.

Example 4.2 Writing Symbols for Ions

Give the symbol for each of the following ions.

a. The ion formed when a barium atom loses two electrons.
b. The ion formed when a phosphorus atom gains three electrons.

Solution

a. A neutral barium atom contains 56 protons and 56 electrons because barium has an atomic number of 56. The barium ion formed by the loss of 2 electrons would still contain 56 protons but would have only 54 electrons because 2 electrons were lost.

56 protons = 56 + charges
54 electrons = 54 − charges

Net charge = 2+

The symbol of the barium ion is thus Ba^{2+}.

b. The atomic number of phosphorus is 15. Thus 15 protons and 15 electrons are present in a neutral phosphorus atom. A gain of 3 electrons raises the electron count to 18.

$$\begin{array}{r} 15 \text{ protons} = 15 + \text{charges} \\ 18 \text{ electrons} = 18 - \text{charges} \\ \hline \text{Net charge} = 3- \end{array}$$

The symbol for the ion is P^{3-}.

The chemical properties of a particle (atom or ion) depend on the particle's electron arrangement. Because an ion has a different electron configuration (fewer or more electrons) from the atom from which it was formed, it has different chemical properties as well. For example, the drug many people call lithium, which is used to treat mental illness (manic-depressive symptoms), does not involve lithium (Li, the element) but rather lithium ions (Li^+). The element lithium, if ingested, would be poisonous and possibly fatal. The lithium ion, ingested in the form of lithium carbonate, has entirely different effects on the human body.

Practice Problems and Questions

4.15 Give the correct notation for each of the following ions.
a. An oxygen atom that has gained two electrons
b. A magnesium atom that has lost two electrons
c. A fluorine atom that has gained one electron
d. An aluminum atom that has lost three electrons

4.16 What would the notation be for an ion with each of the following numbers of protons and electrons?
a. 20 protons and 18 electrons b. 8 protons and 10 electrons
c. 11 protons and 10 electrons d. 13 protons and 10 electrons

4.17 Calculate the number of protons and electrons in each of the following ions.
a. P^{3-} b. S^{2-} c. Mg^{2+} d. Li^+

Learning Focus

On the basis of the octet rule, write the symbols for simple ions of the representative elements.

4.5 Ions and the Octet Rule

The octet rule provides a very simple and straightforward explanation for the charge magnitude associated with ions of the representative elements. *Atoms tend to gain or lose electrons until they have obtained an electron configuration that is the same as that of a noble gas.* The element sodium has the electron configuration

$$1s^22s^22p^63s^1$$

One valence electron is present. Sodium can attain a noble-gas electron configuration by losing this valence electron (to give it the electron configuration of neon) or by gaining seven electrons (to give it the electron configuration of argon).

$$\text{Na } (1s^22s^22p^63s^1) \xrightarrow{\text{Loss of } 1\ e^-} \text{Na}^+ \quad (1s^22s^22p^6) \text{ Electron configuration of neon}$$

$$\text{Na } (1s^22s^22p^63s^1) \xrightarrow{\text{Gain of } 7\ e^-} \text{Na}^{7-} \quad (1s^22s^22p^63s^23p^6) \text{ Electron configuration of argon}$$

The electron loss or gain that involves the fewest electrons will always be the more favorable process, from an energy standpoint, and will be the process that occurs. Thus for sodium, the loss of one electron to form the Na^+ ion is the process that occurs.

The element chlorine has the electron configuration

$$1s^22s^22p^63s^23p^5$$

Seven valence electrons are present. Chlorine can attain a noble-gas electron configuration by losing seven electrons (to give it the electron configuration of neon) or by gaining one electron (to give it the electron configuration of argon). The latter occurs for the reason we previously cited.

Cl ($1s^22s^22p^63s^23p^5$) —Loss of 7 e^-→ Cl^{7+} ($1s^22s^22p^6$)
Electron configuration of neon

Cl ($1s^22s^22p^63s^23p^5$) —Gain of 1 e^-→ Cl^- ($1s^22s^22p^63s^23p^6$)
Electron configuration of argon

The considerations we have just applied to sodium and chlorine lead to the following generalizations:

1. Metal atoms containing one, two, or three valence electrons (the metals in Groups IA, IIA, and IIIA of the periodic table) tend to lose electrons to acquire a noble-gas electron configuration. The noble gas involved is the one preceding the metal in the periodic table.
   Group IA metals form 1^+ ions.
   Group IIA metals form 2^+ ions.
   Group IIIA metals form 3^+ ions.
2. Nonmetal atoms containing five, six, or seven valence electrons (the nonmetals in Groups VA, VIA, and VIIA of the periodic table) tend to gain electrons to acquire a noble-gas electron configuration. The noble gas involved is the one following the nonmetal in the periodic table.
   Group VIIA nonmetals form 1^- ions.
   Group VIA nonmetals form 2^- ions.
   Group VA nonmetals form 3^- ions.

▶ The positive charge on metal ions from Groups IA, IIA, and IIIA has a magnitude equal to the metal's periodic-table group number.

▶ Nonmetals from Groups VA, VIA, and VIIA form negative ions whose charge is equal to the group number minus 8. For example, S, in Group VIA, forms S^{2-} ions ($6 - 8 = -2$).

Elements in Group IVA occupy unique positions relative to the noble gases. They would have to gain or lose four electrons to attain a noble-gas structure. Theoretically, ions with charges of +4 or −4 could be formed by elements in this group, but in most cases, these elements form covalent bonds instead, which are discussed later in this chapter.

Practice Problems and Questions

4.18 State the number of electrons that must be lost by atoms of each of the following elements in order for them to acquire a noble-gas electron configuration.
a. Sodium b. Magnesium c. Aluminum d. Silicon

4.19 State the number of electrons that must be gained by atoms of each of the following elements in order for them to acquire a noble-gas electron configuration.
a. Carbon b. Nitrogen c. Oxygen d. Fluorine

4.20 Indicate whether each of the following elements is more likely to form a positive ion or a negative ion.
a. ${}_{20}Ca$ b. ${}_{16}S$ c. ${}_{19}K$ d. ${}_{17}Cl$

4.21 What is the charge on the simple ion formed by each of the following representative elements?
a. ${}_{4}Be$ b. ${}_{35}Br$ c. ${}_{15}P$ d. ${}_{13}Al$

4.22 What noble-gas element has the same number of electrons as each of the following ions?
a. O^{2-} b. P^{3-} c. Ca^{2+} d. K^+

4.23 In what group in the periodic table are representative elements that form ions with the following charges most likely to be found?
a. +2 b. −2 c. −3 d. +1

Learning Focus

Use Lewis structures to describe the bonding in simple ionic compounds.

▶ No atom can lose electrons unless another atom is available to accept them.

4.6 Ionic Compound Formation

Ion formation, through the loss or gain of electrons by atoms, is not an isolated, singular process. In reality, electron loss and electron gain are always partner processes; if one occurs, the other also occurs. Ion formation requires the presence of two elements: a metal that can donate electrons and a nonmetal that can accept electrons. The electrons lost by the metal are the same ones gained by the nonmetal. The positive and negative ions simultaneously formed from such *electron transfer* attract one another. The result is the formation of an ionic compound.

Lewis structures (Section 4.2) are helpful in visualizing the formation of simple ionic compounds. The reaction between the element sodium (with one valence electron) and chlorine (with seven valence electrons) is represented as follows with such structures:

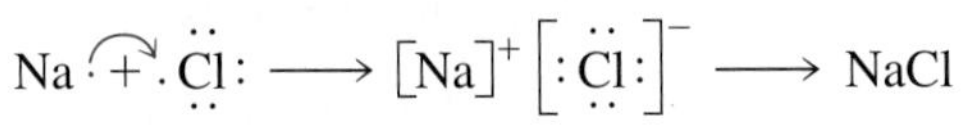

$$\text{Na}\cdot + \cdot\ddot{\underset{..}{\text{Cl}}}: \longrightarrow [\text{Na}]^+\left[:\ddot{\underset{..}{\text{Cl}}}:\right]^- \longrightarrow \text{NaCl}$$

The loss of an electron by sodium empties its valence shell. The next inner shell, which contains eight electrons (a noble-gas configuration), then becomes the valence shell. After the valence shell of chlorine gains one electron, it has the needed eight valence electrons.

When sodium, which has one valence electron, combines with oxygen, which has six valence electrons, the oxygen atom requires the presence of two sodium atoms to acquire two additional electrons.

$$\begin{matrix}\text{Na}\cdot \\ \text{Na}\cdot\end{matrix} + :\dot{\underset{..}{\text{O}}}: \longrightarrow \begin{matrix}[\text{Na}]^+ \\ [\text{Na}]^+\end{matrix}\left[:\ddot{\underset{..}{\text{O}}}:\right]^{2-} \longrightarrow \text{Na}_2\text{O}$$

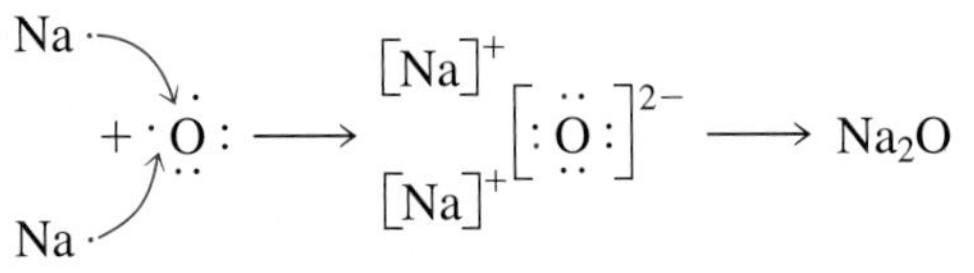

Note that because oxygen has room for two additional electrons, two sodium atoms are required per oxygen atom—hence the formula Na_2O.

An opposite situation occurs in the reaction between calcium, which has two valence electrons, and chlorine, which has seven valence electrons. Here, two chlorine atoms are required to accommodate electrons transferred from one calcium atom, because a chlorine atom can accept only one electron. (It has seven valence electrons and needs only one more.)

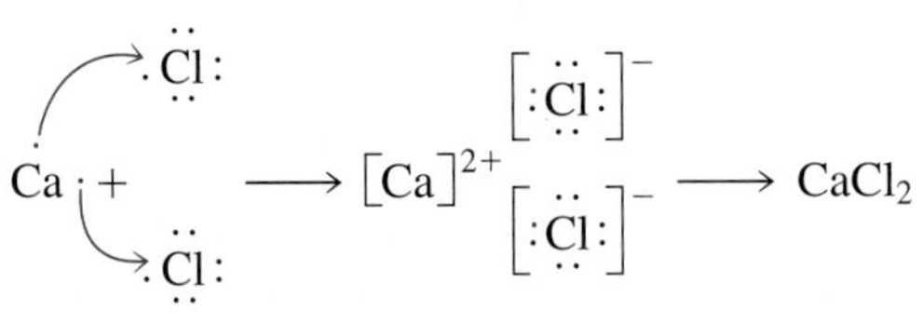

$$\text{Ca}: + \begin{matrix}\cdot\ddot{\underset{..}{\text{Cl}}}: \\ \cdot\ddot{\underset{..}{\text{Cl}}}:\end{matrix} \longrightarrow [\text{Ca}]^{2+}\begin{matrix}\left[:\ddot{\underset{..}{\text{Cl}}}:\right]^- \\ \left[:\ddot{\underset{..}{\text{Cl}}}:\right]^-\end{matrix} \longrightarrow \text{CaCl}_2$$

Example 4.3 Using Lewis Structures to Depict Ionic Compound Formation

Show the formation of the following ionic compounds using Lewis structures.

a. Na_3N **b.** MgO **c.** Al_2S_3

Solution

a. Sodium (a Group IA element) has one valence electron, which it would "like" to lose. Nitrogen (a Group VA element) has five valence electrons and would thus "like" to acquire three more. Three sodium atoms are needed to supply enough electrons for one nitrogen atom.

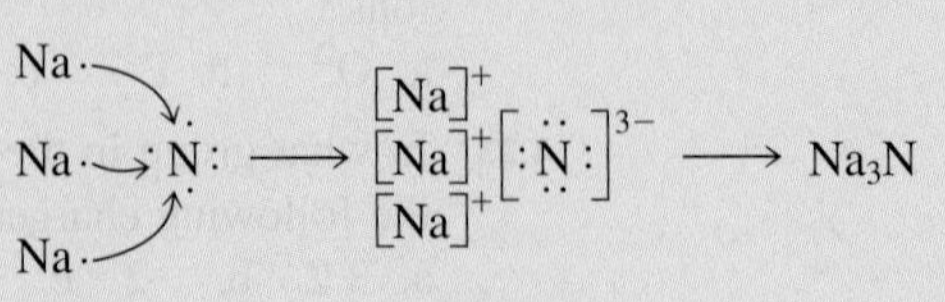

$$\begin{matrix}\text{Na}\cdot \\ \text{Na}\cdot \\ \text{Na}\cdot\end{matrix}\ \dot{\underset{\cdot}{\text{N}}}: \longrightarrow \begin{matrix}[\text{Na}]^+ \\ [\text{Na}]^+ \\ [\text{Na}]^+\end{matrix}\left[:\ddot{\underset{..}{\text{N}}}:\right]^{3-} \longrightarrow \text{Na}_3\text{N}$$

b. Magnesium (a Group IIA element) has two valence electrons, and oxygen (a Group VIA element) has six valence electrons. The transfer of the two magnesium valence electrons to an oxygen atom results in each atom having a noble-gas electron configuration. Thus these two elements combine in a one-to-one ratio.

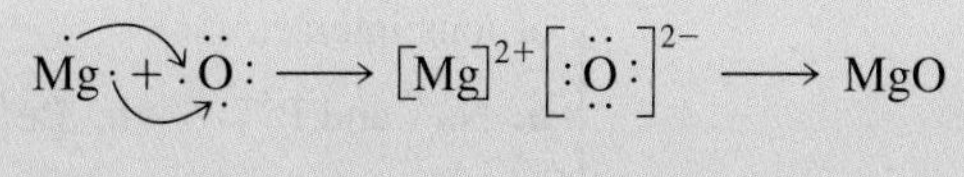

$$Mg\!: + \cdot\ddot{O}: \longrightarrow [Mg]^{2+}[:\ddot{O}:]^{2-} \longrightarrow MgO$$

c. Aluminum (a Group IIIA element) has three valence electrons, all of which need to be lost through electron transfer. Sulfur (a Group VIA element) has six valence electrons and thus needs to acquire two more. Three sulfur atoms are needed to accommodate the electrons given up by two aluminum atoms.

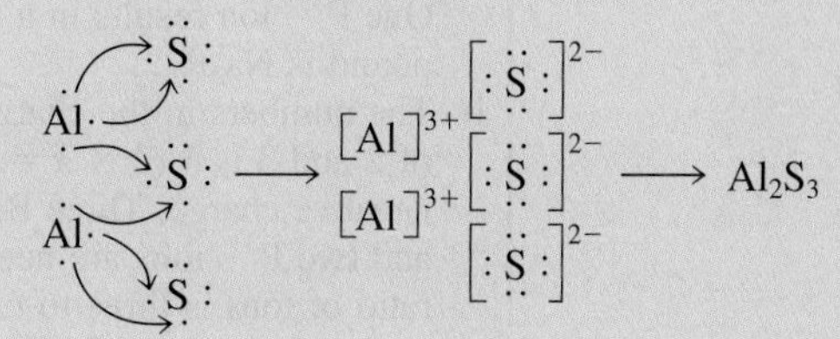

$$\begin{matrix} Al\cdot \\ Al\cdot \end{matrix} + \begin{matrix} \cdot\ddot{S}: \\ \cdot\ddot{S}: \\ \cdot\ddot{S}: \end{matrix} \longrightarrow \begin{matrix} [Al]^{3+} \\ [Al]^{3+} \end{matrix} \begin{matrix} [:\ddot{S}:]^{2-} \\ [:\ddot{S}:]^{2-} \\ [:\ddot{S}:]^{2-} \end{matrix} \longrightarrow Al_2S_3$$

Practice Problems and Questions

4.24 What happens to a metal atom when it is involved in an electron-transfer reaction?

4.25 What happens to a nonmetal atom when it is involved in an electron-transfer reaction?

4.26 Would each of the following atoms be expected to lose or to gain electrons in an electron-transfer reaction?
a. $_{17}Cl$ b. $_{8}O$ c. $_{20}Ca$ d. $_{13}Al$

4.27 Using Lewis structures, show how ionic compounds are formed from atoms of
a. Be and O b. Mg and S c. K and N d. F and Ca

Learning Focus

Using the concept of charge balance and the charges on the ions present, write the correct formula for an ionic compound.

4.7 Formulas for Ionic Compounds

Electron loss always equals electron gain in an electron transfer process. Consequently, ionic compounds are always neutral; no net charge is present. The total positive charge present on the ions that have lost electrons always is exactly counterbalanced by the total negative charge on the ions that have gained electrons. Thus *the ratio in which positive and negative ions combine is the ratio that achieves charge neutrality for the resulting compound.* This generalization can be used, instead of Lewis structures, to determine ionic compound formulas. Ions are combined in the ratio that causes the positive and negative charges to add to zero.

The correct combining ratio when K^+ ions and S^{2-} ions combine is two to one. Two K^+ ions (each of +1 charge) will be required to balance the charge on a single S^{2-} ion.

$$\begin{aligned} 2(K^+)&: \quad (2 \text{ ions}) \times (\text{charge of } +1) = +2 \\ S^{2-}&: \quad (1 \text{ ion}) \times (\text{charge of } -2) = -2 \\ &\qquad\qquad\qquad \text{Net charge} = \ \ 0 \end{aligned}$$

The formula of the compound formed is thus K_2S.

There are three rules to remember when writing formulas for all ionic compounds.

1. The symbol for the positive ions is always written first.
2. The charges on the ions that are present are *not* shown in the formula. You need to know the charges to determine the formula; however, the charges are not explicitly shown in the formula.
3. The numbers in the formula (the subscripts) give the combining ratio for the ions.

Example 4.4 **Using Ionic Charges to Determine the Formula of an Ionic Compound**

Determine the formula for the compound that is formed when each of the following pairs of ions interact.

a. Na^+ and P^{3-} **b.** Be^{2+} and P^{3-}

Solution

a. The Na^+ and P^{3-} ions combine in a three-to-one ratio because this combination causes the charges to add to zero. Three Na^+ ions give a total positive charge of 3. One P^{3-} ion results in a total negative charge of 3. Thus the formula for the compound is Na_3P.

b. The numbers in the charges for these ions are 2 and 3. The lowest common multiple of 2 and 3 is 6 ($2 \times 3 = 6$). Thus we need 6 units of positive charge and 6 units of negative charge. Three Be^{2+} ions are needed to give the 6 units of positive charge, and two P^{3-} ions are needed to give the 6 units of negative charge. The combining ratio of ions is three to two, and the formula is Be_3P_2.

The strategy of finding the lowest common multiple of the numbers in the charges of the ions always works, and it saves you the inconvenience of drawing the Lewis structures.

Practice Problems and Questions

4.28 Write the formula of the ionic compounds formed from Ba^{2+} ions and each of the following ions.
a. Cl^- b. Br^- c. N^{3-} d. O^{2-}

4.29 Write the formula of the ionic compounds formed from F^- ions and each of the following ions.
a. Mg^{2+} b. Be^{2+} c. Li^+ d. Al^{3+}

4.30 Write the formula of the ionic compounds formed from each of the following pairs of ions.
a. Na^+ and S^{2-} b. Ca^{2+} and I^- c. Li^+ and N^{3-} d. Br^- and Al^{3+}

4.31 Why is the formula for the ionic compound sodium chloride always written as NaCl rather than as ClNa?

Learning Focus

Understand the structural characteristics of ionic compounds, including the concept of a *formula unit.*

4.8 The Structure of Ionic Compounds

An ionic compound, in the solid state, consists of positive and negative ions arranged in such a way that each ion is surrounded by nearest neighbors of the opposite charge. Any given ion is bonded by electrostatic (positive – negative) attractions to all the other ions of opposite charge immediately surrounding it. Figure 4.4 shows a two-dimensional cross section and a three-dimensional view of the arrangement of ions in the ionic compound sodium chloride (NaCl). Note in these structural representations that no given ion has a single partner. A given sodium ion has six immediate neighbors (chloride ions) that are equidistant from it. A chloride ion in turn has six immediate sodium ion neighbors.

The alternating array of positive and negative ions present in an ionic compound means that discrete molecules do not exist in such compounds. Therefore, the formulas of ionic compounds cannot represent the composition of molecules of these substances. Instead, such formulas represent the simplest combining ratio for the ions present. The formula for sodium chloride, NaCl, indicates that sodium and chloride ions are present

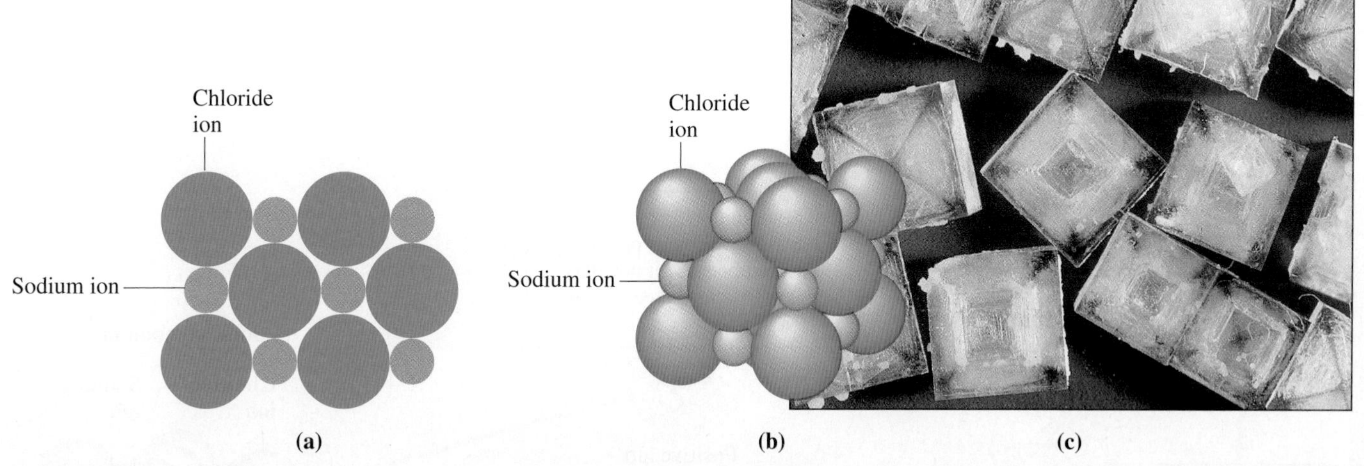

Figure 4.4 (a, b) A two-dimensional cross section and a three-dimensional view of sodium chloride (NaCl), an ionic solid. Both views show an alternating array of positive and negative ions. (c) Sodium chloride crystals.

in a one-to-one ratio in this compound. Chemists use the term *formula unit*, rather than molecule, to refer to the smallest unit of an ionic compound. A **formula unit** *is the smallest whole-number repeating ratio of ions present in an ionic compound that results in charge neutrality.* A formula unit is "hypothetical" because it does not exist as a separate entity; it is only a part of the extended array of ions that constitute an ionic solid (see Figure 4.5).

▶ In Section 1.9 the molecule was described as the smallest unit of a pure substance that is capable of a stable, independent existence. Ionic compounds, with their formula units, are exceptions to this generalization.

Although the formulas for ionic compounds represent only ratios, they are used in equations and chemical calculations in the same way as the formulas for molecular species. Remember, however, that they cannot be interpreted as indicating that molecules exist for these substances. They merely represent the simplest ratio of ions.

Chemical Portraits 6 profiles three relatively common ionic compounds—sodium chloride, calcium oxide, and aluminum oxide—in terms of their occurrence and uses.

Practice Problems and Question

4.32 Describe the general structure of a solid-state ionic compound.

4.33 What is a *formula unit* of an ionic compound?

4.34 In general terms, how many *formula units* are present in a crystal of an ionic compound?

The accompanying Chemistry at a Glance reviews the general concepts we have considered so far about ionic compounds.

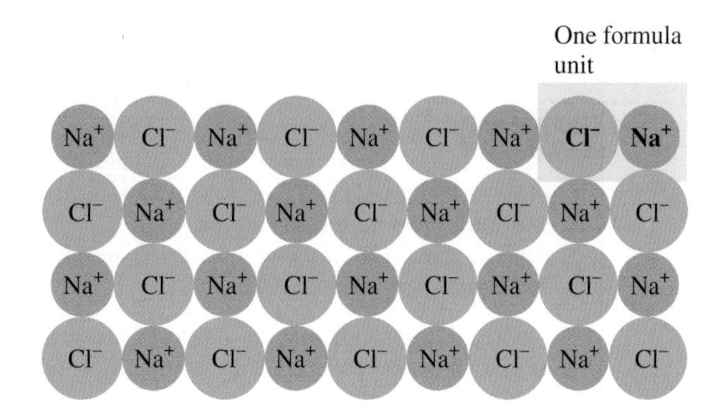

Figure 4.5 Two-dimensional cross section of an ionic solid (NaCl). No molecule can be distinguished in this structure. Instead, we can recognize a basic formula unit that is repeated indefinitely.

Chemistry *at a Glance*

Ionic Bonds and Ionic Compounds

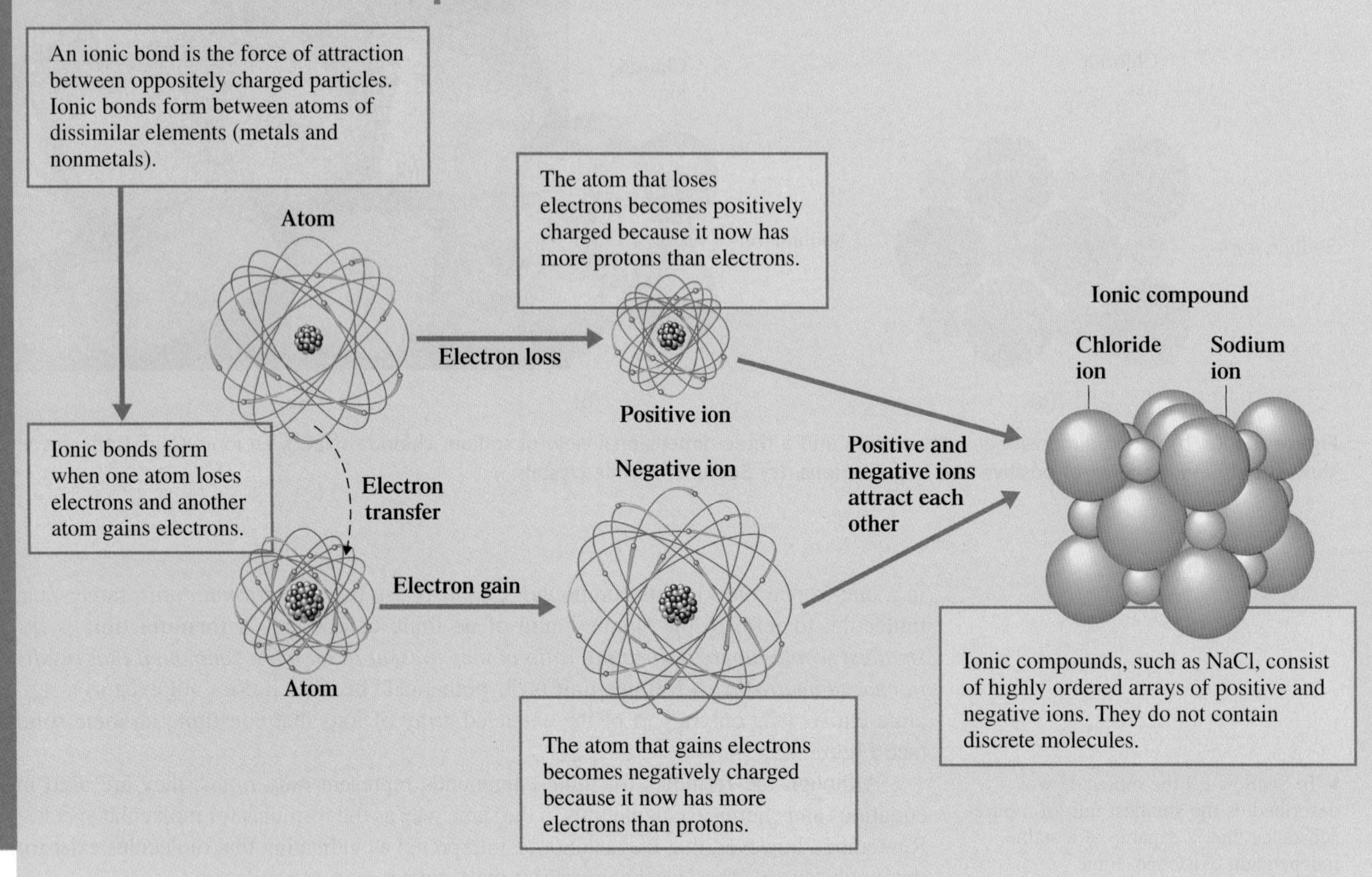

An ionic bond is the force of attraction between oppositely charged particles. Ionic bonds form between atoms of dissimilar elements (metals and nonmetals).
Atom
Ionic bonds form when one atom loses electrons and another atom gains electrons.
Electron transfer
Atom
Electron loss
The atom that loses electrons becomes positively charged because it now has more protons than electrons.
Positive ion
Negative ion
Electron gain
The atom that gains electrons becomes negatively charged because it now has more electrons than protons.
Positive and negative ions attract each other
Ionic compound
Chloride ion
Sodium ion
Ionic compounds, such as NaCl, consist of highly ordered arrays of positive and negative ions. They do not contain discrete molecules.

Learning Focus

On the basis of their chemical formulas, be able to recognize which binary compounds are ionic. Write the name of a binary ionic compound given its formula, or vice versa.

4.9 Recognizing and Naming Binary Ionic Compounds

The term *binary* means "two." A **binary ionic compound** *is an ionic compound in which only two elements are present.* In all binary ionic compounds, one element is a metal and the other element is a nonmetal. The metal is always present as the positive ion, the nonmetal as the negative ion. The joint presence of a metal and a nonmetal in a binary compound is the "recognition key" that the compound is an ionic compound.

Example 4.5 **Classifying a Binary Compound as Ionic on the Basis of Its Chemical Formula**

Which of the following binary compounds is most likely to be an ionic compound?

a. Al_2S_3 **b.** H_2O **c.** KF **d.** NH_3

Solution

a. Ionic; a metal (Al) and a nonmetal (S) are present
b. Not ionic; two nonmetals are present
c. Ionic; a metal (K) and a nonmetal (F) are present
d. Not ionic; two nonmetals are present

The two compounds that are not ionic are *molecular* compounds (Section 4.1). Section 4.10 begins an extended discussion of molecular compounds. In general, molecular compounds contain just nonmetals.

Chemical Portraits 6 Selected Simple Ionic Compounds

Sodium Chloride (NaCl) (Na^+ and Cl^-)

Profile: When finely divided, NaCl is a white crystalline solid; large crystals, however are colorless. It occurs in large underground deposits (rock salt) or can be obtained by solar evaporation of seawater (sea salt). Also known as *table salt,* NaCl is the world's most common food additive. Table salt use tends to enhance other flavors, probably by suppressing the bitter flavors.

Biochemical considerations: Physiologically, NaCl is the major source for Na^+ ions and Cl^- ions, the two most abundant ions in blood plasma and in interstitial fluid (the fluid outside cells). Sodium ions are needed for nerve transmission and muscle contraction. Choride ions are essential to acid-base balance within the body and help regulate fluid flow.

How is table salt commercially produced and what additives does it contain?

Calcium Oxide (CaO) (Ca^{2+} and O^{2-})

Profile: Calcium oxide, often called *lime,* is a white solid in its pure state. It is produced by strongly heating *limestone* ($CaCO_3$), which decomposes, yielding gaseous carbon dioxide (CO_2), and solid lime.

Uses: Lime is a high-volume industrial chemical. Its major use is in steel making, where it is added to a molten charge of iron that contains impurities. The lime interacts with the impurities to form a glassy waste material (*slag*), which floats to the top of the molten iron and is drawn off. Another use for lime is as a component of mortar mixes (dry sand and lime). When water is added to a mortar mix, and as CO_2 is absorbed from the air, the mortar sets as solid $CaCO_3$. A small, but important, use of lime involves the control of soil acidity; when added to soil, it reduces soil acidity. Excess soil acidity can negatively affect plant health.

What is the function for the sand present in a mortar mix?

Aluminum Oxide (Al_2O_3) (Al^{3+} and O^{2-})

Profile: Al_2O_3 occurs as the mineral *corundum* and as *emery*, a granular form of corundum, both of which are very hard substances. This hardness leads to many applications for Al_2O_3 in the area of abrasives. Larger crystals of corundum, when colored with transition metal impurities have gemstone value; both rubies and sapphires are Al_2O_3 crystals; Cr^{3+} ions give red color to rubies and Fe^{2+} and Fe^{3+} ions impart blue color to sapphires.

Uses: Al_2O_3 is used as an abrasive in grinding wheels, sandpaper, and toothpaste. Synthetically produced Al_2O_3 gemstones (artificial rubies) find use as bearings ("jewels") in expensive watches and instruments and as "stones" in inexpensive jewelry.

What is the procedure by which "artificial rubies" are produced?

See the text web site at **www.cengage.com/chemistry/stoker** for answers to the above questions and for further information.

Binary ionic compounds are named using the following rule: *The full name of the metallic element is given first, followed by a separate word containing the stem of the nonmetallic element name and the suffix* -ide. Thus, in order to name the compound NaF, we start with the name of the metal (sodium), follow it with the stem of the name of the nonmetal (fluor-), and then add the suffix *-ide.* The name becomes *sodium fluoride.*

The stem of the name of the nonmetal is the name of the nonmetal with its ending chopped off. Table 4.1 gives the stem part of the name for each of the most common nonmetallic elements. The name of the metal ion is always exactly the same as the name of the metal itself; the metal's name is never shortened. Example 4.6 illustrates the use of the rule for naming binary ionic compounds.

Table 4.1 Names of Selected Common Nonmetallic Ions

Element	Stem	Name of ion	Formula of ion
bromine	brom-	bromide	Br^-
carbon	carb-	carbide	C^{4-}
chlorine	chlor-	chloride	Cl^-
fluorine	fluor-	fluoride	F^-
hydrogen	hydr-	hydride	H^-
iodine	iod-	iodide	I^-
nitrogen	nitr-	nitride	N^{3-}
oxygen	ox-	oxide	O^{2-}
phosphorus	phosph-	phosphide	P^{3-}
sulfur	sulf-	sulfide	S^{2-}

Example 4.6 Naming Binary Ionic Compounds

Name the following binary ionic compounds.

a. MgO **b.** Al_2S_3 **c.** K_3N **d.** $CaCl_2$

Solution

The general pattern for naming binary ionic compounds is

Name of metal + stem of name of nonmetal + *-ide*

a. The metal is magnesium and the nonmetal is oxygen. Thus the compound's name is *magnesium oxide.*

b. The metal is aluminum and the nonmetal is sulfur; the compound's name is *aluminum sulfide.* Note that no mention is made of the subscripts present in the formula—the 2 and the 3. The name of an ionic compound never contains any reference to formula subscript numbers. There is only one ratio in which aluminum and sulfur atoms combine. Thus, just telling the names of the elements present in the compound is adequate nomenclature.

c. Potassium (K) and nitrogen (N) are present in the compound, and its name is *potassium nitride.*

d. The compound's name is *calcium chloride.*

▶ All the inner transition elements (*f* area of the periodic table), most of the transition elements (*d* area), and a few representative metals (*p* area) exhibit variable ionic charge behavior.

▶ An older method for indicating the charge on metal ions uses the suffixes *-ic* and *-ous* rather than the Roman numeral system. It is mentioned here because it is still sometimes encountered. In this system, when a metal has two common ionic charges, the suffix *-ous* is used for the ion of lower charge and the suffix *-ic* for the ion of higher charge. The metal's Latin name is also used. In this older system, iron(II) ion is called ferrous ion, and iron(III) ion is called ferric ion.

Thus far in our discussion of ionic compounds, it has been assumed that the only behavior allowable for an element is that predicted by the octet rule. This is a good assumption for nonmetals and for most representative element metals. However, there are many other metals that exhibit a less predictable behavior because they are able to form more than one type of ion. For example, iron forms both Fe^{2+} ions and Fe^{3+} ions, depending on chemical circumstances.

When we name compounds that contain metals with variable ionic charges, the charge on the metal ion must be incorporated into the name. This is done by using Roman numerals. For example, the chlorides of Fe^{2+} and Fe^{3+} ($FeCl_2$ and $FeCl_3$, respectively) are named iron(II) chloride and iron(III) chloride (Figure 4.6). Likewise, CuO is named copper(II) oxide. If you are uncertain about the charge on the metal ion in an ionic compound, use the charge on the nonmetal ion (which does not vary) to calculate it. For example, in order to determine the charge on the copper ion in CuO, you can note that the oxide ion carries a -2 charge because oxygen is in Group VIA. This means that the copper ion must have a $+2$ charge to counterbalance the -2 charge.

Figure 4.6 (a) Copper(II) oxide (CuO) is black, whereas copper(I) oxide (Cu_2O) is reddish brown. (b) Iron(II) chloride ($FeCl_2$) is green, whereas iron(III) chloride ($FeCl_3$) is bright yellow.

Example 4.7 **Using Roman Numerals in the Naming of Binary Ionic Compounds**

Name the following binary ionic compounds, each of which contains a metal whose ionic charge can vary.

a. AuCl **b.** Fe_2O_3

Solution

We will need to indicate the magnitude of the charge on the metal ion in the name of each of these compounds by means of a Roman numeral.

a. To calculate the metal ion charge, use the fact that total ionic charge (both positive and negative) must add to zero.

$$(\text{Gold charge}) + (\text{chlorine charge}) = 0$$

The chloride ion has a -1 charge (Section 4.5). Therefore,

$$(\text{Gold charge}) + (-1) = 0$$

Thus,

$$\text{Gold charge} = +1$$

Therefore, the gold ion present is Au^+, and the name of the compound is *gold(I) chloride.*

b. For charge balance in this compound we have the equation

$$2(\text{iron charge}) + 3(\text{oxygen charge}) = 0$$

Note that we have to take into account the number of each kind of ion present (2 and 3 in this case). Oxide ions carry a -2 charge (Section 4.5). Therefore,

$$2(\text{iron charge}) + 3(-2) = 0$$

$$2(\text{iron charge}) = +6$$

$$\text{Iron charge} = +3$$

Here, we are interested in the charge on a single iron ion (+3) and not in the total positive charge present (+6). The compound is named *iron(III) oxide* because Fe^{3+} ions are present. As is the case for all ionic compounds, the name does not contain any reference to the numerical subscripts in the compound's formula.

▶ The fixed-charge metals are those in Group IA (+1 ionic charge), those in Group IIA (+2 ionic charge), and five others (Al^{3+}, Ga^{3+}, Zn^{2+}, Cd^{2+}, and Ag^+).

In order to know when to use Roman numerals in binary ionic compound names, you must know which metals exhibit variable ionic charge and which have a fixed ionic charge. There are many more of the former (Roman numeral required) than of the latter (no Roman numeral required). Thus you should learn the identity of the metals that have a fixed ionic charge (the short list); any metal not on the short list must exhibit variable charge. Figure 4.7 shows the metals that always form a single type of ion in ionic compound formation. Ionic compounds that contain these metals are the only ones without Roman numerals in their names.

Practice Problems and Questions

4.35 Which of the following pairs of elements would be expected to form a binary ionic compound?

a. Sodium and oxygen b. Magnesium and sulfur
c. Nitrogen and chlorine d. Copper and fluorine

4.36 Name each of the following binary ionic compounds, each of which contains a fixed-charge metal.

a. KI b. BeO c. AlF_3 d. Na_3P

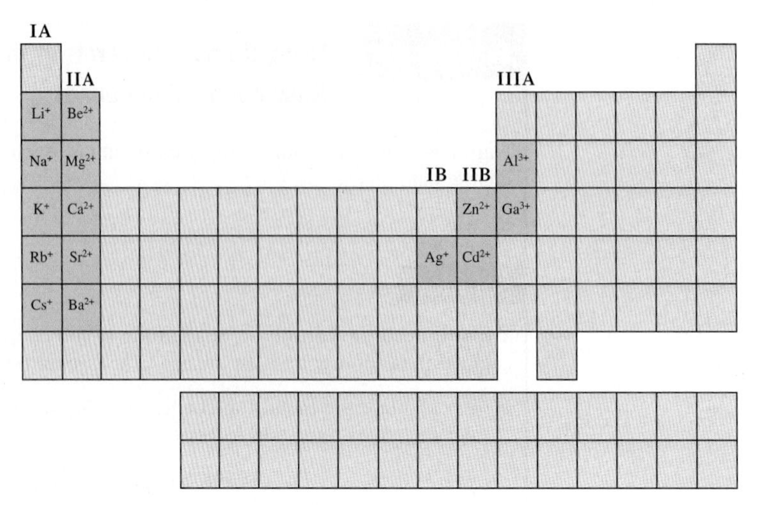

Figure 4.7 A periodic table in which the metallic elements that exhibit a fixed ionic charge are highlighted.

4.37 Calculate the charge on the metal ion in the following binary ionic compounds, each of which contains a variable-charge metal.
a. Ag_2O b. CuO c. SnO_2 d. SnO

4.38 Name the following binary ionic compounds, each of which contains a variable-charge metal.
a. FeO b. Au_2O_3 c. CuS d. $CoBr_2$

4.39 Name each of the following binary ionic compounds.
a. AuCl b. KCl c. AgCl d. $CuCl_2$

4.40 Write formulas for the following binary ionic compounds.
a. Potassium bromide b. Silver fluoride c. Beryllium sulfide d. Zinc chloride

4.41 Write formulas for the following binary ionic compounds.
a. Cobalt(II) oxide b. Cobalt(III) oxide c. Tin(IV) iodide d. Lead(II) nitride

4.10 The Covalent Bond Model

Learning Focus

Be able to explain the major differences between the ionic bond model and the covalent bond model.

The forces that hold atoms in compounds together as a unit are of two general types: (1) ionic bonds (which involve electron transfer) and (2) covalent bonds (which involve electron sharing). The ionic bond model was the subject of Sections 4.4 through 4.9. We now consider the covalent bond model.

We begin our discussion of covalent bonding and the molecular compounds that result from such bonding by listing three key differences between ionic and covalent bonding.

1. Ionic bonds form between atoms of dissimilar elements (a metal and a nonmetal). Covalent bond formation occurs between *similar* or even *identical* atoms. Most often two nonmetals are involved.
2. Electron transfer is the mechanism by which ionic bond formation occurs. Covalent bond formation involves *electron sharing.*
3. Ionic compounds do not contain discrete molecules. Instead, such compounds consist of an extended array of alternating positive and negative ions. In covalently bonded compounds, the basic structural unit is a molecule. Indeed, such compounds are called molecular compounds.

▶ Among the millions of compounds that are known, those that have covalent bonds are dominant. Almost all compounds encountered in the fields of organic chemistry and biochemistry contain covalent bonds.

Consideration of the hydrogen molecule (H_2), the simplest of all molecules, provides initial insights into the nature of the covalent bond and its formation. When two hydrogen atoms, each with a single electron, are brought together, the orbitals that contain the valence electrons *overlap* to create an orbital common to both atoms. This overlapping is

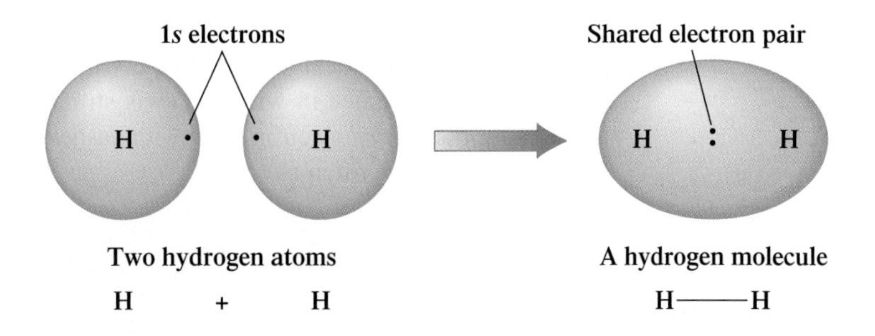

Figure 4.8 Electron sharing can occur only when electron orbitals from two different atoms overlap.

▶ Covalent bonds result from a common attraction of two nuclei for one or more shared pairs of electrons.

shown in Figure 4.8. The two electrons, one from each H atom, now move throughout this new orbital and are said to be *shared* by the two nuclei.

Once two orbitals overlap, the most favorable location for the shared electrons is the area directly between the two nuclei. Here the two electrons can simultaneously interact with (be attracted to) both nuclei, a situation that produces increased stability. This concept of increased stability can be explained by using an analogy. Consider the nuclei of the two hydrogen atoms in H_2 to be "old potbellied stoves" and the two electrons to be running around each of the stoves trying to keep warm. When the two nuclei are together (an H_2 molecule) the electrons have two sources of heat. In particular, in the region between the nuclei (the overlap region) the electrons can keep both front and back warm at the same time. This is a better situation than when each electron has only one "stove" (nucleus) as a source of heat.

In terms of Lewis notation, this sharing of electrons by the two hydrogen atoms is diagrammed as follows:

Shared electron pair

$$\text{H}\cdot \; \cdot\text{H} \longrightarrow \text{H}:\text{H}$$

The two shared electrons do double duty, helping each hydrogen atom achieve a helium noble-gas configuration.

Practice Problems and Questions

4.42 Contrast the types of atoms involved in ionic and covalent bonds.

4.43 Contrast the mechanisms by which ionic and covalent bonds form.

4.44 Contrast the basic structural units in ionic and molecular compounds.

4.11 Lewis Structures for Molecular Compounds

Learning Focus

Use Lewis structures to describe the bonding in simple covalent molecules. Distinguish between nonbonding and bonding electrons in a given Lewis structure.

Using the octet rule, which applies to both electron transfer and electron sharing (Section 4.3), and Lewis structures (Section 4.2), let us now consider the formation of selected simple covalently bonded molecules that contain the element fluorine. Fluorine, located in Group VIIA of the periodic table, has seven valence electrons. Its Lewis structure is

$$\cdot\ddot{\underset{..}{\text{F}}}:$$

Fluorine needs only one electron to achieve the octet of electrons that enables it to have a noble-gas electron configuration. When fluorine bonds to other nonmetals, the octet of electrons is completed by means of electron sharing. The molecules HF, F_2, and BrF, whose electron-dot structures follow, are representative of this situation.

$$\text{H}\cdot \; \cdot\ddot{\underset{..}{\text{F}}}: \longrightarrow \text{H}:\ddot{\underset{..}{\text{F}}}:$$

$$:\ddot{\underset{..}{\text{F}}}\cdot \; \cdot\ddot{\underset{..}{\text{F}}}: \longrightarrow :\ddot{\underset{..}{\text{F}}}:\ddot{\underset{..}{\text{F}}}:$$

$$:\ddot{\underset{..}{\text{Br}}}\cdot \; \cdot\ddot{\underset{..}{\text{F}}}: \longrightarrow :\ddot{\underset{..}{\text{Br}}}:\ddot{\underset{..}{\text{F}}}:$$

The HF and BrF molecules illustrate the point that the two atoms involved in a covalent bond need not be identical (as is the case with H_2 and F_2).

A common practice in writing Lewis structures for covalently bonded molecules is to represent the *shared* electron pairs with dashes. Using this notation, the H_2, HF, F_2, and BrF molecules are written as

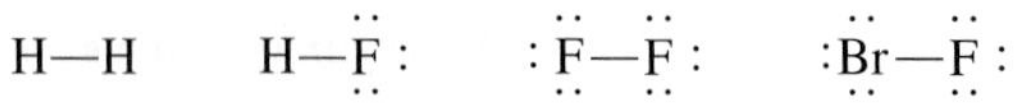

H—H   H—F: :F—F: :Br—F:

The atoms in covalently bonded molecules often possess both *bonding* and *nonbonding* electrons. **Bonding electrons** *are pairs of valence electrons that are shared between atoms in a covalent bond.* Each of the fluorine atoms in the molecules HF, F_2, and BrF possesses one pair of bonding electrons. **Nonbonding electrons** *are pairs of valence electrons that are not involved in electron sharing.* Each of the fluorine atoms in HF, F_2, and BrF possesses three pairs of nonbonding electrons, as does the bromine atom in BrF.

▶ Nonbonding electron pairs are often also referred to as *unshared electron pairs* or *lone electron pairs* (or simply *lone pairs*).

▶ In Section 4.14 we will learn that nonbonding electrons play an important role in determining the shape (geometry) of molecules when three or more atoms are present.

H:H   H:F: :F:F: :Br:F:

Bonding electrons

Nonbonding electrons

Bonding electrons (black)
Nonbonding electrons (blue)

The preceding four examples of Lewis structures involved diatomic molecules, the simplest type of molecule. The "thinking pattern" used to draw these diatomic Lewis structures easily extends to triatomic and larger molecules. Consider the molecules H_2O, NH_3, and CH_4, molecules in which two, three, and four hydrogen atoms are attached, respectively, to the O, N, and C atoms. The hydrogen content of these molecules directly correlates with the fact that oxygen, nitrogen, and carbon have six, five, and four valence electrons, respectively, and therefore need to gain two, three, and four electrons, respectively, through electron sharing in order for the octet rule to be obeyed. The electron-sharing patterns and Lewis structures for these three molecules are as follows:

Oxygen has 6 valence electrons and gains 2 more through sharing.

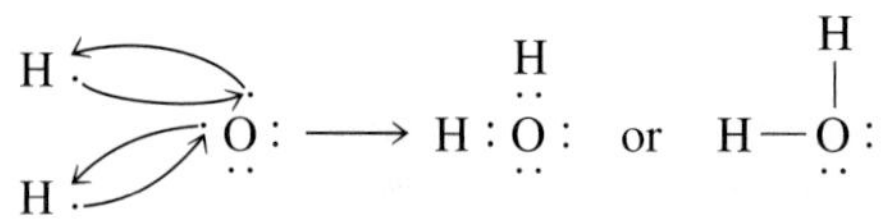

Nitrogen has 5 valence electrons and gains 3 more through sharing.

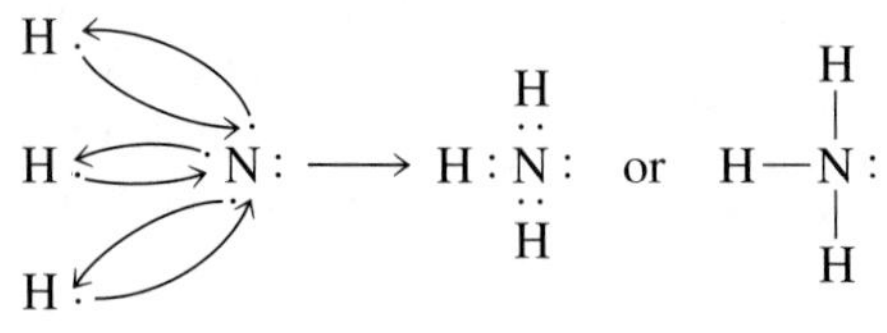

Carbon has 4 valence electrons and gains 4 more through sharing.

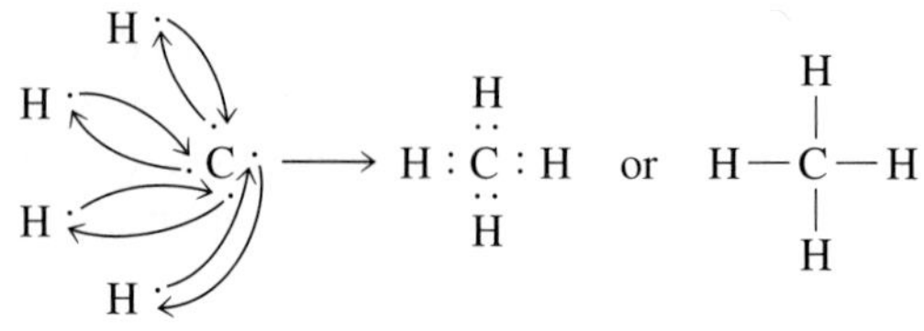

Thus we see here that just as the octet rule was useful in determining the ratio of ions in ionic compounds (Section 4.7), it can be used to predict formulas in molecular compounds. Example 4.8 further illustrates the use of the octet rule to determine formulas for molecular compounds.

Example 4.8 Using the Octet Rule to Predict the Formulas of Simple Molecular Compounds

Draw Lewis structures for the simplest binary compounds that can be formed from the following pairs of nonmetals.

a. Nitrogen and iodine **b.** Sulfur and hydrogen

Solution

a. Nitrogen is in Group VA of the periodic table and has five valence electrons. It will want to form three covalent bonds. Iodine, in Group VIIA of the periodic table, has seven valence electrons and will want to form only one covalent bond. Therefore, three iodine atoms will be needed to meet the needs of one nitrogen atom. The Lewis structure for this molecule is

:I: + :I: + :I: + N: ⟶ :I:N: (with :I: above and below N) or :I—N: (with I bonded above and below N)

Each atom in NI_3 has an octet of electrons, which is circled in the following diagram.

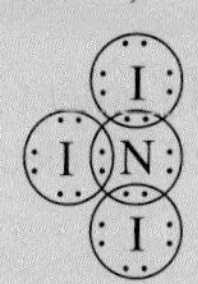

b. Sulfur has six valence electrons and hydrogen has one valence electron. Thus, sulfur will form two covalent bonds (6 + 2 = 8) and hydrogen will form one covalent bond (1 + 1 = 2). Remember that for hydrogen, an "octet" is two electrons; the noble gas that hydrogen mimics is helium, which has only two valence electrons.

H· + H· + S: ⟶ H:S: (with H below S) ⟶ H—S: (with H bonded below S)

Practice Problems and Questions

4.45 Draw Lewis structures to illustrate the covalent bonding in the following diatomic molecules.

a. Br_2 b. HI c. IBr d. BrF

4.46 Draw Lewis structures for the simplest molecular compounds formed between the following pairs of nonmetals.

a. Sulfur and fluorine b. Carbon and iodine
c. Nitrogen and bromine d. Selenium and hydrogen

4.47 What would be the predicted chemical formula for the simplest molecular compound formed between each of the following pairs of nonmetals?

a. Hydrogen and oxygen b. Carbon and bromine
c. Phosphorus and iodine d. Silicon and hydrogen

4.48 How many nonbonding electron pairs are present in each of the following Lewis structures?

a. H:O:O:H b. H:O:H c. H:H d. H:O:Cl:

4.49 Contrast the meaning of the terms *nonbonding electron pair, unshared electron pair,* and *lone electron pair.*

Learning Focus

Understand how *single, double,* and *triple* covalent bonds differ, and be able to recognize such bond types in Lewis structures.

4.12 Single, Double, and Triple Covalent Bonds

A **single covalent bond** *is a bond in which two atoms share one pair of electrons.* All of the molecules considered in the previous section contain *single* covalent bonds.

Single covalent bonds are not adequate to explain covalent bonding in all molecules. Sometimes two atoms must share two or three pairs of electrons in order to provide a complete octet of electrons for each atom involved in the bonding. Such bonds are called *double* covalent bonds and *triple* covalent bonds. A **double covalent bond** *is a bond in which two atoms share two pairs of electrons.* A double covalent bond between two atoms is approximately twice as strong as a single covalent bond between the same two atoms; that is, it takes approximately twice as much energy to break the double bond as it does the single bond. A **triple covalent bond** *is a bond in which two atoms share three pairs of electrons.* A triple covalent bond is approximately three times as strong as a single covalent bond between the same two atoms. The term *multiple covalent bond* is a designation that applies to both double and triple covalent bonds.

One of the simplest molecules possessing a multiple covalent bond is the N_2 molecule, which has a triple covalent bond. A nitrogen atom has five valence electrons and needs three additional electrons to complete its octet.

$$\cdot\ddot{\text{N}}\cdot$$

In order to acquire a noble-gas electron configuration, each nitrogen atom must share three of its electrons with the other nitrogen atom.

:N· + ·N: ⟶ :N:::N: or :N≡N:

▶ A single line (dash) is used to denote a single covalent bond, two lines to denote a double covalent bond, and three lines to denote a triple covalent bond.

Note that all three shared electron pairs are placed in the area between the two nitrogen atoms in this bonding diagram. Just as one line is used to denote a single covalent bond, three lines are used to denote a triple covalent bond.

When you are "counting" electrons in an electron-dot structure to make sure that all atoms in the molecule have achieved their octet of electrons, *all* electrons in a double or triple bond are considered to belong to *both* of the atoms involved in that bond. The "counting" for the N_2 molecule would be

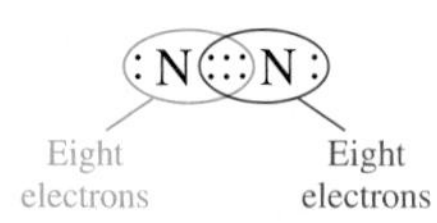

Each of the circles around a nitrogen atom contains eight valence electrons. Circles are never drawn to include just some of the electrons in a double or triple bond.

A slightly more complicated molecule containing a triple covalent bond is the molecule C_2H_2 (acetylene). A carbon–carbon triple covalent bond is present as well as two carbon–hydrogen single bonds. The arrangement of valence electrons in C_2H_2 is as follows:

H· ·C· + ·C· ·H ⟶ H:C:::C:H or H—C≡C—H

The two atoms in a triple covalent bond are commonly the same element. However, they do not have to be. The molecule hydrogen cyanide (HCN) contains a heteroatomic triple covalent bond.

H:C:::N: or H—C≡N:

A common molecule that contains a double covalent bond is carbon dioxide (CO_2). In fact, there are two carbon–oxygen double covalent bonds present in CO_2.

:Ö· ·Ċ· ·Ö: ⟶ :Ö::C::Ö: or :Ö=C=Ö:

Note in the following diagram how the circles are drawn for the octet of electrons about each of the atoms in carbon dioxide.

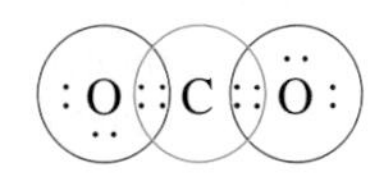

Chemical Portraits 7 profiles the previously mentioned C_2H_2 and HCN molecules as well as the diatomic CO. Molecules of these three compounds have a common bonding feature, the presence of a triple covalent bond.

Practice Problems and Questions

4.50 Contrast the strengths of single, double, and triple covalent bonds.

4.51 Specify the number of single, double, and triple covalent bonds present in molecules represented by the following Lewis structures.
a. :C:::O: b. H:C:::N:
c. :Ö::C::Ö: d. :C̈l:C̈l:

Chemical Portraits 7 Simple Molecules that Contain a Triple Covalent Bond

Carbon Monoxide (CO)
:C≡O:

Profile: Carbon monoxide is a well-known, *toxic,* gaseous air pollutant present in automobile exhaust and cigarette smoke. Automobile catalytic convertors do not prevent CO formation as gasoline burns. Rather they convert the CO formed into carbon dioxide (CO_2), which then enters the atmosphere.

Biochemical considerations: Because CO is colorless, odorless, and tasteless, it gives humans no warning of its presence. It is these properties that make CO so dangerous. Other air pollutants have odors that warn of their presence. CO impairs human health by combining with hemoglobin in red blood cells in a manner that reduces the blood's oxygen-carrying capacity. The effects of CO poisoning are actually the effects of oxygen-deprivation in the body.

How does the level of CO in urban air at present compare with that of 30 years ago?

Hydrogen Cyanide (HCN)
H—C≡N:

Profile: Hydrogen cyanide is a colorless, toxic gas with a faint odor of almonds. The almond smell is not, however, strong enough to be a good warning signal for the presence of the gas. Despite its toxicity, HCN is an important industrial chemical used in the production of nylon and acrylic fibers and in metallurgy in the extraction of silver and gold.

Biological considerations: HCN is toxic to almost all forms of life. Because it can penetrate even into insect eggs, it is used as a fumigate for storage bins and cargo ship holds. In humans, gaseous HCN quickly dissolves in the blood, where it forms cyanide ions (CN^-). These ions stop the process by which cellular energy production occurs, causing death. A few tenths of 1% by volume of HCN in air can cause death within minutes.

What are the symptoms of hydrogen cyanide poisoning?

Acetylene (C_2H_2)
H—C≡C—H

Profile: Acetylene is a colorless, flammable gas with a faint garlic-like odor. It burns brilliantly in air. Early cars had headlights that produced acetylene by the action of slowly dripping water on calcium carbide (CaC_2). This same type of lamp, which was also once used by miners, is still often used by spelunkers (cave explorers).

Uses: Most commercially produced C_2H_2 is converted to chemical intermediates that are used to make plastics, fibers, and resins. About 10% of C_2H_2 production is consumed in working with metals. When C_2H_2 is burned with oxygen in an oxyacetylene welding torch, a very high temperature is produced (3000°C). Such torches are used extensively by welders for cutting and welding steel.

For humans and animals, is acetylene a toxic or nontoxic gas?

See the text web site at **www.cengage.com/chemistry/stoker** for answers to the above questions and for further information.

4.52 Convert each of the following Lewis structures into the form in which lines are used to denote shared electron pairs. Include nonbonding electron pairs in the rewritten structures.

```
                         ..     ..
a. :N:::N:         b. :N::N::O:

c. H:C:H           d. H:C::C:H
     ..                 ..  ..
    :O                  H   H
     ..
```

4.53 Show how each of the following Lewis structures for molecules has electron arrangements such that each atom obeys the octet rule by using circles to "enclose" each octet of electrons. Remember that an "octet" for hydrogen is two electrons.

```
                          ..  ..
a. :C:::O:         b. H:N:N:H
                          ..  ..
                          H   H
    ..   ..
c. :O:S:O:         d. H:C:::C:H
    .. ..  ..
      :O
       ..
```

Figure 4.9 The sulfur dioxide (SO_2) molecule. A computer-generated model.

Learning Focus

Using systematic procedures, draw Lewis structures for molecular compounds given the arrangement of atoms within the compounds.

4.13 A Systematic Method for Drawing Lewis Structures

Drawing Lewis structures for diatomic molecules is usually a straightforward and uncomplicated process. However, with triatomic and larger molecules, simplicity rapidly disappears, and the "bookkeeping" on the electrons present can become complicated unless systematic procedures are followed.

This section presents a stepwise method for distributing valence electrons as bonding and nonbonding pairs in a Lewis structure—a method that overcomes many of the pitfalls associated with drawing Lewis structures. There are six steps in this systematic method for drawing Lewis structures.

Let us apply this stepwise procedure to the molecule SO_2, a molecule in which two oxygen atoms are bonded to a central sulfur atom (see Figure 4.9).

Step 1: *Calculate the total number of valence electrons available in the molecule by adding together the valence electron counts for all atoms in the molecule.* The periodic table is a useful guide for determining this number.

An SO_2 molecule has 18 valence electrons available for bonding. Sulfur (Group VIA) has 6 valence electrons, and each oxygen (also Group VIA) has 6 valence electrons. The total number is therefore $6 + 2(6) = 18$.

Step 2: *Write the symbols of the atoms in the molecule in the order in which they are bonded to one another, and then place a single covalent bond, involving two electrons, between each pair of bonded atoms.* For SO_2, the S atom is the central atom. Thus we have

O : S : O

Determining which atom is the *central atom*—that is, which atom has the most other atoms bonded to it—is the key to determining the arrangement of atoms in a molecule or ion. Most other atoms present will be bonded to the central atom. For common binary molecular compounds, the molecular formula can help us determine the identity of the central atom. The central atom is the atom that appears only once in the formula; for example, S is the central atom in SO_3, O is the central atom in H_2O, and P is the central atom in PF_3. In molecular compounds containing hydrogen, oxygen, and an additional element, that additional element is the central atom; for example, N is the central atom in HNO_3, and S is the central atom in H_2SO_4. In compounds of this type, the oxygen atoms are bonded to the central atom, and the hydrogen atoms are bonded to the oxygens. Carbon is the central atom in nearly all carbon-containing compounds. Neither hydrogen nor fluorine is ever the central atom.

Step 3: *Add nonbonding electron pairs to the structure such that each atom bonded to the central atom has an octet of electrons. Remember that for hydrogen, an "octet" is only 2 electrons.*

For SO_2, addition of the nonbonding electrons gives

$$\mathrm{:\ddot{\underset{..}{O}}:S:\ddot{\underset{..}{O}}:}$$

At this point, 16 of the 18 available electrons have been used.

Step 4: *Place any remaining electrons on the central atom of the structure.*

Placing the two remaining electrons on the S atom gives

$$\mathrm{:\ddot{\underset{..}{O}}:\ddot{S}:\ddot{\underset{..}{O}}:}$$

Step 5: *If there are not enough electrons to give the central atom an octet, then use one or more pairs of nonbonding electrons on the atoms bonded to the central atom to form double or triple bonds.*

The S atom has only 6 electrons. Thus a nonbonding electron pair from an O atom is used to form a sulfur–oxygen double bond.

$$\mathrm{:\ddot{\underset{..}{O}}:\ddot{S}:\ddot{\underset{..}{O}}: \longrightarrow :\ddot{\underset{..}{O}}:\ddot{S}::\ddot{O}:}$$

This structure now obeys the octet rule.

Step 6: *Count the total number of electrons in the completed Lewis structure to make sure it is equal to the total number of valence electrons available for bonding, as calculated in Step 1.* This step serves as a "double-check" on the correctness of the Lewis structure.

For SO_2, there are 18 valence electrons in the Lewis structure of Step 5, the same number we calculated in Step 1.

Example 4.9 Drawing a Lewis Structure Using Systematic Procedures

Draw Lewis structures for the following molecules.

a. PF_3, a molecule in which P is the central atom and all F atoms are bonded to it (see Figure 4.10)

b. HCN, a molecule in which C is the central atom (see Figure 4.11)

Solution

Part a.

Step 1: Phosphorus (Group VA) has 5 valence electrons, and each of the fluorine atoms (Group VIIA) has 7 valence electrons. The total electron count is 5 + 3(7) = 26.

Step 2: Drawing the molecular skeleton with single covalent bonds (two electrons) placed between all bonded atoms gives

$$\begin{array}{c}\mathrm{F:P:F}\\ \mathrm{\cdot\cdot}\\ \mathrm{F}\end{array}$$

Step 3: Adding nonbonding electrons to the structure to complete the octets of all atoms bonded to the central atom gives

$$\begin{array}{c}\mathrm{:\ddot{\underset{..}{F}}:\underset{..}{P}:\ddot{\underset{..}{F}}:}\\ \mathrm{:\underset{..}{F}:}\end{array}$$

At this point, we have used 24 of the 26 available electrons.

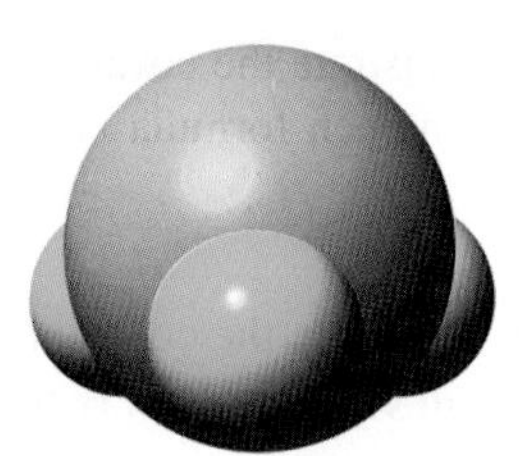

Figure 4.10 The phosphorus trifluoride (PF_3) molecule. A computer-generated model.

Step 4: The central P atom has only 6 electrons; it needs 2 more. The 2 remaining available electrons are placed on the P atom, completing its octet. All atoms now have an octet of electrons.

$$\begin{array}{c} :\ddot{F}:\ddot{P}:\ddot{F}: \\ :\ddot{F}: \end{array}$$

Step 5: This step is not needed; the central atom already has an octet of electrons.

Step 6: There are 26 electrons in the Lewis structure, the same number of electrons we calculated in Step 1.

Part b.

Step 1: Hydrogen (Group IA) has 1 valence electron, carbon (Group IVA) has 4 valence electrons, and nitrogen (Group VA) has 5 valence electrons. The total number of electrons is 10.

Step 2: Drawing the molecular skeleton with single covalent bonds between bonded atoms gives

$$H:C:N$$

Step 3: Adding nonbonding electron pairs to the structure such that the atoms bonded to the central atom have "octets" gives

$$H:C:\ddot{N}:$$

Remember that hydrogen needs only 2 electrons.

Step 4: The structure in Step 3 has 10 valence electrons, the total number available. Thus there are no additional electrons available to place on the carbon atom to give it an octet of electrons.

Step 5: To give the central carbon atom its octet, 2 nonbonding electron pairs on the nitrogen atom are used to form a carbon–nitrogen triple bond.

$$H:C:\ddot{N}: \longrightarrow H:C:::N:$$

Step 6: The Lewis structure has 10 electrons, as calculated in Step 1.

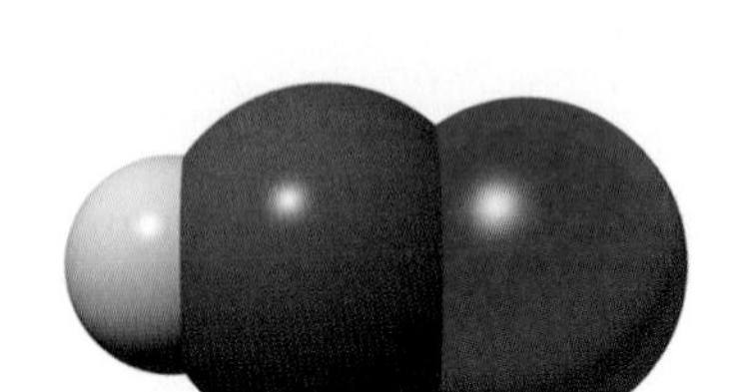

Figure 4.11 The hydrogen cyanide (HCN) molecule. A computer-generated model.

Practice Problems and Questions

4.54 Without actually drawing the Lewis structure, determine the total number of "dots" present in the Lewis structure of each of the following molecules. That is, determine the total number of valence electrons available for bonding in each of the molecules.
a. Cl_2O b. H_2S c. NH_3 d. SO_3

4.55 Using systematic procedures, draw Lewis structures to illustrate the covalent bonding in each of the following molecules. The first atom in each formula is the central atom to which all other atoms are attached.
a. PH_3 b. PCl_3 c. $SiBr_4$ d. OF_2

4.56 Using systematic procedures, draw Lewis structures to illustrate the covalent bonding in each of the following molecules. The first atom in each formula is the central atom to which all other atoms are attached.
a. AsH_3 b. $AsCl_3$ c. CBr_4 d. SCl_2

4.57 Using systematic procedures, draw Lewis structures to illustrate the bonding in the following molecules. In each case, there will be at least one multiple bond present in the molecule.
a. $COCl_2$; both chlorine atoms and the oxygen atom are bonded to the carbon atom.
b. N_2F_2; the two nitrogen atoms are bonded to one another, and each nitrogen atom also has a fluorine atom bonded to it.

c. C_2H_4; the two carbon atoms are bonded to one another, and each carbon atom also has two hydrogen atoms bonded to it.
d. C_2H_2; the two carbon atoms are bonded to one another, and each carbon atom also has a hydrogen atom bonded to it.

Learning Focus

Given a Lewis structure for a molecule, determine its molecular geometry using VSEPR theory.

4.14 The Shape of Molecules: Molecular Geometry

Lewis structures show the numbers and types of bonds present in molecules. They do not, however, convey any information about molecular shape—that is, molecular geometry. **Molecular geometry** *describes the three-dimensional arrangement of atoms in molecules.* Indeed, Lewis structures falsely imply that all molecules have flat, two-dimensional shapes. This is not the case, as the computer-generated models for the molecules SO_2, PF_3, and HCN illustrate (Figures 4.9 through 4.11).

Molecular shape is an important factor in determining the physical and chemical properties of a substance. Dramatic relationships between shape and properties are often observed in research associated with the development of prescription drugs. A small change in overall molecular shape, caused by the addition or removal of atoms, can enhance drug effectiveness and/or decrease drug side effects. Studies also show that the human senses of taste and smell depend in part on the shapes of molecules.

For molecules that contain only a few atoms, molecular shape can be predicted by using the information present in a molecule's Lewis structure and a procedure called valence shell electron pair repulsion (VSEPR) theory. **VSEPR theory** *is a set of procedures for predicting the three-dimensional shape of a molecule using the information contained in the molecule's Lewis structure.*

The central concept of VSEPR theory is that electron pairs in the valence shell of an atom adopt an arrangement in space that minimizes the repulsions between the like-charged (all negative) electron pairs. The specific arrangement adopted by the electron pairs depends on the number of electron pairs present. The electron pair arrangements about a *central atom* in the cases of two, three, and four electron pairs are as follows:

1. Two electron pairs, to be as far apart as possible from one another, are found on opposite sides of a nucleus—that is, at 180° angles to one another (Figure 4.12a). Such an electron pair arrangement is said to be *linear.*
2. Three electron pairs are as far apart as possible when they are found at the corners of an equilateral triangle. In such an arrangement, they are separated by 120° angles, giving a *trigonal planar* arrangement of electron pairs (Figure 4.12b).
3. A *tetrahedral* arrangement of electron pairs minimizes repulsions among four sets of electron pairs (Figure 4.12c). A tetrahedron is a four-sided solid in which all four sides are identical equilateral triangles. The angle between any two electron pairs is 109°.

► The preferred arrangement of a given number of valence electron pairs about a central atom is the one that maximizes the separation among them. Such an arrangement minimizes repulsions between electron pairs.

For actual molecules, the number of valence electron pairs about a *central* atom is determined by using the following VSEPR theory conventions:

1. No distinction is made between bonding and nonbonding electron pairs. Both are counted.
2. Single, double, and triple bonds are all counted equally as "one pair," because each takes up only one region of space about the central atom.

Let us now apply VSEPR theory to molecules in which two, three, and four VSEPR electron pairs are present about a central atom. Our operational rules will be

► The acronym VSEPR is pronounced "vesper."

1. Draw a Lewis structure for the molecule and identify the specific atom for which geometrical information is desired. (This atom will usually be the central atom in the molecule.)
2. Count the number of VSEPR electron pairs present about the atom of interest.

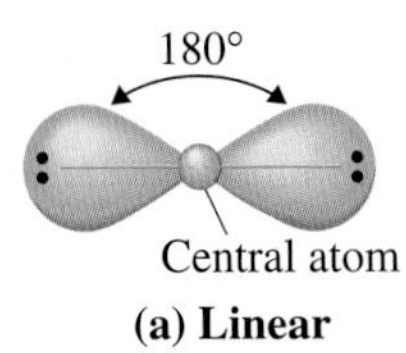

(a) Linear

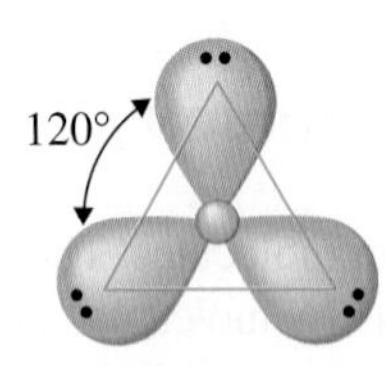

(b) Trigonal planar

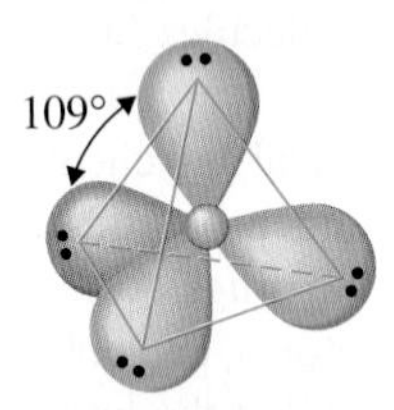

(c) Tetrahedral

Figure 4.12 Arrangements of valence electron pairs about a central atom that minimize repulsions between the pairs.

3. Predict the VSEPR electron pair arrangement about the atom by assuming that the electron pairs orient themselves in a manner that minimizes repulsions (see Figure 4.12).

Molecules with Two VSEPR Electron Pairs

All molecules with two VSEPR electron pairs are *linear.* Two common molecules with two VSEPR electron pairs are carbon dioxide (CO_2) and hydrogen cyanide (HCN), whose Lewis structures are

$$:\ddot{O}=C=\ddot{O}: \qquad H-C\equiv N:$$

In CO_2, the central carbon atom's two VSEPR pairs are the two double bonds. In HCN, the central carbon atom's two VSEPR pairs are a single bond and a triple bond. In both molecules, the VSEPR electron pairs arrange themselves on opposite sides of the carbon atom, which produces a linear molecule.

Molecules with Three VSEPR Electron Pairs

Molecules with three VSEPR electron pairs have two possible molecular structures: *trigonal planar* and *angular.* The former occurs when all three VSEPR pairs are bonding and the latter when one of the three VSEPR pairs is nonbonding. The molecules H_2CO (formaldehyde) and SO_2 (sulfur dioxide) illustrate these two possibilities. Their Lewis structures are

Nonbonding electron pair

Trigonal planar Angular

In both molecules, the VSEPR electron pairs are found at the corners of an equilateral triangle.

The shape of the SO_2 molecule is described as *angular* rather than *trigonal planar*, because molecular shape describes only *atom positions.* The positions of nonbonding electron pairs are not taken into account in describing molecular shape. Do not interpret this to mean that nonbonding electron pairs are unimportant in molecular shape determinations; indeed, in the case of SO_2, it is the presence of the nonbonding electron pair that makes the molecule angular rather than linear.

▶ VSEPR electron pair arrangement and molecular shape are not the same when a central atom possesses nonbonding electron pairs. The word used to describe the shape in such cases does not include the positions of the nonbonding electron pairs.

Molecules with Four VSEPR Electron Pairs

Molecules with four VSEPR electron pairs have three possible molecular shapes: *tetrahedral* (no nonbonding electron pairs present), *trigonal pyramidal* (one nonbonding electron pair present), and *angular* (two nonbonding electron pairs present). The molecules CH_4 (methane), NH_3 (ammonia), and H_2O (water) illustrate this sequence of molecular shapes.

Nonbonding electron pair   Nonbonding electron pairs

Tetrahedral   Trigonal pyramidal   Angular

In all three molecules, the VSEPR electron pairs arrange themselves at the corners of a tetrahedron. Again, note that the word used to describe the shape of the molecule does not take into account the positioning of nonbonding electron pairs.

What we have considered about molecular geometry in this section is summarized in the accompanying Chemistry at a Glance.

▶ "Dotted line" and "wedge" bonds can be used to indicate the directionality of bonds, as shown below.

Bond behind page   Bonds in the plane of the page   Bond in front of page

The Shape (Geometry) of Molecules

PREDICTING MOLECULAR GEOMETRY USING VSEPR THEORY	**Operational rules** 1. Draw a Lewis structure for the molecule. 2. Count the number of VSEPR electron pairs about the central atom in the Lewis structure. 3. Assign a geometry based on minimizing repulsions between electron pairs.

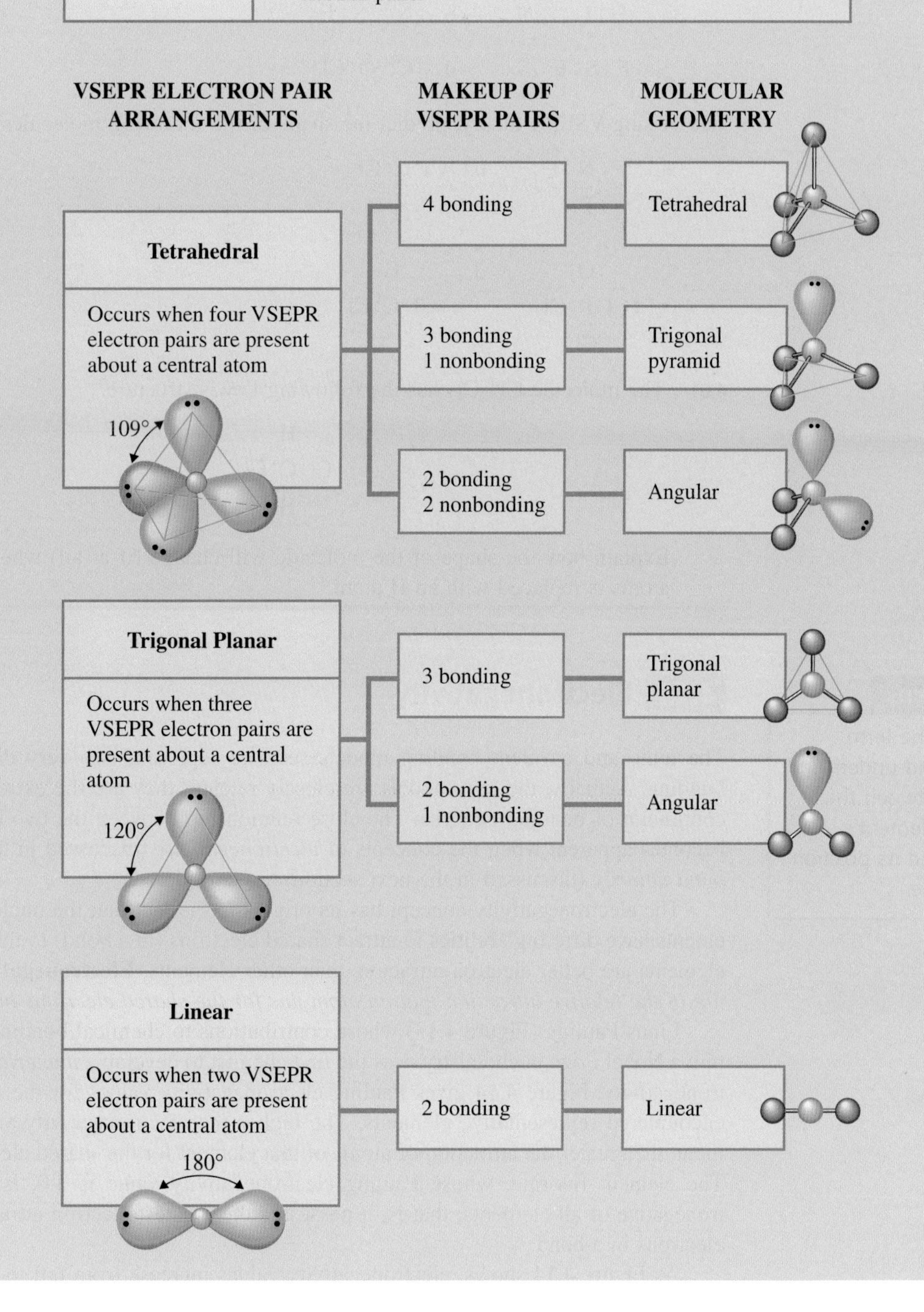

Practice Problems and Questions

4.58 Predict the arrangement in space of the following numbers of VSEPR pairs of electrons about a central atom.

a. Two b. Three c. Four

4.59 Using VSEPR theory, predict whether each of the following triatomic molecules is linear or angular (bent).

```
                              ..  ..  ..
a. :H:C:::N:            b. :N::S:F:
                                      ..

     ..  ..  ..                ..  ..  ..
c. :F:S:F:              d. :Cl:O:Cl:
     ..  ..  ..                ..  ..  ..
```

4.60 Using VSEPR theory, predict the shape of the following molecules.

```
     ..  ..  ..             ..      ..
a. :F:N:F:          b. :Cl:C:Cl:
     ..  ..  ..             ..  ::  ..
        :F:                     :O
         ..                      ..

          ..
c.       :O:           d.       H
      ..  ..  ..              ..  ..  ..
     :Cl:P:Cl:               :Cl:C:Cl:
      ..  ..  ..              ..  ..  ..
        :Cl:                    :Cl:
         ..                      ..
```

4.61 The molecule CH_2Cl_2 has the following Lewis structure.

```
       H
  ..   ..   ..
 :Cl:C:Cl:
  ..   ..   ..
       H
```

Explain how the shape of the molecule will change (if at all) when one of the Cl atoms is replaced with an H atom.

Learning Focus

Be able to define the term *electronegativity* and understand the relationship between the magnitude of an element's electronegativity and its position in the periodic table.

4.15 Electronegativity

The ionic and covalent bonding models seem to represent two very distinct forms of bonding. Actually, the two models are closely related; they are the extremes of a broad continuum of bonding patterns. The close relationship between the two bonding models becomes apparent when the concepts of *electronegativity* (discussed in this section) and *bond polarity* (discussed in the next section) are considered.

The electronegativity concept has its origins in the fact that the nuclei of various elements have differing abilities to attract shared electrons (in a bond) to themselves. Some elements are better electron attractors than other elements. **Electronegativity** *is a measure of the relative attraction that an atom has for the shared electrons in a bond.*

Linus Pauling (Figure 4.13) whose contributions to chemical bonding theory earned him a Nobel Prize in chemistry, was the first chemist to develop a *numerical* scale of electronegativity. Figure 4.14 gives Pauling electronegativity values for the more frequently encountered representative elements. The higher the electronegativity value for an element, the greater the attraction of atoms of that element for the shared electrons in bonds. The element fluorine, whose Pauling electronegativity value is 4.0, is the most electronegative of all elements; that is, it possesses the greatest electron-attracting ability for electrons in a bond.

As Figure 4.14 shows, electronegativity values increase from left to right across periods and from bottom to top within groups of the periodic table. These two trends result in nonmetals generally having higher electronegativities than metals. This fact is consistent with our previous generalization (Section 4.5) that metals tend to lose electrons and nonmetals tend to gain electrons when an ionic bond is formed. Metals (low electronegativities, poor electron attractors) give up electrons to nonmetals (high electronegativities, good electron attractors).

Figure 4.13 Linus Carl Pauling (1901–1994). Pauling received the Nobel Prize in chemistry in 1954 for his work on the nature of the chemical bond. In 1962 he received the Nobel Peace Prize in recognition of his efforts to end nuclear weapons testing.

Note that the electronegativity for an element is not a directly measurable quantity. Rather, electronegativity values are calculated from bond energy information and other related experimental data. Values differ from element to element because of differences in atom size, nuclear charge, and number of inner-shell (nonvalence) electrons.

Practice Problems and Questions

4.62 Using a periodic table, but not a table of electronegativities, identify the atom that has the higher electronegativity in each of the following pairs.
a. N and O b. Cl and Br c. Li and F d. Na and C

4.63 By what constant amount do the electronegativity values for sequential Period 2 elements differ?

4.64 Use the information in Figure 4.14 as a basis for answering the following questions.
a. Which elements have electronegativity values that exceed that of the element carbon?
b. Which elements have electronegativity values of 1.0 or less?
c. What are the four most electronegative elements listed in Figure 4.14?
d. How does the electronegativity of hydrogen compare to that of the Period 2 elements?

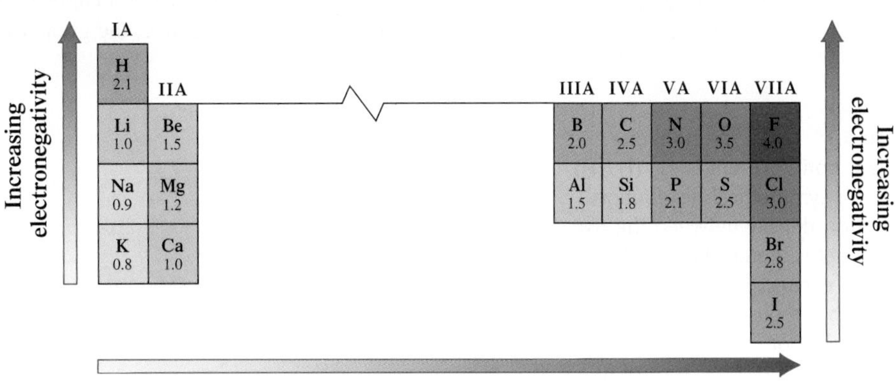

Figure 4.14 Abbreviated periodic table showing Pauling electronegativity values for selected representative elements.

Learning Focus

With the help of a table of electronegativities, be able to classify a bond as *nonpolar covalent, polar covalent,* or *ionic.*

4.16 Bond Polarity

When two atoms of equal electronegativity share one or more pairs of electrons, each atom exerts the same attraction for the electrons, which results in the electrons being *equally* shared. This type of bond is called a nonpolar covalent bond. A **nonpolar covalent bond** *is a covalent bond in which there is equal sharing of electrons between two atoms.*

When the two atoms involved in a covalent bond have different electronegativities, the electron-sharing situation is more complex. The atom that has the higher electronegativity attracts the electrons more strongly than the other atom, which results in an *unequal* sharing of electrons. This type of covalent bond is called a polar covalent bond. A **polar covalent bond** *is a covalent bond in which there is unequal sharing of electrons between two atoms.* Figure 4.15 pictorially contrasts a nonpolar covalent bond and a polar covalent bond using the molecules H_2 and HCl.

The significance of unequal sharing of electrons in a polar covalent bond is that it creates fractional positive and negative charges on atoms. Although both atoms involved in a polar covalent bond are initially uncharged, the unequal sharing means that the electrons spend more time near the more electronegative atom of the bond (producing a fractional negative charge) and less time near the less electronegative atom of the bond (producing a fractional positive charge). The presence of such fractional charges on atoms within a molecule often significantly affects molecular properties (Section 4.17).

▶ The δ^+ and δ^- symbols are pronounced "delta plus" and "delta minus." Whatever the magnitude of δ^+, it must be the same as that of δ^- because the sum of δ^+ and δ^- must be zero.

The fractional charges associated with atoms involved in a polar covalent bond are always values less than 1 because complete electron transfer does not occur. Complete electron transfer, which produces an ionic bond, would produce charges of +1 and −1. A notation that involves the lower-case Greek letter delta (δ) is used to denote fractional charge. The symbol δ^-, meaning "fractional negative charge," is placed above the more electronegative atom of the bond, and the symbol δ^+, meaning "fractional positive charge," is placed above the less electronegative atom of the bond.

With delta notation, the direction of polarity of the bond in hydrogen chloride (HCl) is depicted as

$$\overset{\delta^+}{\mathrm{H}}-\overset{\delta^-}{\ddot{\underset{\cdot\cdot}{\mathrm{Cl}}}}\!:$$

Chlorine is the more electronegative of the two elements; it dominates the electronsharing process and draws the shared electrons closer to itself. Hence the chlorine end of the bond has the δ^- designation (the more electronegative element always has the δ^- designation).

An extension of the reasoning used in characterizing the covalent bond in the HCl molecule as polar leads to the generalization that most chemical bonds are not 100% covalent (equal sharing) or 100% ionic (no sharing). Instead, most bonds are somewhere in between (unequal sharing).

▶ Prediction of bond type, on the basis of electronegativity differences, is as follows:
Nonpolar covalent: zero difference
Polar covalent: greater than 0 but less than 2.0
Ionic: 2.0 or greater

Bond polarity *is a measure of the degree of inequality in the sharing of electrons between two atoms in a chemical bond.* The numerical value of the electronegativity difference between two bonded atoms gives an approximate measure of the polarity of the bond. The greater the numerical difference, the greater the inequality of electron sharing and the

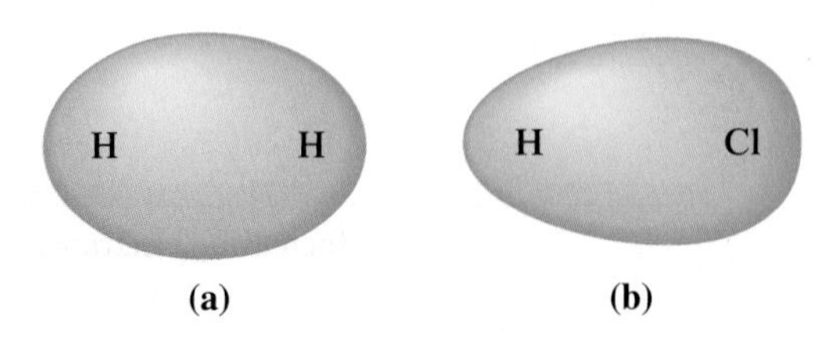

Figure 4.15 (a) In the nonpolar covalent bond present in H_2 (H—H), there is a symmetrical distribution of electron density between the two atoms; that is, equal sharing of electrons occurs. (b) In the polar covalent bond present in HCl (H—Cl), electron density is displaced toward the Cl atom because of its greater electronegativity; that is, unequal sharing of electrons occurs.

greater the polarity of the bond. For electronegativity differences of 2.0 or greater, inequality in electron sharing is sufficiently large that the bond is considered ionic (electron transfer).

Example 4.10 **Using Electronegativity Difference to Predict Bond Polarity and Bond Type**

Consider the following bonds

N—N Ca—F C—O B—H N—O

a. Rank the bonds in order of increasing polarity.
b. Determine the direction of polarity for each bond.
c. Classify each bond as nonpolar covalent, polar covalent, or ionic.

Solution

Let us first calculate the electronegativity difference for each of the bonds by using the electronegativity values in Figure 4.14

N—N: $3.0 - 3.0 = 0.0$

Ca—F: $4.0 - 1.0 = 3.0$

C—O: $3.5 - 2.5 = 1.0$

B—H: $2.1 - 2.0 = 0.1$

N—O: $3.5 - 3.0 = 0.5$

a. Bond polarity increases as electronegativity difference increases. Using the mathematical symbol <, which means "is less than," we can rank the bonds in terms of increasing bond polarity as follows:

N—N < B—H < N—O < C—O < Ca—F
0.0 0.1 0.5 1.0 3.0

b. The direction of bond polarity is from the least electronegative atom to the most electronegative atom. The more electronegative atom bears the partial negative charge (δ^-).

$\overset{}{\text{N—N}}$ $\overset{\delta^+\ \delta^-}{\text{B—H}}$ $\overset{\delta^+\ \delta^-}{\text{N—O}}$ $\overset{\delta^+\ \delta^-}{\text{C—O}}$ $\overset{\delta^+\ \delta^-}{\text{Ca—F}}$

c. Nonpolar covalent bonds require zero difference in electronegativity, and ionic bonds require an electronegativity difference of 2.0 or greater. The in-between region characterizes polar covalent bonds.

Nonpolar covalent: N—N
Polar covalent: B—H, N—O, and C—O
Ionic: Ca—F

The Chemistry at a Glance that follows summarizes important concepts about covalent bonds and electron sharing that we have considered in Sections 4.10 through 4.16.

Practice Problems and Questions

4.65 Place the symbols for partial positive charge (δ^+) and partial negative charge (δ^-) above the appropriate atoms in the following polar covalent bonds. Try to answer this question by using the periodic table rather than Figure 4.14.
a. B—N b. Cl—F c. N—C d. F—O

4.66 Rank the following bonds in order of increasing polarity on the basis of electronegativity differences.
a. H—Cl, H—O, H—Br b. O—F, P—O, Al—O
c. H—Cl, Br—Br, B—N d. P—N, S—O, Br—F

Chemistry *at a Glance*

Covalent Bonds and Molecular Compounds

COVALENT BONDS AND MOLECULAR COMPOUNDS

- A covalent bond results from the sharing of one or more pairs of electrons between atoms.
- A molecule is the basic structural unit in a covalently bonded compound.
- Covalent bonds form between similar or identical atoms—most often between nonmetals.
- Covalent bonds form by the sharing of electrons through an overlap of electron orbitals.

H• + •H ⟶ H:H

Hydrogen atoms — Shared electrons of covalent bond

TYPES OF COVALENT BONDS

CLASSIFICATION BASIS

Total number of shared electrons in the bond

- **Single Covalent Bond** — One shared electron pair — :X:X: Single bond
- **Double Covalent Bond** — Two shared electron pairs — :X::X: Double bond
- **Triple Covalent Bond** — Three shared electron pairs — :X:::X: Triple bond

Electronegativity difference between atoms in bond

- **Nonpolar Covalent Bond** — Equal sharing of electrons occurs because atoms are of the same electronegativity. 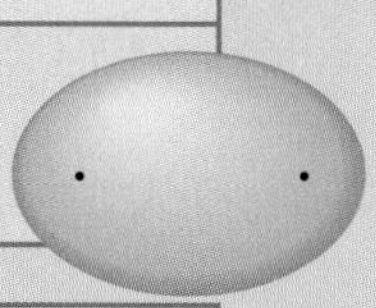
- **Polar Covalent Bond** — Unequal sharing of electrons occurs because atoms have different electronegativities. (δ^+, δ^-)

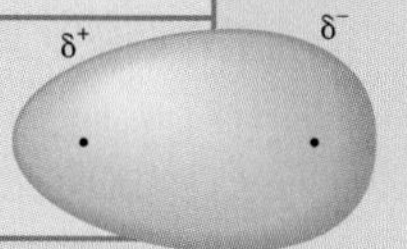

4.67 Characterize each of the following types of bonds in terms of electronegativity difference between bonded atoms.
a. Nonpolar covalent b. Polar covalent c. Ionic

4.68 Classify each of the following bonds as nonpolar covalent, polar covalent, or ionic on the basis of electronegativity difference between bonded atoms.
a. C—O b. Na—Cl c. C—Br d. Ca—S

Learning Focus

Given its molecular geometry, predict whether a given molecule is polar or nonpolar.

4.17 Molecular Polarity

Molecules, as well as bonds (Section 4.16), can have polarity. **Molecular polarity** *is a measure of the degree of inequality in the attraction of bonding electrons to various locations within a molecule.* In terms of electron attraction, if one part of a molecule is favored over other parts, then the molecule is *polar.* A **polar molecule** *is a molecule in which there is an unsymmetrical distribution of electronic charge.* In a polar molecule, bonding electrons are more attracted to one part of the molecule than to other parts. A **nonpolar molecule** *is a molecule in which there is a symmetrical distribution of electronic charge.* Attraction for bonding electrons is the same in all parts of a nonpolar molecule.

▶ A prerequisite for determining molecular polarity is a knowledge of molecular geometry

Molecular polarity depends on two factors: (1) bond polarities and (2) molecular geometry (Section 4.14). In molecules that are symmetrical, the effects of polar bonds may cancel each other, resulting in the molecule as a whole having no polarity.

Determining the molecular polarity of a diatomic molecule is simple because only one bond is present. If that bond is nonpolar, then the molecule is nonpolar; if the bond is polar, then the molecule is polar.

Determining molecular polarity for triatomic molecules is more complicated. Two different molecular geometries are possible: linear and angular. In addition, the symmetrical nature of the molecule must be considered. Let us consider the polarities of three specific triatomic molecules: CO_2 (linear), H_2O (angular), and HCN (linear).

▶ Molecules in which all bonds are polar can be nonpolar if the bonds are so oriented in space that the polarity effects cancel each other.

In the linear CO_2 molecule, both bonds are polar (oxygen is more electronegative than carbon). Despite the presence of these polar bonds, CO_2 molecules are *nonpolar.* The effects of the two polar bonds are canceled as a result of the oxygen atoms being arranged symmetrically around the carbon atom. The shift of electronic charge toward one oxygen atom is exactly compensated for by the shift of electronic charge toward the other oxygen atom. Thus one end of the molecule is not negatively charged relative to the other end (a requirement for polarity), and the molecule is nonpolar. This cancellation of individual bond polarities, with crossed arrows used to denote the polarities, is diagrammed as follows:

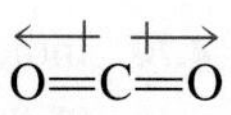

$$\overset{\longleftarrow\!+\;\;+\!\longrightarrow}{\text{O=C=O}}$$

The nonlinear (angular) triatomic H_2O molecule is polar. The bond polarities associated with the two hydrogen–oxygen bonds do not cancel one another because of the nonlinearity of the molecule.

$$\text{H}\overset{\nearrow}{\diagup}\text{O}\overset{\nwarrow}{\diagdown}\text{H}$$

As a result of their orientation, both bonds contribute to an accumulation of negative charge on the oxygen atom. The two bond polarities are equal in magnitude but are not opposite in direction.

The generalization that linear triatomic molecules are nonpolar and nonlinear triatomic molecules are polar, which you might be tempted to make on the basis of our discussion of CO_2 and H_2O molecular polarities, is not valid. The linear molecule HCN, which is polar, invalidates this statement. Both bond polarities contribute to nitrogen's acquiring a partial negative charge relative to hydrogen in HCN.

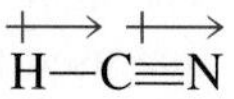

$$\overset{+\!\longrightarrow\;+\!\longrightarrow}{\text{H—C}\equiv\text{N}}$$

(The two polarity arrows point in the same direction because nitrogen is more electronegative than carbon, and carbon is more electronegative than hydrogen.)

Molecules that contain four and five atoms commonly have trigonal planar and tetrahedral geometries, respectively. Such molecules in which all of the atoms attached to the central atom are identical, such as SO_3 (trigonal planar) and CH_4 (tetrahedral), are *nonpolar.* The individual bond polarities cancel as a result of the highly symmetrical arrangement of atoms around the central atom.

If two or more kinds of atoms are attached to the central atom in a trigonal planar or tetrahedral molecule, the molecule is polar. The high degree of symmetry required for cancellation of the individual bond polarities is no longer present. For example, if one of the hydrogen atoms in CH_4 (a nonpolar molecule) is replaced by a chlorine atom, then a polar molecule results, even though the resulting CH_3Cl is still a tetrahedral molecule. A carbon–chlorine bond has a greater polarity than a carbon–hydrogen bond; chlorine has an electronegativity of 3.0 and hydrogen has an electronegativity of only 2.1 Figure 4.16 contrasts the polar CH_3Cl and nonpolar CH_4 molecules. Note that the direction of polarity of the carbon–chlorine bond is opposite to that of the carbon–hydrogen bonds.

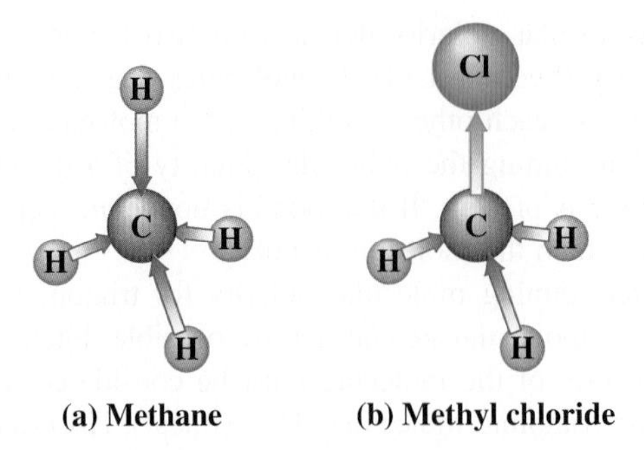

Figure 4.16 (a) Methane (CH_4) is a nonpolar tetrahedral molecule. (b) Methyl chloride (CH_3Cl) is a polar tetrahedral molecule. Bond polarities cancel in the first case, but not in the second.

Practice Problems and Questions

4.69 Indicate whether each of the following triatomic molecules is *polar* or *nonpolar.* The molecular geometry is given in parentheses.
a. CS_2 (linear with C in the center position)
b. H_2Se (angular with Se in the center position)
c. NO_2 (angular with N in the center position)
d. N_2O (linear with N in the center position)

4.70 Indicate whether each of the following hypothetical triatomic molecules is *polar* or *nonpolar.* Assume that A, X, and Y represent elements with different electronegativities.
a. A linear X—A—X molecule
b. A linear X—X—A molecule
c. An angular A—X—Y molecule
d. An angular X—A—Y molecule

4.71 Indicate whether each of the following molecules is *polar* or *nonpolar.* The molecular geometry is given in parentheses.
a. SO_3 (trigonal planar with S in the center position)
b. NCl_3 (trigonal pyramid with N at the apex)
c. CCl_4 (tetrahedral with C in the center position)
d. $CHCl_3$ (tetrahedral with C in the center position)

4.18 Naming Binary Molecular Compounds

Learning Focus

Write the name of a binary molecular compound given its formula, or vice versa.

▶ Numerical prefixes are used in naming binary molecular compounds. They are *never* used, however, in naming binary ionic compounds.

The names of binary molecular compounds are derived by using a rule very similar to that used for naming binary ionic compounds (Section 4.9). However, one major difference exists. Names for binary molecular compounds always contain numerical prefixes that give the number of each type of atom present in addition to the names of the elements present. This is in direct contrast to binary ionic compound nomenclature, where formula subscripts are never mentioned in the names.

Here is the basic rule to use when constructing the name of a binary molecular compound: *The full name of the nonmetal of lower electronegativity is given first, followed by a separate word containing the stem of the name of the more electronegative nonmetal and the suffix* -ide. *Numerical prefixes, giving numbers of atoms, precede the names of both nonmetals.* Thus the compounds N_2O, N_2O_3, and N_2O_4 are dinitrogen monoxide, dinitrogen trioxide, and dinitrogen tetroxide, respectively.

▶ When an element name begins with a vowel, an *a* or *o* at the end of the Greek prefix is dropped for phonetic reasons, as in pentoxide instead of pentaoxide.

Prefixes are necessary because several different compounds exist for most pairs of nonmetals. For example, all of the following nitrogen–oxygen compounds exist: NO, NO_2, N_2O, N_2O_3, N_2O_4, and N_2O_5. Such diverse behavior between two elements is related to the fact that single, double, and triple covalent bonds exist. The prefixes used are the standard numerical prefixes, which are given for the numbers 1 through 10 in Table 4.2. Example 4.11 shows how these prefixes are used in nomenclature for binary covalent compounds.

Table 4.2
Prefixes for 1 Through 10

Prefix	Number
mono-	1
di-	2
tri-	3
tetra-	4
penta-	5
hexa-	6
hepta-	7
octa-	8
ennea-	9
deca-	10

▶ In Section 9.3, we will learn that placing hydrogen first in a formula conveys the message that the compound behaves as an acid in aqueous solution.

▶ Classification of a compound as ionic or molecular determines which set of nomenclature rules is used. For *nomenclature purposes*, binary compounds in which a metal and a nonmetal are present are considered ionic, and binary compounds that contain two nonmetals are considered covalent. Electronegativity differences are *not* used in classifying a compound as ionic or molecular for nomenclature purposes.

Table 4.3
Selected Binary Molecular Compounds That Have Common Names

Compound formula	Accepted common name
H_2O	water
H_2O_2	hydrogen peroxide
NH_3	ammonia
N_2H_4	hydrazine
CH_4	methane
C_2H_6	ethane
PH_3	phosphine
AsH_3	arsine

Example 4.11 Naming Binary Molecular Compounds

Name the following binary molecular compounds.

a. S_2Cl_2 **b.** CS_2 **c.** P_4O_{10} **d.** CBr_4

Solution

The names of each of these compounds will consist of two words. These words will have the following general formats:

$$\text{First word:} \quad (\text{prefix}) + \begin{pmatrix}\text{full name of least} \\ \text{electronegative nonmetal}\end{pmatrix}$$

$$\text{Second word:} \quad (\text{prefix}) + \begin{pmatrix}\text{stem of name of more} \\ \text{electronegative nonmetal}\end{pmatrix} + (\text{ide})$$

a. The elements present are sulfur and chlorine. The two portions of the name (including prefixes) are *disulfur and dichloride*, which are combined to give the name *disulfur dichloride*.
b. When only one atom of the first nonmetal is present, it is customary to omit the initial prefix *mono-*. Thus the name of this compound is *carbon disulfide*.
c. The prefix for four atoms is *tetra-* and for ten atoms is *deca-*. This compound has the name *tetraphosphorus decoxide*.
d. Omitting the initial *mono-* (see part **b**), we name this compound *carbon tetrabromide*.

There is one standard exception to the use of numerical prefixes when naming binary molecular compounds. Compounds in which hydrogen is the first listed element in the formula are named without numerical prefixes. Thus the compounds H_2S and HCl are hydrogen sulfide and hydrogen chloride, respectively.

A few binary molecular compounds have names that are completely unrelated to the rules we have been discussing. They have common names that were coined prior to the development of systematic rules. At one time, in the early history of chemistry, all compounds had common names. With the advent of systematic nomenclature, most common names were discontinued. A few, however, have persisted and are now officially accepted. The most "famous" example is the compound H_2O, which has the systematic name hydrogen oxide, a name that is never used. The compound H_2O is *water*, a name that will never change. Table 4.3 lists other compounds for which common names are used in preference to systematic names.

Practice Problems and Questions

4.72 Name the following binary molecular compounds.
a. SF_4 b. P_4O_6 c. ClO_2 d. H_2S

4.73 Name the following binary molecular compounds.
a. Cl_2O b. CO c. PI_3 d. HI

4.74 Write formulas for the following binary molecular compounds.
a. Iodine monochloride b. Dinitrogen monoxide
c. Sulfur trioxide d. Dioxygen difluoride

4.75 Write formulas for the following binary molecular compounds given their *common* names.
a. Hydrogen peroxide b. Ammonia c. Methane d. Hydrazine

4.19 Polyatomic Ions

Learning Focus

Recognize formulas of common polyatomic ions, write formulas for compounds with polyatomic ions, and name such compounds.

We now revisit the topic of ions, a subject considered in the earlier sections of this chapter. There are two categories of ions: monatomic and polyatomic. A **monatomic ion** *is formed from a single atom that has lost or gained electrons and thus acquired a*

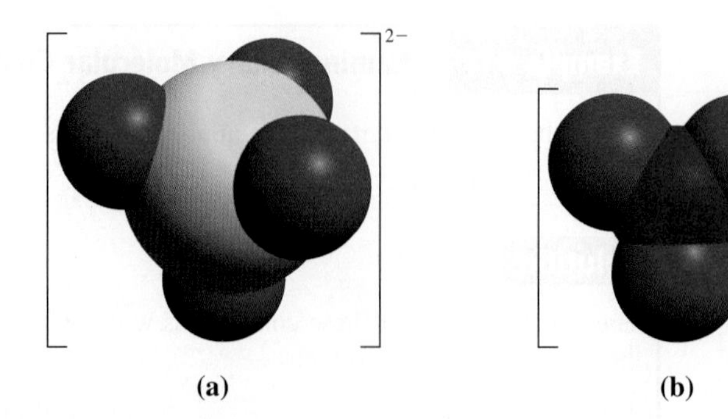

Figure 4.17 Models of (a) a sulfate ion (SO_4^{2-}) and (b) a nitrate ion (NO_3^-).

charge. All of the ions we have discussed so far have been monatomic (Cl^-, Na^+, Ca^{2+}, N^{3-}, and so on).

A **polyatomic ion** *is formed from a group of atoms, held together by covalent bonds, that has acquired a charge.* An example of a polyatomic ion is the sulfate ion, SO_4^{2-} (Figure 4.17a). This ion contains four oxygen atoms and one sulfur atom, and the whole group of five atoms has acquired a -2 charge. The whole sulfate group is the ion rather than any one atom within the group. Covalent bonding, holds the sulfur and oxygen atoms together.

There are numerous ionic compounds in which the positive or negative ion (sometimes both) is polyatomic. Polyatomic ions are very stable and generally maintain their identity during chemical reactions.

Note that polyatomic ions are not molecules. They never occur alone as molecules do. Instead, they are always found associated with ions of opposite charge. Polyatomic ions are *charged pieces* of compounds, not compounds. Ionic compounds require the presence of both positive and negative ions and are neutral overall.

Table 4.4 lists the names and formulas of some of the more common polyatomic ions. The following generalizations concerning polyatomic ion names and charges emerge from consideration of the ions listed in Table 4.4.

Table 4.4
Formulas and Names of Some Common Polyatomic Ions

Key element present	Formula	Name of ion
nitrogen	NO_3^-	nitrate
	NO_2^-	nitrite
	NH_4^+	ammonium
	N_3^-	azide
sulfur	SO_4^{2-}	sulfate
	HSO_4^-	bisulfate or hydrogen sulfate
	SO_3^{2-}	sulfite
	HSO_3^-	bisulfite or hydrogen sulfite
	$S_2O_3^{2-}$	thiosulfate
phosphorus	PO_4^{3-}	phosphate
	HPO_4^{2-}	hydrogen phosphate
	$H_2PO_4^-$	dihydrogen phosphate
	PO_3^{3-}	phosphite
carbon	CO_3^{2-}	carbonate
	HCO_3^-	bicarbonate or hydrogen carbonate
	$C_2O_4^{2-}$	oxalate
	$C_2H_3O_2^-$	acetate
	CN^-	cyanide
chlorine	ClO_4^-	perchlorate
	ClO_3^-	chlorate
	ClO_2^-	chlorite
	ClO^-	hypochlorite
hydrogen	H_3O^+	hydronium
	OH^-	hydroxide

1. Most of the polyatomic ions have a negative charge, which can vary from −1 to −3. Only two positive ions are listed in the table: NH_4^+ (ammonium) and H_3O^+ (hydronium).
2. Two of the negatively charged polyatomic ions, OH^- (hydroxide) and CN^- (cyanide), have names ending in *-ide*, and the rest of them have names ending in either *-ate* or *-ite.*
3. A number of *-ate*, *-ite* pairs of ions exist, as in SO_4^{2-} (sulfate) and SO_3^{2-} (sulfite). The *-ate* ion always has one more oxygen atom than the *-ite* ion. Both the *-ate* and *-ite* ions of a pair carry the same charge.
4. A number of pairs of ions exist wherein one member of the pair differs from the other by having a hydrogen atom present, as in CO_3^{2-} (carbonate) and HCO_3^- (hydrogen carbonate or bicarbonate). In such pairs, the charge on the ion that contains hydrogen is always 1 less than that on the other ion.

▶ Learning the names of the common polyatomic ions is a memorization project. There is no shortcut. The charges and formulas for the various polyatomic ions cannot be easily related to the periodic table, as was the case for many of the monatomic ions.

Writing Formulas for Compounds That Contain Polyatomic Ions

Compounds that contain polyatomic ions offer an interesting combination of both ionic and covalent bonding; covalent bonding occurs *within* the polyatomic ion, and ionic bonding occurs *between* the polyatomic ion and the ion (of opposite charge) that must be present.

Formulas for ionic compounds that contain polyatomic ions are determined in the same way as those for ionic compounds that contain monatomic ions (Section 4.7). The positive and negative charges present must add to zero.

Two conventions not encountered previously in formula writing often arise when we write formulas containing polyatomic ions.

1. When more than one polyatomic ion of a given kind is required in a formula, the polyatomic ion is enclosed in parentheses, and a subscript, placed outside the parentheses, is used to indicate the number of polyatomic ions needed. An example is $Fe(OH)_3$.
2. So that the identity of polyatomic ions is preserved, the same elemental symbol may be used more than once in a formula. An example is the formula NH_4NO_3, where N appears in two locations.

Example 4.12 Illustrates the use of both of these new conventions.

Example 4.12 Writing Formulas for Ionic Compounds Containing Polyatomic Ions

Determine the formulas for the ionic compounds that contain these pairs of ions.

a. Na^+ and SO_4^{2-} **b.** Mg^{2+} and NO_3^- **c.** NH_4^+ and CN^-

Solution

a. In order to equalize the total positive and negative charge, we need two sodium ions (+1 charge) for each sulfate ion (−2 charge). We indicate the presence of two Na^+ ions with the subscript 2 following the symbol of this ion. The formula of the compound is Na_2SO_4. The convention that the positive ion is always written first in the formula still holds when polyatomic ions are present.

b. Two nitrate ions (−1 charge) are required to balance the charge on one magnesium ion (+2 charge). Because more than one polyatomic ion is needed, the formula contains parentheses, $Mg(NO_3)_2$. The subscript 2 outside the parentheses indicates two of what is inside the parentheses. If parentheses were not used, the formula would appear to be $MgNO_{32}$, which is not intended and conveys false information.

c. In this compound, both ions are polyatomic, which is a perfectly legal situation. Because the ions have equal but opposite charges, they combine in a one-to-one ratio. Thus the formula is NH_4CN. No parentheses are necessary because we need only one polyatomic ion of each type in a formula unit. The appearance of the symbol for the element nitrogen (N) at two locations in the formula could be prevented by combining the two nitrogens, resulting in N_2H_4C. But the formula N_2H_4C does not convey the message that NH_4^+ and CN^- ions are present. Thus, when writing formulas that contain polyatomic ions, we always maintain the identities of these ions, even if it means having the same elemental symbol at more than one location in the formula.

Naming Compounds That Contain Polyatomic Ions

The names of ionic compounds containing polyatomic ions are derived in a manner similar to that for binary ionic compounds (Section 4.9). The rule for naming binary ionic compounds is as follows: Give the name of the metallic element first (including, when needed, a Roman numeral indicating ion charge), and then give a separate word containing the stem of the nonmetallic name and the suffix *-ide.*

For our present situation, *if the polyatomic ion is positive, its name is substituted for that of the metal. If the polyatomic ion is negative, its name is substituted for the nonmetal stem plus* -ide. Where both positive and negative ions are polyatomic, dual substitution occurs, and the resulting name includes just the names of the polyatomic ions.

Example 4.13 **Naming Ionic Compounds in Which Polyatomic Ions Are Present**

Name the following compounds, which contain one or more polyatomic ions.

a. $Ca_3(PO_4)_2$ **b.** $Fe_2(SO_4)_3$ **c.** $(NH_4)_2CO_3$

Solution

a. The positive ion present is the calcium ion (Ca^{2+}). We will not need a Roman numeral to specify the charge on a Ca^{2+} ion because it is always +2. The negative ion is the polyatomic phosphate ion (PO_4^{3-}). The name of the compound is *calcium phosphate.* As in naming binary ionic compounds, subscripts in the formula are not incorporated into the name.

b. The positive ion present is iron(III). The negative ion is the polyatomic sulfate ion (SO_4^{2-}). The name of the compound is *iron(III) sulfate.* The determination that iron is present as iron(III) involves the following calculation dealing with charge balance:

$$2(\text{iron charge}) + 3(\text{sulfate charge}) = 0$$

The sulfate charge is −2. (You had to memorize that.) Therefore,

$$2(\text{iron charge}) + 3(-2) = 0$$
$$2(\text{iron charge}) = +6$$
$$\text{Iron charge} = +3$$

c. Both the positive and the negative ions in this compound are polyatomic—the ammonium ion (NH_4^+) and the carbonate ion (CO_3^{2-}). The name of the compound is simply the combination of the names of the two polyatomic ions: *ammonium carbonate.*

It is possible to have ionic compounds that contain more than one type of positive ion, more than one type of negative ion, or more than one type of both positive and negative ions. In such compounds, charge neutrality is still required; total positive charge must equal total negative charge. "Complex" ionic compounds of this type usually have common names (Section 4.18). Chemical Portraits 8 gives three examples of naturally occurring substances in which three types of ions are present.

Chemical Portraits 8 — Complex Ionic Compounds: Three or More Types of Ions

Hydroxyapatite [$Ca_{10}(PO_4)_6(OH)_2$]

Profile: Structurally, bones and the hard outer covering of teeth (enamel) have two parts: a solid mineral phase and a second phase made up primarily of fibrous protein. The mineral phase is *hydroxyapatite,* a substance containing calcium ions (Ca^{2+}), phosphate ions (PO_4^{3-}), and hydroxide ions (OH^-) in a 10:6:2 ratio. The fibrous protein phase is dispersed in the spaces between the ions.

Biochemical considerations: Hydroxyapatite is continually dissolving (demineralization) and reforming (mineralization) in tooth enamel and bones. Tooth decay results when chemical factors within the mouth, mainly acidity, cause demineralization rates to exceed mineralization rates. In most individuals, after age 40, bone demineralization rates begin to exceed bone mineralization rates; more bone material is lost than formed, a condition called *osteoporosis.* In general, osteoporosis occurs at a faster rate in women than men. It is recommended that persons over 40, especially women, include a supplemental source of calcium ion in their diet.

Why does fluoride-ion-containing tooth paste help prevent tooth decay?

Chrysotile [$Mg_3(OH)_4(Si_2O_5)$]

Profile: Asbestos is the family name for a group of fibrous minerals that are resistant to heat, flame, and acids. *Chrysotile*, often called *white asbestos*, is the most abundant and most used of these minerals. It contains magnesium ions (Mg^{2+}), hydroxide ions (OH^-) and disilicate ions ($Si_2O_5^{2-}$) in a 3:4:1 ratio. The disilicate ions present interact among themselves giving the asbestos a layer-like structure; often several layers of the structure can be "peeled off" as a unit.

Uses: Into the 1970s chrysotile and other asbestos minerals were used as thermal insulators in buildings. It is now known that prolonged exposure to *airborne* asbestos fibers (dusts) can damage lung tissue. Use of asbestos-containing insulation in buildings has stopped, and existent asbestos insulation is being replaced in some situations. In other situations, it is safer to leave the asbestos where it is than try to remove it.

What steps has the U.S. government taken to reduce a person's exposure to asbestos products?

Beryl [$Be_3Al_2(Si_6O_{18})$]

Profile: The mineral *beryl* contains the positive ions Be^{2+} and Al^{3+} and the large polyatomic negative ion $Si_6O_{18}^{12-}$; the ion ratio is 3:2:1. High grade beryl crystals are valued as gemstones. Although *pure* beryl crystals are colorless, most beryl crystals contain transition metal impurities that give them distinctive colors. *Emeralds* are intensely green gemstone-quality beryl crystals in which Cr^{3+} ions have replaced up to 2% of the Al^{3+} ions. *Aquamarine* gemstones are beryl crystals in which Fe^{3+} ions are the transition metal impurity; they have a light bluish-green color.

Uses: Only two beryllium minerals, beryl and bertrandite, are of commercial importance; beryl contains about 4% Be and bertrandite contains less than 1% Be. Both are used to obtain the light metal beryllium (element #4). Bertrandite is the principal beryllium mineral mined in the United States and beryl is the principle mineral used in the rest of the world.

What is the chemical formulation for bertrandite in terms of the ions present?

See the text web site at **www.cengage.com/chemistry/stoker** for answers to the above questions and for further information.

Practice Problems and Questions

4.76 With the help of Table 4.4, write formulas for the following polyatomic ions.
a. Hydroxide b. Cyanide c. Ammonium d. Phosphate

4.77 With the help of Table 4.4, write formulas for the following pairs of polyatomic ions.
a. Sulfate and sulfite b. Nitrate and nitrite
c. Carbonate and bicarbonate d. Hydrogen phosphate and dihydrogen phosphate

4.78 Write formulas for the compounds formed between the following positive and negative ions.
a. Mg^{2+} and SO_4^{2-} b. Fe^{3+} and PO_4^{3-} c. Na^+ and CO_3^{2-} d. K^+ and CN^-

4.79 Write formulas for the compounds formed between the following positive and negative ions.
a. Fe^{3+} and OH^- b. Be^{2+} and NO_3^- c. NH_4^+ and S^{2-} d. NH_4^+ and PO_4^{3-}

4.80 Name the following compounds, each of which contains a polyatomic ion and a fixed-charge metal ion.
a. $ZnSO_4$ b. $MgCO_3$ c. K_2CO_3 d. AgOH

4.81 Name the following compounds, each of which contains a polyatomic ion and a variable-charge metal ion.
a. Cu_3PO_4 b. $Fe(NO_3)_3$ c. $FeSO_4$ d. AuCN

CONCEPTS TO REMEMBER

Chemical bonds. Chemical bonds are the attractive forces that hold atoms together in more complex units. Chemical bonds result from the transfer of valence electrons between atoms (ionic bond) or from the sharing of electrons between atoms (covalent bond).

Valence electrons. Valence electrons, for representative elements, are the electrons in the outermost electron shell, which is the shell with the highest shell number. These electrons are particularly important in determining the bonding characteristics of a given atom.

Octet rule. In compound formation, atoms of representative elements lose, gain, or share electrons in such a way that their electron configurations become identical to those of the noble gas nearest them in the periodic table.

Ionic compounds. Ionic compounds usually involve a metal atom and a nonmetal atom. Metal atoms lose one or more electrons, producing positive ions. Nonmetal atoms acquire the electrons lost by the metal atoms, producing negative ions. The oppositely charged ions attract one another, creating ionic bonds.

Charge magnitude for ions. Metal atoms containing one, two, or three valence electrons tend to lose such electrons, producing ions of $+1$, $+2$, and $+3$ charge, respectively. Nonmetal atoms containing five, six, or seven valence electrons tend to gain electrons, producing ions of -3, -2, and -1 charge, respectively.

Formulas for ionic compounds. The ratio in which positive and negative ions combine is the ratio that causes the total amount of positive and negative charges to add up to zero.

Structure of ionic compounds. Ionic solids consist of positive and negative ions arranged in such a way that each ion is surrounded by ions of the opposite charge.

Binary ionic compound nomenclature. Binary ionic compounds are named by giving the full name of the metallic element first, followed by a separate word containing the stem of the nonmetallic element name and the suffix *-ide.* A Roman numeral specifying ionic charge is appended to the name of the metallic element if it is a metal that exhibits variable ionic charge.

Molecular compounds. Molecular compounds usually involve two or more nonmetals. The covalent bonds within molecular compounds involve electron sharing between atoms. The covalent bond results from the common attraction of the two nuclei for the shared electrons.

Bonding and nonbonding electron pairs. Bonding electrons are pairs of valence electrons that are shared between atoms in a covalent bond. Nonbonding electrons are pairs of valence electrons about an atom that are not involved in electron sharing.

Types of covalent bonds. One shared pair of electrons constitutes a single covalent bond. Two or three pairs of electrons may be shared between atoms to give double and triple covalent bonds.

Molecular geometry. Molecular geometry describes the way atoms in a molecule are arranged in space relative to one another. VSEPR theory is a set of procedures used to predict molecular geometry from a compound's Lewis structure. VSEPR theory is based on the concept that valence shell electron pairs about an atom (bonding and nonbonding) orient themselves as far away from one another as possible (to minimize repulsions).

Electronegativity. Electronegativity is a measure of the relative attraction that an atom has for the shared electrons in a bond. Electronegativity values are useful in predicting the type of bond that forms (ionic or covalent).

Bond polarity. When atoms of like electronegativity participate in a bond, the bonding electrons are equally shared and the bond is nonpolar. When atoms of differing electronegativity participate in a bond, the bonding electrons are unequally shared and the bond is polar. In a polar bond, the more electronegative atom dominates the sharing process. The greater the electronegativity difference between two bonded atoms, the greater the polarity of the bond.

Molecular polarity. Molecules as a whole can have polarity. If individual bond polarities do not cancel because of the symmetrical nature of a molecule, then the molecule as a whole is polar.

Binary molecular compound nomenclature. Along with the names of the elements present, names for binary molecular compounds usually contain Greek numerical prefixes that give the number of each type of atom present per molecule.

Polyatomic ions. A polyatomic ion is a group of covalently bonded atoms that has acquired a charge through the loss or gain of electrons. Polyatomic ions are very stable entities that generally maintain their identity during chemical reactions.

KEY REACTIONS AND EQUATIONS

1. Number of valence electrons for representative elements (Section 4.2)

   Number of valence electrons = periodic-table group number

2. Charges on metallic monatomic ions (Section 4.5)

   Group IA metals form 1^+ ions.

   Group IIA metals form 2^+ ions.

   Group IIIA metals form 3^+ ions.

3. Charges on nonmetallic monatomic ions (Section 4.5)

   Group VA nonmetals form 3^- ions.

   Group VIA nonmetals form 2^- ions.

   Group VIIA nonmetals form 1^- ions.

4. Molecular geometry and central atom VSEPR electron pair count (Section 4.14)

   Four VSEPR electron pairs none of which is nonbonding = tetrahedral geometry

   Four VSEPR electron pairs one of which is nonbonding = trigonal pyramidal geometry

   Four VSEPR electron pairs two of which are nonbonding = angular geometry

   Three VSEPR electron pairs none of which is nonbonding = trigonal planar geometry

   Three VSEPR electron pairs one of which is nonbonding = angular geometry

   Two VSEPR electron pairs none of which is nonbonding = linear geometry

5. Bond characterization and electronegativity difference (Section 4.16)

   Electronegativity difference of 2.0 or greater = ionic bond

   Electronegativity difference greater than 0 but less than 2.0 = polar covalent bond

   Electronegativity difference of 0 = nonpolar covalent bond

KEY TERMS

Binary ionic compound (4.11)
Bond polarity (4.16)
Bonding electrons (4.11)
Chemical bond (4.1)
Covalent bond (4.1)
Double covalent bond (4.12)
Electronegativity (4.15)
Formula unit (4.8)
Ion (4.4)
Ionic bond (4.1)
Lewis structure (4.2)
Molecular geometry (4.14)
Molecular polarity (4.17)
Monatomic ion (4.19)
Nonbonding electrons (4.11)
Nonpolar covalent bond (4.17)
Nonpolar molecule (4.17)
Octet rule (4.3)
Polar covalent bond (4.17)
Polar molecule (4.17)
Polyatomic ion (4.19)
Single covalent bond (4.12)
Triple covalent bond (4.12)
Valence electrons (4.2)
VSEPR theory (4.14)

ADDITIONAL PROBLEMS

4.82 Write out the electron configurations for each of the following pairs of atoms and/or ions.
a. Al atom and Al^{3+} ion b. N atom and N^{3-} ion
c. Na^+ ion and Ne atom d. O^{2-} ion and Mg^{2+} ion

4.83 Write out the electron configuration for each of the following representative elements.
a. Period 2 element with four valence electrons
b. Period 2 element with seven valence electrons
c. Period 3 element with two valence electrons
d. Period 3 element with five valence electrons

4.84 What would be the charge on each of the following ions?
a. A sodium ion with ten electrons
b. A fluorine ion with ten electrons
c. A sulfur ion with two fewer protons than electrons
d. A calcium ion with two more protons than electrons

4.85 Write the chemical formula of the ionic compound that could form from the elements X and Z if
a. X has two valence electrons and Z has seven valence electrons
b. X has one valence electron and Z has six valence electrons
c. X has three valence electrons and Z has five valence electrons
d. X has six valence electrons and Z has two valence electrons

4.86 Write formulas (symbol and charge) for both kinds of ions present in each of the following compounds.
a. KCl b. CaS c. BeF_2 d. Al_2S_3

4.87 Classify each of the following compounds as ionic or molecular.
a. Li_2O b. CO_2 c. Na_3N d. NF_3

4.88 What would be the predicted simplest chemical formula for the compound formed from each of the following pairs of elements?
a. Sodium and oxygen b. Magnesium and oxygen
c. Hydrogen and oxygen d. Fluorine and oxygen

4.89 Name each of the following binary molecular or binary ionic compounds.
a. $BeCl_2$ b. NCl_3 c. $AlCl_3$ d. $FeCl_3$

4.90 The element lead forms the ions lead(II) and lead(IV).
a. What is the symbol for each of the ions?
b. How many protons and electrons are present in each ion?
c. What is the formula of lead(II) sulfate?
d. What is the formula of lead(IV) nitrate?

4.91 Successive substitution of F atoms for H atoms in the molecule CH_4 produces the molecules CH_3F, CH_2F_2, CHF_3, and CF_4
a. Draw the Lewis structure for each of the five molecules.
b. Using VSEPR theory, predict the molecular geometry of each of the five molecules.
c. Give the molecular polarity (polar or nonpolar) of each of the five molecules.

4.92 Classify each of the following molecules as polar or nonpolar, or indicate that no such classification is possible because of insufficient information.
a. A molecule in which all bonds are polar
b. A molecule in which all bonds are nonpolar
c. A molecule with two bonds, both of which are polar
d. A molecule with two bonds, one of which is polar and one of which is nonpolar

4.93 Four hypothetical elements A, B, C, and D have electronegativities A = 3.8, B = 3.3, C = 2.8, and D = 1.3 These elements form the diatomic compounds BA, DA, DB, and CA.
a. Characterize the bond in each of the four compounds as nonpolar covalent, polar covalent, or ionic.
b. Arrange the four compounds in order of increasing bond polarity.
c. Arrange the four compounds in order of increasing molecular polarity.

PRACTICE TEST ▸ True/False

4.94 The octet rule is based on the observation that certain electron arrangements are more stable than others.

4.95 The maximum number of valence electrons that a representative or noble-gas element can have is eight.

4.96 Representative elements in the same group in the periodic table have the same number of valence electrons.

4.97 The mechanism for formation of ions is proton loss or proton gain.

4.98 Group VA nonmetals form ions with a −2 charge.

4.99 In an ionic compound, the total charge on the positive ions must equal the total charge on the negative ions.

4.100 Greek numerical prefixes are used in naming ionic compounds.

4.101 A covalent bond results from a common attraction by two nuclei for one or more shared pairs of electrons.

4.102 All atoms in a molecular compound possess both shared and unshared pairs of valence electrons.

4.103 A triple covalent bond is approximately three times as strong as a single covalent bond between the same two atoms.

4.104 A trigonal planar structure minimizes repulsions among four sets of VSEPR electron pairs.

4.105 A molecule with two bonding and two nonbonding VSEPR electron pairs about the central atom would have an angular molecular geometry.

4.106 The element fluorine has the greatest electronegativity of all the elements.

4.107 When the electronegativity difference between two bonded atoms is 2.0 or greater, the bond is considered to be ionic.

4.108 Most, but not all, common polyatomic ions carry a positive charge.

PRACTICE TEST ▶ Multiple Choice

4.109 How many electrons appear in the Lewis structure for an element whose electron configuration is $1s^2 2s^2 2p^4$?
a. Two b. Four c. Six d. Eight

4.110 The correct formula for the ionic compound containing Al^{3+} and S^{2-} ions is
a. AlS b. Al_3S_2 c. Al_2S_3 d. AlS_2

4.111 The correct formula for the ionic compound formed between K and N is
a. K_3N b. KN_3 c. K_2N_3 d. K_3N_2

4.112 The correct name for the ionic compound Fe_2S_3 is
a. iron sulfide b. iron(III) sulfide
c. iron trisulfide d. diiron trisulfide

4.113 Which of the following statements about polyatomic ions is correct?
a. All have names which end in *-ate.*
b. All must contain oxygen.
c. Negative ions are more common than positive ions.
d. All carry a charge of 2.

4.114 Which of the following molecular compounds would have a Lewis structure that contains 10 "electron dots"?
a. HCN b. H_2O c. NH_3 d. CO_2

4.115 Which of the following statements contrasting covalent bonds and ionic bonds is correct?
a. Covalent bonds usually involve two nonmetals, and ionic bonds usually involve two metals.
b. Covalent bonds usually involve two metals, and ionic bonds usually involve a metal and a nonmetal.
c. Covalent bonds usually involve a metal and a nonmetal, and ionic bonds usually involve two nonmetals.
d. Covalent bonds usually involve two nonmetals, and ionic bonds usually involve a metal and a nonmetal.

4.116 Which of the following compounds contains both ionic and covalent bonds?
a. SO_2 b. NaCl c. KOH d. NH_3

4.117 Given the electronegativities K = 0.8, H = 2.1, I = 2.5, indicate which of the following statements is true?
a. H_2, I_2, and HI are nonpolar covalent molecules.
b. HI is a nonpolar covalent compound, and KI is an ionic compound.
c. HI is a nonpolar covalent compound, and I_2 is a polar covalent molecule.
d. H_2 is a nonpolar covalent molecule, and HI is a polar covalent compound.

4.118 Which of the following is a molecular compound that contains four atoms per molecule?
a. Disulfur monoxide
b. Dinitrogen pentoxide
c. Hydrogen peroxide
d. Methane

CHAPTER OUTLINE

CHAPTER FIVE

Chemical Calculations: Formula Masses, Moles, and Chemical Equations

The energy associated with a lightning discharge causes many different chemical reactions to occur within the atmosphere. Each of these chemical reactions can be represented using a chemical equation. Chemical equations is one of the topics considered in this chapter.

In this chapter we discuss "chemical arithmetic," the quantitative relationship between elements and compounds. Anyone who deals with chemical processes needs to understand at least the simpler aspects of this topic. All chemical processes, regardless of where they occur—in the human body, at a steel mill, on top of the kitchen stove, or in a clinical laboratory setting—are governed by the same mathematical rules.

We have already presented some information about chemical formulas (Section 1.10). In this chapter we discuss formulas again, and here we look beyond describing the composition of compounds in terms of constituent atoms. A new unit, the mole, will be introduced and its usefulness discussed. Chemical equations will be considered for the first time. We will learn how to represent chemical reactions by using chemical equations and how to derive quantitative relationships from these equations.

5.1 Formula Masses

Learning Focus

Calculate the formula mass of a substance given its chemical formula and a listing of atomic masses.

Our entry point into the realm of "chemical arithmetic" is a discussion of the quantity called formula mass. The **formula mass** *of a substance is the sum of the atomic masses of all the atoms represented in the chemical formula of the substance*. Formula masses, like the atomic masses from which they are calculated, are relative masses based on the $^{12}_{6}C$ relative-mass scale (Section 3.2). Example 5.1 illustrates how formula masses are calculated.

▶ Many chemists use the term *molecular mass* interchangeably with *formula mass* when dealing with substances that contain discrete molecules. It is incorrect, however, to use the term *molecular mass* when dealing with ionic compounds, because such compounds do not have molecules as their basic structural unit (Section 4.8).

Example 5.1 **Using a Compound's Formula and Atomic Masses to Calculate Formula Mass**

Calculate the formula mass of each of the following substances.

a. SnF_2 (tin(II) fluoride, a toothpaste additive)
b. $Al(OH)_3$ (aluminum hydroxide, a water purification chemical)

Solution

Formula masses are obtained simply by adding the atomic masses of the constituent elements, counting each atomic mass as many times as the symbol for the element occurs in the formula.

a. A formula unit of SnF_2 contains three atoms: one atom of Sn and two atoms of F. The formula mass, the collective mass of these three atoms, is calculated as follows:

$$1 \text{ atom Sn} \times \left(\frac{118.71 \text{ amu}}{1 \text{ atom Sn}}\right) = 118.71 \text{ amu}$$

$$2 \text{ atoms F} \times \left(\frac{19.00 \text{ amu}}{1 \text{ atom F}}\right) = \underline{38.00 \text{ amu}}$$

$$\text{Formula mass} = 156.71 \text{ amu}$$

We derive the conversion factors in the calculation from the atomic masses listed on the inside front cover of the text. Our rules for the use of conversion factors are the same as those discussed in Section 2.7.

Conversion factors are usually not explicitly shown in a formula mass calculation, as they are in our example; the calculation is simplified as follows:

$$\text{Sn:} \quad 1 \times 118.71 \text{ amu} = 118.71 \text{ amu}$$

$$\text{F:} \quad 2 \times 19.00 \text{ amu} = \underline{38.00 \text{ amu}}$$

$$\text{Formula mass} = 156.71 \text{ amu}$$

b. The formula for this compound contains parentheses. Improper interpretation of parentheses (see Section 1.10) is a common error made by students doing formula mass calculations. In the formula $Al(OH)_3$, the subscript 3 outside the parentheses affects both the symbols inside the parentheses. Thus we have

$$\text{Al:} \quad 1 \times 26.98 \text{ amu} = 26.98 \text{ amu}$$

$$\text{O:} \quad 3 \times 16.00 \text{ amu} = 48.00 \text{ amu}$$

$$\text{H:} \quad 3 \times 1.01 \text{ amu} = \underline{3.03 \text{ amu}}$$

$$\text{Formula mass} = 78.01 \text{ amu}$$

In this text, we will always use atomic masses rounded to the hundredths place, as we have done in this example. This rule allows us to use, without rounding, the atomic masses given inside the front cover of the text. A benefit of this approach is that we always use the same atomic mass for a given element and thus become familiar with the atomic masses of the common elements.

Figure 5.1 A basic process in chemical laboratory work is determining the mass of a substance.

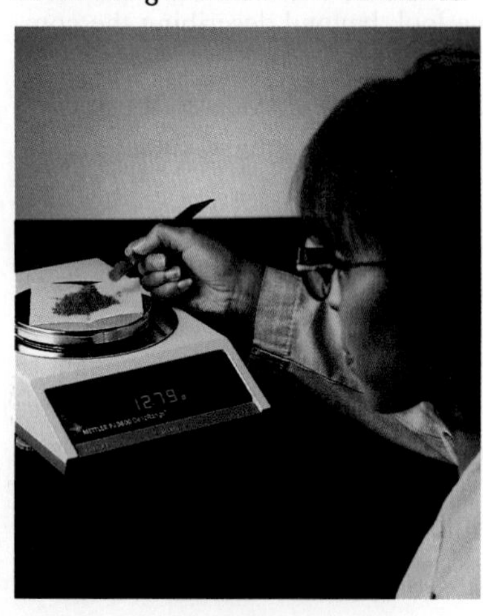

Practice Problems and Questions

5.1 Calculate, to two decimal places, the formula mass of each of the following compounds. Obtain the needed atomic masses from the inside front cover of the text.
a. CH_4 b. SO_3 c. N_2H_4 d. Al_2O_3

5.2 Calculate, to two decimal places, the formula mass of each of the following compounds. Obtain the needed atomic masses from the inside front cover of the text.
a. H_2SO_4 b. $KMnO_4$ c. $COCl_2$ d. Na_3PO_4

5.3 Calculate, to two decimal places, the formula mass of each of the following compounds. Obtain the needed atomic masses from the inside front cover of the text.
a. $Fe(OH)_3$ b. $(NH_4)_2S$ c. $Ba(H_2PO_4)_2$ d. $Ca(C_2H_3O_2)_2$

5.2 The Mole: A Counting Unit for Chemists

Learning Focus

Interpret the mole as a counting unit, and calculate the number of particles in a given number of moles of a chemical substance.

The quantity of material in a sample of a substance can be specified either in terms of units of mass or in terms of units of *amount*. Mass is specified in terms of units such as grams, kilograms, and pounds (Figure 5.1). The amount of a substance is specified by indicating the number of objects present—3, 17, or 437, for instance.

We all use both units of mass and units of amount on a daily basis. For example, when buying oranges at the grocery store, we can decide on quantity in either mass units (4-lb bag or 10-lb bag) or amount units (3 oranges or 8 oranges) (Figure 5.2). In chemistry, as in everyday life, both mass and amount methods of specifying quantity are used. In laboratory work, practicality dictates working with quantities of known mass. Counting out a given number of atoms for a laboratory experiment is impossible because we cannot see individual atoms.

When we perform chemical calculations after the laboratory work has been done, it is often useful and even necessary to think of the quantities of substances present in terms of numbers of atoms or molecules instead of mass. When this is done, very large numbers are always encountered. Any macroscopic-sized sample of a chemical substance contains many trillions of atoms or molecules.

In order to cope with this large-number problem, chemists have found it convenient to use a special unit when counting atoms and molecules. Specialized counting units are used in many areas—for example, a *dozen* eggs or a *ream* (500 sheets) of paper (Figure 5.3).

The chemist's counting unit is the *mole*. What is unusual about the mole is its magnitude. A **mole** *is* 6.02×10^{23} *objects*. The extremely large size of the mole unit is necessitated by the extremely small size of atoms and molecules. To the chemist, *one mole* always means 6.02×10^{23} objects, just as *one dozen* always means 12 objects. Two moles of objects is two times 6.02×10^{23} objects, and five moles of objects is five times 6.02×10^{23} objects.

Avogadro's number *is the name given to the numerical value* 6.02×10^{23}. This designation honors Amedeo Avogadro, an Italian physicist whose pioneering work on gases later proved valuable in determining the number of particles present in given volumes of substances (Figure 5.4). When we solve problems dealing with the number of objects (atoms or molecules) present in a given number of moles of a substance, Avogadro's number becomes part of the conversion factor used to relate the number of objects present to the number of moles present.

► How large is the number 6.02×10^{23}? It would take an ultramodern computer that can count 100 million times a second 190 million years to count 6.02×10^{23} times. If each of the 6 billion people on Earth were made a millionaire (receiving 1 million dollar bills), we would still need 100 million other worlds, each inhabited with the same number of millionaires, in order to have 6.02×10^{23} dollar bills in circulation.

► Why the number 6.02×10^{23}, rather than some other number, was chosen as the counting unit of chemists is discussed in Section 5.3. A more formal definition of the mole will also be presented in that section.

From the definition

$$1 \text{ mole} = 6.02 \times 10^{23} \text{ objects}$$

two conversion factors can be derived:

$$\frac{6.02 \times 10^{23} \text{ objects}}{1 \text{ mole}} \quad \text{and} \quad \frac{1 \text{ mole}}{6.02 \times 10^{23} \text{ objects}}$$

Example 5.2 illustrates the use of these conversion factors in solving problems.

Figure 5.2 Oranges may be bought in units of mass (4-lb bag) or units of amount (3 oranges).

Example 5.2 Calculating the Number of Objects in a Molar Quantity

How many objects are there in each of the following quantities?

a. 0.23 mole of aspirin molecules **b.** 1.6 moles of oxygen atoms

Solution

Dimensional analysis (Section 2.7) will be used to solve each of these problems. Both of the problems are similar in that we are given a certain number of moles of substance and want to find the number of objects present in the given number of moles. We will need Avogadro's number to solve each of these moles-to-particles problems.

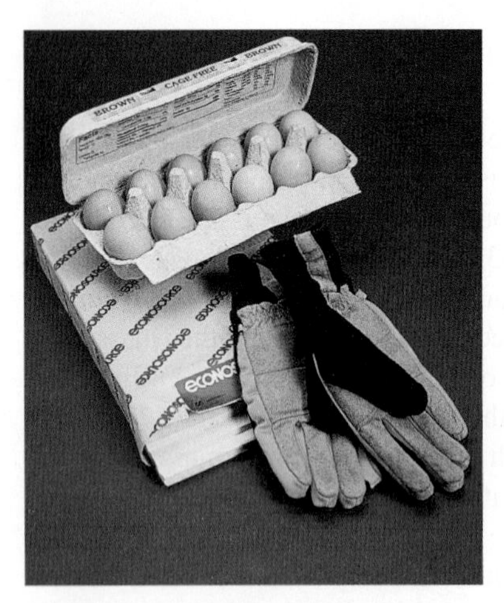

Figure 5.3 Everyday counting units—a dozen, a pair, and a ream.

Moles of Substance —Conversion factor involving Avogadro's number→ Particles of Substance

a. The objects of concern are molecules of aspirin. The given quantity is 0.23 mole of aspirin molecules, and the desired quantity is the number of aspirin molecules.

0.23 mole aspirin molecules = ? aspirin molecules

Applying dimensional analysis here involves the use of a single conversion factor, one that relate moles and molecules.

$$0.23\ \cancel{\text{mole aspirin molecules}} \times \left(\frac{6.02 \times 10^{23}\ \text{aspirin molecules}}{1\ \cancel{\text{mole aspirin molecules}}}\right)$$
$$= 1.4 \times 10^{23}\ \text{aspirin molecules}$$

b. This time we are dealing with atoms instead of molecules. This switch does not change the way we work the problem. We will need the same conversion factor.

The given quantity is 1.6 moles of oxygen atoms, and the desired quantity is the actual number of oxygen atoms present.

1.6 moles oxygen atoms = ? oxygen atoms

The setup is

$$1.6\ \cancel{\text{moles oxygen atoms}} \times \left(\frac{6.02 \times 10^{23}\ \text{oxygen atoms}}{1\ \cancel{\text{mole oxygen atoms}}}\right) = 9.6 \times 10^{23}\ \text{oxygen atoms}$$

Figure 5.4 Amedeo Avogadro (1776–1856) was the first scientist to distinguish between atoms and molecules. His name is associated with the number 6.02×10^{23}, the number of particles (atoms or molecules) in a mole.

Practice Problems and Questions

5.4 Indicate the number of objects present in each of the following molar quantities.
a. The number of apples in 1.00 mole of apples
b. The number of elephants in 1.00 mole of elephants
c. The number of atoms in 1.00 mole of Zn atoms
d. The number of molecules in 1.00 mole of CO_2 molecules

5.5 How many water molecules are present in the following molar quantities of water?
a. 1.00 mole H_2O b. 2.00 moles H_2O c. 0.500 mole H_2O d. 0.621 mole H_2O

5.6 How many atoms are present in the following molar quantities of various elements?
a. 1.50 moles Fe b. 1.50 moles Ni c. 1.50 moles C d. 1.50 moles Ne

5.7 What is the relationship between the number of objects in a mole and Avogadro's number?

5.8 Select the quantity that contains the greater number of atoms in each of the following pairs of substances.
a. 0.100 mole C atoms or 0.200 mole Al atoms
b. Avogadro's number of C atoms or 0.750 mole Al atoms
c. 6.02×10^{23} C atoms or 1.50 moles Al atoms
d. 6.50×10^{23} C atoms or Avogadro's number of Al atoms

Learning Focus

Calculate the mass, in grams, of a given number of moles of a substance, or vice versa.

5.3 The Mass of a Mole

How much does a mole weigh? Are you uncertain about the answer to that question? Let us consider a similar but more familiar question first: "How much does a dozen weigh?" Your response is now immediate: "A dozen what?" The mass of a dozen identical objects obviously depends on the identity of the object. For example, the mass of a dozen elephants is greater than the mass of a dozen peanuts. The mass of a mole, like the mass of a dozen, depends on the identity of the object. Thus the mass of a mole, or *molar mass,* is not a set number; it varies and is different for each chemical substance (see Figure 5.5).

Figure 5.5 The mass of a mole is not a set number of grams; it depends on the substance. For the substances shown, the mass of 1 mole (clockwise from sulfur, the yellow solid) is as follows: sulfur, 32.07 g; zinc, 65.38 g; carbon, 12.0 g; magnesium, 24.30 g; lead, 207.2 g; silicon, 28.09 g; copper, 63.55 g; and in the center, mercury, 200.6 g.

This is in direct contrast to the *molar number,* Avogadro's number, which is the same for all chemical substances.

The **molar mass** *of a substance is a mass in grams that is numerically equal to the substance's formula mass.* For example, the formula mass (atomic mass) of the element sodium is 22.99 amu; therefore, 1 mole of sodium weighs 22.99 g. In Example 5.1, we calculated that the formula mass of tin(II) fluoride is 156.71 amu; therefore, 1 mole of tin(II) fluoride weighs 156.71 g. We can obtain the actual mass in grams of 1 mole of any substance by computing its formula mass (atomic mass for elements) and writing "grams" after it. Thus, when we add atomic masses to get the formula mass (in amu's) of a compound, we are simultaneously finding the mass of 1 mole of that compound (in grams).

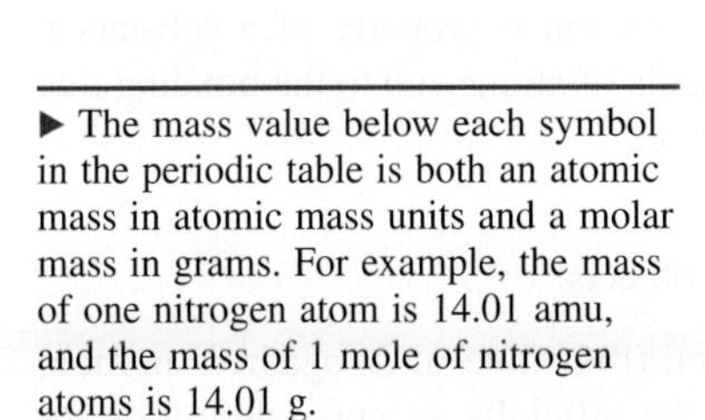

▶ The mass value below each symbol in the periodic table is both an atomic mass in atomic mass units and a molar mass in grams. For example, the mass of one nitrogen atom is 14.01 amu, and the mass of 1 mole of nitrogen atoms is 14.01 g.

It is not a coincidence that the molar mass of a substance and its formula mass or atomic mass match numerically. Avogadro's number has the value that it has in order to cause this relationship to exist. The numerical match between molar mass and atomic or formula mass makes calculating the mass of any given number of moles of a substance a very simple procedure. When you solve problems of this type, the numerical value of the molar mass becomes part of the conversion factor used to convert from moles to grams.

For example, for the compound CO_2, which has a formula mass of 44.01 amu, we can write the equality

$$44.01 \text{ g } CO_2 = 1 \text{ mole } CO_2$$

From this statement (equality), two conversion factors can be written:

$$\frac{44.01 \text{ g } CO_2}{1 \text{ mole } CO_2} \quad \text{and} \quad \frac{1 \text{ mole } CO_2}{44.01 \text{ g } CO_2}$$

Example 5.3 illustrates the use of gram-to-mole conversion factors like these in solving problems.

Example 5.3 Calculating the Mass of a Molar Quantity of Compound

Acetaminophen, the pain-killing ingredient in Tylenol formulations, has the formula $C_8H_9O_2N$. Calculate the mass, in grams, of a 0.30-mole sample of this pain reliever.

Solution

We will use dimensional analysis to solve this problem. The relationship between molar mass and formula mass will serve as a conversion factor in the setup of this problem.

Moles of Substance —(Conversion factor involving molar mass)→ Grams of Substance

The given quantity is 0.30 mole of $C_8H_9O_2N$, and the desired quantity is grams of this same substance.

$$0.30 \text{ mole } C_8H_9O_2N = ? \text{ grams } C_8H_9O_2N$$

The calculated formula mass of $C_8H_9O_2N$ is 151.18 amu. Thus,

$$151.18 \text{ grams } C_8H_9O_2N = 1 \text{ mole } C_8H_9O_2N$$

With this relationship in the form of a conversion factor, the setup for the problem becomes

$$0.30 \text{ mole } C_8H_9O_2N \times \left(\frac{151.18 \text{ g } C_8H_9O_2N}{1 \text{ mole } C_8H_9O_2N} \right) = 45 \text{ g } C_8H_9O_2N$$

The molar mass of *an element* is unique. No two natural elements have the same molar mass. The molar mass of *a compound* lacks uniqueness. More than one compound can have the same molar mass. For example, the compounds carbon dioxide (CO_2), nitrous oxide (N_2O), and propane (C_3H_8) all have a molar mass of 44.0 g. Despite having like molar masses, these compounds have very different chemical properties (see Chemical Portraits 9). Molar mass is a physical rather than a chemical property of a substance. Chemical properties are related to electron arrangements of atoms and to the bonding that results when the atoms interact in compound formation.

In Section 5.2 we defined the mole simply as

$$1 \text{ mole} = 6.02 \times 10^{23} \text{ objects}$$

Although this statement conveys correct information (the value of Avogadro's number to three significant figures is 6.02×10^{23}), it is not the officially accepted definition for Avogadro's number. The official definition, which is based on mass, is as follows: A

Chemical Portraits 9 Compounds of Like Molar Mass

Carbon Dioxide (CO_2) 44.0 g/mole

Profile: Carbon dioxide is a colorless, odorless gas that is nontoxic at normally encountered concentrations. The presence of CO_2 in air is necessary for sustaining human life, in that the growth of plants, a major human food source, depends on *photosynthesis,* a process that requires CO_2. The end products of food metabolism in humans and animals are CO_2 and H_2O.

Uses: Carbon dioxide is used in fighting fires. Nonflammable and about 1.5 times denser than air, CO_2 blankets the fire, thereby diminishing the O_2 supply needed for the fire to burn. When the normal 21% O_2 content of air is reduced to 17% O_2, by dilution with CO_2, most materials cease to burn.

What are the principles of operation for a carbon dioxide fire extinguisher?

Dinitrogen Monoxide (N_2O) 44.0 g/mole

Profile: Dinitrogen monoxide, also known as *nitrous oxide,* is a colorless, nontoxic gas with a slightly sweet taste. N_2O resembles O_2 in that it supports combustion. It does so because it decomposes when heated to form N_2 and O_2.

Biochemical considerations: Nitrous oxide is used as an anesthetic in dental procedures and other minor surgery. It is called "laughing gas" because a person inhaling it becomes somewhat giddy. No satisfactory explanation has yet been proposed for this unusual physiological response. N_2O is also approved for use as a food additive. The largest commercial use of N_2O is as a propellant in whipped-cream dispensers.

What is "unusual" about the structure of a dinitrogen monoxide molecule?

Propane (C_3H_8) 44.0 g/mole

Profile: Propane is a colorless, odorless, nontoxic, highly flammable gas. It is naturally present in crude oil and natural gas deposits from which it is obtained by refinery processing. C_3H_8 is readily converted to a liquid at pressures low enough to be easily maintained in commercial and consumer apparatus, allowing for its easy transport and storage.

Uses: Propane is primarily used as a fuel in homes and businesses located in rural areas, away from natural gas distribution systems. Most home gas barbecues operate on propane. It is easily vaporized and burns with a clean flame to produce CO_2 and H_2O. C_3H_8 is used in limited quantities as a fuel for vehicles.

What are the benefits of using a propane-powered vehicle?

See the text web site at **www.cengage.com/chemistry/stoker** for answers to the above questions and for further information.

mole *is the amount of substance in a system that contains as many elementary particles (atoms, molecules, or formula units) as there are* $^{12}_{6}C$ *atoms in exactly 12 grams of* $^{12}_{6}C$. The value of Avogadro's number is an experimentally determined quantity (the number of atoms in exactly 12 g of $^{12}_{6}C$ atoms) rather than a defined quantity. Its value is not even mentioned in the definition. The most up-to-date experimental value for Avogadro's number is 6.022137×10^{23}, which is consistent with our previous definition (Section 5.2).

Practice Problems and Questions

5.9 How much, in grams, does 1.00 mole of each of the following substances weigh?
a. CO (carbon monoxide) b. CO_2 (carbon dioxide)
c. NaCl (table salt) d. $C_{12}H_{22}O_{11}$ (table sugar)

5.10 What is the mass, in grams, of each of the following quantities of matter?
a. 0.231 mole Ag b. 0.231 mole Au c. 0.231 mole Cu d. 0.231 mole Zn

5.11 How many moles of compound are present in a sample of each of the following substances if each sample weighs 5.00 grams?
a. NH_3 (ammonia) b. H_2O_2 (hydrogen peroxide)
c. SO_2 (sulfur dioxide) d. Zn (zinc)

5.4 The Mole and Chemical Formulas

Learning Focus

Interpret the subscripts in a chemical formula in terms of the number of moles of the various elements present in one mole of the substance.

A chemical formula has two meanings or interpretations: a microscopic-level interpretation and a macroscopic-level interpretation. At a microscopic level, a chemical formula indicates the number of atoms of each element present in one molecule or formula unit of a substance (Section 1.10). *The numerical subscripts in the formula give the number of atoms of the various elements present in one formula unit of the substance.* The formula N_2O_4, interpreted at the microscopic level, conveys the information that two atoms of nitrogen and four atoms of oxygen are present in one molecule of N_2O_4 (Figure 5.6).

Now that the mole concept has been introduced, a macroscopic interpretation of chemical formulas is possible. At a macroscopic level, a chemical formula indicates the number of moles of atoms of each element present in one mole of a substance. *The numerical subscripts in the chemical formula give the number of moles of atoms of the various elements present in 1 mole of the substance.* The designation *macroscopic* is given to this molar interpretation because moles are laboratory-sized quantities of atoms. The formula N_2O_4, interpreted at the macroscopic level, conveys the information that 2 moles of nitrogen atoms and 4 moles of oxygen atoms are present in 1 mole of N_2O_4 molecules. Thus the subscripts in a formula always carry a dual meaning: atoms at the microscopic level and moles of atoms at the macroscopic level.

▶ The molar (macroscopic-level) interpretation of a chemical formula is used in calculations where information about a *particular element within a compound* is needed.

When it is necessary to know the number of moles of a particular element *within* a compound, the subscript of that element's symbol in the chemical formula becomes part of the conversion factor used to convert from moles of compound to moles of element *within* the compound. Using N_2O_4 as our chemical formula, we can write the following conversion factors:

▶ Conversion factors that relate a component of a substance to the substance as a whole are dependent on the formula of the substance. By analogy, the relationship of body parts of an animal to the animal as a whole is dependent on the animal's identity. For example, in 1 mole of elephants there would be 4 moles of elephant legs, 2 moles of elephant ears, 1 mole of elephant tails, and 1 mole of elephant trunks.

$$\text{For N:} \quad \frac{2 \text{ moles N atoms}}{1 \text{ mole } N_2O_4 \text{ molecules}} \quad \text{or} \quad \frac{1 \text{ mole } N_2O_4 \text{ molecules}}{2 \text{ moles N atoms}}$$

$$\text{For O:} \quad \frac{4 \text{ moles O atoms}}{1 \text{ mole } N_2O_4 \text{ molecules}} \quad \text{or} \quad \frac{1 \text{ mole } N_2O_4 \text{ molecules}}{4 \text{ moles O atoms}}$$

Example 5.4 illustrates the use of this type of conversion factor in problem solving.

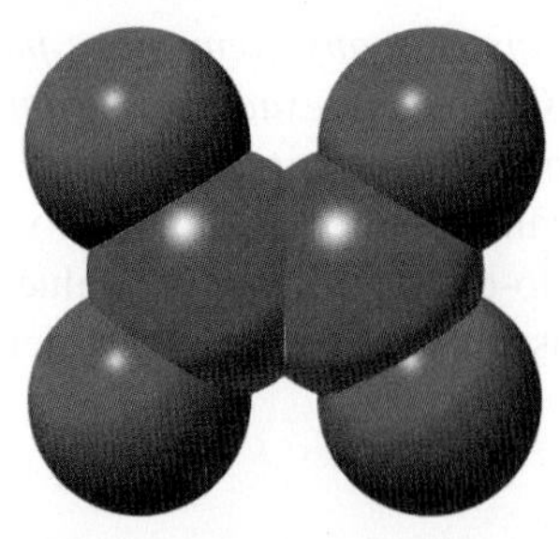

Figure 5.6 A computer-generated model of the molecular structure of N_2O_4. One molecule of N_2O_4 contains two atoms of N and four atoms of O.

Example 5.4 **Calculating Molar Quantities of Compound Components**

Lactic acid, the substance that builds up in muscles and causes them to hurt when they are worked hard, has the formula $C_3H_6O_3$. How many moles of carbon atoms, hydrogen atoms, and oxygen atoms are present in a 1.2-mole sample of lactic acid?

Solution

One mole of $C_3H_6O_3$ contains 3 moles of carbon atoms, 6 moles of hydrogen atoms, and 3 moles of oxygen atoms. We obtain the following conversion factors from this statement:

$$\left(\frac{3 \text{ moles C atoms}}{1 \text{ mole } C_3H_6O_3}\right) \quad \left(\frac{6 \text{ moles H atoms}}{1 \text{ mole } C_3H_6O_3}\right) \quad \left(\frac{3 \text{ moles O atoms}}{1 \text{ mole } C_3H_6O_3}\right)$$

Using the first conversion factor, the moles of carbon atoms present are calculated as follows:

$$1.2 \text{ moles } C_3H_6O_3 \times \left(\frac{3 \text{ moles C atoms}}{1 \text{ mole } C_3H_6O_3}\right) = 3.6 \text{ moles C atoms}$$

Similarly, from the second and third conversion factors, the moles of hydrogen and oxygen atoms present are calculated as follows:

$$1.2 \text{ moles } C_3H_6O_3 \times \left(\frac{6 \text{ moles H atoms}}{1 \text{ mole } C_3H_6O_3}\right) = 7.2 \text{ moles H atoms}$$

$$1.2 \text{ moles } C_3H_6O_3 \times \left(\frac{3 \text{ moles O atoms}}{1 \text{ mole } C_3H_6O_3}\right) = 3.6 \text{ moles O atoms}$$

Practice Problems and Questions

5.12 Write the six mole-to-mole conversion factors that can be derived from each of the following chemical formulas.
a. H_2SO_4 b. $POCl_3$

5.13 List the number of moles of each type of atom that are present in each of the following molar quantities.
a. 2.00 moles SO_2 molecules b. 2.00 moles SO_3 molecules
c. 3.00 moles NH_3 molecules d. 3.00 moles N_2H_4 molecules

5.14 How many total moles of atoms are present in each of the following molar quantities?
a. 4.00 moles SO_3 b. 2.00 moles H_2SO_4
c. 1.00 mole $C_{12}H_{22}O_{11}$ d. 3.00 moles $Mg(OH)_2$

5.5 Calculations Using Molar Mass

Learning Focus

Given information about the moles or mass of a substance, calculate additional information (moles or mass) about the same substance or about a component of the substance.

In Section 5.3, we learned that *molar mass* provides a relationship between the number of grams of a substance and the number of moles of that substance:

Grams of Substance → (Conversion factor involving molar mass) → Moles of Substance

In Section 5.4, we learned that the *molar interpretation of chemical formula subscripts* provides a relationship between the number of moles of a substance and the number of moles of its components:

Moles of Compound → (Conversion factor involving chemical formula subscripts) → Moles of Element within Compound

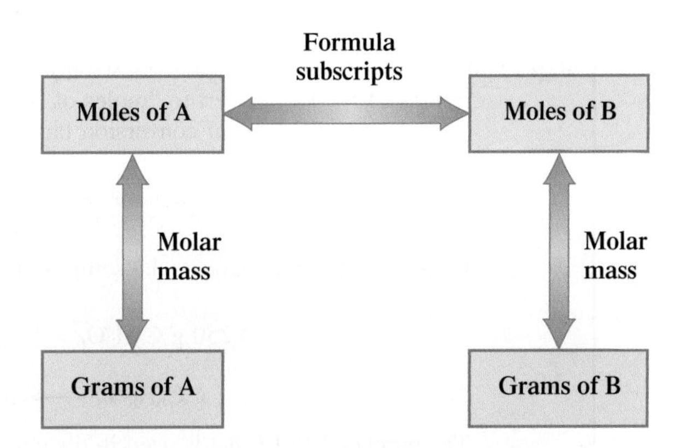

Figure 5.7 In solving chemical-formula-based problems, the only "transitions" allowed are those between quantities (boxes) connected by arrows. Associated with each arrow is the concept on which the required conversion factor is based.

The preceding two concepts can be combined into a single diagram that is very useful in problem solving. This diagram, Figure 5.7, can be viewed as a road map from which conversion factor sequences (pathways) may be obtained. It gives all of the needed relationships for solving two general types of problems:

1. Calculations where information (moles or grams) is given about a particular substance, and additional information (moles or grams) is needed concerning the *same* substance.
2. Calculations where information (moles or grams) is given about a particular substance, and information is needed concerning a *component* of that same substance.

For the first type of problem, only the left side of Figure 5.7 (the "A" boxes) is needed. For problems of the second type, both sides of the diagram (both "A" and "B" boxes) are used.

The thinking pattern needed to use Figure 5.7 is very simple.

1. Determine which box in the diagram represents the *given* quantity in the problem.
2. Locate the box that represents the *desired* quantity.
3. Follow the indicated pathway that takes you from the given quantity to the desired quantity. This involves simply following the arrows. There will always be only one pathway possible for the needed transition.

Examples 5.5 and 5.6 illustrate some of the types of problems that can be solved by using the relationships shown in Figure 5.7.

Example 5.5 **Calculating the Number of Moles in a Given Mass of Compound**

Vitamin C has the formula $C_6H_8O_6$. Calculate the number of moles of vitamin C molecules present in a 0.250-g tablet of vitamin C.

Solution

We will solve this problem by using the three steps of dimensional analysis (Section 2.6) and Figure 5.7.

Step 1: The given quantity is 0.250 g of $C_6H_8O_6$, and the desired quantity is moles of $C_6H_8O_6$.

$$0.250 \text{ g } C_6H_8O_6 = ? \text{ moles } C_6H_8O_6$$

In terms of Figure 5.7, this is a "grams of A" to "moles of A" problem. We are given grams of a substance, A, and desire to find moles of that same substance.

Step 2: Figure 5.7 gives us the pathway we need to solve this problem. Starting with "grams of A," we convert to "moles of A." The arrows between the boxes along our path give the type of conversion factor needed for each step.

$$\boxed{\text{Grams of A}} \xrightarrow[\text{mass}]{\text{Molar}} \boxed{\text{Moles of A}}$$

From dimensional analysis, the setup will involve just one conversion factor.

$$0.250 \text{ g } C_6H_8O_6 \times \left(\frac{1 \text{ mole } C_6H_8O_6}{176.14 \text{ g } C_6H_8O_6} \right)$$

$$\text{g } C_6H_8O_6 \longrightarrow \text{moles } C_6H_8O_6$$

The number 176.14 that is used in the conversion factor is the formula mass of $C_6H_8O_6$. It was not given in the problem but had to be calculated by using atomic masses and the method for calculating formula masses shown in Example 5.1.

Step 3: The solution to the problem, obtained by doing the arithmetic, is

$$\frac{0.250 \times 1}{176.14} \text{ moles } C_6H_8O_6 = 0.00142 \text{ mole } C_6H_8O_6$$

Example 5.6 Calculating the Mass of an Element Present in a Given Mass of Compound

How many grams of nitrogen are present in a 0.10-g sample of caffeine, the stimulant in coffee and tea? The formula of caffeine is $C_8H_{10}N_4O_2$.

Solution

Step 1: There is an important difference between this problem and the preceding one; here we are dealing with not one but two substances, caffeine and nitrogen. The given quantity is grams of caffeine (substance A), and we are asked to find the grams of nitrogen (substance B). This is a "grams of A" to "grams of B" problem.

$$0.10 \text{ g } C_8H_{10}N_4O_2 = ? \text{ g N}$$

Step 2: The appropriate set of conversions for a "grams of A" to "grams of B" problem, from Figure 5.7, is

$$\boxed{\text{Grams of A}} \xrightarrow[\text{mass}]{\text{Molar}} \boxed{\text{Moles of A}} \xrightarrow[\text{subscripts}]{\text{Formula}} \boxed{\text{Moles of B}} \xrightarrow[\text{mass}]{\text{Molar}} \boxed{\text{Grams of B}}$$

The conversion factor setup is

$$0.10 \text{ g } C_8H_{10}N_4O_2 \times \left(\frac{1 \text{ mole } C_8H_{10}N_4O_2}{194.26 \text{ g } C_8H_{10}N_4O_2} \right) \times \left(\frac{4 \text{ moles N}}{1 \text{ mole } C_8H_{10}N_4O_2} \right) \times \left(\frac{14.01 \text{ g N}}{1 \text{ mole N}} \right)$$

The number 194.26 that is used in the first conversion factor is the formula mass for caffeine. The conversion from "moles of A" to "moles of B" (the second conversion factor) is made by using the information contained in the formula $C_8H_{10}N_4O_2$. One mole of caffeine contains 4 moles of nitrogen. The number 14.01 in the final conversion factor is the molar mass of nitrogen.

Step 3: Collecting the numbers from the various conversion factors and doing the arithmetic give us our answer.

$$\left(\frac{0.10 \times 1 \times 4 \times 14.01}{194.26 \times 1 \times 1} \right) \text{ g N} = 0.029 \text{ g N}$$

Practice Problems and Questions

5.15 How much, in grams, does each of the following molar amounts of chemical substance weigh?
a. 0.250 mole S b. 0.250 mole SO_2
c. 0.250 mole SO_3 d. 0.250 mole H_2SO_4

5.16 How many moles of S are present in each of the following gram amounts of chemical substance?
a. 10.0 g S b. 10.0 g SO_2 c. 10.0 g S_2O d. 10.0 g $Na_2S_2O_3$

5.17 How many grams of N are present in each of the following gram amounts of chemical substance?
a. 10.0 g NH_3 b. 10.0 g N_2O_5 c. 10.0 g N_2H_4 d. 10.0 $Al(NO_3)_3$

5.18 How many grams of each of the following compounds is needed to obtain 10.0 g of N?
a. NH_3 b. N_2O_5 c. N_2H_4 d. $Al(NO_3)_3$

Learning Focus

Understand the conventions used in writing chemical equations, recognize whether a chemical equation is balanced, and balance a chemical equation given the formulas of all the reactants and products.

5.6 Writing and Balancing Chemical Equations

A **chemical equation** *is a written statement that uses symbols and formulas instead of words to describe the changes that occur in a chemical reaction.* The following example shows the contrast between a word description of a chemical reaction and a chemical equation for the same reaction.

Word description: Calcium sulfide reacts with water to produce calcium oxide and hydrogen sulfide.
Chemical equation: $CaS + H_2O \longrightarrow CaO + H_2S$

In the same way that chemical symbols are considered the *letters* of chemical language, and formulas are considered the *words* of the language, chemical equations can be considered the *sentences* of chemical language.

Conventions Used in Writing Chemical Equations

Four conventions are used to write chemical equations.

▶ In a chemical equation, the *reactants* (starting materials in a chemical reaction) are always written on the left side of the equation, and the *products* (substances produced in a chemical reaction) are always written on the right side of the equation.

1. *The correct formulas of the* ***reactants*** *(starting materials) are always written on the* ***left*** *side of the equation.*

$$\text{(CaS)} + \text{(H}_2\text{O)} \longrightarrow CaO + H_2S$$

2. *The correct formulas of the* ***products*** *(substances produced) are always written on the* ***right*** *side of the equation.*

$$CaS + H_2O \longrightarrow \text{(CaO)} + \text{(H}_2\text{S)}$$

3. *The reactants and products are separated by an arrow pointing toward the products.*

$$CaS + H_2O \longrightarrow CaO + H_2S$$

This arrow means "to produce."

4. *Plus signs are used to separate different reactants or different products.*

$$CaS \oplus H_2O \longrightarrow CaO \oplus H_2S$$

Plus signs on the reactant side of the equation mean "reacts with," and plus signs on the product side mean "and."

A *valid* chemical equation must satisfy two conditions:

▶ The diatomic elemental gases are the elements whose names end in *-gen* (hydrogen, oxygen, and nitrogen) or *-ine* (fluorine, chlorine, bromine, and iodine).

1. *It must be consistent with experimental facts.* Only the reactants and products that are actually involved in a reaction are shown in an equation. An accurate formula must be used for each of these substances. Elements in solid and liquid states are represented in equations by the chemical symbol for the element. Elements that are gases at room temperature are represented by the molecular formula denoting the form in which they actually occur in nature. The following monatomic, diatomic, and tetratomic elemental gases are known.

Monatomic:	He, Ne, Ar, Kr, Xe
Diatomic:	H_2, O_2, N_2, F_2, Cl_2, Br_2 (vapor), I_2 (vapor)
Tetratomic:	P_4 (vapor), As_4 (vapor)*

2. *There must be the same number of atoms of each kind on both sides of the equation.* Equations that satisfy this condition are said to be balanced. A **balanced chemical equation** *has the same number of atoms of each element involved in the reaction on each side of the equation.* Because the conventions previously listed for writing equations do not guarantee that an equation will be balanced, we now consider procedures for balancing equations.

Guidelines for Balancing Chemical Equations

An unbalanced equation is brought into balance by adding *coefficients* to the equation to adjust the number of reactant or product molecules present. A **coefficient** *is a number that is placed to the left of the chemical formula of a substance and that changes the amount, but not the identity, of the substance.* In the notation $2H_2O$, the 2 on the left is a coefficient; $2H_2O$ means two molecules of H_2O, and $3H_2O$ means three molecules of H_2O. Thus coefficients tell how many formula units of a given substance are present.

▶ The coefficients of a balanced equation represent numbers of molecules or formula units of various species involved in the chemical reaction.

The following is a balanced chemical equation, with the coefficients shown in color.

$$4NH_3 + 3O_2 \longrightarrow 2N_2 + 6H_2O$$

This balanced equation tells us that four NH_3 molecules react with three O_2 molecules to produce two N_2 molecules and six H_2O molecules.

A coefficient of 1 in a balanced equation is not explicity written; it is considered to be understood. Both Na_2SO_4 and Na_2S have "understood coefficients" of 1 in the following balanced equation:

$$Na_2SO_4 + 2C \longrightarrow Na_2S + 2CO_2$$

▶ In balancing a chemical equation, formula subscripts are *never changed.* You must use the formulas just as they are given. *The only thing you can do is add coefficients.*

A coefficient placed in front of a formula applies to the whole formula. By contrast, subscripts, which are also present in formulas, affect only parts of a formula.

Coefficient (affects both H and O)

$$2H_2O$$

Subscript (affects only H)

The preceding notation denotes two molecules of H_2O; it also denotes a total of four H atoms and two O atoms.

Let's look at the mechanics involved in determining the coefficients needed to balance an equation. Suppose we want to balance the chemical equation

$$FeI_2 + Cl_2 \longrightarrow FeCl_3 + I_2$$

*The four elements listed as vapors are not gases at room temperature but vaporize at slightly higher temperatures. The resultant vapors contain molecules with the formulas indicated.

Step 1: *Examine the equation and pick one element to balance first.* It is often convenient to start with the compound that contains the greatest number of atoms, whether a reactant or a product, and key in on the element in that compound that has the greatest number of atoms. Using this guideline, we select $FeCl_3$ and the element chlorine within it.

We note that there are three chlorine atoms on the right side of the equation and two atoms of chlorine on the left (in Cl_2). In order for the chlorine atoms to balance, we will need six on each side; 6 is the lowest number that both 3 and 2 will divide into evenly. In order to obtain six atoms on each side of the equation, we place the coefficient 3 in front of Cl_2 and the coefficient 2 in front of $FeCl_3$.

$$FeI_2 + \textcircled{3}Cl_2 \longrightarrow \textcircled{2}FeCl_3 + I_2$$

We now have six chlorine atoms on each side of the equation.

$$3Cl_2: \quad 3 \times 2 = 6$$
$$2FeCl_3: \quad 2 \times 3 = 6$$

Step 2: *Now pick a second element to balance.* We will balance the iron next. The number of iron atoms on the right side has already been set at 2 by the coefficient previously placed in front of $FeCl_3$. We will need two iron atoms on the reactant side of the equation instead of the one iron atom now present. This is accomplished by placing the coefficient 2 in front of FeI_2.

$$\textcircled{2}FeI_2 + 3Cl_2 \longrightarrow 2FeCl_3 + I_2$$

It is always wise to pick, as the second element to balance, one whose amount is already set on one side of the equation by a previously determined coefficient. If we had chosen iodine as the second element to balance instead of iron, we would have run into problems. Because the coefficient for neither FeI_2 nor I_2 had been determined, we would have had no guidelines for deciding on the amount of iodine needed.

Step 3: *Now pick a third element to balance.* Only one element is left to balance—iodine. The number of iodine atoms on the left side of the equation is already set at four ($2FeI_2$). In order to obtain four iodine atoms on the right side of the equation, we place the coefficient 2 in front of I_2.

$$2FeI_2 + 3Cl_2 \longrightarrow 2FeCl_3 + \textcircled{2}I_2$$

The addition of the coefficient 2 in front of I_2 completes the balancing process; all the coefficients have been determined.

Step 4: *As a final check on the correctness of the balancing procedure, count atoms on each side of the equation.* The following table can be constructed from our balanced equation.

$$2FeI_2 + 3Cl_2 \longrightarrow 2FeCl_3 + 2I_2$$

Atom	Left side	Right side
Fe	$2 \times 1 = 2$	$2 \times 1 = 2$
I	$2 \times 2 = 4$	$2 \times 2 = 4$
Cl	$3 \times 2 = 6$	$2 \times 3 = 6$

All elements are in balance: two iron atoms on each side, four iodine atoms on each side, and six chlorine atoms on each side (see Figure 5.8).

Notice that the elements chlorine and iodine in the equation are written in the form of diatomic molecules (Cl_2 and I_2). This is in accordance with the guideline given at the start of this section on the use of molecular formulas for elements that are gases at room temperature.

$$CaS \longrightarrow Ca + S$$

Figure 5.8 Atoms are neither created nor destroyed in an ordinary chemical reaction. The production of new substances in a reaction results from the rearrangement of the existent groupings of atoms into new groupings. Because only rearrangement occurs, the products always contain the same number of atoms of each kind as do the reactants. This generalization is often referred to as the *law of conservation of mass*. The mass of the reactants and the mass of the products are the same, because both contain exactly the same number of atoms of each kind present.

In Example 5.7 we will balance another chemical equation.

Example 5.7 Balancing a Chemical Equation

Balance the following chemical equation.

$$C_2H_6O + O_2 \longrightarrow CO_2 + H_2O$$

Solution

The element oxygen appears in four different places in this equation. This means we do not want to start the balancing process with the element oxygen. Always start the balancing process with an element that appears only once on both the reactant and product sides of the equation.

Step 1: *Balancing of H atoms.* There are six H atoms on the left and two H atoms on the right. Placing the coefficient 3 in front of H_2O balances the H atoms at six on each side.

$$1C_2H_6O + O_2 \longrightarrow CO_2 + 3H_2O$$

Step 2: *Balancing of C atoms.* An effect of balancing the H atoms at six (Step 1) is the setting of the C atoms on the left side at two; the coefficient in front of C_2H_6O is 1. Placing the coefficient 2 in front of CO_2 causes the carbon atoms to balance at two on each side of the equation.

$$1C_2H_6O + O_2 \longrightarrow 2CO_2 + 3H_2O$$

Step 3: *Balancing of O atoms.* The oxygen content of the right side of the equation is set at seven atoms: four oxygen atoms from $2CO_2$ and three oxygen atoms from $3H_2O$. To obtain seven oxygen atoms on the left side of the equation, we place the coefficient 3 in front of O_2; $3O_2$ gives six oxygen atoms, and there is an additional O in $1C_2H_6O$. The element oxygen is present in all four formulas in the equation.

$$1C_2H_6O + 3O_2 \longrightarrow 2CO_2 + 3H_2O$$

Step 4: *Final check.* The equation is balanced. There are two carbon atoms, six hydrogen atoms, and seven oxygen atoms on each side of the equation.

$$C_2H_6O + 3O_2 \longrightarrow 2CO_2 + 3H_2O$$

Some additional comments and guidelines concerning equations in general, and the process of balancing in particular, are given here.

1. The coefficients in a balanced equation are always the *smallest set of whole numbers* that will balance the equation. We mention this because more than one set of coefficients will balance an equation. Consider the following three equations:

$$2H_2 + O_2 \longrightarrow 2H_2O$$
$$4H_2 + 2O_2 \longrightarrow 4H_2O$$
$$8H_2 + 4O_2 \longrightarrow 8H_2O$$

All three of these equations are mathematically correct; there are equal numbers of hydrogen and oxygen atoms on both sides of the equation. However, the first equation is considered the correct form because the coefficients used there are the smallest set of whole numbers that will balance the equation. The coefficients in the second equation are two times those in the first equations and the third equation has coefficients that are four times those of the first equation.

2. At this point, you are not expected to be able to write down the products for a chemical reaction when given the reactants. After learning how to balance equations, students sometimes get the mistaken idea that they ought to be able to write down equations from scratch. This is not so. You will need more chemical knowledge before attempting this task. At this stage, you should be able to balance simple equations, given *all* of the reactants and *all* of the products.
3. It is often useful to know the physical state of the substances involved in a chemical reaction. We specify physical state by using the symbols (*s*) for solid, (*l*) for liquid, (*g*) for gas, and (*aq*) for aqueous solution (a substance dissolved in water). Two examples of such symbol use in chemical equations are

$$2Fe_2O_3(s) + 3C(s) \longrightarrow 4Fe(s) + 3CO_2(g)$$
$$2HNO_3(aq) + 3H_2S(aq) \longrightarrow 2NO(g) + 3S(s) + 4H_2O(l)$$

Practice Problems and Questions

5.19 Determine whether each of the following equations is balanced or not balanced.
a. $SO_3 + H_2O \longrightarrow H_2SO_4$
b. $CuO + H_2 \longrightarrow Cu + H_2O$
c. $CS_2 + O_2 \longrightarrow CO_2 + SO_2$
d. $AgNO_3 + KCl \longrightarrow KNO_3 + AgCl$

5.20 For each of the following balanced equations, give the number of atoms of each element present on the reactant and product sides of the equation.
a. $2N_2 + 3O_2 \longrightarrow 2N_2O_3$
b. $4NH_3 + 6NO \longrightarrow 5N_2 + 6H_2O$
c. $PCl_3 + 3H_2 \longrightarrow PH_3 + 3HCl$
d. $Al_2O_3 + 6HCl \longrightarrow 2AlCl_3 + 3H_2O$

5.21 What do the symbols in parentheses stand for in the following balanced equations?
a. $CaCO_3(s) \longrightarrow CaO(s) + CO_2(g)$
b. $SO_2(g) + H_2O(l) \longrightarrow H_2SO_3(aq)$

5.22 Balance the following equations.
a. $Cu + O_2 \longrightarrow CuO$
b. $Al + N_2 \longrightarrow AlN$
c. $HgO \longrightarrow Hg + O_2$
d. $H_2O \longrightarrow H_2 + O_2$

5.23 Balance the following equations.
a. $BaCl_2 + Na_2S \longrightarrow BaS + NaCl$
b. $Mg + HBr \longrightarrow MgBr_2 + H_2$
c. $Co + HgCl_2 \longrightarrow CoCl_3 + Hg$
d. $Na + H_2O \longrightarrow NaOH + H_2$

5.24 Balance the following equations.
a. $PbO + NH_3 \longrightarrow Pb + N_2 + H_2O$
b. $NaHCO_3 + H_2SO_4 \longrightarrow Na_2SO_4 + H_2O + CO_2$
c. $TiO_2 + C + Cl_2 \longrightarrow TiCl_4 + CO_2$
d. $NBr_3 + NaOH \longrightarrow N_2 + NaBr + HBrO$

Learning Focus

Interpret coefficients in balanced chemical equations in terms of moles.

5.7 Chemical Equations and the Mole Concept

The coefficients in a balanced chemical equation, like the subscripts in a chemical formula (Section 5.5), have two levels of interpretation—a microscopic level of meaning and a macroscopic level of meaning. The microscopic level of interpretation was used in the previous two sections. *The coefficients in a balanced equation give the numerical relationships among formula units consumed (used up) or produced in the chemical reaction.* Interpreted at the microscopic level, the equation

$$N_2 + 3H_2 \longrightarrow 2NH_3$$

conveys the information that one molecule of N_2 reacts with three molecules of H_2 to produce two molecules of NH_3.

At the macroscopic level of interpretation, chemical equations are used to relate mole-sized quantities of reactants and products to each other. At this level, *the coefficients in a balanced chemical equation give the fixed molar ratios between substances consumed or produced in the chemical reaction.* Interpreted at the macroscopic level, the equation

$$N_2 + 3H_2 \longrightarrow 2NH_3$$

conveys the information that 1 mole of N_2 reacts with 3 moles of H_2 to produce 2 moles of NH_3.

The coefficients in a balanced chemical equation can be used to generate conversion factors to be used in solving problems. Numerous conversion factors are obtainable from a single balanced equation. Consider the following balanced equation:

$$4Fe + 3O_2 \longrightarrow 2Fe_2O_3$$

Three mole-to-mole relationships are obtainable from this equation:

▶ Conversion factors that relate two different substances to one another are valid only for systems governed by the chemical equation from which they were obtained.

4 moles of Fe produces 2 moles of Fe_2O_3.

3 moles of O_2 produces 2 moles of Fe_2O_3.

4 moles of Fe reacts with 3 moles of O_2.

From each of these macroscopic-level relationships, two conversion factors can be written. The conversion factors for the first relationship are

$$\left(\frac{4 \text{ moles Fe}}{2 \text{ moles Fe}_2\text{O}_3}\right) \quad \text{and} \quad \left(\frac{2 \text{ moles Fe}_2\text{O}_3}{4 \text{ moles Fe}}\right)$$

All balanced chemical equations are the source of numerous conversion factors. The more reactants and products there are in the equation, the greater the number of derivable conversion factors. The next section details how conversion factors such as those in the preceding illustration are used in solving problems.

The accompanying Chemistry at a Glance reviews the relationships that involve the mole.

Practice Problems and Questions

5.25 Give an interpretation of the following equations in terms of (1) number of particles and (2) number of moles.

a. $N_2 + 3H_2 \longrightarrow 2NH_3$

b. $CH_4 + 2O_2 \longrightarrow CO_2 + 2H_2O$

5.26 Write the six mole-to-mole conversion factors (three reciprocal pairs) that can be derived from the following balanced equation.

$$2CO + O_2 \longrightarrow 2CO_2$$

5.27 Indicate whether each of the following statements is consistent with or inconsistent with the following balanced equation.

$$3O_2 + CS_2 \longrightarrow CO_2 + 2SO_2$$

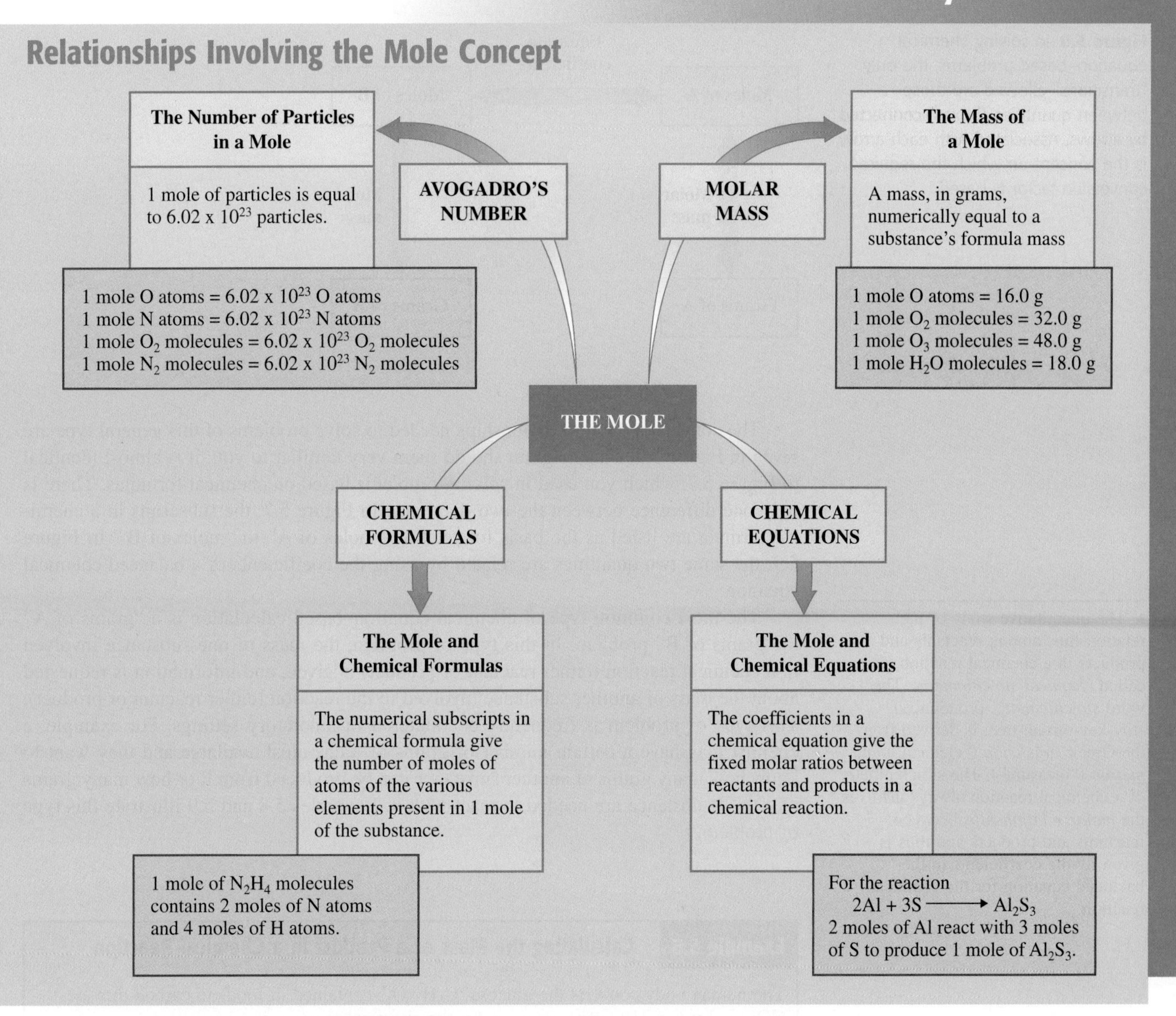

a. 3 moles of O_2 will react with 1.5 moles of CS_2
b. 6 moles of O_2 will react with 2 moles of CS_2
c. The number of moles of CS_2 that reacts and the number of moles of CO_2 produced will always be the same.
d. The molar amounts of the products CO_2 and SO_2 are always in a 2-to-1 ratio.

Learning Focus

Use a balanced chemical equation and other appropriate information to calculate the quantities of reactants consumed or products produced in a chemical reaction.

5.8 Mass Calculations Based on Chemical Equations

When the information contained in a chemical equation is combined with the concept of molar mass (Section 5.3) several useful types of chemical calculations can be carried out. A typical chemical equation–based calculation gives information about one reactant or product of a reaction (number of grams or moles) and requests similar information about another reactant or product of the same reaction. The substances involved in such a calculation may both be reactants or products or may be a reactant and a product.

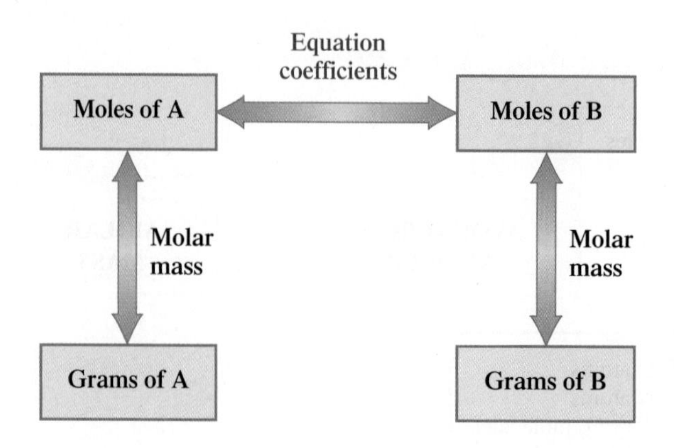

Figure 5.9 In solving chemical equation–based problems, the only "transitions" allowed are those between quantities (boxes) connected by arrows. Associated with each arrow is the concept on which the required conversion factor is based.

The conversion factor relationships needed to solve problems of this general type are given in Figure 5.9. This diagram should seem very familiar to you; it is almost identical to Figure 5.7, which you used in solving problems based on chemical formulas. There is only one difference between the two diagrams. In Figure 5.7, the subscripts in a chemical formula are listed as the basis for relating "moles of A" to "moles of B." In Figure 5.9, the same two quantities are related by using the coefficients of a balanced chemical equation.

The most common type of chemical equation–based calculation is a "grams of A" to "grams of B" problem. In this type of problem, the mass of one substance involved in a chemical reaction (either reactant or product) is given, and information is requested about the mass of another substance involved in the reaction (either reactant or product). This type of problem is frequently encountered in laboratory settings. For example, a chemist may have a certain number of grams of a chemical available and may want to know how many grams of another substance can be produced from it or how many grams of a third substance are needed to react with it. Examples 5.8 and 5.9 illustrate this type of problem.

▶ The quantitative study of the relationships among reactants and products in a chemical reaction is called *chemical stoichiometry*. The word *stoichiometry*, pronounced stoy-key-om-eh-tree, is derived from the Greek *stoicheion* ("element") and *metron* ("measure"). The stoichiometry of a chemical reaction always involves the *molar relationships* between reactants and products and thus is given by the coefficients in the balanced equation for the chemical reaction.

Example 5.8 Calculating the Mass of a Product in a Chemical Reaction

The human body converts the glucose, $C_6H_{12}O_6$, contained in foods to carbon dioxide, CO_2, and water, H_2O. The equation for the chemical reaction is

$$C_6H_{12}O_6 + 6O_2 \longrightarrow 6CO_2 + 6H_2O$$

Assume a person eats a candy bar containing 14.2 g (1/2 oz) of glucose. How many grams of water will the body produce from the ingested glucose, assuming all of the glucose undergoes reaction?

Solution

Step 1: The given quantity is 14.2 g of glucose. The desired quantity is grams of water.

$$14.2 \text{ g } C_6H_{12}O_6 = ? \text{ g } H_2O$$

In terms of Figure 5.9, this is a "grams of A" to "grams of B" problem.

Step 2: Using Figure 5.9 as a road map, we determine that the pathway for this problem is

$$\boxed{\text{Grams of A}} \xrightarrow[\text{mass}]{\text{Molar}} \boxed{\text{Moles of A}} \xrightarrow[\text{coefficients}]{\text{Equation}} \boxed{\text{Moles of B}} \xrightarrow[\text{mass}]{\text{Molar}} \boxed{\text{Grams of B}}$$

The mathematical setup for this problem is

$$14.2 \text{ g } C_6H_{12}O_6 \times \left(\frac{1 \text{ mole } C_6H_{12}O_6}{180.18 \text{ g } C_6H_{12}O_6}\right) \times \left(\frac{6 \text{ moles } H_2O}{1 \text{ mole } C_6H_{12}O_6}\right) \times \left(\frac{18.02 \text{ g } H_2O}{1 \text{ mole } H_2O}\right)$$

$$\text{g } C_6H_{12}O_6 \longrightarrow \text{moles } C_6H_{12}O_6 \longrightarrow \text{moles } H_2O \longrightarrow \text{g } H_2O$$

The 180.18 g in the first conversion factor is the molar mass of glucose, the 6 and 1 in the second conversion factor are the coefficients, respectively, of H_2O and $C_6H_{12}O_6$ in the balanced chemical equation, and the 18.02 g in the third conversion factor is the molar mass of H_2O.

Step 3: The solution to the problem, obtained by doing the arithmetic after all the numerical factors have been collected, is

$$\left(\frac{14.2 \times 1 \times 6 \times 18.02}{180.18 \times 1 \times 1}\right) \text{g } H_2O = 8.52 \text{ g } H_2O$$

Example 5.9 Calculating the Mass of a Substance Taking Part in a Chemical Reaction

The active ingredient in many commercial antacids is magnesium hydroxide, $Mg(OH)_2$, which reacts with stomach acid (HCl) to produce magnesium chloride ($MgCl_2$) and water. The equation for the reaction is

$$Mg(OH)_2 + 2HCl \longrightarrow MgCl_2 + 2H_2O$$

How many grams of $Mg(OH)_2$ are needed to react with 0.30 g of HCl?

Solution

Step 1: This problem, like Example 5.8, is a "grams of A" to "grams of B" problem. It differs from the previous problem in that both the given and the desired quantities involve reactants.

$$0.30 \text{ g HCl} \longrightarrow ? \text{ g } Mg(OH)_2$$

Step 2: The pathway used to solve it will be the same as in Example 5.8.

Grams of A —Molar mass→ Moles of A —Equation coefficients→ Moles of B —Molar mass→ Grams of B

The dimensional-analysis setup is

$$0.30 \text{ g HCl} \times \left(\frac{1 \text{ mole HCl}}{36.46 \text{ g HCl}}\right) \times \left(\frac{1 \text{ mole } Mg(OH)_2}{2 \text{ moles HCl}}\right) \times \left(\frac{58.32 \text{ g } Mg(OH)_2}{1 \text{ mole } Mg(OH)_2}\right)$$

$$\text{g HCl} \longrightarrow \text{moles HCl} \longrightarrow \text{moles } Mg(OH)_2 \longrightarrow \text{g } Mg(OH)_2$$

The balanced chemical equation for the reaction is used as the bridge that enables us to go from HCl to $Mg(OH)_2$. The numbers in the second conversion factor are coefficients from this equation.

Step 3: The solution obtained by combining all of the numbers in the manner indicated in the setup is

$$\left(\frac{0.30 \times 1 \times 1 \times 58.32}{36.46 \times 2 \times 1}\right) \text{g } Mg(OH)_2 = 0.24 \text{ g } Mg(OH)_2$$

To put our answer in perspective, we note that a common brand of antacid tablets has tablets containing 0.10 g of $Mg(OH)_2$.

Practice Problems and Questions

5.28 Ammonia reacts with oxygen to produce nitrogen and water. The balanced chemical equation for the reaction is

$$4NH_3 + 3O_2 \longrightarrow 2N_2 + 6H_2O$$

a. How many grams of NH_3 are needed to produce 2.00 grams of H_2O?
b. How many grams of H_2O can be produced when 10.0 g of O_2 react?
c. How many grams of N_2 are produced at the same time that 20.0 of H_2O is produced?
d. How many moles of NH_3 must react to produce 3.71 moles of N_2?

5.29 The reaction between hydrogen peroxide and hydrogen sulfide produces water and elemental sulfur. The balanced chemical equation for the reaction is

$$H_2O_2 + H_2S \longrightarrow 2H_2O + S$$

a. How many grams of H_2O_2 react at the same time that 40.0 g of H_2S reacts?
b. How many grams of S can be produced when 40.0 g of H_2S reacts?
c. How many moles of H_2S react with 10.0 g of H_2O_2?
d. How many moles of H_2O are produced at the same time that 3.40 moles of S is produced?

CONCEPTS TO REMEMBER

Formula mass. The formula mass of a substance is the sum of the atomic masses of the atoms in its formula.

The mole concept. The mole is the chemist's counting unit. One mole of any substance—element or compound—consists of 6.02×10^{23} formula units of the substance. Avogadro's number is the name given to the numerical value 6.02×10^{23}.

Molar mass. The molar mass of a substance is the mass in grams that is numerically equal to the substance's formula mass. Molar mass is not a set number; it varies and is different for each chemical substance.

The mole and chemical formulas. The numerical subscripts in a chemical formula give the number of moles of atoms of the various elements present in 1 mole of the substance.

Chemical equation. A chemical equation is a written statement that uses symbols and formulas instead of words to represent how reactants undergo transformation into products in a chemical reaction.

Balanced chemical equation. A balanced chemical equation has the same number of atoms of each element involved in the reaction on each side of the equation. An unbalanced equation is brought into balance through the use of coefficients. A coefficient is a number that is placed to the left of the formula of a substance and that changes the amount, but not the identity, of the substance.

The mole and chemical equations. The coefficients in a balanced chemical equation give the molar ratios between substances consumed or produced in the chemical reaction described by the equation.

KEY REACTIONS AND EQUATIONS

1. Calculation of formula mass (Section 5.1)

   Formula mass = sum of atomic masses of all components

2. The mole (Section 5.3)

   $1 \text{ mole} = 6.02 \times 10^{23}$ objects

3. Avogadro's number (Section 5.3)

   Avogadro's number $= 6.02 \times 10^{23}$

4. Mass of a mole (Section 5.4)

   Molar mass = mass, in grams, numerically equal to a substance's formula mass

5. Balanced chemical equation (Section 5.7)

   Balanced chemical equation = same number of atoms of each kind on each side of the equation

KEY TERMS

Avogadro's number (5.2)
Balanced chemical equation (5.6)
Chemical equation (5.6)
Coefficient (5.6)
Formula mass (5.1)
Molar mass (5.3)
Mole (5.2 and 5.3)

ADDITIONAL PROBLEMS

5.30 You are given a sample containing 0.500 mole of a substance.
a. How many atoms are present if the substance is copper metal?
b. How many atoms are present if the substance is iron metal?
c. How many atoms are present if the sample is nitric acid, HNO_3?
d. How many atoms are present if the sample is aspirin, $C_9H_8O_4$?

5.31 The compound 1-propanethiol, which is the eye irritant that is released when fresh onions are chopped up, has a formula mass of 76.18 amu and the chemical formula C_3H_yS. What number does y stand for in this chemical formula?

5.32 Select the quantity that has the greater number of atoms in each of the following pairs of quantities. Make your selection using the periodic table but without performing an actual calculation.
a. 1.00 mole S or 1.00 mole S_8
b. 28.0 g Al or 1.00 mole Al
c. 28.1 g Si or 30.0 g Mg
d. 2.00 g Na or 6.02×10^{23} He atoms

5.33 Select the quantity that has the greater mass in each of the following pairs of quantities. Make your selection using the periodic table but without performing an actual calculation.
a. 1.00 mole Ag or 1.00 mole Au
b. 1.00 mole S or 6.02×10^{23} C atoms
c. 1.00 mole Cl atoms or 1.00 mole Cl_2 molecules
d. 5.00 g He or 6.02×10^{23} Ne atoms

5.34 How many grams of Si would contain the same number of atoms as there are in 2.10 moles of Ar?

5.35 After the following chemical equation was balanced, the name of one of the reactants was substituted for its formula.

$$2 \text{ butyne} + 11O_2 \longrightarrow 8CO_2 + 6H_2O$$

Using only the information found in the chemical equation, determine the chemical formula of butyne.

5.36 What is the Si–N mass ratio needed to obtain an equal number of both types of atoms?

PRACTICE TEST ▶ True/False

5.37 The formula mass for a compound is obtained by adding up the atomic masses of all the atoms present in a formula unit of the compound and dividing by the number of atoms present.

5.38 Because SO_2 and H_2O both contain 3 atoms per molecule, their formula masses will be the same.

5.39 The number of molecules in 1 mole of SO_2 is the same as the number of molecules in 1 mole of SO_3.

5.40 The number of atoms in 1 mole of SO_2 is the same as the number of atoms in 1 mole of SO_3.

5.41 Avogadro's number and the number 6.02×10^{23} are one and the same.

5.42 If the formula mass of a substance is 63.0 amu, then its molar mass will be 63.0 g.

5.43 The numerical subscripts in a chemical formula of a compound give the number of moles of atoms of the various elements present in 1 mole of the substance.

5.44 In 1 mole of the compound H_2SO_4 there are 7 moles of atoms present.

5.45 In a balanced chemical equation, the sums of the coefficients on the two sides of the chemical equation must be equal.

5.46 In a chemical equation, the chemical formulas of the reactants are always written on the left side.

5.47 A coefficient placed to the left of the chemical formula of a substance in a chemical equation changes the amount, but not the identity, of the substance.

5.48 A balanced chemical equation has the same number of atoms of each element involved in the chemical reaction on both sides of the equation.

5.49 The coefficients in a balanced chemical equation give the fixed molar ratios between the substances that are consumed or produced in the chemical reaction described by the chemical equation.

5.50 1 mole of N_2H_4 molecules contains 2 moles of N atoms and 4 moles of H atoms.

5.51 All substances have the same molar mass, which is 6.02×10^{23} atoms.

PRACTICE TEST ▶ Multiple Choice

You will need information found on a periodic table to answer some of the following questions.

5.52 Which of the following statements concerning Avogadro's number is *correct*?
a. It has the value 6.02×10^{-23}.
b. It denotes the number of molecules in 1 mole of any molecular substance.
c. It is the mass, in grams, of 1 mole of any substance.
d. It denotes the number of atoms in 1 mole of any substance.

5.53 Which set of quantities is needed to calculate the mass of a mole of a substance?
a. Chemical formula, Avogadro's number
b. Chemical formula, atomic masses
c. Atomic masses, Avogadro's number
d. Atomic numbers, Avogadro's number

5.54 Which of the following compounds has the largest formula mass?
a. H_2O b. NH_3 c. CO d. BeH_2

5.55 In which of the following molar quantities of sulfur would 8.13×10^{23} atoms of sulfur be present?
a. 1.15 moles S
b. 1.25 moles S
c. 1.35 moles S
d. 1.45 moles S

5.56 Which of the following samples has the largest mass, in grams?

a. 2 moles CO_2
b. 3 moles CO
c. 4 moles H_2O
d. 5 moles H_2

5.57 Which of the following samples contains the greatest number of atoms?

a. 1 mole CO_2
b. 2 moles He
c. 3 moles N_2O
d. 4 moles CO

5.58 Which of the following equations is balanced?

a. $2H_2 + O_2 \longrightarrow H_2O$
b. $2SO_2 + 2O_2 \longrightarrow 3SO_3$
c. $KClO_3 \longrightarrow KCl + 3O_2$
d. $N_2 + 3H_2 \longrightarrow 2NH_3$

5.59 When the equation $C_3H_8 + O_2 \longrightarrow CO_2 + H_2O$ is balanced, the coefficients are

a. 1, 5, 3, 4 b. 1, 3, 4, 5 c. 1, 5, 4, 3 d. 2, 5, 4, 3

5.60 Which of the following is the correct "setup" for the problem "How many grams of S are present in 50.0 g of S_4N_4?"

a. $50.0 \text{ g } S_4N_4 \times \frac{1 \text{ mole } S_4N_4}{184.32 \text{ g } S_4N_4} \times \frac{4 \text{ moles S}}{1 \text{ mole } S_4N_4} \times \frac{32.07 \text{ g S}}{4 \text{ moles S}}$

b. $50.0 \text{ g } S_4N_4 \times \frac{1 \text{ mole } S_4N_4}{184.32 \text{ g } S_4N_4} \times \frac{4 \text{ moles S}}{1 \text{ mole } S_4N_4} \times \frac{32.07 \text{ g S}}{1 \text{ mole S}}$

c. $50.0 \text{ g } S_4N_4 \times \frac{1 \text{ mole } S_4N_4}{184.32 \text{ g } S_4N_4} \times \frac{1 \text{ mole S}}{1 \text{ mole } S_4N_4} \times \frac{32.07 \text{ g S}}{1 \text{ mole S}}$

d. $50.0 \text{ g } S_4N_4 \times \frac{1 \text{ mole } S_4N_4}{184.32 \text{ g } S_4N_4} \times \frac{1 \text{ mole S}}{4 \text{ moles } S_4N_4} \times \frac{32.07 \text{ g S}}{1 \text{ mole S}}$

5.61 Which of the following is the correct "setup" for the problem "How many grams of H_2O form when 3.2 moles of O_2 reacts according to the following reaction?

$$2H_2S + 3O_2 \longrightarrow 2H_2O + 2SO_2$$

a. $3.2 \text{ moles } O_2 \times \frac{18.02 \text{ g } H_2O}{2 \text{ moles } H_2O}$

b. $3.2 \text{ moles } O_2 \times \frac{32.00 \text{ g } O_2}{1 \text{ mole } O_2} \times \frac{18.02 \text{ g } H_2O}{32.00 \text{ g } O_2}$

c. $3.2 \text{ moles } O_2 \times \frac{2 \text{ moles } H_2O}{3 \text{ moles } O_2} \times \frac{18.02 \text{ g } H_2O}{1 \text{ mole } H_2O}$

d. $3.2 \text{ moles } O_2 \times \frac{32.00 \text{ g } O_2}{1 \text{ mole } O_2} \times \frac{2 \text{ moles } H_2O}{3 \text{ moles } O_2}$

CHAPTER OUTLINE

CHAPTER SIX

Gases, Liquids, and Solids

A high-rise building being demolished through the use of explosives. Changes in volume that occur as the explosive (a solid) is converted into gases via a chemical reaction are the basis for the destructive effects of explosives.

In Chapters 3 and 4, we considered the structure of matter from a submicroscopic point of view—in terms of molecules, atoms, protons, neutrons, and electrons. In this chapter, we are concerned with the macroscopic characteristics of matter as represented by the physical states—solid, liquid, and gas. Of particular concern are the properties exhibited by matter in the various physical states and a theory that correlates these properties with molecular behavior.

Learning Focus

List the five statements of kinetic molecular theory and understand the roles that disruptive forces (kinetic energy) and cohesive forces (potential energy) play in determining the physical state of a system.

6.1 The Kinetic Molecular Theory of Matter

Solids, liquids, and gases (Section 1.2) are easily distinguished by using four common physical properties of matter: (1) volume and shape, (2) density, (3) compressibility, and (4) thermal expansion. We discussed the property of density in Section 2.8. **Compressibility** *is a measure of the change in volume resulting from a pressure change.* **Thermal expansion** *is a measure of the volume change resulting from a temperature change.* These distinguishing properties are compared in Table 6.1 for the three states of matter.

The physical characteristics of the solid, liquid, and gaseous states (Table 6.1) can be explained using *kinetic molecular theory,* which is one of the fundamental theories of chemistry.

Table 6.1 Distinguishing Properties of Solids, Liquids, and Gases

Property	Solid state	Liquid state	Gaseous state
volume and shape	definite volume and definite shape	definite volume and indefinite shape; takes the shape of container to the extent that it is filled	indefinite volume and indefinite shape; takes the volume and shape of container it fills
density	high	high, but usually lower than corresponding solid	low
compressibility	small	small, but usually greater than corresponding solid	large
thermal expansion	very small: about 0.01% per °C	small: about 0.10% per °C	moderate: about 0.30% per °C

▶ The word *kinetic* comes from the Greek *kinesis,* which means "movement." The kinetic molecular theory deals with the movement of particles.

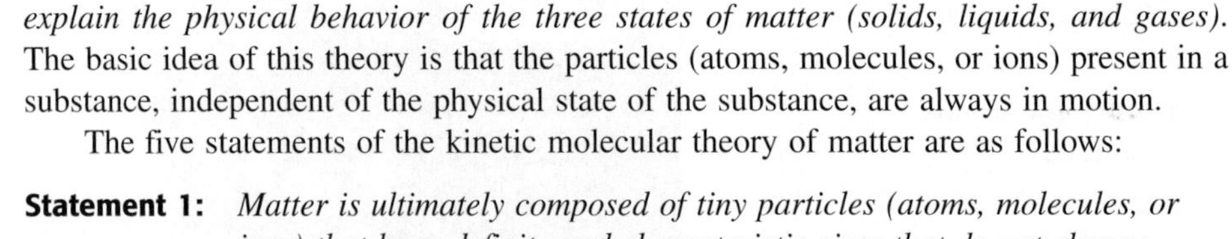

The **kinetic molecular theory of matter** *is a set of five statements that are used to explain the physical behavior of the three states of matter (solids, liquids, and gases).* The basic idea of this theory is that the particles (atoms, molecules, or ions) present in a substance, independent of the physical state of the substance, are always in motion.

The five statements of the kinetic molecular theory of matter are as follows:

Statement 1: *Matter is ultimately composed of tiny particles (atoms, molecules, or ions) that have definite and characteristic sizes that do not change.*

Statement 2: *The particles are in constant random motion and therefore possess kinetic energy.*

Kinetic energy *is energy that matter possesses because of its motion.* An object that is in motion has the ability to transfer its kinetic energy to another object upon collision with that object.

Figure 6.1 The water in the lake behind the dam has potential energy as a result of its position. When the water flows over the dam, its potential energy becomes kinetic energy that can be used to turn the turbines of a hydroelectric plant.

Statement 3: *The particles interact with one another through attractions and repulsions and therefore possess potential energy.*

Potential energy *is energy associated with forces of attraction or repulsion between objects.* It is *stored* energy. Water behind a dam possesses potential energy because of gravitational attraction between the water and Earth (Figure 6.1). The energy released when gasoline burns represents potential energy associated with chemical bonds. The potential energy that is important when we consider the physical characteristics of the three states of matter is that which originates from electrostatic interactions between the particles. **Electrostatic interactions** *are attractions and repulsions that occur between charged particles.* Particles of opposite charge (one positive and the other negative) attract each other, and particles of like charge (both positive or both negative) repel each other.

▶ For gases, the attractions between particles (statement 3) are minimal and as a first approximation are considered to be zero (see Section 6.2).

Statement 4: *The velocity of the particles increases as the temperature is increased.*

The *average* velocity (kinetic energy) of all particles in a system depends on the temperature; velocity increases as temperature increases.

Statement 5: *The particles in a system transfer energy to each other through elastic collisions.*

In an elastic collision, the total kinetic energy remains constant; no kinetic energy is lost. The difference between an *elastic* and an *inelastic* collision is illustrated by comparing the collision of two hard steel spheres with the collision of two masses of putty. The collision of spheres approximates an elastic collision (the spheres bounce off one another and continue moving, as in Figure 6.2); the putty collision has none of these characteristics (the masses "glob" together with no resulting movement).

▶ Two consequences of the elasticity of particle collisions (statement 5) are that (1) the energy of any given particle is continually changing, and (2) particle energies for a system are not all the same; a range of particle energies is always encountered.

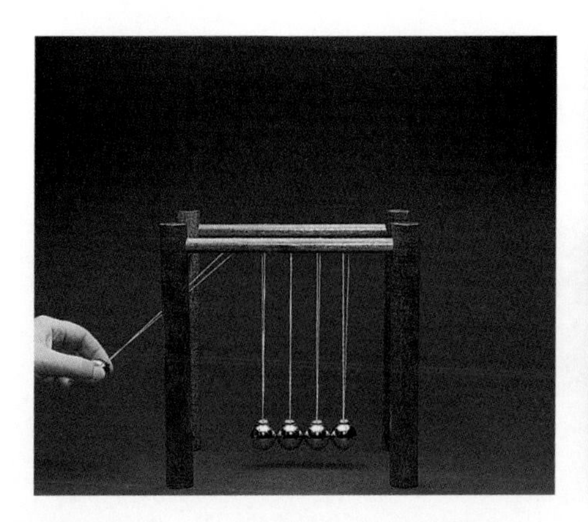
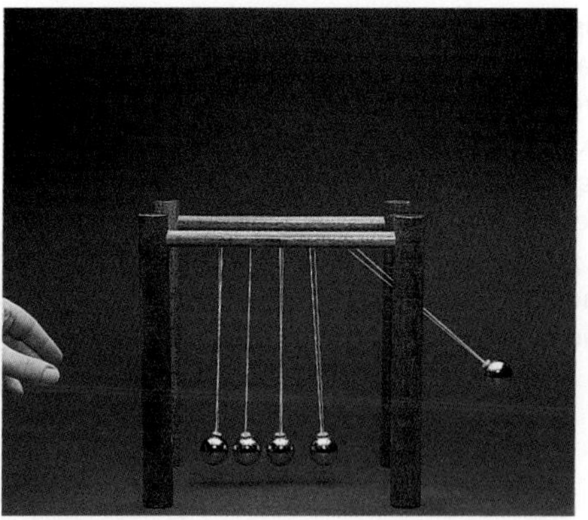

Figure 6.2 Upon release, the steel ball on the left transmits its kinetic energy through a series of elastic collisions to the ball on the right.

The differences among the solid, liquid, and gaseous states of matter can be explained by the relative magnitudes of kinetic energy and potential energy associated with the physical state. Kinetic energy can be considered a *disruptive force* that tends to make the particles of a system increasingly independent of one another. This is because the particles tend to move away from one another as a result of the energy of motion. Potential energy of attraction can be considered a *cohesive force* that tends to cause order and stability among the particles of a system.

How much kinetic energy a chemical system has depends on its temperature. Kinetic energy increases as temperature increases (statement 4 of the kinetic molecular theory of matter). Thus the higher the temperature, the greater the magnitude of disruptive influences within a chemical system. Potential energy magnitude, or cohesive force magnitude, is essentially independent of temperature. The fact that one of the types of forces depends on temperature (disruptive forces) and the other does not (cohesive forces) causes temperature to be the factor that determines in which of the three physical states a given sample is found. We will discuss the reason for this in Section 6.2.

Practice Exercise and Problems

6.1 Indicate what type of energy is related to
a. disruptive forces b. cohesive forces

6.2 Indicate what effect temperature has on the magnitude of
a. disruptive forces b. cohesive forces

6.3 Indicate what effects on a system of particles are created by
a. disruptive forces b. cohesive forces

6.4 What is the relationship between temperature and the average velocity with which particles move?

6.5 How do particles transfer energy from one to another?

Learning Focus

Use kinetic molecular theory to explain the physical characteristics of the solid, liquid, and gaseous states.

6.2 Kinetic Molecular Theory and Physical States

The **solid state** *is the physical state characterized by a dominance of potential energy (cohesive forces) over kinetic energy (disruptive forces).* The particles in a solid are drawn close together in a regular pattern by the strong cohesive forces present (Figure 6.3a). Each particle occupies a fixed position, about which it vibrates because of disruptive kinetic energy.

The **liquid state** *is the physical state characterized by potential energy (cohesive forces) and kinetic energy (disruptive forces) of about the same magnitude.* The liquid state consists of particles that are randomly packed but relatively near one another (Figure 6.3b). The molecules are in constant, random motion; they slide freely over one another but do not move with enough energy to separate. The fact that the particles freely slide over each other indicates the influence of disruptive forces; however, the fact that the particles do not separate indicates fairly strong cohesive forces.

Figure 6.3 (a) In a solid, the particles (atoms, molecules, or ions) are close together and vibrate about fixed sites. (b) The particles in a liquid, though still close together, freely slide over one another. (c) In a gas, the particles are in constant random motion, each particle being independent of the others present.

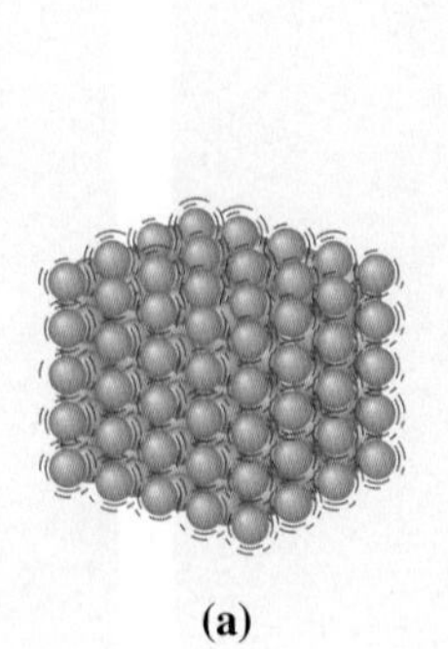

(a)

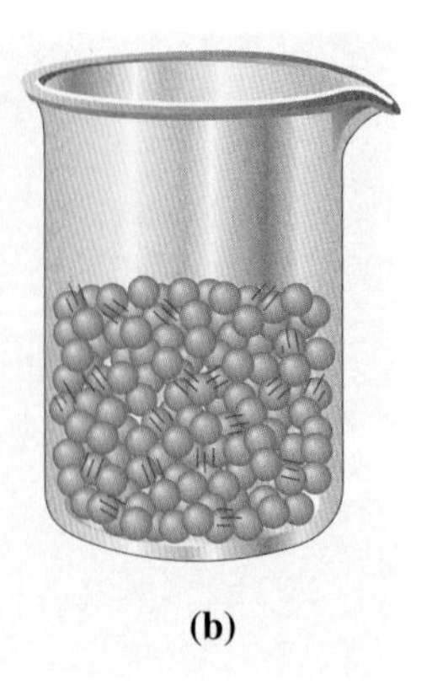

(b)

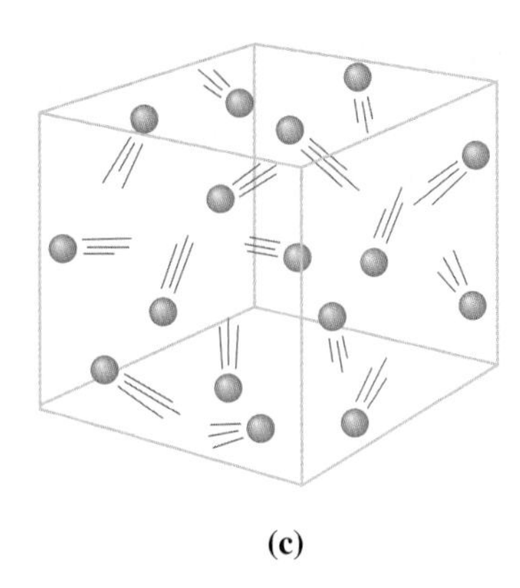

(c)

The **gaseous state** *is the physical state characterized by a complete dominance of kinetic energy (disruptive forces) over potential energy (cohesive forces).* As a result, the particles of a gas move essentially independently of one another, in a totally random manner (Figure 6.3c). Under ordinary pressure, the particles are relatively far apart, except when they collide with one another. In between collisions with one another or the container walls, gas particles travel in straight lines. Particles in a gas are widely separated; essentially, a gas is mostly empty space. When pressure is applied, the particles are easily pushed closer together, decreasing the amount of empty space and the volume of the gas (see Figure 6.4).

Chemical Portraits 10 "Odoriferous" Gaseous Compounds

Ammonia (NH_3)

Profile: Ammonia is a colorless, flammable gas with a *very sharp* and penetrating odor; it is an effective "smelling salt". It is hard to be over-exposed to NH_3; its odor and irritative effects drive a person away from it. The solubility of ammonia in water is greater than that of any other gas; 700 liters of gas dissolve in 1 liter of water at 25°C and 1 atm pressure.

Uses: Direct use as a fertilizer and conversion into other nitrogeneous fertilizers consumes approximately 80% of all NH_3 produced. The high yields of modern agriculture are a direct result of the use of NH_3-based fertilizers. Without such fertilizers, the world's food needs could not be met. Commercial NH_3 production involves combining the elemental gases H_2 and N_2 under high temperature and high pressure conditions, in the presence of a catalyst.

What are the "raw materials" for the industrial production of NH_3?

Hydrogen Sulfide (H_2S)

Profile: Hydrogen sulfide, commonly called "rotten egg gas," is a colorless, very flammable gas, with a strong offensive odor. Small amounts of H_2S are present in unpolluted air. Natural sources include volcanic eruptions and the bacterial decay of organisms. H_2S gas is partially responsible for the unpleasant odor of swamps. Within the atmosphere, chemical reactions convert H_2S to sulfur dioxide (SO_2).

Biochemical considerations: H_2S is more toxic to humans than is hydrogen cyanide (HCN), but its odor warns of its presence; our noses detect it in extremely low, nontoxic concentrations. Not heeding the odor warning is, however, serious; H_2S has an anesthetic effect and your nose rapidly loses its ability to detect it.

What happens to the H_2S naturally present in petroleum and natural gas deposits?

Sulfur Dioxide (SO_2)

Profile: Sulfur dioxide, a colorless, nonflammable gas, is a major air pollutant. It is formed anytime sulfur-containing fuels, such as coal, are burned. Within the atmosphere, SO_2 serves as a precursor for "acid rain."

Biochemical considerations: SO_2 is odorless at typical atmospheric concentrations but has a pungent, choking odor at higher concentrations. Effects of SO_2 on human health include irritation of the respiratory system and irritation of the eyes. Individuals with chronic respiratory diseases are particularly affected. SO_2 is especially toxic to molds and certain bacteria and it is used as a food additive in items such as dried fruits and wines.

What is the mechanism by which SO_2 contributes to the "acid rain" problem?

See the text web site at **www.cengage.com/chemistry/stoker** for answers to the above questions and for further information.

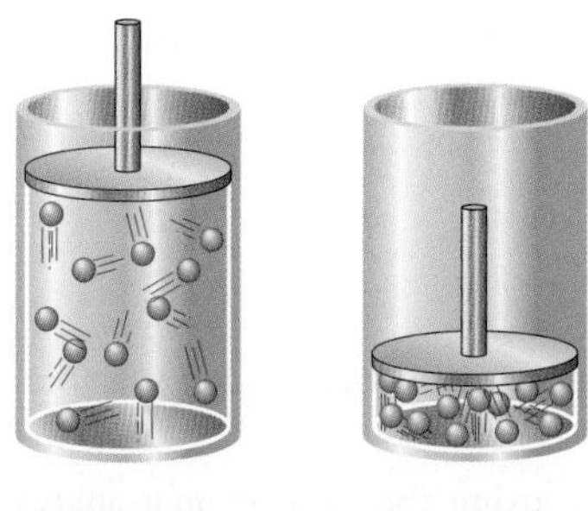

Figure 6.4 When a gas is compressed, the amount of empty space in the container is decreased. The size of the molecules does not change; they simply move closer together.

The freedom of movement that molecules and atoms have in the gaseous state is a prerequisite for the property of "odor" associated with some but not all substances. Whether a substance has an odor depends on whether it can excite the olfactory nerve endings in the nose. Odoriferous substances must be either gases or easily vaporized liquids or solids at room temperature; otherwise, the molecules of such substances would never reach the nose. In addition, "odoriferous" substances must be at least slightly soluble in the mucus that covers nerve endings, and they must be of the correct shape to activate the receptor sites present on the nerve endings. Chemical Portraits 10 profiles three gaseous molecules that meet the prerequisites to be a "smelly" substance.

Practice Problems and Questions

6.6 Indicate in which of the three physical states of matter
a. disruptive forces and cohesive forces are of about the same magnitude
b. disruptive forces are significantly greater than cohesive forces
c. cohesive forces are significantly greater than disruptive forces

6.7 Explain each of the following observations.
a. A container can be half full of a liquid but not half full of a gas.
b. Both liquids and solids are practically incompressible.

Learning Focus

List the units commonly used for each of the gas law variables: amount, pressure, temperature, and volume.

6.3 Gas Law Variables

The behavior of a gas can be described reasonably well by *simple* quantitative relationships called *gas laws*. **Gas laws** *are generalizations that describe in mathematical terms the relationships among the amount, pressure, temperature, and volume of a gas.*

Gas laws involve four variables: amount, pressure, temperature, and volume. Three of these four variables—amount, volume, and temperature—have been discussed previously (Sections 5.2, 2.2 and 2.9, respectively). Amount is usually specified in terms of *moles* of gas present. The unit *liter* or *milliliter* is generally used in specifying gas volume. Only one of the three temperature scales discussed in Section 2.9, the *Kelvin scale,* can be used in gas law calculations if the results are to be valid. We have not yet discussed pressure, the fourth variable. Comments concerning pressure will occupy the remainder of this section.

Pressure *is the force applied per unit area—that is, the total force on a surface divided by the area of that surface.* The mathematical equation for pressure is

$$P \text{ (pressure)} = \frac{F \text{ (force)}}{A \text{ (area)}}$$

For a gas, the force that creates pressure is that which is exerted by the gas molecules or atoms as they constantly collide with the walls of their container. Barometers, manometers, and gauges are the instruments most commonly used to measure gas pressures.

The air that surrounds Earth exerts pressure on every object it touches. A **barometer** *is a device used to measure atmospheric pressure*. The essential components of a simple barometer are shown in Figure 6.5. Atmospheric pressure is expressed in terms of the height of the barometer's mercury column, usually in millimeters of mercury (mm Hg). Another name for millimeters of mercury is *torr,* used in honor of Evangelista Torricelli, the Italian physicist who invented the barometer.

$$1 \text{ mm Hg} = 1 \text{ torr}$$

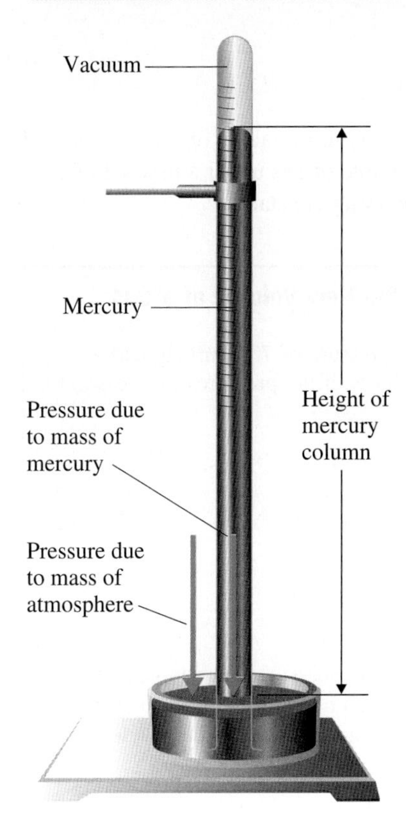

Figure 6.5 The essential components of a mercury barometer are a graduated glass tube, a glass dish, and liquid mercury.

Atmospheric pressure varies with the weather and the altitude. It averages about 760 mm Hg at sea level, and it decreases by approximately 25 mm Hg for every 1000-ft increase in altitude. The pressure unit *atmosphere* (atm) is defined in terms of this average pressure at sea level. By definition,

$$1 \text{ atm} = 760 \text{ mm Hg} = 760 \text{ torr}$$

▶ Millimeters of mercury is the pressure unit most often encountered in clinical work in allied health fields. For example, oxygen and carbon dioxide pressures in respiration are almost always specified in millimeters of mercury.

▶ Blood pressure is measured with the aid of an apparatus known as a sphygmomanometer, which is essentially a barometer tube connected to an inflatable cuff by a hollow tube. A typical blood pressure is 120/80; this ratio means a systolic pressure of 120 mm Hg above atmospheric pressure and a diastolic pressure of 80 mm Hg above atmospheric pressure.

Another commonly used pressure unit is *pounds per square inch* (psi or lb/in.^{2}). One atmosphere is equal to 14.7 psi.

$$1 \text{ atm} = 14.7 \text{ psi}$$

Practice Problems and Questions

6.8 What are the four gas law variables and the common units in which each variable is expressed?

6.9 Carry out the following pressure unit conversions, using the dimensional-analysis method of problem solving.
a. 735 mm Hg to atmospheres
b. 0.530 atm to millimeters of mercury
c. 535 mm Hg to torr
d. 12.0 psi to atmospheres

Learning Focus

State Boyle's law in words and as a mathematical equation; know how to use the law in problem solving.

6.4 Boyle's Law: A Pressure–Volume Relationship

Of the several relationships that exist among gas law variables, the first to be discovered relates gas pressure to gas volume. It was formulated over 300 years ago, in 1662, by the British chemist and physicist Robert Boyle (Figure 6.6). **Boyle's law** states that *the volume of a fixed amount of a gas is* inversely proportional *to the pressure applied to the gas if the temperature is kept constant.* This means that if the pressure on the gas increases, the volume decreases proportionally; conversely, if the pressure decreases, the volume increases. Doubling the pressure cuts the volume in half; tripling the pressure cuts the volume to one-third its original value; quadrupling the pressure cuts the volume to one-fourth; and so on. Figure 6.7 illustrates Boyle's law.

The mathematical equation for Boyle's law is

$$P_1 \times V_1 = P_2 \times V_2$$

where P_1 and V_1 are the pressure and volume of a gas at an initial set of conditions, and P_2 and V_2 are the pressure and volume of the same sample of gas under a new set of conditions, with the temperature and amount of gas remaining constant.

Figure 6.6 Robert Boyle (1627–1691), like most men of the seventeenth century who devoted themselves to science, was self-taught. It was through his efforts that the true value of experimental investigation was first recognized.

Example 6.1 Using Boyle's Law to Calculate the New Volume of a Gas

A sample of O_2 gas occupies a volume of 1.50 L at a pressure of 735 mm Hg and a temperature of 25°C. What volume will it occupy, in liters, if the pressure is increased to 770 mm Hg with no change in temperature?

Solution

A suggested first step in working gas law problems that involve two sets of conditions is to analyze the given data in terms of initial and final conditions.

$P_1 = 735$ mm Hg $P_2 = 770$ mm Hg
$V_1 = 1.50$ L $V_2 = ?$ L

We know three of the four variables in the Boyle's law equation, so we can calculate the fourth, V_2. We will rearrange Boyle's law to isolate V_2 (the quantity to be calculated) on one side of the equation. This is accomplished by dividing both sides of the Boyle's law equation by P_2.

$$P_1V_1 = P_2V_2 \quad \text{(Boyle's law)}$$

$$\frac{P_1V_1}{P_2} = \frac{\cancel{P_2}V_2}{\cancel{P_2}} \quad \text{(Divide each side of the equation by } P_2\text{.)}$$

$$V_2 = V_1 \times \frac{P_1}{P_2}$$

▶ When we know any three of the four quantities in the Boyle's law equation, we can calculate the fourth, which is usually the final pressure, P_2, or the final volume, V_2. The Boyle's law equation is valid only if the temperature and amount of the gas remains constant.

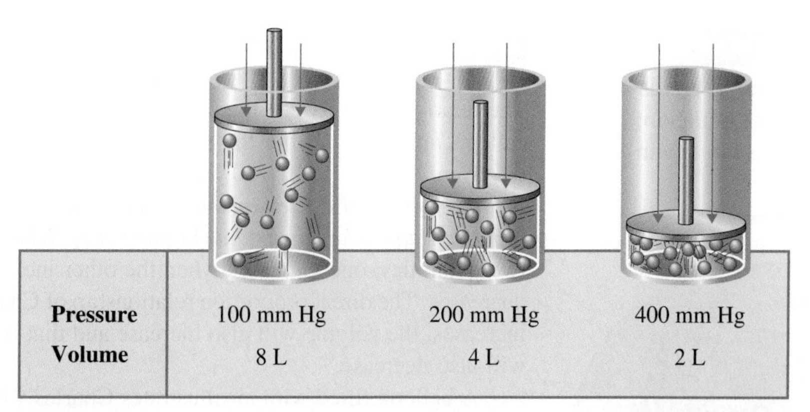

Figure 6.7 Data illustrating the inverse proportionality associated with Boyle's law.

Substituting the given data into the rearranged equation and doing the arithmetic give

$$V_2 = 1.50 \text{ L} \times \left(\frac{735 \cancel{\text{mm Hg}}}{770 \cancel{\text{mm Hg}}}\right) = 1.43 \text{ L}$$

▶ Boyle's law explains the process of breathing. Breathing in occurs when the diaphragm flattens out (contracts). This contraction causes the volume of the thoracic cavity to increase and the pressure within the cavity to drop (Boyle's law) below atmospheric pressure. Air flows into the lungs and expands them, because the pressure is greater outside the lungs than within them. Breathing out occurs when the diaphragm relaxes (moves up), decreasing the volume of the thoracic cavity and increasing the pressure (Boyle's law) within the cavity to a value greater than the external pressure. Air flows out of the lungs. The air flow direction is always from a high-pressure region to a low-pressure region.

Filling a medical syringe with a liquid demonstrates Boyle's law. As the plunger is drawn out of the syringe (see Figure 6.8), the increase in volume inside the syringe chamber results in decreased pressure there. The liquid, which is at atmospheric pressure, flows into this reduced-pressure area. This liquid is then expelled from the chamber by pushing the plunger back in. This ejection of the liquid does not involve Boyle's law; a liquid is incompressible and mechanical force pushes it out.

Figure 6.8 Filling a syringe with a liquid is an application of Boyle's law.

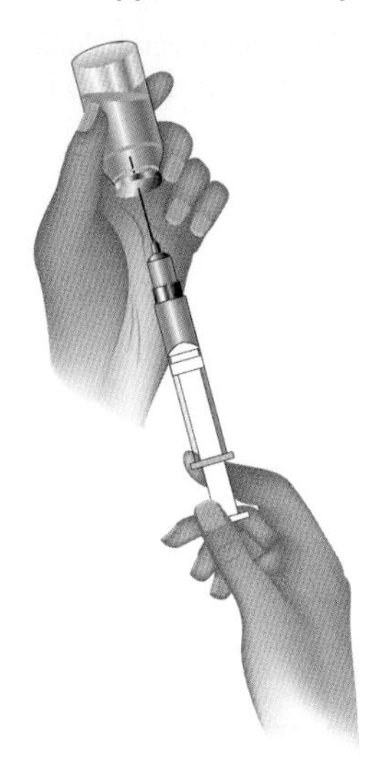

Practice Problems and Questions

6.10 According to Boyle's law, indicate what will happen to the volume of a fixed amount of gas when
a. the pressure on it is increased at constant temperature
b. the pressure on it is decreased at constant temperature

6.11 What is the meaning of the phrase "two variables are inversely proportional"?

6.12 On the basis of Boyle's law, indicate what will happen to the volume of a gas when the pressure on the gas is reduced to
a. one-half its original value
b. one-third its original value
c. one-fourth its original value
d. two-thirds its original value

6.13 Use Boyle's law to calculate the value of the unknown pressure or volume.
a. $V_1 = 2.0$ L, $P_1 = 2.0$ atm, $V_2 = 4.0$ L, $P_2 = ?$
b. $V_1 = 2.0$ L, $P_1 = 2.0$ atm, $V_2 = ?$, $P_2 = 7.0$ atm
c. $V_1 = 2.0$ L, $P_1 = ?$, $V_2 = 8.0$ L, $P_2 = 5.0$ atm
d. $V_1 = ?$, $P_1 = 2.0$ atm, $V_2 = 4.0$ L, $P_2 = 1.0$ atm

6.14 At constant temperature, a sample of 6.0 L of O_2 gas at 3.0 atm pressure is compressed until the volume decreases to 2.5 L. What is the new pressure, in atmospheres.

6.15 A sample of NH_3 gas occupies a volume of 3.00 L at a pressure of 655 mm Hg and a temperature of 25°C. What volume, in liters, will this gas sample occupy at the same temperature if the pressure is increased to 725 mm Hg?

Learning Focus

State Charles's law in words and as a mathematical equation; know how to use the law in problem solving.

6.5 Charles's Law: A Temperature–Volume Relationship

The relationship between the temperature and the volume of a gas at constant pressure is called *Charles's law* after the French scientist Jacques Charles (Figure 6.9). This law was discovered in 1787, over 100 years after the discovery of Boyle's law. **Charles's law** states that *the volume of a fixed amount of gas is* directly proportional *to its Kelvin temperature if the pressure is kept constant* (Figure 6.10). Whenever a *direct* proportion exists between two quantities, one increases when the other increases and one decreases when the other decreases. The direct proportion relationship of Charles's law means that if the temperature increases, the volume will also increase and that if the temperature decreases, the volume will also decrease.

A balloon filled with air illustrates Charles's law. If the balloon is placed near a heat source such as a light bulb that has been on for some time, the heat will cause the balloon to increase visibly in size (volume). Putting the same balloon in the refrigerator will cause it to shrink.

Charles's law, stated mathematically, is

$$\frac{V_1}{T_1} = \frac{V_2}{T_2}$$

where V_1 is the volume of a gas at a given pressure, T_1 is the Kelvin temperature of the gas, and V_2 and T_2 are the volume and Kelvin temperature of the gas under a new set of conditions, with the pressure remaining constant.

Figure 6.9 Jacques Charles (1746–1823), a French physicist, in the process of working with hot air balloons, made the observations that ultimately led to the formulation of what is now known as Charles's law.

▶ When you use the mathematical form of Charles's law, the temperatures used *must be* Kelvin scale temperatures.

▶ Charles's law predicts that gas volume will become smaller and smaller as temperature is reduced, until eventually a temperature is reached at which gas volume becomes zero. This "zero-volume" temperature is calculated to be −273°C and is known as *absolute zero* (see Section 2.8). Absolute zero is the basis for the Kelvin temperature scale. In reality, gas volume never vanishes. As temperature is lowered, at some point before absolute zero, the gas condenses to a liquid, at which point Charles's law is no longer valid.

Example 6.2 Using Charles's Law to Calculate the New Volume of a Gas

A sample of the gaseous anesthetic cyclopropane, with a volume of 425 mL at a temperature of 27°C, is cooled at constant pressure to 20°C. What is the new volume, in milliliters, of the sample?

Solution

First, we will analyze the data in terms of initial and final conditions.

$V_1 = 425\text{ mL}$ $\quad V_2 = ?\text{ mL}$

$T_1 = 27°\text{C} + 273 = 300\text{ K}$ $\quad T_2 = 20°\text{C} + 273 = 293\text{ K}$

Note that both of the given temperatures have been converted to Kelvin scale readings. This change is accomplished by simply adding 273 to the Celsius scale value (Section 2.9).

We know three of the four variables in the Charles's law equation, so we can calculate the fourth, V_2. We will rearrange Charles's law to isolate V_2 (the quantity desired) by multiplying each side of the equation by T_2.

$$\frac{V_1}{T_1} = \frac{V_2}{T_2} \quad \text{(Charles's law)}$$

$$\frac{V_1 T_2}{T_1} = \frac{V_2 \cancel{T_2}}{\cancel{T_2}} \quad \text{(Multiply each side by } T_2.\text{)}$$

$$V_2 = V_1 \times \frac{T_2}{T_1}$$

Substituting the given data into the equation and doing the arithmetic give

$$V_2 = 425\text{ mL} \times \left(\frac{293\text{ }\cancel{K}}{300\text{ }\cancel{K}}\right) = 415\text{ mL}$$

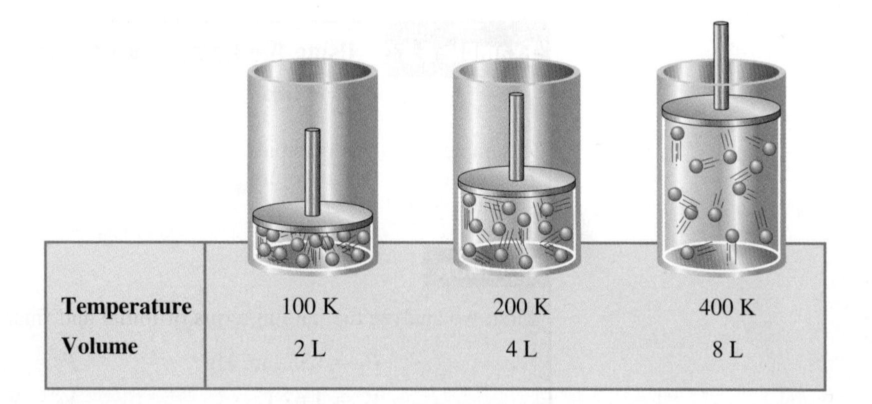

Figure 6.10 Data illustrating the direct proportionality associated with Charles's law.

Charles's law is the principle used in the operation of a convection heater. When air comes in contact with the heating element, it expands (its density becomes less). The hot, less dense air rises, causing continuous circulation of warm air. This same principle has ramifications in closed rooms that lack effective air circulation. The warmer and less dense air stays near the top of the room. This is desirable in the summer but not in the winter.

Practice Problems and Questions

6.16 According to Charles's law, indicate what will happen to the volume of a fixed amount of gas when
a. its temperature is increased at constant pressure
b. its temperature is decreased at constant pressure

6.17 What is the meaning of the phrase "two variables are directly proportional"?

6.18 On the basis of Charles's law, indicate what will happen to the volume of a gas when the Kelvin temperature of the gas is reduced to
a. one-half its original value
b. one-third its original value
c. one-fourth its original value
d. two-thirds its original value

6.19 Use Charles's law to calculate the value of the unknown temperature or volume.
a. $V_1 = 2.00$ L, $T_1 = 327$°C, $V_2 = 4.00$ L, $T_2 = ?$ °C
b. $V_1 = 2.00$ L, $T_1 = 327$°C, $V_2 = ?$ L, $T_2 = 27$°C
c. $V_1 = 2.00$ L, $T_1 = ?$ °C, $V_2 = 8.00$ L, $T_2 = 27$°C
d. $V_1 = ?$ L, $T_1 = 127$°C, $V_2 = 4.00$ L, $T_2 = -73$°C

6.20 At atmospheric pressure, a sample of H_2 gas has a volume of 2.73 L at 27°C. What volume, in liters, will the H_2 gas occupy if the temperature is increased to 127°C and the pressure is held constant?

6.21 A sample of N_2 gas occupies a volume of 375 mL at 25°C and a pressure of 2.0 atm. Determine the temperature, in degrees Celsius, at which the volume of the gas would be 525 mL at the same pressure.

6.6 The Combined Gas Law

Learning Focus

Give the mathematical form of the combined gas law; know how to use the law in problem solving.

The **combined gas law** *is an expression obtained by mathematically combining Boyle's and Charles's laws.* The mathematical equation for the combined gas law is

$$\frac{P_1V_1}{T_1} = \frac{P_2V_2}{T_2}$$

This combined gas law is a much more versatile equation than either of the laws from which it is derived. Using this equation, we can calculate the change in pressure, temperature, or volume that is brought about by changes in the other two variables.

▶ Any time a gas law contains temperature terms, as is the case for both Charles's law and the combined gas law, these temperatures must be specified on the Kelvin temperature scale.

Example 6.3 **Using the Combined Gas Law to Calculate the New Volume of a Gas**

A sample of O_2 gas occupies a volume of 1.62 L at 755 mm Hg pressure and has a temperature of 0°C. What volume, in liters, will this gas sample occupy at 725 mm Hg pressure and 50°C?

Solution

First, we analyze the data in terms of initial and final conditions.

$$P_1 = 755 \text{ mm Hg} \qquad P_2 = 725 \text{ mm Hg}$$
$$V_1 = 1.62 \text{ L} \qquad V_2 = ? \text{ L}$$
$$T_1 = 0°\text{C} + 273 = 273 \text{ K} \qquad T_2 = 50°\text{C} + 273 = 323 \text{ K}$$

We are given five of the six variables in the combined gas law, so we can calculate the sixth one, V_2. Rearranging the combined gas law to isolate the variable V_2 on a side by itself gives

$$V_2 = \frac{V_1 P_1 T_2}{P_2 T_1}$$

Substituting numerical values into this "version" of the combined gas law gives

$$V_2 = 1.62 \text{ L} \times \frac{755 \cancel{\text{mm Hg}}}{725 \cancel{\text{mm Hg}}} \times \frac{323 \cancel{\text{K}}}{273 \cancel{\text{K}}} = 2.00 \text{ L}$$

Practice Problems and Questions

6.22 Rearrange the standard form of the combined gas law so that each of the following variables is by itself on one side of the equation.
a. V_1 b. V_2 c. T_1 d. P_2

6.23 A sample of CO_2 gas has a volume of 15.2 L at a pressure of 1.35 atm and a temperature of 33°C. What is the volume of the gas, in liters, when the pressure and temperature are changed to the following?
a. 1.25 atm and 45°C
b. 1.25 atm and 627°C
c. 3.25 atm and 627°C
d. 6.00 atm and 27°C

6.24 A sample of N_2 gas occupies 65.0 mL at 683 torr and 27°C. Conditions are changed to give a temperature of 33°C and a volume of 85.0 mL. What is the new pressure, in torr, for the N_2 gas?

6.7 The Ideal Gas Law

Learning Focus

Give the mathematical form of the ideal gas law; know how to use the law in problem solving.

The **ideal gas law** *is an equation that includes the quantity of gas in a sample as well as the temperature, pressure, and volume of the sample.* Mathematically, the ideal gas law has the form

$$PV = nRT$$

▶ The ideal gas law is used in calculations when *one* set of conditions is given with one missing variable. The combined gas law (Section 6.6) is used when *two* sets of conditions are given with one missing variable.

In this equation, pressure, temperature, and volume are defined in the same manner as in the gas laws we have already discussed. The symbol n stands for the *number of moles* of gas present in the sample. The symbol R represents the *ideal gas constant,* the proportionality constant that makes the equation valid.

The value of the ideal gas constant (R) varies with the units chosen for pressure and volume. With pressure in atmospheres and volume in liters, R has the value

$$R = \frac{PV}{nT} = 0.0821 \frac{\text{atm} \cdot \text{L}}{\text{mole} \cdot \text{K}}$$

The value of R is the same for all gases under normally encountered conditions of temperature, pressure, and volume.

If three of the four variables in the ideal gas law equation are known, then the fourth can be calculated using the equation. Example 6.4 illustrates the use of the ideal gas law.

Example 6.4 **Using the Ideal Gas Law to Calculate the Volume of a Gas**

The colorless, odorless, tasteless gas carbon monoxide, CO, is a by-product of incomplete combustion of any material that contains the element carbon. Calculate the volume, in liters, occupied by 1.52 moles of this gas at 0.992 atm pressure and a temperature of 65°C.

Solution

This problem deals with only one set of conditions, so the ideal gas equation is applicable. Three of the four variables in the ideal gas equation (P, n, and T) are known, and the fourth (V) is to be calculated.

$$P = 0.992 \text{ atm} \qquad n = 1.52 \text{ moles}$$

$$V = ?\ \text{L} \qquad T = 65°\text{C} = 338 \text{ K}$$

Rearranging the ideal gas equation to isolate V on the left side of the equation gives

$$V = \frac{nRT}{P}$$

Because the pressure is given in atmospheres and the volume unit is liters, the R value 0.0821 is valid. Substituting known numerical values into the equation gives

$$V = \frac{(1.52 \cancel{\text{moles}}) \times \left(0.0821 \frac{\cancel{\text{atm}} \cdot \text{L}}{\cancel{\text{mole}} \cdot \cancel{\text{K}}}\right)(338 \cancel{\text{K}})}{0.992 \cancel{\text{atm}}}$$

Note that all the parts of the ideal gas constant unit cancel except for one, the volume part.

Doing the arithmetic yields the volume of CO.

$$V = \left(\frac{1.52 \times 0.0821 \times 338}{0.992}\right) \text{L} = 42.5 \text{ L}$$

Practice Problems and Questions

6.25 Calculate the volume, in liters, of 0.100 mole of O_2 gas at 0°C and 2.00 atm pressure.

6.26 Calculate the pressure, in atmospheres, of 0.100 mole of O_2 in a 2.00-L container at a temperature of 75°C.

6.27 Calculate the temperature, in degrees Celsius, of 5.23 moles of O_2 gas confined to a volume of 5.23 L at a pressure of 5.23 atm.

6.28 Calculate the number of moles of O_2 gas present in a 6.00-L container that is at a temperature of 27°C and under a pressure of 1.25 atm.

Learning Focus

Use Dalton's law of partial pressures to calculate the partial or total pressure of mixtures of gaseous substances.

6.8 Dalton's Law of Partial Pressures

In a mixture of gases that do not react with one another, each type of molecule moves around in the container as though the other kinds were not there. This type of behavior is possible because a gas is mostly empty space, and attractions between molecules in the gaseous state are negligible at most temperatures and pressures. Each gas in the mixture occupies the entire volume of the container; that is, it distributes itself uniformly throughout the container. The molecules of each type strike the walls of the container as frequently and with the same energy as if they were the only gas in the mixture. Consequently, the pressure exerted by each gas in a mixture is the same as it would be if the gas were alone in the same container under the same conditions.

The English scientist John Dalton (Figure 6.11) was the first to notice the independent behavior of gases in mixtures. In 1803, he published a summary statement concerning this behavior that is now known as Dalton's law of partial pressures. **Dalton's law of partial pressures** states that *the total pressure exerted by a mixture of gases is the sum of the partial pressures of the individual gases.* A **partial pressure** *is the pressure that a gas in a mixture would exert if it were present alone under the same conditions.*

Figure 6.11 John Dalton (1766–1844) throughout his life had a particular interest in the study of weather. From "weather" he turned his attention to the nature of the atmosphere and then to the study of gases in general.

Expressed mathematically, Dalton's law states that

$$P_{\text{Total}} = P_1 + P_2 + P_3 + \cdots$$

where P_{Total} is the total pressure of a gaseous mixture and P_1, P_2, P_3, and so on are the partial pressures of the individual gaseous components of the mixture.

As an illustration of Dalton's law, consider the four identical gas containers shown in Figure 6.12. Suppose we place amounts of three different gases (represented by A, B, and C) into three of the containers and measure the pressure exerted by each sample. We then place all three samples in the fourth container and measure the pressure exerted by this mixture of gases. We find that

$$P_{\text{Total}} = P_A + P_B + P_C$$

Using the actual gauge pressure values given in Figure 6.12, we see that

$$P_{\text{Total}} = 1 + 3 + 2 = 6$$

Example 6.5 **Using Dalton's Law to Calculate Partial Pressure**

The total pressure exerted by a mixture of the three gases oxygen, nitrogen, and water vapor is 742 mm Hg. The partial pressures of the nitrogen and oxygen in the sample are 581 mm Hg and 143 mm Hg, respectively. What is the partial pressure of the water vapor present in the mixture?

Solution

Dalton's law says that

$$P_{\text{Total}} = P_{N_2} + P_{O_2} + P_{H_2O}$$

The known values for variables in this equation are

$$P_{\text{Total}} = 742 \text{ mm Hg}$$
$$P_{N_2} = 581 \text{ mm Hg}$$
$$P_{O_2} = 143 \text{ mm Hg}$$

Rearranging Dalton's law to isolate P_{H_2O} on the left side of the equation gives

$$P_{H_2O} = P_{\text{Total}} - P_{N_2} - P_{O_2}$$

Substituting the known numerical values into this equation and doing the arithmetic give

$$P_{H_2O} = 742 \text{ mm Hg} - 581 \text{ mm Hg} - 143 \text{ mm Hg} = 18 \text{ mm Hg}$$

The Gas Laws

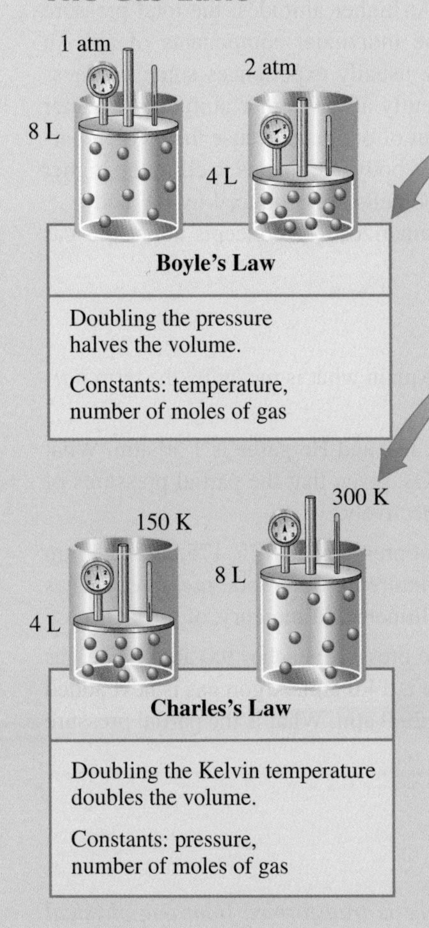
1 atm
2 atm
8 L
4 L
Boyle's Law
Doubling the pressure halves the volume.
Constants: temperature, number of moles of gas
300 K
150 K
8 L
4 L
Charles's Law
Doubling the Kelvin temperature doubles the volume.
Constants: pressure, number of moles of gas

GAS LAW	SYNOPSIS	CONSTANTS	VARIABLES
Boyle's Law $P_1V_1 = P_2V_2$	At constant temperature, the volume of a given gas sample is *inversely proportional* to the pressure applied to it.	temperature, number of moles of gas	pressure, volume
Charles's Law $\frac{V_1}{T_1} = \frac{V_2}{T_2}$	At constant pressure, the volume of a given gas sample is *directly proportional* to its Kelvin temperature.	pressure, number of moles of gas	volume, temperature
Combined Gas Law $\frac{P_1V_1}{T_1} = \frac{P_2V_2}{T_2}$	The product of the pressure and the volume of a given gas sample is *directly proportional* to its Kelvin temperature.	number of moles of gas	pressure, temperature, volume
Ideal Gas Law $PV = nRT$	Relates volume, pressure, temperature, and molar amount of a gas sample under one set of conditions. If three of the four variables are known, the fourth can be calculated from the equation.	$R = 0.0821 \frac{L \cdot atm}{mole \cdot K}$	pressure, volume, temperature, number of moles
Dalton's Law $P_{Total} = P_1 + P_2 + P_3$	The total pressure exerted by a sample that consists of a mixture of gases is equal to the sum of the partial pressures of the individual gases.		

Dalton's Law	P_1	+	P_2	+	P_3	=	P_{Total}
	1	+	3	+	2	=	6

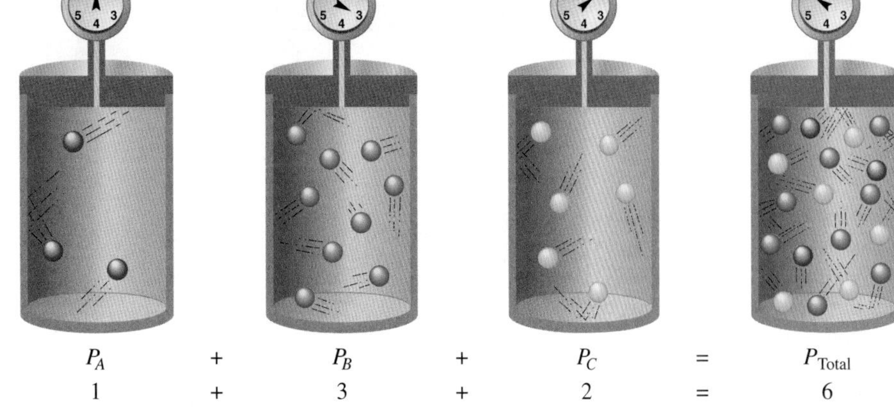
P_A + P_B + P_C = P_{Total}
1 + 3 + 2 = 6

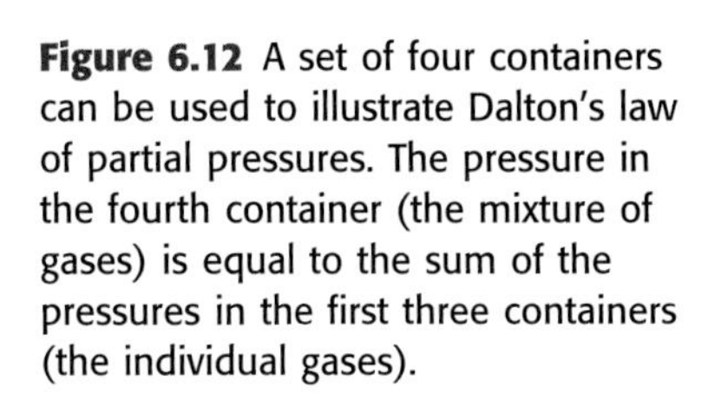
Figure 6.12 A set of four containers can be used to illustrate Dalton's law of partial pressures. The pressure in the fourth container (the mixture of gases) is equal to the sum of the pressures in the first three containers (the individual gases).

Dalton's law of partial pressures is important when we consider the air of our atmosphere, which is a mixture of numerous gases. At higher altitudes, the total pressure of air decreases, as do the partial pressures of the individual components of air. An individual going from sea level to a higher altitude usually experiences some tiredness because his or her body is not functioning as efficiently at the higher altitude. At higher elevation, the red blood cells absorb a smaller amount of oxygen because the oxygen partial pressure at the higher altitude is lower. A person's body acclimates itself to the higher altitude after a period of time as additional red blood cells are produced by the body.

▶ A sample of clean air is the most common example of a mixture of gases that do not react with one another.

The accompanying Chemistry at a Glance summarizes key concepts about the gas laws we have considered in this chapter.

Practice Problems and Questions

6.29 Using air as an example in your explanation, explain what is meant by the term *partial pressure* of a gas.

6.30 The total pressure exerted by a mixture of O_2, N_2, and He gases is 1.50 atm. What is the partial pressure, in atmospheres, of the O_2, given that the partial pressures of the N_2 and the He are 0.75 and 0.33 atm, respectively.

6.31 A gas mixture contains O_2, N_2, and Ar at partial pressures of 125, 175, and 225 mm Hg, respectively. If CO_2 gas is added to the mixture until the total pressure reaches 623 mm Hg, what is the partial pressure, in millimeters of mercury, of the CO_2 gas?

6.32 Helium gas is added to an empty tank until the pressure reaches 9.0 atm. Neon gas is then added to the tank until the total pressure is 14.0 atm. Argon gas is next added to the tank until the total tank pressure reaches 29.0 atm. What is the partial pressure of each gas in the tank?

6.9 Changes of State

Learning Focus

Understand the terminology used to describe various changes of state, and distinguish between exothermic and endothermic changes of state.

A **change of state** *is a process in which a substance is transformed from one physical state to another.* Changes of state are usually accomplished by heating or cooling a substance. Pressure change is also a factor in some systems. Changes of state are examples of physical changes—that is, changes in which chemical composition remains constant. No new substances are ever formed as a result of a change of state.

There are six possible changes of state. Figure 6.13 identifies each of these changes and gives the terminology used to describe them. Four of the six terms used in describ-

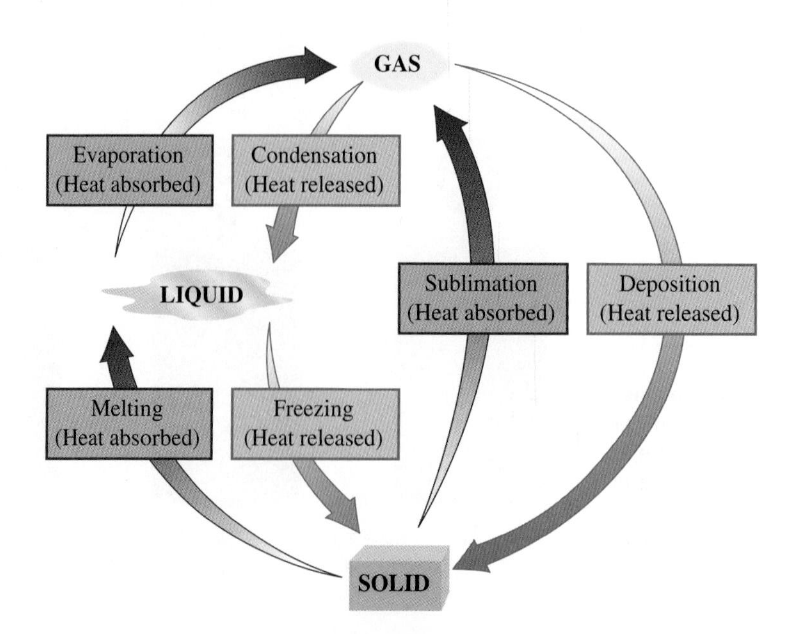

Figure 6.13 There are six changes of state possible for substances. The three endothermic changes, which require the input of heat, are melting, evaporation, and sublimation. The three exothermic changes, which release heat, are freezing, condensation, and deposition.

Figure 6.14 Sublimation and deposition of iodine. (a) The beaker contains iodine crystals, I_2; a dish of ice rests on top of the beaker. (b) Iodine has an appreciable vapor pressure even below its melting point (114°C); thus, when heated carefully, the solid sublimes without melting. The vapor deposits crystals on the cool underside of the dish—the process of deposition.

(a)

(b)

ing state changes are familiar: freezing, melting, evaporation, and condensation. The other two terms—sublimation and deposition—are not so common. *Sublimation* is the direct change from the solid to the gaseous state; *deposition* is the reverse of this, the direct change from the gaseous to the solid state (Figure 6.14).

Although sublimation and deposition are not often encountered, common examples of these processes do exist. Frozen water (ice), dry ice (carbon dioxide), and elemental iodine all readily sublime at normally encountered temperatures and pressures (see Chemical Portraits 11).

Chemical Portraits 11 — Common Substances that Sublime

Water Ice (H_2O)

Profile: Water, the most abundant compound we encounter, is one of the few substances routinely observed in all three physical states: solid ice, liquid water, and gaseous steam. Solid-state water (ice) has an appreciable sublimation rate. When winter temperatures remain below freezing for an extended period of time, ice deposits on streets, sidewalks, etc., "disappear" without melting; sublimation occurs. Ice cubes left in a freezer get smaller with time for the same reason.

Biological considerations: There are three methods by which the human body obtains water, two well-known and one not so well known. Well-known water-sources are the liquids we drink and the foods we eat. The third source is the cells of the human body. Water is a by-product of the metabolic reactions by which cells produce energy from ingested food.

What is the difference between "hard water" and "soft water?"

Dry Ice (CO_2)

Profile: Gaseous CO_2, when cooled to −78°C at atmospheric pressure, changes directly to the solid state producing "dry ice." When warmed, the dry ice sublimes back to the gaseous state.

Uses: Refrigeration is the leading industrial use for CO_2. Industrially, dry ice is obtained by expanding liquid CO_2 from cylinders, producing a "snow" that is compressed into blocks. Dry ice is 1.7 times as dense as water ice and, with a surface temperature of −109.3°F, its net refrigerating effect (on a mass basis) is twice that of water ice. In the large-scale grinding of hamburger, CO_2 snow pellets are added to the meat to absorb processing heat and inhibit bacterial growth and discoloration.

What are some precautions when handling dry ice?

Iodine (I_2)

Profile: Elemental iodine is a violet-black solid that readily sublimes at room temperature to produce violet I_2 vapor. Iodine occurs in seawater and seaweed, from which it can be commercially extracted. A 50% iodine in alcohol solution, tincture of iodine, has use as an antiseptic for treating minor injuries to the skin.

Biological considerations: Iodine is necessary for the proper functioning of the human body; for this reason table salt is "iodized." The hormone thyroxine, produced by the thyroid gland, is an organic molecule that contains 4 iodine atoms. The more thyroxine produced, the higher the basal metabolic rate of a person. The clinical manifestation of iodine deficiency in the body is *goiter,* characterized by an enlarged thyroid gland.

In what form is iodine present in "iodized" salt?

See the text web site at **www.cengage.com/chemistry/stoker** for answers to the above questions and for further information.

Changes of state are classified into two categories based on whether heat (thermal energy) is given up or absorbed during the change process. An **endothermic change of state** *is a change that requires the input (absorption) of heat.* The endothermic changes of state are melting, sublimation, and evaporation. An **exothermic change of state** *is a change that requires heat to be given up (released).* Exothermic changes of state are the reverse of endothermic changes of state; they are freezing, condensation, and deposition.

Practice Problems and Questions

6.33 Indicate whether each of the following is an exothermic or an endothermic change of state.
a. Sublimation b. Melting c. Condensation d. Deposition

6.34 Indicate whether the solid state is involved in each of the following changes of state.
a. Freezing b. Deposition c. Evaporation d. Condensation

6.35 Name the change of state that is the "opposite" of each of the following changes of state.
a. Condensation b. Deposition c. Freezing d. Sublimation

Learning Focus

Explain, on a molecular basis, what happens during the process of evaporation, and list the factors that affect evaporation rates.

6.10 Evaporation of Liquids

Evaporation *is the process by which molecules escape from the liquid phase to the gas phase.* We are all aware that water left in an open container at room temperature slowly disappears by evaporation.

Evaporation can be explained using kinetic molecular theory. Statement 5 of this theory (Section 6.1) indicates that not all the molecules in a liquid (or solid or gas) possess the same kinetic energy. At any given instant, some molecules will have above-average kinetic energies and others will have below-average kinetic energies as a result of collisions between molecules. A given molecule's energy constantly changes as a result of collisions with neighboring molecules. When molecules that happen to be considerably above average in kinetic energy at a given moment are on the liquid surface and are moving in a favorable direction relative to the surface, they can overcome the attractive forces (potential energy) holding them in the liquid and escape.

▶ For a liquid to evaporate, its molecules must gain enough energy to overcome the attractive forces between them.

Evaporation is a surface phenomenon. Surface molecules are subject to fewer attractive forces because they are not completely surrounded by other molecules; thus escape is much more probable. Liquid surface area is an important factor in determining the rate at which evaporation occurs. Increased surface area results in an increased evaporation rate, because a greater fraction of the total molecules are on the surface.

Rate of Evaporation and Temperature

Water evaporates faster from a glass of hot water than from a glass of cold water. A certain minimum kinetic energy is required for molecules to escape from the attractions of neighboring molecules. As the temperature of a liquid increases, a larger fraction of the molecules present acquire this minimum kinetic energy. Consequently, the rate of evaporation always increases as liquid temperature increases.

The escape of high-energy molecules from a liquid during evaporation affects the liquid in two ways: The amount of liquid decreases, and the liquid temperature is lowered. The lower temperature reflects the loss of the most energetic molecules. (Analogously, when all the tall people are removed from a classroom of students, the average height of the remaining students decreases.) A lower average kinetic energy corresponds to a lower temperature (statement 4 of the kinetic molecular theory); hence a cooling effect is produced.

▶ Evaporative cooling is important in many processes. Our own bodies use evaporation to maintain a constant temperature. We perspire in hot weather, and evaporation of the perspiration cools our skin. The cooling effect of evaporation is quite noticeable when one first comes out of an outdoor swimming pool on a hot, breezy day.

The molecules that escape from an evaporating liquid are often collectively referred to as vapor, rather than gas. The term **vapor** *describes gaseous molecules of a substance at a temperature and pressure at which we ordinarily would think of the substance as a* *liquid or solid*. For example, at room temperature and atmospheric pressure, the normal state for water is the liquid state. Molecules that escape (evaporate) from liquid water at these conditions are frequently called *water vapor*.

Practice Problems and Questions

6.36 Indicate whether each of the following changes increases or decreases the rate of evaporation of a liquid.
a. An increase in the temperature of the liquid
b. An increase in the liquid's surface area

6.37 Evaporation of a liquid affects the liquid in two ways. What are these two effects?

6.38 Explain why evaporation is a "cooling process."

6.39 What is the difference between a *gas* and a *vapor*?

6.11 The Vapor Pressure of Liquids

Learning Focus

Know what factors affect a liquid's vapor pressure; be able to relate vapor pressure to the concept of an equilibrium state.

The evaporative behavior of a liquid in a closed container is quite different from its behavior in an open container. Some liquid evaporation occurs in a closed container; this is indicated by a drop in liquid level. However, unlike the liquid level in an open-container system, the liquid level in a closed-container system eventually ceases to drop (becomes constant).

Kinetic molecular theory explains these observations in the following way. The molecules that evaporate in a closed container do not leave the system, as they do in an open container. They find themselves confined in a fixed space immediately above the liquid (see Figure 6.15a). These trapped vapor molecules undergo many random collisions with the container walls, other vapor molecules, and the liquid surface. Molecules that collide with the liquid surface are recaptured by the liquid. Thus two processes, evaporation (escape) and condensation (recapture), take place in a closed container (see Figure 6.15b).

For a short time, the rate of evaporation in a closed container exceeds the rate of condensation, and the liquid level drops. However, as more of the liquid evaporates, the number of vapor molecules increases; the chance of their recapture through striking the liquid surface also increases. Eventually, the rate of condensation becomes equal to the rate of evaporation, and the liquid level stops dropping (see Figure 6.15c). At this point, the number of molecules that escape in a given time is the same as the number recaptured; a steady-state situation has been reached. The amounts of liquid and vapor in the container do not change, even though both evaporation and condensation are still occurring.

▶ Remember that for a system at equilibrium, change at the molecular level is still occurring even though you cannot see it.

This steady-state situation, which will continue as long as the temperature of the system remains constant, is an example of a physical equilibrium state. An **equilibrium state** *is a situation in which two opposite processes take place at equal rates*. For systems in a state of equilibrium, no net macroscopic changes can be detected. However, the system is dynamic; the forward and reverse processes are occurring at equal rates.

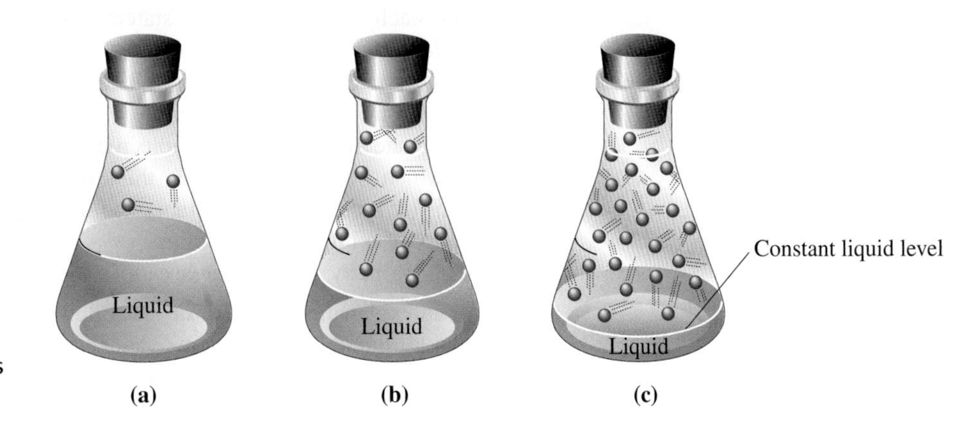

Figure 6.15 In the evaporation of a liquid in a *closed* container (a), the liquid level drops for a time (b) and then becomes constant (ceases to drop). At that point a state of equilibrium has been reached in which the rate of evaporation equals the rate of condensation (c).

Table 6.2
Vapor Pressure of Water at Various Temperatures

Temperature (°C)	Vapor pressure (mm Hg)	Temperature (°C)	Vapor pressure (mm Hg)
0	4.6	50	92.5
10	9.2	60	149.4
20	17.5	70	233.7
25[a]	23.8	80	355.1
30	31.8	90	525.8
37[b]	37.1	100	760.0
40	55.3		

[a]Room temperature
[b]Body temperature

When there is a liquid–vapor equilibrium in a closed container, the vapor in the fixed space immediately above the liquid exerts a constant pressure on both the liquid surface and the walls of the container. This pressure is called the *vapor pressure* of the liquid. **Vapor pressure** *is the pressure exerted by a vapor above a liquid when the liquid and vapor are in equilibrium.*

The magnitude of a vapor pressure depends on the nature and temperature of the liquid. Liquids that have strong attractive forces between molecules have lower vapor pressures than liquids that have weak attractive forces between particles. Substances that have high vapor pressures (weak attractive forces) evaporate readily—that is, they are *volatile.* A **volatile substance** *is a substance that readily evaporates at room temperature because of a high vapor pressure.* Gasoline is a substance whose components are very volatile.

The vapor pressure of all liquids increases with temperature because an increase in temperature results in more molecules having the minimum kinetic energy required for evaporation. Table 6.2 shows the variation in vapor pressure, as temperature increases, of water.

Practice Problems and Questions

6.40 How is the term *equilibrium state* defined?

6.41 Indicate whether each of the following statements concerning a liquid–vapor equilibrium system is *true* or *false.*
a. The pressure exerted by the vapor on the liquid is constant.
b. The rate of evaporation of liquid equals the rate of condensation of vapor.
c. The number of molecules in the vapor state is constant.
d. Molecules are continually entering and leaving the vapor phase of the system.

6.42 Match each of the following statements to the term *vapor, vapor pressure,* or *volatile.*
a. This property can be measured by allowing a liquid and its vapor to reach equilibrium in a closed container.
b. This state involves gaseous molecules of a substance at a temperature and pressure at which we would ordinarily think of the substance as a liquid.
c. This term describes a substance that readily evaporates at room temperature because of a high vapor pressure.
d. This property always increases in magnitude with increasing temperature.

6.43 Give an explanation for each of the following observations.
a. Not all liquids have the same vapor pressure at a given temperature.
b. Increasing the temperature of a liquid increases its vapor pressure.

6.12 Boiling and Boiling Point

Learning Focus

Be able to describe the process of boiling from a molecular viewpoint; understand what is meant by the term *normal boiling point* and know the general relationship between boiling point and external pressure.

In order for a molecule to escape from the liquid state, it usually must be on the surface of the liquid. **Boiling** *is a special form of evaporation where conversion from the liquid state to the vapor state occurs within the body of a liquid through bubble formation.* This phenomenon begins to occur when the vapor pressure of a liquid, which steadily increases as the liquid is heated, reaches a value equal to that of the prevailing external pressure on the liquid; for liquids in open containers, this value is atmospheric pressure. When these two pressures become equal, bubbles of vapor form around any speck of dust or around any irregularity associated with the container surface (Figure 6.16). These vapor bubbles quickly rise to the surface and escape because they are less dense than the liquid itself. We say the liquid is boiling.

Figure 6.16 Bubbles of vapor form within a liquid when the temperature of the liquid reaches the liquid's boiling point.

The **boiling point** *of a liquid is the temperature at which the vapor pressure of a liquid becomes equal to the external (atmospheric) pressure exerted on the liquid.* Because the atmospheric pressure fluctuates from day to day, the boiling point of a liquid does also. Thus, in order for us to compare the boiling points of different liquids, the external pressure must be the same. The boiling point of a liquid that is most often used for comparison and tabulation purposes is called the *normal* boiling point. A liquid's **normal boiling point** *is the temperature at which a liquid boils under a pressure of 760 mm Hg.*

Conditions That Affect Boiling Point

At any given location, the changes in the boiling point of a liquid caused by *natural* variations in atmospheric pressure seldom exceed a few degrees; in the case of water, the maximum is about 2°C. However, variations in boiling points *between* locations at different elevations can be quite striking, as shown in Table 6.3.

The boiling point of a liquid can be increased by increasing the external pressure. This principle is used in the operation of a pressure cooker. Foods cook faster in pressure cookers because the elevated pressure causes water to boil above 100°C. An increase in temperature of only 10°C will cause food to cook in approximately half the normal time (see Figure 6.17).

Liquids that have high normal boiling points or that undergo undesirable chemical reactions at elevated temperatures can be made to boil at low temperatures by reducing the external pressure. This principle is used in the preparation of numerous food products, including frozen fruit juice concentrates. Some of the water in a fruit juice is boiled away at a reduced pressure, thus concentrating the juice without heating it to a high temperature (which spoils the taste of the juice and reduces its nutritional value).

Practice Problems and Questions

6.44 What is meant by the phrase *normal boiling point*?

Table 6.3
Boiling Point of Water at Various Locations That Differ in Elevation

Location	Feet above sea level	P_{atm} (mm Hg)	Boiling point (°C)
top of Mt. Everest, Tibet	29,028	240	70
top of Mt. McKinley, Alaska	20,320	340	79
Leadville, Colorado	10,150	430	89
Salt Lake City, Utah	4,390	650	96
Madison, Wisconsin	900	730	99
New York City, New York	10	760	100
Death Valley, California	−282	770	100.4

Figure 6.17 The converse of the pressure cooker "phenomenon" is that food cooks more slowly at reduced pressures. The pressure reduction associated with higher altitudes, and the accompanying reduction in boiling points of liquids, mean that food cooked over a campfire in the mountains requires longer cooking times.

6.45 Indicate whether each of the following statements about the boiling point of a liquid is *true* or *false*.

a. The boiling point of a liquid heated in an open container is determined by atmospheric pressure.
b. Liquids can be made to boil at temperatures higher than their normal boiling point but never at temperatures lower than their normal boiling point.
c. The boiling point of a liquid increases with increasing external pressure on the liquid.
d. The normal boiling point of a liquid varies from day to day, depending on atmospheric pressure.

6.46 What is the characteristic that distinguishes the process of boiling from the process of evaporation?

Learning Focus

Understand the origins of intermolecular forces and the special significance of hydrogen bonds.

6.13 Intermolecular Forces in Liquids

Boiling points vary greatly among substances. The boiling points of some substances are well below zero; for example, oxygen has a boiling point of −183°C. Numerous other substances do not boil until the temperature is much higher. An explanation of this variation in boiling points involves a consideration of the nature of the intermolecular forces that must be overcome in order for molecules to escape from the liquid state into the vapor state. An **intermolecular force** *is an attractive force that acts between a molecule and another molecule.*

Intermolecular forces are similar in one way to the previously discussed *intramolecular* forces (forces *within* molecules) that are involved in covalent bonding (Sections 4.10 and 4.11); they are electrostatic in origin—that is, they involve positive–negative attractions. A major difference between inter- and intramolecular forces is their strength. Intermolecular forces are weak compared to intramolecular forces (true chemical bonds). Generally, their strength is less than one-tenth that of a single covalent bond. However, intermolecular forces are strong enough to influence the behavior of liquids, and they often do so in very dramatic ways.

There are three main types of intermolecular forces: dipole–dipole interactions, hydrogen bonds, and London forces.

Dipole–Dipole Interactions

A **dipole–dipole interaction** *is an intermolecular force that occurs between polar molecules.* A polar molecule (Section 4.17) has a negative end and a positive end; that is, it has a *dipole* (two poles resulting from opposite charges being separated from one another). As a consequence, the positive end of one molecule attracts the negative end of another molecule, and vice versa. This attraction constitutes a dipole–dipole interaction. The greater the polarity of the molecules, the greater the strength of the dipole–dipole interactions. And the greater the strength of the dipole–dipole interactions, the higher the boiling point of the liquid. Figure 6.18 shows the many dipole–dipole interactions that are possible for a random arrangement of polar chlorine monofluoride (ClF) molecules.

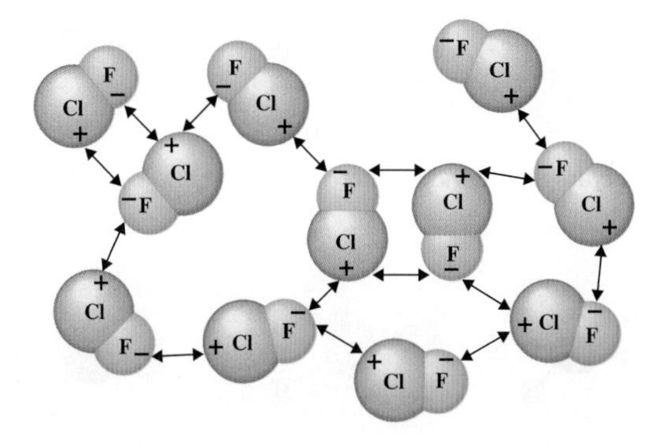

Figure 6.18 There are many dipole–dipole interactions possible between randomly arranged ClF molecules. In each interaction, the positive end of one molecule is attracted to the negative end of a neighboring ClF molecule.

Hydrogen Bonds

Unusually strong dipole–dipole interactions are observed among hydrogen-containing molecules in which hydrogen is covalently bonded to a highly electronegative element of small atomic size (fluorine, oxygen, and nitrogen). Two factors account for the extra strength of these dipole–dipole interactions.

1. The highly electronegative element to which hydrogen is covalently bonded attracts the bonding electrons to such a degree that the hydrogen atom is left with a significant δ^+ charge.

$$\overset{\delta^+}{\text{H}}\text{—}\overset{\delta^-}{\text{F}} \quad \overset{\delta^+}{\text{H}}\text{—}\overset{\delta^-}{\text{O}} \quad \overset{\delta^+}{\text{H}}\text{—}\overset{\delta^-}{\text{N}}$$

   Indeed, the hydrogen atom is essentially a "bare" nucleus, because it has no electrons besides the one attracted to the electronegative element—a unique property of hydrogen.
2. The small size of the hydrogen atom allows the "bare" nucleus to approach closely, and be strongly attracted to, a lone pair of electrons on the electronegative atom of another molecule.

▶ The three elements that have significant hydrogen bonding ability are fluorine, oxygen, and nitrogen. They are all very electronegative elements of small atomic size. Chlorine has the same electronegativity as nitrogen, but its larger atomic size causes it to have little hydrogen bonding ability.

Dipole–dipole interactions of this type are given a special name, hydrogen bonds. A **hydrogen bond** *is an extra-strong dipole–dipole interaction between a hydrogen atom covalently bonded to a small, very electronegative atom (F, O, or N) and a lone pair of electrons on another small, very electronegative atom (F, O, or N).*

Water (H_2O) is the most commonly encountered substance wherein hydrogen bonding is significant. Figure 6.19 depicts the process of hydrogen bonding among water molecules. Note that each oxygen atom in water can participate in two hydrogen bonds—one involving each of its nonbonding electron pairs.

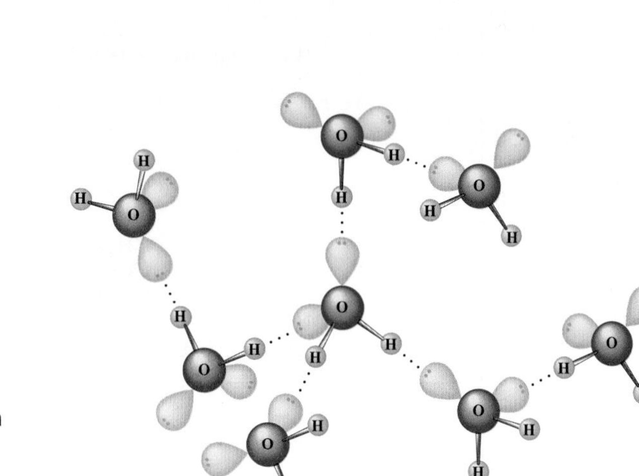

Figure 6.19 Depiction of hydrogen bonding among water molecules. The dotted lines are the hydrogen bonds.

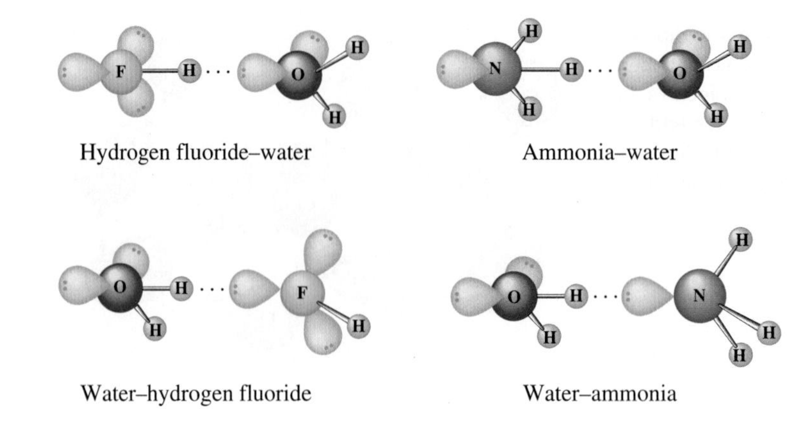

Figure 6.20 Diagrams of hydrogen bonding involving selected simple molecules. The solid lines represent covalent bonds; the dotted lines represent hydrogen bonds.

The two molecules that participate in a hydrogen bond need not be identical. Hydrogen bond formation is possible whenever two molecules, the same or different, have the following characteristics.

▶ A series of dots is used to represent a hydrogen bond, as in the notation
—X—H···Y—
X and Y represent small, highly electronegative elements (fluorine, oxygen, or nitrogen).

1. One molecule has a hydrogen atom attached by a covalent bond to an atom of nitrogen, oxygen, or fluorine.
2. The other molecule has a nitrogen, oxygen, or fluorine atom present that possesses one or more nonbonding electron pairs.

Figure 6.20 gives additional examples of hydrogen bonding involving simple molecules.

The vapor pressures (Section 6.11) of liquids that have significant hydrogen bonding are much lower than those of similar liquids wherein little or no hydrogen bonding occurs. This is because the presence of hydrogen bonds makes it more difficult for molecules to escape from the condensed state; additional energy is needed to overcome the hydrogen bonds. For this reason, boiling points are much higher for liquids in which hydrogen bonding occurs. The effect that hydrogen bonding has on boiling point can be seen by comparing water's boiling point with those of other hydrogen compounds of Group VIA elements—H_2S, H_2Se, and H_2Te (see Figure 6.21). Water is the only compound in this series where significant hydrogen bonding occurs.

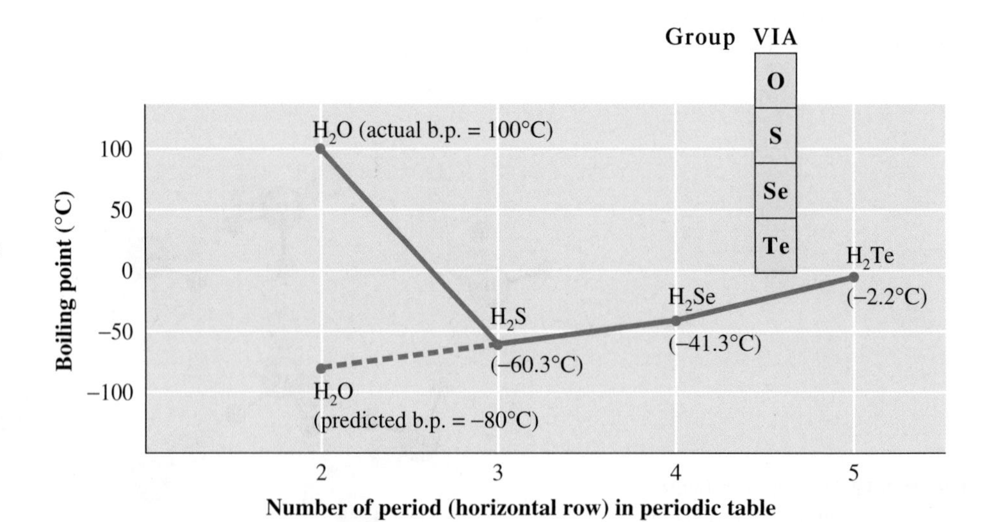

Figure 6.21 If there were no hydrogen bonding between water molecules, the boiling point of water would be approximately −80°C; this value is obtained by extrapolation (extension of the line connecting the three heavier compounds). Because of hydrogen bonding, the actual boiling point of water, 100°C, is nearly 200°C higher than predicted. Indeed, in the absence of hydrogen bonding, water would be a gas at room temperature, and life as we know it on Earth would not be possible.

Intermolecular Forces

INTERMOLECULAR FORCES

- Electrostatic forces that act BETWEEN a molecule and other molecules
- Weaker than chemical bonds (intramolecular forces)
- Strength is generally less than one-tenth that of a single covalent bond

Dipole–Dipole Interactions

- Occur between POLAR molecules
- The positive end of one molecule attracts the negative end of another molecule
- Strength depends on the extent of molecular polarity

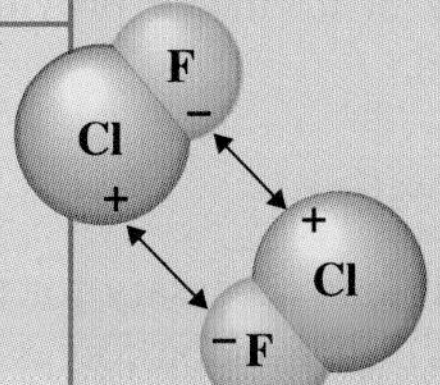

London Forces

- Occur between ALL molecules
- Only type of intermolecular force present between NONPOLAR molecules
- Instantaneous dipole–dipole interactions caused by momentary uneven electron distributions in molecules
- Weakest type of intermolecular force, but important because of their sheer numbers

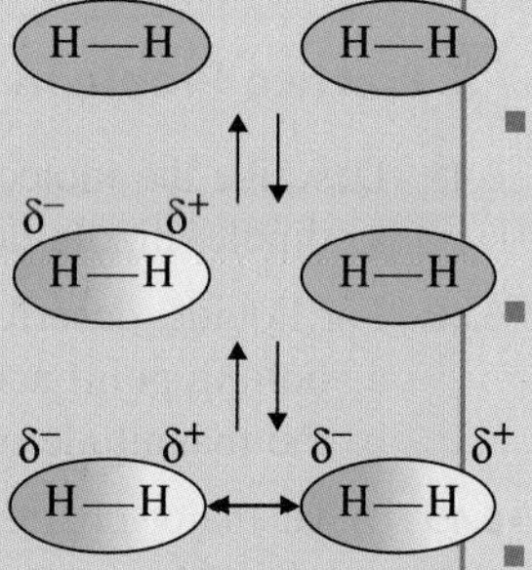

Hydrogen Bonds

- Extra-strong dipole–dipole interactions
- Require the presense of hydrogen covalently bonded to a small very electronegative atom (F, O, or N)
- Interaction is between the H atom and a lone pair of electrons on another small electronegative atom (F, O, or N)

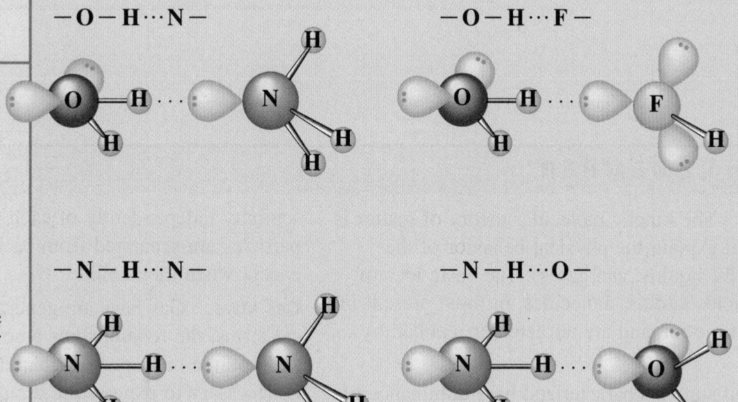

London Forces

The third type of intermolecular force, and the weakest, is the London force, named after the German physicist Fritz London (1900–1954), who first postulated its existence. A **London force** *is a weak temporary intermolecular force that occurs between an atom or molecule (polar or nonpolar) and another atom or molecule (polar or nonpolar).* The origin of London forces is more difficult to visualize than that of dipole–dipole interactions.

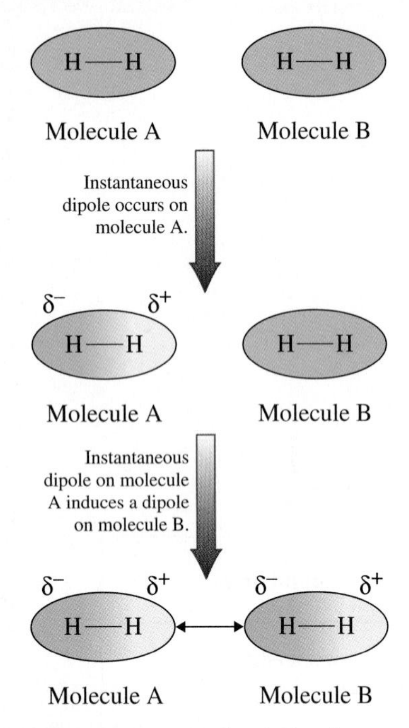

Figure 6.22 Nonpolar molecules such as H_2 can develop instantaneous dipoles and induced dipoles. The attractions between such dipoles, even though they are transitory, create London forces.

London forces result from momentary (temporary) uneven electron distributions in molecules. Most of the time, the electrons in a molecule can be considered to have a predictable distribution determined by their energies and the electronegativities of the atoms present. However, there is a small statistical chance (probability) that the electrons will deviate from their normal pattern. For example, in the case of a nonpolar diatomic molecule, more electron density may temporarily be located on one side of the molecule than on the other. This condition causes the molecule to become polar for an instant. The negative side of this *instantaneously* polar molecule tends to repel electrons of adjoining molecules and causes these molecules also to become polar (*induced polarity*). The original polar molecule and all of the molecules with induced polarity are then attracted to one another. This happens many, many times per second throughout the liquid, resulting in a net attractive force. Figure 6.22 depicts the situation that prevails when London forces exist.

As an analogy for London forces, consider what happens when a bucket filled with water is moved. The water will "slosh" from side to side. This is similar to the movement of electrons. The "sloshing" from side to side is instantaneous; a given "slosh" quickly disappears. "Uneven" electron distribution is likewise a temporary situation.

The accompanying Chemistry at a Glance provides a summary of what we have learned about intermolecular forces.

Practice Problems and Questions

6.47 Describe the molecular conditions necessary for the existence of
a. dipole–dipole interactions b. London forces c. hydrogen bonds

6.48 For liquids, describe the relationship between
a. boiling point and the strength of intermolecular forces
b. the magnitude of vapor pressure and the strength of intermolecular forces

6.49 What major characteristic distinguishes a dipole–dipole interaction from a hydrogen bond?

6.50 What is the difference between an *intra*molecular force and an *inter*molecular force?

6.51 Contrast the strengths of *intra*molecular forces and *inter*molecular forces.

CONCEPTS TO REMEMBER

Kinetic molecular theory. The kinetic molecular theory of matter is a set of five statements that explain the physical behavior of the three states of matter (solids, liquids, and gases). The basic idea of this theory is that the particles (atoms, molecules, or ions) present in a substance are in constant motion and are attracted or repelled by each other.

The solid state. The solid state is characterized by a dominance of potential energy (cohesive forces) over kinetic energy (disruptive forces). As a result, the particles of solids are held in rigid three-dimensional lattices in which the particle's kinetic energy takes the form of vibrations about each lattice site.

The liquid state. The liquid state is one in which neither potential energy (cohesive forces) nor kinetic energy (disruptive forces) dominates. As a result, particles of liquids are randomly arranged but are relatively close to each other and are in constant random motion, sliding freely over each other but without enough kinetic energy to become separated.

The gaseous state. The gaseous state is characterized by a complete dominance of kinetic energy (disruptive forces) over potential energy (cohesive forces). As a result, particles move randomly, essentially independently of each other. Under ordinary pressure, the particles are separated from each other by relatively large distances, except when they collide.

Gas laws. Gas laws are generalizations that describe, in mathematical terms, the relationships among the amount, pressure, temperature, and volume of a specific quantity of gas. When gas laws are used, it is necessary to express the temperature on the Kelvin scale. Pressure is usually expressed in atm, mm Hg, or torr.

Boyle's law. Boyle's law, the pressure–volume law, states that the volume of a sample of a gas is inversely proportional to the pressure applied to the gas if the temperature is kept constant. This means that when the pressure on the gas increases, the volume decreases proportionally; conversely, when the volume decreases, the pressure increases.

Charles's law. Charles's law, the volume–temperature law, states that the volume of a sample of gas is directly proportional to its Kelvin temperature if the pressure is kept constant. This means that when the temperature increases, the volume also increases and that when the temperature decreases, the volume also decreases.

The combined gas law. The combined gas law is an expression obtained by mathematically combining Boyle's and Charles's laws. A change in pressure, temperature, or volume that is brought about by changes in the other two variables can be calculated by using this law.

Ideal gas law. The ideal gas law has the form $PV = nRT$, where R is the ideal gas constant (0.0821 atm · L/mole · K). This equation enables us to calculate any one of the characteristic gas properties (P, V, T, or n), given the other three.

Dalton's law of partial pressures. Dalton's law of partial pressures states that the total pressure exerted by a mixture of gases is the sum of the partial pressures of the individual gases. A partial pressure is the pressure that a gas in a mixture would exert if it were present alone under the same conditions.

Changes of state. Most matter can be changed from one physical state to another by heating, cooling, or changing pressure. The state changes that release heat are called exothermic (condensation, deposition, and freezing), and those that absorb heat are called endothermic (melting, evaporation, and sublimation).

Vapor pressure. The pressure exerted by vapor in equilibrium with its liquid is the vapor pressure of the liquid. Vapor pressure increases as liquid temperature increases.

Boiling and boiling point. Boiling is a special form of evaporation in which bubbles of vapor form within the liquid and rise to the surface. The boiling point of a liquid is the temperature at which the vapor pressure of the liquid becomes equal to the external (atmospheric) pressure exerted on the liquid. The boiling point of a liquid increases or decreases as the prevailing atmospheric pressure increases or decreases.

Intermolecular forces. Intermolecular forces are forces that act between a molecule and another molecule. The three principal types of intermolecular forces in liquids are dipole–dipole interactions, hydrogen bonds, and London forces.

Hydrogen bonds. A hydrogen bond is an extra-strong dipole–dipole interaction between a hydrogen atom covalently bonded to a very electronegative atom (F, O, or N) and a lone pair of electrons on another small, very electronegative atom (F, O, or N).

KEY REACTIONS AND EQUATIONS

1. Boyle's law (Section 6.4)

$$P_1V_1 = P_2V_2 \qquad (n, T \text{ constant})$$

2. Charles's law (Section 6.5)

$$\frac{V_1}{T_1} = \frac{V_2}{T_2} \qquad (n, P \text{ constant})$$

3. Combined gas law (Section 6.6)

$$\frac{P_1V_1}{T_1} = \frac{P_2V_2}{T_2} \qquad (n \text{ constant})$$

4. Ideal gas law (Section 6.7)

$$PV = nRT$$

5. Ideal gas constant (Section 6.7)

$$R = 0.0821 \text{ atm} \cdot \text{L/mole} \cdot \text{K}$$

6. Dalton's law of partial pressures (Section 6.8)

$$P_{\text{Total}} = P_1 + P_2 + P_3 + \cdots$$

KEY TERMS

Barometer (6.3)
Boiling (6.12)
Boiling point (6.12)
Boyle's law (6.4)
Change of state (6.9)
Charles's law (6.5)
Combined gas law (6.6)
Compressability (6.1)
Dalton's law of partial pressures (6.8)
Dipole–dipole interaction (6.13)
Electrostatic interactions (6.1)
Endothermic change of state (6.9)
Equilibrium state (6.11)
Evaporation (6.10)
Exothermic change of state (6.9)
Gas laws (6.3)
Gaseous state (6.2)
Hydrogen bond (6.13)
Ideal gas law (6.7)
Intermolecular force (6.13)
Kinetic energy (6.1)
Kinetic molecular theory of matter (6.1)
Liquid state (6.2)
London force (6.13)
Normal boiling point (6.12)
Partial pressure (6.8)
Potential energy (6.1)
Pressure (6.3)
Solid state (6.2)
Thermal expansion (6.1)
Vapor (6.11)
Vapor pressure (6.11)
Volatile substance (6.11)

ADDITIONAL PROBLEMS

6.52 A sample of He gas is under a pressure of 1.25 atm. Express this pressure in the following units.
a. mm Hg b. torr c. psi d. lb/in.^{2}

6.53 Under which of the following "pressure situations" will a liquid boil?
a. Vapor pressure and atmospheric pressure are equal.
b. Vapor pressure is less than atmospheric pressure.
c. Vapor pressure = 635 mm Hg, atmospheric pressure = 735 mm Hg
d. Vapor pressure = 735 torr, atmospheric pressure = 1.00 atm

6.54 Match the following restrictions on variables to the following gas laws: *Boyle's law, Charles's law*, and *combined gas law*. More than one answer may be correct in a given situation.
a. Number of moles is constant.
b. Pressure is constant.
c. Temperature is constant.
d. Both number of moles and temperature are constant.

6.55 A sample of NO_2 gas in a 575-mL container at a pressure of 1.25 atm and a temperature of 125°C is transferred to a new container with a volume of 825 mL.

a. What is the new pressure, in atmospheres, if no change in temperature occurs?
b. What is the new pressure, in atmospheres, if the temperature changes to 175°C?
c. What is the new temperature, in degrees Celsius, if no change in pressure occurs?
d. What is the new temperature, in degrees Celsius, if the pressure changes to 1.75 atm?

6.56 A gas mixture containing He, Ne, and Ar exerts a pressure of 3.00 atm. What is the partial pressure of each gas present in the mixture under the following conditions?
a. There is an equal number of moles of each gas present.
b. There is an equal number of atoms of each gas present.
c. The partial pressure of He is double that of the other two gases combined.
d. The partial pressure of He is one-half that of Ne and one-third that of Ar.

6.57 The vapor pressure of PBr_3 reaches 400 torr at 150°C. The vapor pressure of PI_3 reaches 400 torr at 57°C.
a. Which substance should evaporate at the slower rate at 100°C?
b. Which substance should have the lower boiling point?
c. Which substance should have the weaker intermolecular forces?
d. Which substance should have the higher vapor pressure at 50°C?

6.58 Match the following statements to the terms *boiling, evaporation, sublimation,* and *condensation*.
a. Bubble formation occurs with a liquid.
b. Molecules from the surface of a liquid become vapor.
c. Molecules from the surface of a solid become vapor.
d. A gas becomes a liquid.

6.59 For which of the following substances, in the pure liquid state, is hydrogen bonding possible?

a. 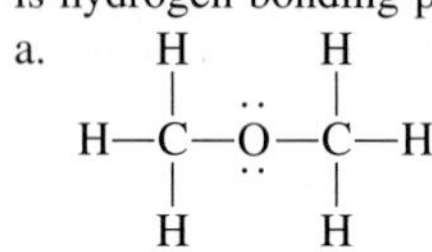

b. 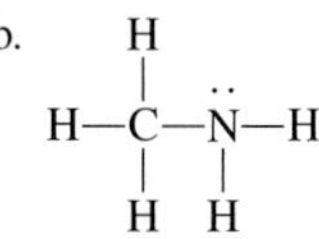

c.
```
   H  H  H
   |  |  |   ..
H—C—C—C—O—H
   |  |  |   ..
   H  H  H
```

d. H—I: (with lone pairs on I)

PRACTICE TEXT ▸ True/False

6.60 In a solid, the particles present have kinetic energy but lack potential energy.

6.61 The average kinetic energy of all particles in a system increases as the temperature of the system increases.

6.62 Torr and millimeters of mercury are two names for the same pressure unit.

6.63 Increasing the pressure on a gas, at constant temperature, causes the volume of the gas to increase.

6.64 A partial pressure is the pressure that a gas in a mixture would exert if it were present alone under the same conditions.

6.65 Sublimation is an endothermic change of state.

6.66 The rate of evaporation of a liquid depends on both liquid surface area and temperature.

6.67 In a liquid–vapor equilibrium system, the rates of evaporation and condensation have both decreased to zero.

6.68 At a given temperature, volatile substances have higher vapor pressures than nonvolatile substances.

6.69 The *normal* boiling point for a liquid is the temperature at which the liquid boils at sea level.

6.70 The boiling point of a liquid can be decreased by increasing the external pressure on the liquid.

6.71 For most compounds in the liquid state, intermolecular forces are stronger than intramolecular forces.

6.72 Dipole–dipole interactions are intermolecular forces that are present in all liquids.

6.73 Liquids with strong intermolecular forces will have higher vapor pressures than liquids with weaker intermolecular forces.

6.74 Water is the most commonly encountered compound in which hydrogen bonding is significant.

PRACTICE TEST ▸ Multiple Choice

6.75 In which of the following groupings of terms are the three terms closely related?
a. Kinetic energy, energy of motion, cohesive forces
b. Potential energy, energy of attraction, disruptive forces
c. Kinetic energy, electrostatic interactions, disruptive forces
d. Potential energy, electrostatic interactions, cohesive forces

6.76 The phrases "particles close together and held in fixed positions" and "completely fills the container" apply, respectively, to
a. liquids and solids
b. solids and gases
c. gases and liquids
d. liquids and gases

6.77 Indicate what the missing words are in the following statement of Charles's law: At constant pressure, the volume of a gas sample is —— proportional to its —— temperature.
a. directly, Celsius
b. directly, Kelvin
c. inversely, Celsius
d. inversely, Kelvin

6.78 A gas has a volume of 6.00 liters at a temperature of 27°C and a pressure of 1.00 atm. What is the volume of the gas, in liters, at a temperature of 327°C and a pressure of 3.00 atm?
a. 1.00 L b. 4.00 L
c. 28.3 L d. 36.0 L

6.79 The correct form of the equation for the ideal gas law is

a. $PV = nRT$ b. $PT = nRV$
c. $P/V = nRT$ d. $PV = n/RT$

6.80 In which of the following pairs of physical changes are both changes exothermic?
a. Sublimation, evaporation
b. Freezing, melting
c. Freezing, condensation
d. Melting, sublimation

6.81 Which of the following is not a factor in determining the magnitude of the vapor pressure of a liquid?
a. The temperature of the liquid
b. The strength of the attractive forces between molecules of the liquid
c. The size of the container for the liquid
d. The type of forces between molecules within the liquid

6.82 The boiling point of a liquid is
a. the temperature at which the rate of sublimation and the rate of evaporation are equal
b. always 100°C or greater
c. the temperature at which the vapor pressure of the liquid equals the external pressure on the liquid
d. the temperature at which liquid–vapor equilibrium is reached

6.83 A liquid placed in a closed container will become a liquid–vapor equilibrium system when
a. all of the liquid has evaporated
b. the rate of evaporation equals the rate of condensation
c. the vapor pressure reaches 1.00 atm
d. molecules cease to pass from the liquid state to the vapor state

6.84 Which of the following statements about intermolecular forces is *correct*?
a. Dipole–dipole interactions occur only between nonpolar molecules.
b. A hydrogen bond is an extremely weak dipole–dipole interaction.
c. London forces are "instantaneous" intermolecular forces.
d. Hydrogen bonding occurs anytime a hydrogen-containing molecule is present in a liquid.

CHAPTER OUTLINE

▶ CHAPTER SEVEN

Solutions

Ocean water is a solution in which many different substances are dissolved.

Solutions are common in nature, and they represent an abundant form of matter. Solutions carry nutrients to the cells of our bodies and carry away waste products. The ocean is a solution of water, sodium chloride, and many other substances (even gold). A large percentage of all chemical reactions take place in solution, including most of those discussed in later chapters in this text.

7.1 Characteristics of Solutions

Learning Focus

Recognize the general properties of solutions and distinguish between the terms *solute* and *solvent.*

All samples of matter are either *pure substances* or *mixtures* (Section 1.5). Pure substances are of two types: *elements* and *compounds*. Mixtures are of two types: *homogeneous* (uniform properties throughout) and *heterogeneous* (different properties in different regions).

Where do solutions fit in this classification scheme? The term *solution* is just an alternative way of saying *homogeneous mixture*. A **solution** *is a homogeneous combination of two or more substances in which each substance retains its own chemical identity.*

It is often convenient to call one component of a solution the solvent and other components that are present solutes (Figure 7.1). A **solvent** *is the component of a solution that is present in the greatest amount.* A solvent can be thought of as the medium in which the other substances present are dissolved. A **solute** *is a solution component that is present in a small amount relative to that of the solvent.* More than one solute can be present in the same solution. For example, both sugar and salt (two solutes) can be dissolved in a container of water (solvent) to give salty sugar water.

▶ "All solutions are mixtures" is a valid statement. However, the reverse statement, "All mixtures are solutions," is not valid. Only those mixtures that are *homogeneous* are solutions.

Figure 7.1 The colored crystals are the solute, and the clear liquid is the solvent. Stirring produces the solution.

In most of the situations we will encounter, the solutes present in a solution will be of more interest to us than to the solvent. The solutes are the active ingredients in the solution. They are the substances that undergo reaction when solutions are mixed.

The general properties of a solution (homogeneous mixture) were outlined in Section 1.5. These properties, restated using the concepts of solvent and solute, are as follows:

▶ Generally, solutions are *transparent;* that is, you can see through them. A synonym for *transparent* is *clear*. Clear solutions may be colorless or colored. A solution of potassium dichromate is a clear yellow-orange solution.

1. A solution contains two or more components: a solvent (the substance present in the greatest amount) and one or more solutes.
2. A solution has a variable composition; that is, the ratio of solute to solvent may be varied.
3. The properties of a solution change as the ratio of solute to solvent is changed.
4. The dissolved solutes are present as individual particles (molecules, atoms, or ions). Intermingling of components at the particle level is a requirement for homogeneity.
5. The solutes remain uniformly distributed throughout the solution and will not settle out with time. Every part of a solution has exactly the same properties and composition as every other part.
6. The solute(s) generally can be separated from the solvent by physical means such as evaporation.

Solutions used in laboratories and clinical settings are most often liquids, and the solvent is nearly always water. However, gaseous solutions (dry air), solid solutions (metal alloys; see Figure 7.2), and liquid solutions in which water is not the solvent (gasoline) are also possible and are relatively common.

Figure 7.2 Jewelry often involves solid solutions in which one metal has been dissolved in another metal.

Practice Problems and Questions

7.1 Indicate whether each of the following statements about the general properties of solutions is *true* or *false*.

a. A solution may contain more than one solute.
b. All solutions have a variable composition.
c. Every part of a solution has exactly the same properties as every other part.
d. For a solution to form, the solute and solvent must chemically react with each other.

7.2 Identify the *solute* and the *solvent* in solutions composed of the following:

a. 5.00 g of sodium chloride and 50.0 g of water
b. 50.0 g of ethyl alcohol and 40.0 g of water
c. 4.00 g of table sugar and 20.0 g of ethyl alcohol
d. 60.0 mL of methyl alcohol and 20.0 mL of ethyl alcohol

7.3 Explain why the statement "All mixtures are solutions" is incorrect.

7.4 Distinguish between the phrases *clear solution* and *colorless solution*.

Learning Focus

Define *solubility*, and distinguish between saturated and unsaturated solutions, between dilute and concentrated solutions and between aqueous and nonaqueous solutions.

7.2 Solubility

In addition to *solvent* and *solute,* several other terms are used to describe characteristics of solutions. The **solubility** *of a solute is the maximum amount of solute that will dissolve in a given amount of solvent.* Many factors affect the numerical value of a solute's solubility in a given solvent, including the nature of the solvent itself, the temperature, and, in some cases, the pressure and presence of other solutes. Solubility is commonly expressed as grams of solute per 100 g of solvent.

Effect of Temperature on Solubility

Most solids become more soluble in water with increasing temperature. The data in Table 7.1 illustrate this temperature–solubility pattern. Here, the solubilities of selected ionic solids in water are given at three different temperatures.

Table 7.1
Solubilities of Various Compounds in Water at 0°C, 50°C, and 100°C

Solute	Solubility (g solute/100 g H_2O)		
	0°C	*50°C*	*100°C*
lead(II) bromide ($PbBr_2$)	0.455	1.94	4.75
silver sulfate (Ag_2SO_4)	0.573	1.08	1.41
copper(II) sulfate ($CuSO_4$)	14.3	33.3	75.4
sodium chloride (NaCl)	35.7	37.0	39.8
silver nitrate ($AgNO_3$)	122	455	952
cesium chloride (CsCl)	161.4	218.5	270.5

Chemical Portraits 12

Effects of Temperature and Pressure on Gas Solubilities

Carbon Dioxide (CO_2)

Profile: Compared to N_2 and O_2, CO_2 is a relatively soluble gas. At 25°C and 1 atm pressure, its solubility in water is approximately 25 times that of O_2 and 50 times that of N_2.

Uses: Refrigeration and beverage carbonation, in that order, are the two leading uses for CO_2. The "bite" and "sparkle" of carbonated soft drinks results from the dissolution of CO_2. Most of the carbonation is from CO_2 itself, although a very small amount of CO_2 is converted to carbonic acid (H_2CO_3). Unopened carbonated beverages are saturated CO_2 solutions at a pressure slightly greater than 1 atm. The pressure drop associated with opening the container decreases CO_2 solubility. The resulting release of CO_2 from solution is usually rapid enough to cause fizzing.

Why does an opened can of soda pop go "flat" faster at room temperature than in the refrigerator?

Nitrogen (N_2)

Profile: Nitrogen gas is relatively insoluble in water; at 25°C and 1 atm pressure, its aqueous solubility is one-half that of O_2 gas.

Biochemical considerations: Like air, water exerts a pressure on anything immersed in it. Deep-sea divers must contend with this increased pressure in terms of gas solubilities. N_2 gas, not very soluble in blood at normal pressures, becomes more soluble at higher pressures. The deeper a diver goes, the more N_2 that dissolves in the blood. If a diver ascends too quickly, the rapid release of N_2 out of the blood causes severe pain in the limbs and joints, a condition called the "bends." Use of a helium-oxygen breathing mixture instead of compressed air can reduce the incidence of the bends. Helium is less soluble in blood than N_2.

What happens to the N_2 gas taken in the human body during the process of breathing?

Oxygen (O_2)

Profile: The solubility of O_2 in water at 25°C and 1 atm pressure is 0.0012 moles per liter. This value is approximately twice that for N_2 gas and 4% of that for CO_2 gas under similar conditions.

Biochemical considerations: Certain minimum dissolved oxygen (DO) levels are necessary to sustain aquatic life in a body of water. DO levels *decrease* as the temperature of a water body *increases* since O_2 solubility is dependent on temperature. On hot summer days the temperature of *shallow* waters can reach a point where DO levels drop below that necessary to sustain life. The result is suffocation of fish and other life present. If industrial *coolant water* is returned, with added heat, to its original source *thermal pollution* can result; its effect is reduced DO levels, which can adversely affect aquatic life.

What types of solutions exist to alleviate thermal pollution problems?

See the text web site at **www.cengage.com/chemistry/stoker** for answers to the above questions and for further information.

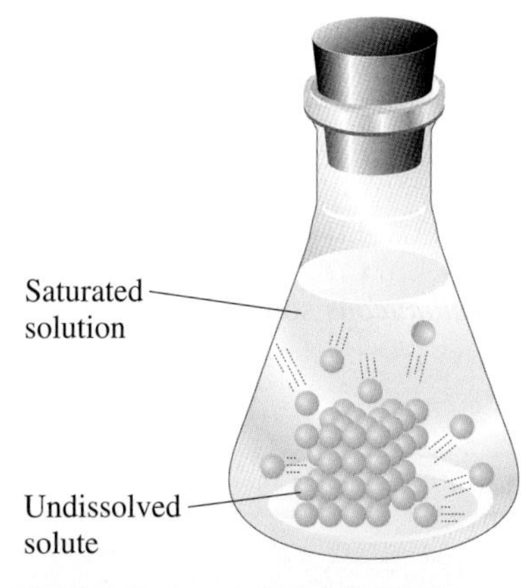

Figure 7.3 In a saturated solution, the dissolved solute is in a dynamic equilibrium with the undissolved solute. Solute enters and leaves the solution at the same rate.

In contrast to the solubilities of solids, gas solubilities in water decrease with increasing temperature. For example, both N_2 and O_2, the major components of air, are less soluble in hot water than in cold water.

Effect of Pressure on Solubility

Pressure has a major effect on the solubility of gases in water. The amount of gas that will dissolve in a liquid at a given temperature is directly related to the pressure of the gas above the liquid. In other words, as the pressure of a gas above a liquid increases, the solubility of the gas increases; conversely, as the pressure of the gas decreases, its solubility decreases. Chemical Portraits 12 profiles important consequences of solubility behavior as a function of temperature and pressure for the gases CO_2, N_2, and O_2. Pressure has little effect on the solubility of solids and liquids in water.

Saturated and Unsaturated Solutions

▶ When the amount of dissolved solute in a solution corresponds to the solute's solubility in the solvent, the solution formed is a saturated solution.

A **saturated solution** *is a solution that contains the maximum amount of solute that can be dissolved under the conditions at which the solution exists.* A saturated solution containing excess undissolved solute is an equilibrium situation where an amount of undissolved solute is continuously dissolving while an equal amount of dissolved solute is continuously crystallizing. Consider the process of adding table sugar (sucrose) to a container of water. Initially, the added sugar dissolves as the solution is stirred. Finally, as we add more sugar, we reach a point where no amount of stirring will cause the added sugar to dissolve. The last-added sugar remains as a solid on the bottom of the container; the solution is saturated. Although it appears to the eye that nothing is happening once the saturation point is reached, this is not the case on the molecular level. Solid sugar from the bottom of the container is continuously dissolving in the water, and an equal amount of sugar is coming out of solution. Accordingly, the net number of sugar molecules in the liquid remains the same. The equilibrium situation in the saturated solution is somewhat similar to the evaporation of a liquid in a closed container (Section 6.11). Figure 7.3 illustrates the dynamic equilibrium process occurring in a saturated solution that contains undissolved excess solute.

▶ When the term *solution* is used, it is generally assumed that "aqueous solution" is meant, unless the context makes it clear that the solvent is not water.

An **unsaturated solution** *is a solution in which less solute than the maximum amount possible is dissolved in the solution.* Most solutions we encounter fall into this category.

The terms *concentrated* and *dilute* are also used to convey qualitative information about the degree of saturation of a solution. A **concentrated solution** *is a solution that contains a large amount of solute relative to the amount that could dissolve.* A concentrated solution does not have to be a saturated solution (see Figure 7.4). A **dilute solution** *is a solution that contains a small amount of solute relative to the amount that could dissolve.*

Figure 7.4 Both solutions contain the same amount of solute. A concentrated solution (left) contains a relatively large amount of solute compared with the amount of solvent. A dilute solution (right) contains a relatively small amount of solute compared with the amount of solvent.

Aqueous and Nonaqueous Solutions

Another set of terms related to solutions is *aqueous* and *nonaqueous*. An **aqueous solution** *is a solution in which water is the solvent.* The presence of water is not a prerequisite for a solution, however. A **nonaqueous solution** *is a solution in which water is not the solvent.* Alcohol-based solutions, rather than water-based solutions, are often encountered in a medical setting.

Practice Problems and Questions

7.5 Using Table 7.1, determine whether each of the following solutions is *saturated* or *unsaturated*.

a. 1.94 g of $PbBr_2$ in 100 g of H_2O at 50°C
b. 34.0 g of NaCl in 100 g of H_2O at 0°C
c. 75.4 g of $CuSO_4$ in 200 g of H_2O at 100°C
d. 0.540 g of Ag_2SO_4 in 50 g of H_2O at 50°C

7.6 Using Table 7.1, determine whether each of the following solutions is *concentrated* or *dilute*.
a. 0.20 g of $CuSO_4$ in 100 g of H_2O at 100°C
b. 1.50 g of $PbBr_2$ in 100 g of H_2O at 50°C
c. 61 g of $AgNO_3$ in 100 g of H_2O at 50°C
d. 0.50 g of Ag_2SO_4 in 100 g of H_2O at 0°C

7.7 Classify each of the following as an *aqueous* or a *nonaqueous* solution.
a. 5.00 g of sodium chloride and 50.0 g of water
b. 50.0 g of ethyl alcohol and 40.0 g of water
c. 40.0 g of ethyl alcohol and 50.0 g of water
d. 4.00 g of table sugar and 20.0 g of ethyl alcohol

7.3 Solution Formation

Learning Focus

Describe, at a molecular level, the solution process as an ionic solute dissolves in water.

In a solution, solute particles are uniformly dispersed throughout the solvent. Considering what happens at the molecular level during the solution process will help us understand how this is accomplished.

In order for a solute to dissolve in a solvent, two types of interparticle attractions must be overcome: (1) attractions between solute particles (solute–solute attractions) and (2) attractions between solvent particles (solvent–solvent attractions). Only when these attractions are overcome can particles in both pure solute and pure solvent separate from one another and begin to intermingle. A new type of interaction, which does not exist prior to solution formation, arises as a result of the mixing of solute and solvent. This new interaction is the attraction between solute and solvent particles (solute–solvent attractions). These attractions are the primary driving force for solution formation.

An important type of solution process is one in which an ionic solid dissolves in water. Let us consider in detail the process of dissolving sodium chloride, a typical ionic solid, in water (Figure 7.5). The polar water molecules become oriented in such a way that the negative oxygen portion points toward positive sodium ions and the positive hydrogen portions point toward negative chloride ions. As the polar water molecules begin to surround ions on the crystal surface, they exert sufficient attraction to cause these ions to break away from the crystal surface. After leaving the crystal, an ion retains its surrounding group of water molecules; it has become a *hydrated ion*. As each hydrated ion leaves the surface, other ions are exposed to the water, and the crystal is picked apart ion by ion. Once in solution, the hydrated ions are uniformly distributed either by stirring or by random collisions with other molecules or ions.

▶ The fact that water molecules are polar is very important in the dissolving of an ionic solid in water.

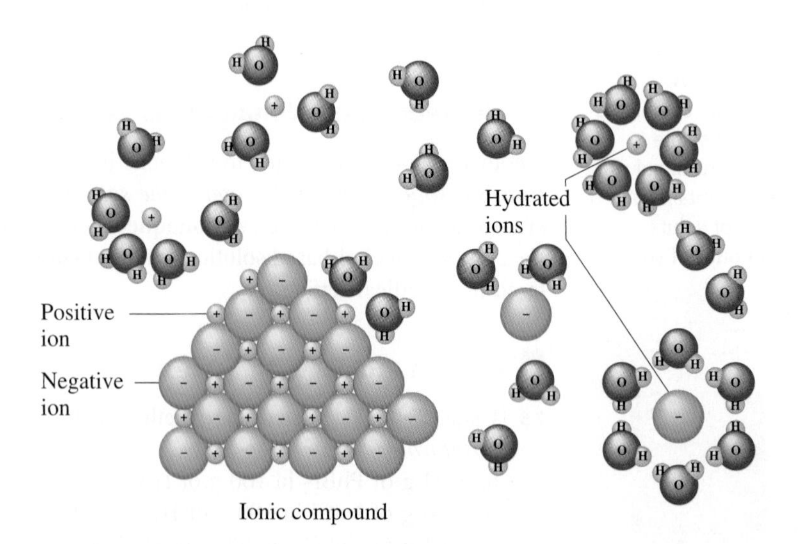

Figure 7.5 When an ionic solid, such as sodium chloride, dissolves in water, the water molecules *hydrate* the ions. The positive ions are bound to the water molecules by their attraction for the partial negative charge on the water's oxygen atom, and the negative ions are bound to the water molecules by their attraction for the partial positive charge on the water's hydrogen atoms.

The random motion of solute ions in solutions causes them to collide with one another, with solvent molecules, and occasionally with the surface of any undissolved solute. Ions undergoing the latter type of collision occasionally stick to the solid surface and thus leave the solution. When the number of ions in solution is low, the chances for collision with the undissolved solute are low. However, as the number of ions in solution increases, so do the chances for collisions, and more ions are recaptured by the undissolved solute. Eventually, the number of ions in solution reaches a level where ions return to the undissolved solute at the same rate at which other ions leave. At this point, the solution is saturated, and the equilibrium process discussed in the previous section is in operation.

Factors That Affect the Rate of Solution Formation

The rate at which a solution forms is governed by how fast the solute particles are distributed throughout the solvent. Three factors that affect the rate of solution formation are

▶ The rate of solution formation reflects how *fast* a given amount of solute dissolves rather than how *much* solute dissolves.

1. *The state of subdivision of the solute.* A crushed aspirin tablet will dissolve in water faster than a whole aspirin tablet. The more compact whole aspirin tablet has less surface area, and thus fewer solute molecules can interact with it at a given time.
2. *The degree of agitation within the mixture.* Stirring a mixture disperses the solute particles more rapidly, increasing the possibilities for solute–solvent interactions and hence boosting the rate of solution formation.
3. *The temperature of the mixture.* Solution formation occurs more rapidly as the temperature is increased. At a higher temperature, both solute and solvent molecules move more rapidly (Section 6.1), so more interactions between them occur within a given time period.

Practice Problems and Questions

7.8 Match the following statements concerning the dissolving of the ionic solid NaCl in water with the terms *hydrated ion, hydrogen atom,* and *oxygen atom.*
a. A Na^+ ion surrounded with water molecules
b. A Cl^- ion surrounded with water molecules
c. The portion of a water molecule attracted to a Na^+ ion
d. The portion of a water molecule attracted to a Cl^- ion

7.9 Indicate whether each of the following actions will *increase* or *decrease* the rate of the dissolving of a sugar cube in water.
a. Cooling the mixture of sugar cube and water
b. Stirring the mixture of sugar cube and water
c. Breaking the sugar cube up into smaller "chunks"
d. Crushing the sugar cube to give a granulated form of sugar

7.4 Solubility Rules

Learning Focus

Explain the solubility rule "like dissolves like," and be able to use the solubility guidelines for common ionic solutes in water.

In this section, we will present some rules for qualitatively predicting solute solubilities. These rules summarize in a concise form the results of thousands of experimental determinations of solute–solvent solubility.

A very useful generalization that relates polarity to solubility is that *substances of like polarity tend to be more soluble in each other than substances that differ in polarity.* This conclusion is often expressed as the simple phrase "*like dissolves like.*" Polar substances, in general, are good solvents for other polar substances but not for nonpolar substances (see Figure 7.6). Similarly, nonpolar substances exhibit greater solubility in nonpolar solvents than in polar solvents.

Table 7.2 Solubility Guidelines for Ionic Compounds in Water

Ion contained in the compound	Solubility	Exceptions
Group IA (Li^+, Na^+, K^+, etc.)	soluble	
ammonium (NH_4^+)	soluble	
nitrates (NO_3^-)	soluble	
chlorides (Cl^-), bromides (Br^-), and iodides (I^-)	soluble	Ag^+, Pb^{2+}, Hg_2^{2+}
sulfates (SO_4^{2-})	soluble	Ca^{2+}, Sr^{2+}, Ba^{2+}, Pb^{2+}
carbonates (CO_3^{2-})	insoluble[a]	Group IA and NH_4^+
phosphates (PO_4^{3-})	insoluble	Group IA and NH_4^+
hydroxides (OH^-)	insoluble	Group IA, Ca^{2+}, Sr^{2+}, Ba^{2+}

[a]All Ionic compounds, even the least soluble ones, dissolve to some slight extent in water. Thus the "insoluble" classification really means ionic compounds that have very limited solubility in water.

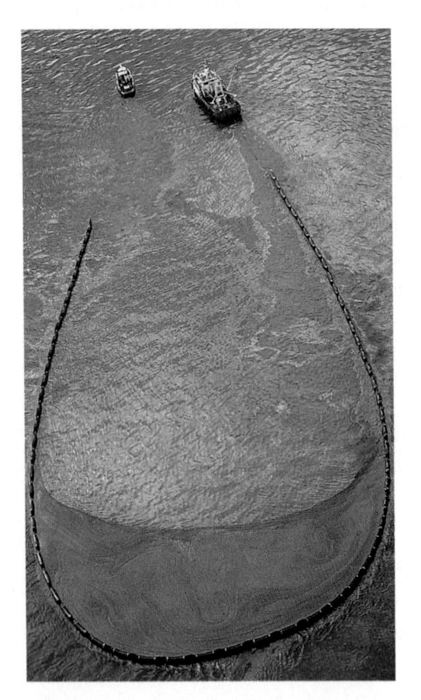

Figure 7.6 Oil spills can be contained to some extent using trawlers and a boom apparatus because oil and water, having different polarities, are relatively insoluble in each other. The oil, which is of lower density, floats on top of the water.

▶ The generalization "like dissolves like" is not adequate for predicting the solubilities of *ionic compounds* in water. More detailed solubility guidelines are needed (see Table 7.2).

The generalization "like dissolves like" is a useful tool for predicting solubility behavior in many, but not all, solute–solvent situations. Results that agree with this generalization are nearly always obtained in the cases of gas-in-liquid and liquid-in-liquid solutions and for solid-in-liquid solutions in which the solute is not an ionic compound. For example, NH_3 gas (a polar gas) is much more soluble in H_2O (a polar liquid) than is O_2 gas (a nonpolar gas).

In the common case of solid-in-liquid solutions in which the solute is an ionic compound, the rule "like dissolves like" is not adequate. Their polar nature would suggest that all ionic compounds are soluble in a polar solvent such as water, but this is not the case. The failure of the generalization for ionic compounds is related to the complexity of the factors involved in determining the magnitude of the solute–solute (ion–ion) and solvent–solute (solvent–ion) interactions. Among other things, both the charge and the size of the ions in the solute must be considered. Changes in these factors affect both types of interactions, but not to the same extent.

Some guidelines concerning the solubility of ionic compounds in water, which should be used in place of "like dissolves like," are given in Table 7.2.

Example 7.1 Predicting Solute Solubility Using Solubility Rules

With the help of Table 7.2, predict the solubility of the following solutes in the solvent indicated.

a. CH_4 (a nonpolar gas) in water
b. Ethyl alcohol (a polar liquid) in chloroform (a polar liquid)
c. AgCl (an ionic solid) in water
d. Na_2SO_4 (an ionic solid) in water
e. $AgNO_3$ (an ionic solid) in water

Solution

a. Insoluble. They are of unlike polarity because water is polar.
b. Soluble. Both substances are polar, so they should be relatively soluble in one another—like dissolves like.
c. Insoluble. Table 7.2 indicates that all chlorides except those of silver, lead, and mercury(I) are soluble. Thus AgCl is one of the exceptions.
d. Soluble. Table 7.2 indicates that all ionic sodium-containing compounds are soluble.
e. Soluble. Table 7.2 indicates that all compounds containing the nitrate ion (NO_3^-) are soluble.

Practice Problems and Questions

7.10 Predict whether the following solutes are very soluble or slightly soluble in water.
a. O_2 (a nonpolar gas) b. CH_3OH (a polar liquid)
c. CBr_4 (a nonpolar liquid) d. NH_3 (a polar gas)

7.11 Assign each of the following types of ionic compounds to the solubility categories *soluble, soluble with exceptions, insoluble,* and *insoluble with exceptions.*
a. Chlorides and sulfates
b. Nitrates and ammonium-ion-containing
c. Carbonates and phosphates
d. Sodium-ion-containing and potassium-ion-containing

7.12 In each of the following sets of ionic compounds, identify, with the help of Table 7.2, the members of the set that are soluble in water.
a. $NaCl$, Na_2SO_4, $NaNO_3$, Na_2CO_3
b. $AgNO_3$, KNO_3, $Ca(NO_3)_2$, $Cu(NO_3)_2$
c. $AgBr$, $CuBr_2$, $PbBr_2$, $NaBr$
d. Ag_2CO_3, $(NH_4)_2CO_3$, $BaCO_3$, $CuCO_3$

Learning Focus

Define and work problems involving the concentration units *percent by mass, percent by volume, mass–volume percent,* and *molarity.*

7.5 Solution Concentration Units

Because solutions are mixtures (Section 7.1), they have a variable composition. Specifying what the composition of a solution is involves specifying solute concentrations. In general, the **concentration** *of a solution is the amount of solute present in a specified amount of solution.* Many methods of expressing concentration exist, and certain methods are better suited for some purposes than others. In this section we consider two methods: *percent concentration* and *molarity.*

Percent Concentration

There are three different ways of representing percent concentration:

1. Percent by mass (or mass–mass percent)
2. Percent by volume (or volume–volume percent)
3. Mass–volume percent

Percent by mass (or mass–mass percent) is the percentage unit most often used in chemical laboratories. **Percent by mass** *is equal to the mass of solute divided by the total mass of solution, multiplied by 100 (to put the value in terms of percentage).*

$$\text{Percent by mass} = \frac{\text{mass of solute}}{\text{mass of solution}} \times 100$$

The solute and solution masses must be measured in the same unit, which is usually grams. The mass of the solution is equal to the mass of the solute plus the mass of the solvent.

$$\text{Mass of solution} = \text{mass of solute} + \text{mass of solvent}$$

▶ The concentration of butterfat in milk is expressed in terms of percent by mass. When you buy 1% milk, you are buying milk that contains 1 g of butterfat per 100 g of milk.

A solution whose mass percent concentration is 5.0% would contain 5.0 g of solute per 100.0 g of solution (5.0 g of solute and 95.0 g of solvent). Thus percent by mass directly gives the number of grams of solute in 100 g of solution. The percent-by-mass concentration unit is often abbreviated as %(m/m).

Example 7.2 Calculating the Percent-by-Mass Concentration of a Solution

What is the percent-by-mass, %(m/m), concentration of sucrose (table sugar) in a solution made by dissolving 7.6 g of sucrose in 83.4 g of water?

Solution

Both the mass of solute and the mass of solvent are known. Substituting these numbers into the percent-by-mass equation

$$\%(\text{m/m}) = \frac{\text{mass of solute}}{\text{mass of solution}} \times 100$$

gives

$$\%(\text{m/m}) = \frac{7.6 \text{ g sucrose}}{7.6 \text{ g sucrose} + 83.4 \text{ g water}} \times 100$$

Remember that the denominator of the preceding equation (mass of solution) is the combined mass of the solute and the solvent.

Doing the mathematics gives

$$\%(\text{m/m}) = \frac{7.6 \text{ g}}{91.0 \text{ g}} \times 100 = 8.4\%$$

Example 7.3 Calculating the Mass of Solute Needed to Produce a Solution of a Given Percent-by-Mass Concentration

How many grams of sucrose must be added to 375 g of water to prepare a 2.75%(m/m) solution of sucrose?

Solution

Often, when a solution concentration is given as part of a problem statement, the concentration information is used in the form of a conversion factor when you solve the problem. That will be the case in this problem.

The given quantity is 375 g of H_2O (grams of solvent), and the desired quantity is grams of sucrose (grams of solute).

$$375 \text{ g } H_2O = ? \text{ g sucrose}$$

The conversion factor relating these two quantities (solvent and solute) is obtained from the given concentration. In a 2.75%-by-mass sucrose solution, there are 2.75 g of sucrose for every 97.25 g of water.

$$100.00 \text{ g solution} - 2.75 \text{ g sucrose} = 97.25 \text{ g } H_2O$$

The relationship between grams of solute and grams of solvent (2.75 to 97.25) gives us the needed conversion factor.

$$\frac{2.75 \text{ g sucrose}}{97.25 \text{ g } H_2O}$$

The problem is set up and solved, using dimensional analysis, as follows:

$$375 \text{ g } \cancel{H_2O} \times \left(\frac{2.75 \text{ g sucrose}}{97.25 \text{ g } \cancel{H_2O}}\right) = 10.6 \text{ g sucrose}$$

The second type of percentage unit, percent by volume (or volume–volume percent), which is abbreviated %(v/v), is used as a concentration unit in situations where the solute and solvent are both liquids or both gases. In these cases, it is more convenient to measure volumes than masses. **Percent by volume** *is equal to the volume of solute divided by the total volume of solution, multiplied by 100.*

$$\text{Percent by volume} = \frac{\text{volume of solute}}{\text{volume of solution}} \times 100$$

Solute and solution volumes must always be expressed in the same units when you use percent by volume.

When the numerical value of a concentration is expressed as a percent by volume, it directly gives the number of milliliters of solute in 100 mL of solution. Thus a 100-mL sample of a 5.0%(v/v) alcohol-in-water solution contains 5.0 mL of alcohol dissolved in enough water to give 100 mL of solution. Note that such a 5.0%(v/v) solution could not be made by adding 5 mL of alcohol to 95 mL of water, because the volumes of two liquids are not usually additive. Differences in the way molecules are packed, as well as differences in distances between molecules, almost always result in the volume of the solution being less than the sum of the volumes of solute and solvent (see Figure 7.7). For example, the final volume resulting from the addition of 50.0 mL of ethyl alcohol to 50.0 mL of water is 96.5 mL of solution (see Figure 7.8). Working problems involving percent by volume entails the same kinds of steps as those used for problems involving percent by mass.

▶ The proof system for specifying the alcoholic content of beverages is twice the percent by volume. Hence 40 proof is 20%(v/v) alcohol; 100 proof is 50%(v/v) alcohol.

The third type of percentage unit in common use is mass–volume percentage, abbreviated %(m/v). This unit, which is often encountered in clinical and hospital settings, is particularly convenient to use when you work with a solid solute, which is easily weighed, and a liquid solvent. Solutions of drugs for internal and external use, intravenous and intramuscular injectables, and reagent solutions for testing are usually labeled in mass–volume percent.

Mass–volume percent *is equal to the mass of solute (in grams) divided by the total volume of solution (in milliliters), multiplied by* 100.

$$\text{Mass–volume percent} = \frac{\text{mass of solute (g)}}{\text{volume of solution (mL)}} \times 100$$

Note that in the definition of mass–volume percent, specific mass and volume units are given. This is necessary because the units do not cancel, as was the case with mass percent and volume percent.

▶ For dilute aqueous solutions, %(m/m) and %(m/v) are almost the same, because mass in grams of the solution equals the volume in milliliters when the density is close to 1.00 g/mL, as it is for pure water and for most dilute solutions.

Mass–volume percent indicates the number of grams of solute dissolved in each 100 mL of solution. Thus a 2.3%(m/v) solution of any solute contains 2.3 g of solute in each 100 mL of solution, and a 5.4%(m/v) solution contains 5.4 g of solute in each 100 mL of solution.

Figure 7.7 When volumes of two different liquids are combined, the volumes are not additive. This process is somewhat analogous to pouring marbles and golf balls together. The marbles can fill in the spaces between the golf balls. This results in the "mixed" volume being less than the sum of the "premixed" volumes.

Figure 7.8 Identical volumetric flasks are filled to the 50.0-mL mark with ethanol and with water. When the two liquids are poured into a 100-mL volumetric flask, the volume is seen to be less than the expected 100.0 mL; it is only 96.5 mL.

Example 7.4 **Calculating the Mass of Solute Needed to Produce a Solution of a Given Mass–Volume Percent Concentration**

Normal saline solution that is used to dissolve drugs for intravenous use is 0.92%(m/v) NaCl in water. How many grams of NaCl are required to prepare 35.0 mL of normal saline solution?

Solution

The given quantity is 35.0 mL of solution, and the desired quantity is grams of NaCl.

$$35.0 \text{ mL solution} = ? \text{ g NaCl}$$

The given concentration, 0.92%(m/v), which means 0.92 g of NaCl per 100 mL of solution, is used as a conversion factor to go from milliliters of solution to grams of NaCl. The setup for the conversion is

$$35.0 \cancel{\text{mL solution}} \times \left(\frac{0.92 \text{ g NaCl}}{100 \cancel{\text{mL solution}}}\right)$$

Doing the arithmetic after canceling the units gives

$$\left(\frac{35.0 \times 0.92}{100}\right) \text{g NaCl} = 0.32 \text{ g NaCl}$$

Molarity

The **molarity** (M) *of a solution is a ratio giving the number of moles of solute per liter of solution.* The mathematical equation for molarity is

$$\text{Molarity (M)} = \frac{\text{moles of solute}}{\text{liters of solution}}$$

A solution containing 1 mole of KBr in 1 L of solution has a molarity of 1 and is said to be a 1 M (1 *molar*) solution.

This concentration unit is often used in laboratories where chemical reactions are being studied. Because chemical reactions occur between molecules and atoms, the mole—a unit that counts particles—is desirable. Equal volumes of two solutions of the same molarity contain the same number of solute molecules.

► In preparing 100 mL of a solution of a specific molarity, enough solvent is added to a weighed amount of solute to give a *final* volume of 100 mL. The weighed solute is not added to a *starting* volume of 100 mL; this would produce a final volume greater than 100 mL, because the solute volume increases the total volume.

In order to find the molarity of a solution, we need to know the solution volume in liters and the number of moles of solute present. An alternative to knowing the number of moles of solute is knowing the number of grams of solute present and the solute's formula mass. The number of moles can be calculated by using these two quantities (Section 5.3).

Example 7.5 Calculating the Molarity of a Solution

Determine the molarities of the following solutions.

a. 4.35 moles of $KMnO_4$ are dissolved in enough water to give 750 mL of solution.
b. 20.0 g of NaOH is dissolved in enough water to give 1.50 L of solution.

Solution

a. The number of moles of solute is given in the problem statement.

$$\text{Moles of solute } (KMnO_4) = 4.35 \text{ moles}$$

The volume of the solution is also given in the problem statement, but not in the right units. Molarity requires liters for the volume units, and we are given milliliters of solution. Making the unit change yields

$$750 \cancel{\text{mL}} \times \left(\frac{10^{-3}\text{ L}}{1 \cancel{\text{mL}}}\right) = 0.750 \text{ L}$$

The molarity of the solution is obtained by substituting the known quantities into the equation

$$M = \frac{\text{moles of solute}}{\text{liters of solution}}$$

which gives

$$M = \frac{4.35 \text{ moles } KMnO_4}{0.750 \text{ L solution}} = 5.80 \frac{\text{moles } KMnO_4}{\text{L solution}}$$

Note that the units for molarity are always moles per liter.

b. This time, the volume of solution is given in liters.

$$\text{Volume of solution} = 1.50 \text{ L}$$

The moles of solute must be calculated from the grams of solute (given) and the solute's molar mass, which is 40.00 g/mole (calculated from atomic masses).

$$20.0 \cancel{\text{g NaOH}} \times \left(\frac{1 \text{ mole NaOH}}{40.00 \cancel{\text{g NaOH}}}\right) = 0.500 \text{ mole NaOH}$$

Substituting the known quantities into the defining equation for molarity gives

$$M = \frac{0.500 \text{ mole NaOH}}{1.50 \text{ L solution}} = 0.333 \frac{\text{mole NaOH}}{\text{L solution}}$$

The mass of solute present in a known volume of solution is an easily calculable quantity if the molarity of the solution is known. When we do such a calculation, molarity serves as a conversion factor that relates liters of solution to moles of solute. In a similar manner, the volume of solution needed to supply a given amount of solute can be calculated by using the solution's molarity as a conversion factor.

Example 7.6 Calculating the Amount of Solute Present in a Given Amount of Solution

How many grams of sucrose (table sugar, $C_{12}H_{22}O_{11}$) are present in 185 mL of a 2.50 M sucrose solution?

Solution

The given quantity is 185 mL of solution, and the desired quantity is grams of $C_{12}H_{22}O_{11}$.

$$185 \text{ mL of solution} = ? \text{ g } C_{12}H_{22}O_{11}$$

The pathway used to solve this problem is

$$\text{mL solution} \longrightarrow \text{L solution} \longrightarrow \text{moles } C_{12}H_{22}O_{11} \longrightarrow \text{g } C_{12}H_{22}O_{11}$$

The given molarity (2.50 M) serves as the conversion factor for the second unit change; the formula mass of sucrose (which is not given and must be calculated) is used to accomplish the third unit change.

The dimensional-analysis setup for this pathway is

$$185 \cancel{\text{mL solution}} \times \left(\frac{10^{-3} \cancel{\text{L solution}}}{1 \cancel{\text{mL solution}}}\right) \times \left(\frac{2.50 \cancel{\text{moles } C_{12}H_{22}O_{11}}}{1 \cancel{\text{L solution}}}\right) \times \left(\frac{342.34 \text{ g } C_{12}H_{22}O_{11}}{1 \cancel{\text{mole } C_{12}H_{22}O_{11}}}\right)$$

Canceling the units and doing the arithmetic, we find that

$$\left(\frac{185 \times 10^{-3} \times 2.50 \times 342.34}{1 \times 1 \times 1}\right) \text{g } C_{12}H_{22}O_{11} = 158 \text{ g } C_{12}H_{22}O_{11}$$

Example 7.7 Calculating the Amount of Solution Needed to Supply a Given Amount of Solute

A typical dose of iron(II) sulfate ($FeSO_4$) used in the treatment of iron-deficiency anemia is 0.35 g. How many milliliters of a 0.10 M iron(II) sulfate solution would be needed to supply this dose?

Solution

The given quantity is 0.35 g of $FeSO_4$, and the desired quantity is milliliters of $FeSO_4$ solution.

$$0.35 \text{ g } FeSO_4 = ? \text{ mL } FeSO_4 \text{ solution}$$

The pathway used to solve this problem is

$$\text{g } FeSO_4 \longrightarrow \text{moles } FeSO_4 \longrightarrow \text{L } FeSO_4 \text{ solution} \longrightarrow \text{mL } FeSO_4 \text{ solution}$$

We accomplish the first unit conversion by using the formula mass of $FeSO_4$ (which must be calculated) as a conversion factor. The second unit conversion involves the use of the given molarity as a conversion factor.

$$0.35 \cancel{\text{g } FeSO_4} \times \left(\frac{1 \cancel{\text{mole } FeSO_4}}{151.92 \cancel{\text{g } FeSO_4}}\right) \times \left(\frac{1 \cancel{\text{L solution}}}{0.10 \cancel{\text{mole } FeSO_4}}\right) \times \left(\frac{1 \text{ mL solution}}{10^{-3} \cancel{\text{L solution}}}\right)$$

Canceling units and doing the arithmetic, we find that

$$\left(\frac{0.35 \times 1 \times 1 \times 1}{151.92 \times 0.10 \times 10^{-3}}\right) \text{mL solution} = 23 \text{ mL solution}$$

See the accompanying Chemistry at a Glance for a review of solutions and the ways in which we represent their concentration.

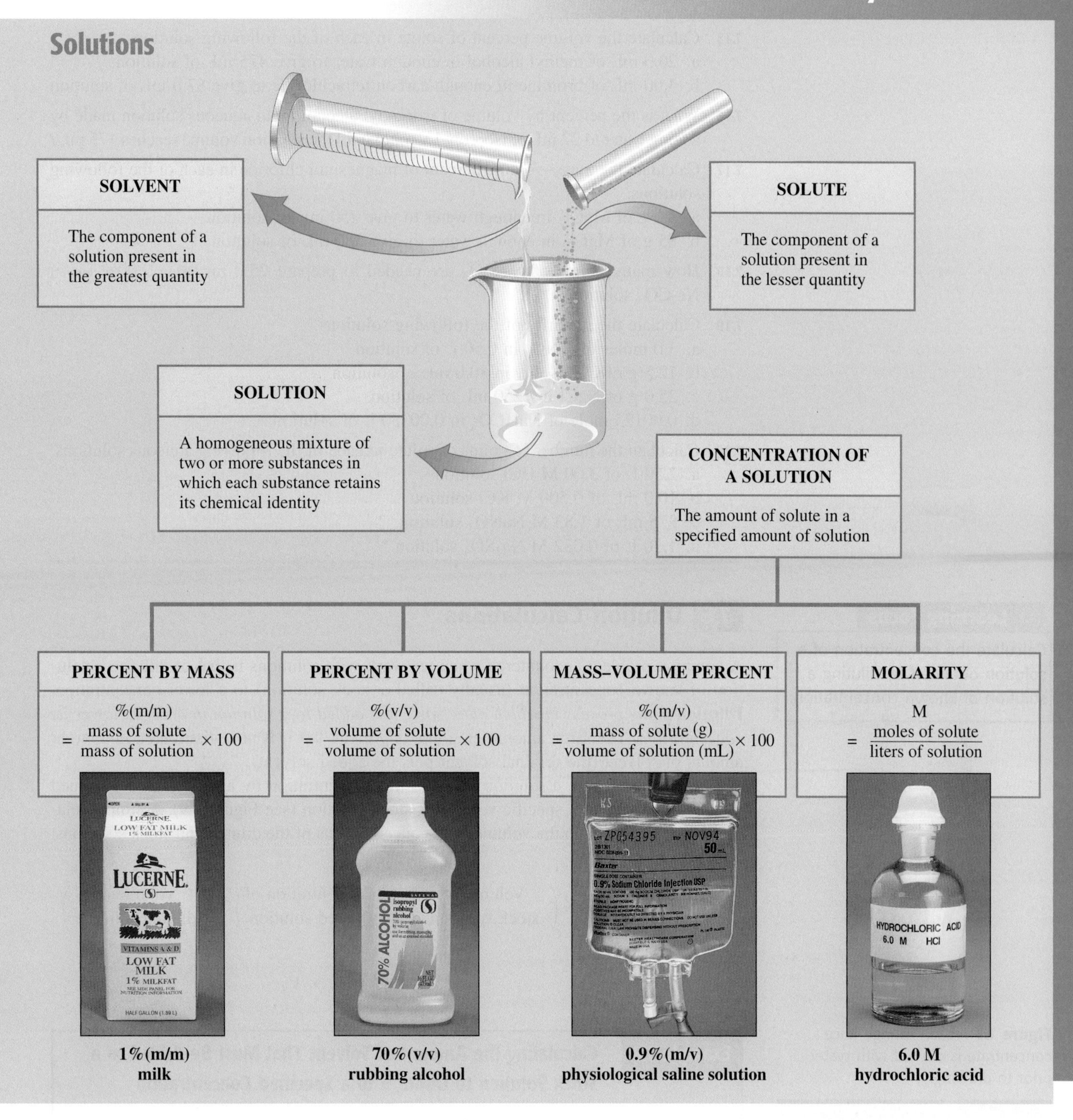

Chemistry at a Glance
Solutions
SOLVENT
The component of a solution present in the greatest quantity
SOLUTE
The component of a solution present in the lesser quantity
SOLUTION
A homogeneous mixture of two or more substances in which each substance retains its chemical identity
CONCENTRATION OF A SOLUTION
The amount of solute in a specified amount of solution
PERCENT BY MASS
%(m/m) = mass of solute / mass of solution × 100
PERCENT BY VOLUME
%(v/v) = volume of solute / volume of solution × 100
MASS–VOLUME PERCENT
%(m/v) = mass of solute (g) / volume of solution (mL) × 100
MOLARITY
M = moles of solute / liters of solution
LUCERNE
LOW FAT MILK
1% MILKFAT
70% ALCOHOL
0.9% Sodium Chloride Injection USP
HYDROCHLORIC ACID
6.0 M HCl
1%(m/m) milk
70%(v/v) rubbing alcohol
0.9%(m/v) physiological saline solution
6.0 M hydrochloric acid

Practice Problems and Questions

7.13 Calculate the mass percent of solute in the following solutions.

a. 6.50 g of NaCl dissolved in 85.0 g of H_2O
b. 2.31 g of LiBr dissolved in 35.0 g of H_2O
c. 12.5 g of KNO_3 dissolved in 125 g of H_2O
d. 0.0032 g of NaOH dissolved in 1.2 g of H_2O

7.14 How many grams of glucose must be added to 275 g of water in order to prepare each of the following percent-by-mass concentrations of aqueous glucose solution?

a. 1.30% b. 5.00% c. 20.0% d. 31.0%

7.15 Calculate the volume percent of solute in each of the following solutions.
a. 20.0 mL of methyl alcohol in enough water to give 475 mL of solution
b. 4.00 mL of bromine in enough carbon tetrachloride to give 87.0 mL of solution

7.16 What is the percent by volume of isopropyl alcohol in an aqueous solution made by adding water to 22 mL of isopropyl alcohol until the solution volume reaches 175 mL?

7.17 Calculate the mass–volume percent of magnesium chloride in each of the following solutions.
a. 5.0 g of $MgCl_2$ in enough water to give 250 mL of solution
b. 85 g of $MgCl_2$ in enough water to give 580 mL of solution

7.18 How many grams of Na_2CO_3 are needed to prepare 25.0 mL of a 2.00% (m/v) Na_2CO_3 solution?

7.19 Calculate the molarity of the following solutions.
a. 3.0 moles of KNO_3 in 0.50 L of solution
b. 12.5 g of $C_{12}H_{22}O_{11}$ in 80.0 mL of solution
c. 25.0 g of NaCl in 1250 mL of solution
d. 0.00125 mole of $NaHCO_3$ in 0.00250 L of solution

7.20 Calculate the number of grams of solute in each of the following aqueous solutions.
a. 2.50 L of 3.00 M HCl solution
b. 10.0 mL of 0.500 M KCl solution
c. 875 mL of 1.83 M $NaNO_3$ solution
d. 1.20 L of 0.032 M Na_2SO_4 solution

7.6 Dilution Calculations

Learning Focus

Calculate the concentration of a solution obtained by diluting a solution of known concentration.

A common problem encountered when we work with solutions is that of diluting a solution of known concentration (usually called a stock solution) to a lower concentration. **Dilution** *is the process in which more solvent is added to a solution in order to lower its concentration.* The same amount of solute is present, but it is now distributed in a larger amount of solvent (the original solvent plus the added solvent).

Often, we prepare a solution of a specific concentration by adding a predetermined volume of solvent to a specific volume of stock solution (see Figure 7.9). A simple relationship exists between the volumes and concentrations of the diluted and stock solutions. This relationship is

$$\begin{pmatrix}\text{Concentration of}\\\text{stock solution}\end{pmatrix} \times \begin{pmatrix}\text{Volume of}\\\text{stock solution}\end{pmatrix} = \begin{pmatrix}\text{Concentration of}\\\text{diluted solution}\end{pmatrix} \times \begin{pmatrix}\text{Volume of}\\\text{diluted solution}\end{pmatrix}$$

or

$$C_s \times V_s = C_d \times V_d$$

Figure 7.9 Frozen orange juice concentrate is diluted with water prior to drinking.

Example 7.8 Calculating the Amount of Solvent That Must Be Added to a Stock Solution to Dilute it to a Specified Concentration

A nurse wants to prepare a 1.0%(m/v) silver nitrate solution from 24 mL of a 3.0%(m/v) stock solution of silver nitrate. How much water should be added to the 24 mL of stock solution?

Solution

The volume of water to be added will be equal to the difference between the final and initial volumes. The initial volume is known (24 mL). The final volume can be calculated by using the equation.

$$C_s \times V_s = C_d \times V_d$$

Once the final volume is known, the difference between the two volumes can be obtained.

Substituting the known quantities into the dilution equation, which has been rearranged to isolate V_d on the left side, gives

$$V_d = \frac{C_s \times V_s}{C_d} = \frac{3.0\%\cancel{(m/v)} \times 24\ \text{mL}}{1.0\%\cancel{(m/v)}} = 72\ \text{mL}$$

The solvent added is

$$V_d - V_s = (72 - 24)\ \text{mL} = 48\ \text{mL}$$

Practice Problems and Questions

7.21 What is the molarity of the solution prepared by diluting 25.0 mL of 0.220 M NaCl to each of the following final volumes?
a. 30.0 mL b. 75.0 mL c. 457 mL d. 2.00 L

7.22 For each of the following solutions, how many milliliters of water should be added to yield a solution that has a concentration of 0.100 M?
a. 50.0 mL of 3.00 M NaCl b. 2.00 mL of 1.00 M NaCl
c. 1.45 L of 6.00 M NaCl d. 75.0 mL of 0.110 M NaCl

7.23 Determine the final concentration of each of the following solutions after 20.0 mL of water has been added.
a. 30.0 mL of 5.0 M NaCl solution b. 30.0 mL of 5.0 M $AgNO_3$ solution
c. 30.0 mL of 7.5 M NaCl solution d. 60.0 mL of 2.0 M $AgNO_3$ solution

Learning Focus

Explain how the presence of a solute in a solution causes vapor-pressure lowering, boiling-point elevation and freezing-point depression, and discuss the important consequences of these colligative properties.

7.7 Colligative Properties of Solutions

Adding a solute to a pure solvent causes its physical properties to change. A special group of physical properties that change when a solute is added are called colligative properties. **Colligative properties** *are the physical properties of a solution that depend only on the number (concentration) of solute particles (molecules or ions) in a given quantity of solvent and not on their chemical identities.* Examples of colligative properties include vapor-pressure lowering, boiling-point elevation, freezing-point depression, and osmotic pressure. The first three of these colligative properties are discussed in this section. The fourth, osmotic pressure, will be considered in Section 7.8.

Adding a nonvolatile solute to a solvent *lowers* the vapor pressure of the resulting solution below that of the pure solvent at the same temperature. (A nonvolatile solute is one that has a low vapor pressure and therefore a low tendency to vaporize.) This lowering of vapor pressure is a direct consequence of some of the solute molecules or ions occupying positions on the surface of the liquid. Their presence decreases the probability of solvent molecules escaping; that is, the number of surface-occupying solvent molecules has been decreased. Figure 7.10 illustrates the decrease in surface concentration of solvent

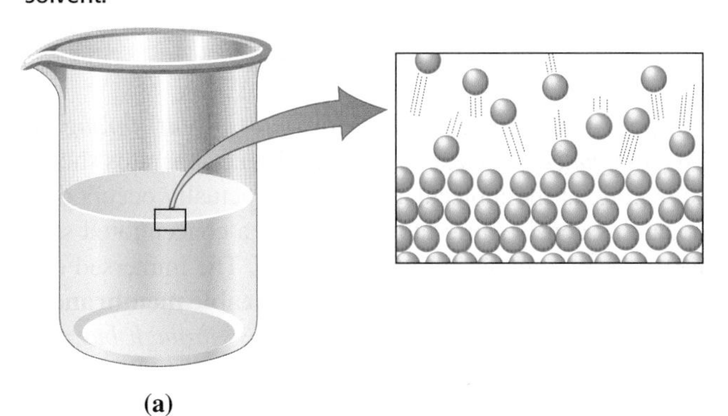
(a)

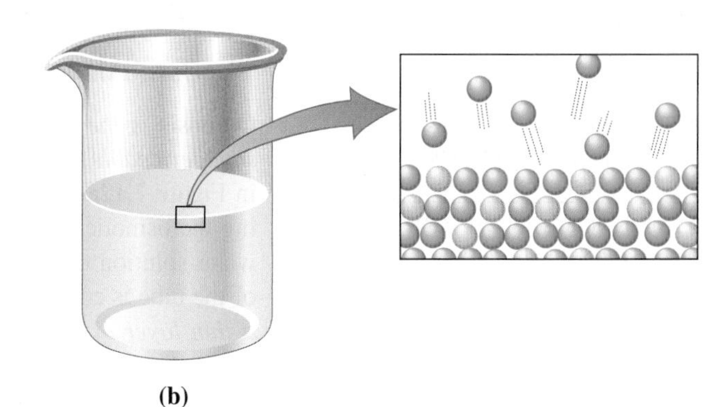
(b)

Figure 7.10 Close-ups of the surface of a liquid solvent (a) before and (b) after solute has been added. There are fewer solvent molecules on the surface of the liquid after solute has been added. This results in a decreased vapor pressure for the solution compared with pure solvent.

Figure 7.11 A water–antifreeze mixture has a higher boiling point and a lower freezing point than pure water.

▶ In the making of homemade ice cream, the function of the rock salt added to the ice is to depress the freezing point of the ice–water mixture surrounding the ice cream mix sufficiently to allow the mix (which contains sugar and other solutes and thus has a freezing point below 0°C) to freeze.

molecules when a solute is added. As the *number* of solute particles increases, the reduction in vapor pressure also increases; thus vapor pressure is a colligative property. What is important is not the identity of the solute molecules but the fact that they take up room on the surface of the liquid.

Adding a nonvolatile solute to a solvent *raises* the boiling point of the resulting solution above that of the pure solvent. This is logical when we remember that the vapor pressure of the solution is lower than that of pure solvent and that the boiling point is dependent on vapor pressure (Section 6.12). A higher temperature will be needed to raise the depressed vapor pressure of the solution to atmospheric pressure; this is the condition required for boiling.

A common application of the phenomenon of boiling point elevation involves automobiles. The coolant ethylene glycol (a nonvolatile solute) is added to car radiators to prevent boilover in hot weather (see Figure 7.11). The engine may not run any cooler, but the coolant–water mixture will not boil until it reaches a temperature well above the normal boiling point of water.

Adding a nonvolatile solute to a solvent *lowers* the freezing point of the resulting solution below that of the pure solvent. The presence of the solute particles within the solution interferes with the tendency of solvent molecules to line up in an organized manner, a condition necessary for the solid state. A lower temperature is necessary before the solvent molecules will form the solid.

Applications of freezing-point depression are even more numerous than those for boiling-point elevation. In climates where the temperature drops below 0°C in the winter, it is necessary to protect water-cooled automobile engines from freezing. This is done by adding antifreeze (usually ethylene glycol) to the radiator. The addition of this nonvolatile material causes the vapor pressure and freezing point of the resulting solution to be much lower than those of pure water. Also in the winter, salt, usually NaCl or $CaCl_2$, is spread on roads and sidewalks to melt ice or prevent it from forming. The salt dissolves in the water to form a solution that will not freeze until the temperature drops much lower than 0°C, the normal freezing point of water.

Practice Problems and Questions

7.24 Why is the vapor pressure of a solution that contains a nonvolatile solute always less than that of the pure solvent?

7.25 How are the boiling point and the freezing point of water affected by the addition of solute?

7.26 Why does seawater evaporate more slowly than fresh water at the same temperature?

7.27 How does the freezing point of seawater compare with that of fresh water?

Learning Focus

Discuss the process of osmosis, osmotic pressure, osmolarity, and the biological importance of osmotic pressure.

7.8 Osmosis and Osmotic Pressure

The process of osmosis and the colligative property of osmotic pressure are extremely important phenomena when we consider biological solutions. These phenomena govern many of the processes important to a functioning human body.

Osmosis

▶ The term *osmosis* comes from the Greek *osmos,* which means "push."

▶ An osmotic semipermeable membrane contains very small pores (holes)—too small to see—that are big enough to let small solvent molecules through but not big enough to let larger solute molecules pass through.

Osmosis *is the passage of a solvent from a dilute solution (or pure solvent) through a semipermeable membrane into a more concentrated solution.* The simple apparatus shown in Figure 7.12a is helpful in explaining, at the molecular level, what actually occurs during the osmotic process. The apparatus consists of a tube containing a concentrated salt-water solution that has been immersed in a dilute salt-water solution. The immersed end of the tube is covered with a semipermeable membrane. A **semipermeable membrane** *is a thin layer of material that allows certain types of molecules to pass through but prohibits the passage of others.* The selectivity of the membrane is based on size differences

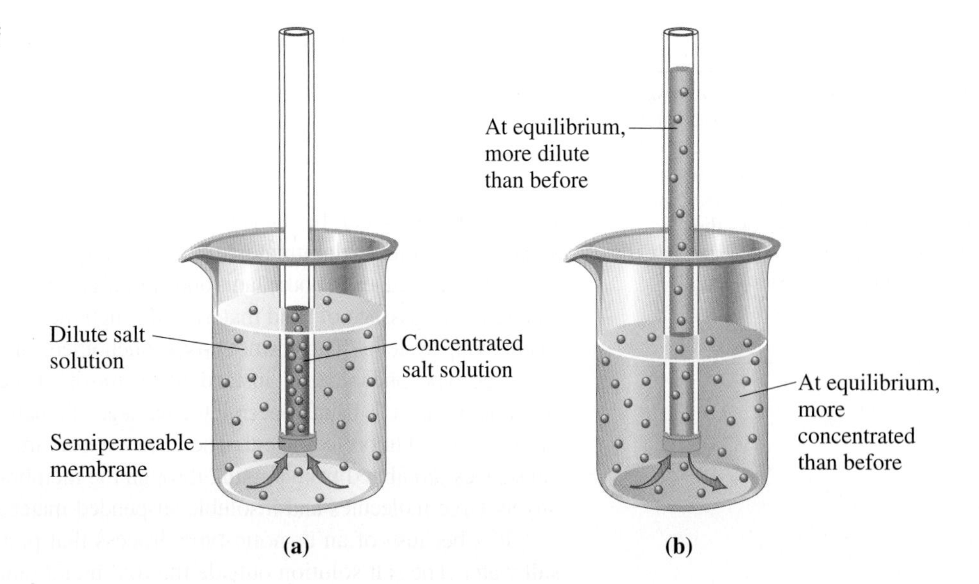

Figure 7.12 (a) Osmosis, the flow of solvent through a semipermeable membrane from a dilute to a more concentrated solution, can be observed with this apparatus. (b) At equilibrium, the molecules move back and forth at equal rates.

between molecules. The particles that are allowed to pass through (usually just solvent molecules like water) are relatively small. Thus, the membrane functions somewhat like a sieve. Using the experimental setup of Figure 7.12a, we can observe a net flow of solvent from the dilute to the concentrated solution over the course of time. This is indicated by a rise in the level of the solution in the tube and a drop in the level of the dilute solution, as shown in Figure 7.12b.

What is actually happening on a molecular level as the process of osmosis occurs? Water is flowing in both directions through the membrane. However, the rate of flow into the concentrated solution is greater than the rate of flow in the other direction (see Figure 7.13). Why? The presence of solute molecules diminishes the ability of water molecules to cross the membrane. The solute molecules literally get in the way; they occupy some of the surface positions next to the membrane. Because there is a greater concentration of solute molecules on one side of the membrane than on the other, the flow rates differ. The flow rate is diminished to a greater extent on the side of the membrane where the greater concentration of solute is present.

The net transfer of solvent across the membrane continues until (1) the concentrations of solute particles on both sides of the membrane become equal or (2) the hydrostatic pressure on the concentrated side of the membrane (from the difference in liquid levels) becomes sufficient to counterbalance the greater escaping tendency of molecules from the dilute side. From here on, there is an equal flow of solvent in both directions across the membrane, and the volume of liquid on each side of the membrane remains constant.

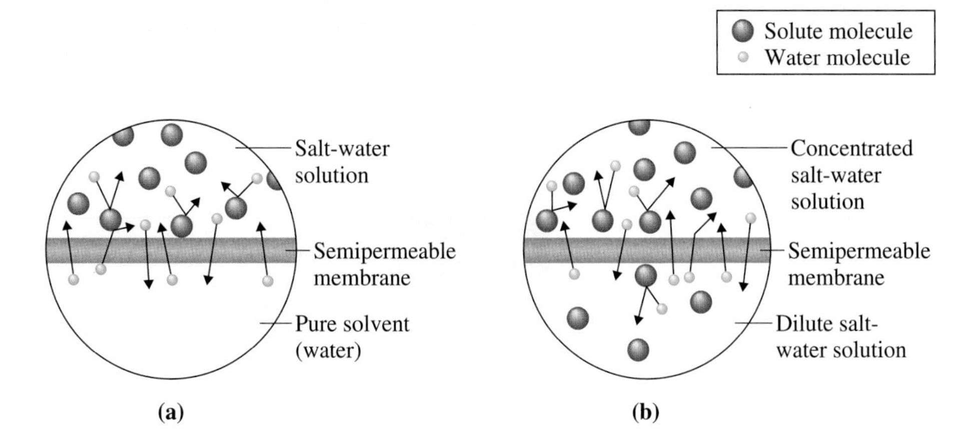

Figure 7.13 Enlarged views of a semipermeable membrane separating (a) pure water and a salt-water solution, and (b) a dilute salt-water solution and a concentrated salt-water solution. In both cases, water moves from the area of lower solute concentration to the area of higher solute concentration.

Osmotic Pressure

▶ A process called *reverse osmosis* is used in the desalination of seawater to make drinking water. Pressure greater than the osmotic pressure is applied on the salt-water side of the membrane to force solvent water across the membrane from the salt-water side to the "pure" water side.

Osmotic pressure *is the amount of pressure that must be applied to prevent the net flow of solvent through a semipermeable membrane from a solution of lower solute concentration to a solution of higher solute concentration.* In terms of Figure 7.12, osmotic pressure is the pressure required to prevent water from rising in the tube. Figure 7.14 shows how this pressure can be measured. The greater the concentration difference between the separated solutions, the greater the magnitude of the osmotic pressure.

Cell membranes in both plants and animals are semipermeable in nature (see Figure 7.15). The selective passage of fluid materials through these membranes governs the balance of fluids in living systems. Thus osmotic-type phenomena are of prime importance for life. We say "osmotic-type phenomena" instead of "osmosis" because the semipermeable membranes found in living cells usually permit the passage of small solute molecules (nutrients and waste products) in addition to solvent. The term *osmosis* implies the passage of solvent only. The substances prohibited from passing through the membrane in osmotic-type processes are relatively large molecules and insoluble suspended materials (see Dialysis, Section 7.9).

It is because of an osmotic-type process that plants will die if they are watered with salt water. The salt solution outside the root membranes is more concentrated than the solution in the root, so water flows out of the roots; then the plant becomes dehydrated and dies. This same principle is the reason for not drinking excessive amounts of salt water, even if you are stranded on a raft in the middle of the ocean. When salt water is taken into the stomach, water flows out of the stomach wall membranes and into the stomach; then the tissues become dehydrated. Drinking seawater will cause greater thirst because the body will lose water rather than absorb it.

Osmolarity

The osmotic pressure of a solution depends on the number of solute particles present. This in turn depends on the solute concentration and on whether the solute forms ions once it is in solution. Note that two factors are involved in determining osmotic pressure.

The fact that some solutes dissociate into ions in solution is of utmost importance in osmotic pressure considerations. For example, the osmotic pressure of a 1 M NaCl solution is twice that of a 1 M glucose solution, despite the fact that both solutions have equal concentrations (1 M). Sodium chloride is an ionic solute, and it dissociates in solution to give two particles (a Na^+ and a Cl^- ion) per formula unit; however, glucose is a molecular solute and does not dissociate. It is the number of particles present that determines osmotic pressure.

The concentration unit osmolarity is used to compare the osmotic pressures of solutions. The **osmolarity** *of a solution is the product of its molarity and the number of particles produced per formula unit when the solute dissociates.* The equation for osmolarity is

$$\text{Osmolarity} = \text{molarity} \times i$$

where i is the number of particles produced from the dissociation of one formula unit of solute. The abbreviation for osmolarity is osmol.

Solutions of equal osmolarity have equal osmotic pressures. If the osmolarity of one solution is three times that of another, then the osmotic pressure of the first solution is three times that of the second solution. A solution with high osmotic pressure will take up more water than a solution of lower osmotic pressure; thus more pressure must be applied to prevent osmosis.

Figure 7.14 Osmotic pressure is the amount of pressure needed to prevent the solution in the tube from rising as a result of the process of osmosis.

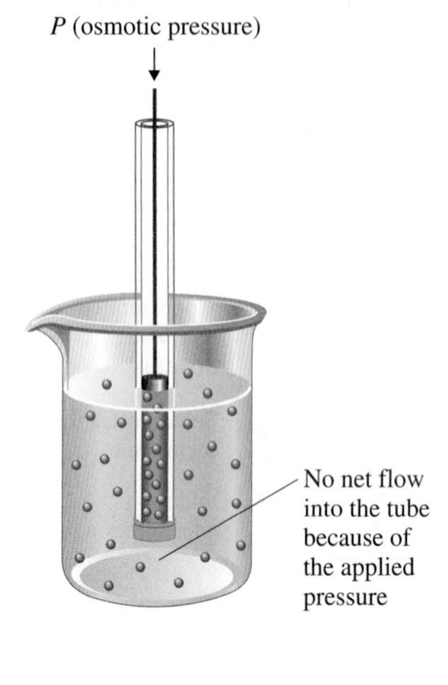
P (osmotic pressure)
No net flow into the tube because of the applied pressure

Example 7.9 Calculating the Osmolarity of Various Solutions

What is the osmolarity of each of the following solutions?

a. 2 M NaCl **b.** 2 M $CaCl_2$ **c.** 2 M glucose
d. 2 M in both NaCl and glucose
e. 2 M in NaCl and 1 M in glucose

Solution

The general equation for osmolarity that was previously given will be applicable in each of the parts of the problem.

$$\text{Osmolarity} = \text{molarity} \times i$$

a. Two particles per dissociation are produced when NaCl dissociates in solution.

$$NaCl \longrightarrow Na^+ + Cl^-$$

The value of i is 2, and the osmolarity is twice the molarity.

$$\text{Osmolarity} = 2\text{ M} \times 2 = 4\text{ osmol}$$

b. For $CaCl_2$, the value of i is 3, because three ions are produced from the dissociation of one $CaCl_2$ unit.

$$CaCl_2 \longrightarrow Ca^{2+} + 2\ Cl^-$$

The osmolarity will therefore be triple the molarity:

$$\text{Osmolarity} = 2\text{ M} \times 3 = 6\text{ osmol}$$

c. Glucose is a nondissociating solute. Thus the value of i is 1, and the molarity and osmolarity will be the same—two molar and two osmolar.

d. With two solutes present, we must consider the collective effects of both solutes. For NaCl, $i = 2$; and for glucose, $i = 1$. The osmolarity is calculated as follows:

$$\text{Osmolarity} = \underbrace{2\text{ M} \times 2}_{\text{NaCl}} + \underbrace{2\text{ M} \times 1}_{\text{glucose}} = 6\text{ osmol}$$

e. This problem differs from the previous one in that the two solutes are not present in equal concentrations. This does not change the way we work the problem. The i values are the same as before, and the osmolarity is

$$\text{Osmolarity} = \underbrace{2\text{ M} \times 2}_{\text{NaCl}} + \underbrace{1\text{ M} \times 1}_{\text{glucose}} = 5\text{ osmol}$$

Figure 7.15 The dissolved substances in tree sap create a more concentrated solution than the surrounding ground water. Water enters membranes in the roots and rises in the tree, creating an osmotic pressure that can exceed 20 atm in extremely tall trees.

Isotonic, Hypertonic, and Hypotonic Solutions

The terms *isotonic solution, hypertonic solution,* and *hypotonic solution* pertain to osmotic-type phenomena in the human body. A consideration of what happens to red blood cells when they are placed in three different liquids will help us understand the differences in meaning of these three terms. The liquids are distilled water, concentrated sodium chloride solution, and physiological saline solution.

When red blood cells are placed in pure water, they swell up (enlarge in size) and finally rupture (burst); this process is called *hemolysis* (Figure 7.16a). Hemolysis is caused by an increase in the amount of water entering the cells compared with the amount of water leaving the cells. This is the result of cell fluid having a greater osmotic pressure than pure water.

When red blood cells are placed in a concentrated sodium chloride solution, a process opposite to hemolysis occurs. This time, water moves from the cells to the solution, causing the cells to shrivel (shrink in size); this process is called *crenation* (Figure 7.16b). Crenation occurs because the osmotic pressure of the concentrated salt solution surrounding the red cells is greater than that of the fluid within the cells. Water always moves in the direction of greater osmotic pressure.

▶ The pickling of cucumbers and salt curing of meat are practical applications of the concept of crenation. A concentrated salt solution (brine) is used to draw water from the cells of the cucumber to produce a pickle. Salt on the surface of the meat preserves the meat by crenation of bacterial cells.

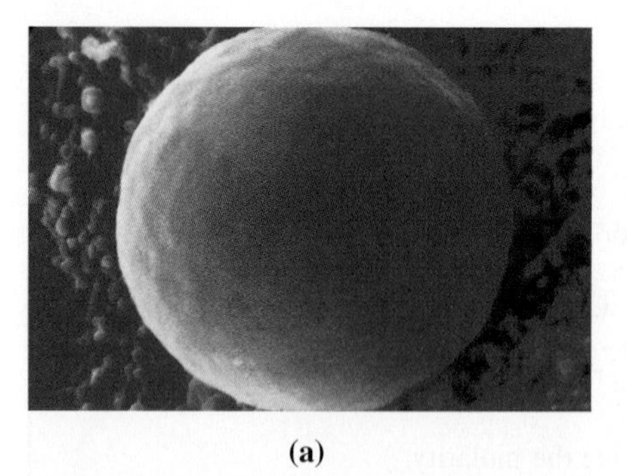
(a)

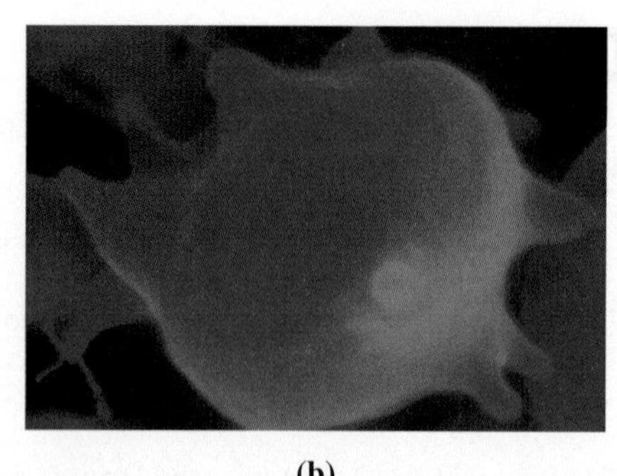
(b)

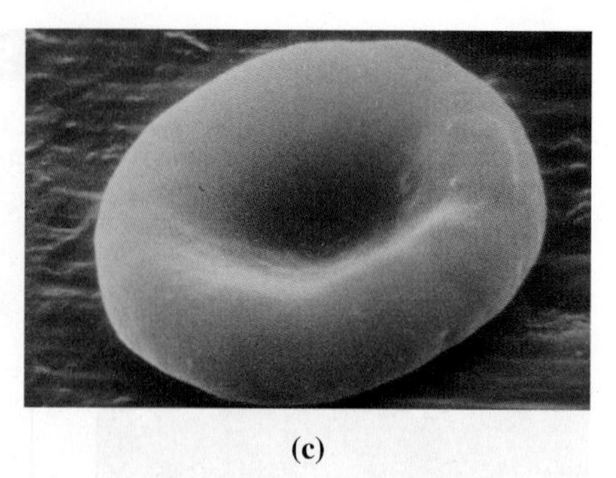
(c)

Figure 7.16 Effects of bathing red blood cells in various types of solutions. (a) Hemolysis in pure water (a hypotonic solution). (b) Crenation in concentrated sodium chloride solution (a hypertonic solution). (c) Cells neither swell nor shrink in a physiological saline solution (an isotonic solution).

▶ The molarity of a 5.0%(m/v) glucose solution is 0.31 M. The molarity of a 0.92%(m/v) NaCl solution is 0.16 M. Despite the differing molarities, these two solutions have the same osmotic pressure. The concept of osmolarity explains why solutions of different concentration can exhibit the same osmotic pressure.

▶ The use of 5%(m/v) glucose solution for intravenous feeding has a shortcoming. A patient can accommodate only about 3 L of water in a day. Three liters of 5%(m/v) glucose water will supply only about 640 kcal of energy, an inadequate amount of energy. A resting patient requires about 1400 kcal/day.

This problem is solved by using solutions that are about 6 times as concentrated as isotonic solutions. They are administered, through a tube, directly into a large blood vessel leading to the heart (the superior vena cava) rather than through a small vein in the arm or leg. The large volume of blood flowing through this vein quickly dilutes the solution to levels that do not upset the osmotic balance in body fluids. Using this technique, patients can be given up to 5000 kcal/day of nourishment.

Finally, when red blood cells are placed in physiological saline solution, a 0.9%(m/v) sodium chloride solution, water flow is balanced and neither hemolysis nor crenation occurs (Figure 7.16c). The osmotic pressure of physiological saline solution is the same as that of red blood cell fluid. Thus the rates of water flow into and out of the red blood cells are the same.

We will now define the terms *isotonic, hypotonic,* and *hypertonic*. An **isotonic solution** *is a solution whose osmotic pressure is equal to that within cells.* Red blood cell fluid, physiological saline solution, and 5%(m/v) glucose water are all isotonic with respect to one another. The processes of replacing body fluids and supplying nutrients to the body intravenously require the use of isotonic solutions such as physiological saline and glucose water. If isotonic solutions were not used, the damaging effects of hemolysis or crenation would occur.

A **hypotonic solution** *is a solution with a lower osmotic pressure than that within cells.* The prefix *hypo-* means "under" or "less than normal." Distilled water is hypotonic with respect to red blood cell fluid, and these cells will hemolyze when placed in it (Figure 7.16a). A **hypertonic solution** *is a solution with a higher osmotic pressure than that within cells.* The prefix *hyper-* means "over" or "more than normal." Concentrated sodium chloride solution is hypertonic with respect to red blood cell fluid, and these cells undergo crenation when placed in it (Figure 7.16b).

It is sometimes necessary to introduce a hypertonic or hypotonic solution, under controlled conditions, into the body to correct an improper "water balance" in a patient. A hypertonic solution will cause the net transfer of water from tissues to blood; then the kidneys will remove the water. Some laxatives, such as Epsom salts, act by forming hypertonic solutions in the intestines. A hypotonic solution can be used to cause water to flow from the blood into surrounding tissue; blood pressure can be decreased in this manner. Table 7.3 summarizes the differences in meaning among the terms *isotonic, hypertonic,* and *hypotonic*.

The accompanying Chemistry at a Glance summarizes the discussion of colligative properties that has been presented in this section and the preceding one.

Table 7.3
Characteristics of Isotonic, Hypertonic, and Hypotonic Solutions

	Type of Solution		
	Isotonic	***Hypertonic***	***Hypotonic***
osmolarity relative to body fluids	equal	greater than	less than
osmotic pressure relative to body fluids	equal	greater than	less than
osmotic effect on cells	equal water flow into and out of cells	net flow of water out of cells	net flow of water into cells

Colligative Properties of Solutions

COLLIGATIVE PROPERTIES OF SOLUTIONS

The physical properties of a solution that depend only on the concentration of solute particles in a given quantity of solute, not on the chemical identity of the particles.

VAPOR-PRESSURE LOWERING

Addition of a nonvolatile solute to a solvent makes the vapor pressure of the solution LOWER than that of the solvent alone.

BOILING-POINT ELEVATION

Addition of a nonvolatile solute to a solvent makes the boiling point of the solution HIGHER than that of the solvent alone.

FREEZING-POINT DEPRESSION

Addition of a nonvolatile solute to a solvent makes the freezing point of the solution LOWER than that of the solvent alone.

OSMOTIC PRESSURE

The pressure required to stop the net flow of water across a semipermeable membrane separating solutions of differing composition.

OSMOLARITY

Osmolarity = molarity × *i*, where *i* = number of particles from the dissociation of one formula unit of solute.

HYPERTONIC SOLUTION

- Solution with an osmotic pressure HIGHER than that in cells.
- Causes cells to crenate (shrink).

ISOTONIC SOLUTION

- Solution with an osmotic pressure EQUAL to that in cells.
- Has no effect on cell size.

HYPOTONIC SOLUTION

- Solution with an osmotic pressure LOWER than that in cells.
- Causes cells to hemolyze (burst).

Practice Problems and Questions

7.28 A semipermeable membrane separates a 4%(m/m) NaCl solution from a 2%(m/m) NaCl solution. Initially, the two solution levels are the same, but over time the levels become different.

a. Which solution level will rise and which will drop?
b. Which solution becomes more concentrated?
c. In which direction, if any, is there a net flow of solvent?
d. In which direction, if any, is there a net flow of solute?

7.29 Indicate whether the osmotic pressure of a 0.1 M NaCl solution will be *less than, the same as,* or *greater than* that of each of the following solutions.

a. 0.1 M NaBr b. 0.050 M $MgCl_2$ c. 0.1 M KNO_3 d. 0.1 M glucose

7.30 Would red blood cells *swell, remain the same size,* or *shrink* when placed in each of the following solutions?
a. 0.9%(m/v) glucose solution
b. 0.9%(m/v) NaCl solution
c. 2.3%(m/v) glucose solution
d. 5.0%(m/v) NaCl solution

7.31 Would red blood cells *crenate, hemolyze,* or *remain unaffected* when placed in each of the following solutions?
a. 0.9%(m/v) glucose solution
b. 0.9%(m/v) NaCl solution
c. 2.3%(m/v) glucose solution
d. 5.0%(m/v) NaCl solution

7.32 Classify each of the following solutions as *isotonic, hypertonic,* or *hypotonic.*
a. 0.9%(m/v) glucose solution
b. 0.9%(m/v) NaCl solution
c. 2.3%(m/v) glucose solution
d. 5.0%(m/v) NaCl solution

7.9 Dialysis

Learning Focus

Understand the differences and similarities between the processes of dialysis and osmosis.

Dialysis is closely related to osmosis. It is the osmotic-type process that occurs in living systems. Osmosis, you recall (Section 7.8), occurs when a solution and a solvent are separated by a semipermeable membrane that allows solvent but not solute to pass through it. There is a net transfer of solvent from the dilute solution (or pure solvent) into the more concentrated solution. **Dialysis** *is the process in which a semipermeable membrane permits the passage of solvent, dissolved ions, and small molecules but blocks the passage of large molecules.* Thus dialysis allows for the separation of small particles from large molecules. Many plant and animal membranes function as dialyzing membranes.

► In *osmosis,* only solvent passes through the membrane. In *dialysis,* both solvent and small solute particles (ions and small molecules) pass through the membrane.

Consider the placement of an aqueous solution of sodium chloride in a dialyzing bag that is surrounded by water (Figure 7.17a). What happens? Sodium ions and chloride ions move through the dialyzing membrane into the water; that is, there is a net movement of ions from a region of high concentration to a region of low concentration. This will occur until both sides of the membrane have equal concentrations of ions (Figure 7.17b).

Dialysis can be used to purify a solution containing suspended protein molecules and solute. The smaller solute molecules pass through the dialyzing membrane and leave the solution. The larger protein molecules remain behind. The result is a purified protein solution (Figure 7.18).

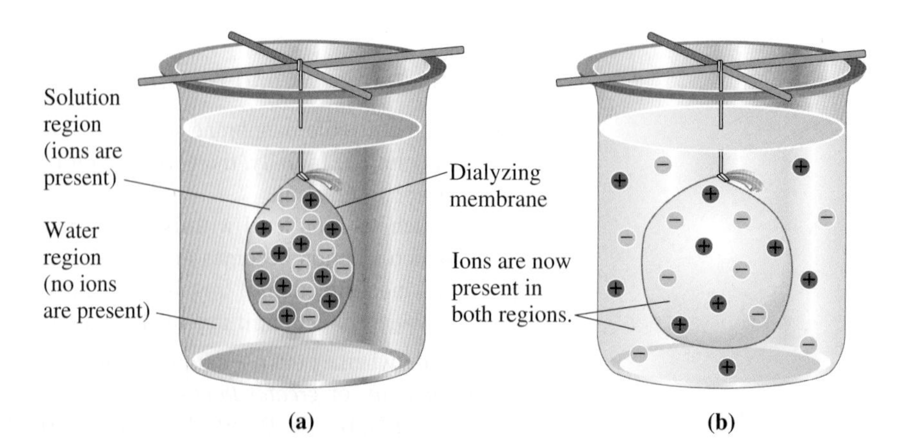

Figure 7.17 In dialysis, there is a net movement of ions from a region of higher concentration to a region of lower concentration. (a) Before dialysis. (b) After dialysis.

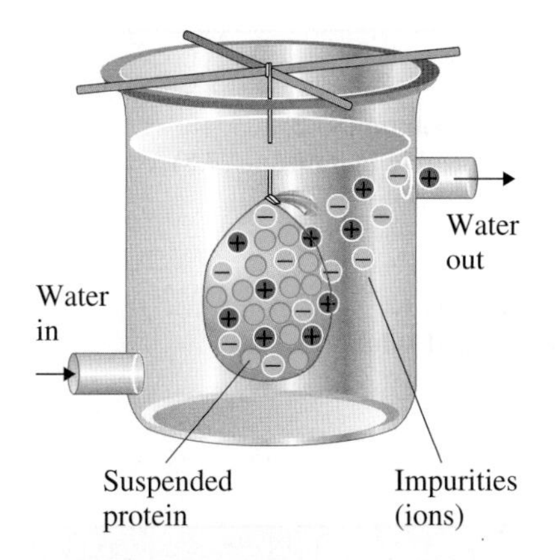

Figure 7.18 Impurities (ions) can be removed from a solution by using a dialysis procedure.

The human kidneys are a complex dialyzing system that is responsible for removing waste products from the blood. The removed products are then eliminated in urine. When the kidneys fail, these waste products build up and eventually poison the body.

When a person goes into shock, there is a sudden increase in the permeability of the membranes of the blood capillaries. Large molecules, such as proteins, leave the bloodstream and leak into the space between cells. This damage disrupts the normal chemistry of the blood. If a patient in shock is left untreated, death can occur.

Practice Problems and Questions

7.33 How do the processes of osmosis and dialysis differ?

7.34 In each of the following situations, which substances, if any, will leave the dialyzing bag?

a. A dialyzing bag containing a dilute solution of sodium chloride is immersed in distilled water.
b. A dialyzing bag containing a dilute solution of sodium chloride is immersed in a concentrated solution of sodium chloride.
c. A dialyzing bag containing a dilute solution of sodium chloride with suspended protein molecules in it is immersed in distilled water.
d. A dialyzing bag containing large protein molecules suspended in water is immersed in distilled water.

CONCEPTS TO REMEMBER

Solution components. The component of a solution that is present in the greatest amount is the *solvent*. A *solute* is a solution component that is present in a small amount relative to the solvent.

Solution characteristics. A solution is a homogeneous (uniform) mixture. Its composition and properties are dependent on the ratio of solute(s) to solvent. Dissolved solutes are present as individual particles (molecules, atoms, or ions).

Solubility. The solubility of a solute is the maximum amount of solute that will dissolve in a given amount of solvent. The extent to which a solute dissolves in a solvent depends on the structure of solute and solvent, the temperature, and the pressure. Molecular polarity is a particularly important factor in determining solubility. A saturated solution contains the maximum amount of solute that can be dissolved under the conditions at which the solution exists.

Solution concentration. Solution concentration is the amount of solute present in a specified amount of solution. Percent solute and molarity are commonly encountered concentration units. Percent concentration units include percent by mass, percent by volume, and mass–volume percent. Molarity gives the moles of solute per liter of solution.

Colligative properties of solutions. Properties of a solution that depend on the number of solute particles in solution, not on their identity, are called colligative properties. Vapor pressure lowering, boiling-point elevation, freezing-point depression, and osmotic pressure are all colligative properties.

Osmosis and osmotic pressure. Osmosis involves the passage of a solvent from a dilute solution (or pure solvent) through a semipermeable membrane into a more concentrated solution. Osmotic pressure is the amount of pressure needed to prevent the net flow of solvent across the membrane in the direction of the more concentrated solution.

Dialysis. Dialysis is the process in which a semipermeable membrane permits the passage of solvent, dissolved ions, and small molecules but blocks the passage of large molecules. Many plant and animal membranes function as dialyzing membranes.

KEY REACTIONS AND EQUATIONS

1. Percent by mass (Section 7.5)

$$\%(\text{m/m}) = \frac{\text{mass of solute}}{\text{mass of solution}} \times 100$$

2. Percent by volume (Section 7.5)

$$\%(\text{v/v}) = \frac{\text{volume of solute}}{\text{volume of solution}} \times 100$$

3. Mass–volume percent (Section 7.5)

$$\%(\text{m/v}) = \frac{\text{mass of solute (g)}}{\text{volume of solution (mL)}} \times 100$$

4. Molarity (Section 7.5)

$$\text{M} = \frac{\text{moles of solute}}{\text{liters of solution}}$$

5. Dilution of stock solution to make less-concentrated solution (Section 7.6)

$$C_s \times V_s = C_d \times V_d$$

6. Osmolarity (Section 7.8)

$$\text{osmol} = \text{M} \times i$$

KEY TERMS

Aqueous solution (7.2)
Colligative properties (7.7)
Concentrated solution (7.2)
Concentration (7.5)
Dialysis (7.9)
Dilute solution (7.2)
Dilution (7.6)
Hypertonic solution (7.8)
Hypotonic solution (7.8)
Isotonic solution (7.8)
Mass–volume percent (7.5)
Molarity (7.5)
Nonaqueous solution (7.2)
Osmolarity (7.8)
Osmosis (7.8)
Osmotic pressure (7.8)
Percent by mass (7.5)
Percent by volume (7.5)
Saturated solution (7.2)
Semipermeable membrane (7.8)
Solubility (7.2)
Solute (7.1)
Solution (7.1)
Solvent (7.1)
Unsaturated solution (7.2)

ADDITIONAL PROBLEMS

7.35 With the help of Table 7.2, determine in which of the following pairs of compounds both members of the pair have like solubility (both soluble or both insoluble).
a. Ag_2SO_4 and $AgNO_3$ b. $ZnCl_2$ and $Mg(OH)_2$
c. $BaBr_2$ and $NiCO_3$ d. NH_4I and $AlPO_4$

7.36 How many grams of solute are dissolved in the following amounts of solution?
a. 134 g of 3.00%(m/m) KNO_3 solution
b. 75.02 g of 9.735%(m/m) NaOH solution
c. 1576 g of 0.800%(m/m) HI solution
d. 1.23 g of 12.0%(m/m) NH_4Cl solution

7.37 What volume of water, in quarts, is contained in 3.50 qt of a 2.00%(v/v) solution of water in acetone?

7.38 How many liters of 0.10 M solution can be prepared from 60.0 g of each of the following solutes?
a. $NaNO_3$ b. HNO_3 c. KOH d. LiCl

7.39 What is the molarity of the solution obtained by concentrating, through evaporation of solvent, 2212 mL of 0.400 M K_2SO_4 solution to each of the following final volumes.
a. 1875 mL b. 1.25 mL c. 853 mL d. 553 mL

7.40 After all the water is evaporated from 10.0 mL of a CsCl solution, 3.75 g of CsCl remains. Express the original concentration of the CsCl solution in each of the following units.
a. Mass–volume percent b. Molarity

7.41 Which of the following aqueous solutions would give rise to a greater osmotic pressure?
a. 8.00 g of NaCl in 375 mL of solution or 4.00 g of NaBr in 155 mL of solution
b. 6.00 g of NaCl in 375 mL of solution or 6.00 g of $MgCl_2$ in 225 mL of solution

PRACTICE TEST ▸ True/False

7.42 A mandatory characteristic for all solutions is the liquid state.

7.43 The terms *saturated solution* and *concentrated solution* have the same meaning.

7.44 When an ionic solute dissolves in water, the individual ions of the solute become *hydrated.*

7.45 The solubility rule "like dissolves like" is not adequate for predicting solubility when the solute is an ionic compound.

7.46 Pulverizing a solid solute before placing it in a solvent should increase the rate at which it dissolves.

7.47 For the solubility rule "like dissolves like," the term *like* refers to molar mass.

7.48 100 mL of a 5.0 M KBr solution has the same concentration as 500 mL of a 1.0 M KBr solution.

7.49 100 mL of a 5.0 M KBr solution contains the same number of moles of solute as 500 mL of a 1.0 M KBr solution.

7.50 The concentration unit *percent by volume* is equal to the volume of solute divided by the volume of solvent, multiplied by 100.

7.51 A 5.0%(m/v) solution contains 5.0 g of solvent per 100 mL of solution.

7.52 When a solution is diluted, the amount of solute present remains constant.

7.53 Adding a nonvolatile solute to a solvent lowers the vapor pressure of the resulting solution below that of the pure solvent at the same temperature.

7.54 The osmolarity of a 0.1 M NaCl solution is twice that of a 0.1 M glucose solution.

7.55 In the process of *hemolysis,* red blood cells increase in size, often to the point where their cell membranes burst.

7.56 An *isotonic* solution is a solution with a higher osmotic pressure than that found within red blood cells.

PRACTICE TEST ▸ Multiple Choice

7.57 In a solution, the *solvent* is
a. the substance being dissolved
b. always a liquid
c. the substance present in the greatest amount
d. always water

7.58 In a *saturated* solution
a. undissolved solute must be present
b. no undissolved solute may be present
c. the solubility limit for the solute has been reached
d. solid crystallizes out if the solution is stirred

7.59 For which of the following types of ionic compounds are most examples *insoluble* in water?
a. Nitrates b. Sulfates c. Phosphates d. Chlorides

7.60 What is the concentration, in mass percent, of a solution that contains 20.0 g of NaCl dissolved in 250.0 g of water?
a. 6.76%(m/m) b. 7.41%(m/m)
c. 8.00%(m/m) d. 8.25%(m/m)

7.61 The defining expression for the *molarity* concentration unit is
a. moles of solute/liters of solution
b. moles of solute/liters of solvent
c. grams of solute/liters of solution
d. grams of solute/liters of solvent

7.62 What volume, in milliliters, of 6.0 M NaOH is needed to prepare 175 mL of 0.20 M NaOH by dilution?
a. 5.3 mL b. 5.8 mL c. 6.2 mL d. 7.1 mL

7.63 Compared to pure water, a 1.0 M sugar–water solution has a
a. lower vapor pressure, lower boiling point, and lower freezing point
b. higher vapor pressure, higher boiling point, and higher freezing point
c. lower vapor pressure, higher boiling point, and lower freezing point
d. lower vapor pressure, lower boiling point, and higher freezing point

7.64 If two sugar solutions whose concentrations are 0.2 M and 0.4 M, respectively, are separated by a semipermeable membrane, during osmosis there is a net flow of
a. sugar molecules from the concentrated to the dilute solution
b. sugar molecules from the dilute to the concentrated solution
c. water molecules from the concentrated to the dilute solution
d. water molecules from the dilute to the concentrated solution

7.65 *Crenation* of red blood cells occurs when the cells are placed in a(n)
a. hypotonic solution
b. isotonic solution
c. hypertonic solution
d. physiological saline solution

7.66 Which of the following solutions is *hypertonic* with respect to red blood cells?
a. 0.5%(m/v) NaCl b. 0.8%(m/v) NaCl
c. 4.0%(m/v) glucose d. 6.0%(m/v) glucose

CHAPTER OUTLINE

CHAPTER EIGHT

Chemical Reactions

Acid-rain-caused corrosion of churches and statues involves a double-replacement reaction that includes sulfuric acid (H_2SO_4). The reaction is $CaCO_3 + H_2SO_4 \longrightarrow CaSO_4 + H_2CO_3$

In the previous two chapters, we considered the properties of matter in various pure and mixed states. Nearly all of the subject matter dealt with interactions and changes of a *physical* nature. We now concern ourselves with the *chemical* changes that occur when various types of matter interact.

We first consider several types of chemical reactions and then discuss important fundamentals common to all chemical changes. Of particular concern to us will be how fast chemical changes occur (chemical reaction rates) and how far chemical changes go (chemical equilibrium).

Learning Focus

Identify a reaction as a combination, decomposition, single-replacement, double-replacement, or combustion reaction.

8.1 Types of Chemical Reactions

A **chemical reaction** *is a process in which at least one new substance is produced as a result of chemical change.* An almost inconceivable number of chemical reactions is possible. The majority of chemical reactions (but not all) fall into five major categories: *combination* reactions, *decomposition* reactions, *single-replacement* reactions, *double-replacement* reactions, and *combustion* reactions.

Combination Reactions

A **combination reaction** *is a reaction in which a single product is produced from two (or more) reactants.* The general equation for a combination reaction involving two reactants is

$$X + Y \longrightarrow XY$$

In such a combination reaction, two substances join together to form a more complicated product (see Figure 8.1). The reactants X and Y can be elements or compounds or an element and a compound. The product of the reaction (XY) is always a compound. Some representative combination reactions that have elements as the reactants are

$$\begin{aligned} Ca + S &\longrightarrow CaS \\ N_2 + 3H_2 &\longrightarrow 2NH_3 \\ 2Na + O_2 &\longrightarrow Na_2O_2 \end{aligned}$$

▶ In organic chemistry (Chapters 10–13), combination reactions are called *addition reactions*. One reactant, usually a small molecule, is considered to be added to a larger reactant molecule to produce a single product.

Some examples of combination reactions in which compounds are involved as reactants are

$$\begin{aligned} SO_3 + H_2O &\longrightarrow H_2SO_4 \\ 2NO + O_2 &\longrightarrow 2NO_2 \\ 2NO_2 + H_2O_2 &\longrightarrow 2HNO_3 \end{aligned}$$

Decomposition Reactions

A **decomposition reaction** *is a reaction in which a single reactant is converted into two (or more) simpler substances (elements or compounds).* Thus a decomposition reaction is the opposite of a combination reaction. The general equation for a decomposition reaction in which there are two products is

$$XY \longrightarrow X + Y$$

Although the products may be elements or compounds, the reactant is *always* a compound.

At sufficiently high temperatures, all compounds can be broken down (decomposed) into their constituent elements. Examples of such reactions include

$$\begin{aligned} 2CuO &\longrightarrow 2Cu + O_2 \\ 2H_2O &\longrightarrow 2H_2 + O_2 \end{aligned}$$

▶ In organic chemistry, decomposition reactions are often called *elimination reactions*. In many reactions, including some metabolic reactions that occur in the human body, either H_2O or CO_2 is eliminated from a molecule (a decomposition).

At lower temperatures, compound decomposition often produces other compounds as products.

$$\begin{aligned} CaCO_3 &\longrightarrow CaO + CO_2 \\ 2KClO_3 &\longrightarrow 2KCl + 3O_2 \\ 4HNO_3 &\longrightarrow 4NO_2 + 2H_2O + O_2 \end{aligned}$$

Decomposition reactions are easy to recognize in that they are the only type of reaction in which there is only one reactant.

Figure 8.1 When a hot nail is stuck into a pile of zinc and sulfur, a fiery combination reaction occurs and zinc sulfide forms.

$$Zn + S \longrightarrow ZnS$$

Single-Replacement Reactions

A **single-replacement reaction** *is a reaction in which an atom or molecule replaces an atom or group of atoms from a compound.* There are always two reactants and two products in a single-replacement reaction. The general equation for a single-replacement reaction is

$$X + YZ \longrightarrow Y + XZ$$

A common type of single-replacement reaction is one in which an element and a compound are reactants and an element and a compound are products. Examples of this type of single-replacement reaction include

$$\begin{aligned} Fe + CuSO_4 &\longrightarrow Cu + FeSO_4 \\ Mg + Ni(NO_3)_2 &\longrightarrow Ni + Mg(NO_3)_2 \\ Cl_2 + NiI_2 &\longrightarrow I_2 + NiCl_2 \\ F_2 + 2NaCl &\longrightarrow Cl_2 + 2NaF \end{aligned}$$

The first two equations illustrate one metal replacing another metal from its compound. The latter two equations illustrate one nonmetal replacing another nonmetal from its compound. A more complicated example of a single-replacement reaction, in which all reactants and products are compounds, is

$$4PH_3 + Ni(CO)_4 \longrightarrow 4CO + Ni(PH_3)_4$$

Double-Replacement Reactions

▶ In organic chemistry, replacement reactions (both single and double) are often called *substitution* reactions. Substitution reactions of the single-replacement type are seldom encountered. However, double-replacement reactions are common in organic chemistry.

A **double-replacement reaction** *is a reaction in which two substances exchange parts with one another and form two different substances.* The general equation for a double-replacement reaction is

$$AX + BY \longrightarrow AY + BX$$

Such reactions can be thought of as involving "partner switching." The AX and BY partnerships are dissolved, and new AY and BX partnerships are formed in their place.

When the reactants in a double-replacement reaction are ionic compounds in solution, the parts exchanged are the positive and negative ions of the compounds present.

$$AgNO_3(aq) + NaCl(aq) \longrightarrow AgCl(s) + NaNO_3(aq)$$
$$2KI(aq) + Pb(NO_3)_2(aq) \longrightarrow 2KNO_3(aq) + PbI_2(s)$$

In most reactions of this type, one of the product compounds is in a different physical state (solid or gas) from that of the reactants (see Figure 8.2). Insoluble solids formed from such a reaction are called *precipitates;* AgCl and PbI_2 are precipitates in the foregoing reactions.

Combustion Reactions

Combustion reactions are a most common type of reaction. A **combustion reaction** *is the reaction of a substance with oxygen (usually from air) that proceeds with the evolution of heat and usually with a flame.* Hydrocarbons, which are binary compounds of carbon and hydrogen (many such compounds exist), are the most common type of compound that undergoes combustion. In hydrocarbon combustion, the carbon of the hydrocarbon combines with the oxygen of air to produce carbon dioxide (CO_2). The hydrogen of the hydrocarbon also interacts with the oxygen of air to give water (H_2O) as a product. The relative amounts of CO_2 and H_2O produced depend on the composition of the hydrocarbon.

$$2C_2H_2 + 5O_2 \longrightarrow 4CO_2 + 2H_2O$$
$$C_3H_8 + 5O_2 \longrightarrow 3CO_2 + 4H_2O$$
$$C_4H_8 + 6O_2 \longrightarrow 4CO_2 + 4H_2O$$

Figure 8.2 A double-replacement reaction involving solutions of potassium iodide and lead(II) nitrate (both colorless solutions) produces yellow, insoluble lead(II) iodide as one of the products.

$2KI(aq) + Pb(NO_3)_2(aq) \longrightarrow 2KNO_3(aq) + PbI_2(s)$

▶ Hydrocarbon combustion reactions are the basis of industrial society, whether it be the burning of gasoline in cars, of natural gas in homes, or of coal in factories. Gasoline, natural gas, and coal all contain hydrocarbons.

Unlike most other chemical reactions, hydrocarbon combustion reactions are carried out for the energy obtained rather than for the material products produced.

Combustion reactions in which oxygen reacts with an element to form a single product can also be categorized as combination reactions. Two such reactions are

$$S + O_2 \longrightarrow SO_2$$
$$2Mg + O_2 \longrightarrow 2MgO$$

The vast majority of human-related air pollutants that enter the atmosphere each year, in terms of both tonnage and kind, are produced through combustion reactions. Chemical Portraits 13 profiles common combustion-generated air pollutants involving the elements nitrogen, sulfur, and carbon.

Example 8.1 Classification of Chemical Reactions

Classify each of the following reactions as as combination, decomposition, single-replacement, double-replacement, or combustion reaction.

a. $2KNO_3 \longrightarrow 2KNO_2 + O_2$
b. $Zn + AgNO_3 \longrightarrow Zn(NO_3)_2 + 2Ag$
c. $Ni(NO_3)_2 + 2NaOH \longrightarrow Ni(OH)_2 + 2NaNO_3$
d. $3Mg + N_2 \longrightarrow Mg_3N_2$

Solution

a. Decomposition. Two substances are produced from a single substance.
b. Single-replacement. An element and a compound are reactants, and an element and a compound are products.
c. Double-replacement. Two compounds exchange parts with each other; the nickel ion (Ni^{2+}) and the sodium ion (Na^+) are "swapping partners."
d. Combination. Two substances combine to form a single substance.

Chemical Portraits 13 Combustion Reactions and Air Pollutants

NO_x Pollution (NO and NO_2)

Profile: All combustion processes generate NO_x pollution. At combustion temperatures, the N_2 and O_2 of air become slightly reactive towards each. The result is production of small amounts of NO and NO_2, regardless of what the fuel is. NO, which is produced at flame temperatures, enters the atmosphere, where it reacts with atmospheric O_2 to produce NO_2.

NO is a colorless and odorless gas. NO_2, in pure form, is a reddish-brown gas that has a strong choking odor. However, at atmospheric concentrations, NO_2 is also odorless and colorless. The presence of NO_x pollution in the atmosphere is the starting point for smog formation.

What chemical reactions link NO_x pollution to smog formation?

SO_x Pollution (SO_2 and SO_3)

Profile: Coal combustion is the major source of SO_x pollution. Sulfur, an impurity in all fossil fuels, is removed from petroleum and natural gas during processing; coal is consumed unprocessed. At flame temperatures, coal's sulfur is converted to SO_2. The SO_2 enters the atmosphere, where it reacts with atmospheric O_2 to produce SO_3.

SO_2, is a colorless, odorless gas at typical atmospheric concentrations. It, however, has a pungent and choking odor at higher concentrations. SO_3 is a reactive colorless gas; once formed, it rapidly interacts with H_2O vapor to produce sulfuric acid (H_2SO_4). SO_x pollution is the major starting point for the formation of acid rain.

What chemical reactions link SO_x pollution to acid rain formation?

CO_x Pollution (CO and CO_2)

Profile: All common fuels are mixtures of carbon-hydrogen compounds. During combustion, fuel C atoms combine with O_2 to produce CO_2, and fuel H atoms combine with O_2 to form H_2O. CO_2 formation is a two-step process with both steps occurring at flame temperatures. First fuel C is converted to CO, which then reacts with more O_2 to produce CO_2. Small amounts of CO product result when inadequate air-fuel mixing occurs; the dominate product is always, however, CO_2.

Atmospheric CO_2 is a "greenhouse" gas; it absorbs heat radiated from Earth as it cools at night. This traps heat in Earth's lower atmosphere and contributes to the phenomenon known as "global warming."

What other greenhouse gases besides CO_2 are present in the atmosphere?

See the text web site at **www.cengage.com/chemistry/stoker** for answers to the above questions and for further information.

Chemistry *at a Glance*

Types of Chemical Reactions

COMBINATION REACTION

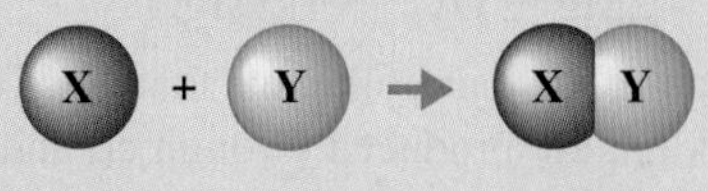

$$2Al + 3I_2 \longrightarrow 2AlI_3$$

Aluminum reacts with iodine to form aluminum iodide.

DECOMPOSITION REACTION

$$2HgO \longrightarrow 2Hg + O_2$$

Mercury (II) oxide decomposes to form mercury and oxygen.

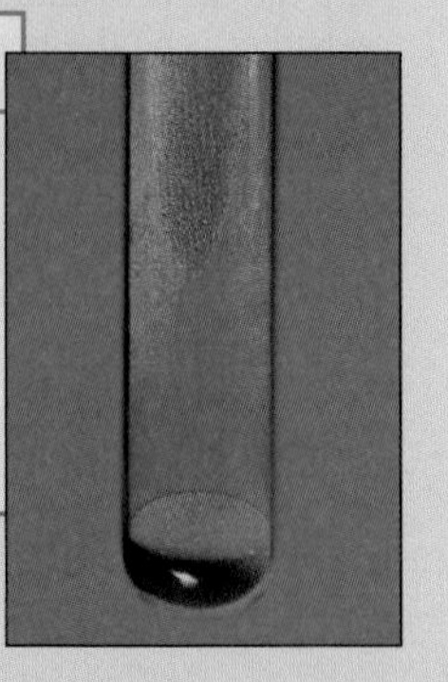

SINGLE-REPLACEMENT REACTION

$$Zn + CuSO_4 \longrightarrow Cu + ZnSO_4$$

Zinc reacts with copper (II) sulfate to form copper and zinc sulfate.

DOUBLE-REPLACEMENT REACTION

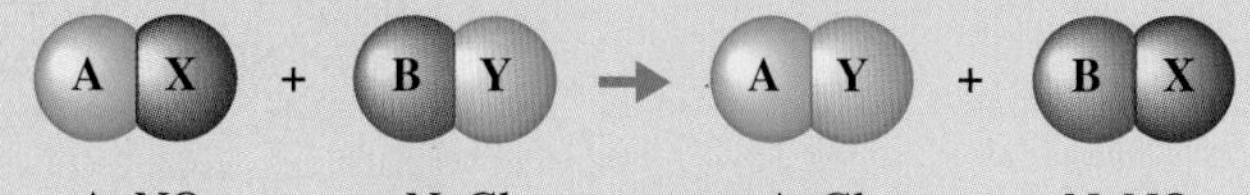

$$AgNO_3 + NaCl \longrightarrow AgCl + NaNO_3$$

Silver nitrate reacts with sodium chloride to form silver chloride and sodium nitrate.

Not all reactions fall into one of the five categories we have discussed in this section. But even though this classification system is not all-inclusive, it is still very useful because of the many reactions it does help correlate. The accompanying Chemistry at a Glance summarizes pictorially much of the material considered in this section.

Practice Problems and Questions

8.1 What is the general chemical equation for each of the following types of reactions?
a. Decomposition b. Combination c. Double-replacement d. Single-replacement

8.2 Classify each of the following reactions as a combination, decomposition, single-replacement, double-replacement, or combustion reaction.
a. $3CuSO_4 + 2Al \longrightarrow Al_2(SO_4)_3 + 3Cu$
b. $K_2CO_3 \longrightarrow K_2O + CO_2$
c. $2C_2H_6 + 7O_2 \longrightarrow 4CO_2 + 6H_2O$
d. $Mg + 2HCl \longrightarrow MgCl_2 + H_2$

8.3 Classify each of the following reactions as a combination, decomposition, single-replacement, double-replacement, or combustion reaction.
a. $2NaHCO_3 \longrightarrow Na_2CO_3 + CO_2 + H_2O$
b. $2Fe + 3Cl_2 \longrightarrow 2FeCl_3$
c. $2AgNO_3 + K_2SO_4 \longrightarrow Ag_2SO_4 + 2KNO_3$
d. $2H_2S + 3O_2 \longrightarrow 2SO_2 + 2H_2O$

8.4 Indicate to which of the following types of reactions each of the statements applies: combination, decomposition, single-replacement, double-replacement, and combustion. More than one answer is possible for a given statement.

a. An element may be a reactant.
b. An element may be a product.
c. A compound may be a reactant.
d. A compound may be a product.

Learning Focus

Know three common ways of defining *oxidation* and *reduction*; be able to recognize oxidation–reduction reactions.

8.2 Oxidation–Reduction Reactions

Many of the reactions considered in the previous section are also characterized as *oxidation–reduction reactions.* Such reactions, known also by the shortened designation *redox reactions,* represent a very important type of chemical reaction. Redox reactions occur all around us and even within us. The bulk of the energy needed for the functioning of the human body is obtained from food via oxidation–reduction processes. Such diverse phenomena as the electricity obtained from a battery to start a car, the use of natural gas to heat a home, the rusting of iron, and the functioning of an antiseptic agent to kill or prevent the growth of bacteria involve oxidation–reduction reactions. Knowledge of this type of reaction is fundamental to understanding many biological and technological processes.

The terms *oxidation* and *reduction* can be defined in a number of ways, three of which we will consider in this section. These three different approaches to *redox reactions* focus on oxygen atoms, hydrogen atoms, and electrons, respectively.

Oxygen-Based Redox Definitions

Historically, the word *oxidation* was first used to describe the reaction of a substance with oxygen. According to this historical definition, each of the following reactions involves oxidation.

$$4Fe + 3O_2 \longrightarrow 2Fe_2O_3$$
$$S + O_2 \longrightarrow SO_2$$
$$CH_4 + 2O_2 \longrightarrow CO_2 + 2H_2O$$

The first-listed reactant in each of these chemical equations is said to have been *oxidized.*

Originally, the term *reduction* referred to processes whereby oxygen was removed from a compound. A particularly common type of reduction reaction, on the basis of this original definition, is the removal of oxygen from a metal oxide to produce the free metal.

$$CuO + H_2 \longrightarrow Cu + H_2O$$
$$2Fe_2O_3 + 3C \longrightarrow 4Fe + 3CO_2$$

The word *reduction* comes from the reduction in mass of the metal-containing species; the metal has a mass less than that of the metal oxide. The metal oxide is said to have been *reduced.*

Our first set of definitions for oxidation and reduction is related to these historical oxygen-based origins.

1. **Oxidation** *is the process whereby a reactant in a chemical reaction gains one or more oxygen atoms.*
2. **Reduction** *is the process whereby a reactant in a chemical reaction loses one or more oxygen atoms.*

Note that these definitions of the oxidation process and the reduction process are "opposites." That is, the definitions involve the gain and loss, respectively, of the same thing. This will always be the case in oxidation–reduction definitions.

The processes of oxidation and reduction are also always complementary processes; that is, they always occur together. You cannot have one without the other. When oxygen atoms are lost by one reactant, they must be gained by another reactant; oxygen atoms cannot simply disappear. In the next-to-last of the preceding example equations,

$$CuO + H_2 \longrightarrow Cu + H_2O$$

note that CuO is oxidized (oxygen loss) and H_2 (the other reactant) is reduced (oxygen gain).

Hydrogen-Based Redox Definitions

Our second set of definitions of oxidation and reduction focus on the loss and gain of hydrogen atoms by reactant molecules.

1. **Oxidation** *is the process whereby a reactant in a chemical reaction loses one or more hydrogen atoms.*
2. **Reduction** *is the process whereby a reactant in a chemical reaction gains one or more hydrogen atoms.*

Let us consider again the reaction involving CH_4 and O_2, this time from a "hydrogen viewpoint."

$$CH_4 + 2O_2 \longrightarrow CO_2 + 2H_2O$$

Methane (CH_4) loses hydrogen atoms and thus is oxidized. The O_2 gains hydrogen atoms (to become H_2O) and thus is reduced. In organic chemistry and biochemistry, the oxidation of carbon-containing molecules often involves the gain of hydrogen atoms.

Electron-Based Redox Definitions

▶ Note that the oxygen-based definition of oxidation involves the word *gain* (of oxygen atoms) and that the hydrogen-based definition of oxidation involves the word *loss* (of hydrogen atoms).

Our third set of definitions of oxidation and reduction focuses on the loss and gain of electrons.

1. **Oxidation** *is the process whereby a reactant in a chemical reaction loses one or more electrons.*
2. **Reduction** *is the process whereby a reactant in a chemical reaction gains one or more electrons.*

These electron-based definitions are easily applied to chemical reactions that involve ionic compounds. Consider the reaction between the elements Mg and S to form the ionic compound MgS. The Mg atoms lose electrons to form Mg^{2+} ions, and the S atoms gain electrons to form S^{2-} ions (Section 4.5).

$$Mg + S \longrightarrow Mg^{2+} + S^{2-} \longrightarrow MgS$$

In this reaction, both oxidation and reduction occur.

$$Mg \longrightarrow Mg^{2+} + 2e^- \quad \text{(oxidation, loss of electrons)}$$

$$S + 2e^- \longrightarrow S^{2-} \quad \text{(reduction, gain of electrons)}$$

Again, as previously noted, the processes of oxidation and reduction occur simultaneously. You cannot have one without the other. The electrons lost by the Mg atoms are the same electrons that are gained by the S atoms.

A slightly more complicated redox equation involving ions is the single-replacement reaction (Section 8.1).

$$Zn + CuSO_4 \longrightarrow ZnSO_4 + Cu$$

Rewriting this equation to show the ions that are present gives

$$Zn + Cu^{2+} + SO_4^{2-} \longrightarrow Zn^{2+} + SO_4^{2-} + Cu$$

Table 8.1
Characteristics of Oxidation and Reduction

Oxidation	
Always involves	***May involve***
Loss of electrons	Gain of oxygen Loss of hydrogen
Reduction	
Always involves	***May involve***
Gain of electrons	Loss of oxygen Gain of hydrogen

Now it is apparent that Zn is oxidized and Cu^{2+} is reduced.

$$Zn \longrightarrow Zn^{2+} + 2e^- \quad \text{(oxidation, loss of electrons)}$$
$$Cu^{2+} + 2e^- \longrightarrow Cu \quad \text{(reduction, gain of electrons)}$$

Universality of Oxidation–Reduction Definitions

Of the three sets of oxidation–reduction definitions that we have considered, the "best" set is the electron-based set. Electron loss and electron gain (electron transfer) are characteristic of all redox reactions. Hence this set of definitions applies to all redox reactions. Obviously, the oxygen-based and hydrogen-based definitions for oxidation and reduction have more limited application. Oxygen-based definitions cannot be applied to redox reaction wherein no oxygen is present. Similarly, hydrogen-based definitions are applicable only when hydrogen is present in reactant molecules. Table 8.1 summarizes the concepts of oxidation and reduction that are contained in the three sets of definitions we have considered in this section.

The transfer of electrons associated with oxidation and reduction is relatively easy to recognize in reactions where ions are present. However, when ions are not present (molecular compounds; Section 4.11), recognizing electron transfer is not so straightforward a process. It is in this situation that the oxygen-based and hydrogen-based definitions become useful. Thus, most often, these latter definitions are used in reactions involving molecular compounds.

Non-oxidation-reduction Reactions

An **oxidation–reduction reaction** *is a chemical reaction in which there is a transfer of electrons between reactants.* Although many reactions are *oxidation–reduction reactions,* there are numerous reactions in which no electron transfer occurs. Such reactions are called *non-oxidation-reduction (nonredox) reactions.* A **non-oxidation-reduction reaction** *is a chemical reaction in which there is no transfer of electrons between reactants.* The following double-replacement reaction (Section 8.1) is a nonredox reaction.

$$AgNO_3(aq) + NaCl(aq) \longrightarrow AgCl(s) + NaNO_3(aq)$$

Rewriting this equation to emphasize the ions present in the various compounds gives

$$Ag^+ + NO_3^- + Na^+ + Cl^- \longrightarrow Ag^+ + Cl^- + Na^+ + NO_3^-$$

The same ions are present on both sides of the equation; no electron transfer has occurred. The driving force for this "partner-switching" reaction (Section 8.1) is the formation of AgCl, an insoluble ionic compound (Section 7.4).

Example 8.2 **Recognizing Oxidation–Reduction Reactions**

Determine whether each of the following reactions is an oxidation–reduction reaction.

a. $4Al + 3O_2 \longrightarrow 2Al_2O_3$
b. $CH_4 + Cl_2 \longrightarrow CH_3Cl + HCl$
c. $2AgNO_3 + BaCl_2 \longrightarrow 2AgCl + Ba(NO_3)_2$
d. $C_3H_8 + 5O_2 \longrightarrow 3CO_2 + 4H_2O$

Solution

a. Oxidation–reduction reaction; Al loses electrons to form Al^{3+} ion.
b. Oxidation–reduction reaction; CH_4 loses a H atom in forming CH_3Cl.
c. Non-oxidation-reduction reaction; the same ions exist on both sides of the equation.
d. Oxidation–reduction reaction; the carbon-containing compound (C_3H_8) gains oxygen.

Practice Exercises and Questions

8.5 Indicate whether each of the following describes the process of oxidation or the process of reduction.
a. Loss of electrons
b. Loss of oxygen atoms
c. Loss of hydrogen atoms
d. Gain of oxygen atoms

8.6 Classify each of the following reactions as a redox reaction or a nonredox reaction.
a. $2Cu + O_2 \longrightarrow 2CuO$
b. $Cl_2 + 2KI \longrightarrow 2KCl + I_2$
c. $Zn + CuCl_2 \longrightarrow ZnCl_2 + Cu$
d. $Zn(NO_3)_2 + CuCl_2 \longrightarrow ZnCl_2 + Cu(NO_3)_2$

8.7 Indicate which substance is oxidized and which substance is reduced in each of the following redox reactions.
a. $2Li + F_2 \longrightarrow 2LiF$
b. $Fe + CuSO_4 \longrightarrow FeSO_4 + Cu$
c. $2C_4H_8 + 12O_2 \longrightarrow 8CO_2 + 8H_2O$
d. $Ca + S \longrightarrow CaS$

8.8 Each of the following "equations" shows only part of a chemical reaction. Indicate whether the reactant shown is being oxidized or reduced.
a. $C_2H_4O \longrightarrow C_2H_4O_2$
b. $SO_3 \longrightarrow S$
c. $Fe^{2+} \longrightarrow Fe^{3+}$
d. $H_2O_2 \longrightarrow H_2O$

8.3 Oxidizing Agents and Reducing Agents

Learning Focus

Define the terms *oxidizing agent* and *reducing agent.*

Oxidizing Agents and Reducing Agents

There are two different ways of looking at the reactants in a redox reaction. First, the reactants can be viewed as being "acted on." From this viewpoint, one reactant is *oxidized* (the one that loses electrons) and one is *reduced* (the one that gains electrons). Second, the reactants can be looked on as "bringing about" the reaction. In this approach, the terms *oxidizing agent* and *reducing agent* are used. An **oxidizing agent** *causes oxidation by accepting electrons from the other reactant.* This acceptance of electrons means that the oxidizing agent itself is reduced. Similarly, the **reducing agent** *causes reduction by providing electrons for the other reactant to accept.* Thus the reducing agent and the substance oxidized are one and the same, as are the oxidizing agent and the substance reduced:

Substance oxidized = reducing agent
Substance reduced = oxidizing agent

▶ The terms *oxidizing agent* and *reducing agent* sometimes cause confusion, because the oxidizing agent is not oxidized (it is reduced) and the reducing agent is not reduced (it is oxidized). A simple analogy is that a travel agent is not the one who takes a trip; he or she is the one who plans (causes) the trip that is taken.

Chemical Portraits 14 profiles three common oxidizing agents which can be used to kill bacteria in selected situations. Each oxidizing agent also has many other uses.

Example 8.3 Identifying the Oxidizing and Reducing Agents in a Redox Reaction

For the redox reaction

$$FeO + CO \longrightarrow Fe + CO_2$$

identify the following species.

a. The substance oxidized **b.** The substance reduced
c. The oxidizing agent **d.** The reducing agent

Solution

We first identify the substances oxidized and reduced. Using the oxygen-based redox definitions, FeO loses oxygen and thus is reduced, and CO gains oxygen and thus is oxidized. Alternatively, using the electron-based redox definitions, the iron in FeO, which is Fe^{2+} ion, is changed to Fe, a process that involves the gain of two electrons; thus FeO is reduced, and CO, the other reactant, must be oxidized.

a. The substance oxidized is CO.
b. The substance reduced is FeO.
c. The oxidizing agent and the substance reduced are always one and the same. Thus FeO is the oxidizing agent.
d. The reducing agent and the substance oxidized are always one and the same. Thus CO is the reducing agent.

Chemical Portraits 14 Commonly Used Oxidizing Agents

Hydrogen Peroxide (H_2O_2)

Profile: Hydrogen peroxide is a colorless, syrupy liquid in its pure state. It readily decomposes producing H_2O and O_2 gas. Since light accelerates H_2O_2 decomposition, H_2O_2 and its solutions are stored in dark bottles. It is primarily used in aqueous solution form.

Biochemical Considerations: A 3% aqueous H_2O_2 solution is used as a mild antiseptic. When applied to minor wounds, *catalase* (a substance in blood) causes the H_2O_2 to decompose. The released O_2, which causes "foaming" at the wound site, is the antiseptic agent. A 6% to 12% H_2O_2 solution is used to bleach hair. The H_2O_2 oxidizes the dark pigment *melanin* in hair to colorless products.

When was H_2O_2 discovered, and how is it produced?

Ozone (O_3)

Profile: Two forms of oxygen exist: ordinary oxygen (O_2) and ozone (O_3). O_3 has a faintly blue color and a sharp, irritating odor. "Good ozone" occurs in the upper atmosphere (the ozone layer) where it screens out ultraviolet light from the sun. "Bad ozone" occurs in the lower atmosphere as a component of smog. It can cause headaches and eye irritation.

Uses: O_3 is an extremely powerful oxidizing agent. Of common oxidizing agents, only F_2 is more powerful. O_3 is sometimes used in place of chlorine to kill bacteria in drinking water. It is also used as a bleach for wood pulp in the papermaking industry.

What are typical concentrations for O_3 in the ozone layer and in smog?

Chlorine (Cl_2)

Profile: Chlorine, a pale, yellow-green gas in the pure state, is prepared commercially from salt (NaCl). It is a toxic gas; inhalation in large amounts can cause extensive lung damage.

Uses: Aqueous solutions of Cl_2 are powerful oxidizing agents. They are used in disinfecting drinking water supplies and swimming pools, where they destroy bacteria and viruses. The sterilizing action of Cl_2 is not entirely due to Cl_2 itself but also to hypochlorous acid (HOCl), which forms when chlorine reacts with water. Household laundry bleaches contain compounds related to HOCl. A faint Cl_2 odor is often noticeable when such bleaches are used.

At what stage in drinking water purification is Cl_2 added?

See the text web site at **www.cengage.com/chemistry/stoker** for answers to the above questions and for further information.

Practice Problems and Questions

8.9 In each of the following statements, choose the word in parentheses that best completes the statement.
a. The substance oxidized in a redox reaction is the (oxidizing, reducing) agent.
b. The oxidizing agent (gains, loses) electrons during a redox reaction.

8.10 Indicate whether each of the following substances loses or gains electrons in a redox reaction.
a. The oxidizing agent
b. The reducing agent
c. The substance undergoing oxidation
d. The substance undergoing reduction

8.11 Identify which substance is the oxidizing agent and which substance is the reducing agent in each of the following redox reactions.
a. $2Li + F_2 \longrightarrow 2LiF$
b. $Fe_2O_3 + 3CO \longrightarrow 2Fe + 3CO_2$
c. $Ca + S \longrightarrow CaS$
d. $Fe + Ag_2SO_4 \longrightarrow FeSO_4 + 2Ag$

Learning Focus

Explain, in terms of collision theory, the conditions necessary for a reaction to take place.

8.4 Collision Theory and Chemical Reactions

What causes a chemical reaction, either redox or nonredox, to take place? A set of three generalizations, developed after study of thousands of different reactions, helps answer this question. Collectively these generalizations are known as collision theory. **Collision theory** *is a set of statements that give the conditions that must be met before a chemical reaction will take place.* Central to collision theory are the concepts of molecular collisions, activation energy, and collision orientation. The statements of collision theory are

1. *Molecular collisions.* Reactant particles must interact (that is, collide) with one another before any reaction can occur.
2. *Activation energy.* Colliding particles must possess a certain minimum total amount of energy, called the activation energy, if the collision is to be effective (that is, result in reaction).
3. *Collision orientation.* Colliding particles must come together in the proper orientation unless the particles involved are single atoms or small, symmetrical molecules.

Let's look at these statements in the context of a reaction between two molecules or ions.

Molecular Collisions

When reactions involve two or more reactants, collision theory assumes (statement 1) that the reactant molecules, ions, or atoms must come in contact (collide) with one another in order for any chemical change to occur. The validity of this statement is fairly obvious. Reactants cannot react if they are separated from each other.

Most reactions are carried out either in liquid solution or in the gaseous phase, wherein reacting particles are more free to move around, and thus it is easier for the reactants to come in contact with one another. Reactions in which reactants are solids can and do occur; however, the conditions for molecular collisions are not as favorable as they are for liquids and gases. Reactions of solids usually take place only on the solid surface and thus include only a small fraction of the total particles present in the solid. As the reaction proceeds and products dissolve, diffuse, or fall from the surface, fresh solid is exposed. In this way, the reaction can eventually consume all of the solid. The rusting of iron is an example of this type of process.

Figure 8.3 Rubbing a match head against a rough surface provides the activation energy needed for the match to ignite.

Activation Energy

The collisions between reactant particles do not always result in the formation of reaction products. Sometimes, reactant particles rebound unchanged from a collision. Statement 2 of collision theory indicates that in order for a reaction to occur, particles must collide with a certain minimum energy; that is, the kinetic energies of the colliding particles must add to a certain minimum value. **Activation energy** *is the minimum combined kinetic energy that reactant particles must possess in order for their collision to result in a reaction.* Every chemical reaction has a different activation energy. In a slow reaction, the activation energy is far above the average energy content of the reacting particles. Only those few particles with above-average energy undergo collisions that result in reaction; this is the reason for the overall slowness of the reaction.

It is sometimes possible to start a reaction by providing activation energy and then have the reaction continue on its own. Once the reaction is started, enough energy is released to activate other molecules and keep the reaction going. The striking of a kitchen match is an example of such a situation (see Figure 8.3). Activation energy is initially provided by rubbing the match head against a rough surface; heat is generated by friction. Once the reaction is started, the match continues to burn.

Collision Orientation

Reaction rates are sometimes very slow because reactant molecules must be oriented in a certain way in order for collisions to lead successfully to products. For nonspherical molecules and nonspherical polyatomic ions, orientation relative to one another at the moment of collision is a factor that determines whether a collision produces a reaction.

As an illustration of the importance of proper collision orientation, consider the chemical reaction between NO_2 and CO to produce NO and CO_2.

▶ Many reactions in the human body do not occur unless specialized proteins called *enzymes* (Chapter 16) are present. One of the functions of these enzymes is to hold reactant molecules in the orientation required for a reaction to occur.

$$NO_2(g) + CO(g) \longrightarrow NO(g) + CO_2(g)$$

In this reaction, an O atom is transferred from an NO_2 molecule to a CO molecule. The collision orientation most favorable for this to occur is one that puts an O atom from NO_2 near a C atom from CO at the moment of collision. Such an orientation is shown in Figure 8.4 (top). In Figure 8.4 (bottom), three undesirable NO_2–CO orientations are shown, where the likelihood of successful reaction is very low.

Figure 8.4 In the reaction of NO_2 with CO to produce NO and CO_2, the most favorable collision orientation is one that puts an O atom from NO_2 in close proximity to the C atom of CO.

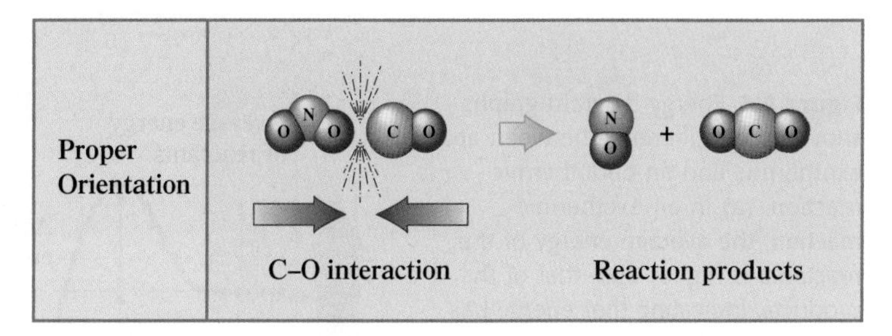

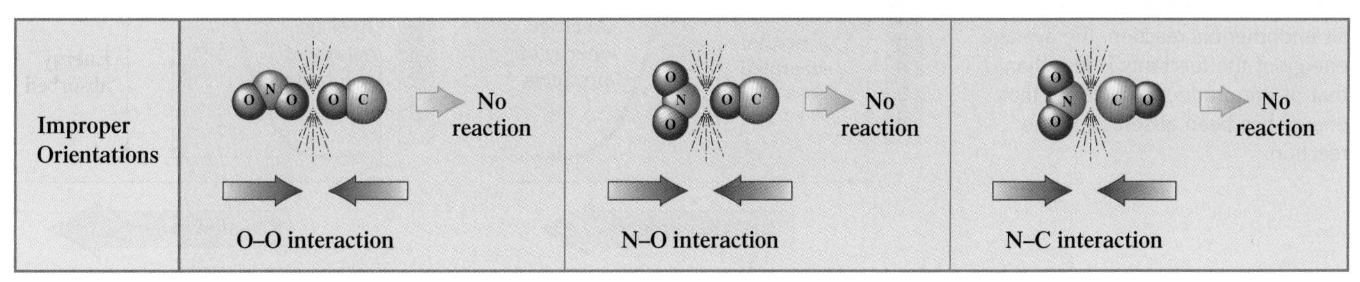

Practice Problems and Questions

8.12 Why are most chemical reactions carried out either in liquid solution or in the gaseous phase?

8.13 The collisions between reactant particles do not always result in the formation of reaction products. What role does each of the following play in this situation?
a. Activation energy b. Collision orientation

8.14 Explain why a reaction with a high activation energy would be expected to be a slow reaction.

8.5 Exothermic and Endothermic Chemical Reactions

Learning Focus

Explain the relationship between reactant energies and product energies for exothermic and endothermic chemical reactions.

▶ *Exothermic* means energy is released; energy is a "product" of the chemical reaction. *Endothermic* means energy is absorbed; energy is a "reactant" in the reaction.

In Section 6.9, the terms *exothermic* and *endothermic* were used to classify changes of state. Melting, sublimation, and evaporation are endothermic changes of state, and freezing, condensation, and deposition are exothermic changes of state. The terms *exothermic* and *endothermic* are also used to classify chemical reactions. An **exothermic chemical reaction** *is one in which energy is released as the reaction occurs.* The burning of a fuel (reaction of the fuel with oxygen) is an exothermic process. An **endothermic chemical reaction** *is one that requires the continuous input of energy as the reaction occurs.* The photosynthesis process that occurs in plants is an example of an endothermic reaction. Light is the energy source for photosynthesis. Light energy must be continuously supplied in order for photosynthesis to occur; a green plant that is kept in the dark will die.

What determines whether a chemical reaction is exothermic or endothermic? The answer to this question is related to the strength of chemical bonds—that is, the energy stored in chemical bonds. Different types of bonds, such as oxygen–hydrogen bonds and fluorine–nitrogen bonds, have different energies associated with them. In a chemical reaction, bonds are broken within reactant molecules, and new bonds are formed within product molecules. The energy balance between this bond-breaking and bond-forming determines whether there is a net loss or a net gain of energy.

An exothermic reaction (release of energy) occurs when the energy required to break bonds in the reactants is less than the energy released by bond formation in the products. The opposite situation applies for an endothermic reaction. There is more energy stored in product molecule bonds than in reactant molecule bonds. The necessary additional energy must be supplied from external sources as the reaction proceeds. Figure 8.5 illustrates the energy relationships associated with exothermic and endothermic chemical reactions. Note that both of these diagrams contain a "hill" or "hump." The height of this "hill" corresponds

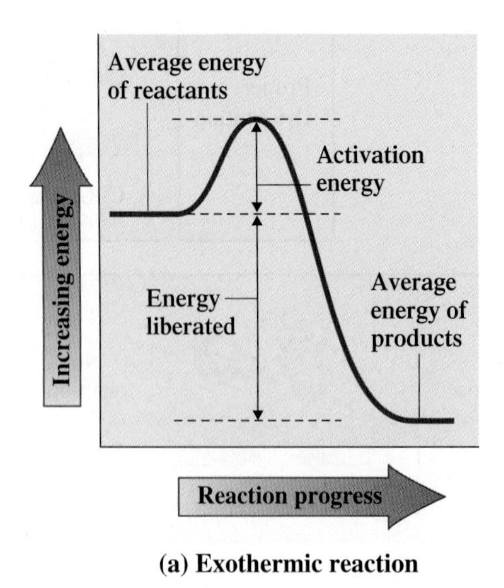

(a) Exothermic reaction

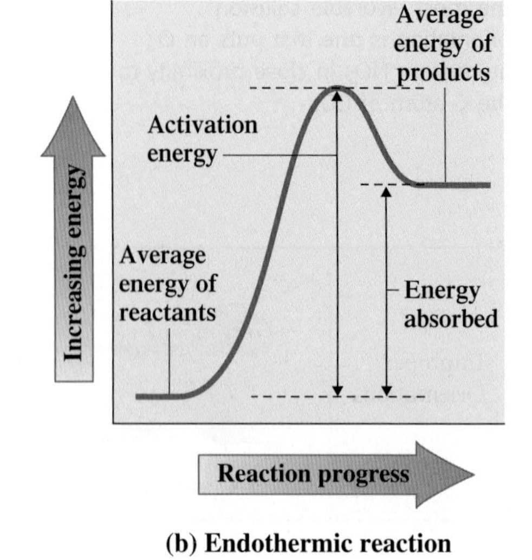

(b) Endothermic reaction

Figure 8.5 Energy diagram graphs showing the difference between an exothermic and an endothermic reaction. (a) In an exothermic reaction, the average energy of the reactants is higher than that of the products, indicating that energy has been released in the reaction. (b) In an endothermic reaction, the average energy of the reactants is less than that of the products, indicating that energy has been absorbed in the reaction.

to the activation energy needed for reaction between molecules to occur. This activation energy is independent of whether a given reaction is exothermic or endothermic.

Practice Problems and Questions

8.15 In a certain chemical reaction, the energy of the products is 35 kcal lower than the energy of the reactants.
a. Is the chemical reaction exothermic or endothermic?
b. Is energy released or absorbed in the chemical reaction?

8.16 Indicate whether each of the following is a characteristic of an endothermic reaction or of an exothermic reaction.
a. There is more energy stored in product molecule bonds than in reactant molecule bonds.
b. The energy required to break bonds in the reactants is less than the energy released by bond formation in the products.

8.17 Classify each of the following chemical reactions as an exothermic reaction or as an endothermic reaction on the basis of the placement of the term *heat* in the equation for the chemical reaction.
a. $C_2H_4 + 3O_2 \longrightarrow 2CO_2 + 2H_2O + \text{heat}$
b. $N_2 + 2O_2 + \text{heat} \longrightarrow 2NO_2$
c. $2H_2O + \text{heat} \longrightarrow 2H_2 + O_2$
d. $2KClO_3 \longrightarrow 2KCl + 3O_2 + \text{heat}$

Learning Focus

List four factors that affect reaction rates and explain how they operate in terms of collision theory.

8.6 Factors That Influence Reaction Rates

The **rate of a chemical reaction** *is the rate at which reactants are consumed or products produced in a given time period.* Natural processes have a wide range of reaction rates (see Figure 8.6). In this section we consider four different factors that affect reaction rate: (1) the physical nature of the reactants, (2) reactant concentrations, (3) reaction temperature, and (4) the presence of catalysts.

(a) (b) (c) (d)

Figure 8.6 Natural processes occur at a wide range of reaction rates. A fire (a) is a much faster reaction than the ripening of fruit (b), which is much faster than the process of rusting (c), which is much faster than the process of aging (d).

Physical Nature of Reactants

The physical nature of reactants includes not only the physical state of each reactant (solid, liquid, or gas) but also the particle size. In reactions where reactants are all in the same physical state, the reaction rate is generally faster between liquid-state reactants than between solid reactants and is fastest between gaseous reactants. Of the three states of matter, the gaseous state is the one where there is the most freedom of movement; hence, collisions between reactants are the most frequent in this state.

▶ For solid-state reactants, reaction rate increases as subdivision of the solid increases.

In the solid state, reactions occur at the boundary surface between reactants. The reaction rate increases as the amount of boundary surface area increases. Subdividing a solid into smaller particles increases surface area and thus increases reaction rate.

When the particle size of a solid is extremely small, reaction rates can be so fast that an explosion results. Although a lump of coal is difficult to ignite, the spontaneous ignition of coal dust is a real threat to underground coal-mining operations.

Reactant Concentrations

An increase in the concentration of a reactant causes an increase in the rate of the reaction. Combustible substances burn much more rapidly in pure oxygen than in air (21% oxygen). A person with a respiratory problem such as pneumonia or emphysema is often given air enriched with oxygen because an increased partial pressure of oxygen facilitates the absorption of oxygen in the alveoli of the lungs and thus expedites all subsequent steps in respiration.

▶ Reaction rate increases as the concentration of reactants increases.

Increasing the concentration of a reactant means that there are more molecules of that reactant present in the reaction mixture; thus collisions between this reactant and other reactant particles are more likely. An analogy can be drawn to the game of billiards. The more billiard balls there are on the table, the greater the probability that a moving cue ball will strike one of them.

When the concentration of reactants is increased, the actual quantitative change in reaction rate is determined by the specific reaction. The rate usually increases, but not to the same extent in all cases. Sometimes the rate doubles with a doubling of concentration, but not always.

Reaction Temperature

The effect of temperature on reaction rates can also be explained by using the molecular-collision concept. An increase in the temperature of a system results in an increase in the average kinetic energy of the reacting molecules. The increased molecular speed causes more collisions to take place in a given time. Because the average energy of the colliding molecules is greater, a larger fraction of the collisions will result in reaction from the point of view of activation energy. As a rule of thumb, chemists have found that for the temperature ranges we normally encounter, the rate of a chemical reaction doubles for every 10°C increase in temperature (see Figure 8.7).

▶ Reaction rate increases as the temperature of the reactants increases.

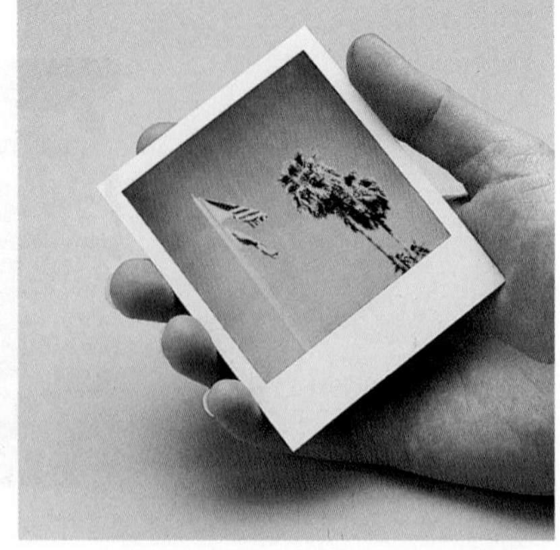

Figure 8.7 You may have noticed that pictures from an "instant camera" develop more rapidly on warm days than on cold days. The chemical reactions involved in the development process occur faster at higher temperatures.

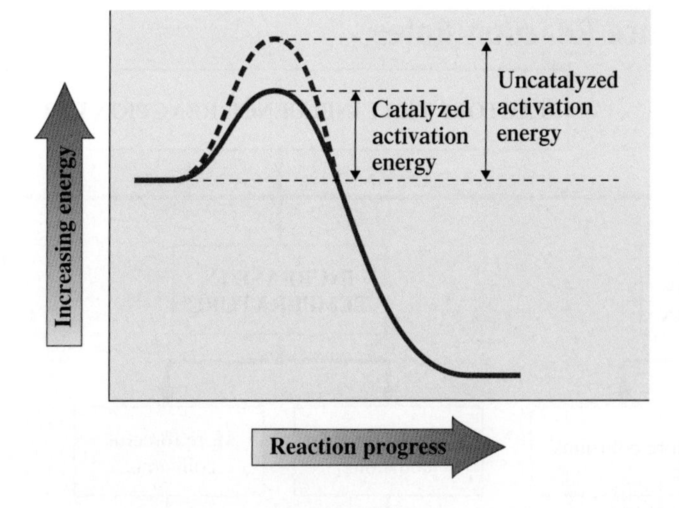

Figure 8.8 Catalysts lower the activation energy for chemical reactions. Reactions proceed more rapidly with the lowered activation energy.

Presence of Catalysts

A **catalyst** *is a substance that increases a reaction rate without being consumed in the reaction.* Catalysts enhance reaction rates by providing alternative reaction pathways that have lower activation energies than the original, uncatalyzed pathway. This lowering of activation energy is diagrammatically shown in Figure 8.8.

▶ Catalysts lower the activation energy for a reaction. Lowered activation energy increases the rate of a reaction.

Catalysts exert their effects in varying ways. Some catalysts provide a lower-energy pathway by entering into a reaction and forming an "intermediate," which then reacts further to produce the desired products and regenerate the catalyst. The following equations, where C is the catalyst, illustrate this concept.

Uncatalyzed reaction: $X + Y \longrightarrow XY$

Catalyzed reaction: *Step 1:* $X + C \longrightarrow XC$

Step 2: $XC + Y \longrightarrow XY + C$

Solid-state catalysts often act by providing a surface to which reactant molecules are physically attracted and on which they are held with a particular orientation. These "held" reactants are sufficiently close to and favorably oriented toward one another that the reaction takes place. The products of the reaction then leave the surface and make it available to catalyze other reactants.

▶ Catalysts are extremely important for the proper functioning of the human body and other biological systems. Enzymes, which are proteins, are the catalysts within the human body (Chapter 16). They cause many reactions to take place rapidly under mild conditions and at body temperature. Without these enzymes, the reactions would proceed very slowly and then only under harsher conditions.

The accompanying Chemistry at a Glance summarizes the factors that influence reaction rates.

Practice Problems and Questions

8.18 Substances burn more rapidly in pure oxygen than in air. Explain why.

8.19 Milk will sour in a couple of days when left at room temperature, yet it can remain unspoiled for 2 weeks when refrigerated. Explain why.

8.20 Will each of the following changes increase or decrease the rate of this chemical reaction?

$$2CO + O_2 \longrightarrow 2CO_2$$

a. Adding some O_2 to the reaction mixture
b. Lowering the temperature of the reaction mixture
c. Adding a catalyst to the reaction mixture
d. Removing some CO from the reaction mixture

8.21 Using collision theory, indicate why each of the following factors influences the rate of a chemical reaction.

a. Temperature of the reactants
b. Presence of a catalyst
c. Physical nature of the reactants
d. Reactant concentrations

Chemistry *at a Glance*

Factors That Influence Reaction Rates

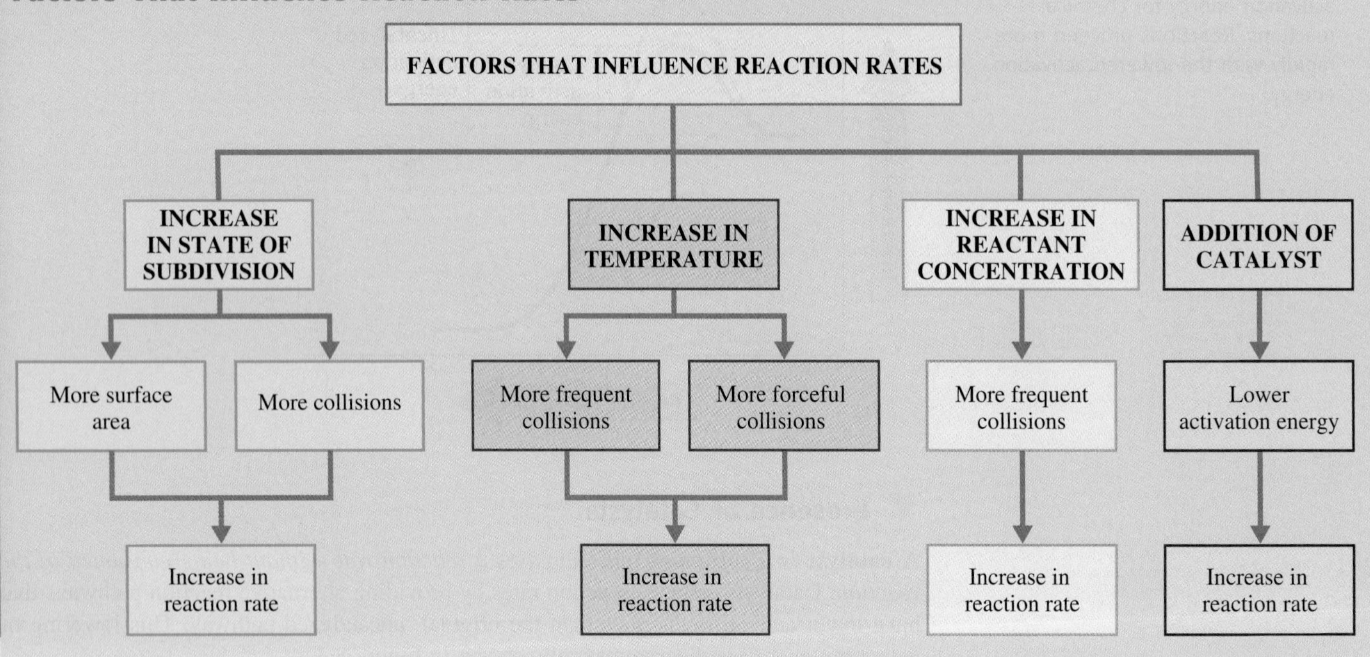

Learning Focus

Explain what the term *chemical equilibrium* means, and discuss the conditions necessary for attainment of this state.

8.7 Chemical Equilibrium

In our discussions of chemical reactions up to this point, we have assumed that chemical reactions go to completion; that is, reactions continue until one or more of the reactants are used up. This assumption is valid as long as product concentrations are not allowed to build up in the reaction mixture. If one or more products are gases that can escape from the reaction mixture or insoluble solids that can be removed from the reaction mixture, no product buildup occurs.

When product buildup does occur, reactions do not go to completion. This is because product molecules begin to react with one another to re-form reactants. With time, a steady-state situation results wherein the rate of formation of products and the rate of re-formation of reactants are equal. At this point, the concentrations of all reactants and all products remain constant, and a state of *chemical equilibrium* is reached. **Chemical equilibrium** *is the process wherein two opposing chemical reactions occur simultaneously at the same rate.* We discussed equilibrium situations in Sections 6.11 (vapor pressure) and 7.2 (saturated solutions), but the previous examples involved physical equilibrium rather than chemical equilibrium. Figure 8.9 illustrates an "everyday" equilibrium situation.

▶ A chemical reaction is in a state of chemical equilibrium when the rates of the forward and reverse reactions are equal. At this point, the concentrations of reactants and products no longer change.

The conditions that exist in a system in a state of chemical equilibrium can best be seen by considering an actual chemical reaction. Suppose equal molar amounts of gaseous H_2 and I_2 are mixed together in a closed container and allowed to react.

$$H_2 + I_2 \longrightarrow 2HI$$

Initially, no HI is present, so the only reaction that can occur is that between H_2 and I_2. However, as the HI concentration increases, some HI molecules collide with one another in a way that causes a reverse reaction to occur:

$$2HI \longrightarrow H_2 + I_2$$

Figure 8.9 These jugglers provide an illustration of equilibrium. Each throws balls to the other at the same rate at which he receives balls from that person. Because balls are thrown continuously in both directions, the number of balls moving in each direction is constant, and the number of balls each juggler has at a given time remains constant.

The initially low concentration of HI makes this reverse reaction slow at first, but as the concentration of HI increases, the reaction rate also increases. At the same time that the reverse-reaction rate is increasing, the forward-reaction rate (production of HI) is decreasing as the reactants are used up. Eventually, the concentrations of H_2, I_2, and HI in the reaction mixture reach a level at which the rates of the forward and reverse reactions become equal. At this point, a state of chemical equilibrium has been reached.

Figure 8.10a illustrates the behavior of reaction rates over time for both the forward and reverse reactions in the H_2–I_2–HI system. Figure 8.10b illustrates the important point that the reactant and product concentrations are usually not equal at the point at which equilibrium is reached.

▶ At chemical equilibrium, forward and reverse reaction rates are equal. Reactant and product concentrations, although constant, do not have to be equal.

The equilibrium involving H_2, I_2, and HI could have been established just as easily by starting with pure HI and allowing it to change into H_2 and I_2 (the reverse reaction). The final position of equilibrium does not depend on the direction from which equilibrium is approached.

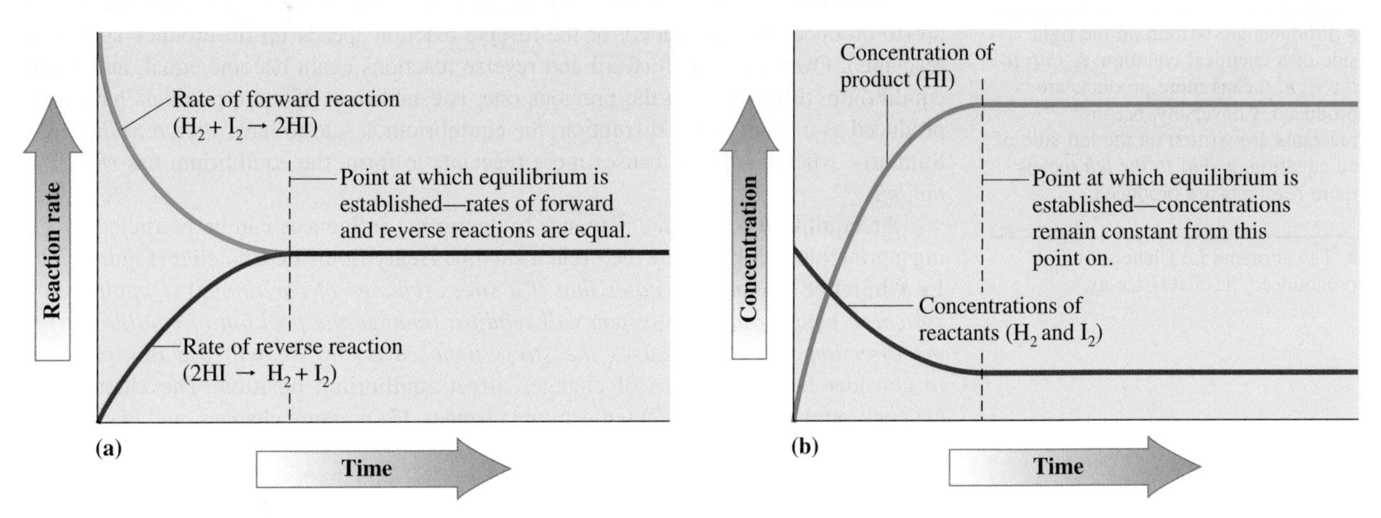

Figure 8.10 Graphs showing how reaction rates and reactant concentrations vary with time for the chemical system H_2–I_2–HI. (a) At equilibrium, rates of reaction are equal. (b) At equilibrium, concentrations of reactants remain constant but are not equal.

It is normal procedure to represent an equilibrium by using a single equation and two half-headed arrows pointing in opposite directions. Thus the reaction between H_2 and I_2 at equilibrium is written as

$$H_2 + I_2 \rightleftharpoons 2HI$$

The half-headed arrows denote a chemical system at equilibrium.

▶ Theoretically, all reactions are reversible (can go in either direction). Sometimes, the reverse reaction is so slight, however, that we say the reaction has "gone to completion" because no detectable reactants remain.

The term *reversible reaction* is often used in discussions of chemical equilibrium. A **reversible reaction** *is a chemical reaction in which the conversion of reactants into products (the forward reaction) and the conversion of products into reactants (the reverse reaction) occur simultaneously.* The half-headed arrow used to denote a system at equilibrium also carries the message that the reaction is reversible.

Practice Problems and Questions

8.22 What notation is used in a chemical equation to denote that the system is in a state of chemical equilibrium?

8.23 At chemical equilibrium, what is the relationship between the rates of the forward and reverse reactions?

8.24 What is a reversible chemical reaction?

8.25 Indicate whether each of the following conditions always applies to a system at chemical equilibrium.
a. The rates of the forward and reverse reactions are equal.
b. The concentrations of the reactants and products are equal.
c. The rates of the forward and reverse reactions are not changing.
d. The concentrations of the reactants and products are not changing.

Learning Focus

Use Le Châtelier's principle to predict the effect that concentration, temperature, and pressure changes will have on an equilibrium system.

8.8 Altering Equilibrium Conditions: Le Châtelier's Principle

A chemical system at equilibrium is very susceptible to disruption from outside forces. A change in temperature or a change in pressure can upset the balance within the equilibrium system. Changes in the concentrations of reactants or products upset an equilibrium also.

Disturbing an equilibrium has one of two results: Either the forward reaction speeds up (to produce more products), or the reverse reaction speeds up (to produce additional reactants). Over time, the forward and reverse reactions again become equal, and a new equilibrium, different from the previous one, is established. If more products have been produced as a result of the disruption, the equilibrium is said to have *shifted to the right.* Similarly, when disruption causes more reactants to form, the equilibrium has *shifted to the left.*

▶ Products are written on the right side of a chemical equation. A *shift to the right* means more products are produced. Conversely, because reactants are written on the left side of an equation, a *shift to the left* means more reactants are produced.

An equilibrium system's response to disrupting influences can be predicted by using a principle introduced by the French chemist Henry Louis Le Châtelier (Figure 8.11). **Le Châtelier's principle** *states that if a stress (change of conditions) is applied to a system in equilibrium, the system will readjust (change the position of equilibrium) in the direction that best reduces the stress imposed on it.* We will use this principle to consider how four types of changes affect equilibrium position. The changes are (1) concentration changes, (2) temperature changes, (3) pressure changes, and (4) addition of catalysts.

▶ The surname Le Châtelier is pronounced "le-SHOT-lee-ay."

Concentration Changes

Adding a reactant or product to or removing it from a reaction mixture at equilibrium always upsets the equilibrium. If an additional amount of any reactant or product has been *added* to the system, the stress is relieved by shifting the equilibrium in the direction that

Figure 8.11 Henri Louis Le Châtelier (1850–1936), although most famous for the principle that bears his name, was amazingly diverse in his interests. He worked on metallurgical processes, cements, glasses, fuels, and explosives and was also noted for his skills in industrial management.

consumes (uses up) some of the added reactant or product. Conversely, if a reactant or product is *removed* from an equilibrium system, the equilibrium shifts in a direction that will *produce* more of the substance that was removed.

Let us consider the effect that concentration changes will have on the gaseous equilibrium

$$N_2(g) + 3H_2(g) \rightleftharpoons 2NH_3(g)$$

Suppose some additional H_2 is added to the equilibrium mixture. The stress of "added H_2" causes the equilibrium to shift to the right; that is, the forward reaction rate increases in order to use up some of the additional H_2.

Stress: Too much H_2
Response: Use up "extra" H_2

$$N_2(g) + 3H_2(g) \rightleftharpoons 2NH_3(g)$$

Shift to the right →
$[N_2]$ decreases, $[H_2]$ decreases, $[NH_3]$ increases

As the H_2 reacts, the amount of N_2 also decreases (it reacts with the H_2) and the amount of NH_3 increases (it is formed as H_2 and N_2 react).

With time, the equilibrium shift to the right caused by the addition of H_2 will cease because a new equilibrium condition (not identical to the original one) has been reached. At this new equilibrium condition, most (but not all) of the added H_2 will have been converted to NH_3. Necessary accompaniments to this change are a decreased N_2 concentration (some of it reacted with the H_2) and an increased NH_3 concentration (that produced from the N_2–H_2 reaction). Figure 8.12 quantifies the changes that occur in the N_2–H_2–NH_3 equilibrium system when it is upset by the addition of H_2 for a specific set of concentrations.

Consider again the reaction between N_2 and H_2 to form NH_3.

$$N_2(g) + 3H_2(g) \rightleftharpoons 2NH_3(g)$$

Le Châtelier's principle applies in the same way to removing a reactant or product from the equilibrium mixture as it does to adding a reactant or product at equilibrium. Suppose that at equilibrium we remove some NH_3. The equilibrium position shifts to the right to replenish the NH_3.

Figure 8.12 Concentration changes that result when H_2 is added to an equilibrium mixture involving the system

$$N_2(g) + 3H_2(g) \rightleftharpoons 2NH_3(g)$$

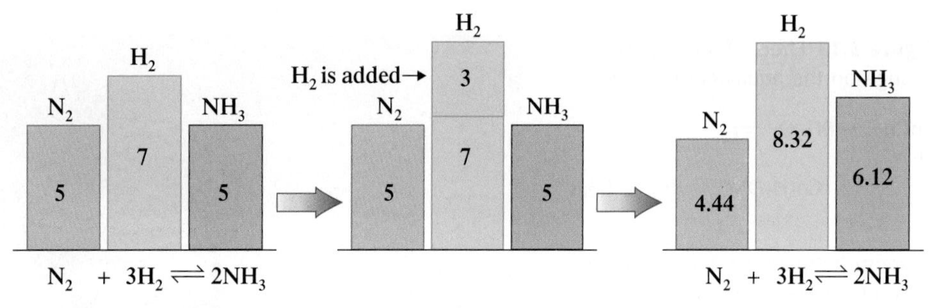

(1) Original equilibrium conditions

(2) Increase in $[H_2]$ upsets equilibrium; reaction shifts to right as more N_2 reacts with the additional H_2.

(3) New equilibrium conditions. Compared with the original equilibrium in (1):
$[N_2]$ has decreased.
$[H_2]$ has increased because of addition. (Note that $[H_2]$ is actually decreased from conditions at (2) because some of it has reacted with N_2 to form more NH_3.)
$[NH_3]$ has increased.

▶ Thousands of chemical equilibria simultaneously exist in biological systems. Many of them are interrelated. When the concentration of one substance changes, many equilibria are affected.

Within the human body, numerous equilibrium situations exist that shift in response to a concentration change. Consider, for example, the equilibrium between glucose in the blood and stored glucose (glycogen) in the liver:

$$\text{Glucose in blood} \rightleftharpoons \text{stored glucose} + H_2O$$

Strenuous exercise or hard work causes our blood glucose level to decrease. Our bodies respond to this stress (not enough glucose in the blood) by the liver converting glycogen into glucose. Conversely, when an excess of glucose is present in the blood (after a meal), the liver converts the excess glucose in the blood to its storage form (glycogen).

Temperature Changes

Le Châtelier's principle can be used to predict the influence of temperature changes on an equilibrium, provided we know whether the reaction is exothermic or endothermic. For *exothermic reactions,* heat can be treated as one of the *products;* for *endothermic reactions,* heat can be treated as one of the *reactants.*

Consider the exothermic reaction

$$H_2(g) + F_2(g) \rightleftharpoons 2HF(g) + \text{heat}$$

Heat is produced when the reaction proceeds to the right. Thus, if we add heat to an exothermic system at equilibrium (by raising the temperature), the system will shift to the left in an "attempt" to decrease the amount of heat present. When equilibrium is reestablished, the concentrations of H_2 and F_2 will be higher and the concentration of HF will have decreased. Lowering the temperature of an exothermic reaction mixture causes the reaction to shift to the right as the system acts to replace the lost heat (Figure 8.13).

The behavior, with temperature change, of an equilibrium system involving an endothermic reaction, such as

$$\text{Heat} + 2CO_2(g) \rightleftharpoons 2CO(g) + O_2(g)$$

is opposite to that of an exothermic reaction, because a shift to the left produces heat. Consequently, an increase in temperature will cause the equilibrium to shift to the right (to decrease the amount of heat present), and a decrease in temperature will produce a shift to the left (to generate more heat).

Figure 8.13 Effect of temperature change on the equilibrium mixture

$$\underset{\text{Blue}}{CoCl_4^{2-}} + 6H_2O \rightleftharpoons \underset{\text{Pink}}{Co(H_2O)_6^{2+}} + 4Cl^- + \text{heat}$$

At room temperature, the equilibrium mixture is blue from $CoCl_4^{2-}$. When cooled by the ice bath, the equilibrium mixture turns pink from $Co(H_2O)_6^{2+}$. The temperature decrease causes the equilibrium position to shift to the right.

Pressure Changes

Pressure changes affect systems at equilibrium only when gases are involved, and then only in cases where the chemical reaction is such that a change in the total number of moles in the gaseous state occurs. This latter point can be illustrated by considering the following two gas-phase reactions:

$$\underbrace{2H_2(g) + O_2(g)}_{\text{3 moles of gas}} \longrightarrow \underbrace{2H_2O(g)}_{\text{2 moles of gas}}$$

$$\underbrace{H_2(g) + Cl_2(g)}_{\text{2 moles of gas}} \longrightarrow \underbrace{2HCl(g)}_{\text{2 moles of gas}}$$

In the first reaction, the total number of moles of gaseous reactants and products decreases as the reaction proceeds to the right. This is because 3 moles of reactants combine to give only 2 moles of products. In the second reaction, there is no change in the total number of moles of gaseous substances present as the reaction proceeds. This is because 2 moles of reactants combine to give 2 moles of products. Thus a pressure change will shift the position of the equilibrium in the first reaction but not in the second reaction.

Pressure changes are usually brought about through volume changes. A pressure increase results from a volume decrease, and a pressure decrease results from a volume increase (Section 6.4). Le Châtelier's principle correctly predicts the direction of the equilibrium position shift resulting from a pressure change only when the pressure change is caused by a change in volume. It does not apply to pressure increases caused by the addition of a nonreactive (inert) gas to the reaction mixture. This addition has no effect on the equilibrium position. The partial pressure (Section 6.8) of each of the gases involved in the reaction remains the same.

▶ Increasing the pressure associated with an equilibrium system by adding an inert gas (a gas that is not a reactant or a product in the reaction) does not affect the position of the equilibrium.

According to Le Châtelier's principle, the stress of increased pressure is relieved by decreasing the number of moles of gaseous substances in the system. This is accomplished by the reaction shifting in the direction of the fewer moles; that is, it shifts to the side of the equation that contains the fewer moles of gaseous substances. For the reaction

$$2NO_2(g) + 7H_2(g) \rightleftharpoons 2NH_3(g) + 4H_2O(g)$$

an increase in pressure would shift the equilibrium position to the right, because there are 9 moles of gaseous reactants and only 6 moles of gaseous products. On the other hand, the stress of decreased pressure causes an equilibrium system to produce more moles of gaseous substances.

Addition of Catalysts

Catalysts cannot change the position of an equilibrium. A catalyst functions by lowering the activation energy for a reaction. It speeds up both the forward and the reverse reactions, so it has no net effect on the position of the equilibrium. However, the lowered activation energy allows equilibrium to be established more quickly than if the catalyst were absent.

Example 8.4 Using Le Châtelier's Principle to Predict How Various Changes Affect an Equilibrium System

How will the gas-phase equilibrium

$$CH_4(g) + 2H_2S(g) + \text{heat} \rightleftharpoons CS_2(g) + 4H_2(g)$$

be affected by the following?

a. The removal of $H_2(g)$
b. The addition of $CS_2(g)$
c. An increase in the temperature
d. An increase in the volume of the container (a decrease in pressure)

Solution

a. The equilibrium will *shift to the right,* according to Le Châtelier's principle, in an "attempt" to replenish the H_2 that was removed.
b. The equilibrium will *shift to the left* in an attempt to use up the extra CS_2 that has been placed in the system.
c. Raising the temperature means that heat energy has been added. In an attempt to minimize the effect of this extra heat, the position of the equilibrium will *shift to the right,* which is a direction that consumes heat; heat is one of the reactants in an endothermic reaction.
d. The system will *shift to the right,* the direction that produces more moles of gaseous substances (an increase of pressure). In this way, the reaction produces 5 moles of gaseous products for every 3 moles of gaseous reactants consumed.

Practice Problems and Questions

8.26 Consider the following chemical system at equilibrium.

$$2Cl_2(g) + 2H_2O(g) \rightleftharpoons 4HCl(g) + O_2(g)$$

For each of the following adjustments of conditions, indicate the effect (shifts left, shifts right, or no effect) on the position of equilibrium.
a. Increasing the Cl_2 concentration
b. Increasing the O_2 concentration
c. Decreasing the H_2O concentration
d. Decreasing the HCl concentration

8.27 Consider the following chemical system at equilibrium.

$$C_6H_6(g) + 3H_2(g) \rightleftharpoons C_6H_{12}(g) + \text{heat}$$

For each of the following adjustments of conditions, indicate the effect (shifts left, shifts right, or no effect) on the position of equilibrium.
a. Increasing the C_6H_{12} concentration
b. Decreasing the C_6H_6 concentration
c. Increasing the temperature
d. Decreasing the pressure by increasing the volume of the container

8.28 Consider the following chemical system at equilibrium.

$$CO(g) + H_2O(g) + \text{heat} \rightleftharpoons CO_2(g) + H_2(g)$$

For each of the following adjustments of conditions, indicate the effect (shifts left, shifts right, or no effect) on the position of equilibrium.
a. Lowering the temperature of the equilibrium mixture
b. Adding a catalyst to the equilibrium mixture
c. Increasing the pressure of the equilibrium mixture by adding a nonreactive (inert) gas
d. Increasing the size of the reaction container

CONCEPTS TO REMEMBER

Chemical reaction. A process in which at least one new substance is produced as a result of chemical change.

Combination reaction. A chemical reaction in which a single product is produced from two or more reactants.

Decomposition reaction. A chemical reaction in which a single reactant is converted into two or more simpler substances (elements or compounds).

Single-replacement reaction. A chemical reaction in which an atom or a molecule replaces an atom or a group of atoms from a compound.

Double-replacement reaction. A chemical reaction in which two substances exchange parts with one another and form two different substances.

Combustion reaction. A chemical reaction in which oxygen (usually from air) reacts with a substance with the evolution of heat and (usually) the presence of a flame.

Oxidation. The process whereby a reactant in a chemical reaction (1) gains one or more oxygen atoms, (2) loses one or more hydrogen atoms, or (3) loses one or more electrons.

Reduction. The process whereby a reactant in a chemical reaction (1) loses one or more oxygen atoms, (2) gains one or more hydrogen atoms, or (3) gains one or more electrons.

Redox reaction. A chemical reaction in which there is a transfer of electrons from one reactant to another reactant.

Nonredox reaction. A chemical reaction in which there is no transfer of electrons from one reactant to another reactant.

Oxidizing and Reducing Agents. An oxidizing agent causes oxidation by accepting electrons from another reactant; it itself is reduced. A reducing agent causes reduction by providing electrons for another reactant; it itself is oxidized.

Collision theory. Collision theory summarizes the conditions required for a chemical reaction to take place. The three basic tenets of collision theory are as follows: (1) Reactant molecules must collide with each other. (2) The collision must involve a certain minimum of energy. (3) In some cases, colliding molecules must be oriented in a specific way if reaction is to occur.

Exothermic and endothermic chemical reactions. An exothermic chemical reaction releases energy as the reaction occurs. An endothermic chemical reaction requires an input of energy as the reaction occurs.

Reaction rates. Reaction rate is the speed at which reactants are converted to products. Four factors affect the rates of all reactions: (1) the physical nature of the reactants, (2) reactant concentrations, (3) reaction temperature, and (4) the presence of catalysts.

Chemical equilibrium. Chemical equilibrium is the state wherein the rate of the forward reaction is equal to the rate of the reverse reaction. Equilibrium is indicated in chemical equations by writing half-headed arrows pointing in both directions between reactants and products.

Le Châtelier's principle. Le Châtelier's principle states that when a stress (change of conditions) is applied to a system in equilibrium, the system will readjust (change the position of the equilibrium) in the direction that best reduces the stress imposed on it. Stresses known to change an equilibrium position include (1) changes in amount of reactants and/or products, (2) changes in temperature, and (3) changes in pressure.

KEY REACTIONS AND EQUATIONS

1. Combination reaction (Section 8.1)

$$X + Y \longrightarrow XY$$

2. Decomposition reaction (Section 8.1)

$$XY \longrightarrow X + Y$$

3. Single-replacement reaction (Section 8.1)

$$X + YZ \longrightarrow Y + XZ$$

4. Double-replacement reaction (Section 8.1)

$$AX + BY \longrightarrow AY + BX$$

KEY TERMS

Activation energy (8.4)
Catalyst (8.6)
Chemical equilibrium (8.7)
Chemical reaction (8.1)
Collision theory (8.4)
Combination reaction (8.1)
Combustion reaction (8.1)
Decomposition reaction (8.1)
Double-replacement reaction (8.1)
Endothermic chemical reaction (8.5)
Exothermic chemical reaction (8.5)
Le Châtelier's principle (8.8)
Non-oxidation-reduction reaction (8.2)
Oxidation (8.2)
Oxidation–reduction reaction (8.2)
Oxidizing agent (8.3)
Rate of chemical reaction (8.6)
Reducing agent (8.3)
Reduction (8.2)
Reversible reaction (8.7)
Single-replacement reaction (8.1)

ADDITIONAL PROBLEMS

8.29 Characterize each of the following reactions using one selection from the choices *redox* and *nonredox* combined with one selection from the choices *combination, decomposition, single-replacement,* and *double-replacement.*

a. $Zn + Cu(NO_3)_2 \longrightarrow Zn(NO_3)_2 + Cu$
b. $2SO_2 + O_2 \longrightarrow 2SO_3$
c. $2CuO \longrightarrow 2Cu + O_2$
d. $NaCl + AgNO_3 \longrightarrow AgCl + NaNO_3$

8.30 In each of the following statements, choose the word in parentheses that best completes the statement.

a. The process of reduction is associated with the (loss, gain) of electrons.
b. The oxidizing agent in a redox reaction is the substance that undergoes (oxidation, reduction).
c. The reducing agent in a redox reaction is the substance that (loses, gains) electrons.
d. The substance that undergoes oxidation in a redox reaction is the (oxidizing, reducing) agent.

8.31 For the redox reaction

$$2Cl_2 + 2H_2O \longrightarrow 4HCl + O_2$$

identify

a. the substance that is oxidized
b. the substance that is reduced
c. the oxidizing agent
d. the reducing agent

8.32 The characteristics of four reactions, each of which involves only two reactants, follow.

Reaction	Activation energy	Temperature	Concentration of reactants
1	low	low	1 mole/L of each
2	high	low	1 mole/L of each
3	low	high	1 mole/L of each
4	low	low	1 mole/L of first reactant and 4 moles/L of second reactant

For each of the following pairs of the preceding reactions, compare the reaction rates when the two reactants are first mixed. Indicate which reaction is faster.
a. 1 and 2 b. 1 and 3 c. 1 and 4 d. 2 and 3

8.33 A characteristic of a chemical equilibrium system is *constant* reactant and product concentrations. What is the difference between *constant* product and reactant concentrations and *equal* product and reactant concentrations?

8.34 The reaction between a particular metal and oxygen to produce the metal's oxide is found (1) to be very slow and (2) to produce heat.
a. Is the reaction exothermic or endothermic?
b. Does the forward reaction have a high or a low activation energy?
c. Is the reaction a redox reaction or a nonredox reaction?
d. Is the reaction a combination reaction or a single-replacement reaction?

PRACTICE TEST True/False

8.35 The presence of only one reactant is a characteristic of all decomposition reactions.

8.36 There are always two reactants and two products in a single-replacement reaction.

8.37 Some combustion reactions are also combination reactions.

8.38 The oxygen-based definition for reduction involves a reactant gaining one or more oxygen atoms.

8.39 In a redox reaction, the substance that undergoes oxidation is also called the reducing agent.

8.40 In an exothermic reaction, the energy of the reactants is lower than the energy of the products.

8.41 In a chemical reaction, collisions between reactant molecules do not always result in the formation of reaction products.

8.42 Most "slow" reactions have "low" activation energies.

8.43 An increase in the concentration of a reactant causes an increase in the rate of the reaction.

8.44 A catalyst is a substance that increases the rate of a reaction as it is consumed in the reaction.

8.45 All cases of chemical equilibrium involve reversible reactions.

8.46 For a system in a state of chemical equilibrium, the rate of formation of products and the rate of re-formation of reactants are equal.

8.47 When the disruption of a system at chemical equilibrium causes more products to form, the system is said to have "shifted to the right."

8.48 Addition of heat to an exothermic system at equilibrium will cause the system to "shift to the left."

8.49 A catalyst functions by lowering the activation energy for a reaction.

PRACTICE TEST Multiple Choice

8.50 Which of the following equations is a general representation for a single-replacement reaction?
a. $X + Y \longrightarrow XY$
b. $XY \longrightarrow X + Y$
c. $X + YZ \longrightarrow Y + XZ$
d. $AX + BY \longrightarrow AY + BX$

8.51 Which substance is oxidized in the following redox reaction?

$$2H_2S + O_2 \longrightarrow 2H_2O + 2S$$

a. H_2S b. O_2 c. H_2O d. S

8.52 In a redox reaction, the substance reduced
a. is also the reducing agent
b. always gains electrons
c. never contains oxygen
d. must contain hydrogen

8.53 Most reactions are carried out in liquid solution or in the gaseous phase, because in such situations,
a. activation energies are higher
b. it is easier for reactants to come in contact with each other
c. kinetic energies of reactants are lower
d. products are less apt to decompose

8.54 The minimum combined kinetic energy that reactant particles must possess in order for their collision to result in a reaction is called the
a. equilibrium energy
b. collision energy
c. activation energy
d. orientation energy

8.55 Which of the following changes would be most likely to *decrease* the rate of a reaction?
a. Increasing the state of subdivision of a reactant
b. Increasing the concentration of a reactant
c. Decreasing the reaction temperature
d. Decreasing the activation energy for the reaction

8.56 Catalysts are correctly characterized by each of the following statements except one. Which statement is *not* true of catalysts?
a. They can be solids, liquids, or gases.
b. They lower the activation energy for the reaction.
c. They do not actively participate in the reaction.
d. They are not consumed in the reaction.

8.57 Which of the following conditions characterizes a system in a state of chemical equilibrium?
a. The concentrations of reactants and products are equal.
b. The rate of the forward reaction has dropped to zero.
c. Reactants are being consumed at the same rate at which they are being produced.
d. Reactant molecules no longer react with each other.

8.58 CO_2 and H_2 are allowed to react until an equilibrium is established as follows:

$$CO_2(g) + H_2(g) \rightleftharpoons H_2O(g) + CO(g)$$

What will be the effect on the equilibrium of removing CO from the equilibrium mixture?
a. The equilibrium will shift to the left.
b. The H_2 concentration will increase.
c. The CO and CO_2 concentrations will increase.
d. The H_2 concentration will decrease and the H_2O concentration will increase.

8.59 Which of the following changes will cause the equilibrium position to shift to the left for this reaction?

$$4NH_3(g) + 3O_2(g) \rightleftharpoons 2N_2(g) + 6H_2O(g) + \text{heat}$$

a. Adding more NH_3
b. Increasing the temperature
c. Adding a catalyst
d. Decreasing the pressure by increasing the volume

CHAPTER OUTLINE

CHAPTER NINE

Acids, Bases, and Salts

Fish are very sensitive to the acidity of the water present in an aquarium.

Acids, bases, and salts are among the most common and important compounds known. In the form of aqueous solutions, these compounds are key materials in both biological systems and the chemical industry. A major ingredient of gastric juice in the stomach is hydrochloric acid. Quantities of lactic acid are produced when the human body is subjected to strenuous exercise. The lye used in making soap contains the base sodium hydroxide. Bases are ingredients in many stomach antacid formulations. The white crystals you sprinkle on your food to make it taste better represent only one of many hundreds of salts that exist.

9.1 Arrhenius Acid–Base Theory

Learning Focus

Using Arrhenius acid–base theory, define an acid and a base.

In 1884, the Swedish chemist Svante August Arrhenius (1859–1927) proposed that acids and bases be defined in terms of the chemical species they form when they dissolve in water. An **Arrhenius acid** *is a hydrogen-containing compound that, in water, produces hydrogen ions* (H^+). The acidic species in Arrhenius theory is thus the hydrogen ion. An **Arrhenius base** *is a hydroxide-containing compound that, in water, produces hydroxide ions* (OH^-). The basic species in Arrhenius theory is thus the hydroxide ion. For this reason, Arrhenius bases are also called *hydroxide bases*.

Two common examples of Arrhenius acids are HNO_3 (nitric acid) and HCl (hydrochloric acid).

$$HNO_3(l) \xrightarrow{H_2O} H^+(aq) + NO_3^-(aq)$$
$$HCl(g) \xrightarrow{H_2O} H^+(aq) + Cl^-(aq)$$

Figure 9.1 The difference between the aqueous solution processes of ionization (Arrhenius acids) and dissociation (Arrhenius bases). Ionization is the production of ions from a *molecular compound* that has been dissolved in solution. Dissociation is the production of ions from an *ionic compound* that has been dissolved in solution.

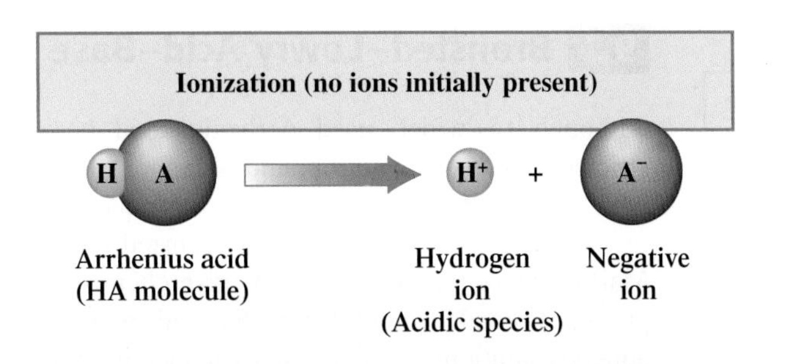

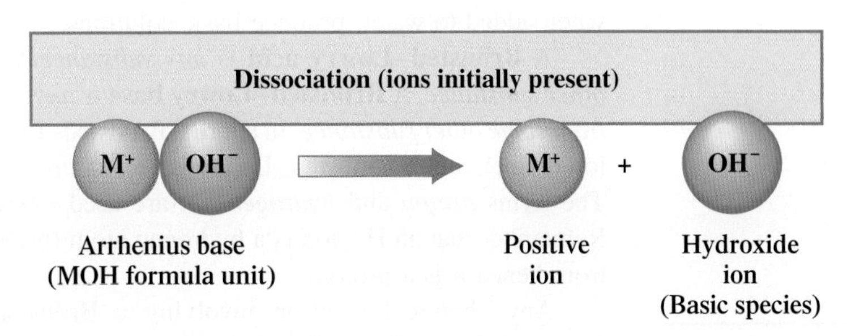

When Arrhenius acids are in the pure state (not in solution), they are covalent compounds; that is, they do not contain H^+ ions. This ion is formed through an interaction between water and the acid when they are mixed. **Ionization** *is the process in which individual positive and negative ions are produced from a molecular compound that is dissolved in solution.*

Two common examples of Arrhenius bases are NaOH (sodium hydroxide) and KOH (potassium hydroxide).

$$NaOH(s) \xrightarrow{H_2O} Na^+(aq) + OH^-(aq)$$

$$KOH(s) \xrightarrow{H_2O} K^+(aq) + OH^-(aq)$$

In direct contrast to acids, Arrhenius bases are ionic compounds in the pure state. When these compounds dissolve in water, the ions separate to yield the OH^- ions. **Dissociation** *is the process in which individual positive and negative ions are released from an ionic compound that is dissolved in solution.* Figure 9.1 contrasts the processes of ionization (acids) and dissociation (bases).

Arrhenius acids have a sour taste, change blue litmus paper to red (see Figure 9.2), and are corrosive to many materials. Arrhenius bases have a bitter taste, change red litmus paper to blue, and are slippery (soapy) to the touch. [The bases themselves are not slippery but react with the fats in the skin to form new slippery (soapy) compounds.]

Figure 9.2 Litmus is a vegetable dye obtained from certain lichens found principally in the Netherlands. Paper treated with this dye turns from blue to red in acids (left) and from red to blue in bases (right).

Practice Problems and Questions

9.1 Classify each of the following as a property of an Arrhenius acid or as a property of an Arrhenius base.

a. Has a sour taste
b. Has a bitter taste
c. Changes blue litmus paper to red
d. Changes red litmus paper to blue

9.2 Write equations depicting the behavior of the following Arrhenius acids and bases in water.

a. HI (hydroiodic acid) b. HClO (hypochlorous acid)
c. LiOH (lithium hydroxide) d. CsOH (cesium hydroxide)

9.3 What word is used to describe the formation of ions, in aqueous solution

a. from a molecular compound b. from an ionic compound

9.4 In Arrhenius acid–base theory, indicate what ion is responsible for the properties of

a. acidic solutions b. basic solutions

Learning Focus

Using Brønsted–Lowry acid–base theory, define an acid and a base.

9.2 Brønsted–Lowry Acid–Base Theory

Although it is widely used, Arrhenius acid–base theory has some shortcomings. It is restricted to aqueous solution, and it cannot explain why compounds like ammonia (NH_3), which do not contain hydroxide ion, produce a basic water solution.

In 1923, Johannes Nicolaus Brønsted (1879–1947), a Danish chemist, and Thomas Martin Lowry (1874–1936), a British chemist, independently and almost simultaneously proposed broadened definitions for acids and bases that applied in both aqueous and nonaqueous solutions and that also explained how some nonhydroxide-containing substances, when added to water, produce basic solutions.

A **Brønsted–Lowry acid** *is any substance that can donate a proton* (H^+) *to some other substance*. A **Brønsted–Lowry base** *is any substance that can accept a proton* (H^+) *from some other substance*. In short, a Brønsted–Lowry acid is a *proton donor* (or hydrogen ion donor), and a Brønsted–Lowry base is a *proton acceptor* (or hydrogen ion acceptor). The terms *proton* and *hydrogen ion* are used interchangeably in acid–base discussions. Remember that an H^+ ion is a hydrogen atom (proton plus electron) that has lost its electron; hence it is a proton.

▶ The terms *hydrogen ion* and *proton* are used synonymously in acid–base discussions. Why? The predominant hydrogen isotope, ${}^1_1H$, is unique in that no neutrons are present; it consists of a proton and an electron. Thus the ion ${}^1_1H^+$, a hydrogen atom that has lost its only electron, is simply a proton.

Any chemical reaction involving a Brønsted–Lowry acid must also involve a Brønsted–Lowry base. You cannot have one without the other. Proton donation (from an acid) cannot occur unless an acceptor (a base) is present.

Brønsted–Lowry acid–base theory also includes the concept that hydrogen ions in an aqueous solution do not exist in the free state but, rather, react with water to form *hydronium ions*. The attraction between a hydrogen ion and polar water molecules is sufficiently strong to bond the hydrogen ion to a water molecule to form a hydronium ion (H_3O^+).

$$H^+ + :\ddot{O}-H \longrightarrow \left[H:\ddot{O}-H \right]^+$$
(with H bonded below O in each)

Hydronium ion

▶ A Brønsted–Lowry base—a proton acceptor—must contain an atom that possesses a pair of nonbonding electrons that can be used in forming a covalent bond to an incoming proton (from a Brønsted–Lowry acid).

When gaseous hydrogen chloride dissolves in water, it forms hydrochloric acid. This is a simple Brønsted–Lowry acid–base reaction. The chemical equation for this process is

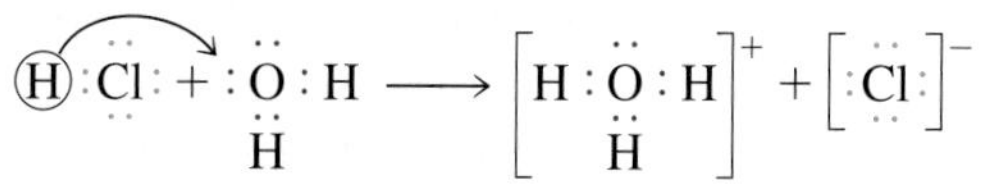

$$\underset{\text{Acid}}{H:\ddot{\underset{..}{Cl}}:} + \underset{\text{Base}}{:\ddot{\underset{..}{O}}:H} \longrightarrow \left[H:\ddot{\underset{..}{O}}:H \right]^+ + \left[:\ddot{\underset{..}{Cl}}: \right]^-$$
(with H bonded below O in each)

The hydrogen chloride behaves as an acid by donating a proton to a water molecule. Because the water molecule accepts the proton, to become H_3O^+, it is the base.

It is not necessary that a water molecule be one of the reactants in a Brønsted–Lowry acid–base reaction; the reaction does not have to take place in the liquid state. Brønsted–Lowry acid–base theory can be used to describe gas-phase reactions. The white solid haze that often covers glassware in a chemistry laboratory results from the gas-phase reaction between HCl and NH_3:

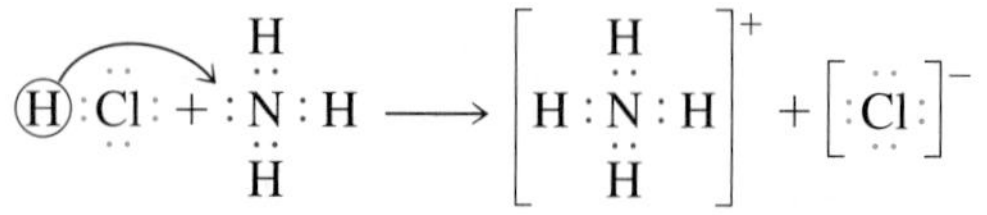

$$\underset{\text{Acid}}{H:\ddot{\underset{..}{Cl}}:} + \underset{\text{Base}}{:\ddot{N}:H} \longrightarrow \left[H:\ddot{\underset{..}{N}}:H \right]^+ + \left[:\ddot{\underset{..}{Cl}}: \right]^-$$
(with H above and below N in each)

This is a Brønsted–Lowry acid–base reaction because the HCl molecules donate protons to the NH_3, forming NH_4^+ and Cl^- ions. These ions instantaneously combine to form the white solid NH_4Cl (see Figure 9.3).

Table 9.1 Summary of Acid–Base Definitions

Arrhenius acid: hydrogen-containing species that produces H^+ ion in aqueous solution
Arrhenius base: hydroxide-containing species that produces OH^- ion in aqueous solution
Brønsted–Lowry acid: proton (H^+) donor
Brønsted–Lowry base: proton (H^+) acceptor

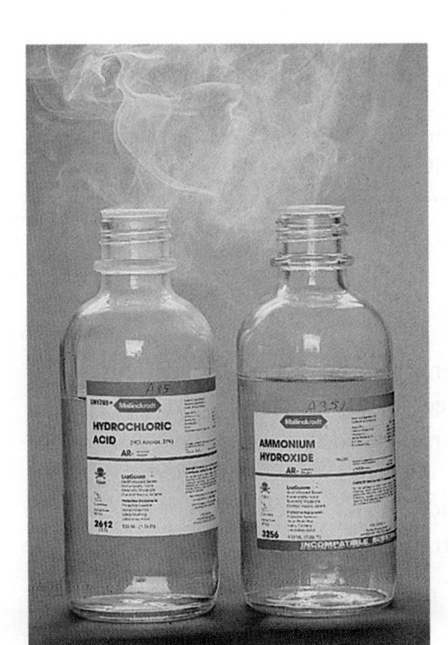

Figure 9.3 A white cloud of finely divided solid NH_4Cl is produced by the acid-base reaction that results when the colorless gases HCl and NH_3 mix. (The gases involved escape from the concentrated solutions of HCl and NH_3.)

All the acids and bases included in the Arrhenius theory are also acids and bases according to the Brønsted–Lowry theory. However, the converse is not true; some substances that are not considered Arrhenius bases are Brønsted–Lowry bases. Table 9.1 summarizes these two types of acid–base definitions.

Conjugate Acid–Base Pairs

For most Brønsted–Lowry acid–base reactions, 100% proton transfer does not occur. Instead, an equilibrium situation (Section 8.7) is reached in which a forward and a reverse reaction occur at the same rate.

The equilibrium mixture for a Brønsted–Lowry acid–base reaction always has two acids and two bases present. Consider the acid–base reaction involving hydrogen fluoride and water:

$$HF(aq) + H_2O(l) \rightleftharpoons H_3O^+(aq) + F^-(aq)$$

For the forward reaction, the HF molecules donate protons to water molecules. Thus the HF is functioning as an acid, and the H_2O is functioning as a base.

$$\underset{\text{Acid}}{HF(aq)} + \underset{\text{Base}}{H_2O(l)} \longrightarrow H_3O^+(aq) + F^-(aq)$$

For the reverse reaction, the one going from right to left, a different picture emerges. Here, H_3O^+ is functioning as an acid (by donating a proton), and F^- behaves as a base (by accepting the proton).

$$\underset{\text{Acid}}{H_3O^+(aq)} + \underset{\text{Base}}{F^-(aq)} \longrightarrow HF(aq) + H_2O(l)$$

▶ *Conjugate* means "coupled" or "joined together" (as in a pair).

The two acids and two bases involved in a Brønsted–Lowry acid–base equilibrium mixture can be grouped into two conjugate acid–base pairs. A **conjugate acid–base pair** *is two species that differ by one proton.* The two conjugate acid–base pairs in our example are

$$\underset{\text{Acid}}{HF(aq)} + \underset{\text{Base}}{H_2O(l)} \rightleftharpoons \underset{\text{Acid}}{H_3O^+(aq)} + \underset{\text{Base}}{F^-(aq)}$$

(Conjugate pair: HF and F^-; Conjugate pair: H_2O and H_3O^+)

▶ Every acid has a conjugate base, and every base has a conjugate acid. In general terms, these relationships can be diagrammed as follows:

$$\underset{\text{Acid}}{HA} + \underset{\text{Base}}{B} \rightleftharpoons \underset{\text{Conjugate acid}}{HB^+} + \underset{\text{Conjugate base}}{A^-}$$

The **conjugate base** *of an acid is the species that remains when an acid loses a proton.* The conjugate base of HF is F^-. The **conjugate acid** *of a base is the species formed when a base accepts a proton.* The H_3O^+ ion is the conjugate acid of H_2O. A slash mark separating an acid (on the left) from its conjugate base (on the right) is often used in representing conjugate acid–base pairs: for example, HF/F^- and H_3O^+/H_2O.

Example 9.1 Determining the Formula of One Member of a Conjugate Acid–Base Pair When Given the Other Member

Write the formula of each of the following.

a. The conjugate base of HSO_4^- **b.** The conjugate acid of NO_3^-
c. The conjugate base of H_3PO_4 **d.** The conjugate acid of $HC_2O_4^-$

Solution

a. A conjugate base can always be found by removing one H^+ from a given acid. Removing one H^+ (both the atom and the charge) from HSO_4^- leaves SO_4^{2-}. Thus SO_4^{2-} is the conjugate base of HSO_4^-.

b. A conjugate acid can always be found by adding one H^+ to a given base. Adding one H^+ (both the atom and the charge) to NO_3^- produces HNO_3. Thus HNO_3 is the conjugate acid of NO_3^-.

c. Proceeding as in part **a,** the removal of a H^+ ion from H_3PO_4 produces the $H_2PO_4^-$ ion. Thus $H_2PO_4^-$ is the conjugate base of H_3PO_4.

d. Proceeding as in part **b,** the addition of a H^+ ion to $HC_2O_4^-$ produces the $H_2C_2O_4$ molecule. Thus $H_2C_2O_4$ is the conjugate acid of $HC_2O_4^-$.

Amphoteric Substances

▶ The term *amphoteric* comes from the Greek *amphoteres,* which means "partly one and partly the other." Just as an amphibian is an animal that lives partly on land and partly in the water, an amphoteric substance is sometimes an acid and sometimes a base.

Some molecules or ions are able to function as either an acid or a base, depending on the kind of substance with which they react. Such molecules are said to be amphoteric. An **amphoteric substance** *can either lose or accept a proton and thus can function as either an acid or a base.*

Water is the most common amphoteric substance. Water functions as a base in the first of the following two reactions and as an acid in the second.

$$\underset{\text{Acid}}{HNO_3(aq)} + \underset{\text{Base}}{H_2O(l)} \rightleftharpoons H_3O^+(aq) + NO_3^-(aq)$$

$$\underset{\text{Base}}{NH_3(aq)} + \underset{\text{Acid}}{H_2O(l)} \rightleftharpoons NH_4^+(aq) + OH^-(aq)$$

Practice Problems and Questions

9.5 In each of the following reactions, decide whether the underlined species is functioning as a Brønsted–Lowry acid or base.

a. $\underline{HF} + H_2O \longrightarrow H_3O^+ + F^-$
b. $H_2O + \underline{S^{2-}} \longrightarrow HS^- + OH^-$
c. $H_2O + \underline{H_2CO_3} \longrightarrow H_3O^+ + HCO_3^-$
d. $\underline{HCO_3^-} + H_2O \longrightarrow H_3O^+ + CO_3^{2-}$

9.6 Write equations to illustrate the acid–base reactions that can take place between the following Brønsted–Lowry acids and bases.

a. Acid: $HClO$; base: H_2O
b. Acid: $HClO_4$; base: NH_3
c. Acid: H_3O^+; base: OH^-
d. Acid: H_3O^+; base: NH_2^-

9.7 Write the formula of each of the following.

a. Conjugate base of H_2SO_3
b. Conjugate acid of CN^-
c. Conjugate base of $HC_2O_4^-$
d. Conjugate acid of HPO_4^{2-}

9.8 For each of the following amphoteric substances, write the two equations needed to describe its behavior in aqueous solution.

a. HS^- b. HPO_4^{2-}

9.9 What term is used interchangeably with *hydrogen ion* in both Arrhenius and Brønsted–Lowry acid–base discussions?

9.10 How does a *hydronium ion* differ from a *hydrogen ion*?

9.11 What advantages does Brønsted–Lowry acid–base theory have over Arrhenius acid–base theory?

9.3 Mono-, Di-, and Triprotic Acids

Learning Focus

On the basis of chemical formula, differentiate among mono-, di-, and triprotic acids, and write equations for the stepwise dissociation of di- and triprotic acids.

Acids can be classified according to the number of hydrogen ions they can transfer per molecule during an acid–base reaction. A **monoprotic acid** *is an acid that transfers one* H^+ *ion (proton) per molecule during an acid–base reaction.* Hydrochloric acid (HCl) and nitric acid (HNO_3) are both monoprotic acids.

A **diprotic acid** *is an acid that can transfer two* H^+ *ions (two protons) per molecule during an acid–base reaction.* Carbonic acid (H_2CO_3) is a diprotic acid. The transfer of protons for a diprotic acid always occurs in steps. For H_2CO_3, the two steps are

$$H_2CO_3(aq) + H_2O(l) \rightleftharpoons H_3O^+(aq) + HCO_3^-(aq)$$
$$HCO_3^-(aq) + H_2O(l) \rightleftharpoons H_3O^+(aq) + CO_3^{2-}(aq)$$

▶ If the double arrows in the equation for a system at equilibrium are of unequal length, the longer arrow indicates the direction in which the equilibrium is displaced.
$\leftharpoondown\!\!\!\rightharpoonup$ (longer left arrow) Equilibrium displaced toward reactants
$\rightleftharpoons$ (longer right arrow) Equilibrium displaced toward products

A few triprotic acids exist. A **triprotic acid** *is an acid that can transfer three* H^+ *ions (three protons) per molecule during an acid–base reaction.* Phosphoric acid, H_3PO_4, is the most common triprotic acid. The three proton-transfer steps for this acid are

$$H_3PO_4(aq) + H_2O(l) \rightleftharpoons H_3O^+(aq) + H_2PO_4^-(aq)$$
$$H_2PO_4^-(aq) + H_2O(l) \rightleftharpoons H_3O^+(aq) + HPO_4^{2-}(aq)$$
$$HPO_4^{2-}(aq) + H_2O(l) \rightleftharpoons H_3O^+(aq) + PO_4^{3-}(aq)$$

The general term **polyprotic acid** *describes acids that can transfer two or more* H^+ *ions (protons) during an acid–base reaction.*

The number of hydrogen atoms present in one molecule of an acid *cannot* always be used to classify the acid as mono-, di-, or triprotic. For example, a molecule of acetic acid contains four hydrogen atoms, and yet it is a monoprotic acid. Only one of the hydrogen atoms in acetic acid is *acidic;* that is, only one of the hydrogen atoms leaves the molecule when it is in solution.

Whether a hydrogen atom is acidic is related to its location in a molecule—that is, to which other atom it is bonded. From a structural viewpoint, the acidic behavior of acetic acid can be represented by the equation

$$\text{H–}\underset{\text{H}}{\overset{\text{H}}{\text{C}}}\text{–}\overset{\text{O}}{\overset{\|}{\text{C}}}\text{–O–H} + H_2O \rightleftharpoons H_3O^+ + \left[\text{H–}\underset{\text{H}}{\overset{\text{H}}{\text{C}}}\text{–}\overset{\text{O}}{\overset{\|}{\text{C}}}\text{–O}\right]^-$$

Note that one hydrogen atom is bonded to an oxygen atom and the other three hydrogen atoms are bonded to a carbon atom. The hydrogen atom bonded to the oxygen atom is the only acidic hydrogen; the hydrogen atoms that are bonded to carbon atoms are too tightly held to be removed by reaction with water molecules. Water has very little effect on a carbon–hydrogen bond, because that bond is only slightly polar. On the other hand, the hydrogen bonded to oxygen is involved in a very polar bond because of oxygen's large electronegativity (Section 4.15). Water, which is a polar molecule, readily attacks this bond.

Writing the formula for acetic acid as $HC_2H_3O_2$ instead of $C_2H_4O_2$ emphasizes that there are two different kinds of hydrogen. One of the hydrogen atoms is acidic and the other three are not. When some hydrogen atoms are acidic and others are not, we write the acidic hydrogens first, thus separating them from the other hydrogen atoms in the formula. Citric acid, the principal acid in citrus fruits, is another example of an acid that contains both acidic and nonacidic hydrogens (see Figure 9.4). Its formula, $H_3C_6H_5O_7$, indicates that three of the eight hydrogen atoms present in a molecule are acidic.

Figure 9.4 The sour taste of limes and other citrus fruit is due to the citric acid present in the fruit juice.

Practice Problems and Questions

9.12 Classify the following acids as monoprotic, diprotic, or triprotic.

a. $HClO_4$ (perchloric acid) b. $H_2C_2O_4$ (oxalic acid)
c. $HC_2H_3O_2$ (acetic acid) d. H_2SO_4 (sulfuric acid)

9.13 How many acidic and nonacidic hydrogen atoms are present in each of the following molecules?
a. HNO_3 (nitric acid)
b. $H_2C_2H_4O_4$ (succinic acid)
c. $HC_4H_7O_2$ (butyric acid)
d. CH_4 (methane)

9.14 Write equations showing all steps in the dissociation of citric acid ($H_3C_6H_5O_7$).

9.15 The formula for lactic acid is preferably written as $HC_3H_5O_3$ rather than as $C_3H_6O_3$. Explain why.

9.16 Pyruvic acid, which is produced in metabolic reactions, has the structure

```
   H  O  O
   |  ‖  ‖
H—C—C—C—O—H
   |
   H
```

Would you predict that this acid is a mono-, di-, tri-, or tetraprotic acid? Give your reasoning.

Learning Focus

Differentiate between and give examples of strong and weak acids and of strong and weak bases.

9.4 Strength of Acids and Bases

Brønsted–Lowry acids vary in their ability to transfer protons and produce hydronium ions in aqueous solution. Acids can be classified as strong or weak on the basis of the extent to which proton transfer occurs in aqueous solution. A **strong acid** *is a substance that transfers 100%, or very nearly 100%, of its protons to water.* Thus if an acid is strong, nearly all of the acid molecules present give up protons to water. This extensive transfer of protons produces many hydronium ions (the acidic species) within the solution. A **weak acid** *is a substance that transfers only a small percentage of its protons to water.* The extent of proton transfer for weak acids is usually less than 5%.

The vast majority of acids are weak rather than strong. The six most commonly encountered strong acids are listed in Table 9.2.

▶ Learn the names and formulas of the six commonly encountered strong acids, and then assume that all other acids you encounter are weak unless you are told otherwise.

The extent to which an acid transfers protons to water depends on the molecular structure of the acid; molecular polarity and the strength and polarity of individual bonds are particularly important factors in determining whether an acid is strong or weak.

The difference between a strong acid and a weak acid can also be stated in terms of the position of equilibrium (Section 8.7). Consider the reaction wherein HA represents the acid and H_3O^+ and A^- are the products from the proton transfer to H_2O. For strong acids, the equilibrium lies far to the right (100% or almost 100% proton transfer).

$$HA + H_2O \rightleftharpoons H_3O^+ + A^-$$

For weak acids, the equilibrium position lies far to the left (less than 5% proton transfer):

$$HA + H_2O \rightleftharpoons H_3O^+ + A^-$$

▶ It is important not to confuse the terms *strong* and *weak* with the terms *concentrated* and *dilute*. *Strong* and *weak* apply to the *extent of proton transfer,* not to the concentration of acid or base. *Concentrated* and *dilute* are relative concentration terms. Stomach acid (gastric juice) is a dilute (not weak) solution of a strong acid (HCl); it is 5% by mass hydrochloric acid.

Thus, in solutions of strong acids, the predominant species are H_3O^+ and A^-. In solutions of weak acids, the predominant species is HA; very little proton transfer has occurred. The differences between strong and weak acids, in terms of species present in solution, are illustrated in Figure 9.5.

The strong acids most likely to be encountered in a laboratory setting are sulfuric acid, nitric acid, and hydrochloric acid. These same three acids are "heavy duty" participants in the chemical industry. Chemical Portraits 15 profiles the "big three" of the strong acids.

Just as there are strong acids and weak acids, there are also strong bases and weak bases. As with acids, there are only a few strong bases. Strong bases are limited to the hydroxides of Groups IA and IIA listed in Table 9.3. Of the strong bases, only NaOH and KOH are commonly encountered in a chemical laboratory.

Table 9.2
Commonly Encountered Strong Acids

HCl	hydrochloric acid
HBr	hydrobromic acid
HI	hydroiodic acid
HNO_3	nitric acid
$HClO_4$	perchloric acid
H_2SO_4	sulfuric acid

Figure 9.5 A comparison of the number of H_3O^+ ions (the acidic species) present in strong acid and weak acid solutions of the same concentration.

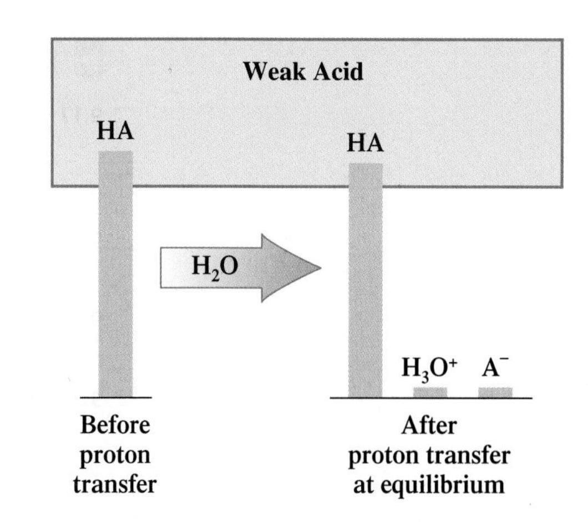

Table 9.3
Commonly Encountered Strong Hydroxide Bases

Group IA hydroxides	Group IIA hydroxides
LiOH	
NaOH	
KOH	$Ca(OH)_2$
RbOH	$Sr(OH)_2$
CsOH	$Ba(OH)_2$

Only one of the many weak bases that exist is fairly common—aqueous ammonia. In this solution of ammonia gas (NH_3) in water, small amounts of OH^- ions are produced through the reaction of NH_3 molecules with water.

$$NH_3(g) + H_2O(l) \rightleftharpoons NH_4^+(aq) + OH^-(aq)$$

A solution of aqueous ammonia is sometimes erroneously called ammonium hydroxide. Aqueous ammonia is the preferred designation because most of the NH_3 present has not reacted with water; the equilibrium position lies far to the left. Only a few ammonium ions (NH_4^+) and hydroxide ions (OH^-) are present.

Chemical Portraits 15 The Three Most Used Strong Acids

Sulfuric Acid (H_2SO_4)

Profile: Sulfuric acid is the world's most widely used industrial chemical. In the U.S. its annual production amount is twice that of any other substance. In the pure state, H_2SO_4 is a clear, colorless, odorless, oily liquid; H_2SO_4 of commerce is 93–98% acid—the remainder is water. H_2SO_4 has a great affinity for water, abstracting it from air and also from many organic substances; it chars wood and sugar.

Uses: Most U.S. sulfuric acid production is consumed in the phosphate fertilizer industry where it is used to convert insoluble phosphate rock into soluble phosphate forms. Other than in a chemical laboratory, the closest encounter most humans have with H_2SO_4 is in automobile batteries, which contain 38% (m/m) H_2SO_4 solution.

Why is sulfuric acid of biological importance for sea slugs?

Nitric Acid (HNO_3)

Profile: Nitric acid was once called "aqua fortis," Latin for "strong water;" it dissolves several metals, including Ag and Cu, which are not attacked by other acids. Colorless concentrated NHO_3 solution cannot be stored in the presence of light without photodecomposition occuring, which produces a yellow-brown discoloration. The yellow color is due to the presence of NO_2 gas. HNO_3 stains proteins, including woolen fabrics and human skin, a bright yellow.

Uses: The largest use for HNO_3 is in nitrate fertilizer production; it is the source for nitrate ions. HNO_3 is also important in the production of explosives, including nitroglycerin and trinitrotoluene (TNT).

What is the process by which HNO_3 is commercially produced?

Hydrochloric Acid (HCl)

Profile: Hydrochloric acid is a solution of hydrogen chloride gas (HCl) in water. HCl's largest single use is in metal surface cleaning (such as removing rust from steel), a process called "pickling." HCl, sold in home improvement stores under the name *muriatic acid,* is used to clean bathtubs and to clean mortar from brick and stone surfaces.

Biochemical considerations: HCl is present in the gastric juice of the stomach; its function is to activate protein-digesting enzymes. Gastric juice HCl concentration is about 5% (m/m); this creates an acidity effect about the same as that of lemon juice (citric acid). Gastric juice HCl causes the sensation of heartburn if it is refluxed into the esophagus.

What role does HCl play in the formation of stomach ulcers?

See the text web site at **www.cengage.com/chemistry/stoker** for answers to the above questions and for further information.

Practice Problems and Questions

9.17 What is the difference between a strong acid and a weak acid in terms of extent of proton transfer to water molecules?

9.18 What are the formulas and names of the six common strong acids.

9.19 Classify each of the following acids as a *strong acid* or a *weak acid.*
a. $HClO_4$ b. HNO_3 c. HCN d. $H_2C_2O_4$

9.20 Classify each of the following acids as a *strong acid* or a *weak acid.*
a. Sulfuric acid b. Acetic acid c. Citric acid d. Hydrochloric acid

9.21 Classify each of the following bases as a *strong base* or a *weak base.*
a. Sodium hydroxide b. Potassium hydroxide
c. Aqueous ammonia d. Methylamine

Learning Focus

Distinguish between compounds that are classified as acids, bases, and salts.

9.5 Salts

To a nonscientist, the term *salt* denotes a white granular substance that is used as a seasoning for food. To the chemist, the term *salt* has a much broader meaning; sodium chloride (table salt) is only one of thousands of salts known to a chemist. From a chemical viewpoint, a **salt** *is an ionic compound containing a metal or polyatomic ion as the positive ion and a nonmetal or polyatomic ion (except hydroxide) as the negative ion.* (Ionic compounds that contain hydroxide ion are bases rather than salts.)

Much information about salts has been presented in previous chapters, although the term *salt* was not explicitly used in these discussions. Formula writing and nomenclature for binary ionic compounds (salts) were covered in Sections 4.7 and 4.9. Many salts contain polyatomic ions such as nitrate and sulfate; these ions were discussed in Section 4.19. The solubility of ionic compounds (salts) in water was the topic of Section 7.4.

All common soluble salts are *completely* dissociated into ions in solution (Section 7.3). Even if a salt is only slightly soluble, the small amount that does dissolve completely dissociates. Thus the terms *weak* and *strong,* which are used to denote qualitatively the percent dissociation of acids and bases, are not applicable to salts. We do not use the terms *weak salt* and *strong salt.*

Acids, bases, and salts are related in that a salt is one of the products that results from the reaction of an acid with a hydroxide base. This particular type of reaction will be discussed in Section 9.6.

Practice Problems and Questions

9.22 Classify each of the following substances as an acid, a base, or a salt.
a. HBr b. NaI c. KOH d. KCl

9.23 Classify each of the following substances as an acid, a base, or a salt.
a. $Ba(OH)_2$ b. $(NH_4)_2SO_4$ c. H_2CO_3 d. $Al(NO_3)_3$

9.24 Write a balanced equation for the dissociation into ions of each of the following soluble salts.
a. CaS b. $MgSO_4$ c. $CaCl_2$ d. Na_2CO_3

Learning Focus

Define the term *neutralization* and be able to write chemical equations for various acid–base neutralization processes.

9.6 Acid–Base Neutralization Reactions

When acids and hydroxide bases are mixed, they react with one another and their acidic and basic properties disappear; we say they have neutralized each other. **Neutralization** *is the reaction between an acid and a hydroxide base to form a salt and water.* The neutralization process can be viewed as either a double-replacement reaction or a proton transfer reaction.

Figure 9.6 The acid–base reaction between sulfuric acid (H_2SO_4) and barium hydroxide [$Ba(OH)_2$] produces the insoluble salt $BaSO_4$. Note that the salt contains the positive ion from the base and the negative ion from the acid, a situation that will always be the case.

From a *double-replacement viewpoint* (Section 8.1),

$$AX + BY \longrightarrow AY + BX$$

we have, for the HCl–KOH neutralization,

$$\underset{\text{Acid}}{HCl} + \underset{\text{Base}}{KOH} \longrightarrow \underset{\text{Water}}{HOH} + \underset{\text{Salt}}{KCl}$$

The salt that is formed contains the negative ion from the acid ionization and the positive ion from the base dissociation (see Figure 9.6).

From a *proton transfer viewpoint,* the formation of water results from the transfer of protons from H_3O^+ ions (the acidic species in aqueous solution) to OH^- ions (the basic species) (Figure 9.7).

Any time an acid is completely reacted with a base, neutralization occurs. It does not matter whether the acid and base are strong or weak. Sodium hydroxide (a strong base) and nitric acid (a strong acid) react as follows:

$$HNO_3 + NaOH \longrightarrow NaNO_3 + H_2O$$

The equation for the reaction of potassium hydroxide (a strong base) with hydrocyanic acid (a weak acid) is

$$HCN + KOH \longrightarrow KCN + H_2O$$

Note that in both reactions, the products are a salt ($NaNO_3$ in the first reaction and KCN in the second) and water.

▶ Hydrochloric acid (HCl), which is necessary for proper digestion of food, is present in the gastric juices of the human stomach. Overeating and emotional factors can cause the stomach to produce too much hydrochloric acid, a condition often called acid indigestion or heartburn. Substances known as *antacids* provide symptomatic relief from this condition. Over-the-counter antacids such as Maalox, Tums, and Alka-Seltzer contain one or more *basic* substances that are capable of neutralizing the hydrochloric acid in gastric juice.

Balancing Acid–Base Neutralization Equations

In any acid–base neutralization reaction, the amounts of H^+ ion and OH^- ion that react are equal. These two ions always react in a one-to-one ratio to form water. This constant reaction ratio between the two ions enables us quickly to balance chemical equations for neutralization reactions.

Let us consider the neutralization reaction between H_2SO_4 and KOH.

$$H_2SO_4 + KOH \longrightarrow \text{salt} + H_2O$$

Because the acid H_2SO_4 is diprotic and the base KOH contains only one OH^- ion, we will need twice as many base molecules as acid molecules. Thus we place the coefficient 2 in front of the formula for KOH in the chemical equation; this gives two H^+ ions reacting with two OH^- ions to produce two H_2O molecules.

$$H_2SO_4 + 2KOH \longrightarrow \text{salt} + 2H_2O$$

The salt formed is K_2SO_4; there are two K^+ ions and one SO_4^{2-} ion on the left side of the equation, which combine to give the salt. The balanced equation is

$$H_2SO_4 + 2KOH \longrightarrow K_2SO_4 + 2H_2O$$

Figure 9.7 Formation of water by the transfer of protons from H_3O^+ ions to OH^- ions.

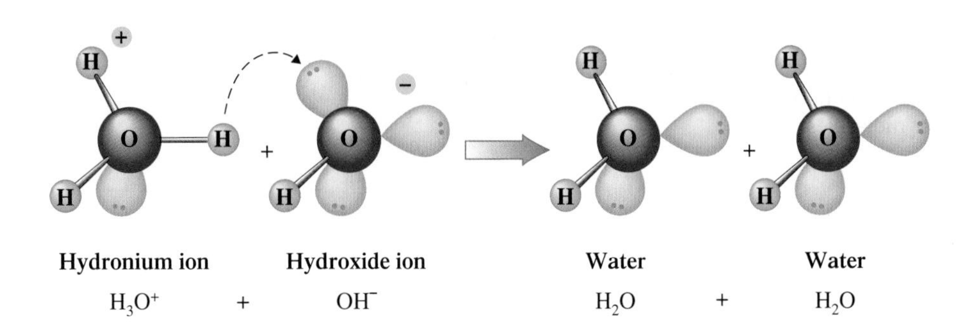

Practice Problems and Questions

9.25 Indicate whether each of the following reactions is an acid–base neutralization reaction.
a. $NaCl + AgNO_3 \longrightarrow AgCl + NaNO_3$
b. $HNO_3 + NaOH \longrightarrow NaNO_3 + H_2O$
c. $HBr + KOH \longrightarrow KBr + H_2O$
d. $H_2SO_4 + Pb(NO_3)_2 \longrightarrow PbSO_4 + 2HNO_3$

9.26 For neutralization, what is the molecular ratio in which each of the following acid–base pairs will react?
a. HNO_3 and NaOH
b. H_2SO_4 and NaOH
c. H_2SO_4 and $Ba(OH)_2$
d. HNO_3 and $Ba(OH)_2$

9.27 Write a balanced equation for each of the following acid–base neutralization reactions.
a. $HNO_3 + KOH \longrightarrow$ salt + H_2O
b. $HNO_3 + LiOH \longrightarrow$ salt + H_2O
c. $HNO_3 + Ba(OH)_2 \longrightarrow$ salt + H_2O
d. $H_2SO_4 + Ba(OH)_2 \longrightarrow$ salt + H_2O

9.28 Write a balanced equation for each of the following acid–base neutralization reactions.
a. HCl and LiOH
b. H_3PO_4 and NaOH
c. H_2CO_3 and $Ca(OH)_2$
d. H_3PO_4 and $Ca(OH)_2$

9.7 Self-Ionization of Water

Learning Focus

Use the ion product for water to calculate the $[H_3O^+]$ and $[OH^-]$ in a solution; tell whether a solution is acidic, basic, or neutral when given its $[H_3O^+]$ or $[OH^-]$.

Although we usually think of water as a covalent substance, experiments show that an *extremely small* percentage of water molecules in pure water interact with one another to form ions, a process that is called *self-ionization* (Figure 9.8). This interaction can be thought of as the transfer of protons between water molecules (Brønsted–Lowry theory, Section 9.2):

$$H_2O + H_2O \rightleftharpoons H_3O^+ + OH^-$$

The net effect of this transfer is the formation of *equal amounts* of hydronium and hydroxide ion. Such behavior for water should not seem surprising; we have already discussed the fact that water is an amphoteric substance (Section 9.2)—one that can either gain or lose protons. We have already seen several reactions in which H_2O acts as an acid and others where it acts as a base.

At any given time, the number of H_3O^+ and OH^- ions present in a sample of pure water is always extremely small. At equilibrium and 24°C, the H_3O^+ and OH^- concentrations are 1.00×10^{-7} M (0.000000100 M).

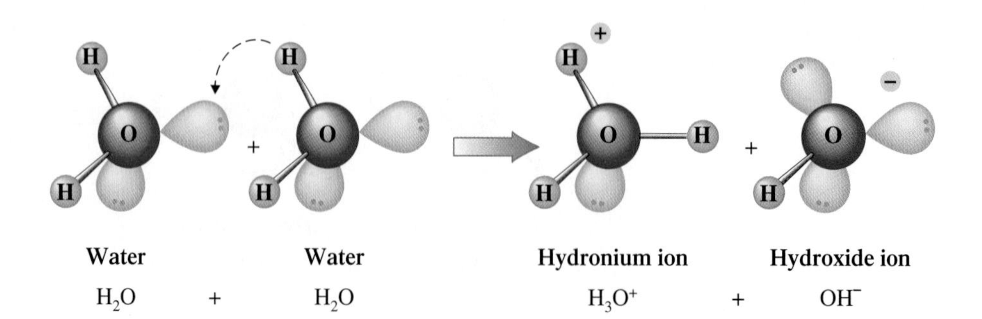

Figure 9.8 Self-ionization of water through proton transfer between water molecules.

Ion Product Constant for Water

The constant concentration of H_3O^+ and OH^- ions present in pure water at 24°C can be used to calculate a very useful number called the ion product constant for water. The **ion product constant for water** *is the numerical value* 1.00×10^{-14}, *obtained by multiplying together the molar concentrations of* H_3O^+ *ion and* OH^- *ion present in pure water.* We have the following equation for the ion product constant for water:

$$\begin{aligned}\text{Ion product constant for water} &= [H_3O^+] \times [OH^-] \\ &= (1.00 \times 10^{-7}) \times (1.00 \times 10^{-7}) \\ &= 1.00 \times 10^{-14}\end{aligned}$$

Remember that square brackets mean concentration in moles per liter (molarity).

The ion product constant expression for water is valid not only in pure water but also in water with solutes present. At all times, the product of the hydronium ion and hydroxide ion molarities in an aqueous solution at 24°C must equal 1.00×10^{-14}. Thus, if $[H_3O^+]$ is increased by the addition of an acidic solute, then $[OH^-]$ must decrease so that their product will still be 1.00×10^{-14}. Similarly, if additional OH^- ions are added to the water, then $[H_3O^+]$ must correspondingly decrease.

We can easily calculate the concentration of either H_3O^+ ion or OH^- ion present in an aqueous solution, if we know the concentration of the other ion, by simply rearranging the ion product expression $[H_3O^+] \times [OH^-] = 1.00 \times 10^{-14}$.

$$[H_3O^+] = \frac{1.00 \times 10^{-14}}{[OH^-]} \quad \text{or} \quad [OH^-] = \frac{1.00 \times 10^{-14}}{[H_3O^+]}$$

▶ If we know $[H_3O^+]$, we can always calculate $[OH^-]$, and vice versa, because of the ion product constant for water:

$[H_3O^+] \times [OH^-] = 1.00 \times 10^{-14}$

Example 9.2 Calculating the Hydroxide Ion Concentration of a Solution from a Given Hydronium Ion Concentration

Sufficient acidic solute is added to a quantity of water to produce $[H_3O^+] = 4.0 \times 10^{-3}$. What is $[OH^-]$ in this solution?

Solution

$[OH^-]$ can be calculated by using the ion product expression for water, rearranged in the form

$$[OH^-] = \frac{1.00 \times 10^{-14}}{[H_3O^+]}$$

Substituting into this expression the known $[H_3O^+]$ and doing the arithmetic give

$$[OH^-] = \frac{1.00 \times 10^{-14}}{4.0 \times 10^{-3}} = 2.5 \times 10^{-12}$$

▶ Neither $[H_3O^+]$ nor $[OH^-]$ is ever zero in an aqueous solution.

The relationship between $[H_3O^+]$ and $[OH^-]$ is that of an inverse proportion; when one increases, the other decreases. If $[H_3O^+]$ increases by a factor of 10^2, then $[OH^-]$ decreases by the same factor, 10^2. A graphical portrayal of this increase–decrease relationship for $[H_3O^+]$ and $[OH^-]$ is given in Figure 9.9.

Acidic, Basic, and Neutral Solutions

Since a small amount of H_3O^+ ion and OH^- ion are present in all aqueous solutions, what determines whether a given solution is acidic or basic? It is the relative amounts of these two ions present. An **acidic solution** *is one in which the concentration of* H_3O^+ *ion is higher than that of* OH^- *ion.* A **basic solution** *is one in which the concentration of the* OH^- *ion is higher than that of the* H_3O^+ *ion.* A basic solution is also often referred to as an *alkaline solution.* It is possible to have an aqueous solution that is neither acidic nor

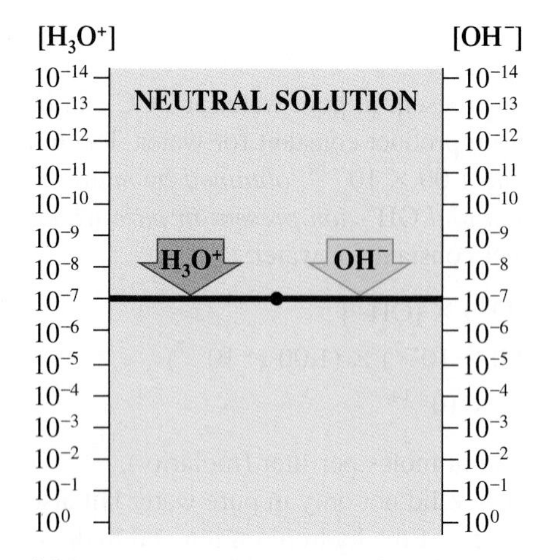

(a) In pure water the concentration of hydronium ions, $[H_3O^+]$, and that of hydroxide ions, $[OH^-]$, are equal. Both are 1.00×10^{-7} M at 24°C.

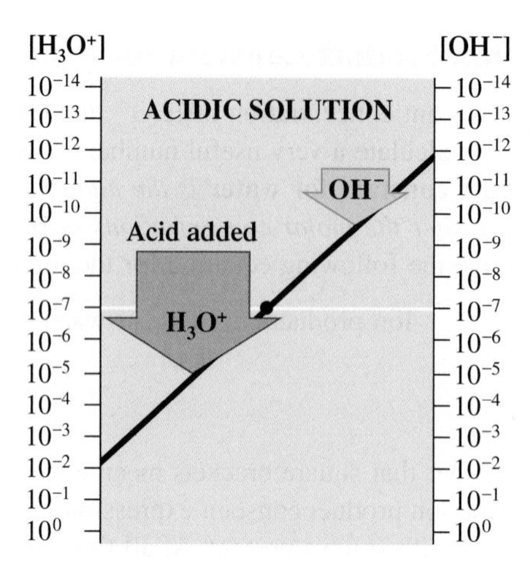

(b) If $[H_3O^+]$ is increased by a factor of 10^5 (from 10^{-7} M to 10^{-2} M), then $[OH^-]$ is decreased by a factor of 10^5 (from 10^{-7} M to 10^{-12} M).

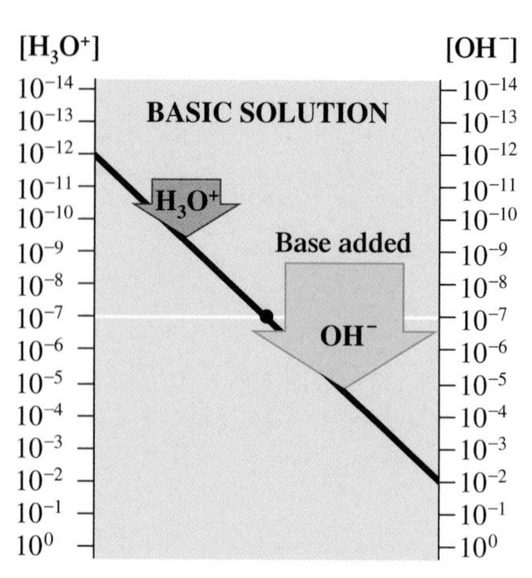

(c) If $[OH^-]$ is increased by a factor of 10^5 (from 10^{-7} M to 10^{-2} M), then $[H_3O^+]$ is decreased by a factor of 10^5 (from 10^{-7} M to 10^{-12} M).

Figure 9.9 The relationship between $[H_3O^+]$ and $[OH^-]$ in aqueous solution is an inverse proportion; when $[H_3O^+]$ is increased, $[OH^-]$ decreases, and vice versa.

basic but is, rather, a neutral solution. A **neutral solution** *is one in which the concentrations of* H_3O^+ *and* OH^- *ions are equal.* Table 9.4 summarizes the relationships between $[H_3O^+]$ and $[OH^-]$ that we have just considered.

Practice Problems and Questions

9.29 *Pure* water contains small amounts of two ions.
a. What is the identity of the two ions?
b. What is the source of the two ions?
c. What are the concentrations of the two ions at 24°C?
d. Why are the concentrations of the two ions equal?

9.30 Calculate the molar H_3O^+ ion concentration for an aqueous solution when the OH^- ion concentration has the following values.
a. 3.0×10^{-4} M b. 6.7×10^{-6} M c. 9.1×10^{-8} M d. 1.2×10^{-11} M

9.31 Calculate the molar OH^- ion concentration for an aqueous solution when the H_3O^+ ion concentration has the following values.
a. 4.0×10^{-5} M b. 7.7×10^{-7} M c. 8.1×10^{-10} M d. 1.2×10^{-3} M

9.32 How does the concentration of H_3O^+ ion compare with that of OH^- ion in each of the following?
a. Acidic solution b. Basic solution c. Neutral solution

9.33 Indicate whether solutions with the following hydronium ion concentrations are acidic, basic, or neutral.
a. $[H_3O^+] = 1.0 \times 10^{-3}$ b. $[H_3O^+] = 2.0 \times 10^{-5}$
c. $[H_3O^+] = 3.0 \times 10^{-7}$ d. $[H_3O^+] = 4.0 \times 10^{-9}$

9.34 Indicate whether solutions with the following hydroxide ion concentrations are acidic, basic, or neutral.
a. $[OH^-] = 1.0 \times 10^{-3}$ b. $[OH^-] = 1.0 \times 10^{-5}$
c. $[OH^-] = 1.0 \times 10^{-7}$ d. $[OH^-] = 1.0 \times 10^{-9}$

Table 9.4
Relationship Between $[H_3O^+]$ and $[OH^-]$ in Neutral, Acidic, and Basic Solutions

neutral solution	$[H_3O^+] = [OH^-] = 1.00 \times 10^{-7}$
acidic solution $[H_3O^+] > [OH^-]$	$[H_3O^+]$ is greater than 1.00×10^{-7} $[OH^-]$ is less than 1.00×10^{-7}
basic solution $[OH^-] > [H_3O^+]$	$[H_3O^+]$ is less than 1.00×10^{-7} $[OH^-]$ is greater than 1.00×10^{-7}

Learning Focus

Calculate pH from hydronium ion concentration, or vice versa; given the pH, tell whether a solution is acidic, basic, or neutral.

9.8 The pH Concept

Hydronium ion concentrations in aqueous solution range from relatively high values (10 M) to extremely small ones (10^{-14} M). It is inconvenient to work with numbers that extend over such a wide range; a hydronium ion concentration of 10 M is 1000 trillion times larger than a hydronium ion concentration of 10^{-14} M. The pH scale was developed as a more practical way to handle such a wide range of numbers. The **pH scale** *is a scale of small numbers that is used to specify molar hydronium ion concentration in an aqueous solution.*

▶ The pH scale is a compact method for representing solution acidity.

The calculation of pH scale values involves the use of logarithms. The **pH** *of a solution is the negative logarithm of the solution's molar hydronium ion concentration.* Expressed mathematically, the definition of pH is

$$pH = -\log [H_3O^+]$$

(The letter *p*, as in pH, has been chosen to mean "negative logarithm of.")

▶ The *p* in pH comes from the German word *potenz,* which means "power," as in "power of 10."

Integral pH Values

For any hydronium ion concentration expressed in exponential notation in which the coefficient is 1.0, the pH is given directly by the negative of the exponent value of the power of 10:

$$[H_3O^+] = 1.0 \times 10^{-x}$$
$$pH = x$$

Thus, if the hydronium ion concentration is 1.0×10^{-9}, then the pH will be 9.00. This simple relationship between pH and hydronium ion concentration is valid only when the coefficient in the exponential notation expression for the hydronium ion concentration is 1.0.

▶ The rule for the number of significant figures in a logarithm is: The number of digits after the decimal place in a logarithm is equal to the number of significant figures in the original number.

$$[H_3O^+] = \underbrace{6.3}_{\text{Two significant figures}} \times 10^{-5}$$
$$pH = 4.\underbrace{20}_{\text{Two digits}}$$

Example 9.3 Calculating the pH of a Solution When Given its Hydronium Ion or Hydroxide Ion Concentration

Calculate the pH for each of the following solutions.

a. $[H_3O^+] = 1.0 \times 10^{-6}$ **b.** $[OH^-] = 1.0 \times 10^{-6}$

Solution

a. Because the coefficient in the exponential expression for the molar hydronium ion concentration is 1.0, the pH can be obtained from the relationships

$$[H_3O^+] = 1.0 \times 10^{-x}$$
$$pH = x$$

The power of 10 is -6 in this case, so the pH will be 6.00.

b. The given quantity involves hydroxide ion rather than hydronium ion. Thus we must calculate the hydronium ion concentration first and then calculate the pH.

$$[H_3O^+] = \frac{1.00 \times 10^{-14}}{1.0 \times 10^{-6}} = 1.0 \times 10^{-8}$$

A solution with a hydronium ion concentration of 1.0×10^{-8} M will have a pH of 8.00.

Nonintegral pH Values

If the coefficient in the exponential expression for the molar hydronium ion concentration is *not* 1.0, then the pH will have a nonintegral value; that is, it will not be a whole number. For example, consider the following nonintegral pH values.

$$[H_3O^+] = 6.3 \times 10^{-5} \qquad pH = 4.20$$
$$[H_3O^+] = 5.3 \times 10^{-5} \qquad pH = 4.28$$
$$[H_3O^+] = 2.2 \times 10^{-4} \qquad pH = 3.66$$

Figure 9.10 gives nonintegral pH value ranges for selected fruits and vegetables.

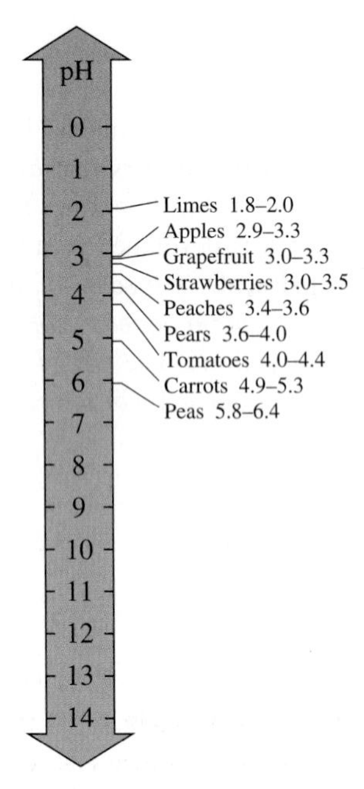

Figure 9.10 Most fruits and vegetables are acidic. Tart or sour taste is an indication that such is the case. Nonintegral pH values for selected foods are as shown here.

The easiest way to obtain nonintegral pH values such as these involves using an electronic calculator that allows for the input of exponential numbers and that has a base-10 logarithm key (LOG).

In using such an electronic calculator, you can obtain logarithm values simply by pressing the LOG key after having entered the number whose log is desired. For pH, you must remember that after obtaining the log value, you must change signs because of the negative sign in the defining equation for pH.

Example 9.4 **Calculating the pH of a Solution When Given Its Hydronium Ion Concentration**

Calculate the pH for each of the following solutions.

a. $[H_3O^+] = 7.23 \times 10^{-8}$ **b.** $[H_3O^+] = 5.70 \times 10^{-3}$

Solution

a. Using an electronic calculator, first enter the number 7.23×10^{-8} into the calculator. Then use the LOG key to obtain the logarithm value, -7.1408617. Changing the sign of this number (because of the minus sign in the definition of pH) and adjusting for significant figures yield a pH value of 7.141.

b. Entering the number 5.70×10^{-3} into the calculator and then using the LOG key give a logarithm value of -2.2441251. This value translates into a pH value, after rounding, of 2.244.

pH Values and Hydronium Ion Concentration

It is often necessary to calculate the hydronium ion concentration for a solution from its pH value. This type of calculation, which is the reverse of that illustrated in Examples 9.3 and 9.4, is shown in Example 9.5.

Figure 9.11 Relationships among pH values, $[H_3O^+]$, and $[OH^-]$ at 24°C.

$[H_3O^+]$	pH	$[OH^-]$	
10^{-0}	0	10^{-14}	Acidic
10^{-1}	1	10^{-13}	
10^{-2}	2	10^{-12}	
10^{-3}	3	10^{-11}	
10^{-4}	4	10^{-10}	
10^{-5}	5	10^{-9}	
10^{-6}	6	10^{-8}	
10^{-7}	7	10^{-7}	Neutral
10^{-8}	8	10^{-6}	
10^{-9}	9	10^{-5}	
10^{-10}	10	10^{-4}	
10^{-11}	11	10^{-3}	
10^{-12}	12	10^{-2}	
10^{-13}	13	10^{-1}	
10^{-14}	14	10^{-0}	Basic

Example 9.5 **Calculating the Molar Hydronium Ion Concentration of a Solution from the Solution's pH**

The pH of a solution is 6.80. What is the molar hydronium ion concentration for this solution?

Solution

From the defining equation for pH, we have

$$\mathrm{pH} = -\log [H_3O^+] = 6.80$$
$$\log [H_3O^+] = -6.80$$

To find $[H_3O^+]$, we need to determine the *antilog* of -6.80.

How an antilog is obtained using a calculator depends on the type of calculator. Many calculators have an antilog function (sometimes labeled INV log) that performs this operation. If this key is present, then

1. Enter the number -6.80. Note that it is the *negative* of the pH that is entered into the calculator.
2. Press the INV log key (or an inverse key and then a log key). The result is the desired hydronium ion concentration.

$$\log [H_3O^+] = -6.80$$
$$\mathrm{antilog}\ [H_3O^+] = 1.5848931 \times 10^{-7}$$

Rounded off, this value translates into a hydronium ion concentration of 1.6×10^{-7} M.

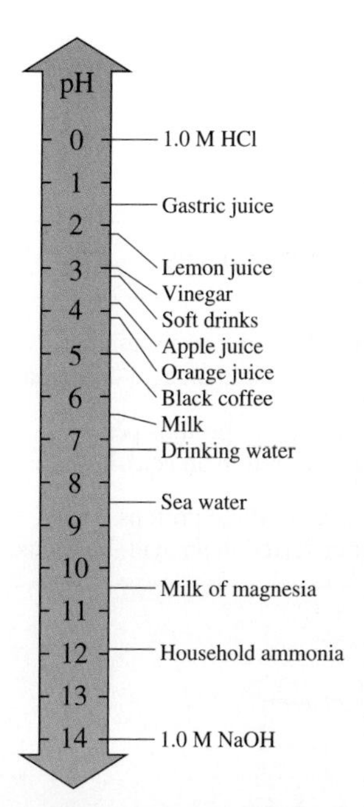

Figure 9.12 The pH values of selected common liquids. The lower the numerical value of the pH, the more acidic the substance.

Some calculators use a 10^x key to perform the antilog operation. Use of this key is based on the mathematical identity

$$\text{antilog } x = 10^x$$

In our case, this means

$$\text{antilog of } -6.80 = 10^{-6.80}$$

If the 10^x key is present, then

1. Enter the number -6.80 (the negative of the pH).
2. Press the function key 10^x. The result is the desired hydronium ion concentration.

$$[H_3O^+] = 10^{-6.80} = 1.6 \times 10^{-7}$$

▶ Solutions of low pH are more acidic than solutions of high pH; conversely, solutions of high pH are more basic than solutions of low pH.

Interpreting pH Values

The pH is simply a way of expressing hydronium ion concentration and thus identifies a solution as acidic, basic, or neutral. At 24°C, a neutral solution has a pH value of 7.0. Values of pH that are less than 7.0 correspond to acidic solutions, and values of pH that are greater than 7.0 are associated with basic solutions. The relationships among $[H_3O^+]$, $[OH^-]$, and pH are summarized in Figure 9.11. Note the following trends from the information presented in this figure.

1. The higher the concentration of hydronium ion, the lower the pH value. Another statement of this same trend is that lowering the pH always corresponds to increasing the hydronium ion concentration.
2. A change of 1 unit in pH always corresponds to a tenfold change in hydronium ion concentration. For example,

$$\text{Difference of 1} \begin{cases} \text{pH} = 1.0, \text{ then } [H_3O^+] = 0.1\,\text{M} \\ \text{pH} = 2.0, \text{ then } [H_3O^+] = 0.01\,\text{M} \end{cases} \text{tenfold difference}$$

Table 9.5
The Normal pH Range of Selected Body Fluids

Type of fluid	pH value
bile	6.8–7.0
blood plasma	7.3–7.5
gastric juices	1.0–3.0
milk	6.6–7.6
saliva	6.5–7.5
spinal fluid	7.3–7.5
urine	4.8–8.4

In a laboratory, solutions of any pH can be created. The range of pH values that are displayed by natural solutions is more limited than that of prepared solutions, but solutions corresponding to most pH values can be found (see Figure 9.12). A pH meter (Figure 9.13) helps chemists determine accurate pH values.

The pH values of several human body fluids are given in Table 9.5. Most human body fluids except gastric juices and urine have pH values within one unit of neutrality. Both blood plasma and spinal fluid are always slightly basic.

The accompanying Chemistry at a Glance summarizes what we have said about acids and acidity.

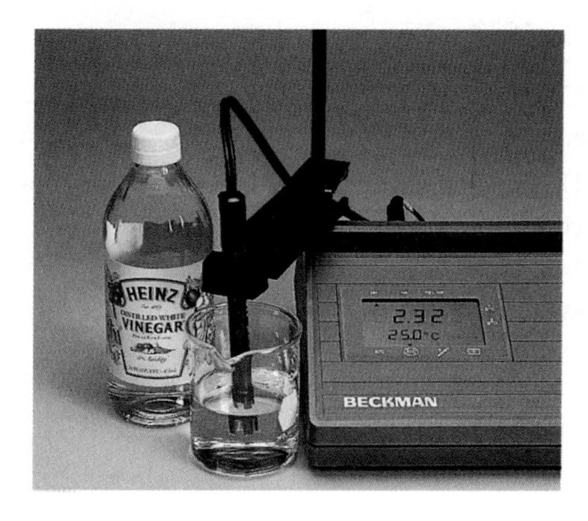

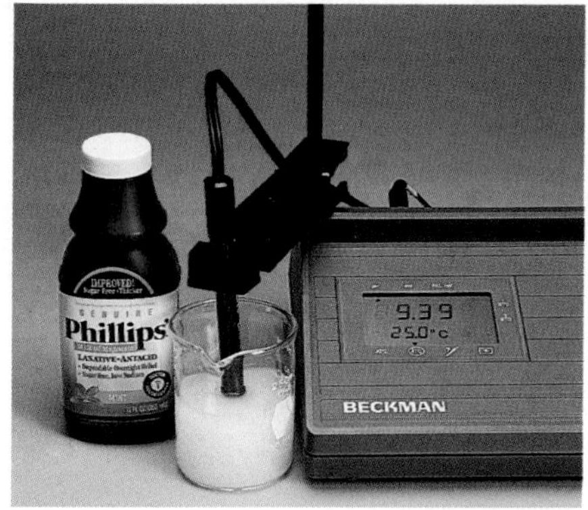

Figure 9.13 A pH meter gives an accurate measurement of pH values. The pH of vinegar is 2.32 (left). The pH of milk of magnesia in water is 9.39 (right).

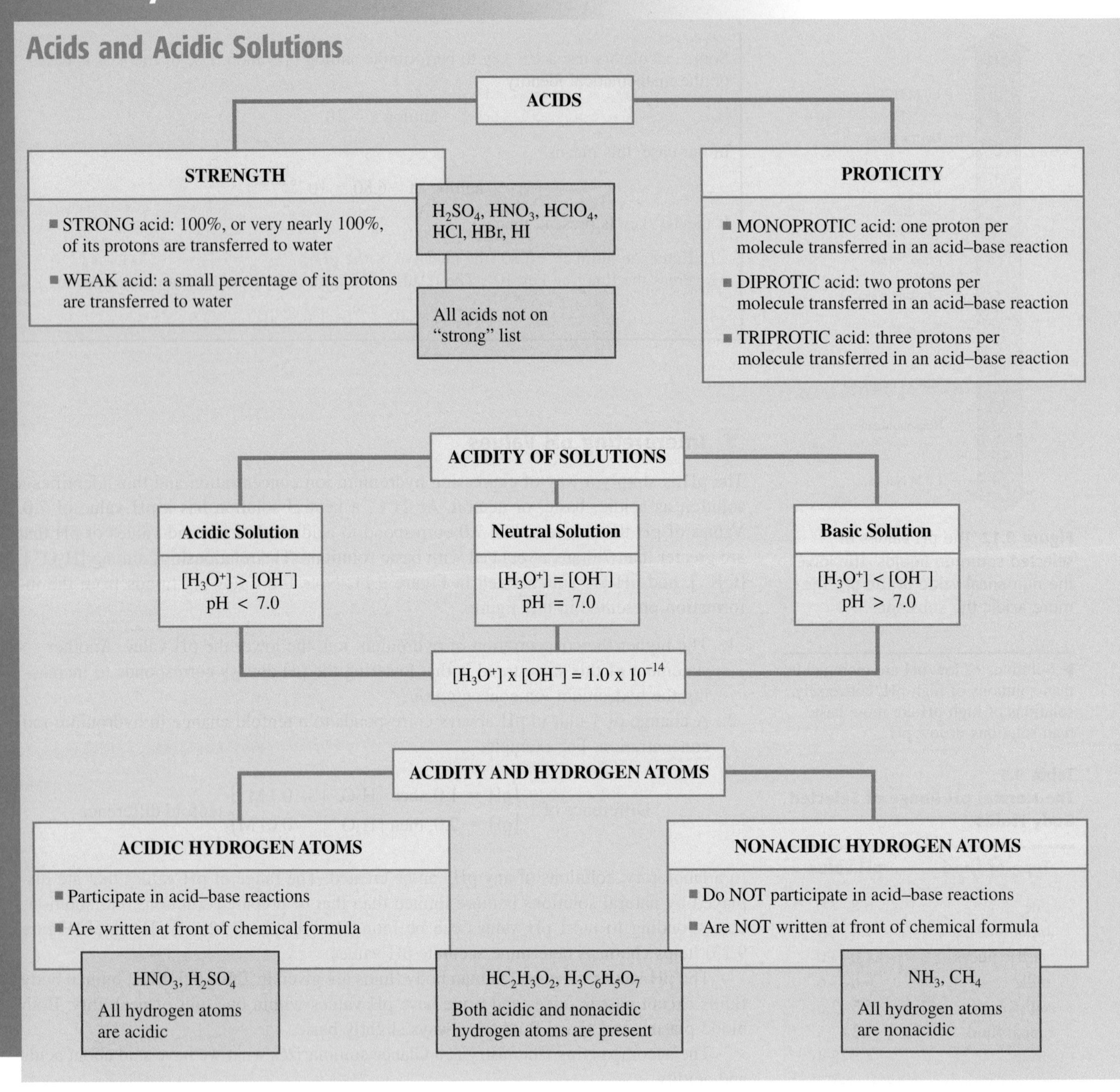

Practice Problems and Questions

9.35 Calculate the pH of solutions with the following hydronium ion concentrations.
a. 1.0×10^{-4} M b. 1.0×10^{-11} M c. 1.0×10^{-7} M d. 1.0×10^{-5} M

9.36 Use a calculator to determine the pH of solutions with the following hydronium ion concentrations.
a. 2.1×10^{-8} M b. 4.0×10^{-8} M c. 6.7×10^{-2} M d. 1.2×10^{-10} M

9.37 What is the molar hydronium ion concentration in solutions with each of the following pH values?
a. 2.00 b. 4.00 c. 7.00 d. 9.00

9.38 Use a calculator to determine the molar hydronium ion concentration in solutions with each of the following pH values?
a. 3.67 b. 5.09 c. 7.35 d. 12.45

9.39 Indicate whether solutions with the following pH values are *acidic, basic,* or *neutral.*
a. 6.47 b. 7.20 c. 11.25 d. 3.20

9.40 Complete the following table.

	$[H_3O^+]$	$[OH^-]$	pH	Acidic, basic, or neutral?
a.	______	1.00×10^{-7}	___	______
b.	______	______	5.00	______
c.	1.00×10^{-3}	______	___	______
d.	1.00×10^{-10}	______	___	______

9.9 Buffers

Learning Focus

Describe how a buffer system causes a solution to be resistant to change in pH.

A **buffer** *is a solution that resists major changes in pH when small amounts of acid or base are added to it.* Buffers are used in a laboratory setting to maintain optimum pH conditions for chemical reactions. Many commercial products contain buffers, which are needed to maintain optimum pH conditions for product behavior. Examples include buffered aspirin (Bufferin) and pH-controlled hair shampoos. Most human body fluids are highly buffered. For example, a buffer system maintains blood's pH at a value close to 7.4, an optimum pH for oxygen transport.

▶ A less common type of buffer involves a weak base and its conjugate acid. We will not consider this type of buffer here.

Buffers contain two chemical species: (1) a substance to react with and remove added base, and (2) a substance to react with and remove added acid. Typically, a buffer system is composed of a weak acid *and* its conjugate base—that is, a conjugate acid–base pair (Section 9.2). Such acid–base pairs that are commonly employed as buffers include $HC_2H_3O_2/C_2H_3O_2^-$, $H_2PO_4^-/HPO_4^{2-}$, and H_2CO_3/HCO_3^-.

Example 9.6 Recognizing Pairs of Chemical Substances That Can Function as a Buffer in Aqueous Solution

Predict whether each of the following pairs of substances could function as a buffer system in aqueous solution.

a. HCl and NaCl **b.** HCN and KCN
c. HCl and HCN **d.** NaCN and KCN

Solution

Buffer solutions contain either a weak acid and a salt of that weak acid or a weak base and a salt of that weak base.

a. No. We have an acid and the salt of that acid. However, the acid is a strong acid rather than a weak acid.
b. Yes. HCN is a weak acid, and KCN is a salt of that weak acid.
c. No. Both HCl and HCN are acids. No salt is present.
d. No. Both NaCN and KCN are salts. No weak acid is present.

As an illustration of buffer action, consider a buffer solution containing approximately equal concentrations of acetic acid (a weak acid) and sodium acetate (a salt of this weak acid). This solution resists pH change by the following mechanisms:

1. When a small amount of a strong acid such as HCl is added to the solution, the newly added H_3O^+ ions react with the acetate ions from the sodium acetate to give acetic acid.

$$H_3O^+ + C_2H_3O_2^- \longrightarrow HC_2H_3O_2 + H_2O$$

Most of the added H_3O^+ ions are tied up in acetic acid molecules, and the pH changes very little.

2. When a small amount of a strong base such as NaOH is added to the solution, the newly added OH^- ions react with the acetic acid (neutralization) to give acetate ions and water.

$$OH^- + HC_2H_3O_2 \longrightarrow C_2H_3O_2^- + H_2O$$

Most of the added OH^- ions are converted to water, and the pH changes only slightly.

▶ To resist both increases and decreases in pH effectively, a weak acid buffer must contain significant amounts of both the weak acid and its conjugate base. If a solution has a large amount of weak acid but very little conjugate base, it will be unable to consume much added acid. Consequently, the pH tends to drop significantly when acid is added. Conversely, a solution that contains a large amount of conjugate base but very little weak acid will provide very little protection against added base. Addition of just a little base will cause a big change in pH.

The reactions that are responsible for the buffering action in the acetic acid/acetate ion system can be summarized as follows:

$$C_2H_3O_2^- \underset{OH^-}{\overset{H_3O^+}{\rightleftharpoons}} HC_2H_3O_2$$

Note that one member of the buffer pair (acetate ion) removes excess H_3O^+ ion and that the other (acetic acid) removes excess OH^- ion. The buffering action always results in the active species being converted to its partner species.

Example 9.7 Writing Equations for Reactions That Occur in a Buffered Solution

Write an equation for each of the following buffering actions.

a. The response of $H_2PO_4^-/HPO_4^{2-}$ buffer to the addition of H_3O^+ ions
b. The response of HCN/CN^- buffer to the addition of OH^- ions

Solution

a. The base in a conjugate acid–base pair is the species that responds to the addition of acid. (Recall, from Section 9.2, that the base in a conjugate acid–base pair always has one less hydrogen than the acid.) The base for this reaction is HPO_4^{2-}. The equation for the buffering action is

$$H_3O^+ + HPO_4^{2-} \longrightarrow H_2PO_4^- + H_2O$$

In the buffering response, the base is always converted into its conjugate acid.

b. The acid in a conjugate acid–base pair is the species that responds to the addition of base. The acid for this reaction is HCN. The equation for the buffering action is

$$HCN + OH^- \longrightarrow CN^- + H_2O$$

Water will always be one of the products of buffering action.

▶ A common misconception about buffers is that a buffered solution is always a neutral (pH 7.0) solution. This is false. One can buffer a solution at any desired pH. A pH 7.4 buffer will hold the pH of the solution near pH 7.4, whereas a pH 9.3 buffer will tend to hold the pH of a solution near pH 9.3. The pH of a buffer is determined by the degree of weakness of the weak acid used and by the concentrations of the acid and its conjugate base.

A false notion about buffers is that they will hold the pH of a solution *absolutely* constant. The addition of even small amounts of a strong acid or a strong base to any solution, buffered or not, will lead to a change in pH. The important concept is that the shift in pH will be much less when an effective buffer is present (see Table 9.6).

Table 9.6
A Comparison of pH Changes in Buffered and Unbuffered Solutions

Unbuffered Solution	
1 liter water	pH = 7.0
1 liter water + 0.01 mole strong base (NaOH)	pH = 12.0
1 liter water + 0.01 mole strong acid (HCl)	pH = 2.0
Buffered Solution	
1 liter buffer[a]	pH = 7.2
1 liter buffer[a] + 0.01 mole strong base (NaOH)	pH = 7.3
1 liter buffer[a] + 0.01 mole strong acid (HCl)	pH = 7.1

[a]Buffer = equal amounts of 0.1 M HPO_4^{2-} and 0.1 M $H_2PO_4^-$

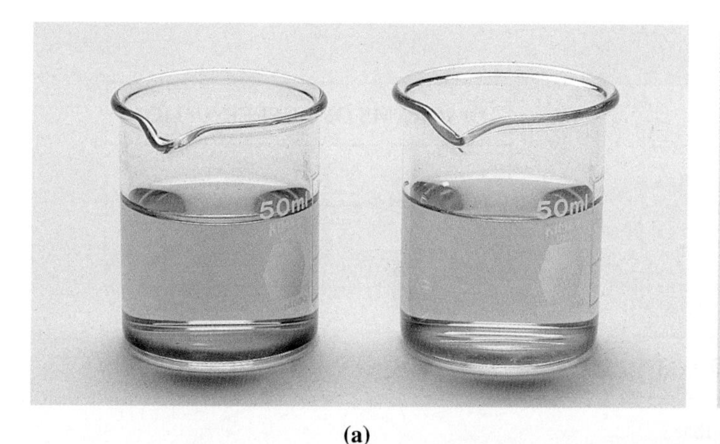

(a)

(b)

Figure 9.14 (a) The buffered solution on the left and the unbuffered solution on the right have the same pH (pH 8). They are basic solutions. (b) After the addition of 1 mL of a 0.01 M HCl solution, the pH of the buffered solution has not perceptibly changed, but the unbuffered solution has become acidic, as indicated by the change in the color of the acid–base indicator present.

Buffer systems have their limits. If large amounts of H_3O^+ or OH^- are added to a buffer, the buffer capacity can be exceeded; then the buffer system is overwhelmed and the pH changes (Figure 9.14). For example, if large amounts of H_3O^+ were added to the acetate/acetic acid buffer previously discussed, the H_3O^+ ion would react with acetate ion until the acetate was depleted. Then the pH would begin to drop as free H_3O^+ ions accumulated in the solution.

Additional insights into the workings of buffer systems are obtained by considering buffer action within the framework of Le Châtelier's principle and an equilibrium system. Let us again consider an acetic acid/acetate ion buffer system. An equilibrium is established in solution between the acetic acid and the acetate ion.

$$HC_2H_3O_2(aq) + H_2O(l) \rightleftharpoons H_3O^+(aq) + C_2H_3O_2^-(aq)$$

This equilibrium system functions in accordance with *Le Châtelier's principle* (Section 8.8), which states that an equilibrium system, when stressed, will shift its position in such a way as to counteract the stress. Stresses for the buffer will be (1) addition of base (hydroxide ion) and (2) addition of acid (hydronium ion). Further details concerning these two stress situations are as follows.

Addition of base [OH^- ion] *to the buffer.* The addition of base causes the following changes to occur in the solution:

1. The added OH^- ion reacts with H_3O^+ ion, producing water (neutralization).
2. The neutralization reaction produces the stress of *not enough* H_3O^+ ion, because H_3O^+ ion was consumed in the neutralization.
3. The equilibrium shifts to the right, in accordance with Le Châtelier's principle, to produce more H_3O^+ ion, which maintains the pH close to its original level.

Addition of acid [H_3O^+ ion] *to the buffer.* The addition of acid causes the following changes to occur in the solution:

1. The added H_3O^+ ion increases the overall amount of H_3O^+ ion present.
2. The stress on the system is *too much* H_3O^+ ion.
3. The equilibrium shifts to the left in accordance with Le Châtelier's principle, consuming most of the excess H_3O^+ ion and resulting in a pH close to the original level.

The accompanying Chemistry at a Glance reviews some of what we have said about buffer systems.

Chemistry at a Glance

Buffer Systems

BUFFER SOLUTION

- A solution that resists major change in pH when small amounts of strong acid or strong base are added
- A typical buffer system contains a weak acid and its conjugate base

Common biological buffer systems are
H_2CO_3/HCO_3^-
$H_2PO_4^-/HPO_4^{2-}$

Weak Acid

- The buffer component that reacts with added base
- Reaction converts it into its conjugate base

Conjugate Base of Weak Acid

- The buffer component that reacts with added acid
- Reaction converts it into its parent acid

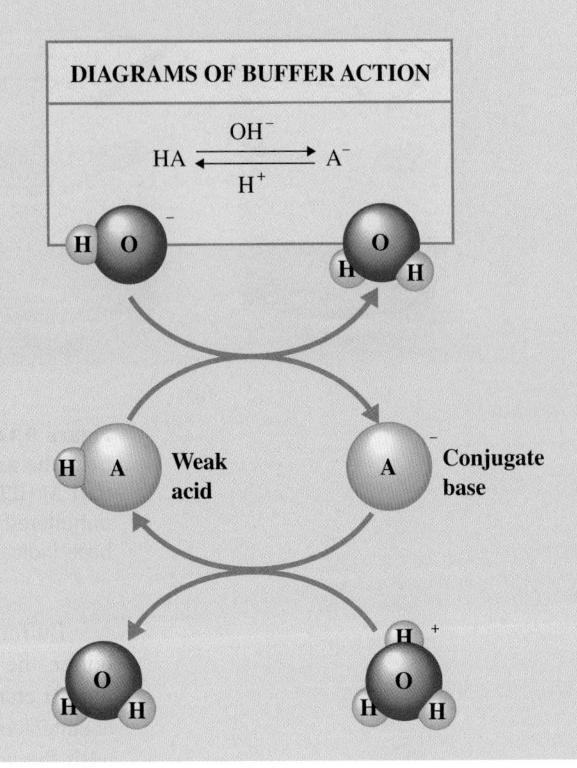

Practice Problems and Questions

9.41 Predict whether each of the following pairs of substances could function as a buffer system in aqueous solution.
a. HNO_3 and $NaNO_3$ b. HF and NaF c. KCl and KCN d. H_2CO_3 and $NaHCO_3$

9.42 Identify the two "active species" in each of the following buffer systems.
a. HCN and KCN b. H_3PO_4 and NaH_2PO_4
c. H_2CO_3 and $KHCO_3$ d. $NaHCO_3$ and K_2CO_3

9.43 Write an equation for each of the following buffering actions.
a. The response of a HF/F^- buffer to the addition of H_3O^+ ions.
b. The response of a H_2CO_3/HCO_3^- buffer to the addition of OH^- ions.
c. The response of a HCO_3^-/CO_3^{2-} buffer to the addition of H_3O^+ ions.
d. The response of a $H_3PO_4/H_2PO_4^-$ buffer to the addition of OH^- ions.

9.10 Acid–Base Titrations

Learning Focus

Calculate the molarity of an acid or base from titration information.

The analysis of solutions to determine the concentration of acid or base present is performed regularly in many laboratories. Such activity is different from determining a solution's pH value. The pH of a solution gives information about the *concentration* of hydronium ions in solution. Only ionized molecules influence the pH value. The concentration of an acid or a base gives information about the *total number* of acid or base molecules present; both dissociated and undissociated molecules are counted. Thus acid or base concentration is a measure of total acidity or total basicity.

The procedure most frequently used to determine the concentration of an acid or a base solution is an acid–base titration. In an **acid–base titration,** *a measured volume of an acid or a base of known concentration is exactly reacted with a measured volume of a base or an acid of unknown concentration.*

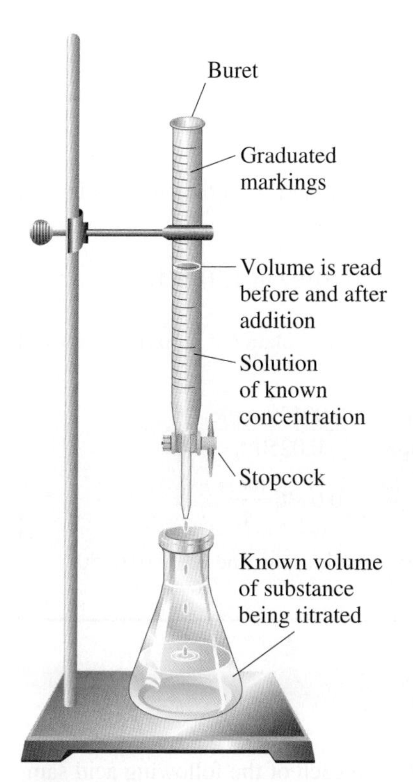

Figure 9.15 A schematic diagram showing the setup used for titration procedures.

Suppose we want to determine the concentration of an acid solution by means of titration. We first measure out a known volume of the acid solution into a beaker or flask. Then we slowly add a solution of base of known concentration to the flask or beaker by means of a buret (Figure 9.15). We continue to add base until all the acid has completely reacted with the added base. The volume of base needed to reach this point is obtained from the buret readings. When we know the original volume of acid, the concentration of the base, and the volume of added base, we can calculate the concentration of the acid, as will be shown in Example 9.8.

In order to complete a titration successfully, we must be able to detect when the reaction between acid and base is complete. Neither the acid nor the base gives any outward sign that the reaction is complete. Thus an indicator is always added to the reaction mixture (Figure 9.16). An **indicator** *is a compound that exhibits different colors depending on the* pH *of its surroundings*. Typically, an indicator is one color in basic solutions and another color in acidic solutions. An indicator is selected that changes color at a pH that corresponds as nearly as possible to the pH of the solution when the titration is complete.

Example 9.8 shows how titration data are used to calculate the molarity of an acid solution of unknown concentration.

Example 9.8 Calculating an Unknown Molarity from Acid–Base Titration Data

In a sulfuric acid (H_2SO_4)–sodium hydroxide (NaOH) acid–base titration, 17.3 mL of 0.126 M NaOH is needed to neutralize 25.0 mL of H_2SO_4 of unknown concentration. Find the molarity of the H_2SO_4 solution, given that the neutralization reaction that occurs is

$$H_2SO_4(aq) + 2NaOH(aq) \longrightarrow Na_2SO_4(aq) + 2H_2O(l)$$

Solution

First, we calculate the number of moles of H_2SO_4 that reacted with the NaOH. The pathway for this calculation, using dimensional analysis (Section 5.8), is

$$\text{mL of NaOH} \longrightarrow \text{L of NaOH} \longrightarrow \text{moles of NaOH} \longrightarrow \text{moles of } H_2SO_4$$

The sequence of conversion factors that effects this series of unit changes is

$$17.3\ \cancel{\text{mL NaOH}} \times \left(\frac{10^{-3}\ \cancel{\text{L NaOH}}}{1\ \cancel{\text{mL NaOH}}}\right) \times \left(\frac{0.126\ \cancel{\text{mole NaOH}}}{1\ \cancel{\text{L NaOH}}}\right) \times \left(\frac{1\ \text{mole } H_2SO_4}{2\ \cancel{\text{moles NaOH}}}\right)$$

Figure 9.16 An acid–base titration using an indicator that is yellow in acidic solution and red in basic solution.

The first conversion factor derives from the definition of a milliliter, the second conversion factor derives from the definition of molarity (Section 7.5), and the third conversion factor uses the coefficients in the balanced chemical equation for the titration reaction (Section 5.7).

The number of moles of H_2SO_4 that react is obtained by combining all the numbers in the dimensional analysis setup in the manner indicated.

$$\left(\frac{17.3 \times 10^{-3} \times 0.126 \times 1}{1 \times 1 \times 2}\right) \text{mole } H_2SO_4 = 0.00109 \text{ mole } H_2SO_4$$

Now that we know how many moles of H_2SO_4 reacted, we calculate the molarity of the H_2SO_4 solution using the definition for molarity.

$$\text{Molarity } H_2SO_4 = \frac{\text{moles } H_2SO_4}{\text{L } H_2SO_4 \text{ solution}} = \frac{0.00109 \text{ mole}}{0.0250 \text{ L}}$$

$$= 0.0436 \frac{\text{mole}}{\text{L}}$$

Note that the units in the denominator of the molarity equation must be liters (0.0250) rather than milliliters (25.0).

Practice Problems and Questions

9.44 It requires 34.5 mL of 0.102 M NaOH to neutralize each of the following acid samples. What is the molarity of each of the acid samples?

a. 25.0 mL of H_2SO_4
b. 20.0 mL of HClO
c. 20.0 mL of H_3PO_4
d. 10.0 mL of HNO_3

9.45 What volume, in milliliters, of a 0.100 M NaOH solution would be needed to neutralize each of the following acid samples?

a. 10.00 mL of 0.175 M H_2SO_4
b. 50.00 mL of 0.500 M H_3PO_4
c. 5.00 mL of 0.500 M HNO_3
d. 75.00 mL of 0.00030 M HCl

CONCEPTS TO REMEMBER

Arrhenius acid–base theory. An Arrhenius acid is a hydrogen-containing compound that, in water, produces hydrogen ions. An Arrhenius base is a hydroxide-containing compound that, in water, produces hydroxide ions.

Brønsted–Lowry acid–base theory. A Brønsted–Lowry acid is any substance that can donate a proton (H^+) to some other substance. A Brønsted–Lowry base is any substance that can accept a proton from some other substance. Proton donation (from an acid) cannot occur unless an acceptor (a base) is present.

Conjugate acids and bases. A conjugate acid–base pair is two species that differ by one proton. The conjugate base of an acid is the species that remains when the acid loses a proton. The conjugate acid of a base is the species formed when the base accepts a proton.

Polyprotic acids. Polyprotic acids are acids that can transfer two or more hydrogen ions during an acid–base reaction.

Strengths of acids and bases. Acids can be classified as strong or weak in terms of the extent to which proton transfer occurs in aqueous solution. A strong acid completely transfers its protons to water. A weak acid transfers only a small percentage of its protons to water.

Salts. Salts are ionic compounds containing a metal or polyatomic ion as the positive ion and a nonmetal or polyatomic ion (except hydroxide ion) as the negative ion. Ionic compounds containing hydroxide ion are bases rather than salts.

Neutralization. Neutralization is the reaction between an acid and a hydroxide base to form a salt and water.

Self-ionization of water. In pure water, a small number of water molecules (1.0×10^{-7} mole/L) donate protons to other water molecules to produce small concentrations (1.0×10^{-7} mole/L) of hydronium and hydroxide ions.

The pH scale. The pH scale is a scale of small numbers that are used to specify molar hydronium ion concentration in an aqueous solution. Mathematically, the pH is the negative logarithm of the hydronium ion concentration. Solutions with a pH lower than 7.0 are acidic, those with a pH higher than 7.0 are basic, and those with a pH equal to 7.0 are neutral.

Buffer solutions. A buffer solution is a solution that resists pH change when small amounts of acid or base are added to it. The resistance to pH change in most buffers is caused by the presence of a weak acid and a salt of its conjugate base.

Acid–base titrations. Titration is a procedure in which an acid–base neutralization reaction is used in determining an unknown concentration. A measured volume of an acid or a base of known concentration is exactly reacted with a measured volume of a base or an acid of unknown concentration.

KEY REACTIONS AND EQUATIONS

1. Ion product constant for water (Section 9.7)

$$[H_3O^+][OH^-] = 1.0 \times 10^{-14}$$

2. Relationship between $[H_3O^-]$ and pH (Section 9.8)

$$pH = -\log [H_3O^+]$$

KEY TERMS

Acid–base titration (9.10)
Acidic solution (9.7)
Amphoteric substance (9.2)
Arrhenius acid (9.1)
Arrhenius base (9.1)
Basic solution (9.7)
Brønsted–Lowry acid (9.2)
Brønsted–Lowry base (9.2)
Buffer (9.9)
Conjugate acid (9.2)
Conjugate acid–base pair (9.2)
Conjugate base (9.2)
Diprotic acid (9.3)
Dissociation (9.1)
Indicator (9.10)
Ion product constant for water (9.7)
Ionization (9.1)
Monoprotic acid (9.3)
Neutral solution (9.7)
Neutralization (9.6)
pH (9.8)
pH scale (9.8)
Polyprotic acid (9.3)
Salt (9.5)
Strong acid (9.4)
Triprotic acid (9.3)
Weak acid (9.4)

ADDITIONAL PROBLEMS

9.46 In which of the following pairs of substances do the two members of the pair constitute a conjugate acid–base pair?
a. HN_3 and N_3^- b. H_2SO_4 and SO_4^{2-}
c. H_2CO_3 and $HClO_3$ d. NH_4^+ and NH_3

9.47 For which of the following pairs of acids are both members of the pair of "like strength"—that is, both strong or both weak?
a. HNO_3 and HNO_2 b. HCl and HBr
c. H_3PO_4 and $HClO_4$ d. H_2CO_3 and $H_2C_2O_4$

9.48 Solution A has a pH of 3.20, solution B a pH of 4.20, solution C a pH of 11.20, and solution D a pH of 7.20. Arrange the four solutions in order of
a. decreasing acidity
b. increasing $[H_3O^+]$
c. decreasing $[OH^-]$
d. increasing basicity

9.49 Classify solutions with the following characteristics as *acidic, basic,* or *neutral.*
a. Hydronium ion and hydroxide ion concentrations are equal.
b. Hydronium ion concentration is twice the hydroxide ion concentration.
c. Hydroxide ion concentration is twice the hydronium ion concentration.
d. Hydroxide ion concentration is three times the hydronium ion concentration.

9.50 Write a balanced chemical equation for the preparation of each of the following salts via an acid–base neutralization reaction.
a. NaCl b. Li_2SO_4 c. KNO_3 d. K_3PO_4

9.51 Calculate the pH of each of the following solutions
a. $[H_3O^+] = 1.0 \times 10^{-4}$ b. $[H_3O^+] = 1.0 \times 10^{-6}$
c. $[OH^-] = 1.0 \times 10^{-3}$ d. $[OH^-] = 1.0 \times 10^{-11}$

9.52 Identify the buffer system(s) (conjugate acid–base pair(s)) present in a solution that contains equal molar amounts of HCN, KCN, NaCN, and NaCl.

9.53 Arrange the following 0.1 M solutions in order of increasing pH: HCl (strong acid), HCN (weak acid), NaOH (strong base), NaCl (salt).

PRACTICE TEST ▶ True/False

9.54 In Arrhenius acid–base theory, the basic species is the OH^- ion.

9.55 In Brønsted–Lowry acid–base theory, a base is any substance that can donate a proton to another substance.

9.56 Any chemical reaction involving a Brønsted–Lowry acid must also involve a Brønsted–Lowry base.

9.57 The conjugate base of a Brønsted–Lowry acid always contains one more hydrogen atom than the Brønsted–Lowry acid.

9.58 Hydrochloric acid is a strong diprotic acid.

9.59 All hydrogen-containing compounds behave as Arrhenius acids in aqueous solution.

9.60 All ionic compounds that contain polyatomic ions are salts.

9.61 In an acid–base neutralization reaction, the compounds H_2SO_4 and NaOH would react in a two-to-one ratio.

9.62 Small amounts of both hydronium and hydroxide ions are present in *pure* water.

9.63 A solution with a hydroxide ion concentration of 1.0×10^{-8} M would be classified as basic.

9.64 A solution with a pH of 7.00 is considered neither acidic nor basic.

9.65 A solution with a pH of 6.40 is more acidic than a solution with a pH of 6.80.

9.66 If the hydronium ion concentration of a solution is 1.00×10^{-8} M, then the hydroxide ion concentration of the same solution is 1.00×10^{-6} M.

9.67 Most common buffers contain a weak acid and its conjugate base.

9.68 Buffering action always results in the active species of the buffer being converted into its partner species.

PRACTICE TEST ▸ Multiple Choice

9.69 In an Arrhenius acid–base context, the compounds HCl, HNO_3, and NaOH, when dissolved in water, are, respectively, a(n)
a. acid, acid, and base
b. base, base, and acid
c. base, acid, and acid
d. acid, base, and acid

9.70 In the reaction $N_3^- + H_2O \longrightarrow HN_3 + OH^-$, the Brønsted–Lowry base is
a. N_3^- b. H_2O c. HN_3 d. OH^-

9.71 In which of the following pairs of acids are both members of the pair strong acids?
a. H_3PO_4 and H_2SO_4 b. HCl and HNO_3
c. H_2CO_3 and HBr d. $HC_2H_3O_2$ and HI

9.72 Which of the following is produced in the *first step* of the dissociation of the polyprotic acid H_3PO_4?
a. H_3PO_3 b. $H_2PO_4^+$ c. $H_2PO_4^-$ d. PO_4^{3-}

9.73 In the neutralization reaction between equal molar concentrations of $Al(OH)_3$ and HNO_3, the acid and base will react in a molar ratio of
a. one-to-one b. one-to-three
c. three-to-one d. three-to-three

9.74 Which of the following hydronium ion concentrations would produce an *acidic* solution?
a. 1.00×10^{-5} M b. 1.00×10^{-7} M
c. 1.00×10^{-9} M d. 1.00×10^{-11} M

9.75 A solution with a pH of 12.0 is correctly described as being
a. strongly acidic b. weakly acidic
c. strongly basic d. weakly basic

9.76 The pH of a solution for which $[OH^-]$ is 1.0×10^{-6} is
a. 1.00 b. 8.00 c. 6.00 d. −6.00

9.77 Changing the pH of a solution from 4.00 to 5.00 corresponds to
a. increasing the $[H^+]$ by a factor of 2
b. decreasing the $[H^+]$ by a factor of 2
c. increasing the $[H^+]$ by a factor of 10
d. decreasing the $[H^+]$ by a factor of 10

9.78 A buffer solution could be prepared from which of the following pairs of substances?
a. NaCl and HCl b. NaOH and HCl
c. KCN and HCN d. NaCN and KCN

CHAPTER OUTLINE

CHAPTER TEN

Saturated Hydrocarbons

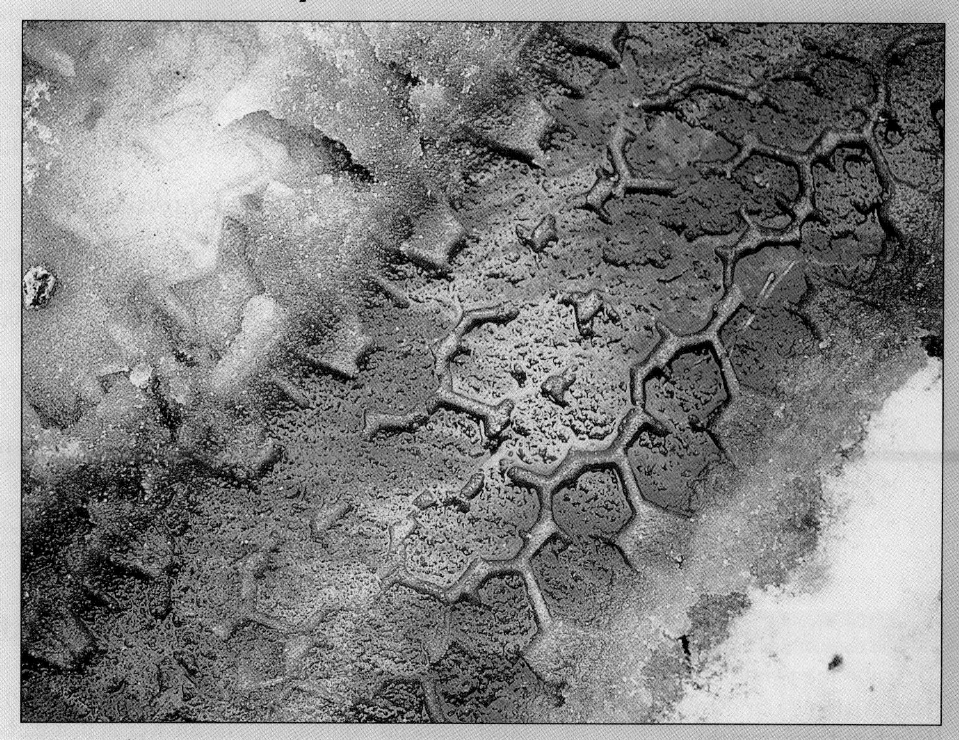

A tire tread in snow with oil (hydrocarbons) causing the iridescent pattern.

This chapter is the first of four that deal with the subject of organic chemistry and organic compounds. Organic compounds are the chemical basis for life itself, as well as an important component of the basis for our current high standard of living. Organic compounds are found in natural gas, petroleum, coal, gasoline, and many synthetic materials such as dyes, plastics, and clothing fibers. Proteins, carbohydrates, enzymes, and hormones are complex organic molecules.

10.1 Organic and Inorganic Compounds

Learning Focus

Understand the historical and modern definitions of the term *organic chemistry.*

During the latter part of the eighteenth century and the early part of the nineteenth century, chemists began to categorize compounds into two types: organic and inorganic. Compounds obtained from living organisms were called *organic* compounds, and compounds obtained from mineral constituents of the earth were called *inorganic* compounds.

During this early period, chemists believed that a special "vital force," supplied by a living organism, was necessary for the formation of an organic compound. This concept was proved incorrect in 1828 by the German chemist Friedrick Wöhler. Wöhler heated an aqueous solution of two inorganic compounds, ammonium chloride and silver cyanate, and obtained urea (a component of urine).

$$NH_4Cl + AgNCO \longrightarrow \underset{\text{urea}}{(NH_2)_2CO} + AgCl$$

▶ The historical origins of the terms *organic* and *inorganic* involve the following conceptual pairings:
*org*anic—living *org*anisms
*in*organic—*in*animate materials

Soon other chemists had successfully synthesized organic compounds from inorganic starting materials. As a result, the vital force theory was completely abandoned. The terms *organic* and *inorganic* continue to be used, but the definitions of these terms have changed.

Today, **organic chemistry** *is defined as the study of hydrocarbons (compounds of hydrogen and carbon) and their derivatives.* Nearly all compounds found in living organisms are still classified as organic compounds, as are many compounds that have been synthesized in the laboratory that have never been found in a living organism.

In essence, organic chemistry is the study of the compounds of one element (carbon), and inorganic chemistry is the study of the compounds of the other 112 elements. This unequal partitioning occurs because there are approximately 7 million organic compounds and only an estimated 1.5 million inorganic compounds (see Figure 10.1). This is an approximately 5:1 ratio between organic and inorganic compounds.

▶ Some textbooks define organic chemistry as the study of carbon-containing compounds. Almost all carbon-containing compounds qualify as organic compounds. However, the oxides of carbon, carbonates, cyanides, and metallic carbides are classified as inorganic rather than organic compounds. Inorganic carbon compounds involve carbon atoms that are not bonded to hydrogen atoms (CO, CO_2, Na_2CO_3, and so on).

Practice Problems and Questions

10.1 Indicate whether each of the following statements is *true* or *false*.

a. Most, but not all, compounds found in living organisms are organic compounds.

b. The number of organic compounds exceeds the number of inorganic compounds by a factor of 2.

c. Chemists now believe that a special "vital force" is needed to form an organic compound.

d. Organic compounds cannot be prepared from inorganic starting materials.

10.2 What are the historical origins for the terms *organic* and *inorganic*?

10.3 What is the modern definition for the term *organic chemistry*?

10.2 Bonding Characteristics of the Carbon Atom

Learning Focus

Describe the ways in which carbon atoms can meet their four-bond requirement.

What is unique about carbon that causes it to form 5 times as many compounds as all the other elements combined? The answer is that carbon atoms can be covalently bonded to other carbon atoms and to atoms of other elements in a wide variety of ways. Many of the bonding patterns result in molecules that contain chains or rings of carbon atoms. Sometimes, both chains and rings are present in the same molecule.

The variety of covalent bonding "behaviors" possible for carbon atoms is related to carbon's electron configuration. Carbon is a member of Group IVA of the periodic table, so carbon atoms possess four valence electrons (Section 4.2). In compound formation, four additional valence electrons are needed to give carbon atoms an octet of valence electrons (the octet rule, Section 4.3). These additional electrons are obtained by electron sharing (covalent bond formation). The sharing of *four* valence electrons requires the formation of *four* covalent bonds.

Carbon can meet this four-bond requirement in three different ways:

1. *By bonding to four other atoms.* This situation requires the presence of four single bonds.

—C— (with bonds above and below)

Four single bonds

2. *By bonding to three other atoms.* This situation requires the presence of two single bonds and one double bond.

—C= (with a bond above)

Two single bonds and one double bond

Figure 10.1 Sheer numbers is one reason why organic chemistry is a separate field of chemical study. Approximately 7 million organic compounds are known, compared to "just" 1.5 million inorganic compounds.

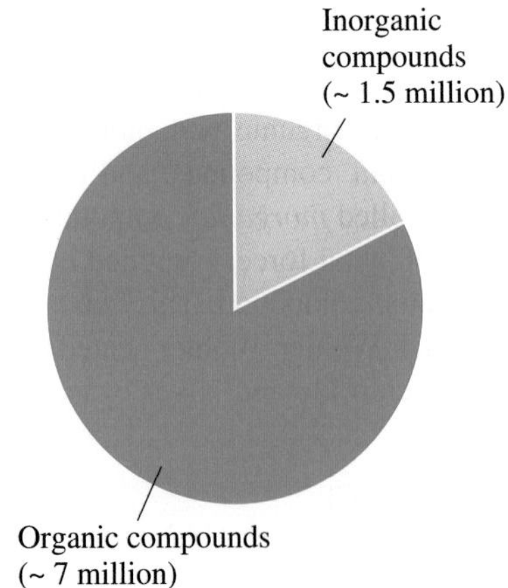

▶ Carbon atoms in organic compounds, in accordance with the octet rule, always form four covalent bonds.

3. *By bonding to two other atoms.* This situation requires the presence of either two double bonds or a triple bond and a single bond.

=C=	—C≡
Two double bonds	One triple bond and one single bond

Practice Problems and Questions

10.4 In organic compounds, carbon atoms always form four bonds. Why?

10.5 Indicate whether each of the following *meets* or *does not meet* the four-bond requirement for carbon atoms.
a. Two single bonds and a double bond
b. A single bond and two double bonds
c. A single bond and a triple bond
d. A double bond and a triple bond

Learning Focus

Define the terms *hydrocarbon* and *hydrocarbon derivative* and list the two major classes of hydrocarbons.

10.3 Hydrocarbons and Hydrocarbon Derivatives

The field of organic chemistry encompasses the study of hydrocarbons and hydrocarbon derivatives (Section 10.1). A **hydrocarbon** *is a compound that contains only carbon and hydrogen atoms.* Thousands of hydrocarbons are known. A **hydrocarbon derivative** *is a compound that contains carbon and hydrogen and one or more additional elements.* Additional elements commonly found in hydrocarbon derivatives include O, N, S, P, F, Cl, and Br. Millions of hydrocarbon derivatives are known.

Hydrocarbons may be divided into two large classes: saturated and unsaturated. A **saturated hydrocarbon** *is a hydrocarbon in which all carbon–carbon bonds are single bonds.* Saturated hydrocarbons are the simplest type of organic compound. An **unsaturated hydrocarbon** *is a hydrocarbon that contains one or more carbon–carbon multiple bonds: double bonds, triple bonds, or both.* In general, saturated and unsaturated hydrocarbons undergo distinctly different chemical reactions.

Saturated hydrocarbons are the subject of this chapter. Unsaturated hydrocarbons are considered in the next chapter. Figure 10.2 summarizes the terminology presented in this section.

▶ The term *saturated* has the general meaning that there is no more room for something. Its use with hydrocarbons comes from early studies in which chemists tried to add hydrogen atoms to various hydrocarbon molecules. Compounds to which no more hydrogen atoms could be added (because they already contained the maximum number) were called saturated, and those to which hydrogen could be added were called unsaturated.

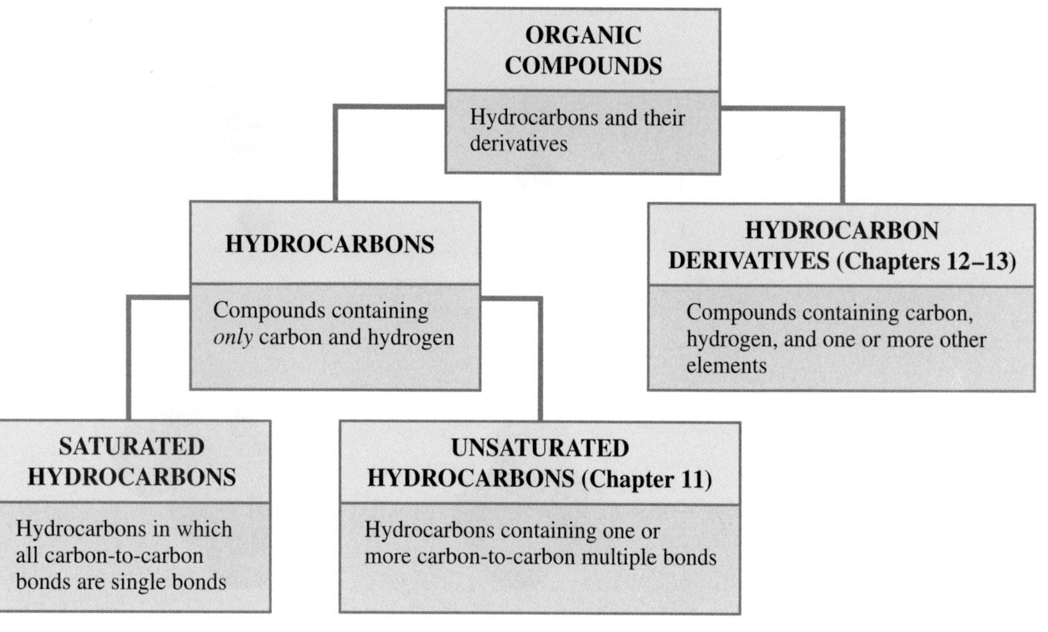

Figure 10.2 A summary of classification terms for organic compounds.

Practice Problems and Questions

10.6 What is the difference between a *saturated* and an *unsaturated* hydrocarbon?

10.7 What is the difference between a *hydrocarbon* and a *hydrocarbon derivative*?

10.8 Contrast *hydrocarbons* and *hydrocarbon derivatives* in terms of the number of compounds that are known.

Learning Focus

Understand the structural characteristics of simple alkanes.

10.4 Alkanes: The Simplest Saturated Hydrocarbons

In a saturated hydrocarbon, the carbon atom arrangement may be *acyclic* or *cyclic*. The term *acyclic* means "not cyclic." Examples of acyclic and cyclic carbon atom arrangements are

```
                         C
                       /   \
                      C     C
C—C—C—C—C—C           |     |
                      C     C
                       \   /
                         C
```

Acyclic arrangement of carbon atoms — Cyclic arrangement of carbon atoms

In this section we consider acyclic saturated hydrocarbons. Cyclic saturated hydrocarbons are considered in Section 10.9.

An **alkane** *is a saturated hydrocarbon in which the carbon atom arrangement is acyclic.* Such compounds are considered the simplest type of hydrocarbon. The molecular formulas of all alkanes fit the general formula C_nH_{2n+2}, where *n* is the number of carbon atoms present. The number of hydrogen atoms present in an alkane is always twice the number of carbon atoms plus two more, as in C_4H_{10}, C_5H_{12}, and C_8H_{18}.

The three simplest alkanes are methane (CH_4), ethane (C_2H_6), and propane (C_3H_8). Ball-and-stick and space-filling models showing the molecular structures of these three alkanes are given in Figure 10.3. Note, from these models, how each carbon atom in each of the models participates in four bonds (Section 10.2). Note also that the geometrical

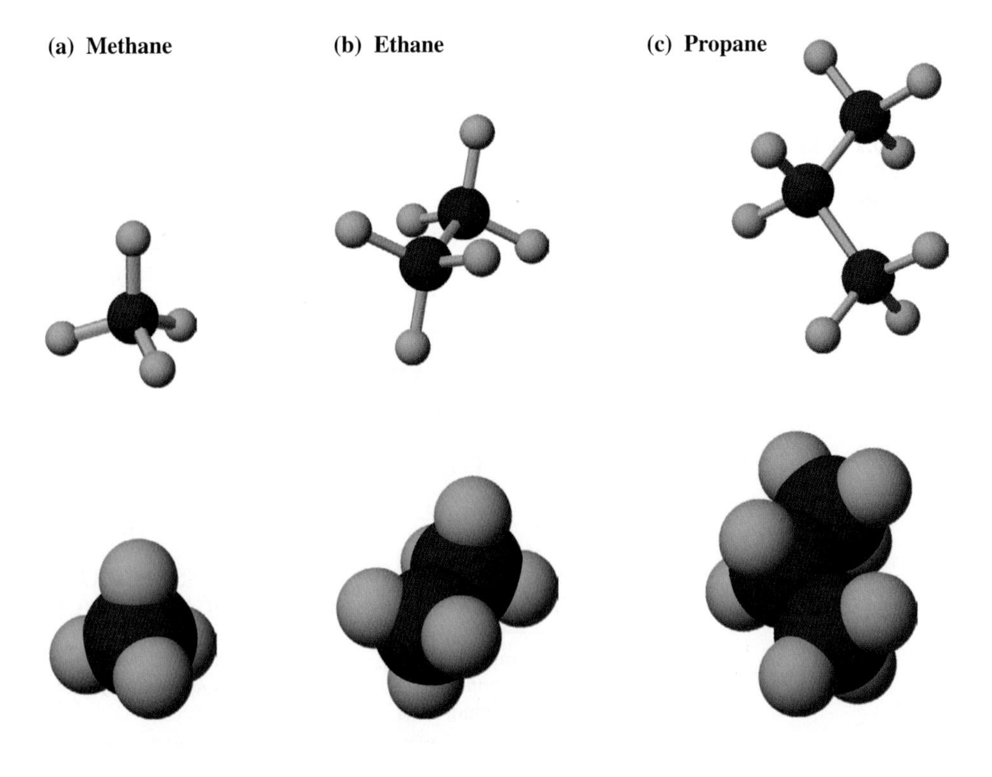

Figure 10.3 Ball-and-stick and space-filling models showing the molecular structures of (a) methane, (b) ethane, and (c) propane, the three simplest alkanes.

arrangement of atoms about each carbon atom is tetrahedral, an arrangement consistent with the principles of VSEPR theory (Section 4.14). The tetrahedral arrangement of the atoms bonded to alkane carbon atoms is fundamental to understanding the structural aspects of organic chemistry.

Practice Problems and Questions

10.9 How many carbon atoms are present in each of the following alkanes?
a. Ethane b. Methane c. Propane

10.10 How many carbon–carbon single bonds and carbon–hydrogen bonds are present in each of the following alkanes?
a. Methane b. Propane c. Ethane

10.11 How many carbon–carbon single bonds and carbon–hydrogen bonds are present in alkanes with each of the following chemical formulas?
a. CH_4 b. C_2H_6 c. C_3H_8

10.12 How many hydrogen atoms would be present in alkanes with the following numbers of carbon atoms?
a. 4 b. 6 c. 7 d. 10

10.13 How many carbon atoms would be present in alkanes with the following numbers of hydrogen atoms?
a. 6 b. 10 c. 16 d. 22

Learning Focus

Be able to write *expanded* and *condensed* structural formulas for simple alkanes.

10.5 Structural Formulas

The structures of alkanes and other types of organic compounds are generally represented in two dimensions rather than three (Figure 10.3) because of the difficulty in drawing the latter. These two-dimensional structural representations make no attempt to portray accurately the bond angles or molecular geometry of molecules. Their purpose is to convey information about which atoms in a molecule are bonded to which other atoms.

Two-dimensional structural representations for organic molecules are of two types: expanded structural formulas and condensed structural formulas. An **expanded structural formula** *shows, in two dimensions, all atoms in a molecule and all the bonds connecting them.* When written out, expanded structural formulas generally occupy a lot of space, and condensed structural formulas represent a shorthand method for conveying the same information. A **condensed structural formula** *uses groupings of atoms, in which central atoms and the atoms connected to them are written as a group, to convey molecular structural information.* The expanded and condensed structural formulas for methane, ethane, and propane follow.

▶ Structural formulas, whether expanded or condensed, do not show the geometry (shape) of the molecule. That information can be conveyed only by 3-D drawings or models such as those in Figure 10.3.

	Methane	Ethane	Propane
Expanded structural formula	$\begin{array}{c} \text{H} \\ | \\ \text{H—C—H} \\ | \\ \text{H} \end{array}$	$\begin{array}{c} \text{H H} \\ | \;\; | \\ \text{H—C—C—H} \\ | \;\; | \\ \text{H H} \end{array}$	$\begin{array}{c} \text{H H H} \\ | \;\; | \;\; | \\ \text{H—C—C—C—H} \\ | \;\; | \;\; | \\ \text{H H H} \end{array}$
Condensed structural formula	CH_4	$CH_3—CH_3$	$CH_3—CH_2—CH_3$

The condensed structural formulas of hydrocarbons in which a long chain of carbon atoms is present are often condensed even more. The formula

$$CH_3—CH_2—CH_2—CH_2—CH_2—CH_2—CH_2—CH_3$$

can be further abbreviated as

$$CH_3—(CH_2)_6—CH_3$$

in which parentheses and a subscript are used to denote the number of $—CH_2—$ groups in the chain.

▶ Only two types of bonds may be present in an alkane: carbon–carbon single bonds and carbon–hydrogen bonds.

In situations where the focus is on the arrangement of carbon atoms in a molecule, *skeletal formulas* that omit the hydrogen atoms are often used.

C—C—C—C—C means the same as $CH_3—CH_2—CH—CH_2—CH_3$

Skeletal formula — Condensed structural formula

The skeletal formula still represents a unique compound, because we know that each carbon atom shown must have enough hydrogen atoms attached to it to give the carbon four bonds.

Practice Problems and Questions

10.14 Propane, the three-carbon alkane, has the formula C_3H_8. For propane, write each of the following.
a. Expanded structural formula
b. Condensed structural formula

10.15 Butane is the alkane that has four carbon atoms in a row. For butane, write each of the following.
a. Expanded structural formula
b. Condensed structural formula
c. Molecular formula

10.16 Convert each of the following alkane skeletal formulas into a condensed structural formula.
a. C—C—C—C—C—C b. C—C—C—C—C—C—C—C

10.17 Convert each of the following alkane condensed structural formulas into an expanded structural formula.
a. $CH_3—CH_2—CH_2—CH_2—CH_3$ b. $CH_3—(CH_2)_3—CH_3$

Learning Focus

Understand the concept of structural isomerism and the difference between a straight-chain alkane and a branched-chain alkane.

10.6 Structural Isomerism

The molecular formulas CH_4, C_2H_6, and C_3H_8 represent the alkanes methane, ethane, and propane, respectively. Next in the alkane molecular formula sequence is C_4H_{10}, which would be expected to be the molecular formula of the four-carbon alkane. A new phenomenon arises, however, when an alkane has four or more carbon atoms. There is more than one structural formula that is consistent with the molecular formula. Consequently, more than one compound is consistent with the molecular formula. This situation brings us to the topic of structural isomerism.

▶ The word *isomer* comes from the Greek *isos,* which means "the same," and *meros,* which means "parts." Isomers have the same parts put together in different ways.

Structural isomers *are compounds with the same molecular formula but different structural formulas—that is, different bonding arrangements between atoms.* Structural isomers, with their differing structural formulas, always have different properties and are different compounds.

There are two four-carbon alkane structural isomers, both with the molecular formula C_4H_{10}.

$CH_3—CH_2—CH_2—CH_3$

Isomer I

$CH_3—CH—CH_3$ with CH_3 attached to the central CH

Isomer II

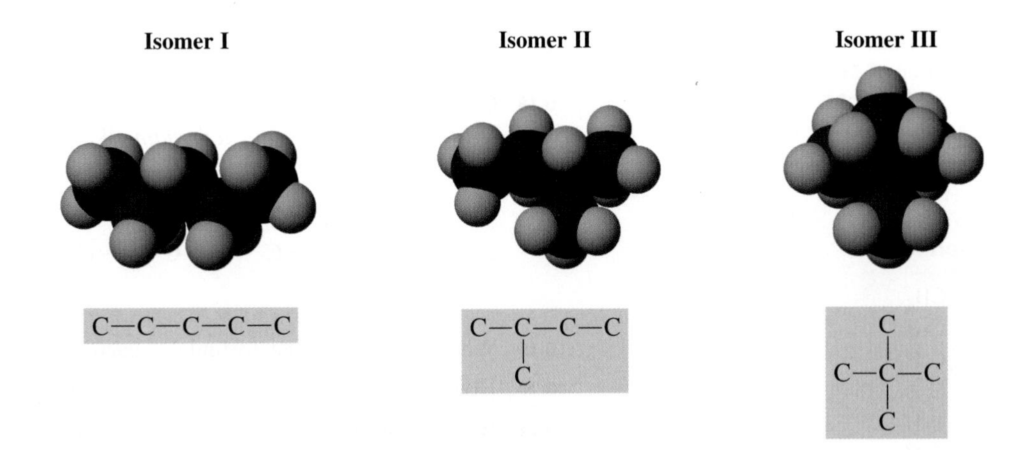

Figure 10.4 Space-filling models and skeletal formulas for the three isomeric C_5H_{12} alkanes.

▶ It is the existence of structural isomers that necessitates the use of structural formulas in organic chemistry. Structural isomers always have the same molecular formula and different structural formulas.

The first C_4H_{10} isomer has a chain of four carbon atoms. It is an example of a continuous-chain alkane. A **continuous-chain alkane** *is an alkane in which all carbon atoms are connected in a continuous nonbranching chain.* The second C_4H_{10} isomer has a chain of three carbon atoms with the fourth carbon attached as a branch on the middle carbon of the three-carbon chain. It is an example of a branched-chain alkane. A **branched-chain alkane** *is an alkane with one or more branches (of carbon atoms) attached to a continuous chain of carbon atoms.*

There are three structural isomers for alkanes with five carbon atoms (C_5H_{12}):

▶ Structural isomers of the type we are now considering are also called *skeletal isomers.* Such isomers differ in carbon atom arrangements; that is, the carbon skeletons differ. For alkanes, skeletal isomers are possible whenever more than three carbon atoms are present.

$CH_3—CH_2—CH_2—CH_2—CH_3$ (Isomer I)

$CH_3—CH(CH_3)—CH_2—CH_3$ (Isomer II)

$CH_3—C(CH_3)_2—CH_3$ (Isomer III)

Figure 10.4 shows space-filling models for the three isometric C_5 alkanes. Note how isomer III, the most branched isomer, has the most compact, most spherical three-dimensional shape.

Table 10.1 contrasts the differing physical properties of the two C_4 alkanes and the three C_5 alkanes. Section 10.8 considers how such structural isomers are named.

The number of possible alkane structural isomers increases dramatically with increasing number of carbon atoms in the alkane, as shown in Table 10.2. Structural isomerism is one of the major reasons for the existence of so many organic compounds.

Table 10.1
Selected Physical Properties of C_4 and C_5 Alkanes

Alkane	Boiling point	Density (at 20°C)
C_4H_{10} alkanes		
Isomer I	−0.5°C	0.579 g/mL
Isomer II	−11.6°C	0.549 g/mL
C_5H_{12} alkanes		
Isomer I	36.1°C	0.626 g/mL
Isomer II	27.8°C	0.620 g/mL
Isomer III	9.5°C	0.614 g/mL

Table 10.2
Number of Structural Isomers Possible for Alkanes of Various Carbon-Chain Lengths

Molecular formula	Possible number of structural isomers
CH_4	1
C_2H_6	1
C_3H_8	1
C_4H_{10}	2
C_5H_{12}	3
C_6H_{14}	5
C_7H_{16}	9
C_8H_{18}	18
C_9H_{20}	35
$C_{10}H_{22}$	75
$C_{15}H_{32}$	4,347
$C_{20}H_{42}$	336,319
$C_{30}H_{62}$	4,111,846,763

Practice Problems and Questions

10.18 Classify each of the following alkanes, represented using skeletal formulas, as a *straight-chain* or a *branched-chain* alkane.

```
a. C—C—C—C                 b. C—C—C—C
                                 |
                                 C

c. C—C—C—C—C               d. C—C—C—C—C—C
     |   |
     C   C
```

10.19 What is the molecular formula of each of the alkanes in Problem 10.18?

10.20 How do the molecular formulas of structural isomers compare?

10.21 How do the physical properties of structural isomers compare?

10.22 Indicate whether the members of each of the following pairs of alkanes, represented using skeletal formulas, are structural isomers.

```
a. C—C—C—C and C—C—C—C—C

b. C—C—C—C and C—C—C—C
                   |
                   C

c. C—C—C—C—C and C—C—C—C
     |               |
     C               C

d. C—C—C—C—C—C and C—C—C—C
                     |  |
                     C  C
```

Learning Focus

Recognize structural formulas that represent different conformations of the same alkane.

10.7 Conformations for Alkanes

Conformations *are differing orientations of a molecule made possible by rotations about single bonds.* Rotation about carbon–carbon single bonds is an important property of alkane molecules. Two groups of atoms in an alkane connected by a carbon–carbon single bond can rotate with respect to one another around that bond, much as a wheel rotates around an axle.

```
   CH3                      CH3
    |       CH3              |         H
    C         |              C         |
   / \       C       →      / \        C
  H   H     / \            H   H      / \
           H   H                   CH3   H
```

As a result of rotation around single bonds, alkane molecules (except for methane) can exist in infinite numbers of orientations, or conformations.

The following skeletal formulas represent four different conformations for a continuous-chain, six-carbon alkane molecule:

```
                                                          C
                                                          |
C—C—C—C—C—C   C—C—C—C—C   C—C—C—C   C—C—C—C
              |           |     |   |
              C           C     C   C
```

All four skeletal formulas represent the same molecule; that is, they are different conformations of the same molecule. In all four cases, a continuous chain of six carbon atoms is present. In all except the first case, the chain is "bent," but bends do not disrupt the continuity of the chain.

```
C—C—C—C—C—C     C—C—C—C—C     C—C—C—C     C—C—C—C
                |             |     |     |     |
                C             C     C     C     C
```

▶ You should learn to recognize molecules drawn in several different ways (conformations). Like friends, they can be recognized whether they are sitting, reclining, or standing.

Note that the structures

```
C—C—C—C—C     and     C—C—C—C—C
|                       |
C                       C
```

are representations for two different alkanes. The first structure involves a continuous chain of six carbon atoms, and the second structure involves a continuous chain of five carbon atoms to which a branch is attached. There is no way that you can get a chain of six carbon atoms out of the second structure without "back-tracking"—and "back-tracking" is not allowed.

Example 10.1 Recognizing Different Conformations of a Molecule and Structural Isomers

Determine whether the members of each of the following pairs of structural formulas represent (1) different conformations of the same molecule, (2) different compounds that are structural isomers, or (3) different compounds that are not structural isomers.

a. $CH_3—CH_2—CH_2—CH_3$ and

```
CH2—CH2
|    |
CH3  CH3
```

b.

```
CH2—CH2—CH3     and     CH2—CH2—CH2—CH3
|                       |
CH3                     CH3
```

c.

```
CH3—CH—CH3     and     CH3—CH2—CH2
    |                          |
    CH3                        CH3
```

Solution

a. Both molecules have the molecular formula C_4H_{10}. The connectivity of carbon atoms is the same for both molecules; a continuous chain of four carbon atoms. For the second structural formula, we need to go around two corners to get a four-carbon-atom chain, which is fine because of the free rotation associated with single bonds in alkanes.

```
C—C—C—C     C—C
            |  |
            C  C
```

With the same molecular formula and the same connectivity of atoms, these two structural formulas are conformations of the same molecule.

b. The molecular formula of the first compound is C_4H_{10}, and that of the second compound is C_5H_{12}. Thus the two structural formulas represent different compounds that are not structural isomers.

c. Both molecules have the same molecular formula, C_4H_{10}. The connectivity of atoms is different. In the first case, we have a chain of three carbon atoms with a branch off the chain. In the second case, a continuous chain of four carbon atoms is present.

```
C—C—C     C—C—C
  |           |
  C           C
```

These two structural formulas are those of structural isomers.

Practice Problems and Questions

10.23 For each of the following arrangements of carbon atoms, determine the number of carbon atoms present in the longest continuous chain of carbon atoms.

```
a.              C               b. C—C—C—C—C—C
                |                    |       |
    C—C—C—C—C—C—C                    C       C
        |
        C

c.          C—C—C               d. C—C—C—C—C—C—C
            |                          |   |
    C—C—C—C                            C   C
            |                          |   |
            C—C—C—C                    C   C
                                       |   |
                                       C   C
```

10.24 Determine whether the members of each of the following pairs of structural formulas represent: (1) different conformations of the same molecule, (2) different compounds that are structural isomers, or (3) different compounds that are not structural isomers.

```
a. CH3—CH2—CH2—CH2—CH3      and    CH3—CH2
                                        |
                                        CH2—CH2
                                            |
                                            CH3

b. CH3—CH—CH2—CH3           and    CH3—CH—CH2
       |                               |   |
       CH3                             CH3 CH3

c. CH3—CH—CH2—CH3           and    CH2—CH2—CH2
       |                           |       |
       CH3                         CH3     CH3
```

10.8 IUPAC Nomenclature for Alkanes

> **Learning Focus**
> Given their structures, name alkanes using IUPAC rules and vice versa.

▶ IUPAC is pronounced "eye-you-pack."

▶ Continuous-chain alkanes are also frequently called *straight-chain alkanes* and *normal-chain alkanes.*

▶ You need to memorize the prefixes in column 2 of Table 10.3. This is the way to count from 1 to 10 in "organic chemistry language."

When relatively few organic compounds were known, chemists arbitrarily named them using what today are called *common names.* These common names gave no information about the structures of the compounds they described. However, as more organic compounds became known, this nonsystematic approach to naming compounds became unwieldy.

Today, formal systematic rules exist for generating names for organic compounds. These rules, which were formulated, and are updated periodically, by the International Union of Pure and Applied Chemistry (IUPAC), are known as *IUPAC rules.* The advantage of the IUPAC naming system is that it assigns each compound a name that not only identifies it but also enables one to draw its structural formula.

IUPAC names for the first 10 *continuous-chain* alkanes are given in Table 10.3. Note that all of these names end in *-ane*, the characteristic ending for all alkane names. Note also that beginning with the five-carbon alkane, Greek numerical prefixes are used to denote the actual number of carbon atoms in the continuous chain.

To name *branched-chain* alkanes, we must be able to name the branch or branches that are attached to the main carbon chain. These branches are formally called substituents. A **substituent** *is an atom or group of atoms attached to a chain (or ring) of carbon atoms.*

For branched-chain alkanes, the substituents are specifically called *alkyl groups.* An **alkyl group** *is the group of atoms that would be obtained by removing a hydrogen atom from an alkane.*

The four most commonly encountered alkyl groups are the four simplest:

```
—CH3          —CH2—CH3          —CH2—CH2—CH3          —CH—CH3
                                                        |
                                                        CH3
Methyl group  Ethyl group       Propyl group          Isopropyl group
```

Table 10.3 IUPAC Names for the First Ten Continuous-Chain Alkanes[a]

Molecular formula	IUPAC prefix	IUPAC name	Structural formula
CH_4	meth-	methane	CH_4
C_2H_6	eth-	ethane	$CH_3—CH_3$
C_3H_8	prop-	propane	$CH_3—CH_2—CH_3$
C_4H_{10}	but-	butane	$CH_3—CH_2—CH_2—CH_3$
C_5H_{12}	pent-	pentane	$CH_3—CH_2—CH_2—CH_2—CH_3$
C_6H_{14}	hex-	hexane	$CH_3—CH_2—CH_2—CH_2—CH_2—CH_3$
C_7H_{16}	hept-	heptane	$CH_3—CH_2—CH_2—CH_2—CH_2—CH_2—CH_3$
C_8H_{18}	oct-	octane	$CH_3—CH_2—CH_2—CH_2—CH_2—CH_2—CH_2—CH_3$
C_9H_{20}	non-	nonane	$CH_3—CH_2—CH_2—CH_2—CH_2—CH_2—CH_2—CH_2—CH_3$
$C_{10}H_{22}$	dec-	decane	$CH_3—CH_2—CH_2—CH_2—CH_2—CH_2—CH_2—CH_2—CH_2—CH_3$

[a]The IUPAC naming system also includes prefixes for naming continuous-chain alkanes that have more than 10 carbon atoms, but we will not consider them in this text.

The extra-long bond in these formulas (on the left) denotes the point of attachment of the group to the carbon chain. Note that there are two three-carbon alkyl groups that differ in the point of attachment to the carbon chain. A propyl group attaches to the carbon chain through an "end" carbon atom; an isopropyl group attaches to a carbon chain through the "middle" carbon atom of the three.

Alkyl groups do not lead a stable, independent existence; that is, they are not molecules. They are "parts" of a compound, always found attached to another entity (usually a carbon chain).

We are now ready for the IUPAC rules for naming branched-chain alkanes.

Rule 1: *Identify the longest continuous carbon chain (the parent chain), which may or may not be shown in a straight line, and name the chain.*

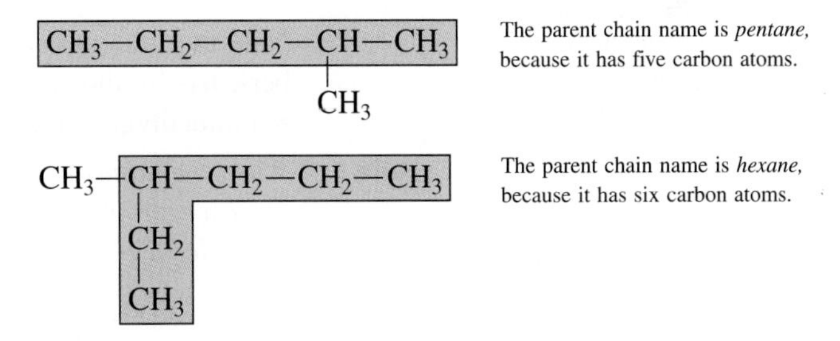

The parent chain name is *pentane*, because it has five carbon atoms.

The parent chain name is *hexane*, because it has six carbon atoms.

▶ Additional guidelines for numbering carbon atom chains:

1. If each end of the chain has a substituent the same distance in, number from the end closest to the second-encountered substituent.
2. If there are substituents equidistant from each end of the chain and there is no third substituent to use as the "tie-breaker," begin numbering nearest the substituent that has alphabetical priority—that is, the substituent whose name occurs first in the alphabet.

Rule 2: *Number the carbon atoms in the parent chain from the end of the chain nearest a substituent (alkyl group).*

There always are two ways to number the chain (either from left to right or from right to left). This rule gives the first-encountered alkyl group the lowest possible number.

$\overset{5}{C}H_3—\overset{4}{C}H_2—\overset{3}{C}H_2—\overset{2}{C}H(CH_3)—\overset{1}{C}H_3$

Right-to-left numbering system

$CH_3—\overset{3}{C}H(\overset{2}{C}H_2—\overset{1}{C}H_3)—\overset{4}{C}H_2—\overset{5}{C}H_2—\overset{6}{C}H_3$

Left-to-right numbering system

Rule 3: *If only one alkyl group is present, name and locate it (by number), and attach the number and name to that of the parent carbon chain.*

```
 5     4     3     2    1
CH3—CH2—CH2—CH—CH3
                  |
                 CH3
```

2-Methylpentane

```
      3     4     5     6
CH3—CH—CH2—CH2—CH3
     |
    2CH2
     |
    1CH3
```

3-Methylhexane

Note that the name is written as one word, with a hyphen between the number (location) and the name of the alkyl group.

Rule 4: *If two or more of the* same kind *of alkyl group are present in a molecule, indicate the number with a Greek numerical prefix (*di-, tri-, tetra-, penta-, *and so forth). In addition, a number specifying the location of each identical group must be included. These position numbers, separated by commas, precede the numerical prefix. Numbers are separated from words by hyphens.*

```
 1    2    3     4    5
CH3—CH—CH2—CH—CH3
     |         |
    CH3       CH3
```

2,4-Dimethylpentane

```
           CH3
 1     2    3|   4     5
CH3—CH2—C—CH2—CH3
            |
           CH3
```

3,3-Dimethylpentane

▶ There must be as many numbers as there are alkyl groups in the IUPAC name of a branched-chain alkane.

Note that the numerical prefix *di-* must always be accompanied by two numbers, *tri-* by three, and so on, even if the same number is written twice, as in 3,3-dimethylpentane.

Rule 5: *When two kinds of alkyl groups are present on the same carbon chain, number each group separately, and list the names of the alkyl groups in alphabetical order.*

```
 5     4     3    2    1
CH3—CH2—CH—CH—CH3
           |    |
          CH2  CH3
           |
          CH3
```

3-Ethyl-2-methylpentane

Note that ethyl is named first in accordance with the alphabetical rule.

```
 1     2    3    4    5     6     7     8
CH3—CH2—CH—CH—CH—CH2—CH2—CH3
          |    |    |
         CH2  CH2  CH2
          |    |    |
         CH3  CH2  CH2
               |    |
              CH3  CH3
```

3-Ethyl-4,5-dipropyloctane

▶ Numerical prefixes that designate numbers of alkyl groups, such as *di-*, *tri-*, and *tetra-*, are not considered when determining alphabetical priority for alkyl groups.

Note that the prefix *di-* does not affect the alphabetical order; ethyl precedes propyl.

Rule 6: *Follow IUPAC punctuation rules, which include the following: (1) Separate numbers from each other by commas. (2) Separate numbers from letters by hyphens. (3) Do not add a hyphen or space between the last-named substituent and the name of the parent alkane that follows.*

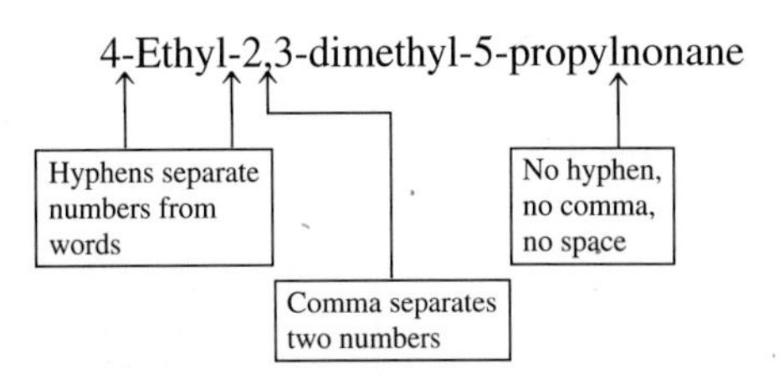

Example 10.2 Determining IUPAC Names for Branched-Chain Alkanes

Give the IUPAC name for each of the following branched-chain alkanes.

a. $CH_3-CH-CH-CH_3$ (with CH_2-CH_3 on the first CH and CH_3 on the second CH)

b. $CH_3-CH-CH_2-CH_2-CH-CH_2-CH-CH_3$ (with CH_3 on the first CH, CH_2-CH_3 on the second CH, and CH_3 on the third CH)

Solution

a. The longest carbon chain possesses five carbon atoms. Thus the parent chain name is pentane.

$CH_3-CH-CH-CH_3$ (with CH_2-CH_3 on the first CH and CH_3 on the second CH; parent chain boxed)

This parent chain is numbered from right to left, because an alkyl substituent is closer to the right end of the chain than to the left end.

$\underset{}{(CH_3)}-\overset{3}{C}H-\overset{2}{C}H-\overset{1}{C}H_3$ (with $\overset{4}{C}H_2-\overset{5}{C}H_3$ on C-3 and (CH_3) on C-2; methyl groups circled)

There are two methyl group substituents (circled). One methyl group is located on carbon 2 and the other on carbon 3. The IUPAC name for the compound is 2,3-dimethylpentane.

b. There are eight carbon atoms in the longest carbon chain, so the parent name is octane. There are three alkyl groups present (circled).

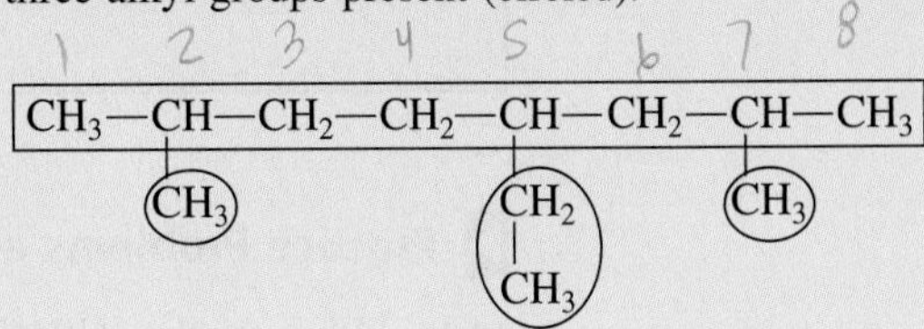

Selection of the numbering system to be used cannot be made based on the "first-encountered-alkyl-group rule" because an alkyl group is equidistant from each end of the chain. Thus the second-encountered alkyl group is used as the "tie-breaker." It is closer to the right end of the parent chain (carbon 4) than to the left end (carbon 5). Thus we use the right-to-left numbering system.

$$\overset{8}{CH_3}-\overset{7}{CH}(CH_3)-\overset{6}{CH_2}-\overset{5}{CH_2}-\overset{4}{CH}(CH_2-CH_3)-\overset{3}{CH_2}-\overset{2}{CH}(CH_3)-\overset{1}{CH_3}$$

Two different kinds of alkyl groups are present: ethyl and methyl. Ethyl has alphabetical priority over methyl and precedes methyl in the IUPAC name. The IUPAC name is 4-ethyl-2,7-dimethyloctane.

▶ Always compare the total number of carbon atoms in the name with the number in the structure to make sure they match. The name 4-ethyl-2,7-dimethyloctane indicates the presence of 2 + 2(1) + 8 = 12 carbon atoms. The structure does have 12 carbon atoms.

After you learn the rules for naming alkanes, it is relatively easy to reverse the procedure and translate the name of an alkane into a structural formula. Example 10.3 shows how this is done.

Example 10.3 Generating the Structural Formula of an Alkane from its IUPAC Name

Draw the condensed structural formula for 3-ethyl-2,3-dimethylpentane.

Solution

Step 1: The name of this compound ends in *pentane,* so the longest continuous chain has five carbon atoms. Draw this chain of five carbon atoms and number it.

$$\overset{1}{C}-\overset{2}{C}-\overset{3}{C}-\overset{4}{C}-\overset{5}{C}$$

Step 2: Complete the carbon skeleton by attaching alkyl groups as they are specified in the name. An ethyl group goes on carbon 3, and methyl groups are attached to carbons 2 and 3.

$$\overset{1}{C}-\overset{2}{C}(C)-\overset{3}{C}(C)(C-C)-\overset{4}{C}-\overset{5}{C}$$

Step 3: Add hydrogen atoms to the carbon skeleton so that each carbon atom has four bonds.

$$\overset{1}{CH_3}-\overset{2}{CH}(CH_3)-\overset{3}{C}(CH_3)(CH_2-CH_3)-\overset{4}{CH_2}-\overset{5}{CH_3}$$

Practice Problems and Questions

10.25 What are the IUPAC names for the two four-carbon alkane structural isomers (see Section 10.7 for their structures).

10.26 What are the IUPAC names for the three five-carbon alkane structural isomers (see Section 10.7 for their structures).

10.27 Assign an IUPAC name to each of the following alkanes.

a.
```
CH3—CH2—CH2—CH—CH3
             |
             CH3
```

b.
```
CH3—CH—CH—CH2—CH—CH3
    |   |       |
    CH2 CH3     CH3
    |
    CH3
```

c.
```
           CH3
           |
CH3—CH—C—CH2—CH3
    |   |
    CH3 CH2
        |
        CH3
```

d.
```
CH3—CH—CH—CH—CH2—CH3
    |   |   |
    CH3 CH2 CH3
        |
        CH3
```

10.28 Draw a condensed structural formula for each of the following alkanes.
a. 2-Methylbutane
b. 3,4-Dimethylhexane
c. 3-Ethyl-3-methylpentane
d. 2,3,4,5-Tetramethylheptane

10.29 Draw a condensed structural formula for each of the following alkanes.
a. 4-Methylheptane
b. 4-Ethylheptane
c. 4-Propylheptane
d. 4-Isopropylheptane

10.30 Without actually drawing a structural formula, determine the number of carbon atoms present in each of the following alkanes.
a. 3-Methylhexane
b. 2,2-Dimethylhexane
c. 5-Propyldecane
d. 3,3,4,4-Tetramethylheptane

10.31 Without actually drawing a structural formula, determine the number of alkyl groups present in each of the alkanes of Problem 10.30.

10.32 Without actually drawing a structural formula, determine the number of substituents present in each of the alkanes of Problem 10.30.

10.9 Cycloalkanes

Learning Focus

Draw structure formulas, including line-angle drawings, for cycloalkanes.

A **cycloalkane** *is a saturated hydrocarbon in which the carbon atoms are connected to one another in a cyclic (ring) arrangement.* The simplest cycloalkane is cyclopropane, which contains a cyclic arrangement of three carbon atoms. Figure 10.5 shows a three-dimensional model of cyclopropane's structure and those of the four-, five-, and six-carbon cycloalkanes.

▶ It takes a minimum of three carbon atoms to form a cyclic arrangement of carbon atoms.

Cyclopropane's three carbon atoms lie in a flat ring. In all other cycloalkane molecules, some puckering of the ring occurs; that is, the ring systems are nonplanar, as shown in Figure 10.5.

Figure 10.5 Three-dimensional representations of the structures of simple cycloalkanes.

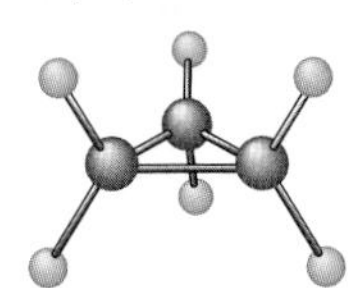

(a) Cyclopropane, C_3H_6

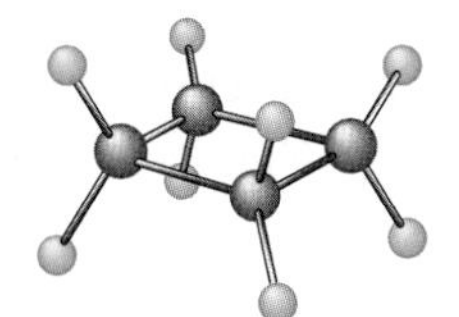

(b) Cyclobutane, C_4H_8

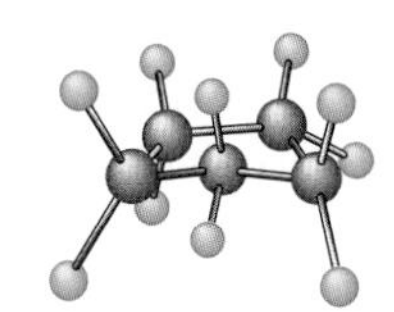

(c) Cyclopentane, C_5H_{10}

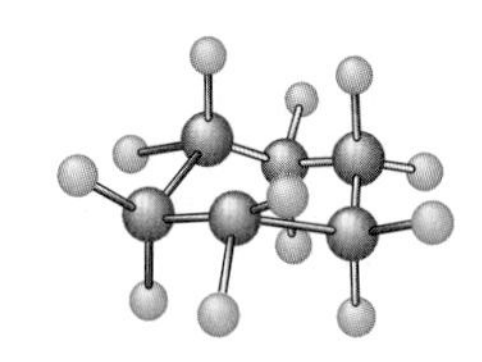

(d) Cyclohexane, C_6H_{12}

The general formula for cycloalkanes is C_nH_{2n}. Thus a given cycloalkane contains two fewer hydrogen atoms than an alkane with the same number of hydrogen atoms (C_nH_{2n+2}). Butane (C_4H_{10}) and cyclobutane (C_4H_8) are not structural isomers; structural isomers must have the same molecular formula (Section 10.6).

Line–angle drawings are often used to represent cycloalkane structures. A **line–angle drawing** *is an abbreviated structural formula representation in which an angle represents a carbon atom and a line represents a bond.* The line–angle drawing for cyclopropane is a triangle, that for cyclobutane a square, that for cyclopentane a pentagon, and that for cyclohexane a hexagon.

Cyclopropane Cyclobutane Cyclopentane Cyclohexane

Alkyl groups may be attached to cycloalkane rings in a manner similar to that which occurs with alkanes. Examples of such compounds include the following:

CH_3 CH_2-CH_3 CH_3 CH_2-CH_3 CH_3 CH_3

Practice Problems and Questions

10.33 Indicate whether each of the following molecular formulas could be that of a cycloalkane.
a. C_4H_8 b. C_5H_{12} c. C_3H_6 d. C_2H_2

10.34 The alkanes propane and cyclopropane are not structural isomers. Explain why.

10.35 What is the molecular formula for each of the following cycloalkanes?
a. Cyclopropane
b. Cyclooctane
c. Cyclobutane to which a methyl group is attached
d. Cyclopentane to which two methyl groups are attached

Learning Focus

Given their structures, name cycloalkanes using IUPAC rules, and vice versa.

10.10 IUPAC Nomenclature for Cycloalkanes

IUPAC naming procedures for cycloalkanes are similar to those for alkanes. The ring portion of a cycloalkane molecule serves as the name base, and the prefix *cyclo-* is used to indicate the presence of the ring. Alkyl group substituents are named in the same manner as in alkanes. Numbering conventions used in locating substituents include the following:

1. If there is just one ring substituent, it is not necessary to locate it by number.
2. When two ring substituents are present, the carbon atoms in the ring are numbered beginning with the substituent of higher alphabetical priority and proceeding in the direction (clockwise or counterclockwise) that gives the other substituent the lower number.
3. When three or more ring substituents are present, ring numbering begins at the substituent that leads to the lowest set of location numbers. When two or more equivalent numbering sets exist, alphabetical priority among substituents determines the set used.

Example 10.4 illustrates the use of the ring-numbering guidelines.

▶ Cycloalkanes of ring sizes ranging from 3 to over 30 are found in nature, and in principle, there is no limit to ring size. Five-membered rings (cyclopentanes) and six-membered rings (cyclohexanes) are especially abundant in nature.

Example 10.4 Determining IUPAC Names for Cycloalkanes

Assign IUPAC names to each of the following cycloalkanes.

a. 

b. 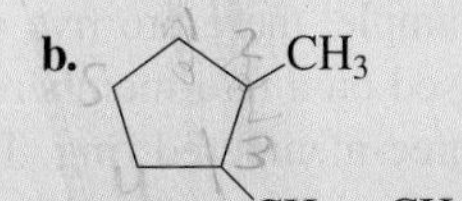

c. CH_3, CH_3, $CH_2—CH_2—CH_3$ 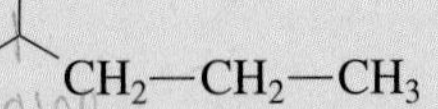

Solution

a. This molecule is a cyclobutane (four-carbon ring) with a methyl substituent. The IUPAC name is simply methylcyclobutane. No number is needed to locate the methyl group, because all four ring positions are equivalent.

b. This molecule is a cyclopentane with ethyl and methyl substituents. The numbers for the carbon atoms that bear the substituents are 1 and 2. On the basis of alphabetical priority, the number 1 is assigned to the carbon atom that bears the ethyl group. The IUPAC name for the compound is 1-ethyl-2-methylcyclopentane.

c. This molecule is a dimethylpropylcyclohexane. Two different 1,2,3 numbering systems exist for locating the substituents.

On the basis of alphabetical priority, we use the first numbering system. This system has carbon 1 bearing a methyl group, methyl has alphabetical priority over propyl. Thus the compound name is 1,2-dimethyl-3-propylcyclohexane.

Practice Problems and Questions

10.36 Assign an IUPAC name to each of the following cycloalkanes.

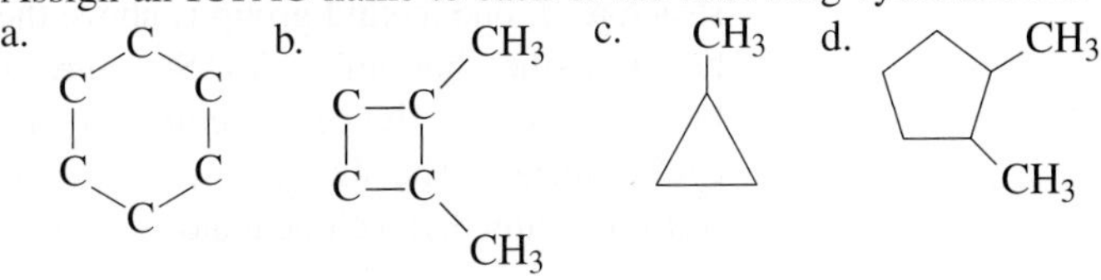

10.37 Assign an IUPAC name to each of the following cycloalkanes.

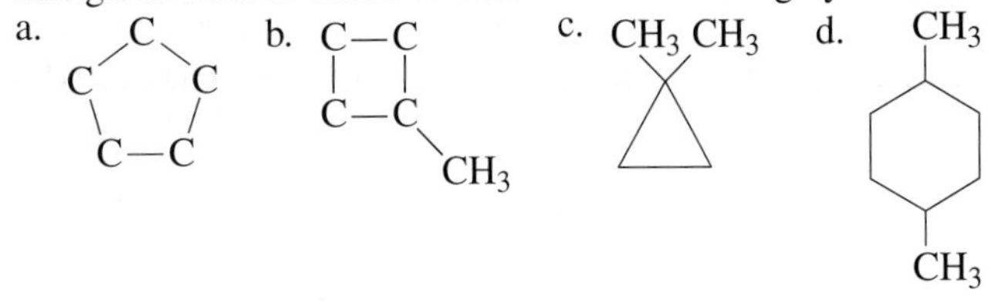

10.38 Draw structural formulas, with line–angle drawings denoting ring structure, for the following cycloalkanes.

a. Ethylcyclobutane b. 1,2-Diethylcyclobutane
c. 1,1-Diethylcyclobutane d. 1,3-Diethylcyclopentane

10.39 Draw structural formulas, with line–angle drawings denoting ring structure, for the following cycloalkanes.

a. Methylcylohexane b. Ethylcyclohexane
c. Propylcyclohexane d. Isopropylcyclohexane

Learning Focus

Be able to recognize structural and geometrical isomerism forms of cycloalkanes.

10.11 Isomerism in Cycloalkanes

Structural isomers are possible for cycloalkanes that contain four or more carbon atoms. For example, there are five cycloalkane structural isomers that have the formula C_5H_{10}: one based on a five-membered ring, one based on a four-membered ring, and three based on a three-membered ring. These isomers are

CH_3 CH_3 CH_3 CH_3CH_3 $CH_2—CH_3$

▶ *Cis–trans isomers* have the same molecular formula and the same structural formula. The only difference between them is the orientation of atoms in space. *Structural isomers* have the same molecular formula but different structural formulas.

A second type of isomerism is possible for some *substituted* cycloalkanes. This new type of isomerism is *cis–trans* isomerism. ***Cis–trans* isomers** *are compounds that have the same molecular and structural formulas but different arrangements of atoms in space because of restricted rotation around bonds.*

In alkanes, there is free rotation about all carbon–carbon bonds (Section 10.7). In cycloalkanes, the ring structure restricts rotation for the carbon atoms in the ring. The consequence of this lack of rotation in a cycloalkane is the creation of "top" and "bottom" positions for the two attachments on each of the ring carbon atoms. This "top–bottom" situation may lead to *cis–trans* isomerism in cycloalkanes with two or more substituents on the ring.

Consider the following two structures for the molecule 1,2-dimethylcyclopentane.

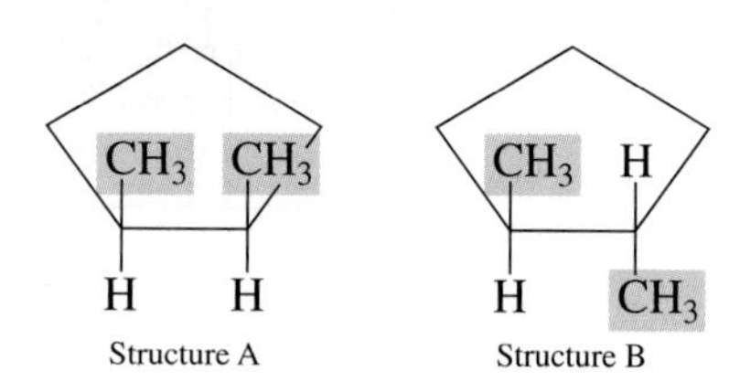

Structure A Structure B

In structure A, both methyl groups are above the plane of the ring (the "top" side). In structure B, one methyl group is above the plane of the ring (the "top" side) and the other below it (the "bottom" side). Structure A cannot be converted into structure B without breaking bonds. Hence structures A and B are isomers; there are two 1,2-dimethylcyclopentanes. The first isomer is called *cis*-1,2-dimethylcyclopentane and the second *trans*-1,2-dimethylcyclopentane.

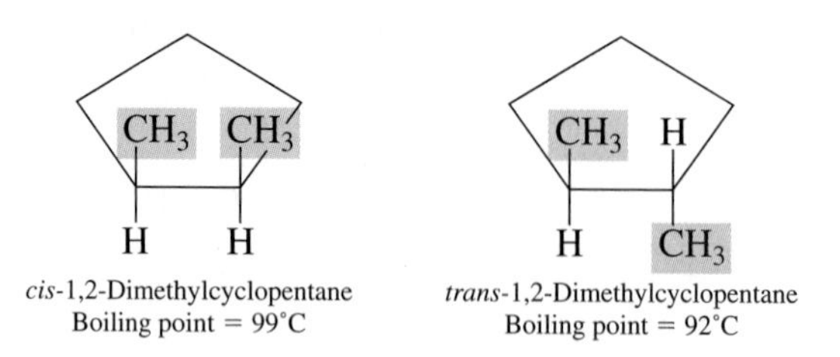

cis-1,2-Dimethylcyclopentane
Boiling point = 99°C

trans-1,2-Dimethylcyclopentane
Boiling point = 92°C

▶ The Latin *cis* means "on the same side," and the Latin *trans* means "across." Consider the use of the prefix *trans*- in the phrase "transatlantic voyage."

A ***cis* isomer** *is an isomer in which two atoms or groups are on the same side of a restricted rotation "barrier" in a molecule.* In our current discussion, the restricted rotation barrier is the ring of carbon atoms. A ***trans* isomer** *is an isomer in which two atoms or groups are on different sides of a restricted rotation "barrier" in a molecule.*

Cis–trans isomerism can occur in rings of all sizes. The presence of a substituent on each of two carbon atoms in the ring is the minimum requirement for its occurrence. In biochemistry, we will find that the human body often selectively distinguishes between the *cis* and *trans* isomers of a compound. One isomer will be active in the body and the other inactive.

▶ In cycloalkanes, *cis–trans* isomerism can also be denoted by using wedges and dotted lines. A heavy wedge-shaped bond to a ring structure indicates a bond *above* the plane of the ring, and a broken dotted line indicates a bond *below* the plane of the ring.

CH_3, CH_3 or CH_3, CH_3, H, H

cis-1,2-Dimethylcyclopropane

CH_3, CH_3 or CH_3, H, H, CH_3

trans-1,2-Dimethylcyclopropane

Example 10.5 Identifying and Naming Cycloalkane *Cis–Trans* Isomers

Determine whether *cis–trans* isomerism is possible for each of the following cycloalkanes. If so, then draw structural formulas for the *cis* and *trans* isomers.

a. Methylcyclohexane **b.** 1,1-Dimethylcyclohexane
c. 1,3-Dimethylcyclobutane **d.** 1-Ethyl-2-methylcyclobutane

Solution

a. *Cis–trans* isomerism is not possible because we do not have two substituents on the ring.

b. *Cis–trans* isomerism is not possible. We have two substituents on the ring, but they are on the same carbon atom. Each of two different carbons must bear substituents.

c. *Cis–trans* isomerism does exist.

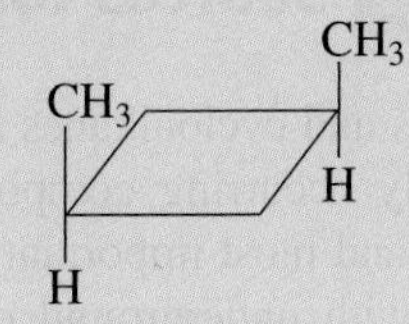

cis-1,3-Dimethylcyclobutane

CH_3, H, H, CH_3

trans-1,3-Dimethylcyclobutane

d. *Cis–trans* isomerism does exist.

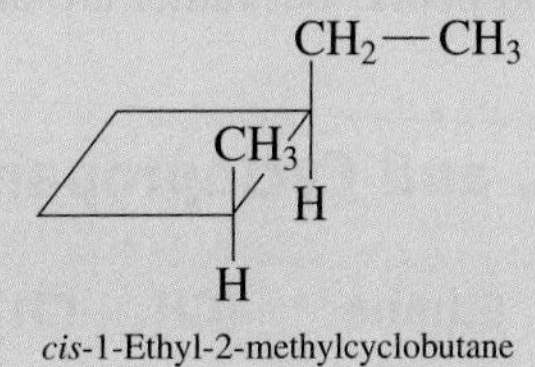

cis-1-Ethyl-2-methylcyclobutane

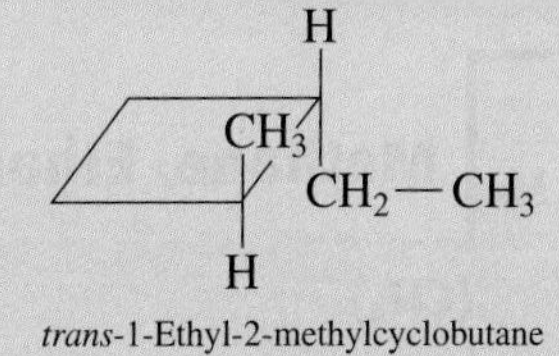

trans-1-Ethyl-2-methylcyclobutane

Practice Questions and Problems

10.40 Determine whether the members of each of the following pairs of structural formulas represent (1) different orientations of the same molecule, (2) different compounds that are structural isomers, or (3) different compounds that are not structural isomers.

a.

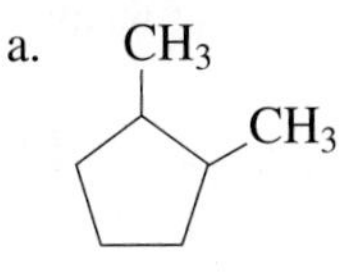

and CH_3, CH_3

b.

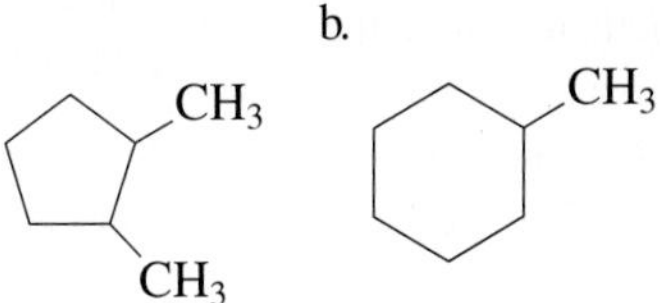

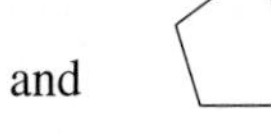

c.

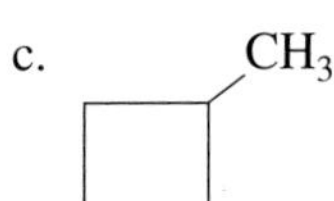

and

d.

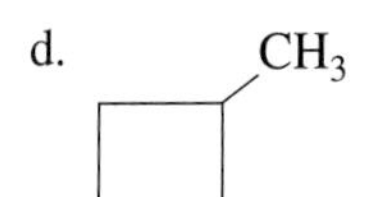

and CH_3

10.41 Determine whether *cis–trans* isomerism is possible for each of the following cycloalkanes. If so, then draw structural formulas for the *cis* and *trans* isomers.

a. 1-Ethyl-1-methylcyclopentane
b. Ethylcyclohexane
c. 1,4-Diethylcyclohexane
d. 1,1-Dimethylcyclooctane

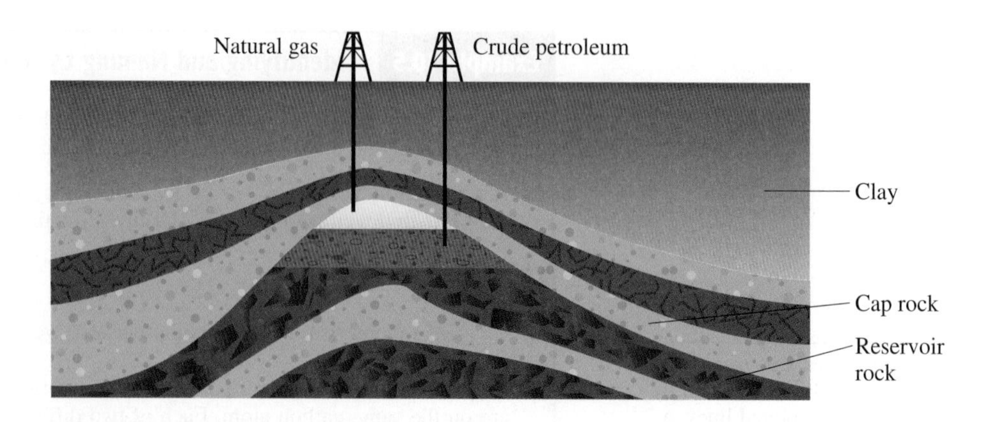

Figure 10.6 A rock formation such as this is necessary for the accumulation of petroleum and natural gas.

10.12 Sources for Alkanes and Cycloalkanes

Learning Focus

List the major natural sources for alkanes and cycloalkanes.

▶ The word *petroleum* comes from the Latin *petra,* which means "rock," and *oleum,* which means "oil."

Alkanes and cycloalkanes are not "laboratory curiosities" but rather extremely important naturally occurring compounds. Natural gas and petroleum (crude oil) constitute their largest and most important natural source. Deposits of these substances are usually associated with underground dome-shaped rock formations (Figure 10.6). When a hole is drilled into such a rock formation, it is possible to recover some of the trapped hydrocarbons (Figure 10.7). Petroleum and natural gas do not occur in the earth in the form of "liquid pools" but rather are dispersed throughout a porous rock formation.

Chemical Portraits 16

Methane, Ethane, and Cyclopropane

Methane (CH_4)

Profile: CH_4, the simplest hydrocarbon, is a colorless, odorless, non-poisonous, flammable gas. Its major sources are natural gas and petroleum. It is primarily used as a fuel for heating, lighting, and cooking. Commercial natural gas used in home heating is approximately 90%(v/v) CH_4. It is a major constituent of the atmospheres of the Sun's outer planets (Jupiter, Saturn, Uranus, and Pluto).

Biochemical considerations: Often called *marsh gas,* methane enters the atmosphere in significant amounts from the decomposition of animal and plant matter in an oxygen-deficient environment. It is a *greenhouse gas* that contributes to global warming. Bacteria found in the digestive tracts of termites and plant-eating animals have the ability to produce CH_4 from plant materials. The CH_4 output of a large cow can reach 20 liters per day.

As greenhouse gases, how do the contributions of carbon dioxide and methane compare?

Ethane ($CH_3—CH_3$)

Profile: Ethane, the two carbon alkane, is a colorless, odorless, flammable gas obtained directly from natural gas and petroleum. Commercial natural gas is about 10% ethane. Because methane and ethane are both odorless, a small amount of a "smelly" sulfur-containing compound is added to natural gas so consumers can detect a gas leak. The major non-fuel use for ethane is in preparing ethylene, which is the starting material for the production of many types of plastics.

Biochemical considerations: Chloroethane, an ethane molecule in which a chlorine atom has replaced a hydrogen atom, is a local anesthetic. When applied to the skin, its rapid evaporation lowers the skin temperature, deadening the local nerve endings for a brief period of time.

What is the relationship between ethane and polyethylene plastics?

Cyclopropane

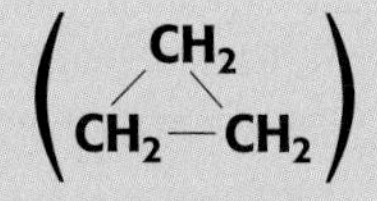

Profile: Cyclopropane, the simplest cycloalkane, is a colorless, flammable gas with a characteristic but not unpleasant odor resembling that of petroleum ether.

Biochemical considerations: Cyclopropane has seen extensive use as a general inhalation anesthetic. It is quick acting and the safety margin between anesthetic and lethal doses is wider than for most other anesthetics. The chief disadvantage in its use is its flammability; extra safety precautions must be taken. Nonflammable anesthetics, many of which are halogenated hydrocarbons, have now largely replaced cyclopropane. Introduction of halogen atoms into a hydrocarbon usually reduces its flammability.

Contrast the properties of cyclopropane with those of propane.

See the text web site at **www.cengage.com/chemistry/stoker** for answers to the above questions and for further information.

Figure 10.7 An oil rig pumping oil from an underground rock formation.

Unprocessed natural gas contains 50%–90% methane, 1%–10% ethane, and up to 8% higher-molecular-mass alkanes (predominantly propane and butanes). The higher alkanes found in crude natural gas are removed prior to release of the gas into the pipeline distribution systems. Because the removed alkanes can be liquefied by the use of moderate pressure, they are stored as liquids under pressure in steel cylinders and are marketed as bottled gas.

Crude petroleum is a complex mixture of hydrocarbons (both cyclic and acyclic) that can be separated into useful fractions through refining. During refining, the physical separation of the crude into component fractions is accomplished by fractional distillation, a process that takes advantage of boiling-point differences between the components of the crude petroleum. Each fraction contains hydrocarbons within a specific boiling-point range. The fractions obtained from a typical fractionation process are shown in Figure 10.8.

Chemical Portraits 16 profiles the two simplest alkanes, methane and ethane and also cyclopropane, the simplest cycloalkane.

Practice Problems and Questions

10.42 Indicate the largest natural sources for
a. alkanes b. cycloalkanes

10.43 What physical characteristic is associated with the rock formations in which natural gas and petroleum collect?

10.44 The physical separation of crude petroleum into its component fractions is accomplished by taking advantage of differences in what physical property of the various hydrocarbons present?

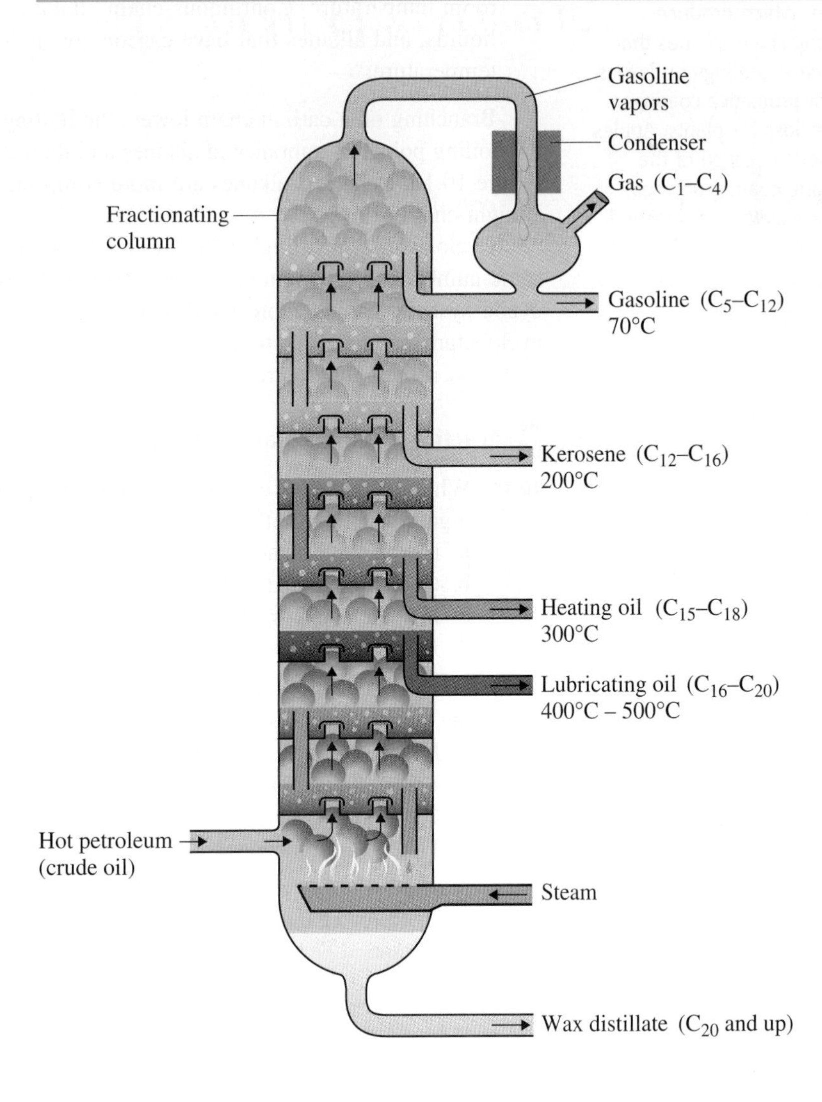

Figure 10.8 The complex hydrocarbon mixture present in petroleum is separated into simpler mixtures by means of a fractionating column.

Learning Focus

List generalizations about common physical properties of alkanes and cycloalkanes.

10.13 Physical Properties of Alkanes and Cycloalkanes

In this section, we consider a number of generalizations about the physical properties of alkanes and cycloalkanes.

1. *Alkanes and cycloalkanes are insoluble in water.* Water molecules are polar, and alkane and cycloalkane molecules are nonpolar. Molecules of unlike polarity have limited solubility in one another (Section 7.4). The water insolubility of alkanes makes them good preservatives for metals. They prevent water from reaching the metal surface and causing corrosion. They also have biological functions as protective coatings (see Figure 10.9).
2. *Alkanes and cycloalkanes have densities lower than that of water.* Alkane and cycloalkane densities fall in the range 0.6 g/mL to 0.8 g/mL, compared with water's density of 1.0 g/mL. When alkanes and cycloalkanes are mixed with water, two layers form (because of insolubility), with the hydrocarbon layer on top (because of its lower density). This density difference between alkanes/cycloalkanes and water explains why oil spills in aqueous environments spread so quickly. The *floating* oil follows the movement of the water.
3. *The boiling points of continuous-chain alkanes and cycloalkanes increase with an increase in carbon chain length or ring size.* For continuous-chain alkanes, the boiling point increases roughly 30°C for every carbon atom added to the chain. This trend, shown in Figure 10.10, is the result of increasing London force strength (Section 6.13). London forces become stronger as molecular surface area increases. Short, continuous-chain alkanes (1 to 4 carbon atoms) are gases at room temperature. Continuous-chain alkanes containing 5 to 17 carbon atoms are liquids, and alkanes that have carbon chains longer than this are solids at room temperature.

Figure 10.9 The insolubility of alkanes in water is used to advantage by many plants, which produce unbranched long-chain alkanes that serve as protective coatings on leaves and fruits. Such protective coatings minimize water loss for plants. Apples can be "polished" because of the long-chain alkane coating on their skin, which involves the unbranched alkanes $C_{27}H_{56}$ and $C_{29}H_{60}$. The leaf wax of cabbage and broccoli is mainly unbranched $C_{29}H_{60}$.

Branching on a carbon chain lowers the boiling point of an alkane. A comparison of the boiling points of unbranched alkanes and their 2-methyl-branched isomers is given in Figure 10.10. Branched alkanes are more compact, with smaller surface areas than their straight-chain isomers.

Cycloalkanes have higher boiling points than their noncyclic counterparts with the same number of carbon atoms (Figure 10.10). These differences are due in large part to cyclic systems having more rigid and more symmetrical structures. Cyclopropane and cyclobutane are gases at room temperature, and cyclopentane through cyclooctane are liquids at room temperature.

Practice Problems and Questions

10.45 Which member of each of the following pairs of saturated hydrocarbons has the higher boiling point?

a. Hexane and octane
b. Cyclobutane and cyclopentane
c. Pentane and 2-methylbutane
d. Pentane and cyclopentane

10.46 For which of the following pairs of compounds do both members of the pair have the same physical state (solid, liquid, or gas) at room temperature and pressure?

a. Ethane and hexane
b. Cyclopropane and butane
c. Cyclobutane and cyclopentane
d. Pentane and octane

10.47 Indicate whether each of the following saturated hydrocarbons has a density less than or greater than that of water.

a. Butane b. Cyclobutane
c. Methane d. 3-Methylpentane

Figure 10.10 Trends in normal boiling points for continuous-chain alkanes, 2-methyl branched alkanes, and unsubstituted cycloalkanes as a function of the number of carbon atoms present. Although you are not expected to memorize specific boiling points and specific melting points of hydrocarbons, you should develop a feel for general trends in these values. For example, in a series of alkanes, melting point increases as carbon chain length increases.

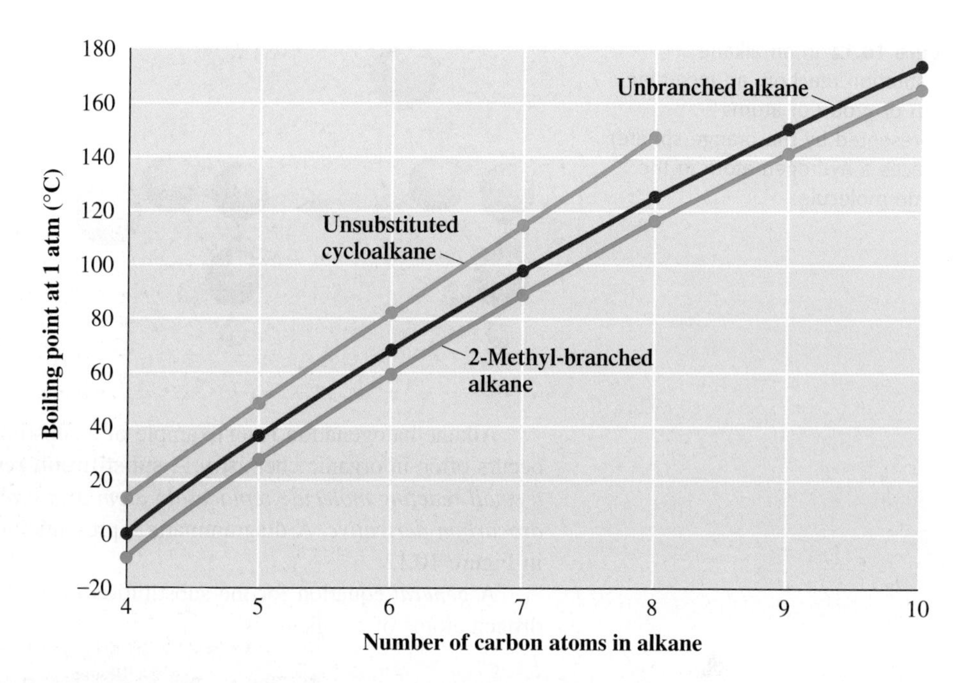

Learning Focus

Predict the products from alkane and cycloalkane combustion and halogenation reactions.

10.14 Chemical Properties of Alkanes and Cycloalkanes

Alkanes are the least reactive type of organic compound. They can be heated for long periods of time in strong acids and bases with no appreciable reaction. Strong oxidizing agents and reducing agents have little effect on alkanes.

Alkanes are not absolutely unreactive. Two important reactions that they undergo are combustion, which is reaction with oxygen, and halogenation, which is reaction with halogens.

▶ The term *paraffins* is an older name for the alkane family of compounds. This name comes from the Latin *parum affinis,* which means "little activity." That is a good summary of the general chemical properties of alkanes.

Combustion

A **combustion reaction** *is a reaction between a substance and oxygen (usually from air) that proceeds with the evolution of heat and light (usually as a flame).* Alkanes readily undergo combustion when ignited. When sufficient oxygen is present to support total combustion, carbon dioxide and water are the products.

$$CH_4 + 2O_2 \longrightarrow CO_2 + 2H_2O + \text{energy}$$

$$2C_6H_{14} + 19O_2 \longrightarrow 12CO_2 + 14H_2O + \text{energy}$$

The exothermic nature (Section 8.5) of alkane combustion reactions explains the extensive use of alkanes as fuels. Natural gas is predominantly methane. Propane is used for home heating and in gas barbecue units (see Figure 10.11). Butane fuels portable camping stoves. Gasoline is a complex mixture of many alkanes and other types of hydrocarbons.

Figure 10.11 Propane fuel tank on a home barbecue unit.

Halogenation

The halogens are the elements in Group VIIA of the periodic table: fluorine (F_2), chlorine (Cl_2), bromine (Br_2), and iodine (I_2) (Section 3.4). A **halogenation reaction** *is a reaction between a substance and a halogen in which one or more halogen atoms are incorporated into molecules of the substance.*

Halogenation of an alkane produces a hydrocarbon derivative in which one or more halogen atoms have been substituted for hydrogen atoms. An example of an alkane halogenation reaction is

$$\text{H-}\underset{\text{H}}{\overset{\text{H}}{\text{C}}}\text{-}\underset{\text{H}}{\overset{\text{H}}{\text{C}}}\text{-H} + Br_2 \xrightarrow[\text{light}]{\text{Heat or}} \text{H-}\underset{\text{H}}{\overset{\text{H}}{\text{C}}}\text{-}\underset{\text{H}}{\overset{\text{H}}{\text{C}}}\text{-Br} + HBr$$

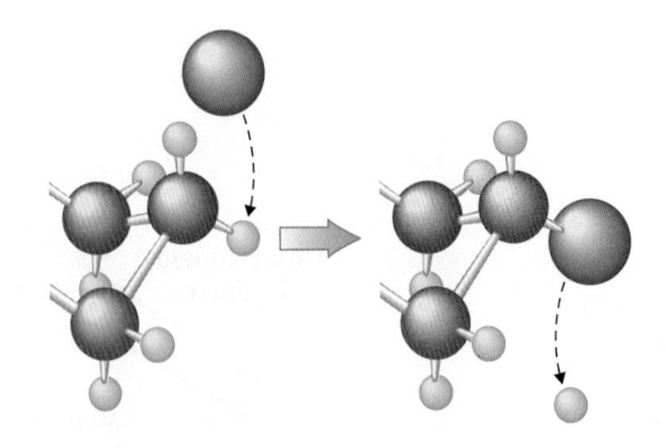

Figure 10.12 In an alkane substitution reaction, an incoming atom or group of atoms (represented by the orange sphere) replaces a hydrogen atom in the alkane molecule.

Alkane halogenation is an example of a substitution reaction, a type of reaction that occurs often in organic chemistry. A **substitution reaction** *is a reaction in which part of a small reacting molecule replaces an atom or a group of atoms on a hydrocarbon or hydrocarbon derivative.* A diagrammatic representation of a substitution reaction is shown in Figure 10.12.

A *general* equation for the substitution of a single halogen atom for one of the hydrogen atoms of an alkane is

$$\underset{\text{Alkane}}{R{-}H} + \underset{\text{Halogen}}{X_2} \xrightarrow[\text{light}]{\text{Heat or}} \underset{\text{Halogenated alkane}}{R{-}X} + \underset{\text{Hydrogen halide}}{H{-}X}$$

▶ Occasionally, it is useful to represent alkyl groups in a nonspecific way. The symbol R is used for this purpose. Just as *city* is a generic term for Chicago, New York, or San Francisco, the symbol R is a generic designation for any alkyl group. The symbol R comes from the German word *radikal,* which means, in a chemical context, "molecular fragment."

Note the following features of this general equation:

1. The notation R—H is a general formula for an alkane. R— is the general symbol for an alkyl group. Addition of a hydrogen atom to an alkyl group produces the parent hydrocarbon of the alkyl group.
2. The notation R—X on the product side is the general formula for a halogenated alkane. X is the general symbol for a halogen atom.
3. Reaction conditions are noted by placing these conditions on the equation arrow that separates reactants from products. Halogenation of an alkane requires the presence of heat or light.

In halogenation of an alkane, the alkane is said to undergo *fluorination, chlorination, bromination,* or *iodination,* depending on the identity of the halogen reactant. Chlorination and bromination are the two widely used alkane halogenation reactions. Fluorination reactions generally proceed too quickly to be useful, and iodination reactions go too slowly.

Halogenation usually results in the formation of a mixture of products rather than a single product. More than one product results because more than one hydrogen atom on an alkane can be replaced with halogen atoms. To illustrate this concept, let us consider the chlorination of methane, the simplest alkane.

Methane and chlorine, when heated to a high temperature or in the presence of light, react as follows:

$$CH_4 + Cl_2 \xrightarrow[\text{light}]{\text{Heat or}} CH_3Cl + HCl$$

The reaction does not stop at this stage, however, because the chlorinated methane product can react with additional chlorine to produce polychlorinated products.

$$CH_3Cl + Cl_2 \xrightarrow[\text{light}]{\text{Heat or}} CH_2Cl_2 + HCl$$

$$CH_2Cl_2 + Cl_2 \xrightarrow[\text{light}]{\text{Heat or}} CHCl_3 + HCl$$

$$CHCl_3 + Cl_2 \xrightarrow[\text{light}]{\text{Heat or}} CCl_4 + HCl$$

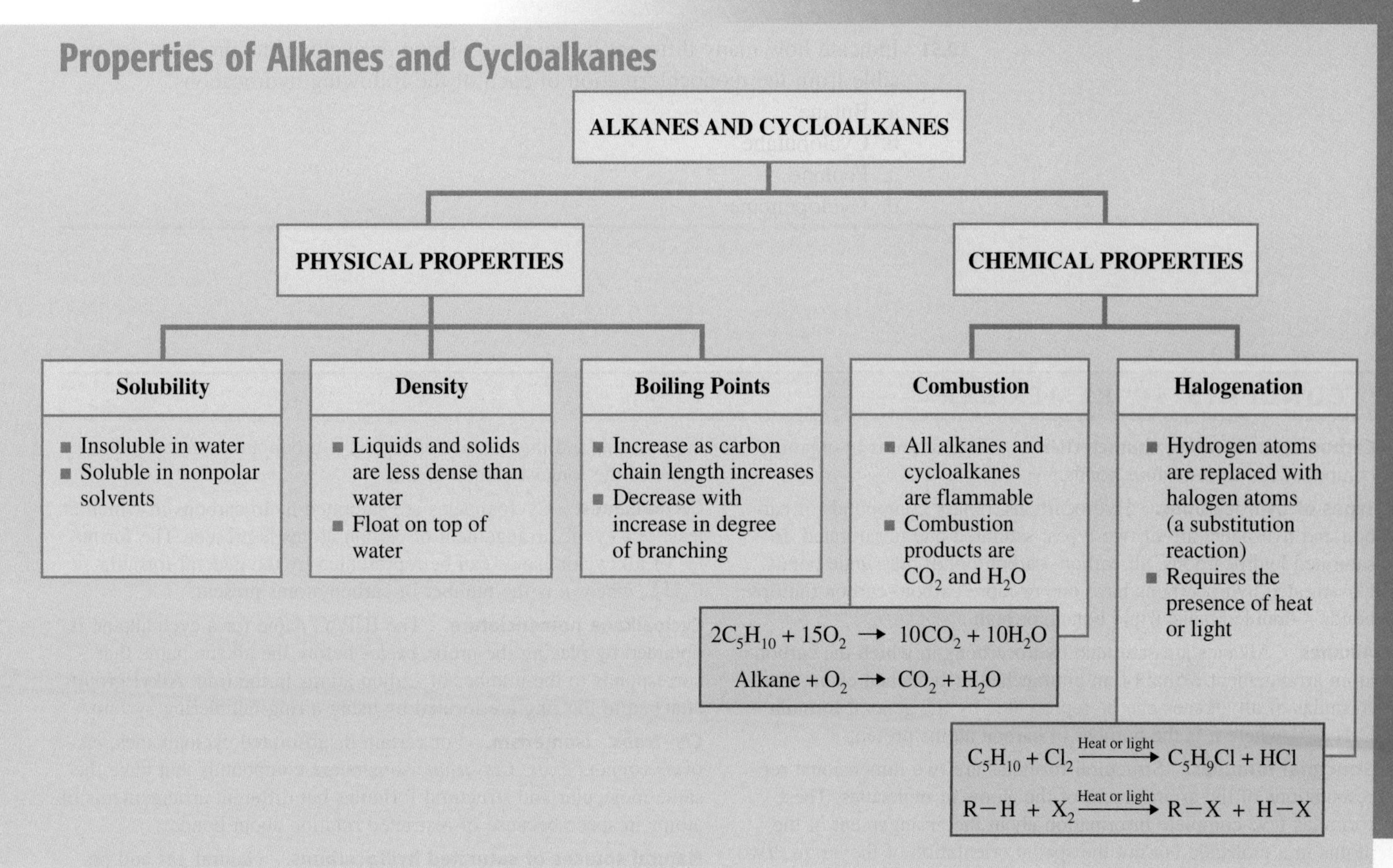

By controlling the reaction conditions and the ratio of chlorine to methane, it is possible to *favor* formation of one or another of the possible chlorinated methane products.

The chemical properties of cycloalkanes are similar to those of alkanes. Cycloalkanes readily undergo combustion as well as chlorination and bromination. With unsubstituted cycloalkanes, monohalogenation produces a single product, because all hydrogen atoms present in the cycloalkane are equivalent to one another.

$$\text{(cyclohexane)} + Br_2 \xrightarrow[\text{light}]{\text{Heat or}} \text{(cyclohexane–Br)} + HBr$$

The subject of halogenated hydrocarbons will occupy our attention again in Chapter 12. At that time, IUPAC nomenclature, uses, and properties of this type of hydrocarbon derivative will be considered.

The accompanying Chemistry at a Glance summarizes the properties (both physical and chemical) of alkanes and cycloalkanes.

Practice Problems and Questions

10.48 Write the formulas for the two products from the complete combustion of each of the following alkanes or cycloalkanes.

a. C_3H_8 b. $CH_3{—}(CH_2)_7{—}CH_3$ c. Cyclopentane d. 3-Methyloctane

10.49 Write molecular formulas for all possible halogenated products from the bromination of methane.

10.50 The monochlorination of propane produces two halogenated alkanes. Write structural formulas for these two compounds.

10.51 Indicate how many different halogenated alkane or cycloalkane products are possible from the monochlorination of each of the following hydrocabons.
a. Butane
b. Cyclobutane
c. Pentane
d. Cyclopentane

CONCEPTS TO REMEMBER

Carbon atom bonding characteristics. Carbon atoms in organic compounds must have four bonds.

Types of hydrocarbons. Hydrocarbons, binary compounds of carbon and hydrogen, are of two types: saturated and unsaturated. In saturated hydrocarbons, all carbon–carbon bonds are single bonds. Unsaturated hydrocarbons have one or more carbon–carbon multiple bonds—double bonds, triple bonds, or both.

Alkanes. Alkanes are saturated hydrocarbons in which the carbon atom arrangement is that of an unbranched or branched chain. The formulas of all alkanes can be represented by the general formula C_nH_{2n+2}, where n is the number of carbon atoms present.

Structural formulas. Structural formulas are two-dimensional representations of the arrangement of the atoms in molecules. These formulas give complete information about the arrangement of the atoms in a molecule but not the spatial orientation of the atoms. Two types of structural formulas are commonly encountered: expanded and condensed.

Structural isomerism. Structural isomers are two or more compounds that have the same molecular formula but different structural formulas—that is, different arrangements of atoms within the molecule.

Conformations. Conformations are differing orientations of the same molecule made possible by free rotation about single bonds in the molecule.

Alkane nomenclature. The IUPAC name for an alkane is based on the longest continuous chain of carbon atoms in the molecule. A group of carbon atoms attached to the chain is an alkyl group. Both the position and the identity of the alkyl group are prefixed to the name of the longest carbon chain.

Cycloalkanes. Cycloalkanes are saturated hydrocarbons in which at least one cyclic arrangement of carbon atoms is present. The formulas of all cycloalkanes can be represented by the general formula C_nH_{2n}, where n is the number of carbon atoms present.

Cycloalkane nomenclature. The IUPAC name for a cycloalkane is obtained by placing the prefix *cyclo-* before the alkane name that corresponds to the number of carbon atoms in the ring. Alkyl groups attached to the ring are located by using a ring-numbering system.

***Cis–trans* isomerism.** For certain disubstituted cycloalkanes, *cis–trans* isomers exist. *Cis–trans* isomers are compounds that have the same molecular and structural formulas but different arrangements of atoms in space because of restricted rotation about bonds.

Natural sources of saturated hydrocarbons. Natural gas and petroleum are the largest and most important natural sources of both alkanes and cycloalkanes.

Physical properties of saturated hydrocarbons. Saturated hydrocarbons are not soluble in water and have lower densities than water. Melting and boiling points increase with increasing carbon chain length or ring size.

Chemical properties of saturated hydrocarbons. Two important reactions that saturated hydrocarbons undergo are combustion and halogenation. In combustion, saturated hydrocarbons burn in air to produce CO_2 and H_2O. Halogenation is a substitution reaction in which one or more hydrogen atoms of the hydrocarbon are replaced by halogen atoms.

KEY REACTIONS AND EQUATIONS

1. Combustion (rapid reaction with O_2) of alkanes (Section 10.14)

$$\text{Alkane} + O_2 \longrightarrow CO_2 + H_2O$$

2. Halogenation of alkanes (Section 10.14)

$$R{-}H + X_2 \xrightarrow[\text{light}]{\text{Heat or}} R{-}X + H{-}X$$

KEY TERMS

Alkane (10.4)
Alkyl group (10.8)
Branched-chain alkane (10.6)
***Cis* isomer** (10.11)
***Cis–trans* isomers** (10.11)
Combustion reactions (10.14)
Condensed structural formula (10.5)
Conformations (10.7)
Continuous-chain alkane (10.6)
Cycloalkane (10.9)
Expanded structural formula (10.5)
Halogenation reaction (10.14)
Hydrocarbon (10.3)
Hydrocarbon derivative (10.3)
Line–angle drawing (10.9)
Organic chemistry (10.1)
Saturated hydrocarbon (10.3)
Structural isomers (10.6)
Substituent (10.8)
Substitution reaction (10.14)
***Trans* isomer** (10.11)
Unsaturated hydrocarbon (10.3)

ADDITIONAL PROBLEMS

10.52 On the basis of the general formula for an alkane, determine the following for specific alkanes.

a. Number of hydrogen atoms present when 8 carbon atoms are present
b. Number of carbon atoms present when 10 hydrogen atoms are present
c. Number of carbon atoms present when 41 total atoms are present
d. Total number of covalent bonds present in the molecule when 7 carbon atoms are present

10.53 On the basis of the general formula for a cycloalkane, determine the following for specific cycloalkanes.

a. Number of hydrogen atoms present when 8 carbon atoms are present
b. Number of carbon atoms present when 10 hydrogen atoms are present
c. Number of carbon atoms present when 42 total atoms are present
d. Total number of covalent bonds present in the molecule when 7 carbon atoms are present

10.54 Convert each of the following expanded structural formulas into a condensed structural formula.

a.
```
  H H H H H H
  | | | | | |
H-C-C-C-C-C-C-H
  | | | | | |
  H H H H H H
```

b.
```
  H H H H
  | | | |
H-C-C-C-C-H
  | | | |
  H | H H
H-C-H
  |
  H
```

c.
```
  H H   H   H H
  | |   |   | |
H-C-C---C---C-C-H
  | |   |   | |
  H |   H   | H
H-C-H     H-C-H
  |         |
  H         H
```

d.
```
  H H H H H H
  | | | | | |
H-C-C-C-C-C-C-H
  | | | | | |
  H H | H H H
    H-C-H
      |
    H-C-H
      |
      H
```

10.55 Draw the indicated type of formula for the following alkanes.

a. The expanded structural formula for a continuous-chain alkane with four carbon atoms
b. The expanded structural formula for $CH_3—(CH_2)_3—CH_3$
c. The condensed structural formula, using parentheses for the $—CH_2—$ groups, for the straight-chain alkane $C_{12}H_{26}$
d. The molecular formula for the alkane $CH_3—(CH_2)_6—CH_3$

10.56 Explain why the following alkane names are *not* correct IUPAC names. Then give the correct IUPAC name.

a. 4-Methylpentane
b. 2-Ethyl-2-methylpropane
c. 2,3,3-Trimethylbutane
d. 2-Methyl-4-methylhexane

10.57 What is wrong with each of the following attempts to name a cycloalkane using IUPAC rules?

a. Dimethylcyclohexane
b. 3,4-Dimethylcyclobutane
c. 1-Ethylcyclobutane
d. 2-Ethyl-1-methylcyclopentane

10.58 Indicate whether the members of each of the following pairs of compounds are structural isomers.

a. Hexane and 2-methylhexane
b. Hexane and 2,2-dimethylbutane
c. Hexane and methylcyclopentane
d. Hexane and cyclohexane

10.59 There are four dimethylcyclohexane structural isomers. Give the IUPAC names of the four isomers.

10.60 There are five alkane structural isomers with the formula C_6H_{14}. Give the IUPAC names of the five isomers.

10.61 Answer the following questions about the unbranched alkane with seven carbon atoms.

a. How many hydrogen atoms are present?
b. How many carbon–carbon bonds are present?
c. How many carbon atoms have two hydrogen atoms bonded to them?
d. How many total covalent bonds are present?
e. Is it a solid, a liquid, or a gas at room temperature?
f. Is it less dense or more dense than water?
g. Is it soluble or insoluble in water?
h. Is it flammable or inflammable in air?

PRACTICE TEST ▶ True/False

10.62 In organic compounds, carbon atoms always form four bonds.

10.63 In a saturated hydrocarbon, the arrangement of carbon atoms may be acyclic or cyclic.

10.64 The general molecular formula for all alkanes and cycloalkanes is C_nH_{2n+2}.

10.65 All of the bonds present in methane are carbon–hydrogen bonds.

10.66 Structural isomers have the same structural formula but different molecular formulas.

10.67 The number of alkane structural isomers possible increases with increasing number of carbon atoms in the alkane.

10.68 Hexane and octane are both branched-chain alkanes.

10.69 The compound 4-ethyl-2,7-dimethyloctane contains 13 carbon atoms.

10.70 The simplest cycloalkane is cycloethane.

10.71 Cycloalkanes with alkyl groups as substituents are not known.

10.72 *Cis–trans* isomers have the same molecular formula and the same structural formula.

10.73 Natural gas and petroleum are important natural sources of both alkanes and cycloalkanes.

10.74 "Liquid pools" of natural gas and petroleum occur in certain locations within the earth.

10.75 Alkanes are not soluble in water, but cycloalkanes are.

10.76 Halogenation of an alkane is an example of a substitution reaction.

PRACTICE TEST ▸ Multiple Choice

10.77 Which of the following bonding "behaviors" is *not* possible for a carbon atom in an organic compound?
a. Formation of four single bonds
b. Formation of two double bonds
c. Formation of a single bond and a triple bond
d. Formation of a double bond and a triple bond

10.78 The classification *saturated* hydrocarbon includes
a. alkanes but not cycloalkanes
b. cycloalkanes but not alkanes
c. both alkanes and cycloalkanes
d. neither alkanes nor cycloalkanes

10.79 All of the bonds within an alkane are explicitly shown in
a. its condensed structural formula
b. its expanded structural formula
c. its molecular formula
d. both its condensed structural formula and its expanded structural formula

10.80 Which of the following is a *correct* alkane structural formula?
a. $CH_3-CH_2-CH_3$ b. $CH_3-CH_3-CH_2$
c. $CH_2-CH_2-CH_2$ d. $CH_2-CH_3-CH_2$

10.81 Which of the following alkanes has the IUPAC name 2-methylbutane?
a. $CH_3-\underset{\underset{\displaystyle CH_3}{|}}{CH}-CH_3$
b. $CH_3-CH_2-\underset{\underset{\displaystyle CH_3}{|}}{CH}-CH_3$
c. $CH_3-\underset{\underset{\displaystyle CH_3}{|}}{CH}-\underset{\underset{\displaystyle CH_3}{|}}{CH}-CH_3$
d. $CH_3-\underset{\underset{\displaystyle CH_3}{|}}{CH}-CH_2-CH_2-CH_3$

10.82 How many hydrogen atoms are present in the molecule 3,5-dimethyloctane?
a. 18 b. 20 c. 22 d. 24

10.83 In which of the following pairs of alkanes are the two members of the pair structural isomers?
a. Hexane and 2-methylhexane
b. 2,3-Dimethylhexane and 2,3-dimethylheptane
c. Butane and pentane
d. 2-Methylpentane and 2,3-dimethylbutane

10.84 When the molecular formulas for cyclic and noncyclic alkanes with the same number of carbon atoms are compared, it is always found that the cycloalkane has
a. two more hydrogen atoms
b. the same number of hydrogen atoms as the noncyclic alkane
c. two fewer hydrogen atoms
d. four fewer hydrogen atoms

10.85 For which of the following cycloalkanes is *cis–trans* isomerism possible?
a. Isopropylcyclobutane
b. Propylcyclopropane
c. 1-Methyl-1-propylcyclopentane
d. 1-Ethyl-4-methylcyclohexane

10.86 Which of the following is a general physical property of alkanes?
a. Not soluble in water
b. More dense than water
c. Liquid at room temperature
d. Very polar molecules

CHAPTER OUTLINE

CHAPTER ELEVEN

Unsaturated Hydrocarbons

When acetylene, an unsaturated hydrocarbon, is burned with oxygen in an oxyacetylene welding torch, a temperature high enough to cut metals is produced.

Two general types of hydrocarbons exist: *saturated and unsaturated.* Saturated hydrocarbons, which were discussed in Chapter 10, include the *alkanes* and *cycloalkanes*. All bonds in saturated hydrocarbons are single bonds. Unsaturated hydrocarbons, the other type of hydrocarbon, are the subject of this chapter. One or more carbon–carbon multiple bonds are present in such compounds. There are three classes of unsaturated hydrocarbons: the *alkenes,* the *alkynes,* and the *aromatic hydrocarbons,* all of which we will consider.

11.1 Unsaturated Hydrocarbons

Learning Focus

Define the term *unsaturated hydrocarbon* and identify the functional groups present in these compounds.

An **unsaturated hydrocarbon** *is a hydrocarbon that contains one or more carbon–carbon multiple bonds: double bonds, triple bonds, or both.* Unsaturated hydrocarbons have *physical* properties similar to those of saturated hydrocarbons. However, their *chemical* properties are much different. Unsaturated hydrocarbons are chemically more reactive than their saturated counterparts. The increased reactivity of unsaturated hydrocarbons is related to the presence of the carbon–carbon multiple bond(s) in such compounds. These multiple bonds serve as locations where chemical reactions can occur.

Whenever a specific portion of a molecule governs its chemical properties, that portion of the molecule is called a functional group. A **functional group** *is the part of a molecule where most of its chemical reactions occur.* Carbon–carbon multiple bonds are the functional group for an unsaturated hydrocarbon.

▶ The field of organic chemistry is organized in terms of functional groups.

▶ Alkanes and cycloalkanes (Chapter 10) lack functional groups; as a result, they are relatively unreactive.

The study of various functional groups and their respective reactions provides the organizational structure for organic chemistry. Each of the organic chemistry chapters that follow introduces new functional groups that characterize families of hydrocarbon derivatives.

Unsaturated hydrocarbons are subdivided into three groups on the basis of the type of multiple bond(s) present: (1) *alkenes,* which contain one or more carbon–carbon double bonds, (2) *alkynes,* which contain one or more carbon–carbon triple bonds, and (3) *aromatic hydrocarbons,* which exhibit a special type of "delocalized" bonding that involves a six-membered carbon ring (to be discussed in Section 11.8).

We begin our consideration of unsaturated hydrocarbons with a discussion of alkenes. Information about alkynes and aromatic hydrocarbons then follows.

Practice Problems and Questions

11.1 Indicate whether each of the following are *similar* or *different.*
a. Physical properties of saturated and unsaturated hydrocarbons
b. Chemical properties of saturated and unsaturated hydrocarbons

11.2 What is a *functional group*?

11.3 What is the functional group present in each of the following types of hydrocarbons?
a. Alkene b. Alkyne c. Aromatic hydrocarbon d. Alkane

11.2 Characteristics of Alkenes and Cycloalkenes

Learning Focus

Define the terms *alkene* and *cycloalkene,* and give the general formulas for alkenes and cycloalkenes that contain only one double bond.

An **alkene** *is an acyclic unsaturated hydrocarbon that contains one or more carbon–carbon double bonds.* Note the close similarity between the family names *alkene* and *alkane* (Section 10.4), they differ only in their endings: *-ene* versus *-ane.* The *-ene* ending means a double bond is present.

The simplest type of alkene contains only one carbon–carbon double bond. Such compounds have the general formula C_nH_{2n}. Thus alkenes with one double bond have two fewer hydrogen atoms than do alkanes (C_nH_{2n+2}).

The two simplest alkenes are ethene (C_2H_4) and propene (C_3H_6).

$$CH_2{=}CH_2 \qquad CH_2{=}CH{-}CH_3$$

Ethene Propene

▶ The general formula for an alkene with one double bond, C_nH_{2n}, is the same as that for a cycloalkane (Section 10.9). Thus such alkenes and cycloalkanes with the same number of carbon atoms are isomeric with one another.

▶ An older but still widely used name for alkenes is *olefins,* pronounced "oh-la-fins." The term *olefin* means "oil-forming." Many alkenes react with Cl_2 to form "oily" compounds.

Comparing the geometrical shape of ethene with that of methane (the simplest alkane) reveals a major difference. The arrangement of bonds about the carbon atom in methane is tetrahedral (Section 10.4), whereas the carbon atoms in ethene have a trigonal planar arrangement of bonds; that is, they form a flat, triangle-shaped arrangement (see Figure 11.1). The two carbon atoms participating in a double bond and the four other atoms attached to these two carbon atoms always lie in a plane with a trigonal planar arrangement of atoms about each carbon atom of the double bond. Such an arrangement of atoms is consistent with the principles of VSEPR theory (Section 4.14).

A **cycloalkene** *is a cyclic unsaturated hydrocarbon that contains one or more carbon–carbon double bonds within the ring system.* Cycloalkenes in which there is only one double bond have the general formula C_nH_{2n-2}. This general formula reflects the loss of four hydrogen atoms from that of an alkane (C_nH_{2n+2}). Note that two hydrogen atoms are lost because of the double bond and two because of the ring structure.

The simplest cycloalkene is the compound cyclopropene (C_3H_4), a three-membered carbon ring system containing one double bond.

H H H H C C C or

Cyclopropene

Figure 11.1 Three-dimensional representations of the structures of ethene and methane. In ethene, the atoms are in a flat (planar) rather than a tetrahedral arrangement. Bond angles are 120°.

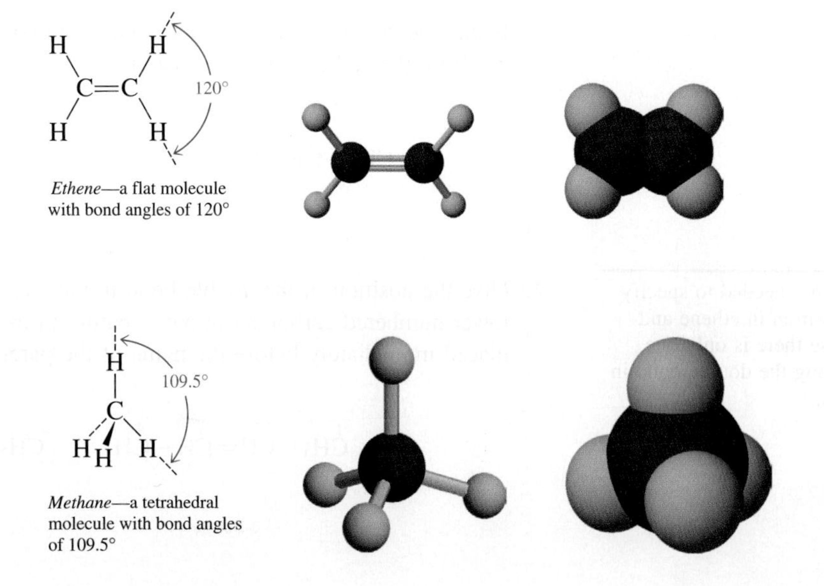

Alkenes with more than one carbon–carbon double bond are relatively common. When two double bonds are present, the compounds are often called *dienes*. Cycloalkenes that contain more than one double bond are possible but are not common.

Practice Problems and Questions

11.4 The general formula for an alkane is C_nH_{2n+2}. What is the general formula for an alkene with one double bond?

11.5 The general formula for a cycloalkane is C_nH_{2n}. What is the general formula for a cycloalkene with one double bond?

11.6 Describe the arrangement of atoms, in terms of bond angles, for ethene, the simplest alkene.

11.7 What is the relationship between the terms *olefin* and *alkene*?

Learning Focus

Given their structures, name alkenes and cycloalkenes using IUPAC rules, and vice versa.

11.3 Names for Alkenes and Cycloalkenes

The rules previously presented for naming alkanes and cycloalkanes (Sections 10.8 and 10.10) can be used, with some modification, to name alkenes and cycloalkenes.

1. Replace the alkane suffix *-ane* with the suffix *-ene,* which is used to indicate the presence of a carbon–carbon double bond.
2. Select as the parent carbon chain the longest continuous chain of carbon atoms that *contains both carbon atoms of the double bond.* For example, select

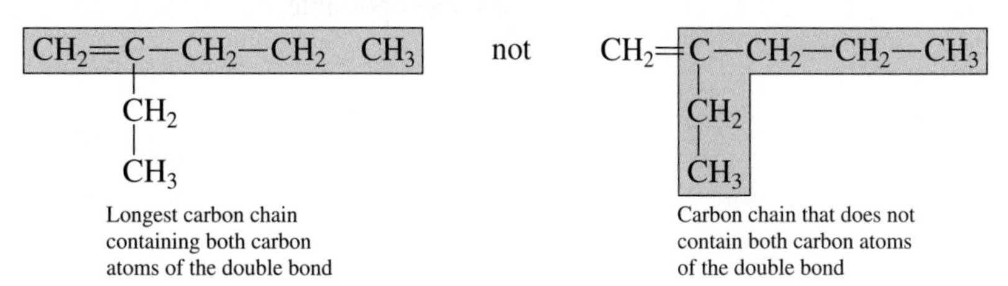

▶ Carbon–carbon double bonds take precedence over alkyl groups in determining the direction in which the parent carbon chain is numbered.

3. Number the parent carbon chain beginning at the end nearest the double bond.

$$\overset{1}{C}H_3-\overset{2}{C}H=\overset{3}{C}H-\overset{4}{C}H_2-\overset{5}{C}H_3 \quad \text{not} \quad \overset{5}{C}H_3-\overset{4}{C}H=\overset{3}{C}H-\overset{2}{C}H_2-\overset{1}{C}H_3$$

If the double bond is equidistant from both ends of the parent chain, begin numbering from the end closer to a substituent.

$$\overset{4}{C}H_3-\overset{3}{C}H=\overset{2}{C}H-\underset{\substack{|\\ Cl}}{\overset{1}{C}H_2} \quad \text{not} \quad \overset{1}{C}H_3-\overset{2}{C}H=\overset{3}{C}H-\underset{\substack{|\\ Cl}}{\overset{4}{C}H_2}$$

▶ A number is not needed to specify double bond position in ethene and propene, because there is only one way of positioning the double bond in these molecules.

4. Give the position of the double bond in the chain as a *single* number, which is the lower-numbered carbon atom participating in the double bond. This number is placed immediately before the name of the parent carbon chain.

$$\overset{1}{C}H_3-\overset{2}{C}H=\overset{3}{C}H-\overset{4}{C}H_3 \qquad \overset{1}{C}H_2=\overset{2}{C}H-\underset{\substack{|\\ CH_3}}{\overset{3}{C}H}-\overset{4}{C}H_3$$

2-Butene 3-Methyl-1-butene

5. Use the suffixes *-diene, -triene, -tetrene,* and so on when more than one double bond is present in the molecule. A separate number must be used to locate each double bond.

$$\overset{1}{C}H_2=\overset{2}{C}H-\overset{3}{C}H=\overset{4}{C}H_2 \qquad \overset{1}{C}H_2=\overset{2}{C}H-\underset{\substack{|\\ CH_3}}{\overset{3}{C}H}-\overset{4}{C}H=\overset{5}{C}H_2$$

1,3-Butadiene 3-Methyl-1,4-pentadiene

6. Do not use a number to locate the double bond in unsubstituted cycloalkenes with only one double bond, because that bond is assumed to be between carbons 1 and 2.
7. In substituted cycloalkenes with only one double bond, the double-bonded carbon atoms are numbered 1 and 2 in the direction (clockwise or counterclockwise) that gives the first-encountered substituent the lower number. Again, no number is used in the name to locate the double bond.

Cyclohexene 4-Methylcyclohexene

8. In cyclic hydrocarbon systems with more than one double bond within the ring, assign one double bond the numbers 1 and 2 and the other double bonds the lowest numbers possible.

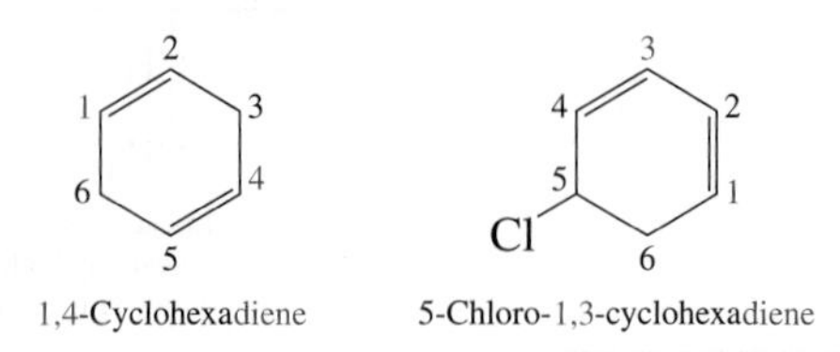

1,4-Cyclohexadiene 5-Chloro-1,3-cyclohexadiene

Example 11.1 **Assigning IUPAC Names to Alkenes and Cycloalkenes**

Assign IUPAC names to the following alkenes and cycloalkenes.

a. $CH_3—CH{=}CH—CH_2—CH_2—CH_3$

b. $CH_3—CH_2—C{=}CH_2$ with $CH_2—CH_3$ on C

c. (cyclobutene ring with CH_3)

d. (cyclopentadiene ring with CH_3)

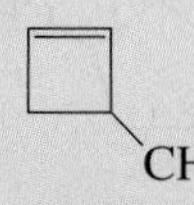

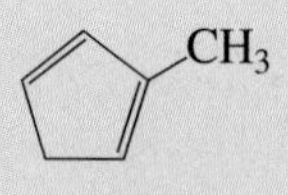

Solution

a. The carbon chain in this hexene is numbered from the end closest to the double bond.

$$\overset{1}{C}H_3—\overset{2}{C}H{=}\overset{3}{C}H—\overset{4}{C}H_2—\overset{5}{C}H_2—\overset{6}{C}H_3$$

The complete IUPAC name is 2-hexene.

b. The longest carbon chain containing *both* carbons of the double bond has four carbon atoms. Thus we have a butene.

$$\boxed{CH_3—CH_2—C{=}CH_2}$$ with $CH_2—CH_3$ on C

The chain is numbered from the end closest to the double bond. The complete IUPAC name is 2-ethyl-1-butene.

c. This compound is a methylcyclobutene. The numbers 1 and 2 are assigned to the carbon atoms of the double bond, and the ring is numbered clockwise, which results in a carbon 3 location for the methyl group. (Counterclockwise numbering would have placed the methyl group on carbon 4.) The complete IUPAC name of the cycloalkene is 3-methylcyclobutene. The double bond is understood to involve carbons 1 and 2.

d. A ring system containing five carbon atoms, two double bonds, and a methyl substituent on the ring is called a methylcyclopentadiene. Two different numbering systems produce the same locations (carbons 1 and 3) for the double bonds.

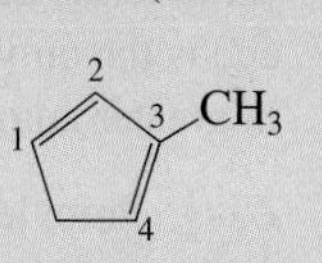

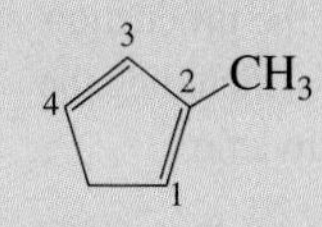

(ring numbered 1, 2, 3 (CH_3), 4) (ring numbered 1, 2 (CH_3), 3, 4)

The counterclockwise numbering system assigns the lower number to the methyl group. The complete IUPAC name of the compound is 2-methyl-1,3-cyclopentadiene.

Common Names

The simpler members of most families of organic compounds have common names in addition to IUPAC names. Often these common names predate the advent of systematic IUPAC rules for naming organic compounds. It would be nice if such common names did not exist, but they do. Hence there are two names for many simple organic compounds, and we must be familiar with both of them.

The two simplest alkenes, ethene and propene, have common names that you should be familiar with. They are, respectively, ethylene and propylene.

$CH_2{=}CH_2$
Ethylene
(IUPAC name: ethene)

$CH_2{=}CH—CH_3$
Propylene
(IUPAC name: propene)

Chemical Portraits 17 profiles ethylene, propylene, and the five carbon diene isoprene (2-methyl-1,3-butadiene).

Practice Problems and Questions

11.8 Assign an IUPAC name to each of the following alkenes.

a. $CH_3—CH{=}CH—CH_3$

b. $CH_3—CH_2—CH{=}CH—CH_3$

c. $CH_3—CH_2—C(CH_3){=}CH—CH_3$

d. $CH_3—C(CH_3){=}CH—CH(CH_3)—CH_3$

11.9 Assign an IUPAC name to each of the following cycloalkenes.

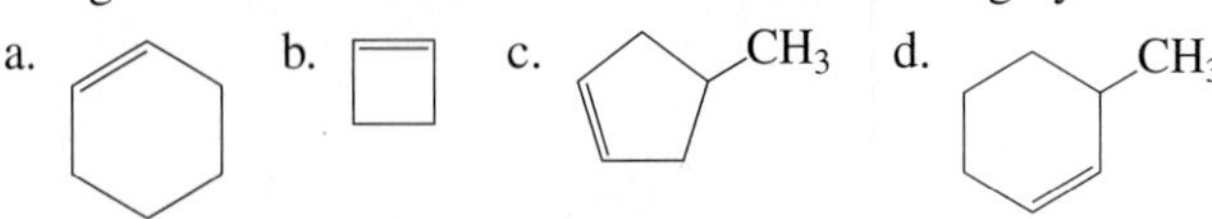

11.10 Assign an IUPAC name to each of the following dienes.

a. $CH_2{=}CH—CH{=}CH_2$

b. $CH_2{=}C(CH_3)—CH{=}CH_2$

c. $CH_2{=}C(CH_3)—C(CH_3){=}CH_2$

d. $CH(CH_3){=}CH—CH{=}CH_2$

11.11 Assign an IUPAC name to each of the following cyclodienes.

a. b. c. CH_3 d. CH_3

Chemical Portraits 17 Simple Alkenes with One or Two Double Bonds

Ethylene (Ethene) ($CH_2{=}CH_2$)

Profile: Ethylene, the simplest unsaturated hydrocarbon, is a colorless, flammable gas with a slightly sweet odor. It is a petrochemical, that is, it is made from petroleum products. Industrially produced C_2H_4 is the starting material for the production of many plastics and fibers, including the well-known plastic polyethylene.

Biochemical considerations: Ethylene occurs naturally in small amounts (a few parts per million) in plants where it functions as a plant hormone. It stimulates the ripening of fruit. In the commercial fruit industry, bananas, tomatoes, and some citrus fruits are picked green to prevent spoiling and bruising during transportation to markets. At their destination, the fruits and vegetables are exposed to C_2H_4 to stimulate ripening.

What is the process for the production of polyethylene from ethylene?

Propylene (Propene) ($CH_2{=}CH—CH_3$)

Profile: Propylene, a colorless, flammable gas under normal conditions, exhibits mild anesthetic properties. It is the second-most important petrochemical in terms of amount produced; ethylene is the only petrochemical produced in greater amounts.

Uses: Propylene's most important use is as the starting material for the production of the plastic polypropylene, a substance that can be formulated to be much harder than polyethylene. The diverse uses for this plastic include indoor-outdoor carpet fibers, bottles, and molded parts (including car battery cases, replacements for metal parts in automobiles, and even heart valve components).

What is the basis for propylene's classification as a petrochemical?

Isoprene (2-methyl-1,3-butadiene)

Profile: Isoprene, a five-carbon diene, is a volatile, fragrant, naturally-occurring liquid at normal conditions. It is the "building block" for numerous natural products, called *terpenes*, which give plants their characteristic fragrances. Terpenes are formed by joining isoprene units in a "head-to-tail" manner. The number of carbon atoms present in a terpene is always a multiple of the number 5.

Biochemical considerations: The terpene called beta-carotene, the molecule responsible for the yellow-orange color of carrots, apricots, and yams, is an eight-unit isoprene molecule. In the human body, beta-carotene serves as a precursor of vitamin A; splitting a beta-carotene molecule in half produces two vitamin A molecules. Beta-carotene also has antioxidant functions in the human body.

What is the structure of limonene (oil of orange), a compound containing two isoprene units?

See the text web site at **www.cengage.com/chemistry/stoker** for answers to the above questions and for further information.

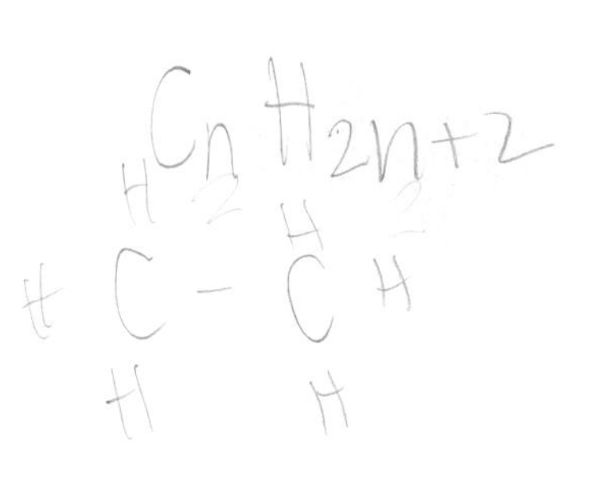

11.12 Draw a condensed structural formula for each of the following unsaturated hydrocarbons.

a. Ethene b. Propene c. Ethylene d. Propylene

11.13 Draw a condensed structural formula for each of the following unsaturated hydrocarbons.

a. 1-Pentene b. 2-Pentene c. 1,3-Pentadiene d. 1,4-Pentadiene

11.14 Draw a condensed structural formula for each of the following unsaturated hydrocarbons.

a. 3-Methyl-1-pentene b. 3-Methylcyclopentene
c. 3-Ethyl-1,4-hexadiene d. 3-Ethyl-1,4-cyclohexadiene

Learning Focus

Recognize, identify, and name alkene *cis–trans* isomers.

▶ Structural isomers that differ only in the location of the functional group are often called *positional isomers*. The compounds 1-butene and 2-butene are positional isomers.

$CH_2{=}CH{-}CH_2{-}CH_3$
1-Butene

$CH_3{-}CH{=}CH{-}CH_3$
2-Butene

11.4 Isomerism in Alkenes

Structural isomerism is possible for alkenes, just as it was for alkanes (Section 10.6). Table 11.1 compares structural isomer possibilities for four- and five-carbon alkane and alkene systems. The potential for structural isomers is greater for alkenes than for alkanes, because there is more than one location where the double bond can be placed in systems containing four or more carbon atoms.

Cis–trans isomerism (Section 10.11) is possible for some alkenes. Such isomerism results from structural rigidity associated with carbon–carbon double bonds. Just as a carbon ring prevents free rotations of atoms (Section 10.11), so does a carbon–carbon double bond.

To determine whether an alkene has *cis* and *trans* isomers, draw the alkene structure in a manner that emphasizes the four attachments to the double-bonded carbon atoms.

$$\rangle C{=}C\langle$$

Table 11.1 A Comparison of Structural Isomerism Possibilities for Four- and Five-Carbon Alkane and Alkene Systems

Four-carbon alkanes (two isomers)	Four-carbon alkenes (three isomers)	
$CH_3{-}CH_2{-}CH_2{-}CH_3$ Butane	$CH_2{=}CH{-}CH_2{-}CH_3$ 1-Butene	$CH_2{=}C{-}CH_3$ CH_3 2-Methyl-1-propene
$CH_3{-}CH{-}CH_3$ CH_3 2-Methylpropane	$CH_3{-}CH{=}CH{-}CH_3$ 2-Butene	
Five-carbon alkanes (three isomers)	**Five-carbon alkenes (five isomers)**	
$CH_3{-}CH_2{-}CH_2{-}CH_2{-}CH_3$ Pentane	$CH_2{=}CH{-}CH_2{-}CH_2{-}CH_3$ 1-Pentene	$CH_2{=}CH{-}CH{-}CH_3$ CH_3 3-Methyl-1-butene
$CH_3{-}CH{-}CH_2{-}CH_3$ CH_3 2-Methylbutane	$CH_3{-}CH{=}CH{-}CH_2{-}CH_3$ 2-Pentene	$CH_3{-}C{=}CH{-}CH_3$ CH_3 2-Methyl-2-butene
CH_3 $CH_3{-}C{-}CH_3$ CH_3 2,2-Dimethylpropane	$CH_2{=}C{-}CH_2{-}CH_3$ CH_3 2-Methyl-1-butene	

Figure 11.2 *Cis–trans* isomers: Different representations of the *cis* and *trans* isomers of 2-butene.

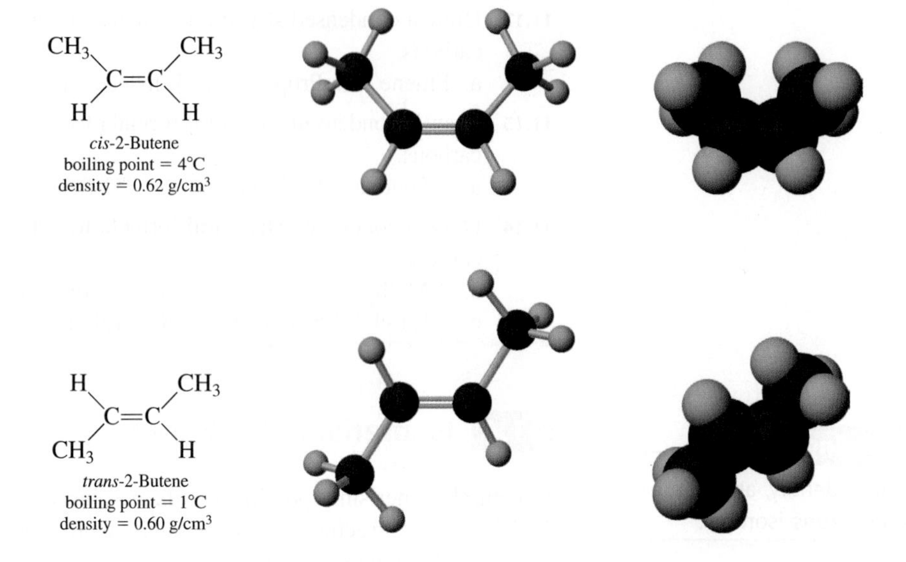

▶ The double bond of alkenes, like the ring of cycloalkanes, imposes rotational restrictions.

If *each* of the two carbons of the double bond has two *different* groups attached to it, *cis* and *trans* isomers exist.

Two groups are different → CH_3, H; C=C; $CH_2—CH_3$, H ← Two groups are different

The simplest alkene for which *cis* and *trans* isomers exist is 2-butene.

▶ Recall, from Section 10.11, that *cis* means "on the same side" and *trans* means "across."

CH_3 CH_3 / C=C / H H — Structure A (*cis*-2-butene)

CH_3 H / C=C / H CH_3 — Structure B (*trans*-2-butene)

In structure A, the *cis* isomer, both methyl groups are on the same side of the double bond. In structure B, the *trans* isomer, the methyl groups are on opposite sides of the double bond. The only way to convert structure A to structure B is to break the double bond. At room temperature, such bond breaking does not occur. Hence these two structures represent two different compounds (*cis–trans* isomers) that differ in boiling point, density, and so on. Figure 11.2 shows three-dimensional representations of the *cis* and *trans* isomers of 2-butene.

Cis–trans isomerism is not possible when one of the double-bonded carbons bears two identical groups. Thus neither 1-butene nor 2-methylpropene is capable of existing in *cis* and *trans* forms.

Two identical groups {H, H}; C=C; $CH_2—CH_3$, H — 1-Butene

Two identical groups {H, H}; C=C; {CH_3, CH_3} Two identical groups — 2-Methylpropene

Practice Problems and Questions

11.15 Structurally, how does 1-butene differ from 2-butene?

11.16 Indicate whether each of the following pairs of alkenes are structural isomers.

a. 1-Butene and 2-butene
b. 1-Butene and 1-pentene
c. 2-Methyl-1-butene and 2-methyl-2-butene
d. 2-Methyl-1-butene and 1-pentene

11.17 Structurally, how does *cis*-2-butene differ from *trans*-2-butene?

11.18 Indicate whether each of the following structures is (1) a *cis isomer,* (2) a *trans isomer,* or (3) neither a *cis* isomer nor a *trans* isomer.

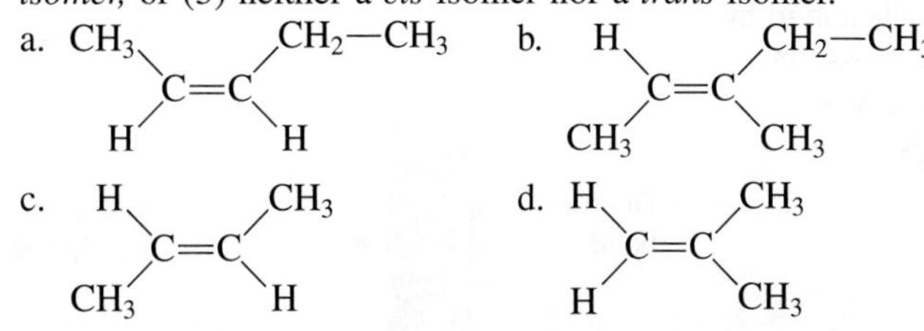

11.19 Assign an IUPAC name to each of the molecules in Problem 11.18.

11.20 For each of the following compounds, tell whether *cis–trans* isomers exist.

a. $CH_3—CH_2—CH{=}CH_2$
b. $CH_3—CH_2—C(CH_3){=}CH_2$
c. $CH_2{=}CH—CH_3$
d. $CH_3—C(CH_3){=}CH—CH_3$

11.21 Draw a structural formula for each of the following compounds.

a. *trans*-3-Methyl-3-hexene
b. *cis*-2-Pentene
c. *trans*-5-Methyl-2-heptene
d. *cis*-4-Methyl-2-pentene

Learning Focus

Describe the general physical properties of alkenes, and write equations and draw structures of products in both symmetrical and unsymmetrical addition reactions of alkenes, using Markovnikov's rule when necessary.

11.5 Physical and Chemical Properties of Alkenes

The general physical properties of alkenes include insolubility in water, solubility in nonpolar solvents, and densities lower than that of water. Thus alkenes have physical properties similar to those of alkanes (Section 10.13). The melting point of an alkene is usually lower than that of the alkane with the same number of carbon atoms.

Alkenes with 2 to 4 carbon atoms are gases at room temperature. Unsubstituted alkenes with 5 to 17 carbon atoms and one double bond are liquids, and those with still more carbon atoms are solids.

Alkenes, like alkanes, are very flammable. The combustion products, as with any hydrocarbon, are carbon dioxide and water.

$$\underset{\text{Ethene}}{C_2H_4} + 3O_2 \longrightarrow 2CO_2 + 2H_2O$$

Pure alkenes are, however, too expensive to be used as fuel.

Aside from combustion, almost all other reactions of alkenes take place at the carbon–carbon double bond(s). Such reactions are called *addition reactions* because a substance is *added* to the double bond. This behavior contrasts with that of alkanes, where the most common reaction type, aside from combustion, is *substitution* (Section 10.14).

An **addition reaction** *is a reaction in which atoms or groups of atoms are added to each carbon atom of a carbon–carbon multiple bond.* A general equation for an alkene addition reaction is

$$>C{=}C< \; + \; A—B \longrightarrow —\underset{A}{\overset{|}{\underset{|}{C}}}—\underset{B}{\overset{|}{\underset{|}{C}}}—$$

In this reaction, the A part of the reactant A—B becomes attached to one carbon atom of the double bond, and the B part to the other carbon atom (see Figure 11.3). As this occurs, the carbon–carbon double bond simultaneously becomes a carbon–carbon single bond.

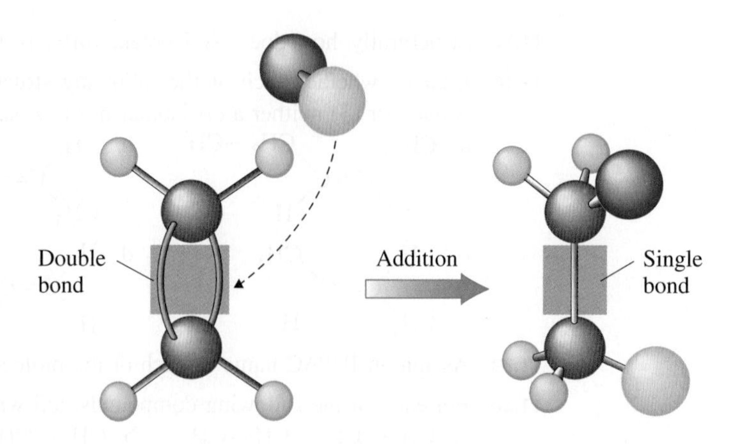

Figure 11.3 In an alkene addition reaction, the atoms provided by an incoming molecule are attached to the carbon atoms originally joined by a double bond. In the process, the double bond becomes a single bond.

▶ The following word associations are important to remember:
alkane—substitution reaction
alkene—addition reaction
An analogy can be drawn to a basketball team. When a *substitution* is made, one player leaves the game as another enters. The number of players on the court remains at five per team. If *addition* were allowed during a basketball game, two players could enter the game and no one would leave; there would be seven players per team on the court rather than five.

Addition reactions can be classified as symmetrical or unsymmetrical. In a **symmetrical addition reaction,** *identical atoms (or groups of atoms) are added to each carbon of the multiple bond.* In an **unsymmetrical addition reaction,** *different atoms (or groups of atoms) are added to the carbon atoms of the multiple bond.*

Symmetrical Addition Reactions

The two most common examples of symmetrical addition reactions are hydrogenation and halogenation.

Hydrogenation of an alkene involves the addition of a hydrogen atom to each carbon of the double bond. Hydrogenation is usually accomplished by heating the alkene and H_2 in the presence of a catalyst.

$$\underset{\text{Propene}}{CH_2{=}CH{-}CH_3} + H_2 \xrightarrow[\substack{150°C \\ 12\text{–}15\ \text{atm} \\ \text{pressure}}]{\text{Ni or Pt}} \underset{\text{Propane}}{\underset{|}{\overset{H}{CH_2}}{-}\underset{|}{\overset{H}{CH}}{-}CH_3}$$

▶ Hydrogenation of an alkene requires a catalyst. No reaction occurs if the catalyst is not present.

The identity of the catalyst used in hydrogenation is specified by writing it above the arrow in the chemical equation. In this case, Ni or Pt is commonly used as the catalyst. In general terms, hydrogenation of an alkene can be written as

$$\underset{\text{Alkene}}{>C{=}C<} + H_2 \xrightarrow[\substack{\text{Heat,} \\ \text{pressure}}]{\text{Ni or Pt}} \underset{\text{Alkane}}{-\overset{H}{\underset{|}{\overset{|}{C}}}-\overset{H}{\underset{|}{\overset{|}{C}}}-}$$

The hydrogenation of vegetable oils is a very important commercial process today. Vegetable oils from sources such as soybeans and cottonseeds are composed of long-chain organic molecules that contain several double bonds. When these oils are hydrogenated, they are converted to low-melting solids that are used in margarines and shortenings.

Halogenation of an alkene involves the addition of a halogen atom to each carbon of the double bond. Halogenation of alkenes most often involves Cl_2 or Br_2. No catalyst is needed. The product is always a dihalogenated alkene derivative.

$$\underset{\text{2-Butene}}{CH_3{-}CH{=}CH{-}CH_3} + Cl_2 \longrightarrow \underset{\text{Dihalogenated alkane}}{CH_3{-}\overset{Cl}{\overset{|}{CH}}{-}\overset{Cl}{\overset{|}{CH}}{-}CH_3}$$

Figure 11.4 A bromine in water solution is reddish brown (left). When a small amount of such a solution is added to an unsaturated hydrocarbon, the added solution is decolorized as the bromine adds to the hydrocarbon to form colorless dibromo compounds (right).

In general terms, halogenation of an alkene can be written as

$$\underset{\text{Alkene}}{>C=C<} + \underset{\text{Halogen}}{X_2} \longrightarrow \underset{\text{Dihalogenated alkane}}{-\overset{X}{\overset{|}{\underset{|}{C}}}-\overset{X}{\overset{|}{\underset{|}{C}}}-} \quad (X = Cl, Br)$$

Bromination is often used to test for the presence of carbon–carbon double bonds in organic substances. Bromine in water or carbon tetrachloride is reddish brown. The dibromo compound(s) formed from the symmetrical addition of bromine to an organic compound is (are) colorless. Thus the decolorization of a Br_2 solution indicates the presence of carbon–carbon double bonds (see Figure 11.4).

Unsymmetrical Addition Reactions

Two important types of unsymmetrical addition reactions are hydrohalogenation and hydration.

Hydrohalogenation of an alkene involves the addition of the reactant HCl, HBr, or HI to a carbon–carbon double bond. Hydrohalogenation reactions require no catalyst. For *symmetrical* alkenes, such as ethene, only one product results from hydrohalogenation.

$$\underset{\text{Ethene}}{CH_2{=}CH_2} + H{-}Cl \longrightarrow \underset{\text{A halogenated hydrocarbon}}{\overset{H}{\overset{|}{CH_2}}{-}\overset{Cl}{\overset{|}{CH_2}}}$$

Hydration of an alkene involves the addition of the reactant H_2O to a carbon–carbon double bond. Hydration requires a small amount of H_2SO_4 (sulfuric acid) as a catalyst. For *symmetrical* alkenes, only one product results from hydration.

▶ The addition of water to carbon–carbon double bonds occurs in many biochemical reactions that take place in the human body.

$$\underset{\text{Ethene}}{CH_2{=}CH_2} + H{-}OH \xrightarrow{H_2SO_4} \underset{\text{An alcohol}}{\overset{H}{\overset{|}{CH_2}}{-}\overset{OH}{\overset{|}{CH_2}}}$$

In this equation, the water (H_2O) is written as H—OH to emphasize how this molecule adds to the double bond (the H goes to one carbon atom and the OH to the other carbon atom). Note also that the product of this hydration reaction contains an —OH group. Hydrocarbon derivatives of this type are called *alcohols*. Such compounds are considered in Chapter 12.

When the alkene involved in a hydrohalogenation or hydration reaction is itself *unsymmetrical,* more than one product is possible. (An unsymmetrical alkene is one in which the two carbon atoms of the double bond are not equivalently substituted.) For example, the addition of HCl to propene (an unsymmetrical alkene) could produce either 1-chloropropane or 2-chloropropane, depending on whether the H from the HCl attaches itself to carbon 2 or carbon 1.

$$\underset{\text{Propene}}{CH_2{=}CH{-}CH_3} + HCl \longrightarrow \underset{\text{1-Chloropropane}}{\overset{Cl}{\overset{|}{C}}H_2{-}\overset{H}{\overset{|}{C}}H{-}CH_3}$$

or

$$\underset{\text{Propene}}{CH_2{=}CH{-}CH_3} + HCl \longrightarrow \underset{\text{2-Chloropropane}}{\overset{H}{\overset{|}{C}}H_2{-}\overset{Cl}{\overset{|}{C}}H{-}CH_3}$$

When two isomeric products are possible, one product often predominates. The dominant product can be predicted by using Markovnikov's rule, named after the Russian chemist Vladimir V. Markovnikov (see Figure 11.5). **Markovnikov's rule** states that *when an unsymmetrical molecule of the form* HQ *adds to an unsymmetrical alkene, the hydrogen atom from the* HQ *becomes attached to the unsaturated carbon atom that already has the most hydrogen atoms.* Thus the major product in our example involving propene is 2-chloropropane.

▶ Two catchy summaries of Markovnikov's rule are "Hydrogen goes where hydrogen is" and "The rich get richer" (in terms of hydrogen).

Example 11.2 Predicting Products in Alkene Addition Reactions Using Markovnikov's Rule

Using Markovnikov's rule, predict the predominant product in each of the following addition reactions.

a. $CH_3{-}CH_2{-}CH_2{-}CH{=}CH_2 + HBr \rightarrow$

b. (1-methylcyclopentene ring)$-CH_3 + HCl \rightarrow$

c. $CH_3{-}CH{=}CH{-}CH_2{-}CH_3 + HBr \rightarrow$

Solution

a. The hydrogen atom will add to carbon 1, because carbon 1 already contains more hydrogen atoms than carbon 2. The predominant product of the addition will be 2-bromopentane.

$$CH_3{-}CH_2{-}CH_2{-}\overset{②}{C}H{=}\overset{①}{C}H_2 + HBr \longrightarrow CH_3{-}CH_2{-}CH_2{-}\overset{Br}{\overset{|}{C}}H{-}\overset{H}{\overset{|}{C}}H_2$$

b. Carbon 1 of the double bond does not have any H atoms directly attached to it. Carbon 2 of the double bond has one H atom (H atoms are not shown in the structure but are implied) attached to it. The H atom from the HCl will add to carbon 2, giving 1-chloro-1-methylcyclopentane as the product.

(1-methylcyclopentene, with carbon ① bearing CH_3 and carbon ② bearing H) $+ HCl \longrightarrow$ (cyclopentane ring with CH_3 and Cl on carbon 1, H and H on carbon 2)

Figure 11.5 Vladimir Vasilevich Markovnikov (1837–1904). A professor of chemistry at several Russian universities, Markovnikov (pronounced Mar-cove-na-coff) synthesized rings containing four carbon atoms and seven carbon atoms, thereby disproving the notion of the day that carbon could form only five- and six-membered rings.

c. Each carbon atom of the double bond in this molecule has one hydrogen atom. Thus Markovnikov's rule does not favor either carbon atom. The result is two isomeric products that are formed in almost equal quantities.

$$\underset{\text{2-Bromopentane}}{CH_3-\underset{\displaystyle Br}{\underset{|}{CH}}-CH_2-CH_2-CH_3} \quad \text{and} \quad \underset{\text{3-Bromopentane}}{CH_3-CH_2-\underset{\displaystyle Br}{\underset{|}{CH}}-CH_2-CH_3}$$

In compounds that contain more than one carbon–carbon double bond, such as dienes and trienes, addition can occur at each of the double bonds. In the complete hydrogenation of a diene and in that of a triene, the amount of hydrogen needed is twice as much and three times as much, respectively, as that needed for the hydrogenation of an alkene with one double bond.

$$CH_2{=}CH-CH_2-CH_2-CH_2-CH_3 + 1H_2 \xrightarrow{Ni} CH_3-(CH_2)_4-CH$$

$$CH_2{=}CH-CH{=}CH-CH_2-CH_3 + 2H_2 \xrightarrow{Ni} CH_3-(CH_2)_4-CH_3$$

$$CH_2{=}CH-CH{=}CH-CH{=}CH_2 + 3H_2 \xrightarrow{Ni} CH_3-(CH_2)_4-CH_3$$

Example 11.3 Predicting Reactants and Products in Alkene Addition Reactions

Supply the structural formula of the missing substance in each of the following addition reactions.

a. $CH_3-CH_2-CH{=}CH_2 + H_2O \xrightarrow{H_2SO_4} ?$

b. $? + Br_2 \rightarrow$ (cyclohexane ring with Br, Br, CH_3)

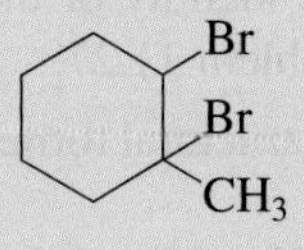

c. 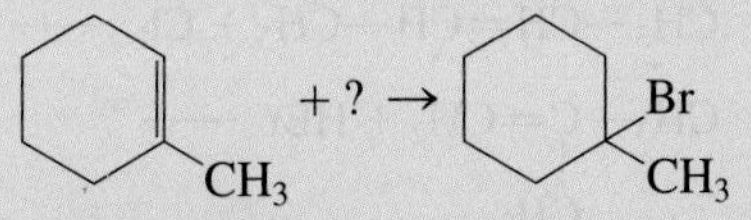 (cyclohexene ring with CH_3) $+ ? \rightarrow$ (cyclohexane ring with Br, CH_3)

d. $CH_3-CH{=}CH-CH{=}CH_2 + 2H_2 \xrightarrow{Ni} ?$

Solution

a. This is a hydration reaction. Using Markovnikov's rule, we determine that the H will become attached to carbon 1, which has more hydrogen atoms than carbon 2, and that the —OH group will be attached to carbon 2.

$$CH_3-CH_2-\overset{\displaystyle OH}{\overset{|}{CH}}-CH_3$$

b. The reactant alkene will have to have a double bond between the two carbon atoms that bromine atoms are attached to in the product.

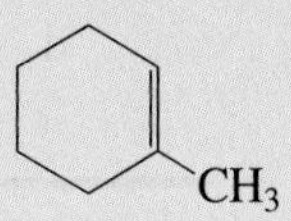

(cyclohexene ring with CH_3)

c. The small reactant molecule that adds to the double bond is HBr. The added Br atom from the HBr is explicitly shown in the product's structural formula, but the added H atom is not shown.

d. Hydrogen will add at each of the double bonds. The product hydrocarbon is pentane.

$$CH_3{-}CH_2{-}CH_2{-}CH_2{-}CH_3$$

Practice Problems and Questions

11.22 Indicate whether each of the following alkenes would be expected to be a solid, a liquid, or a gas at room temperature and pressure.
a. Ethene b. Propene c. 1-Hexene d. 1-Octene

11.23 What is the difference between an addition reaction and a substitution reaction?

11.24 Which of the following are addition reactions?
a. $C_4H_8 + Cl_2 \longrightarrow C_4H_8Cl_2$
b. $C_6H_6 + Cl_2 \longrightarrow C_6H_5Cl + HCl$
c. $C_3H_6 + HCl \longrightarrow C_3H_7Cl$
d. $C_7H_{16} \longrightarrow C_7H_8 + 4H_2$

11.25 When alkenes are hydrogenated, to which carbon atoms do the incoming hydrogen atoms become attached?

11.26 Why is Markovnikov's rule sometimes needed to predict the major product in the hydration of an alkene but never needed to predict the major product in the hydrogenation of an alkene?

11.27 Draw the structure of the major organic product when propene reacts with each of the following substances.
a. H_2 b. Cl_2 c. H_2O d. HBr

11.28 Indicate the identity of the catalyst needed, if any, to carry out each of the reactions in Problem 11.27.

11.29 Supply the structural formula of the product in each of the following alkene addition reactions.
a. $CH_3{-}CH{=}CH{-}CH_3 + Cl_2 \longrightarrow ?$
b. $CH_3{-}\underset{\displaystyle CH_3}{\underset{|}{C}}{=}CH_2 + HBr \longrightarrow ?$
c. $CH_3{-}CH_2{-}CH{=}CH_2 + HCl \longrightarrow ?$
d. (cyclopentene) $+ H_2 \xrightarrow[\text{catalyst}]{\text{Ni}} ?$

11.30 How many molecules of H_2 gas will react with one molecule of each of the following unsaturated hydrocarbons in a complete hydrogenation reaction?
a. $CH_3{-}CH{=}CH{-}CH{=}CH{-}CH_3$
b. (cyclobutene ring with CH_3 substituent)
c. (cyclohexene ring with $CH{=}CH_2$ substituent)
d. $CH_3{-}CH{=}C{=}\underset{\displaystyle CH_3}{\underset{|}{C}}{-}CH{=}CH_2$

Learning Focus

List common polymers of alkenes and the alkenes monomers from which they are formed.

11.6 Polymers of Alkenes

A **polymer** *is a large molecule formed by the repetitive bonding together of many smaller molecules*. The smaller repeating units of the polymer are called *monomers*. A **monomer** *is the small molecule that is repeated many times in a polymer*. The process by which a polymer is made is called *polymerization*. A **polymerization reaction** *is a reaction in which the repetitious combining of many small molecules (monomers) produces a very large molecule (the polymer)*. With appropriate catalysts, simple alkenes and simple substituted alkenes readily undergo polymerization.

The type of polymer that alkenes and substituted alkenes form is an *addition polymer*. An **addition polymer** *is a polymer in which the monomers simply "add together" with no other products formed besides the polymer*. Addition polymerization is similar to the addition reactions described in Section 11.5 except that there is no reactant other than the alkene or substituted alkene.

▶ The word *polymer* comes from the Greek *poly*, which means "many," and *meros*, which means "parts."

The simplest alkene addition polymer has ethylene (ethene) as the monomer. With appropriate catalysts, ethylene readily adds to itself to produce polyethylene.

$$5\ \mathrm{H_2C{=}CH_2} \ (\text{drawn as five } \mathrm{C{=}C} \text{ units, each C bearing two H}) \xrightarrow{\text{Catalyst}} -\mathrm{CH_2}-\mathrm{CH_2}-\mathrm{CH_2}-\mathrm{CH_2}-\mathrm{CH_2}-\mathrm{CH_2}-\mathrm{CH_2}-\mathrm{CH_2}-\mathrm{CH_2}-\mathrm{CH_2}-$$

Polyethylene

An *exact* formula for a polymer such as polyethylene cannot be written, because the length of the carbon chain varies from polymer molecule to polymer molecule. In recognition of this "inexactness" of formula, the notation used for denoting polymer formulas is independent of carbon chain length. We write the formula of the simplest repeating unit (the monomer with the double bond changed to a single bond) in parentheses and then add the subscript *n* after the parentheses, with *n* being understood to represent a very large number. Using this notation, we have, for the formula of polyethylene,

$$\left(-\mathrm{CH_2}-\mathrm{CH_2}- \right)_n$$

This notation clearly identifies the basic repeating unit found in the polymer.

Many substituted alkenes undergo polymerization in a manner similar to that for ethene when treated with the proper catalyst. For a monosubstituted-ethene monomer, the general polymerization equation is

$$\mathrm{H_2C{=}CH{-}Z} \xrightarrow{\text{Polymerization}} \left(\mathrm{CH_2}-\mathrm{CH(Z)} \right)_n$$

Variation in the substituent group Z can change polymer properties dramatically, as is shown by the entries in Table 11.2, a listing of monomers for and uses for the ethene-based polymers polyethylene, polypropylene, poly(vinyl chloride) [PVC], Teflon, and polystyrene. Figure 11.6 depicts the preparation of polystyrene.

Figure 11.6 When styrene, $C_6H_5{-}CH{=}CH_2$, is heated with a catalyst (benzoyl peroxide), it yields a viscous liquid. After some time, this liquid sets to a hard plastic (sample shown at left).

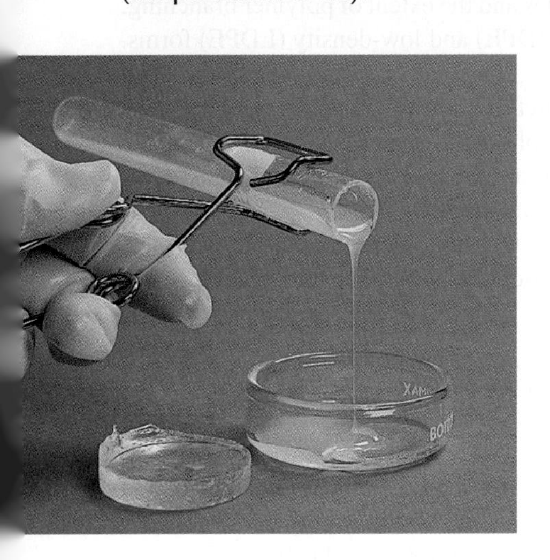

Table 11.2 Some Common Polymers Obtained from Ethene-Based Monomers

Polymer formula and name	Monomer formula and name	Uses of polymer
$-\left(CH_2-CH_2\right)_n-$ polyethylene	$CH_2=CH_2$ ethylene	bottles, plastic bags, toys, electrical insulation
$-\left(CH_2-CH(CH_3)\right)_n-$ polypropylene	$CH_2=CH(CH_3)$ propylene	indoor-outdoor carpeting, bottles, molded parts (including heart valves)
$-\left(CH_2-CHCl\right)_n-$ poly(vinyl chloride) (PVC)	$CH_2=CHCl$ vinyl chloride	plastic wrap, bags for intravenous drugs, garden hose, plastic pipe, simulated leather (Naugahyde)
$-\left(CF_2-CF_2\right)_n-$ Teflon	$CF_2=CF_2$ tetrafluoroethylene	cooking utensil coverings, electrical insulation, component of artificial joints in body parts replacement
$-\left(CH_2-CH(C_6H_5)\right)_n-$ polystyrene	$CH_2=CH(C_6H_5)$ styrene	toys, styrofoam packaging, cups, simulated wood furniture

The properties of an ethene-based polymer depend not only on monomer identity but also on the average size (length) of polymer molecules and the extent of polymer branching. For example, polyethylene has both high-density (HDPE) and low-density (LDPE) forms. HDPE, which has nonbranched polymer molecules, is a rigid material used in threaded bottle caps, toys, bottles, and milk jugs. LDPE, which has highly branched polymer molecules, is a flexible material used in plastic bags, plastic film, and squeeze bottles (see Figure 11.7). Objects made of HDPE hold their shape in boiling water, whereas those made of LDPE become severely deformed at this temperature.

When dienes such as 1,3-butadiene are used as the monomers in addition polymerization reactions, the resulting polymers contain double bonds and are thus still unsaturated.

$$\underset{\text{1,3-Butadiene}}{CH_2=CH-CH=CH_2} \xrightarrow{\text{Polymerization}} \underset{\text{Polybutadiene}}{-\!\left(CH_2-CH=CH-CH_2\right)_n\!-}$$

In general, unsaturated polymers are much more flexible than the ethene-based saturated polymers listed in Table 11.2. Natural rubber is a flexible addition polymer whose repeating unit is isoprene—that is, 2-methyl-1,3-butadiene (see Figure 11.8).

Figure 11.7 Objects made of polyethylene. Those objects with a rigid structure contain HDPE (high-density polyethylene), and those objects with flexible characteristics contain LDPE (low-density polyethylene).

$$CH_2{=}\underset{\underset{CH_3}{|}}{C}{-}CH{=}CH_2 \xrightarrow{\text{Polymerization}} \left(CH_2{-}\underset{\underset{CH_3}{|}}{C}{=}CH{-}CH_2\right)_n$$

Isoprene (2-methyl-1,3-butadiene) → Polyisoprene (natural rubber)

Saran Wrap is a polymer in which two different monomers are present: chloroethene (vinyl chloride) and 1,1-dichloroethene.

$$\underset{\text{Vinyl chloride}}{CHCl{=}CH_2} + \underset{\text{1,1-Dichloroethene}}{CCl_2{=}CH_2} \xrightarrow{\text{Polymerization}} \underset{\text{Saran Wrap}}{\left(\overbrace{CH_2{-}CHCl}^{\text{1st monomer}}{-}\overbrace{CCl_2{-}CH_2}^{\text{2nd monomer}}\right)_n}$$

Such a polymer is an example of a *copolymer*. A **copolymer** *is a polymer in which two different monomers are present.* Another important copolymer is styrene–butadiene rubber, the leading synthetic rubber in use today. It contains the monomers 1,3-butadiene and styrene in a 3:1 ratio and is a major ingredient in automobile tires.

The accompanying Chemistry at a Glance summarizes the reaction chemistry of alkenes presented in this and the previous section.

Figure 11.8 Natural rubber being harvested in Malaysia.

Practice Problems and Questions

11.31 What is the relationship between a monomer and its polymer?

11.32 What is a copolymer?

11.33 Draw the structural formula of the monomer from which each of the following addition polymers is made.

a. $\left(CF_2{-}CF_2\right)_n$ b. $\left(CH_2{-}CCl{=}CH{-}CH_2\right)_n$ c. $\left(CH_2{-}CHCl\right)_n$ d. $\left(CH_2{-}CH(C_6H_5)\right)_n$

11.34 With the help of Table 11.2, draw the "start" (the first three repeating units) of the structural formula of the addition polymers made from the following monomers.

a. Ethylene b. Vinyl chloride c. Tetrafluoroethylene d. Propylene

Chemistry at a Glance

Chemical Reactions of Alkenes

ALKENE

COMBUSTION
- Products are CO_2 and H_2O
- Does not involve the process of addition

ADDITION
- Addition of a small symmetrical molecule
- Addition of a small unsymmetrical molecule

POLYMERIZATION
- Alkene molecules undergo an addition reaction with one another
- Specific catalysts are needed

H_2 Hydrogenation	Br_2 or Cl_2 Halogenation	HBr or HCl Hydrohalogenation	H_2O (H—OH) Hydration
H_2 + >C=C< → (Ni or Pt catalyst) —C(H)—C(H)—	X_2 + >C=C< → —C(X)—C(X)—	HX + >C=C< → —C(H)—C(X)—	H—OH + >C=C< → (H_2SO_4 catalyst) —C(H)—C(OH)—
Alkane	Dihaloalkane	Monohaloalkane*	Alcohol*

*Markovnikov's rule is needed to predict the product's exact structure if the alkene is unsymmetrical.

Learning Focus

Know the general formula and structural characteristics for an alkyne; given their structure, name alkynes using IUPAC rules, and vice versa.

11.7 Alkynes

Alkynes represent a second type of unsaturated hydrocarbon. An **alkyne** *is an acyclic unsaturated hydrocarbon in which one or more carbon–carbon triple bonds are present.* As the family name *alkyne* indicates, the characteristic "ending" associated with a triple bond is *-yne*.

The general formula for an alkyne with one triple bond is C_nH_{2n-2}. Thus the simplest member of this type of alkyne has the formula C_2H_2, and the next member, with $n = 3$, has the formula C_3H_4.

$$CH{\equiv}CH \qquad CH{\equiv}C{-}CH_3$$

Ethyne Propyne

The presence of a carbon–carbon triple bond in a molecule always results in a linear arrangement for the two atoms attached to the carbons of the triple bond. Thus, ethyne is a linear molecule (see Figure 11.9).

Because of the linearity (180° angles) about the triple bond, *cis–trans* isomerism, like that found for 2-butene (CH_3—CH═CH—CH_3), is not possible for alkynes such as 2-butyne (CH_3—C≡C—CH_3).

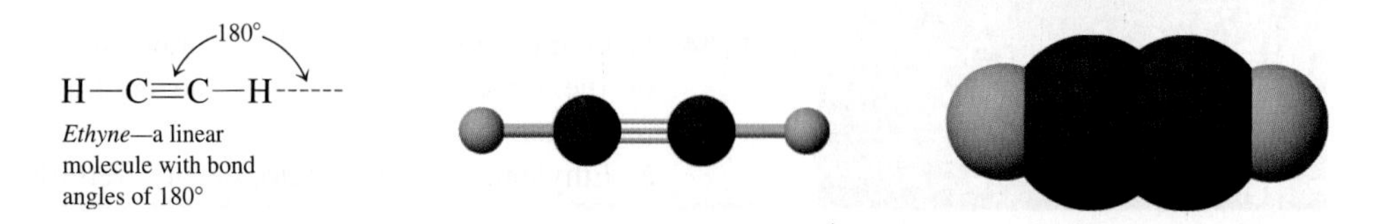

Figure 11.9 Structural representations of ethyne (acetylene), the simplest alkyne. The molecule is linear—that is, all four atoms lie in a straight line.

▶ *Cycloalkynes,* molecules that contain a triple bond as part of a ring structure, are known, but they are not common. Because of the 180° angle associated with a triple bond, a ring system containing a triple bond has to be quite large. The smallest cycloalkyne that has been isolated is cyclooctyne.

The simplest alkyne, ethyne (C_2H_2), is the most important alkyne from an industrial standpoint. A colorless gas, it goes by the common name *acetylene* and is used in oxy-acetylene torches, which are high-temperature torches used for cutting and welding metals.

IUPAC Nomenclature for Alkynes

The rules for naming alkynes are identical to those used to name alkenes (Section 11.3), except the ending *-yne* is used instead of *-ene*. Consider the following structures and their IUPAC names.

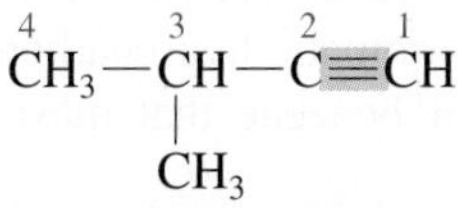

$$\overset{4}{CH_3}-\overset{3}{\underset{\displaystyle CH_3}{\underset{|}{CH}}}-\overset{2}{C}\equiv\overset{1}{CH}$$

3-Methyl-1-butyne

$$\overset{1}{CH_3}-\overset{2}{CH_2}-\overset{3}{C}\equiv\overset{4}{C}-\overset{5}{CH_2}-\overset{\displaystyle CH_3}{\overset{|}{\underset{\displaystyle CH_3}{\underset{|}{\overset{6}{C}}}}}-\overset{7}{CH_3}$$

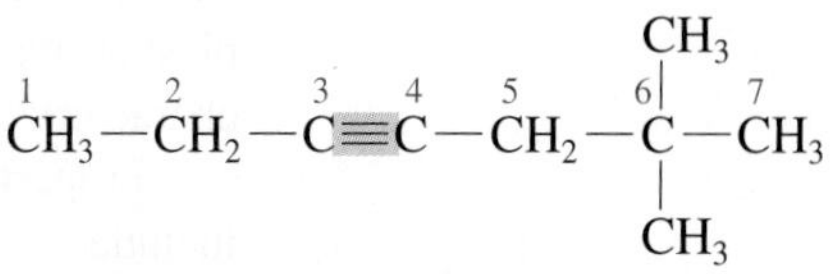

6,6-Dimethyl-3-heptyne

$$\overset{1}{CH}\equiv\overset{2}{C}-\overset{3}{CH_2}-\overset{4}{CH_2}-\overset{5}{CH_2}-\overset{6}{C}\equiv\overset{7}{CH}$$

1,6-Heptadiyne

Physical and Chemical Properties of Alkynes

The physical properties of alkynes are similar to those of alkenes and alkanes. In general, alkynes are insoluble in water but soluble in organic solvents, have low densities, and have boiling points that increase with molecular mass. Low-molecular-mass alkynes are gases at room temperature.

The chemical reactions of alkynes are similar to those of alkenes. The same substances that add to double bonds (H_2, HCl, Cl_2, and so on) can add to triple bonds. However, two molecules of a specific reactant can add to the triple bond. The first molecule converts the triple bond to a double bond, and the second molecule then converts the double bond to a single bond. For example, propyne reacts with H_2 to form propene first and then to form propane.

$$\underset{\text{An alkyne (propyne)}}{CH\equiv C-CH_3} \xrightarrow[Ni]{H_2} \underset{\text{An alkene (propene)}}{CH_2=CH-CH_3} \xrightarrow[Ni]{H_2} \underset{\text{An alkane (propane)}}{CH_3-CH_2-CH_3}$$

Practice Problems and Questions

11.35 What is the functional group for an alkyne?

11.36 Contrast alkynes and alkenes in terms of
a. general physical properties b. general chemical properties

11.37 Assign an IUPAC name to each of the following alkynes.

a. $CH_3-CH_2-CH_2-CH_2-C\equiv CH$

b. $CH_3-C\equiv C-CH_3$

c. 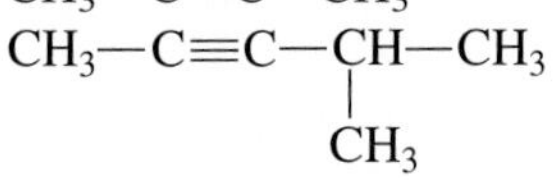 $CH_3-C\equiv C-\underset{\displaystyle CH_3}{\underset{|}{CH}}-CH_3$

d. 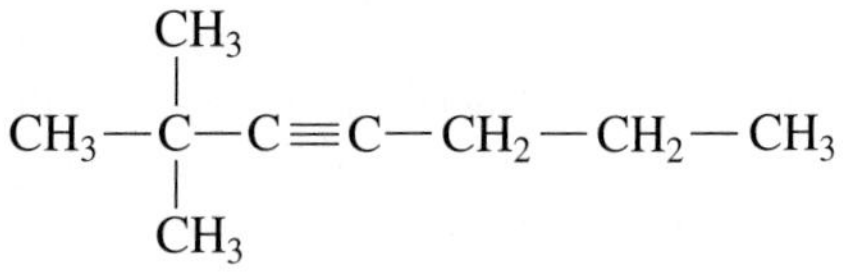 $CH_3-\overset{\displaystyle CH_3}{\overset{|}{\underset{\displaystyle CH_3}{\underset{|}{C}}}}-C\equiv C-CH_2-CH_2-CH_3$

11.38 Draw the structural formula for each of the following alkynes.
a. 2-Pentyne b. 3-Hexyne c. 3-Methyl-1-butyne d. 4-Methyl-2-pentyne

11.39 Supply the structural formula of the product in each of the following alkyne addition reactions.

a. $CH\equiv CH + 2H_2 \xrightarrow{Ni} ?$ b. $CH_3-C\equiv CH + 2Br_2 \longrightarrow ?$

c. $CH_3-C\equiv CH + 2HBr \longrightarrow ?$ d. $CH\equiv CH + 1HCl \longrightarrow ?$

Learning Focus

Define the term *aromatic hydrocarbon* and explain how the bonding in such compounds differs from that in other unsaturated hydrocarbons.

11.8 Aromatic Hydrocarbons

The *aromatic hydrocarbons* are the third class of unsaturated hydrocarbons—alkenes and alkynes being the other two classes. An **aromatic hydrocarbon** *is an unsaturated cyclic hydrocarbon that does not readily undergo addition reactions.* This reaction behavior, which is almost opposite to that of alkenes and alkynes, explains the separate classification for aromatic hydrocarbons.

Aromatic hydrocarbons contain a bonding feature different from that of carbon–carbon double and triple bonds. A discussion of the bonding present in *benzene,* the simplest aromatic hydrocarbon, will be used to illustrate this new bonding feature and to also characterize the aromatic hydrocarbon functional group.

Properties of benzene that must be accounted for through bonding considerations include

1. A molecular structure that involves a six-membered ring of carbon atoms in which all carbon–carbon bonds are identical.
2. A molecular structure in which each carbon atom has one hydrogen atom attached to it, with all hydrogen atoms equivalent to one another.
3. A molecular formula of C_6H_6, indicating a high degree of unsaturation because of the low hydrogen-to-carbon ratio.
4. A chemical reactivity different from that normally associated with an *unsaturated* compound. Substitution reactions, rather than addition reactions, occur.

▶ The name Kekulé is pronounced "Keck-u-la."

In 1865 the German chemist August Kekulé (see Figure 11.10) proposed a cyclic, alternating single- and double-bonded structure for benzene.

```
           H
           |
   H     C     H
     \ //  \  /
      C      C
      |      ||
      C      C
    /  \\  /   \
   H     C      H
         |
         H
```

or

Kekulé further proposed that the double and single bonds in the carbon ring of benzene were oscillating rapidly around the ring. In order to indicate the oscillation of bonds around the ring, Kekulé drew two equivalent structures for benzene that differed only in the location of the double bonds (1,3,5 positions versus 2,4,6 positions) and connected the two structures by using a double-headed arrow.

Figure 11.10 Friedrich August Kekulé (1829–1896) originally entered the German University of Giessen to study architecture but switched to chemistry after taking a chemistry course. In 1890, on the 25th anniversary of his proposal on the cyclic structure of benzene, he gave a speech in which he stated that the idea for benzene's structure came as a result of his dozing off in front of a fire while working on a textbook. He dreamed of chains of carbon atoms twisting and turning in a snake-like motion, when suddenly the head of one snake seized hold of its own tail and formed a spinning ring. This concept—that chains of carbon atoms could have cyclic structures—was a new idea at the time.

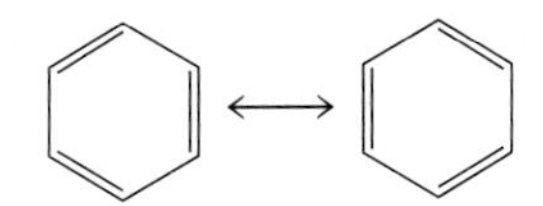

▶ We know that neither of the Kekulé structures for benzene is, by itself, correct, because if it were, benzene would have three separate double bonds and would undergo addition reactions. Benzene does not undergo addition reactions.

Such bond oscilliation would make all carbon–carbon bonds equivalent, the "real" benzene structure being an "average" of the two Kekulé structures; that is, the bonds present in the ring would be an average of a single bond and a double bond.

An alternative notation for denoting the bonding in benzene—a notation that involves a single structure—is

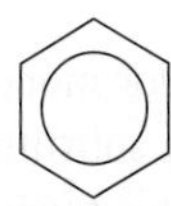

The interpretation of this "circle-in-the-ring" structure is that carbon atoms form three "normal" bonds and one "delocalized" bond (the circle). In the delocalized bond, the electrons are free to move about the ring rather than being tied down to a specific location. This "delocalized" bond phenomenon is what causes benzene and other aromatic hydrocarbons to be resistant to addition reactions. Addition reactions would require breaking up the delocalized bonding system.

The structure represented by the notation

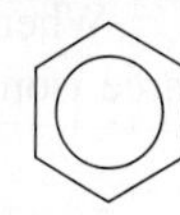

is called an *aromatic ring* and is the functional group present in aromatic hydrocarbons.

Practice Problems and Questions

11.40 Draw a structural representation for the functional group present in an aromatic hydrocarbon.

11.41 A circle (ring) within a hexagon is often used to represent an aromatic ring. What does the circle represent?

11.42 Contrast aromatic hydrocarbons with alkenes and alkynes in terms of the types of reactions that they undergo.

Learning Focus

Given their structures, name aromatic hydrocarbons using IUPAC rules, and vice versa.

11.9 Names for Aromatic Hydrocarbons

Replacement of one or more of the hydrogen atoms on benzene with other groups produces benzene derivatives. The IUPAC system of naming monosubstituted benzene derivatives uses the name of the substituent as a prefix to the name *benzene*. Examples of this type of nomenclature include

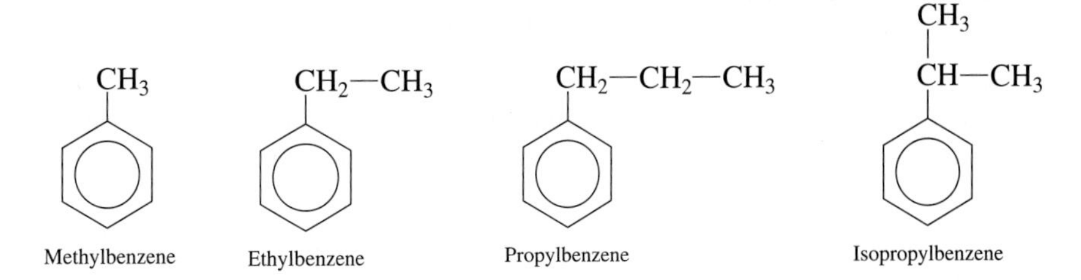

Methylbenzene also goes by the name *toluene;* in fact, *toluene* is the name usually encountered for this compound.

Monosubstituted benzene structures are often drawn with the substituent at the "12 o'clock" position, as in the previous structures. However, because all the hydrogen atoms in benzene are equivalent, it does not matter at which carbon of the ring the substituted group is located. Each of the following formulas represents toluene.

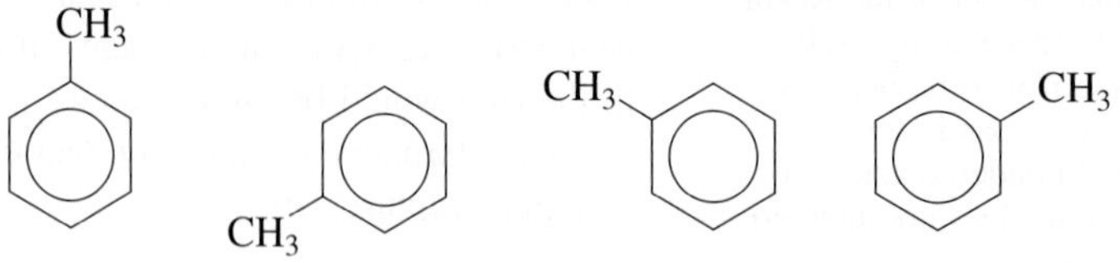

CH_3 CH_3 CH_3 CH_3

▶ The word *phenyl* comes from "phene," a European term used during the 1800s for benzene. The word is pronounced *fen*-nil.

For monosubstituted benzene rings that have a group attached that is not easily named as a substituent, the benzene ring is often treated as a group attached to this substituent. In this reversed approach, the benzene ring attachment is called a *phenyl* group, and the compound is named according to the rules for naming alkanes, alkenes, and alkynes.

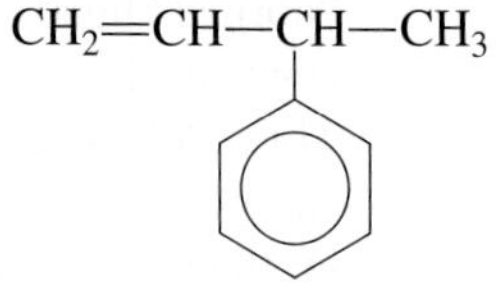

$CH_2{=}CH{-}CH{-}CH_3$

3-Phenyl-1-butene

When two substituents, either the same or different, are attached to a benzene ring, three isomeric structures are possible.

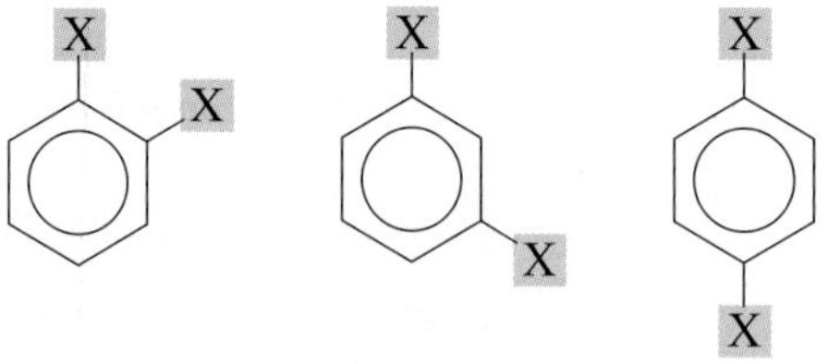

X X X X X X

▶ *Cis–trans* isomerism is not possible for disubstituted benzenes. All 12 atoms of benzene are in the same plane—that is, benzene is a flat molecule. When a substituent group replaces an H atom, the atom that bonds the group to the ring is also in the plane of the ring.

To distinguish among these three isomers, we must specify the positions of the substituents relative to one another. This can be done in either of two ways: by using numbers and by using nonnumerical prefixes.

When numbers are used, the three isomeric dimethylbenzenes have the first-listed set of names:

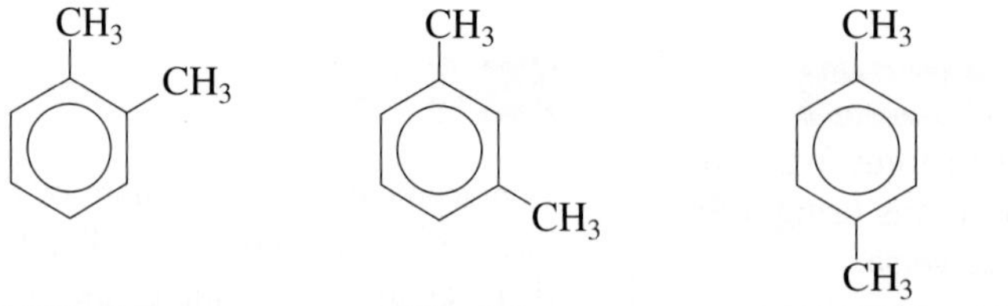

CH_3 CH_3 CH_3 CH_3 CH_3 CH_3

1,2-Dimethylbenzene (*ortho*-dimethylbenzene) 1,3-Dimethylbenzene (*meta*-dimethylbenzene) 1,4-Dimethylbenzene (*para*-dimethylbenzene)

▶ Learn the meaning of the prefixes *ortho-*, *meta-*, and *para-*. These prefixes are extensively used in naming disubstituted benzenes.

X

← *ortho* to X

← *meta* to X

para to X

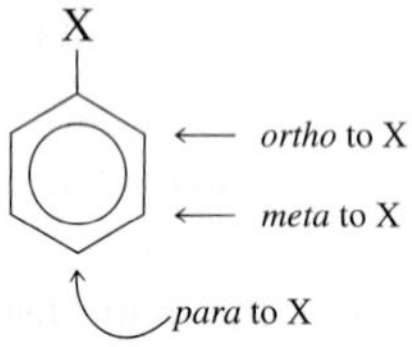

The prefix system uses the prefixes *ortho-*, *meta-*, and *para-* (abbreviated *o-*, *m-*, and *p-*).

Ortho- means 1,2 disubstitution; the substituents are on adjacent carbon atoms.

Meta- means 1,3 disubstitution; the substituents are one carbon removed from each other.

Para- means 1,4 disubstitution; the substituents are two carbons removed from each other (on opposite sides of the ring).

When prefixes are used, the three isomeric dimethylbenzenes have the second-listed set of names above.

▶ The use of *ortho-*, *meta-*, and *para-* in place of position numbers is reserved exclusively for disubstituted benzenes. The system is never used with cyclohexanes or other ring systems.

When more than two groups are present on the benzene ring, their positions are indicated with *numbers*. The ring is numbered so as to obtain the lowest possible numbers for the carbon atoms that have substituents. If there is a choice of numbering systems (two systems give the same lowest set), then the group that comes first alphabetically is given the lower number.

CH_3 H_3C CH_3

1,2,4-Trimethylbenzene

$CH_2—CH_3$ H_3C CH_3

1-Ethyl-3,5-dimethylbenzene

Benzene and its substituted derivatives are not the only type of aromatic hydrocarbon that exists. Another large class of aromatic hydrocarbons is the fused-ring aromatic hydrocarbons. A **fused-ring aromatic hydrocarbon** *is an aromatic compound whose structure contains two or more rings fused together.* Two carbon rings that share a pair of carbon atoms are said to be *fused.*

The two simplest fused-ring aromatic compounds are naphthalene and anthracene. Both are solids at room temperature.

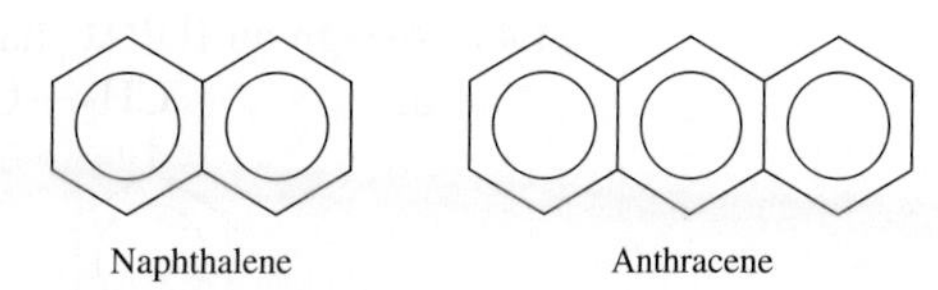

Naphthalene Anthracene

Chemical Portraits 18 profiles the single ring aromatic compounds benzene and toluene, and the fused-ring aromatic compound naphthalene.

Chemical Portraits 18 Single- and Double-Ring Aromatic Hydrocarbons

Benzene (C_6H_6)

Profile: Benzene is a clear, colorless, volatile, highly flammable liquid that was once highly used as a solvent for organic compounds. Its solvent use has been greatly reduced as inhalation exposure can cause nausea and respiratory problems. Many important larger organic molecules have benzene rings present as part of their structures. In such situations, the properties of benzene are not manifest.

Biochemical considerations: Gasoline contains about 2% benzene. Inhalation of gasoline vapors, while refueling an automobile, constitutes low-level exposure to benzene vapors. Being around a cigarette smoker also involves benzene exposure; benzene is a combustion product present in cigarette smoke.

What is known about the effects of long-term benzene exposure on humans?

Toluene (C_7H_8)

Profile: Toluene, also called methylbenzene, is a clear, colorless, flammable liquid with a benzene-like odor. The name toluene comes from *Tolu balsam*, the yellow-brown, pleasant-smelling gum of the South American tree *Toluifera balsamum*, which was the original source for toluene. Today, its source is processes associated with petroleum refining.

Uses: Toluene is used as an octane-enhancer in gasoline, and as a raw material to produce other chemicals. Benzoic acid, a preservative for foods, beverages, and cosmetics is produced from toluene. Makers of explosives use toluene to make trinitrotoluene, commonly called TNT. Paint manufacturers use it as a lacquer solvent.

How does the structure of TNT compare with that of toluene itself?

Naphthalene ($C_{10}H_8$)

Profile: Naphthalene is a volatile, white crystalline solid that readily sublimes at room temperature. Obtained from coal tar, naphthalene's major industrial uses are in making resins and plastics. Another use is that of an insecticidal fumigant for areas such as cargo holds of ships.

Biochemical considerations: Naphthalene vapor has an odor that we recognize as "mothballs;" indeed, naphthalene crystals have been extensively used as a moth repellent. Naphthalene crystals are "old" mothballs. A "newer" version of mothballs involves the compound para-dichlorobenzene, $C_6H_4Cl_2$, another white solid that readily sublimes at room temperature. Most "mothballs" are now para-dichlorobenzene.

Are there any health risks associated with breathing "mothball" vapor?

See the text web site at **www.cengage.com/chemistry/stoker** for answers to the above questions and for further information.

Practice Problems and Questions

11.43 Assign an IUPAC name to each of the following disubstituted benzenes. Use numbers rather than prefixes to locate the substituents on the benzene ring.

a.
CH₃
CH₂—CH₂—CH₃

b.
CH₂—CH₂—CH₃
CH—CH₃
CH₃

c.
CH₂—CH₃
CH₂—CH₃

d.
CH₃
CH₂—CH₃

11.44 Assign to each of the compounds in Problem 11.43 an IUPAC name in which the substituents on the benzene ring are located using the *ortho-*, *meta-*, *para-* prefix system.

11.45 Assign an IUPAC name to each of the following substituted benzenes.

a.
CH₂—CH₃
H₃C
CH₂—CH₃

b.
CH₂—CH₂—CH₃
CH—CH₃
CH₃
CH₃

c.
CH₂—CH₂—CH₃
CH₃
CH₃

d.
CH₃
CH₃
H₃C
CH₃

11.46 Assign an IUPAC name to each of the following compounds, in which the benzene ring is treated as a substituent—that is, as a phenyl group.

a.
CH₃—CH—CH₂—CH₃

b.
CH₃—CH—CH=CH₂

c.
CH₃—CH—CH₂—CH₂
CH₃

d.
CH₃—CH—CH₂—CH—CH₃

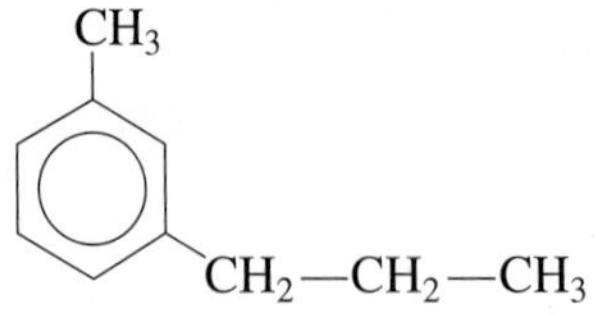

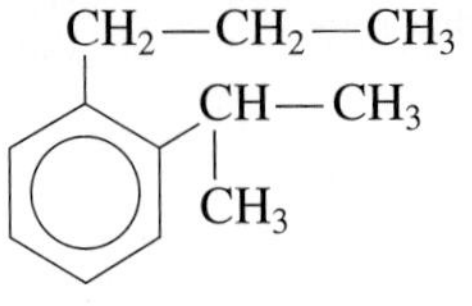

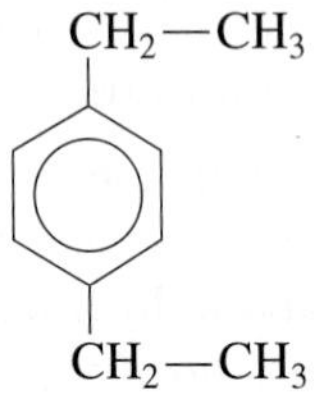

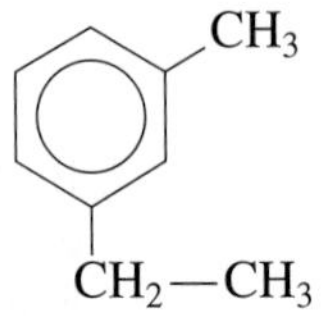

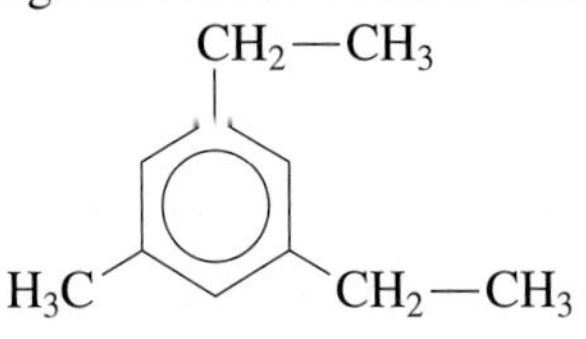

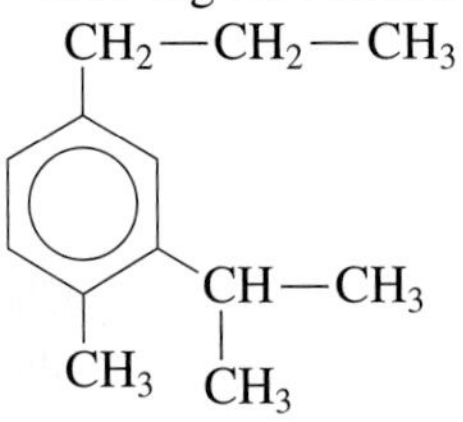

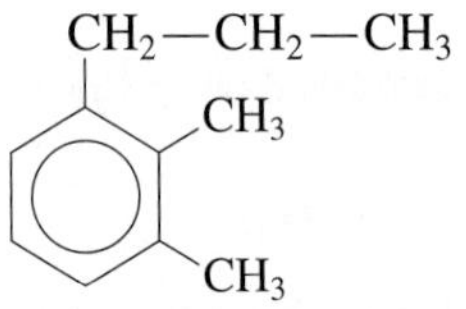

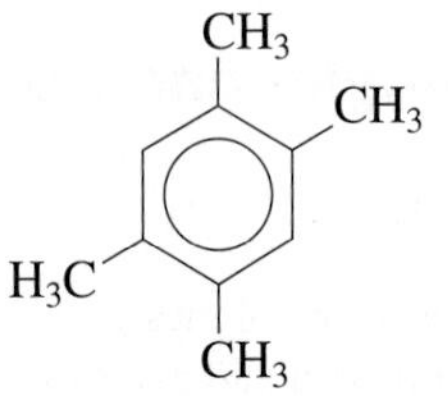

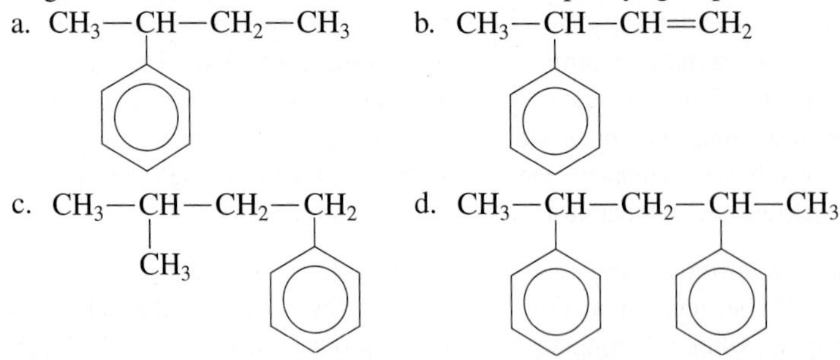

Learning Focus

Describe the general physical properties of aromatic hydrocarbons; write equations for the aromatic substitution reactions involving alkylation and halogenation.

11.10 Properties and Reactions of Aromatic Hydrocarbons

In general, aromatic hydrocarbons resemble other hydrocarbons in physical properties. They are insoluble in water, are good solvents for other nonpolar materials, and are less dense than water.

Benzene, monosubstituted benzenes, and many disubstituted benzenes are liquids at room temperature. Benzene itself is a colorless, flammable liquid that burns with a sooty flame because of incomplete combustion.

At one time, coal tar was the main source of aromatic hydrocarbons. Petroleum is now the primary source of such compounds. At high temperatures, with special catalysts, saturated hydrocarbons obtained from petroleum can be converted to aromatic hydrocarbons. The production of toluene from heptane is representative of such a conversion.

$$CH_3-CH_2-CH_2-CH_2-CH_2-CH_2-CH_3 \xrightarrow[\text{High temperature}]{\text{Catalyst}} C_6H_5-CH_3 + 4H_2$$

Aromatic Substitution Reactions

We have noted that aromatic hydrocarbons undergo substitution reactions rather than addition reactions (Section 11.8). As you recall from Section 10.14, substitution reactions are characterized by different atoms or groups of atoms replacing hydrogen atoms in a hydrocarbon molecule. Two important types of substitution reactions for benzene are alkylation and halogenation.

1. *Alkylation:* An alkyl group (R—) from an alkyl chloride (R—Cl) substitutes for a hydrogen atom on the benzene ring. A catalyst, $AlCl_3$, is needed for alkylation.

$$C_6H_6 + CH_3-CH_2-Cl \xrightarrow{AlCl_3} C_6H_5-CH_2-CH_3 + HCl$$

Benzene Chloroethane Ethylbenzene

In general terms, the alkylation of benzene can be written as

$$C_6H_6 + R-Cl \xrightarrow{AlCl_3} C_6H_5-R + HCl$$

Alkylation is the most important industrial reaction of benzene.

2. *Halogenation* (bromination or chlorination): A hydrogen atom on a benzene ring can be replaced by bromine or chlorine if benzene is treated with Br_2 or Cl_2 in the presence of a catalyst. The catalyst is usually $FeBr_3$ for bromination and $FeCl_3$ for chlorination.

$$C_6H_6 + Br_2 \xrightarrow{FeBr_3} C_6H_5-Br + HBr$$

$$C_6H_6 + Cl_2 \xrightarrow{FeCl_3} C_6H_5-Cl + HCl$$

Practice Problems and Questions

11.47 Indicate whether each of the following is a general property of aromatic hydrocarbons.

a. Soluble in water
b. Less dense than water
c. Usually a gas at room temperature
d. Flammable in air

11.48 Complete the following reaction equations by supplying the formula of the missing reactant or product.

a. benzene + ? $\xrightarrow{FeBr_3}$ bromobenzene (benzene ring with Br) + HBr

b. benzene + Cl_2 $\xrightarrow{FeCl_3}$ chlorobenzene (benzene ring with Cl) + ?

c. benzene + $CH_3—CH(CH_3)—Cl$ $\xrightarrow{AlCl_3}$? + HCl

d. benzene + ? $\xrightarrow{AlBr_3}$ ethylbenzene (benzene ring with $CH_2—CH_3$) + HBr

CONCEPTS TO REMEMBER

Unsaturated hydrocarbons An unsaturated hydrocarbon is a hydrocarbon that contains one or more carbon–carbon multiple bonds. Three main classes of unsaturated hydrocarbons exist: alkenes, alkynes, and aromatic hydrocarbons.

Alkenes and cycloalkenes. An alkene is an acyclic unsaturated hydrocarbon in which one or more carbon–carbon double bonds are present. A cycloalkene is a cyclic unsaturated hydrocarbon that contains one or more carbon–carbon double bonds within the ring system.

Alkene nomenclature. Alkenes and cycloalkenes are given IUPAC names using rules similar to those for alkanes and cycloalkanes, except that the ending *-ene* is used. Also, the double bond takes precedence both in selecting and in numbering the main chain or ring.

***Cis–trans* isomers.** Because rotation about a carbon–carbon double bond is restricted, some alkenes exist in two isomeric (*cis–trans*) forms. *Cis–trans* isomerism is possible when each carbon of the double bond is attached to two different groups.

Physical properties of alkenes. Alkenes and alkanes have similar physical properties. They are nonpolar, insoluble in water, less dense than water, and soluble in nonpolar solvents.

Addition reactions of alkenes. Numerous substances, including H_2, Cl_2, Br_2, HCl, HBr, and H_2O, add to an alkene carbon–carbon double bond. When both the alkene and the reactant are unsymmetrical, the addition proceeds according to Markovnikov's rule: The carbon atom of the double bond that already has the greater number of H atoms gets one more.

Addition polymers. Addition polymers are formed from alkene monomers that undergo repeated addition reactions with each other. Many familiar and widely used materials, such as fibers and plastics, are addition polymers.

Alkynes and cycloalkynes. Alkynes and cycloalkynes are unsaturated hydrocarbons that contain one or more carbon–carbon triple bonds. They are named in the same way as alkenes and cycloalkenes, except that their parent names end in *-yne*. Like alkenes, alkynes undergo addition reactions. These occur in two steps, an alkene forming first and then an alkane.

Aromatic hydrocarbons. Benzene, the simplest aromatic hydrocarbon, and other members of this family of compounds contain a six-membered ring with a cyclic, delocalized bond. This aromatic ring is often drawn as a hexagon containing a circle.

Nomenclature of aromatic hydrocarbons. Monosubstituted benzene compounds are named by adding the substituent name to the word *benzene*. Positions of substituents in disubstituted benzenes are indicated by using a numbering system or the *ortho-* (1,2), *meta-* (1,3), and *para-* (1,4) prefix system.

Reactions of aromatic hydrocarbons. Aromatic hydrocarbons undergo substitution reactions rather than addition reactions. Important substitution reactions are alkylation and halogenation.

KEY REACTIONS AND EQUATIONS

1. Halogenation of an alkene (Section 11.5)

$$\text{>C=C<} + Br—Br \longrightarrow —\underset{Br}{\overset{|}{C}}—\underset{Br}{\overset{|}{C}}—$$

2. Hydrogenation of an alkene (Section 11.5)

$$\text{>C=C<} + H—H \xrightarrow{Ni} —\underset{H}{\overset{|}{C}}—\underset{H}{\overset{|}{C}}—$$

3. Hydrohalogenation of an alkene (Section 11.5)

$$\text{>C=C<} + H—Cl \longrightarrow —\underset{H}{\overset{|}{C}}—\underset{Cl}{\overset{|}{C}}—$$

4. Hydration of an alkene (Section 11.5)

$$\text{>C=C<} + H—OH \xrightarrow{H_2SO_4} —\underset{H}{\overset{|}{C}}—\underset{OH}{\overset{|}{C}}—$$

5. Hydrogenation of an alkyne (Section 11.7)

$$-C\equiv C- + H_2 \xrightarrow{Ni} -\underset{}{\overset{H}{\overset{|}{C}}}=\overset{H}{\overset{|}{C}}- \xrightarrow[Ni]{H_2} -\underset{\underset{H}{|}}{\overset{H}{\overset{|}{C}}}-\underset{\underset{H}{|}}{\overset{H}{\overset{|}{C}}}-$$

6. Halogenation of an alkyne (Section 11.7)

$$-C\equiv C- + Br_2 \longrightarrow -\overset{Br}{\overset{|}{C}}=\overset{Br}{\overset{|}{C}}- \xrightarrow{Br_2} -\underset{\underset{Br}{|}}{\overset{Br}{\overset{|}{C}}}-\underset{\underset{Br}{|}}{\overset{Br}{\overset{|}{C}}}-$$

7. Hydrohalogenation of an alkyne (Section 11.7)

$$-C\equiv C- + HBr \longrightarrow -\overset{H}{\overset{|}{C}}=\overset{Br}{\overset{|}{C}}- \xrightarrow{HBr} -\underset{\underset{H}{|}}{\overset{H}{\overset{|}{C}}}-\underset{\underset{Br}{|}}{\overset{Br}{\overset{|}{C}}}-$$

8. Alkylation of benzene (Section 11.10)

$$C_6H_6 + R-Cl \xrightarrow{AlCl_3} C_6H_5-R + HCl$$

9. Halogenation of benzene (Section 11.10)

$$C_6H_6 + Br_2 \xrightarrow{FeBr_3} C_6H_5-Br + HBr$$

KEY TERMS

Addition polymer (11.6)
Addition reaction (11.5)
Alkene (11.2)
Alkyne (11.7)
Aromatic hydrocarbon (11.8)
Copolymer (11.6)
Cycloalkene (11.2)
Functional group (11.1)
Fused-ring aromatic hydrocarbon (11.9)
Markovnikov's rule (11.5)
Monomer (11.6)
Polymer (11.6)
Polymerization reaction (11.6)
Symmetrical addition reaction (11.5)
Unsaturated hydrocarbon (11.1)
Unsymmetrical addition reaction (11.5)

ADDITIONAL PROBLEMS

11.49 Identify the functional group present in each of the following types of hydrocarbons.
a. Alkene
b. Cycloalkene
c. Alkyne
d. Aromatic hydrocarbon

11.50 What is the molecular formula for each of the following?
a. The simplest alkene
b. The simplest cycloalkene
c. The simplest alkyne
d. The simplest alkane

11.51 Indicate whether the first-listed hydrocarbon in each of the following pairs of hydrocarbons contains (1) more hydrogen atoms, (2) the same number of hydrogen atoms, or (3) fewer hydrogen atoms than the second-listed hydrocarbon.
a. Propane and propene
b. Propene and propyne
c. Propene and cyclopropene
d. Propyne and cyclopropene

11.52 Draw a condensed structural formula for each of the following unsaturated hydrocarbons.
a. 5-Methyl-2-hexyne
b. 2-Methyl-2-butene
c. 1,6-Heptadiene
d. 3-Methyl-1,4-pentadiyne

11.53 How many molecules of H_2 will react with one molecule of each of the compounds in Problem 11.52 when the appropriate catalyst is present?

11.54 How many different compounds fit each of the following descriptions?
a. Pentadiene
b. Cyclopentene
c. Dimethylbenzene
d. Trimethylbenzene

11.55 Indicate whether each of the following pairs of structural formulas represent (1) structural isomers, (2) *cis–trans* isomers, (3) identical structures, or (4) different compounds that are not isomers.
a. $CH_2{=}CH{-}CH_3$ and $CH_3{-}CH{=}CH_2$
b.
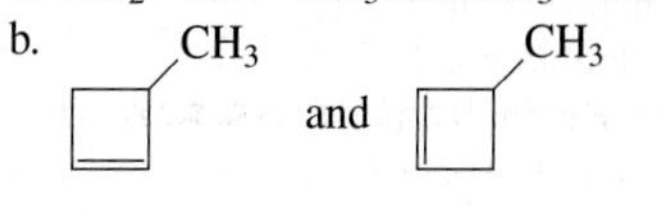

c.
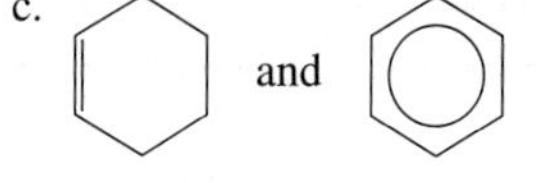

d.
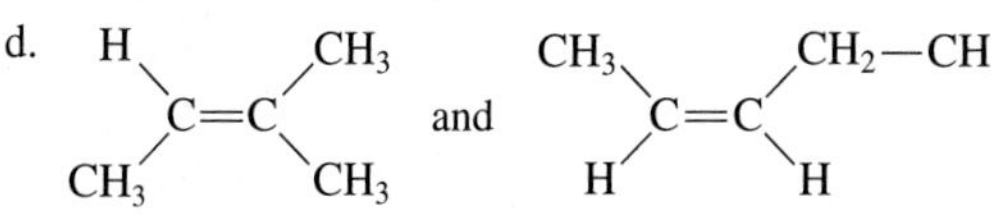

11.56 Assign an IUPAC name to each of the compounds in Problem 11.55.

11.57 Indicate whether each of the following pairs of hydrocarbons are structural isomers?
a. Propene and cyclopropene
b. 1-Pentene and 2-pentene
c. *cis*-2-Butene and *trans*-2-butene
d. Cyclobutene and 2-butyne

11.58 Contrast the compounds cyclohexane, cyclohexene, and benzene in terms of
a. the number of carbon atoms present
b. the number of hydrogen atoms present
c. whether they undergo substitution or addition reactions
d. whether they are a solid, a liquid, or a gas at room temperature and pressure

11.59 Give an acceptable alternative name for each of the following hydrocarbons.
a. Acetylene
b. Propylene
c. Ethylene
d. Toluene

PRACTICE TEST ▶ True/False

11.60 The three classes of *unsaturated* hydrocarbons are alkanes, alkenes, and alkynes.

11.61 The general formula for an alkene with one double bond is C_nH_{2n-2}.

11.62 Carbon–carbon double bonds take precedence over alkyl groups in determining the numbering system for a chain of carbon atoms.

11.63 *Cis–trans* isomerism is possible for both 1-butene and 2-butene.

11.64 Both alkenes and alkynes have limited water solubility.

11.65 Alkenes undergo substitution reactions, and alkynes and aromatic compounds undergo addition reactions.

11.66 Markovnikov's rule is often needed in predicting the major organic product in the hydrogenation of an alkene.

11.67 PVC and Teflon are addition polymers in which the monomers are substituted ethenes.

11.68 A copolymer is a polymer in which two different monomers are present.

11.69 *Cis–trans* isomerism is not possible for alkynes regardless of the number of substituents present.

11.70 Ethylene and acetylene are, respectively, the simplest alkene and the simplest alkyne.

11.71 The "circle-in-the-ring" in the structure of benzene represents a "delocalized" bond.

11.72 Ortho-dimethylbenzene and 1,3-dimethylbenzene are two names for the same compound.

11.73 Halogenation is a reaction that all three types of unsaturated hydrocarbons undergo.

11.74 The functional group for all unsaturated hydrocarbons is a carbon–carbon double bond.

PRACTICE TEST ▶ Multiple Choice

11.75 All of the following compounds are unsaturated hydrocarbons except one. The exception is
a. 2-butene b. 3-heptyne
c. cyclopropane d. 1,3-dimethylbenzene

11.76 The correct IUPAC name for the compound $CH_2{=}CH{-}CH{=}CH_2$ is
a. 1,4-butene b. 1,3-butadiene
c. 1,3-dibutene d. 1,4-dibutene

11.77 Which of the following types of unsaturated hydrocarbons does *not* have the general formula C_nH_{2n-2}?
a. Alkenes with one double bond
b. Cycloalkenes with one double bond
c. Alkenes with two double bonds
d. Alkynes with one triple bond

11.78 Which of the following is a structural isomer of 1-pentene?
a. 1-Butene
b. Cyclopentene
c. 2-Methyl-2-butene
d. 1,3-Pentadiene

11.79 *Cis–trans* isomerism is possible for which of the following alkenes?
a. $CH_2{=}CH{-}CH_2{-}CH_2{-}CH_3$
b. $CH_3{-}CH{=}CH{-}CH_2{-}CH_3$
c. $CH_3{-}CH{=}CH_2$
d. $CH_2{=}CH{-}CH_2{-}CH_3$

11.80 Which of the following reactions can be used to convert an alkene to an alkane?
a. Hydrogenation b. Halogenation
c. Hydrohalogenation d. Hydration

11.81 Markovnikov's rule is needed to predict the major organic product in the reaction between HCl and which of the following hydrocarbons.
a. $CH_2{=}CH_2$
b. $CH_3{-}CH_3$
c. $CH_2{=}CH{-}CH_3$
d. $CH_3{-}CH{=}CH{-}CH_3$

11.82 Which of the following addition polymers contains only carbon and hydrogen?
a. Polystyrene b. PVC c. Teflon d. Saran

11.83 Which of the following is a correct pairing of "prefix" and "numbers"?
a. *para-* and 1,2-
b. *ortho-* and 1,4-
c. *meta-* and 1,3-
d. *iso-* and 2,3-

11.84 Which of the following compounds is *not* an aromatic compound?
a. Ethylbenzene
b. 3-Methylcyclopropene
c. 2-Phenylbutane
d. Toluene

CHAPTER OUTLINE

CHAPTER TWELVE

Hydrocarbon Derivatives I: Carbon–Heteroatom Single Bonds

The physiological effects of poison ivy are caused by certain organic compounds (hydrocarbon derivatives) that contain —OH functional groups that are present in the leaves.

This chapter is the first of two on the subject of hydrocarbon derivatives. In this chapter we consider hydrocarbon derivatives whose functional group involves a heteroatom single-bonded to a carbon atom (halogenated hydrocarbons, alcohols, ethers, thiols, and amines). The next chapter focuses on derivatives whose functional group contains a carbon–oxygen double bond (aldehydes, ketones, carboxylic acids, esters, and amides).

Learning Focus

Recognize the functional groups present in halogenated hydrocarbons, alcohols, ethers, thiols, and amines.

12.1 Functional Groups Containing Carbon–Heteroatom Single Bonds

Hydrocarbon derivatives, by definition, must contain one or more atoms of some other element than carbon and hydrogen. These additional atoms are called *heteroatoms*. A **heteroatom** *is an atom, other than carbon or hydrogen, present in a hydrocarbon derivative.*

The classes of hydrocarbon derivatives considered in this chapter share the common feature of having heteroatoms that are single-bonded to a carbon atom. These classes are

1. **Halogenated hydrocarbons.** One or more of four different heteroatoms can be present in a halogenated hydrocarbon. These atoms are the halogens (Section 3.4), which are fluorine, chlorine, bromine, and iodine. In halogenated hydrocarbons, the heteroatom (halogen atom) itself is the functional group. Using the general designation for a halogen atom, which is X (Section 10.14), the generalized formula for

a halogenated hydrocarbon is R—X. [Remember that the symbol R designates the carbon–hydrocarbon portion of the molecule (Section 10.14).]

Halogenated Hydrocarbon: R—X

2. **Alcohols.** The heteroatom in an alcohol is an oxygen atom. The O atom itself is only part of the functional group rather than the entire functional group as was the case with the halogens. The alcohol functional group is the —OH group.

Alcohol: R—OH

3. **Ethers.** Ethers also have oxygen as the heteroatom present. The O atom is, however, bonded to two carbon atoms rather than to a carbon atom and a hydrogen atom as in alcohols. The functional group present is —O—, and the generalized formula for an ether is R—O—R.

Ether: R—O—R

4. **Thiols.** Thiols are sulfur-containing hydrocarbon derivatives. The functional group present is —SH, and the generalized designation for a thiol is R—SH.

Thiol: R—SH

5. **Amines.** A nitrogen atom is the heteroatom present in amines. The functional group itself contains three atoms, one nitrogen, and two hydrogens; its designation is $—NH_2$. The overall designation for an amine is $R—NH_2$.

Amine: $R—NH_2$

Practice Problems and Questions

12.1 Identify the heteroatom present in each of the following types of hydrocarbon derivatives.
a. Alcohol b. Ether c. Amine d. Thiol

12.2 Identify the complete functional group present in each of the following types of hydrocarbon derivatives.
a. Ether b. Amine c. Thiol d. Alcohol

12.3 Identify the class of hydrocarbon derivative associated with each of the following generalized formulas.
a. R—X b. R—OH c. R—SH d. R—O—R

12.2 Halogenated Hydrocarbons

Learning Focus

Know the general structural characteristics of halogenated hydrocarbons and be able to name such compounds given their structural formulas, and vice versa.

A **halogenated hydrocarbon** *is a hydrocarbon derivative in which one or more halogen atoms are present.* Such compounds were briefly mentioned in each of the previous two chapters on organic chemistry in the sections dealing with reactions of hydrocarbons. Alkanes, alkenes, alkynes, and aromatic hydrocarbons all undergo halogenation reactions to produce halogenated hydrocarbons. For alkanes and aromatic hydrocarbons, halogenation is a substitution reaction (Sections 10.14 and 11.10). For alkenes and alkynes, halogenation involves an addition reaction (Sections 11.5 and 11.7). The functional groups present in halogenated hydrocarbons are the halogen atoms themselves.

Naming of Halogenated Hydrocarbons

The IUPAC rules for naming halogenated hydrocarbons are similar to those for naming hydrocarbons themselves, with the following modifications:

1. Halogen atoms, treated as substituents on a carbon chain, are called *fluoro-*, *chloro-*, *bromo-*, and *iodo-*.

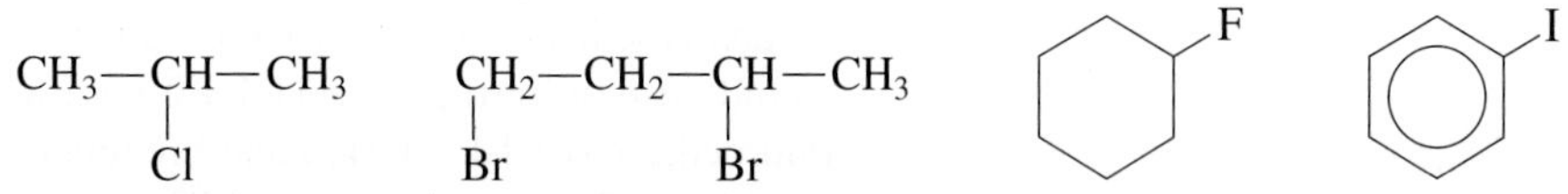

$CH_3—CH(Cl)—CH_3$ 2-Chloropropane

$CH_2(Br)—CH_2—CH(Br)—CH_3$ 1,3-Dibromobutane

Fluorocyclohexane

Iodobenzene

2. When a carbon chain bears both a halogen and an alkyl substituent, the two substituents are considered of equal rank in determining the numbering system for the chain. The chain is numbered from the end closer to a substituent. Alphabetical priority determines the order in which all substituents present are listed and also selects between equivalent numbering systems.

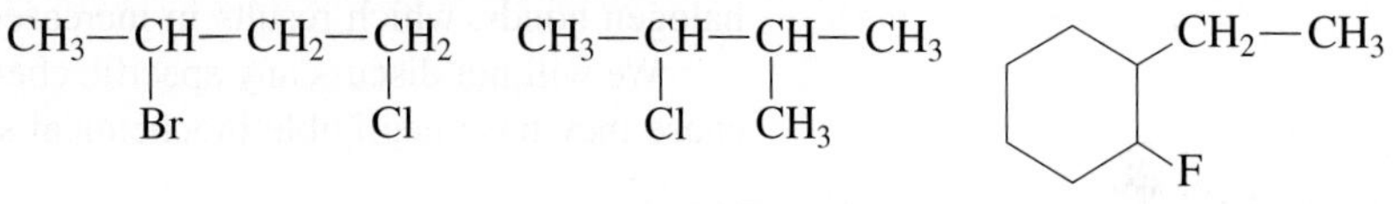

$CH_3—CH(Br)—CH_2—CH_2(Cl)$ 3-Bromo-1-chlorobutane

$CH_3—CH(Cl)—CH(CH_3)—CH_3$ 2-Chloro-3-methylbutane

1-Ethyl-2-fluorocyclohexane

3. When a carbon chain contains a multiple bond, the chain is numbered from the end closest to the multiple bond, regardless of where alkyl and halogen substituents are located.

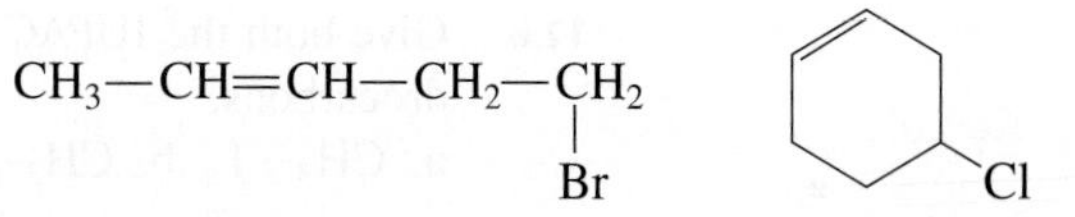

$CH_3—CH=CH—CH_2—CH_2(Br)$ 5-Bromo-2-pentene

4-Chlorocyclohexene

▶ An alternative designation for a halogenated alkane is *alkyl halide*.

▶ The contrast between IUPAC and common names for halogenated hydrocarbons is as follows:
IUPAC (one word)
haloalkane
chloromethane
Common (two words)
alkyl halide
methyl chloride

A common (non-IUPAC) system also exists for naming simple halogenated hydrocarbons; it is used particularly with simple halogenated alkanes. These common names have two parts. The first part is the name of the hydrocarbon portion of the molecule (the alkyl group). The second part (as a separate word) identifies the halogen portion, which is named as though it were an ion (chloride, bromide, and so on), even though no ions are present (all bonds are covalent bonds). The following examples contrast the IUPAC names and the common names (in parentheses).

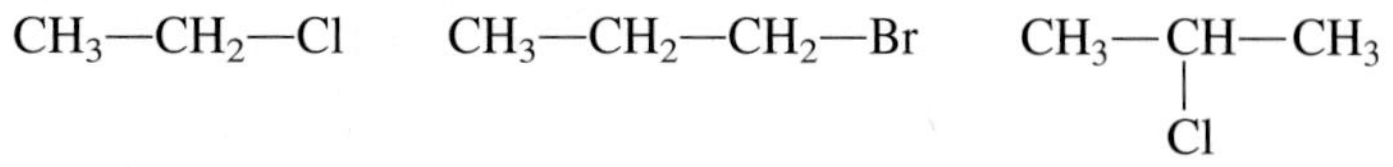

$CH_3—CH_2—Cl$ Ethyl chloride (chloroethane)

$CH_3—CH_2—CH_2—Br$ Propyl bromide (1-bromopropane)

$CH_3—CH(Cl)—CH_3$ Isopropyl chloride (2-chloropropane)

A number of polyhalogenated methanes have acquired additional common names that are usually not clearly related to their structures. Table 12.1 gives five important examples of this additional nomenclature. The compounds Freon-11 and Freon-12, the last two entries in Table 12.1, are examples of chlorofluorocarbons (CFCs). CFCs are synthetic compounds that have been heavily used as refrigerants and as air conditioning chemicals. We now know that CFCs are factors in the destruction of stratospheric (high-altitude) ozone.

Table 12.1 Additional Common Names for Selected Polyhalogenated Methanes

Formula	Name
CH_2Cl_2	methylene chloride
$CHCl_3$	chloroform
CCl_4	carbon tetrachloride
CCl_3F	Freon-11
CCl_2F_2	Freon-12

Uses, Properties, and Reactions of Halogenated Hydrocarbons

No *large* quantities of halogenated hydrocarbons occur naturally, unlike the case for alkanes (petroleum and natural gas). However, *small* amounts of several different halogenated hydrocarbons are found in nature. Most are chlorine- and bromine-containing substances isolated from various species that live in the oceans and seas—sponges, mollusks, and other aquatic creatures. Dissolved chlorine-, bromine-, and iodine-containing ionic species are present in ocean water. Halogenated hydrocarbons used in bulk as chemical intermediates in industry are produced synthetically.

Low-molar-mass halogenated hydrocarbons are excellent solvents and find use as cleaners and degreasers. Carbon tetrachloride (CCl_4) was the first major dry-cleaning solvent; because of toxicity it is no longer used. Replacement solvents include 1,1,1-trichloroethane ($CCl_3—CH_3$), 1,1,2-trichloroethene ($CCl_2═CHCl$), and 1,1,2,2-tetrachloroethene ($CCl_2═CCl_2$). Other uses for simple halogenated hydrocarbons include use as refrigerants [1,1,1,2-tetrafluoroethane ($CF_3—CH_2F$) and 1,1-dichloro-1-fluoroethane ($CCl_2F—CH_3$)] and fire retardants [bromotrifluoromethane (CF_3Br) and bromochlorodifluoromethane ($CBrClF_2$)].

Some halogenated alkanes have densities that are greater than that of water, a situation not common for organic compounds. Chloroalkanes containing two or more chlorine atoms, bromoalkanes, and iodoalkanes are all more dense than water; this is because of the high atomic masses of these elements compared to that of carbon.

The boiling points of halogenated alkanes are generally higher than those of the corresponding alkanes. An important factor contributing to this effect is the polarity of carbon–halogen bonds, which results in increased dipole–dipole interactions (Section 6.13).

We will not discuss any specific chemical reactions of halogenated hydrocarbons because they have negligible biochemical significance.

Practice Problems and Questions

12.4 What is the identity of the functional group present in a monohalogenated alkane?

12.5 There are two functional groups present in a monohalogenated alkene. Identify these two functional groups by name.

12.6 Give both the IUPAC and the common names for these halogenated saturated hydrocarbons.

a. $CH_3—I$ b. $CH_3—CH_2—CH_2—Cl$ c. $CH_3—\underset{\displaystyle F}{\underset{|}{CH}}—CH_3$ d. 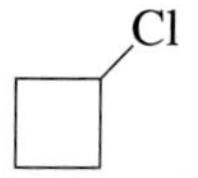

12.7 Give the IUPAC name for each of the following halogenated unsaturated hydrocarbons.

a. $CH_3—\underset{\displaystyle Cl}{\underset{|}{C}}═CH—CH_3$ b. $CH_3—CH═CH—\underset{\displaystyle Br}{\underset{|}{CH}}—CH═CH_2$

c. 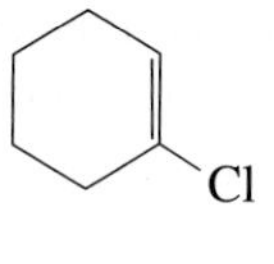

d. 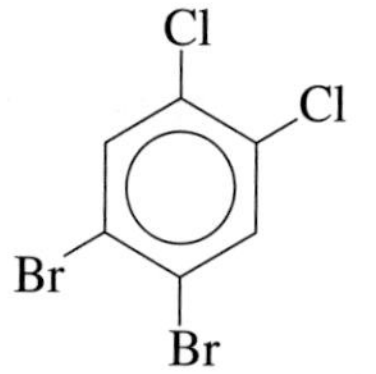

12.8 Draw structural formulas for the following halogenated hydrocarbons.
a. Trichloromethane
b. 1,2-Dichloro-1,1,2,2-tetrafluoroethane
c. Isopropyl bromide
d. *trans*-1-Bromo-3-chlorocyclopentane

12.9 Draw structural formulas for the following halogenated hydrocarbons.
a. 3-Chloro-1-pentene
b. 1,4-Dibromo-1,3-butadiene
c. 2,3-Dibromocyclohexene
d. p-Dichlorobenzene

12.10 There are four structurally isomeric difluoropropanes. What are the IUPAC names for these four compounds?

12.11 Indicate whether each of the following statements is true or false.
a. Several halogenated alkanes are relatively abundant naturally occurring substances.
b. All halogenated alkanes are more dense than water.
c. There are no naturally occurring halogenated hydrocarbons.
d. Several halogenated hydrocarbons are important industrial chemicals.

12.12 Which of the following halogenated alkanes would be predicted to have a density greater than that of water?
a. Fluoromethane b. Chloromethane c. Dichloromethane d. Bromomethane

12.13 What does the notation CFC stand for and what is the environmental problem associated with the use of CFCs?

Learning Focus

Know the general structural characteristics of alcohols and phenols; be able to name such compounds given their structural formulas, or vice versa.

12.3 Structural Features and Naming of Alcohols and Phenols

Alcohols and phenols (a special type of alcohol) contain the same functional group, a *hydroxyl group*. A **hydroxyl group** *is the —OH functional group*. Thus oxygen is the heteroatom present in alcohols and phenols, and it is bonded to a carbon atom (of a carbon chain or ring) and also a hydrogen atom (of the functional group).

We should note that the normal bonding behavior for an oxygen atom in an organic compound is the formation of two covalent bonds. Oxygen is a member of Group VIA of the periodic table and thus possesses six valence electrons. To complete its octet by electron sharing (Section 4.11), an oxygen atom must form two covalent bonds. Thus, in organic chemistry, carbon forms four bonds, hydrogen and halogens form one bond, and oxygen forms two bonds.

An **alcohol** *is a hydrocarbon derivative in which a hydroxyl group (—OH) is attached to a saturated carbon atom.* Even though alcohols contain an —OH group, they are not hydroxides. A *hydroxide* is a compound in which the polyatomic OH^- ion (Section 4.19) is present. Alcohols are not ionic compounds. In an alcohol, the —OH group is *covalently* bonded to a carbon atom.

Saturated carbon atom → —C—OH ← Alcohol functional group

The generalized formula for an alcohol is R—OH, where R represents the hydrocarbon portion of the molecule. Examples of structural formulas for simple alcohols include

$$CH_3—OH \qquad CH_3—CH_2—OH \qquad CH_3—CH_2—CH_2—OH$$

Figure 12.1 shows space-filling models for these three same alcohols, which are the simplest unbranched-chain alcohols.

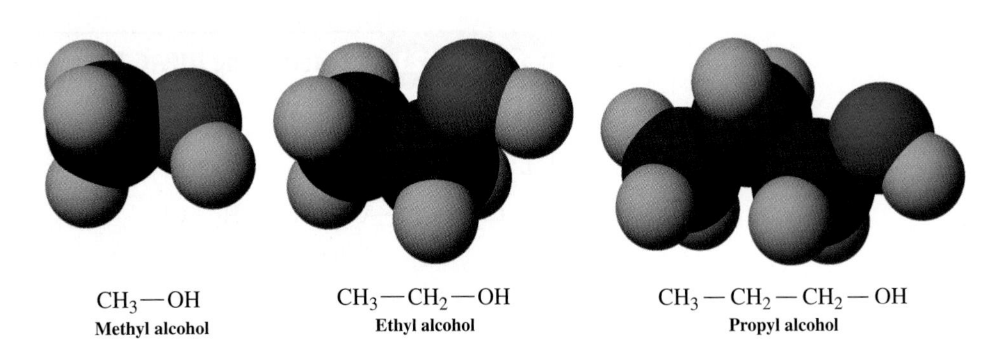

Figure 12.1 Space-filling models for the three simplest unbranched-chain alcohols: methyl alcohol, ethyl alcohol, and propyl alcohol.

▶ The generic term *aryl group* (Ar) is the aromatic counterpart of the nonaromatic generic term *alkyl group* (R).

A **phenol** *is a compound in which a hydroxyl group (—OH) is attached to a carbon atom that is part of an aromatic carbon ring system.* The general formula for phenols is Ar—OH, where Ar represents an aryl group. An **aryl group** *is an aromatic carbon ring system from which one hydrogen atom has been removed.*

Aromatic ring system → [benzene ring] —OH ← Hydroxyl functional group

A hydroxyl group is the functional group for both phenols and alcohols.

The following compounds are examples of simple phenols.

[benzene ring]—OH   Cl—[benzene ring]—OH CH_3, CH_3—[benzene ring]—OH

Naming of Alcohols

Common names exist for alcohols with simple (generally C_1 through C_4) alkyl groups. The word *alcohol,* as a separate word, is placed after the name of the alkyl or cycloalkyl group present.

$CH_3—OH$ Methyl alcohol

$CH_3—CH_2—OH$ Ethyl alcohol

$CH_3—CH_2—CH_2—OH$ Propyl alcohol

$CH_3—CH(CH_3)—OH$ Isopropyl alcohol

[cyclobutane]—OH Cyclobutyl alcohol

IUPAC rules for naming alcohols that contain a single hydroxyl group follow.

Rule 1: *Name the longest carbon chain to which the hydroxyl group is attached.* The chain name is obtained by dropping the final *-e* from the alkane name and adding the suffix *-ol.*

Rule 2: *Number the chain starting at the end nearest the hydroxyl group, and use the appropriate number to indicate the position of the —OH group.* (In numbering of the longest carbon chain, the hydroxyl group has priority over double and triple bonds, as well as over alkyl, cycloalkyl, and halogen substituents.)

Rule 3: *Name and locate any other substituents present.*

Rule 4: *In alcohols where the —OH group is attached to a carbon atom in a ring, the hydroxyl group is assumed to be on carbon 1.*

Example 12.1 Determining IUPAC Names for Alcohols

Name the following alcohols, utilizing IUPAC nomenclature rules.

a. $CH_3—CH_2—C(CH_3)(OH)—CH_2—CH_2—CH_3$

b. $CH_3—CH_2—CH(CH_2—OH)—CH_2—CH_3$

c. (cyclohexane ring bearing CH_3, CH_3, and $-OH$ substituents)

Solution

a. The longest carbon chain that contains the alcohol functional group has six carbons. When we change the *-e* to *-ol,* hexane becomes *hexanol.* Numbering the chain from the end nearest the —OH group identifies carbon number 3 as the location of both the —OH group and a methyl group. The complete name is 3-methyl-3-hexanol.

$$\overset{1}{CH_3}-\overset{2}{CH_2}-\overset{3}{\underset{\underset{OH}{|}}{\overset{\overset{CH_3}{|}}{C}}}-\overset{4}{CH_2}-\overset{5}{CH_2}-\overset{6}{CH_3}$$

b. The longest carbon chain containing the —OH group has four carbon atoms. It is numbered from the end closest to the —OH group as follows:

$$CH_3-CH_2-\overset{2}{\underset{\underset{\overset{1}{CH_2}-OH}{|}}{CH}}-\overset{3}{CH_2}-\overset{4}{CH_3}$$

The base name is 1-butanol. The complete name is 2-ethyl-1-butanol.

c. This alcohol is a cyclohexanol. The carbon to which the —OH group is attached is assumed to be carbon number 1. The complete name for this alcohol is 3,4-dimethylcyclohexanol. Note that the number 1 is not part of the name.

(cyclohexanol ring numbered 1–4: —OH on C1, CH_3 on C3 and C4)

▶ The contrast between IUPAC and common names for alcohols is as follows:

IUPAC (one word)
alkanol
ethanol
Common (two words)
alkyl alcohol
ethyl alcohol

Polyhydroxy alcohols—alcohols that possess more than one hydroxyl group—can be named with only a slight modification of the preceding IUPAC rules. An alcohol in which two hydroxyl groups are present is named as a *diol,* one containing three hydroxyl groups is named as a *triol,* and so on. In these names for diols, triols, and so forth, the final *-e* of the parent alkane name is retained for pronunciation reasons.

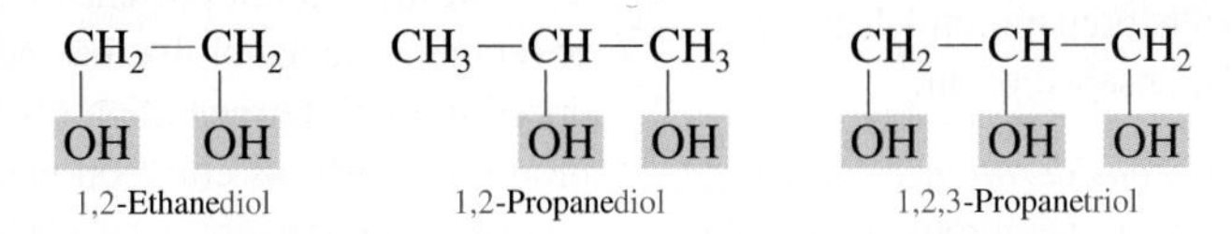

Each of the preceding three polyhydroxy alcohols has a common name with which most people are familiar. The common names are, respectively, ethylene glycol, propylene glycol, and glycerin. Ethylene glycol and propylene glycol, both colorless, odorless, high-boiling liquids, are the main ingredients in automobile "year-round" antifreeze and airplane "de-icers." Glycerin, a clear, thick liquid that has the consistency of honey, has a great affinity for water. It is used as a moisturizer in skin lotions and soap; it is also a good lubricant. Glycerin is normally present in the human body because it is a product of fat metabolism.

Chemical Portraits 19 profiles the two simplest alcohols—methanol and ethanol—as well as the naturally occurring substituted cyclohexanol called menthol (common name). Note the closeness of the names methanol and menthol.

Naming of Phenols

Besides being the name for a family of compounds, *phenol* is also the IUPAC-approved name of the simplest member of the phenol family of compounds

OH

Phenol

Substituted phenols are named as derivatives of phenol. Ring numbering always begins with the hydroxyl group and proceeds in the direction that gives the lower number to the next carbon atom bearing a substituent.

OH, Cl — 3-Chlorophenol (or *meta*-chlorophenol)

OH, CH_3, $CH_2—CH_3$ — 4-Ethyl-2-methylphenol

OH, Br, Br — 2,5-Dibromophenol

Chemically, the behavior of phenols differs considerably from that of alkyl alcohols. This is the reason for considering them a group separate from the other alcohols.

Chemical Portraits 19 Some Commonly Encountered Alcohols

Methanol (Methyl Alcohol) ($CH_3—OH$)

Profile: Methanol, a clear, colorless, flammable, poisonous liquid, has excellent solvent properties. It is the solvent of choice for paints, shellacs, and varnishes. For safety reasons, it is the fuel for race cars; methanol fires are easier to put out than gasoline fires. Methanol is sometimes called *wood alcohol,* terminology drawing attention to an early method for its preparation—heating wood in the absence of air.

Biochemical considerations: Drinking methanol is very dangerous; its metabolism in the human body produces formaldehyde and formic acid. Formaldehyde is toxic to the eye and as little as 1 oz can cause optic nerve damage and blindness. Formic acid causes acidosis.

What is the current method for the industrial production of methanol?

Ethanol (Ethyl Alcohol) ($CH_3—CH_2—OH$)

Profile: Ethanol is a clear, colorless, flammable liquid which—like methanol—is a good industrial solvent. Industrial ethanol is most often *denatured,* that is, it contains an added substance that makes it unfit to drink. It is also used as an octane-booster for gasoline.

Biochemical considerations: Drinking alcohol, the alcohol present in alcoholic beverages, is ethanol that has been produced through fermentation of grain products; hence its common name *grain alcohol.* Pure ethanol is tasteless; the taste of alcoholic beverages comes from other fermentation products present. A 70% ethanol solution is used as an antiseptic to cleanse an area of skin before giving an injection or taking a blood sample; in this function, ethanol destroys bacteria by coagulating their protein.

What are the "pros and cons" of the use of ethanol as a gasoline alternative?

Menthol (5-methyl-2-isopropylcyclohexanol)

Profile: Menthol, a substituted cyclohexanol, has a peppermint taste and odor. It occurs naturally in peppermint oil and can be made synthetically. In its pure state, menthol is a white crystalline solid. It is only slightly soluble in water.

Biochemical considerations: Topical application of menthol to the skin causes a refreshing, cooling sensation followed by a slight burning-and-prickling sensation. At the same time as cooling is perceived, it depresses the nerves for pain reception. The cooling sensation occurs independent of body temperature. Products that often contain menthol include throat sprays and lozenges, cough drops, chest-rub preparations, and aftershave lotions.

Why is menthol often added to toothpastes and mouthwashes?

See the text web site at **www.cengage.com/chemistry/stoker** for answers to the above questions and for further information.

Practice Problems and Questions

12.14 In organic compounds, how many covalent bonds do the following types of atoms form?
a. Oxygen b. Hydrogen c. Halogens d. Carbon

12.15 What are the name and chemical composition of the functional group common to both alcohols and phenols?

12.16 Characterize each of the following structural notations as representing an alcohol, a phenol, or neither an alcohol nor a phenol.
a. R—OH b. R—O—R c. Ar—OH d. Ar—O—Ar

12.17 Characterize each of the following structural formulas as representing an alcohol, a phenol, or neither an alcohol nor a phenol.
a. CH_3—OH b. CH_3—O—CH_3 c. CH_3—CH_2—OH d. CH_3—CH_2—O—CH_3

12.18 Indicate whether each of the following alcohol names is a common name or an IUPAC name.
a. Propyl alcohol b. 2-Butanol c. 2-Methyl-1-pentanol d. Cyclopropyl alcohol

12.19 What is the IUPAC name for each of the following alcohols?
a. Methyl alcohol b. Propyl alcohol
c. Isopropyl alcohol d. Cyclobutyl alcohol

12.20 Assign an IUPAC name to each of the following alcohols.

a. 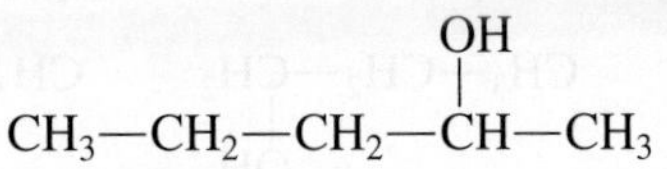

b. CH_3—CH_2—OH

c. 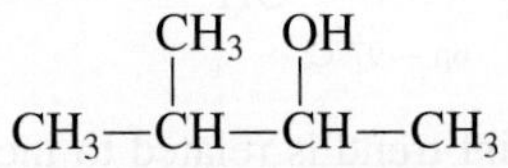

d. 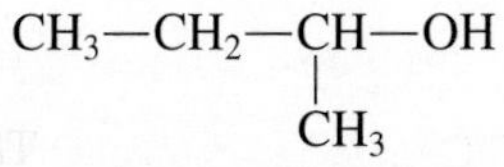

12.21 Write a structural formula for each of the following alcohols.
a. 3-Pentanol
b. 3-Ethyl-3-hexanol
c. 2-Methyl-1-propanol
d. 2-Methylcyclobutanol

12.22 Assign an IUPAC name to each of the following alcohols.

a. 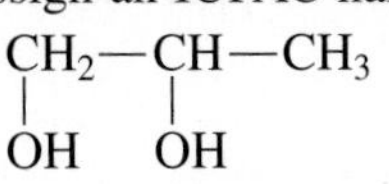

b. 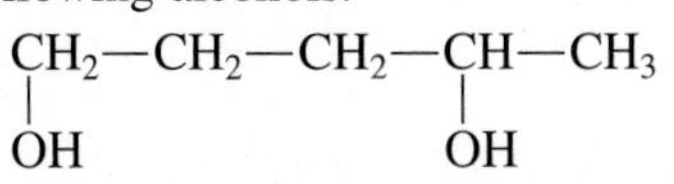

c. 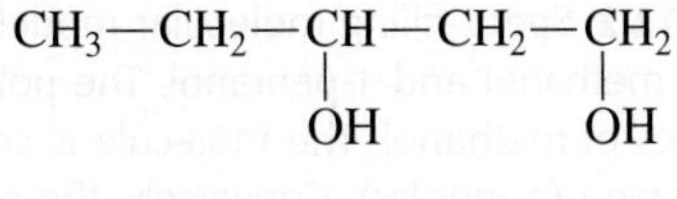

d. 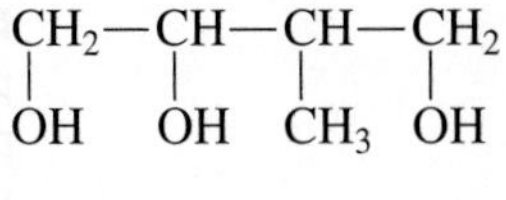

12.23 Name the following phenols.

a. 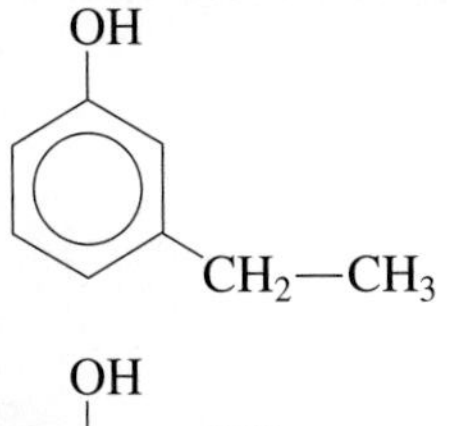

b. 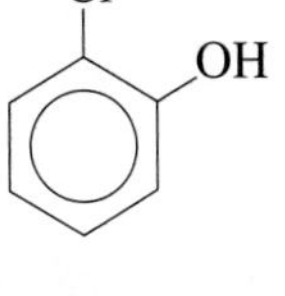

c. 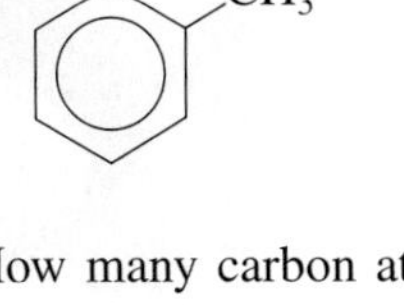

d. 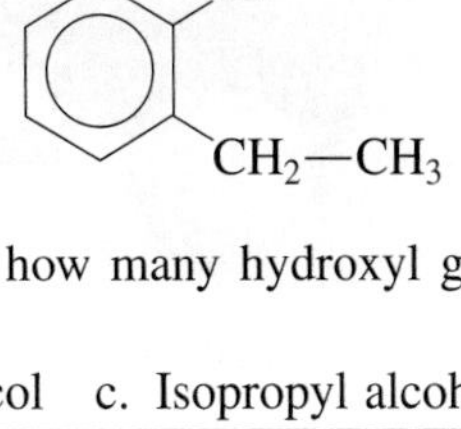

12.24 How many carbon atoms and how many hydroxyl groups are present in each of the following compounds?
a. Glycerin b. Ethylene glycol c. Isopropyl alcohol d. Phenol

Learning Focus

List general physical properties of alcohols and phenols and tell how hydrogen bonding influences them; discuss products from the combustion, dehydration, oxidation, and halogenation of alcohols and phenols.

12.4 Properties and Reactions of Alcohols and Phenols

Alcohol and phenol molecules have both polar and nonpolar character. The hydroxyl group or groups present are polar, and the alkyl (R) portion or aromatic (Ar) portion of the molecule is nonpolar.

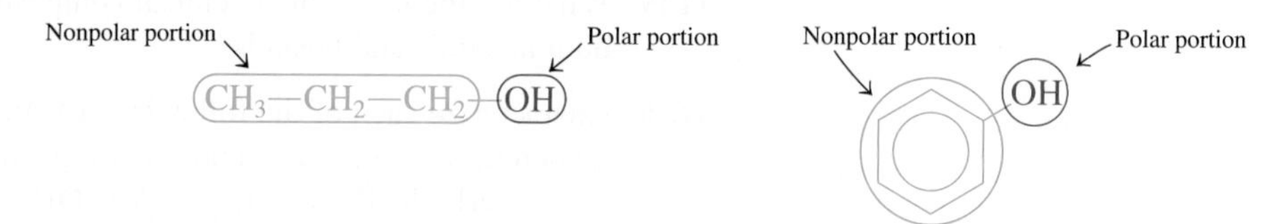

The physical properties of an alcohol or phenol depend on whether the polar or the nonpolar portion of its structure "dominates." For alcohols, factors that determine this include the *length* of the nonpolar carbon chain present and the *number* of polar hydroxyl groups present (see Figure 12.2).

Boiling Points and Water Solubilities

Figure 12.3a shows that the boiling point for unbranched-chain alcohols with an —OH group on an end carbon, 1-alcohols, increases of the length of the carbon chain increases. This trend results from increasing London forces (Section 7.13) with increasing carbon chain length. Alcohols with more than one hydroxyl group present have significantly higher boiling points (bp) than their monohydroxy counterparts.

$CH_3—CH_2—CH_2—OH$ (OH on the third carbon) bp = 97°C

$CH_3—CH(OH)—CH_2(OH)$ bp = 188°C

$CH_2(OH)—CH(OH)—CH_2(OH)$ bp = 290°C

This boiling-point trend is related to increased hydrogen bonding between alcohol molecules (to be discussed shortly).

Small monohydroxy alcohols are soluble in water in all proportions. As carbon chain length increases beyond three carbons, solubility in water rapidly decreases (Figure 12.3b) because of the increasingly nonpolar character of the alcohol. Alcohols with two —OH groups present are more soluble in water than their counterparts with only one —OH group. Increased hydrogen bonding is responsible for this. Diols containing as many as seven carbon atoms show appreciable solubility in water.

Figure 12.2 Space-filling molecular models showing the nonpolar (green) and polar (pink) parts of methanol and 1-pentanol. The polar hydroxyl functional group dominates the physical properties of methanol. The molecule is completely soluble in water (polar) but only partially so in hexane (nonpolar). Conversely, the nonpolar portion of 1-pentanol dominates its physical properties; it is infinitely soluble in hexane and has limited solubility in water.

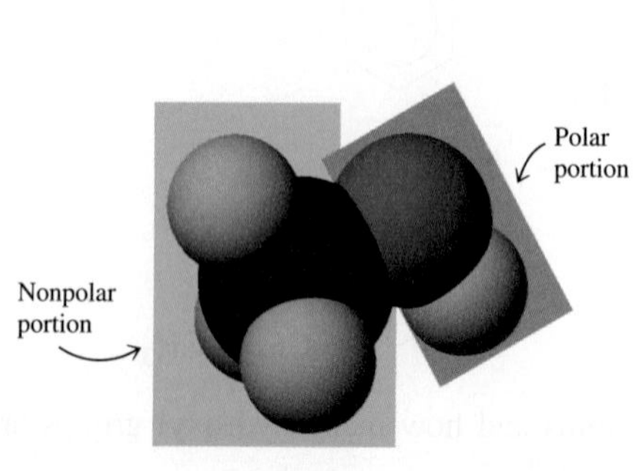

Methanol

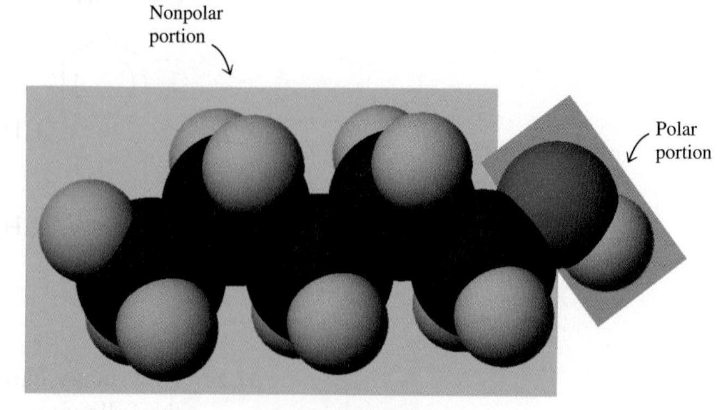

1-Pentanol

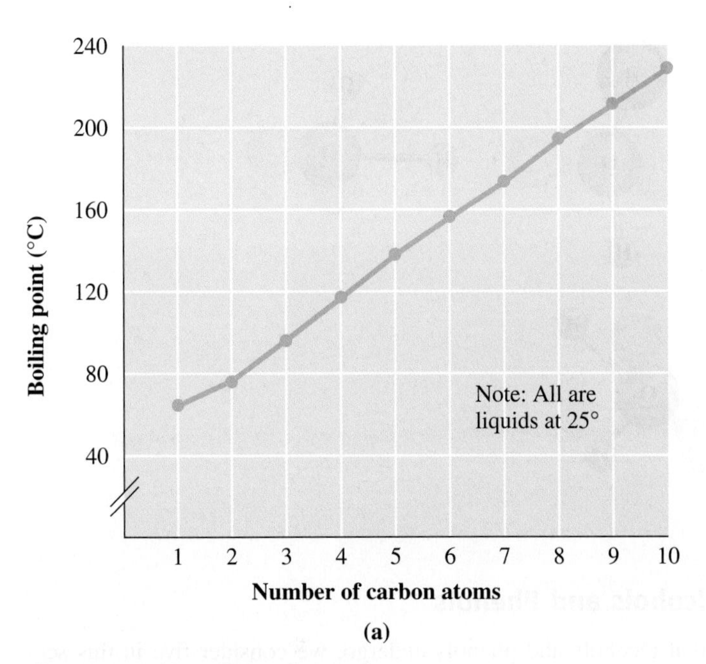

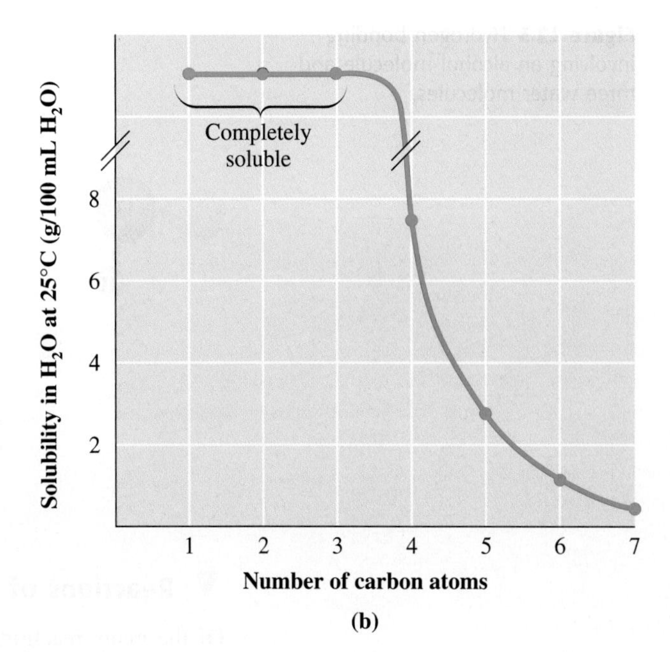

Figure 12.3 (a) Boiling points and (b) solubilities in water of selected 1-alcohols.

A comparison of the properties of alcohols with their alkane counterparts shows that

1. Alcohols have *higher* boiling points than alkanes of similar molecular mass.
2. Alcohols have much *higher* solubility in water than alkanes of similar molecular mass.

The differences in physical properties between alcohols and alkanes are related to hydrogen bonding. Because of their hydroxyl group(s), alcohols can participate in hydrogen bonding, whereas alkanes cannot. Hydrogen bonding between alcohol molecules (see Figure 12.4) is similar to that which occurs between water molecules (Section 6.13).

Extra energy is needed to overcome alcohol–alcohol hydrogen bonds before alcohol molecules can enter the vapor phase. Hence alcohol boiling points are higher than those for the corresponding alkanes (where no hydrogen bonds are present).

Alcohol molecules can also hydrogen-bond to water molecules (see Figure 12.5). The formation of such hydrogen bonds explains the solubility of small alcohol molecules in water. As the alcohol chain length increases, alcohols become more alkane-like (nonpolar), and solubility decreases.

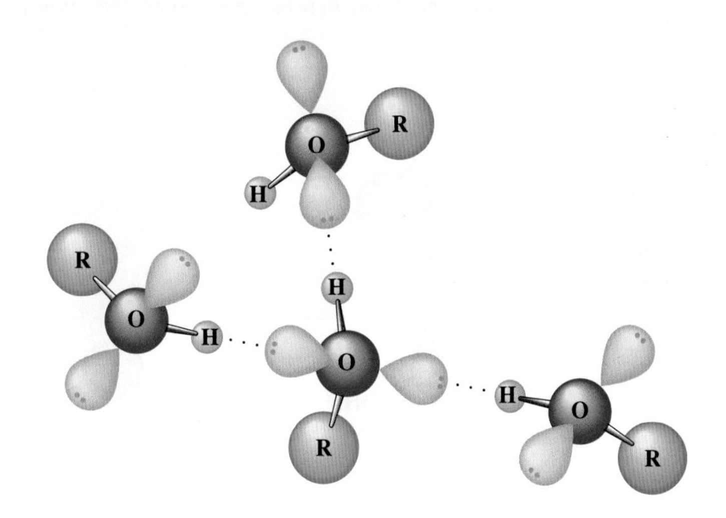

Figure 12.4 Hydrogen bonding among alcohol molecules.

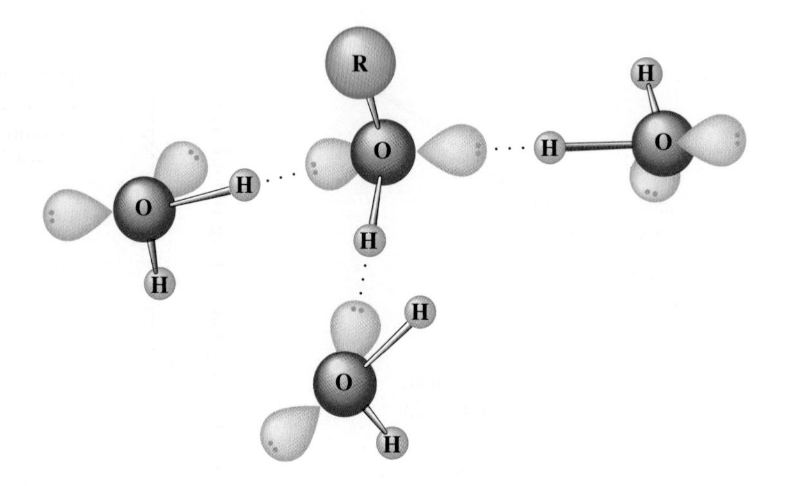

Figure 12.5 Hydrogen bonding involving an alcohol molecule and three water molecules.

Reactions of Alcohols and Phenols

Of the many reactions that alcohols and phenols undergo, we consider five in this section: (1) combustion, (2) dehydration, (3) oxidation, (4) halogenation, and (5) acid–base reactions.

Combustion

As discussed in the previous two chapters, hydrocarbons of all types undergo combustion in air to produce carbon dioxide and water. Alcohols and phenols are also flammable; the combustion products are also carbon dioxide and water. Methyl alcohol is a fuel for racing cars. Oxygenated gasoline, used in the winter in many areas of the United States because it burns "cleaner," contains ethyl alcohol as one of the "oxygenates."

Dehydration

Dehydration is a reaction of alcohols but not of phenols; phenols cannot be dehydrated. A **dehydration reaction** *is a reaction in which the components of water (H and OH) are removed from a single reactant or from two reactants (H from one and OH from the other).* In the alcohol dehydrations that we consider, both water components are removed from the same molecule. Such an alcohol dehydration is an example of an *elimination reaction* (see Figure 12.6), as contrasted to a substitution reaction (Figure 10.12) or an addition reaction (Figure 11.3). An **elimination reaction** *is a reaction in which two groups or two atoms on neighboring carbon atoms are removed, or eliminated, from a molecule, leaving a multiple bond between the two carbon atoms.*

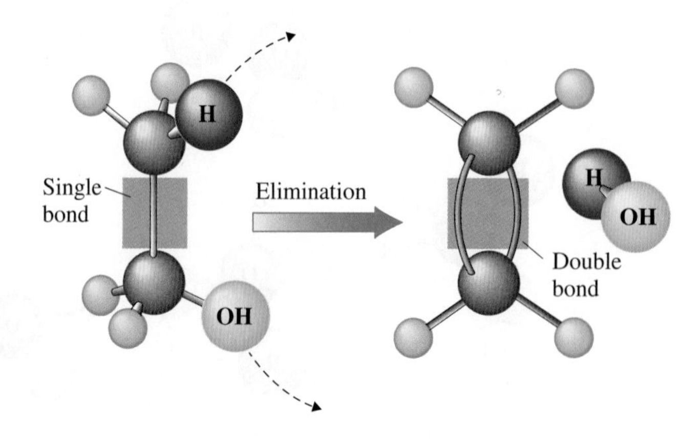

Figure 12.6 In an intramolecular alcohol dehydration (elimination reaction), the components of water (H and OH) are removed from neighboring carbon atoms with the resultant introduction of a double bond into the molecule.

Reaction conditions for the dehydration of an alcohol are a temperature of 180°C and the presence of sulfuric acid (H_2SO_4) as a catalyst. The dehydration product is an alkene.

$$-\underset{\substack{|\\H}}{\overset{|}{C}}-\underset{\substack{|\\OH}}{\overset{|}{C}}- \xrightarrow[180°C]{H_2SO_4} \;\rangle C=C\langle \; + H-OH$$

$$CH_3-\underset{\substack{|\\H}}{CH}-\underset{\substack{|\\OH}}{CH_2} \xrightarrow[180°C]{H_2SO_4} CH_3-CH=CH_2 + H_2O$$

▶ Dehydration of alcohols to form carbon–carbon double bonds occurs in several metabolic pathways in living systems, including the citric acid cycle (Section 18.5). In these biochemical dehydrations, enzymes serve as catalysts instead of acids, and the reaction temperature is 37°C instead of the elevated temperatures required in the laboratory.

In some alcohol molecules there may be more than one carbon atom adjacent to the hydroxyl-bearing carbon from which a hydrogen atom can be removed. In such cases, the adjacent carbon atom bonded to the *fewest* hydrogen atoms loses a hydrogen atom.

$$\underset{\substack{|\\H}}{CH_2}-\underset{\substack{|\\OH}}{CH}-\underset{\substack{|\\H}}{CH}-CH_3 \xrightarrow[180°C]{H_2SO_4} CH_3-CH=CH-CH_3 + H_2O$$

2-Butanol → 2-Butene

Removal (of the H on the first carbon) produces 1-butene; removal (of the H on the third carbon) produces 2-butene.

Dehydration and hydration are "opposite" processes. In Section 11.5 we considered the *hydration* of an alkene to produce an alcohol. *Dehydration* of an alcohol to produce an alkene, our current topic, is the reverse of this.

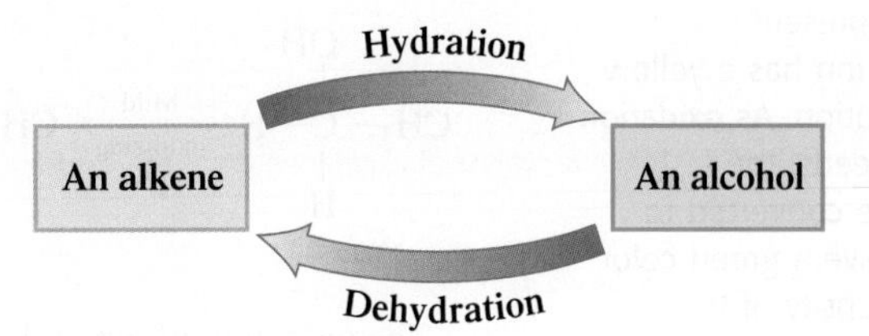

Oxidation

Some alcohols can be oxidized by mild oxidizing agents; others cannot be oxidized by the same oxidizing agents. An alcohol classification system is used to "sort out" whether oxidation will occur or not.

Alcohols are classified as primary (1°), secondary (2°), or tertiary (3°), depending on the number of carbon atoms bonded to the carbon atom that bears the hydroxyl group. A **primary alcohol** *is an alcohol in which the hydroxyl-bearing carbon atom is attached to only one other carbon atom.* A **secondary alcohol** *is an alcohol in which the hydroxyl-bearing carbon atom is attached to two other carbon atoms.* A **tertiary alcohol** *is an alcohol in which the hydroxyl-bearing carbon atom is attached to three other carbon atoms.* Chemical reactions of alcohols often depend on alcohol class (1°, 2°, or 3°).

▶ Pronounce 1° as "primary," 2° as "secondary," and 3° as "tertiary."

▶ Methyl alcohol, CH_3—OH, an alcohol in which the hydroxyl-bearing carbon atom is attached to three hydrogen atoms, does not fit any of the alcohol classification definitions. It is usually grouped with the primary alcohols because its reactions are similar to those of these alcohols.

$$CH_3-\underset{\substack{|\\H}}{\overset{\substack{H\\|}}{C}}-OH \qquad CH_3-\underset{\substack{|\\H}}{\overset{\substack{CH_3\\|}}{C}}-OH \qquad CH_3-\underset{\substack{|\\CH_3}}{\overset{\substack{CH_3\\|}}{C}}-OH$$

1° Alcohol 2° Alcohol 3° Alcohol

Both 1° and 2° alcohols are oxidized by common oxidizing agents. Tertiary (3°) alcohols and phenols do not react with the oxidizing agents that cause 1° and 2° alcohol oxidization.

In Section 8.2 we saw that oxidation can be defined in several ways. The definition that most directly applies to the situation now under consideration is that oxidation can be recognized by a reactant's loss of two hydrogen atoms. In alcohol oxidation, the hydroxyl group loses its hydrogen atom, and a second hydrogen atom is lost from the hydroxyl–bearing carbon atom. The net result is the formation of a carbon–oxygen double bond.

$$\underset{\text{An alcohol}}{-\overset{\overset{\textstyle \text{O—H}}{|}}{\underset{|}{\text{C}}}\text{—H}} \xrightarrow[\text{agent}]{\text{Mild oxidizing}} \underset{\text{Compound containing a carbon–oxygen double bond}}{-\overset{\overset{\textstyle \text{O}}{\|}}{\underset{|}{\text{C}}}} + 2\text{H}$$

The two "removed" hydrogen atoms combine with oxygen supplied by the oxidizing agent to give H_2O.

Primary and secondary alcohols, the two types of oxidizable alcohols, yield different products upon oxidation; a 1° alcohol produces an *aldehyde,* and a 2° alcohol produces a *ketone*. Generalized structures for these two types of product compounds are

$$\underset{\text{Aldehyde}}{\text{R—}\overset{\overset{\textstyle \text{O}}{\|}}{\text{C}}\text{—H}} \qquad \underset{\text{Ketone}}{\text{R—}\overset{\overset{\textstyle \text{O}}{\|}}{\text{C}}\text{—R}}$$

Note that aldehydes and ketones, both of which will be discussed in detail in the next chapter, contain a carbon–oxygen double bond. Specific examples of 1° and 2° alcohol oxidations are

$$\underset{1^\circ\text{ alcohol}}{\text{CH}_3\text{—}\overset{\overset{\textstyle \text{OH}}{|}}{\underset{\underset{\textstyle \text{H}}{|}}{\text{C}}}\text{—H}} \xrightarrow[\text{oxidizing agent}]{\text{Mild}} \underset{\text{Aldehyde}}{\text{CH}_3\text{—}\overset{\overset{\textstyle \text{O}}{\|}}{\text{C}}\text{—H}} \qquad \underset{2^\circ\text{ alcohol}}{\text{CH}_3\text{—}\overset{\overset{\textstyle \text{OH}}{|}}{\underset{\underset{\textstyle \text{H}}{|}}{\text{C}}}\text{—CH}_3} \xrightarrow[\text{oxidizing agent}]{\text{Mild}} \underset{\text{Ketone}}{\text{CH}_3\text{—}\overset{\overset{\textstyle \text{O}}{\|}}{\text{C}}\text{—CH}_3}$$

The oxidation of ethanol, a 1° alcohol, is the basis for the "breathalyzer test" that law enforcement officers use to determine whether an individual suspected of driving under the influence (DUI) has a blood alcohol level exceeding legal limits (see Figure 12.7).

Figure 12.7 "Breathalyzer test" used by police officers to determine whether a DUI suspect is legally drunk.

The DUI suspect is required to breathe into an apparatus containing a solution of potassium dichromate ($K_2Cr_2O_7$). The unmetabolized alcohol in the person's breath is oxidized by the dichromate ion ($Cr_2O_7^{2-}$), and the extent of the reaction gives a measure of the amount of alcohol present.

The dichromate ion has a yellow-orange color in solution. As oxidation of the alcohol proceeds, the dichromate ions are converted to Cr^{3+} ions, which have a green color in solution. The intensity of the green color that develops is measured and is proportional to the amount of ethanol in the suspect's breath, which in turn has been shown to be proportional to the person's blood alcohol level.

Halogenation

Alcohols and phenols undergo halogenation reactions in which the hydroxyl group is replaced by a halogen atom. Using such a substitution reaction to produce alkyl halides is more convenient than halogenation of an alkane (Section 10.14), because mixtures of products are *not* obtained. In the product molecules, the halogen is found only where the —OH groups were originally located.

Acid–Base Reactions

One of the most important properties of phenols is their acidity. The hydroxyl hydrogen atom of a phenol can be removed from the phenol to form a phenoxide anion. Alcohols do not exhibit this acidity property. When phenol itself is reacted with sodium hydroxide (a base), the hydroxyl hydrogen atom is removed, producing the salt sodium phenoxide.

$$\underset{\text{Phenol}}{\text{C}_6\text{H}_5\text{—OH}} + \text{NaOH}(aq) \longrightarrow \underset{\text{Sodium phenoxide}}{\text{C}_6\text{H}_5\text{—O}^-\text{Na}^+} + \text{H}_2\text{O}$$

Summary of Reactions Involving Alcohols

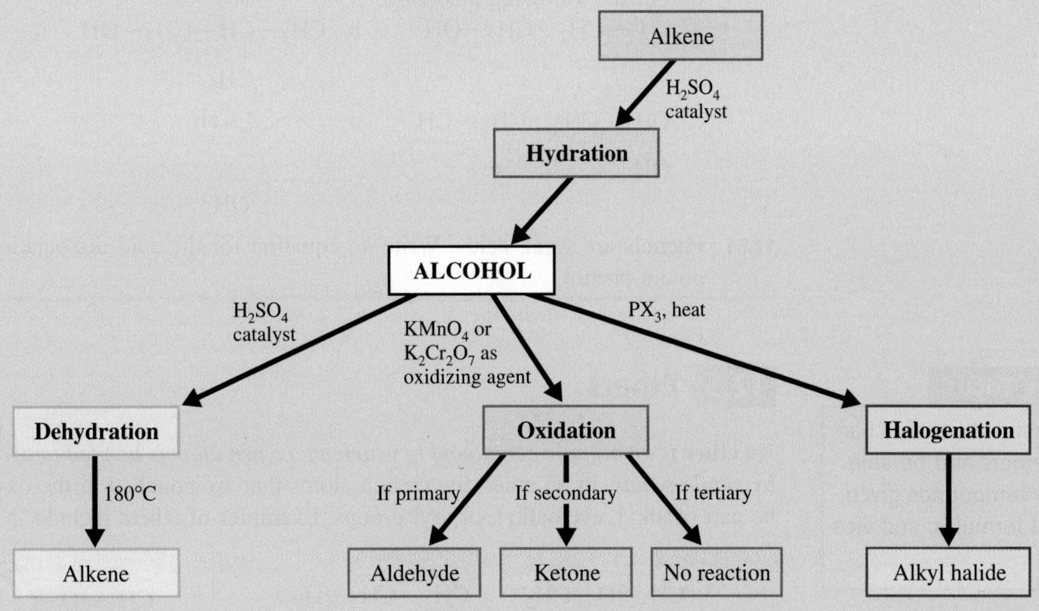

The accompanying Chemistry at a Glance summarizes important chemical reactions that alcohols undergo.

Practice Problems and Questions

12.25 Ethanol, the two-carbon alcohol, is soluble in water; ethane, the two-carbon alkane, is not soluble in water. Explain why.

12.26 1-Propanol is soluble in water; 1-octanol is not soluble in water. Explain why.

12.27 Which member of each of the following pairs of compounds would you expect to have the higher boiling point?

a. 1-Butanol and 1-heptanol
b. Ethanol and 1,2-ethanediol

12.28 Draw the condensed structural formula of the alkene produced by the dehydration of each of the following alcohols.

a. $CH_3{-}\underset{\displaystyle OH}{\underset{|}{CH}}{-}CH_3$ b. $CH_3{-}\underset{\displaystyle CH_3}{\underset{|}{CH}}{-}CH_2{-}OH$

c. $CH_3{-}\underset{\displaystyle OH}{\underset{|}{CH}}{-}CH_2{-}CH_3$ d. $CH_3{-}CH_2{-}CH_2{-}OH$

12.29 Draw the condensed structural formula of the aldehyde or ketone produced in each of the following oxidation reactions. If no oxidation occurs, write "no reaction."

a. $CH_3{-}CH_2{-}CH_2{-}OH$ b. $CH_3{-}\underset{\displaystyle CH_3}{\underset{|}{CH}}{-}CH_2{-}OH$

c. $CH_3{-}\underset{\displaystyle OH}{\underset{|}{CH}}{-}CH_2{-}CH_3$ d. $CH_3{-}\underset{\displaystyle CH_3}{\underset{|}{CH}}{-}OH$

12.30 Draw the condensed structural formula of the alkyl halide produced by halogenation of the following alcohols.

a. $CH_3—CH_2—CH_2—OH$

b. $CH_3—CH(CH_3)—CH_2—OH$

c. $CH_2(OH)—CH_2—CH_2—CH_3$

d. 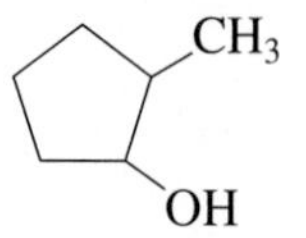 cyclopentane ring with CH_3 and OH on adjacent carbons

12.31 Phenols are weak acids. Write an equation for the acid dissociation of the compound phenol.

Learning Focus

Know the general structural characteristics of ethers and be able to name such compounds given their structural formulas, and vice versa.

12.5 Ethers

An **ether** *is an organic compound in which an oxygen atom is bonded to two carbon atoms by single bonds.* In an ether, the carbon atoms that are attached to the oxygen atom can be part of alkyl, cycloalkyl, or aryl groups. Examples of ethers include

$CH_3—O—CH_3$ $CH_3—CH_2—O—$(cyclopentyl) $CH_3—O—$(phenyl)

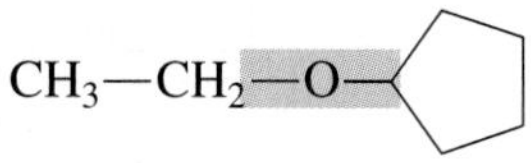

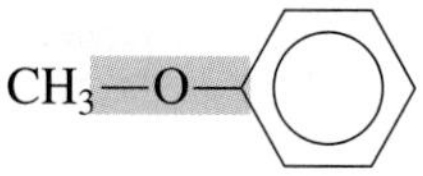

The two groups attached to the oxygen atom of an ether can be the same (first structure), but they need not be so (second and third structures).

All ethers contain a C—O—C unit, which is the ether functional group.

Ether functional group

---C—O—C---

Generalized formulas for ethers, which depend on the types of groups attached to the oxygen atom (alkyl or aryl), include R—O—R, R—O—R′ (where R′ is an alkyl group different from R), R—O—Ar, and Ar—O—Ar. Note that unlike alcohols and phenols, ethers do not possess a hydroxyl (—OH) group.

Names of Ethers

Common names for ethers are formed by naming the two hydrocarbon groups attached to the oxygen atom and adding the word *ether*. The hydrocarbon groups are listed in alphabetical order. When both hydrocarbon groups are the same, the prefix *di-* is placed before the name of the hydrocarbon group. In this system, ether names consist of two or three separate words.

$CH_3—O—CH_2—CH_3$ — Ethyl methyl ether

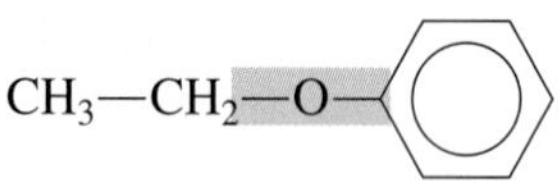

$CH_3—CH_2—O—$(phenyl) — Ethyl phenyl ether

$CH_3—O—CH_3$ — Dimethyl ether

In the IUPAC nomenclature system, ethers are named as substituted hydrocarbons. The smaller hydrocarbon attachment and the oxygen atom are called an *alkoxy group,* and this group is considered a substituent on the larger hydrocarbon group. An **alkoxy group** *is an alkyl group to which an oxygen atom has been added.* Simple alkoxy groups include the following:

$CH_3—O—$ — Methoxy group

$CH_3—CH_2—O—$ — Ethoxy group

$CH_3—CH_2—CH_2—O—$ — Propoxy group

The general symbol for an alkoxy group is —O—R (or —OR).

▶ The contrast between IUPAC and common names for ethers is as follows:

IUPAC (one word)
alkoxyalkane
2-methoxybutane
Common (three or two words)
alkyl alkyl ether
ethyl methyl ether
or
dialkyl ether
dipropyl ether

The following examples illustrate IUPAC ether nomenclature. In each structure, the alkoxy groups present have been highlighted.

$CH_3—O—CH_2—CH_2—CH_2—CH_3$

1-Methoxybutane

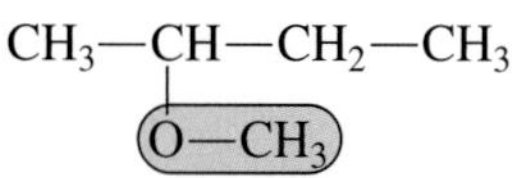

Methoxycyclohexane

$CH_3—CH_2—CH_2—O—CH_2—CH_2—CH_3$

1-Propoxypropane

$CH_3—CH(O—CH_3)—CH_2—CH_3$

2-Methoxybutane

$CH_3—CH(CH_3)—CH_2—O—CH_2—CH_3$

1-Ethoxy-2-methylpropane

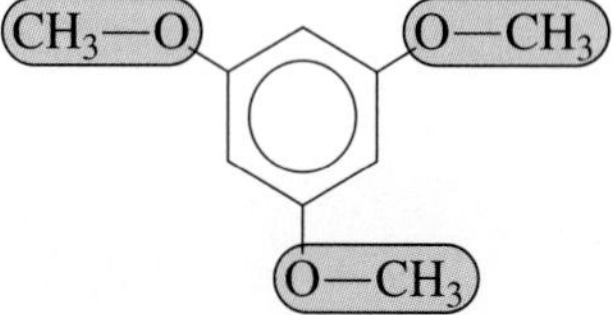

1,3,5-Trimethoxybenzene

Uses, Properties and Reactions of Ethers

For many people, the word *ether* evokes thoughts of hospital operating rooms and anesthesia. This response derives from the *former* large-scale use of diethyl ether as a general anesthetic. Diethyl ether, which is a highly flammable compound, has been replaced by nonflammable general anesthetics that became widely available in the early 1960s. In general, nonflammability was achieved by incorporating halogen atoms into anesthetic molecules. Some of the "new" anesthetics are not ethers; others are halogenated ethers. Two of the most commonly used anesthetics are the hexahalogenated ethers enflurane and isoflurane.

$H—CFCl—CF_2—O—CF_2—H$

Enflurane

$F—CF_2—CHCl—O—CF_2—H$

Isoflurane

The boiling points of ethers are similar to those of alkanes of comparable molecular mass and are much lower than those of alcohols of comparable molecular mass. The much higher boiling point of the alcohol results from hydrogen bonding between alcohol molecules. Ether molecules, like alkanes, cannot hydrogen-bond to one another. Ether oxygen atoms have no hydrogen atom attached directly to them.

Ethers, in general, are more soluble in water than are alkanes of similar molecular mass, because ether molecules are able to form hydrogen bonds with water (Figure 12.8). Ethers have water solubilities similar to those of alcohols of the same molecular mass. For example, diethyl ether and butyl alcohol have the same solubility in water. Because ethers can also hydrogen-bond to alcohols, alcohols and ethers tend to be mutually soluble. Nonpolar substances tend to be more soluble in ethers than in alcohols because ethers have no hydrogen-bonding network that has to be broken up for solubility to occur.

Two chemical properties of ethers are especially important.

1. *Ethers are flammable.* Special care must be exercised in laboratories where ethers are used. Diethyl ether, whose boiling point of 35°C is only a few degrees above room temperature, is a particular flash-fire hazard.
2. *Ethers react slowly with oxygen from the air to form unstable hydroperoxides and peroxides.*

$R—O—O—H$ Hydroperoxide

$R—O—O—R$ Peroxide

Such compounds, when concentrated, represent an explosion hazard and must be removed before *stored* ethers are used.

▶ The term *ether* comes from the Latin *aether,* which means "to ignite." This name is given to these compounds because of their high vapor pressure at room temperature, which makes them very flammable.

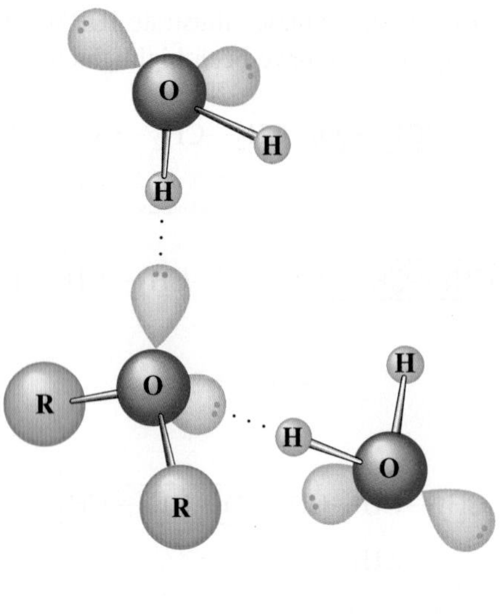

Figure 12.8 Hydrogen bonding between an ether molecule and two water molecules.

Ethers are unreactive toward acids, bases, and oxidizing agents. Like alkanes, they do undergo halogenation reactions.

The general chemical unreactivity of ethers, coupled with the fact that most organic compounds are ether-soluble, makes ethers excellent solvents in which to carry out organic reactions. Their relatively low boiling points simplify their separation from the reaction products.

Practice Problems and Questions

12.32 Indicate whether or not each of the following structural notations represents an ether.

a. R—O—R b. R—O—R′ c. Ar—O—R d. Ar—O—Ar

12.33 What is the difference in meaning associated with each of the following pairs of notations?

a. R—O—R and R—O—R′ b. Ar—O—R and Ar—O—Ar

12.34 Assign an IUPAC name to each of the following ethers.

a. $CH_3—O—CH_2—CH_2—CH_3$ b. $CH_3—CH_2—CH_2—O—CH_2—CH_3$

c. $CH_3—CH—CH_3$ with $O—CH_3$ on the middle carbon d. $CH_3—CH—O—CH_3$ with CH_3 on the CH carbon

12.35 Assign an IUPAC name to each of the following ethers.

a. 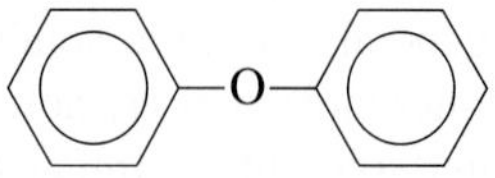

b. 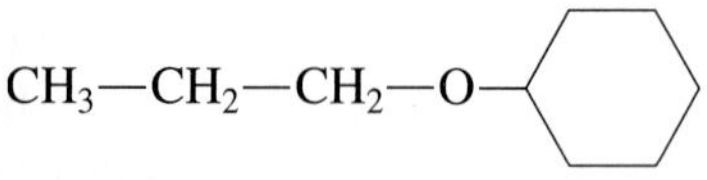

c. 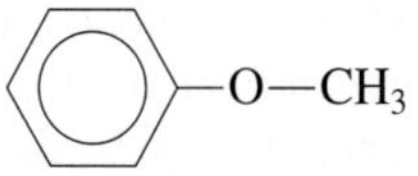

d. 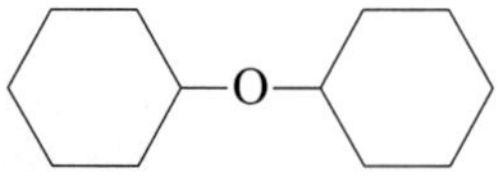

12.36 Assign a common name to each of the ethers in Problem 12.34.

12.37 Assign a common name to each of the ethers in Problem 12.35.

12.38 Draw the condensed structural formula for each of the following ethers.

a. Isopropyl propyl ether b. Ethyl phenyl ether

c. 2-Ethoxypentane d. Ethoxycyclobutane

12.39 Dimethyl ether and ethyl alcohol have the same molecular mass. Dimethyl ether is a gas at room temperature, and ethyl alcohol is a liquid at room temperature. Explain these observations.

12.40 Compare the solubility in water of ethers and alcohols that have similar molecular masses.

12.41 What are the two chemical hazards associated with ether use?

Learning Focus

Know the general structural characteristics of thiols and be able to name such compounds given their structural formulas, and vice versa.

12.6 Thiols

Many organic compounds containing oxygen have sulfur analogs, in which a sulfur atom has replaced an oxygen atom. Because sulfur is in the same group of the periodic table as oxygen, such a replacement proceeds without problems as far as electrons and bonding are concerned. Sulfur, like oxygen (Section 12.3), always forms two bonds in organic compounds.

Thiols, the sulfur analogs of alcohols, contain —SH functional groups instead of —OH functional groups. The thiol functional group is called a *sulfhydryl* group. A **sulfhydryl group** *is the —SH functional group.*

Hydroxyl group ↙	Sulfhydryl group ↙
R—OH	R—SH
An alcohol	A thiol

A **thiol** *is a hydrocarbon derivative in which a sulfhydryl group (—SH) is present.*

Naming of Thiols

Thiols are named in the same way as alcohols in the IUPAC system, except that the *-ol* becomes *-thiol*. The prefix *thio-* indicates the substitution of a sulfur atom for an oxygen atom in a compound.

$$CH_3—\underset{\displaystyle OH}{\underset{|}{CH}}—CH_2—CH_3 \qquad CH_3—\underset{\displaystyle SH}{\underset{|}{CH}}—CH_2—CH_3$$

2-Butanol 2-Butanethiol

As in the case of diols and triols, the *-e* at the end of the alkane name is also retained for thiols.

Properties and Reactions of Thiols

The most obvious characteristic of thiols is their strong, often disagreeable, odors. Skunk scent (see Figure 12.9), for example, is caused primarily by the following two simple thiols:

$$\begin{matrix} H_3C & & H \\ & C{=}C & \\ H & & CH_2—SH \end{matrix} \qquad CH_3—\underset{\displaystyle CH_3}{\underset{|}{CH}}—CH_2—CH_2—SH$$

trans-2-Butenethiol 3-Methyl-1-butanethiol

▶ Bacteria in the mouth interact with saliva and leftover food to produce such compounds as hydrogen sulfide, and methanethiol. These compounds, which have odors detectable in air at concentrations of parts per billion, are responsible for "morning breath."

Figure 12.9 Thiols are responsible for the strong odor of "essence of skunk." Their odor is an effective defense mechanism.

Table 12.2
Selected Thiols That Are Present in Foods

Flavor or odor of	Structures of compounds involved
oysters	$CH_3—SH$
cheddar cheese	$CH_3—SH$
freshly chopped onions	$CH_3—CH_2—CH_2—SH$
garlic	$CH_2{=}CH—CH_2—SH$

The familiar odor of natural gas results from the addition of thiols to the gas. The exceptionally low threshold of detection for thiols enables consumers to smell a gas leak long before the gas reaches dangerous levels. Many of the strong odors associated with foods are caused by small thiol molecules (Table 12.2).

In general, thiols are more reactive than their alcohol counterparts. The larger size of a sulfur atom compared to an oxygen atom results in a carbon–sulfur covalent bond that is weaker than a carbon–oxygen covalent bond. An added factor is that sulfur's electronegativity (2.5) is significantly lower than that of oxygen (3.5).

Thiols are easily oxidized but yield different products than their alcohol analogs. Thiols form *disulfides*. Each of two thiol groups loses a hydrogen atom, thus linking the two sulfur atoms together via a disulfide group, —S—S—.

$$R—SH + HS—R \xrightarrow{\text{Oxidation}} \underset{\text{A disulfide}}{R—S—S—R} + 2H$$

Reversal of this reaction, a reduction process, is also readily accomplished. Breaking of the disulfide bond regenerates two thiol molecules. These two "opposite reactions" are of biochemical importance in protein chemistry (Chapter 16).

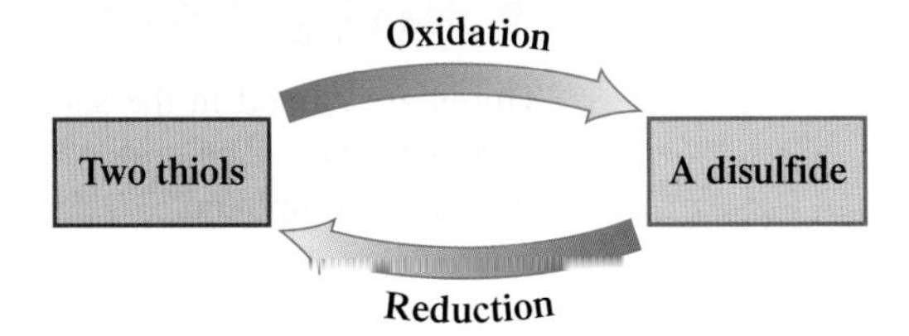

Practice Problems and Questions

12.42 What is the difference between a hydroxyl group and a sulfhydryl group?

12.43 Draw a condensed structural formula for each of the following thiols.
a. Methanethiol b. 2-Propanethiol
c. Cyclopentanethiol d. 1,2-Ethanedithiol

12.44 Give the IUPAC name for each of the following thiols.
a. $CH_3—CH_2—SH$ b. $CH_3—CH(CH_3)—SH$
c. (cyclohexane ring)—SH d. $CH_3—CH{=}CH—SH$

12.45 What is the "dominating" characteristic of thiols?

12.46 Contrast the organic products that result from the oxidation of an alcohol with those that result from the oxidation of a thiol.

12.47 Write the formulas for the sulfur-containing organic products of the following reactions.
a. $2CH_3—CH_2—SH \xrightarrow[\text{agent}]{\text{Oxidizing}}$
b. $CH_3—CH_2—S—S—CH_2—CH_3 \xrightarrow[\text{agent}]{\text{Reducing}}$

Learning Focus

Know the general structural characteristics of amines; be able to name such compounds given their structural formulas, or vice versa.

12.7 Structural Features and Naming of Amines

Nitrogen is the heteroatom in amines. The normal bonding behavior for a nitrogen atom in an organic compound is the formation of three covalent bonds. Nitrogen is a member of Group VA of the periodic table and thus possesses five valence electrons. To complete its octet by electron sharing (Section 4.11), a nitrogen atom must form three covalent bonds. Thus, in organic chemistry, carbon forms four bonds, hydrogen and halogens form one bond, oxygen forms two bonds, and nitrogen forms three bonds.

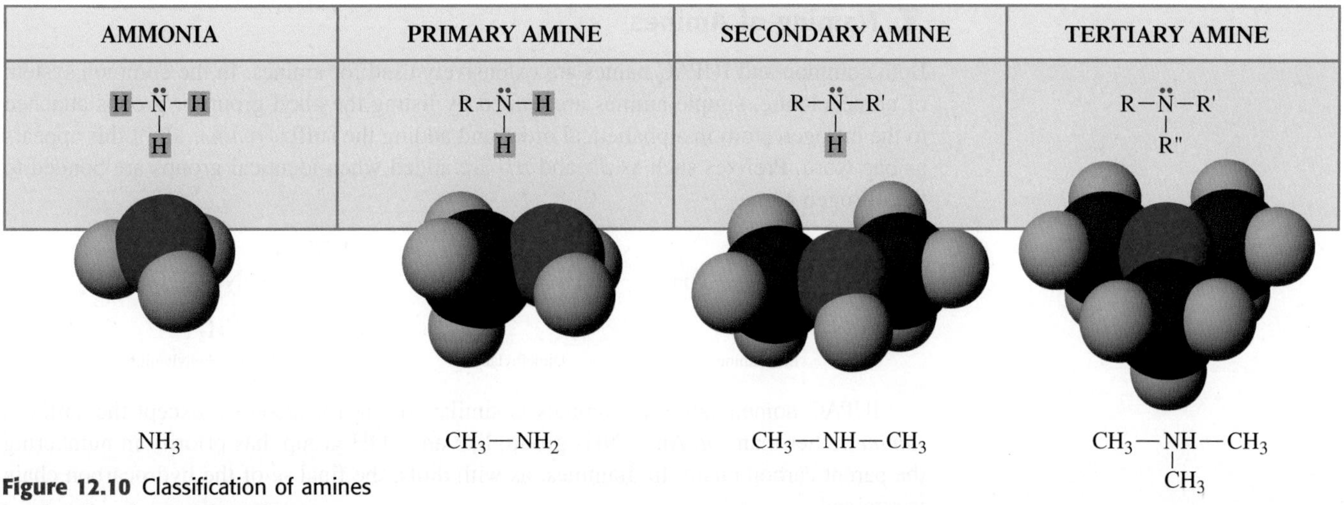

AMMONIA	PRIMARY AMINE	SECONDARY AMINE	TERTIARY AMINE
H—N̈—H, with H below N	R—N̈—H, with H below N	R—N̈—R', with H below N	R—N̈—R', with R'' below N
NH_3	$CH_3—NH_2$	$CH_3—NH—CH_3$	$CH_3—N—CH_3$, with CH_3 below N

Figure 12.10 Classification of amines is related to the number of R groups attached to the nitrogen atom.

Amines are often thought of as being organic derivatives of the inorganic molecule ammonia (NH_3). An **amine** *is an organic derivative of ammonia (NH_3) in which one or more hydrogen atoms on the nitrogen atom of ammonia have been replaced by an alkyl, a cycloalkyl, or an aryl group.* Amines can be classified as primary (1°), secondary (2°), or tertiary (3°), depending on the number of hydrogen atoms of ammonia that have been replaced by hydrocarbon groups (Figure 12.10).

The functional group present in a primary amine, the $—NH_2$ group, is called an *amino* group. An **amino group** *is the $—NH_2$ functional group.* Secondary and tertiary amines possess substituted amino groups.

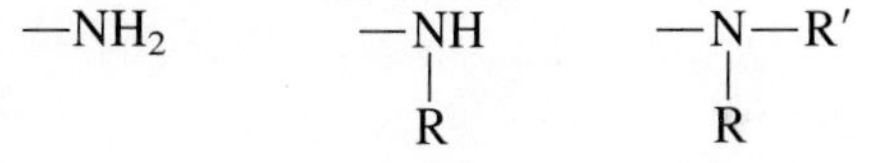

- $—NH_2$ — Amino group
- —NH, with R below N — Monosubstituted amino group
- —N—R', with R below N — Disubstituted amino group

Example 12.2 Classifying Amines as Primary, Secondary, or Tertiary

Classify each of the following amines as a primary, secondary, or tertiary amine.

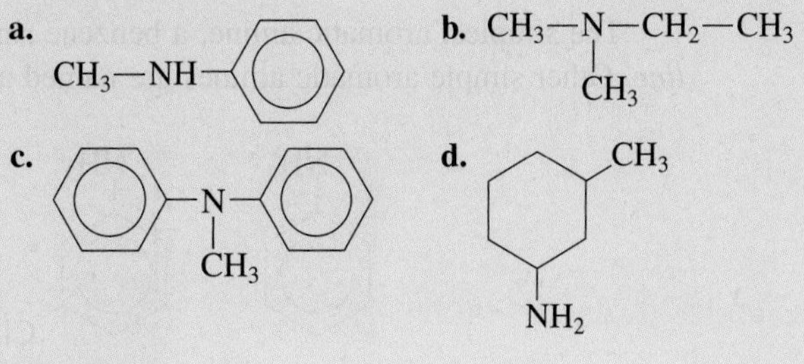

a. $CH_3—NH—$ attached to a benzene ring

b. $CH_3—N—CH_2—CH_3$, with CH_3 below N

c. N bonded to two benzene rings, with CH_3 below N

d. A cyclohexane ring with CH_3 and NH_2 substituents

Solution

The number of carbon atoms directly bonded to the nitrogen atom determines the amine classification.

a. This is a secondary amine, because the nitrogen is bonded to both a methyl group and a phenyl group.
b. Here we have a tertiary amine, because the nitrogen atom is bonded to two methyl groups and an ethyl group.
c. This is also a tertiary amine; the nitrogen atom is bonded to two phenyl groups and a methyl group.
d. This is a primary amine. The nitrogen atom is bonded to only one carbon atom.

Naming of Amines

Both common and IUPAC names are extensively used for amines. In the common system of nomenclature, simple amines are named by listing the alkyl group or groups attached to the nitrogen atom in alphabetical order and adding the suffix *-amine;* all of this appears as one word. Prefixes such as *di-* and *tri-* are added when identical groups are bonded to the nitrogen atom.

$CH_3-CH_2-NH_2$

Ethylamine

$CH_3-NH-CH_3$

Dimethylamine

$C_6H_5-N(CH_3)-CH_2-CH_3$

Ethylmethylphenylamine

IUPAC nomenclature for amines is similar to that for alcohols, except the suffix is *-amine* rather than *-ol*. An $-NH_2$ group, like an $-OH$ group, has priority in numbering the parent carbon chain. In diamines, as with diols, the final *-e* of the hydrocarbon chain is retained.

$CH_3-CH_2-CH_2-CH_2-NH_2$

1-Butanamine

$CH_3-CH(NH_2)-CH_2-CH_3$

2-Butanamine

$CH_3-CH(CH_3)-CH_2-CH_2-NH_2$

3-Methyl-1-butanamine

$H_2N-CH_2-CH_2-CH_2-CH_2-NH_2$

1,4-Butanediamine

In secondary and tertiary amines, the prefix *N-* is used for each substituent on the nitrogen atom.

$CH_3-CH_2-CH(NH-CH_3)-CH_3$

N-methyl-2-butanamine

$CH_3-CH_2-N(CH_3)-CH_2-CH_2-CH_3$

N-ethyl-*N*-methyl-1-propanamine

$CH_3-CH(CH_3)-CH(NH-CH_3)-CH_2-CH_3$

2,*N*-dimethyl-3-pentanamine

The simplest aromatic amine, a benzene ring bearing an amino group, is called *aniline*. Other simple aromatic amines are named as derivatives of aniline.

Aniline

m-Chloroaniline

2,3-Dichloroaniline

In secondary and tertiary aromatic amines, the additional group or groups attached to the nitrogen atom are located using the prefix *N-*.

N-ethylaniline

N,N-dimethylaniline

3,*N*-dimethylaniline

▶ The contrast between IUPAC names and common names for primary, secondary, and tertiary amines is as follows:

Primary Amines
IUPAC (one word)
alkanamine
Common (one word)
alkylamine

Secondary Amines
IUPAC (one word)
N-alkylalkanamine
Common (one word)
alkylalkylamine

Tertiary Amines
IUPAC (one word)
N-alkyl-*N*-alkylalkanamine
Common (one word)
alkylalkylalkylamine

Example 12.3 Determining IUPAC Names for Amines

Assign IUPAC names to each of the following amines.

a. $CH_3—CH_2—NH—(CH_2)_4—CH_3$ **b.** 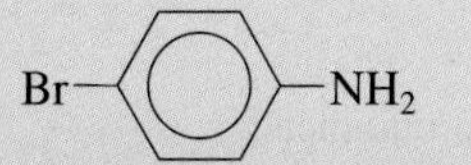 Br—⟨benzene ring⟩—NH_2

c. $H_2N—CH_2—CH_2—NH_2$ **d.** 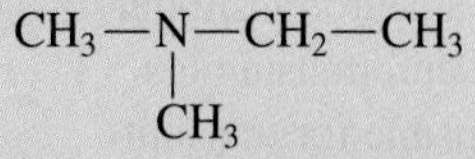 $CH_3—\underset{\displaystyle CH_3}{\underset{|}{N}}—CH_2—CH_3$

Solution

a. The longest carbon chain has five carbons. The name of the compound is *N-ethyl-1-pentanamine*.
b. This compound is named as a derivative of aniline: *4-bromoaniline* (or *p*-bromoaniline). The carbon in the ring to which the $—NH_2$ is attached is carbon 1.
c. Two $—NH_2$ groups are present in this molecule. The name is *1,2-ethanediamine*.
d. This is a tertiary amine in which the longest carbon chain has two carbons (ethane). The base name is thus *ethanamine*. We also have two methyl groups attached to the nitrogen atom. The name of the compound is *N,N-dimethylethanamine*.

Several amines have important functions in the human body. They include dopamine (a brain chemical) and epinephrine (a central nervous system stimulant). These two amines as well as the amine ephedrin, a component of many "dietary supplements," are profiled in Chemical Portraits 20. All three of these amines have structures that contain hydroxyl (alcohol) groups in addition to amine groups, a situation commonly encountered for biologically important amines. Compounds like this with more than one functional group are called *polyfunctional* organic compounds. A **polyfunctional organic compound** *is an organic compound in which two or more different functional groups are present.*

Practice Problems and Questions

12.48 What is the difference between a primary amine and a tertiary amine?

12.49 Classify each of the following amines as a primary, a secondary, or a tertiary amine.

a. 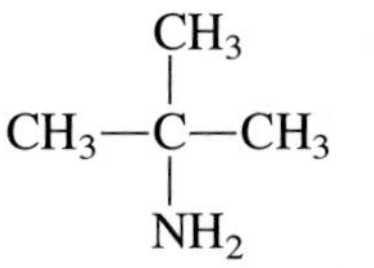 $CH_3—\overset{\displaystyle CH_3}{\overset{|}{\underset{\displaystyle NH_2}{\underset{|}{C}}}}—CH_3$ b. 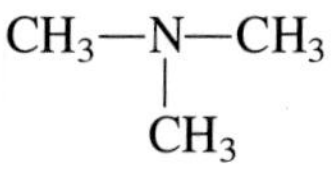 $CH_3—\underset{\displaystyle CH_3}{\underset{|}{N}}—CH_3$

c. $CH_3—NH—CH_3$ d. 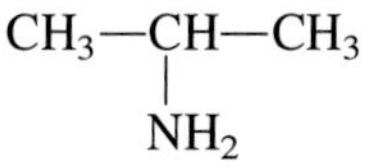 $CH_3—\underset{\displaystyle NH_2}{\underset{|}{CH}}—CH_3$

12.50 Assign a common name to each of the following amines.

a. $CH_3—NH—CH_2—CH_3$ b. $CH_3—CH_2—CH_2—NH_2$

c. 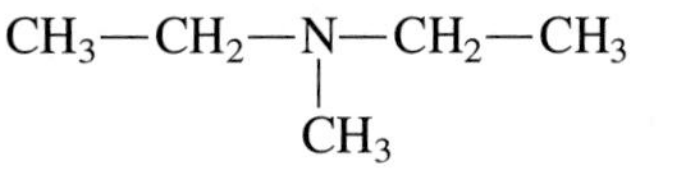 $CH_3—CH_2—\underset{\displaystyle CH_3}{\underset{|}{N}}—CH_2—CH_3$ d. 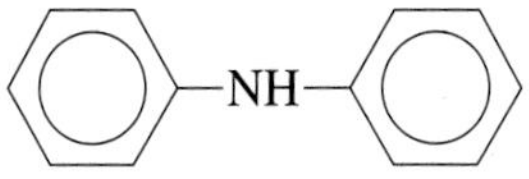 ⟨benzene ring⟩—NH—⟨benzene ring⟩

12.51 Assign an IUPAC name to each of the following amines.

a. 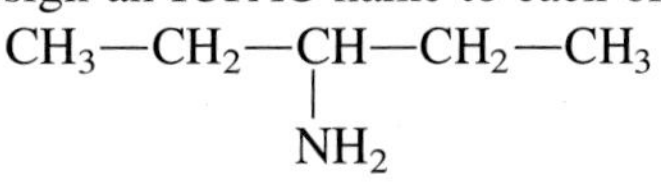 $CH_3—CH_2—\underset{\displaystyle NH_2}{\underset{|}{CH}}—CH_2—CH_3$ b. 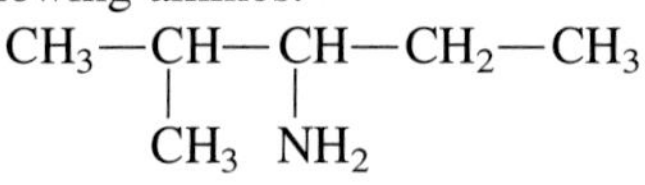 $CH_3—\underset{\displaystyle CH_3}{\underset{|}{CH}}—\underset{\displaystyle NH_2}{\underset{|}{CH}}—CH_2—CH_3$

c. 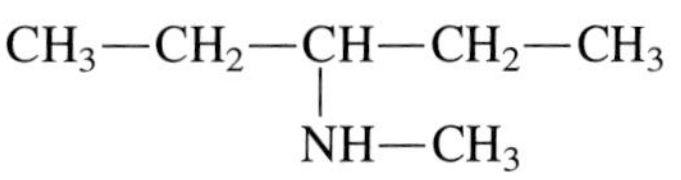 $CH_3—CH_2—\underset{\displaystyle NH—CH_3}{\underset{|}{CH}}—CH_2—CH_3$ d. 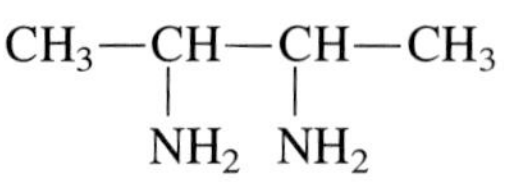 $CH_3—\underset{\displaystyle NH_2}{\underset{|}{CH}}—\underset{\displaystyle NH_2}{\underset{|}{CH}}—CH_3$

Chemical Portraits 20 Biologically Important Hydroxy Amines

Dopamine

Profile: Dopamine, pronounced "DOPE-a-mean", is a neurotransmitter found in the brain. Neurotransmitters are molecules responsible for transmitting signals between various nerve cells in the body.

Biochemical considerations: Abnormally low brain dopamine levels are associated with Parkinson's disease, a neurological disease characterized by tremors, loss of motor function, and speech problems. Administration of dopamine to a patient with Parkinson's disease does not stop the symptoms of the disease, because dopamine in the blood cannot cross the blood-brain barrier. The drug *L-dopa,* which can cross the blood-brain barrier, does give relief from Parkinson's symptoms; inside brain cells, it is converted to dopamine.

What are dopamine agonists and what are dopamine antagonists?

Epinephrine

Profile: Epinephrine, pronounced "ep-in-NEFF-rin", is a hormone produced in the human body by the adrenal glands. It functions both as a neurotransmitter and a central nervous system stimulant, with the latter role being more important. An alternate name for epinephrine is *adrenaline.*

Biochemical considerations: Pain, excitement, and fear trigger the release of large amounts of adrenaline (epinephrine) into the bloodstream. The effect is increased blood glucose levels, which in turn increases blood pressure, rate and force of heart contraction, and muscular strength. These changes cause the body to function at a "higher" level. Adrenaline is often called the "fight or flight" hormone.

What are some medical uses of epinephrine?

Ephedrin

Profile: Ephedrin, pronounced "eh-FEH-drin", is a substance extracted from the Asian plant ephedra that is used in many products sold as weight-loss aids and energy boosters.

Biochemical considerations: Within the body, ephedrin behaves as a stimulant to the heart and the central nervous system. The FDA calls it an "amphetamine-like compound." Currently, its use sends a "mixed-message." It is banned in the Olympics and college sports. It is banned in professional football, but not in professional basketball or baseball. It is sold as a "dietary supplement" at nutrition stores, supermarkets and on the Internet. It is bought by millions of Americans who take billions of doses.

What are the "pros and cons" of ephedrin use?

See the text web site at **www.cengage.com/chemistry/stoker** for answers to the above questions and for further information.

12.52 Name each of the following aromatic amines as a derivative of aniline.

a. $CH_3-N-CH_2-CH_3$ (N bonded to a benzene ring)

b. benzene ring $-N(CH_3)-CH_2-CH_3$

c. benzene ring $-NH-CH(CH_3)-CH_3$

d. benzene ring $-NH_2$ with ortho CH_3

12.53 Draw a condensed structural formula for each of the following amines.

a. Ethylamine b. 2-Methyl-2-butanamine
c. 1,3-Propanediamine d. *N*-Methylaniline

12.8 Properties and Reactions of Amines

Learning Focus

List general properties of amines; understand why amines exhibit basic properties; write equations for the formation of amine salts.

The methylamines (mono-, di-, and tri-) and ethylamine are gases at room temperature and have ammonia-like odors. Most other amines are liquids, and many have odors resembling that of raw fish. A few amines, particularly diamines, have strong, disagreeable odors. The foul odor arising from dead fish and decaying flesh is due to amines released by the bacterial decomposition of protein. Two of these "odoriferous" compounds are the diamines putrescine and cadaverine.

$H_2N—(CH_2)_4—NH_2$ Putrescine (1,4-butanediamine)

$H_2N—(CH_2)_5—NH_2$ Cadaverine (1,5-pentanediamine)

Amines of low molecular mass, with fewer than six carbon atoms, are infinitely soluble in water. This solubility results from hydrogen bonding between the amines and water. Even tertiary amines are water-soluble, because the amine nitrogen atom has a nonbonding electron pair that can form a hydrogen bond with a hydrogen atom of water (see Figure 12.11).

Basicity of Amines

Amines, like ammonia, are weak bases. In Section 9.4 we saw that ammonia's weak-base behavior results from its accepting a proton (H^+) from water to produce ammonium ion (NH_4^+) and hydroxide ion (OH^-).

$$\underset{\text{Ammonia}}{\ddot{N}H_3} + HOH \rightleftharpoons \underset{\text{Ammonium ion}}{NH_4^+} + \underset{\text{Hydroxide ion}}{OH^-}$$

Amines behave in a similar manner.

$$\underset{\text{Methylamine}}{CH_3—\ddot{N}H_2} + HOH \rightleftharpoons \underset{\text{Methylammonium ion}}{[CH_3—\overset{+}{N}H_3]} + \underset{\text{Hydroxide ion}}{OH^-}$$

The result of the interaction of an amine with water is a basic solution containing substituted ammonium ions and hydroxide ions. A **substituted ammonium ion** *is an ammonium ion in which one or more alkyl, cycloalkyl, or aryl groups have been substituted for hydrogen atoms.*

Three important generalizations apply to substituted ammonium ions.

▶ Amines, like ammonia, have a pair of unshared electrons on the nitrogen atom present. These unshared electrons can accept a hydrogen ion from water. Thus both amines and ammonia produce basic aqueous solutions.

1. *Substituted ammonium ions are always charged species rather than neutral molecules.*
2. *The nitrogen atom in an ammonium ion or a substituted ammonium ion participates in four bonds.* In a neutral compound, nitrogen atoms form only three bonds. Four bonds about a nitrogen atom are possible, however, when the species is a positive ion.
3. *Substituted ammonium ions have common names derived from the names of the "parent" amines.* Replacement of the word *amine* in the name of the "parent" amine with the words *ammonium ion* generates the name of the substituted ammonium ion. The following two examples illustrate this nomenclature pattern.

▶ Substituted ammonium ions always contain one more hydrogen atom than their "parent" amine. They also always carry a +1 charge, whereas the "parent" amine is a neutral molecule.

$$\underset{\text{Ethylamine}}{CH_3—CH_2—NH_2} \xrightarrow{H_2O} \underset{\text{Ethylammonium ion}}{CH_3—CH_2—\overset{+}{N}H_3} + OH^-$$

$$\underset{\text{Ethyldimethylamine}}{CH_3—\underset{\substack{|\\CH_3}}{N}—CH_2—CH_3} \xrightarrow{H_2O} \underset{\text{Ethyldimethylammonium ion}}{CH_3—\underset{\substack{|\\CH_3}}{\overset{+}{N}H}—CH_2—CH_3} + OH^-$$

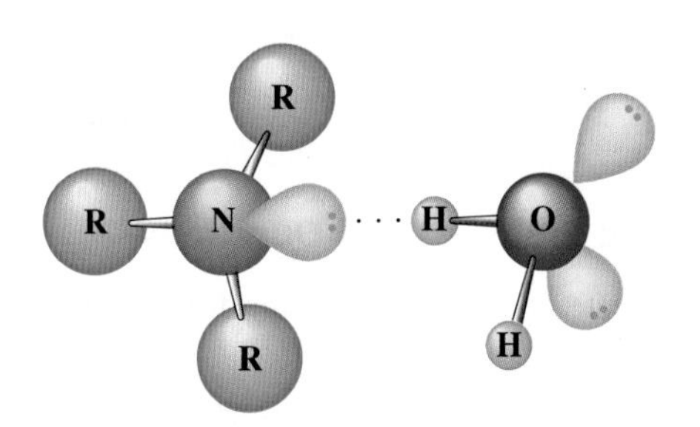

Figure 12.11 Hydrogen bonding between a tertiary amine molecule and a water molecule.

Formation of Amine Salts

The reaction of an acid with a base (neutralization) produces a salt (Section 9.6). Because amines are bases, their reaction with an acid produces a salt, an amine salt.

$$\underset{\text{Amine}}{CH_3\text{—}\ddot{N}H_2} + \underset{\text{Acid}}{H\text{—}Cl} \longrightarrow \underset{\text{Amine salt}}{CH_3\text{—}\overset{+}{N}H_3\ Cl^-}$$

Aromatic amines react with acids in a similar manner.

$$\underset{\text{Amine}}{C_6H_5\text{—}\ddot{N}H\text{—}CH_3} + \underset{\text{Acid}}{H\text{—}Cl} \longrightarrow \underset{\text{Amine salt}}{C_6H_5\text{—}\overset{+}{N}H_2\text{—}CH_3\ Cl^-}$$

An **amine salt** *is an ionic compound in which the positive ion is a mono-, di-, or trisubstituted ammonium ion (RNH_3^+, $R_2NH_2^+$, or R_3NH^+) and the negative ion comes from the acid.* Amine salts can be obtained in crystalline form (odorless, white crystals) by evaporating the water from the acidic solutions in which amine salts are prepared.

Amine salts are named using standard nomenclature procedures for ionic compounds (Section 4.9). The name of the positive ion, the substituted ammonium or anilinium ion, is given first and is followed by a separate word for the name of the negative ion.

$$\underset{\text{Ethylammonium chloride}}{CH_3\text{—}CH_2\text{—}\overset{+}{N}H_3\ Cl^-} \qquad \underset{\text{Dimethylammonium bromide}}{CH_3\text{—}\overset{+}{N}H_2\text{—}CH_3\ Br^-}$$

Many higher-molecular-mass amines are water-insoluble; however, virtually all amine salts are water-soluble. Thus amine salt formation provides a means for converting water-insoluble compounds into water-soluble compounds. Many drugs that contain amine functional groups are administered to patients in the form of amine salts because of their increased solubility in water in this form.

Many people unknowingly use acids to form amine salts when they put vinegar or lemon juice on fish. Such action converts amines in fish (often smelly compounds) to salts, which are odorless.

The process of forming amine salts with acids is an easily reversed process. Treating an amine salt with a strong base such as NaOH regenerates the "parent" amine.

$$\underset{\text{Amine salt}}{CH_3\text{—}\overset{+}{N}H_3\ Cl^-} + \underset{\text{Base}}{NaOH} \longrightarrow \underset{\text{Amine}}{CH_3\text{—}NH_2} + NaCl + H_2O$$

The "opposite nature" of the formation of an amine salt from an amine and the regeneration of the amine from its amine salt can be diagrammed as follows:

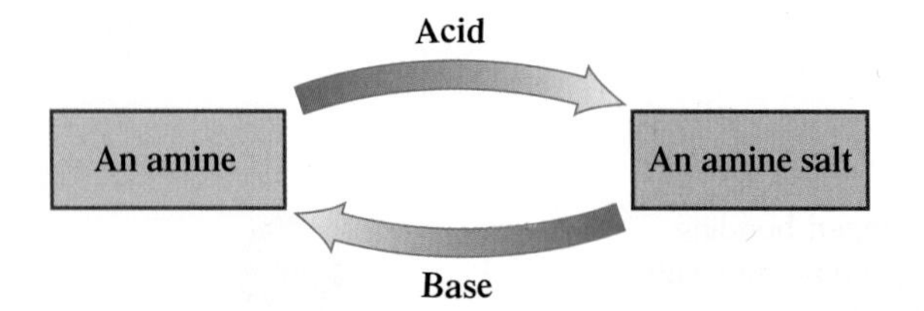

Practice Problems and Questions

12.54 Indicate whether each of the following amines is a liquid or a gas at room temperature.
a. Butylamine b. Dimethylamine c. Ethylamine d. Dibutylamine

12.55 Which compound in each of the following pairs of amines would you expect to be more soluble in water? Justify each answer.

a. $CH_3—CH_2—NH_2$ and $CH_3—CH_2—CH_2—CH_2—CH_2—NH_2$

b. $CH_3—CH_2—CH_2—NH_2$ and $H_2N—CH_2—CH_2—CH_2—NH_2$

12.56 Write an equation showing the basic behavior of each of the following amines in water.

a. $CH_3—NH_2$ b. $CH_3—CH_2—CH_2—NH_2$

c. $CH_3—NH—CH_3$ d. $CH_3—CH_2—NH—CH_2—CH_3$

12.57 Write the condensed structural formula of the salts obtained from the reactions of the amines in Problem 12.56 with hydrochloric acid (HCl).

12.58 Show the structures of the missing substance(s) in the following reactions involving amine salts.

a. $CH_3—CH(CH_3)—\overset{+}{N}H_3\,Cl^- + NaOH \longrightarrow ? + NaCl + H_2O$

b. $? + NaOH \longrightarrow CH_3—NH—CH_3 + NaCl + H_2O$

c. $C_6H_5—\overset{+}{N}H(CH_3)—CH_3\,Br^- + NaOH \longrightarrow ? + ? + H_2O$

d. $CH_3—\overset{+}{N}H_2—CH_3\,Cl^- + NaOH \longrightarrow ? + NaCl + H_2O$

12.59 Name each of the following amine salts.

a. $CH_3—CH_2—CH_2—\overset{+}{N}H_3\,Cl^-$

b. $CH_3—CH_2—CH_2—\overset{+}{N}H_2(CH_3)\,Cl^-$

c. $CH_3—CH_2—\overset{+}{N}H(CH_3)—CH_3\,Br^-$

d. $C_6H_{11}—\overset{+}{N}H(CH_3)—CH_3\,Br^-$ (cyclohexyl)

Learning Focus

List the general characteristics of heterocyclic compounds and heterocylic amines; know what an alkaloid is and be able to give examples of such compounds.

12.9 Heterocyclic Amines

A **heterocyclic organic compound** *is a cyclic organic compound in which one or more of the carbon atoms in the ring have been replaced with heteroatoms.* The most common heteroatoms are oxygen and nitrogen. Examples of heterocyclic ring systems include

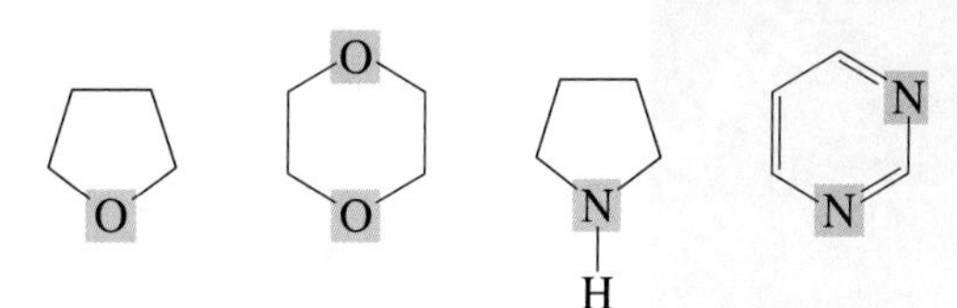

The first two ring systems are examples of heterocyclic ethers, and the latter two ring systems examples of heterocyclic amines. Note that in the ether ring systems, each oxygen atom forms two bonds to carbon atoms, a requirement for ethers (Section 12.5), and in the amine ring systems, each nitrogen atom forms three bonds, at least two of which involve carbon, a requirement for heterocyclic amines.

Heterocyclic amines are the most common heterocyclic organic compounds. These compounds are the "parent" compounds for numerous derivatives that are important in medicinal, agricultural, food, and industrial chemistry, as well as in the functioning of the human body.

Three additional examples of heterocyclic amine structures are

Purine Pyridine Pyrrolidine

(Nearly all heterocyclic compounds go by common names.)

The two most widely used central nervous system (CNS) stimulants in our society, caffeine and nicotine, are heterocyclic amine derivatives. Caffeine's structure is based on a purine ring system. Nicotine's structure contains one pyridine ring and one pyrrolidine ring.

Caffeine Nicotine

Note that caffeine and nicotine are also examples of *polyfunctional* organic compounds (Section 12.7). Caffeine contains amine functional groups, ketone functional groups (Section 13.4), and a carbon–carbon double bond. Nicotine contains amine functional groups and carbon–carbon double bonds.

Caffeine and nicotine are also examples of *alkaloids*. An **alkaloid** *is a physiologically active nitrogen-containing organic compound extracted from plants*. Most alkaloids are heterocyclic amines. Caffeine, the most widely used CNS stimulant today, is found in coffee beans and tea leaves. Nicotine, the second most widely used CNS stimulant in our society, is found in smoking tobacco and chewing tobacco.

An extremely important family of alkaloids is the narcotic painkillers, a class of drugs derived from the resin (opium) of the oriental poppy plant (see Figure 12.12). The most important drugs obtained from opium are morphine and codeine. Synthetic modification of morphine produces the illegal drug heroin. These three compounds have similar chem-

Figure 12.12 Oriental poppy plants, the source of the painkiller morphine.

Morphine Codeine

Heroin

ical structures. The structures differ only in the two attachments on the bottom left and bottom right of the structures. In morphine two hydroxyl groups are present, in codeine a hydroxyl and a methoxy (ether) group are present, and in heroin two ester groups (Section 13.8) are present.

Morphine is one of the most effective painkillers known; its painkilling properties are about a hundred times greater than those of aspirin. Morphine acts by blocking the process in the brain that interprets pain signals coming from the peripheral nervous system. The major drawback to the use of morphine is that it is addictive.

Codeine is a methylmorphine. Almost all codeine used in modern medicine is produced by methylating the more abundant morphine. Codeine is less potent than morphine, having a painkilling effect about one sixth that of morphine.

Heroin is a synthetic compound, produced from morphine. Chemical modification increases painkilling potency; heroin has more than three times the painkilling effect of morphine. However, heroin is so addictive that it has no accepted medical use in the United States. Most often its production from morphine is part of illegal drug operations.

Practice Problems and Questions

12.60 Indicate whether each of the following compounds is a *heterocyclic* compound.

a. NH_2, CH_3 (benzene ring)

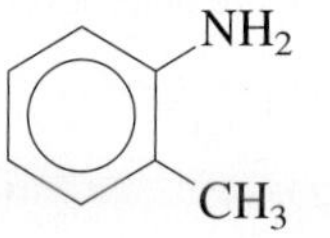

b. N—CH_3 (six-membered ring)

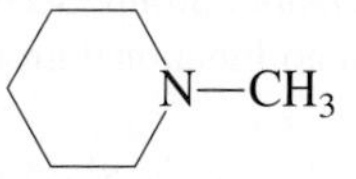

c. NH—CH_3 (benzene ring)

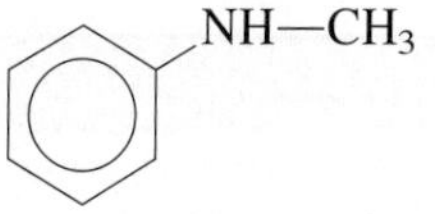

d. O (six-membered ring)

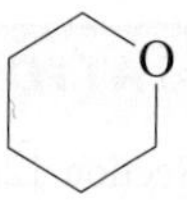

12.61 Indicate whether each of the compounds in Problem 12.60 is a *heterocyclic amine*.

12.62 Classify each of the compounds in Problem 12.60 as a primary amine, a secondary amine, a tertiary amine, or a nonamine.

12.63 Indicate whether each of the compounds in Problem 12.60 is a *polyfunctional* compound.

12.64 What is an alkaloid?

CONCEPTS TO REMEMBER

Hydrocarbon derivatives. Hydrocarbon derivatives contain one or more heteroatoms, atoms other than carbon or hydrogen. Halogenated hydrocarbons have halogen heteroatoms. Oxygen is the heteroatom in alcohols, phenols, and ethers. Thiols contain sulfur as a heteroatom, and nitrogen is the heteroatom in amines.

Halogenated hydrocarbons. Halogenated hydrocarbons are hydrocarbon derivatives in which one or more halogen atoms have replaced hydrogen atoms. Such compounds are named using the rules that apply to branched-chain hydrocarbons, with halogen substituents being treated the same way as alkyl groups.

Alcohols and phenols. Alcohols are hydrocarbon derivatives that contain a *hydroxyl group* (—OH) attached to a saturated carbon atom. The general formula for an alcohol is R—OH. Phenols have the general formula Ar—OH, where Ar represent an aryl group derived from an aromatic compound.

Nomenclature of alcohols and phenols. The IUPAC names of simple alcohols end in *-ol,* and their chains are numbered to give precedence to the location of the —OH group. The common names have the word *alcohol* preceded by the name of the alkyl group. Phenols are named as derivatives of the parent compound phenol.

Physical properties of alcohols. Alcohol molecules hydrogen-bond to each other and to water molecules. They thus have higher-than-normal boiling points, and the low-molecular-mass alcohols are soluble in water.

Classifications of alcohols. Alcohols are classified on the basis of the number of carbon atoms bonded to the carbon attached to the —OH group. In primary alcohols, the —OH group is bonded to a carbon atom bonded to only one other C atom. In secondary alcohols, the —OH-containing C atom is attached to two other C atoms. In tertiary alcohols, it is attached to three other C atoms.

Reactions of alcohols. Alcohols can be dehydrated in the presence of sulfuric acid to form alkenes. Oxidation of primary alcohols produces an aldehyde. Secondary alcohols are oxidized to ketones, and tertiary alcohols are resistant to oxidation.

Ethers. The general formula for an ether is R—O—R′, where R and R′ are alkyl, cycloalkyl, or aryl groups. In the IUPAC system, ethers are named as alkoxy derivatives of alkanes. Common names are obtained by giving the R group names in alphabetical order and adding the word *ether*.

Properties of ethers. Ethers have lower boiling points then alcohols because ether molecules do not hydrogen-bond to each other. Ethers are slightly soluble in water because water forms hydrogen bonds with ethers.

Thiols and disulfides. Thiols are the sulfur analogs of alcohols. They have the general formula R—SH. The —SH group is called the sulfhydryl group. Oxidation of thiols forms disulfides, which have the general formula R—S—S—R. The most distinctive physical property of thiols is their foul odor.

Amines. An amine is an organic derivative of ammonia (NH_3) in which one or more hydrogen atoms on the nitrogen atom of ammonia have been replaced by an alkyl, a cycloalkyl, or an aryl group. Amines can be classified as primary, secondary, or tertiary, depending on the number of hydrogen atoms of NH_3 that have been replaced by hydrocarbon groups. The functional group for a primary amine, the $—NH_2$ group, is called an *amino* group. Secondary and tertiary amines possess substituted amino functional groups.

Nomenclature for amines. Common names for amines are formed by listing the hydrocarbon groups attached to the nitrogen atom in alphabetical order, followed by the suffix *-amine*. In the IUPAC system, the *-e* ending of the name of the longest carbon chain present is changed to *-amine,* and a number is used to locate the position of the amino group. Carbon-chain substituents are given numbers to designate their locations.

Properties of amines. The methylamines and ethylamine are gases at room temperature; amines of higher molecular mass are usually liquids and smell like raw fish. Primary and secondary, but not tertiary, amines can participate in hydrogen bonding to other amine molecules.

Reactions of amines. Amines are weak bases because of the ability of the unshared electron pair on the amine nitrogen atom to accept a proton from water. The reaction of a strong acid with an amine forms an amine salt. Such salts are more soluble in water than is the parent amine.

Heterocyclic amines. In a heterocyclic amine, the nitrogen atoms of amino groups present are part of either an aromatic or a nonaromatic ring system. An *alkaloid* is a physiologically active nitrogen-containing organic compound extracted from plants. Most alkaloids are heterocyclic amines.

KEY REACTIONS AND EQUATIONS

1. Dehydration of alcohols to give alkenes (Section 12.4)

$$\underset{\text{H}\quad\text{OH}}{-\overset{|}{\underset{|}{C}}-\overset{|}{\underset{|}{C}}-} \xrightarrow[180°C]{H_2SO_4} \rangle C{=}C\langle + H_2O$$

2. Oxidation of a primary alcohol to given an aldehyde (Section 12.4)

$$R-\overset{OH}{\underset{H}{\overset{|}{\underset{|}{C}}}}-H \xrightarrow{[O]} \underset{\text{Aldehyde}}{R-\overset{O}{\overset{\|}{C}}-H} + H_2O$$

3. Oxidation of a secondary alcohol to give a ketone (Section 12.4)

$$R-\overset{OH}{\underset{H}{\overset{|}{\underset{|}{C}}}}-R' \xrightarrow{[O]} R-\overset{O}{\overset{\|}{C}}-R' + H_2O$$

4. Attempted oxidation of a tertiary alcohol, which gives no reaction (Section 12.4)

$$R-\overset{OH}{\underset{R''}{\overset{|}{\underset{|}{C}}}}-R' \xrightarrow{[O]} \text{no reaction}$$

5. Oxidation of a thiol to give a disulfide (Section 12.6)

$$R-SH + HS-R \xrightarrow{[O]} R-S-S-R + 2H$$

6. Reduction of a disulfide to give a thiol (Section 12.6)

$$R-S-S-R + 2H \longrightarrow R-SH + HS-R$$

7. Reaction of amines with water to give a basic solution (Section 12.8)

$$R-NH_2 + H_2O \rightleftharpoons R-\overset{+}{N}H_3 + OH^-$$

8. Reaction of amines with acids to produce amine salts (Section 12.8)

$$R-NH_2 + HCl \longrightarrow R-\overset{+}{N}H_3\ Cl^-$$

9. Conversion of an amine salt into an amine (Section 12.8)

$$R-\overset{+}{N}H_3\ Cl^- + NaOH \longrightarrow R-NH_2 + NaCl + H_2O$$

KEY TERMS

Alcohol (12.3)
Alkaloid (12.9)
Alkoxy group (12.5)
Amine (12.7)
Amine salt (12.8)
Amino group (12.7)
Aryl group (12.3)
Dehydration reaction (12.4)
Elimination reaction (12.4)
Ether (12.5)
Halogenated hydrocarbon (12.2)
Heteroatom (12.1)
Heterocyclic organic compound (12.9)
Hydroxyl group (12.3)
Phenol (12.3)
Polyfunctional organic compound (12.7)
Primary alcohol (12.4)
Secondary alcohol (12.4)
Substituted ammonium ion (12.8)
Sulfhydryl group (12.6)
Tertiary alcohol (12.4)
Thiol (12.6)

ADDITIONAL PROBLEMS

12.65 In what type of organic compound would each of the following functional groups be found?
a. Hydroxyl group b. Amino group
c. Sulfhydryl group d. Alkoxy group

12.66 What is the generalized formula for each of the functional groups listed in Problem 12.65?

12.67 What is the minimum number of carbon–heteroatom single bonds that must be present in each of the following types of compounds?
a. Ether b. Alcohol c. Thiol d. Amine

12.68 Indicate whether hydrogen–heteroatom bonds must be present in each of the following types of compounds.
a. Ether b. Alcohol c. Thiol d. Amine

12.69 How many hydrogen–heteroatom bonds are present in each of the following types of compounds?
a. 1° alcohol b. 1° amine c. 2° alcohol d. 2° amine

12.70 What is the molecular formula of each of the following compounds?
a. Ethanol b. Ethanamine
c. Methoxymethane d. Ethanethiol

12.71 Draw the structures of the following ring-containing compounds?
a. Phenol b. Cyclohexanol
c. Aniline d. Cyclohexanamine

12.72 Classify each of the following compounds as an alcohol, an ether, a thiol, or an amine.
a. $CH_3—CH_2—SH$ b. $CH_3—CH_2—OH$
c. $CH_3—CH_2—NH_2$ d. $CH_3—CH_2—O—CH_3$

12.73 Assign an IUPAC name to each of the compounds in Problem 12.72.

12.74 Assign an IUPAC name to each of the following compounds.
a. $HO—CH_2—CH_2—OH$ b. $H_2N—CH_2—CH_2—NH_2$
c. $HS—CH_2—CH_2—SH$ d. $Br—CH_2—CH_2—Br$

12.75 What type of organic compound is produced from each of the following reactions?
a. Oxidation of a primary alcohol
b. Reaction of an amine with a strong acid
c. Dehydration of a secondary alcohol
d. Reduction of a disulfide

12.76 Which of the terms *primary, secondary,* and *tertiary* applies to each of the following alcohols and amines?
a. Propyl alcohol b. Isopropyl alcohol
c. Methyl amine d. Dimethylamine

12.77 How many different saturated molecules exist that fit each of the following descriptions?
a. Three-carbon amine
b. Three-carbon alcohol
c. Three-carbon ether
d. Three-carbon dibromoalkane

PRACTICE TEST ▶ True/False

12.78 Hydrocarbon derivatives must contain at least one atom of an element other than carbon or hydrogen.

12.79 Halogen substituents take precedence over alkyl groups when numbering the carbon chain in a halogenated hydrocarbon.

12.80 CFCs are a group of alcohols added to gasoline to increase its octane rating.

12.81 Alcohols and phenols contain the same functional group.

12.82 Ethylene glycol and 1,2-ethanediol are two names for the same compound.

12.83 A five-carbon unbranched-chain alcohol would be more soluble in water than methanol, the one-carbon alcohol.

12.84 Primary alcohols are resistant to oxidation, but secondary and tertiary alcohols readily undergo oxidation reactions.

12.85 A hydroxyl group and a sulfhydryl group contain the same number of atoms.

12.86 Phenol and aniline are both monosubstituted benzene molecules.

12.87 Two carbon–oxygen single bonds are present in simple ether molecules.

12.88 Halogenated ethers are usually less flammable than nonhalogenated ethers.

12.89 A characteristic of many thiols is a "fishy" odor.

12.90 Oxidation of a thiol produces a disulfide.

12.91 The nitrogen atom in a primary amine participates in two nitrogen–hydrogen bonds.

12.92 Dissolving an amine in water produces an acidic solution.

PRACTICE TEST ▶ Multiple Choice

12.93 In hydrocarbon derivatives, atoms of which of the following elements always forms two bonds?
a. Oxygen b. Nitrogen c. Chlorine d. Bromine

12.94 Which of the following functional groups always contains three atoms?
a. Hydroxyl b. Alkoxy c. Amino d. Sulfhydryl

12.95 Alcohols have higher boiling points than alkanes of similar molecular mass because
a. alcohols are ionic compounds and alkanes are covalent compounds
b. alkane molecules are polar and alcohol molecules are not
c. hydrogen bonding occurs between alcohol molecules but not between alkane molecules
d. alcohols are acidic and alkanes are not

12.96 Which of the following is a secondary alcohol?
a. Ethanol b. 1-Propanol
c. 2-Propanol d. 3-Methyl-1-butanol

12.97 The common name for the ether 2-ethoxypropane is
a. Diethyl ether b. Diisopropyl ether
c. Ethyl propyl ether d. Ethyl isopropyl ether

12.98 The sulfhydryl functional group is found in
a. thiols
b. disulfides
c. both thiols and disulfides
d. neither thiols nor disulfides

12.99 The compound $CH_3—CH_2—CH_2—NH—CH_3$ is an example of a
a. primary amine
b. secondary amine
c. tertiary amine
d. unsubstituted amine

12.100 Many amines, in the liquid state, have odors resembling that of
a. garlic b. raw fish c. roses d. skunks

12.101 The amine salt produced from the reaction of ethyl amine and HCl would have the name
a. ethyl amine chloride b. ethyl chloride
c. ethylammonium chloride d. ammonium chloride

12.102 Heterocyclic amines have structures that require
a. two separate carbon ring systems
b. a fused carbon ring system
c. a ring system with an amino group attachment
d. a ring system in which the nitrogen of an amino or substituted amino group is part of the ring

CHAPTER OUTLINE

CHAPTER THIRTEEN

Hydrocarbon Derivatives II: Carbon–Oxygen Double Bonds

Benzaldehyde, a compound containing a carbon–oxygen double bond, is the main flavor component in almonds. Compounds with carbon–oxygen double bonds are responsible for the odor and taste of numerous nuts and spices.

In the previous chapter we considered five types of hydrocarbon derivatives: halogenated hydrocarbons, alcohols, ethers, thiols, and amines. The structural characteristic that links these diverse types of compounds is the presence of a carbon–heteroatom single bond.

In this chapter we consider five more types of hydrocarbon derivatives: aldehydes, ketones, carboxylic acids, esters, and amides. Again, although the compound types are diverse, a common structural characteristic links them, the presence of a carbon–oxygen double bond.

13.1 Functional Groups Containing the Carbon–Oxygen Double Bond

Learning Focus

List the structural characteristics associated with a carbonyl group and recognize functional groups that contain the carbonyl group.

▶ The word *carbonyl* is pronounced "carbon-EEL."

Hydrocarbon derivatives which contain a carbon–oxygen double bond are said to contain a *carbonyl group*. A **carbonyl group** *consists of a carbon atom and an oxygen atom joined by a double bond.*

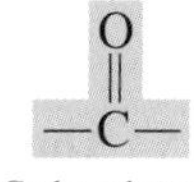

Carbonyl group

The carbon atom of a carbonyl group must form two other bonds (to give it four bonds). The nature of these two additional bonds determines the type of carbonyl-containing compound we are dealing with.

Five major classes of carbonyl-containing hydrocarbon derivatives are considered in this chapter.

1. **Aldehydes.** In an aldehyde, one of the two additional bonds that the carbonyl carbon atom forms must be to a hydrogen atom. The other may be to a hydrogen atom, an alkyl or cycloalkyl group, or an aromatic ring system.

$-\overset{\overset{\large O}{\|}}{C}-H$ $H-\overset{\overset{\large O}{\|}}{C}-H$ $CH_3-CH_2-\overset{\overset{\large O}{\|}}{C}-H$ $C_6H_5-\overset{\overset{\large O}{\|}}{C}-H$

Aldehyde functional group; Simplest aldehyde; Other examples of aldehydes

2. **Ketones.** In a ketone, both of the additional bonds of the carbonyl carbon atom must be to another carbon atom; alkyl, cycloalkyl, or aromatic groups can be involved.

$C-\overset{\overset{\large O}{\|}}{C}-C$ $CH_3-\overset{\overset{\large O}{\|}}{C}-CH_3$ $CH_3-\overset{\overset{\large O}{\|}}{C}-CH_2-CH_3$ $C_5H_9-\overset{\overset{\large O}{\|}}{C}-CH_3$

Ketone functional group; Simplest ketone; Other examples of ketones

3. **Carboxylic acids.** In a carboxylic acid, one of the two additional bonds of the carbonyl carbon atom must be to a hydroxyl group, and the other may be to a hydrogen atom, an alkyl or cycloalkyl group, or an aromatic ring system. The structural parameters for a carboxylic acid are the same as those for an aldehyde except that a hydroxyl group replaces the mandatory hydrogen atom of an aldehyde.

$-\overset{\overset{\large O}{\|}}{C}-OH$ $H-\overset{\overset{\large O}{\|}}{C}-OH$ $CH_3-CH_2-\overset{\overset{\large O}{\|}}{C}-OH$ $C_6H_5-\overset{\overset{\large O}{\|}}{C}-OH$

Carboxylic acid functional group; Simplest carboxylic acid; Other examples of carboxylic acids

4. **Esters.** In an ester, one of the two additional bonds of the carbonyl carbon atom must be to an oxygen atom of an alkyl, cycloalkyl, or aromatic group and the other may be to a hydrogen atom, alkyl or cycloalkyl group, or an aromatic ring system. Parameters for an ester differ from those for a carboxylic acid only in that an —OH group has become an —O—R or O—Ar group.

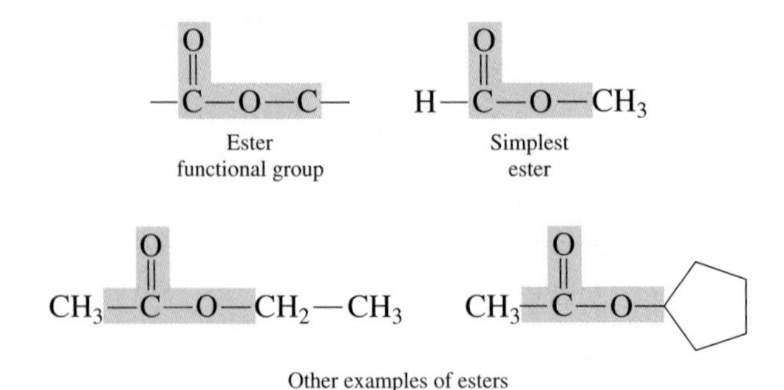

Other examples of esters

5. **Amides.** The previous four types of carbonyl compounds contain the elements carbon, hydrogen, and oxygen. Amides are different from these four types of compounds in that nitrogen, in addition to carbon, hydrogen, and oxygen, is present. In an amide, an amino group (—NH_2) or substituted amino group replaces the —OH group of a carboxylic acid.

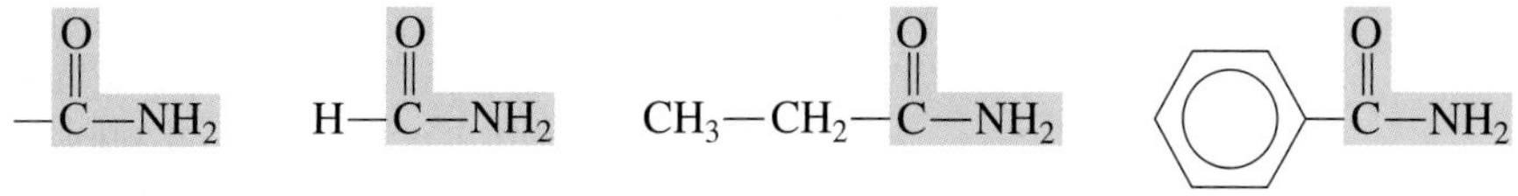

$$-\overset{\overset{\displaystyle O}{\|}}{C}-NH_2 \qquad H-\overset{\overset{\displaystyle O}{\|}}{C}-NH_2 \qquad CH_3-CH_2-\overset{\overset{\displaystyle O}{\|}}{C}-NH_2 \qquad C_6H_5-\overset{\overset{\displaystyle O}{\|}}{C}-NH_2$$

Amide functional group — Simplest amide — Other examples of amides

Practice Problems and Questions

13.1 What is a carbonyl group?

13.2 Indicate whether each of the following types of compounds contains a carbonyl group.
a. Aldehyde b. Ester c. Alcohol d. Carboxylic acid

13.3 What elements are present in each of the following types of hydrocarbon derivatives?
a. Carboxylic acid b. Amide c. Ester d. Ketone

13.4 Identify the type of hydrocarbon derivative associated with each of the following functional group designations.

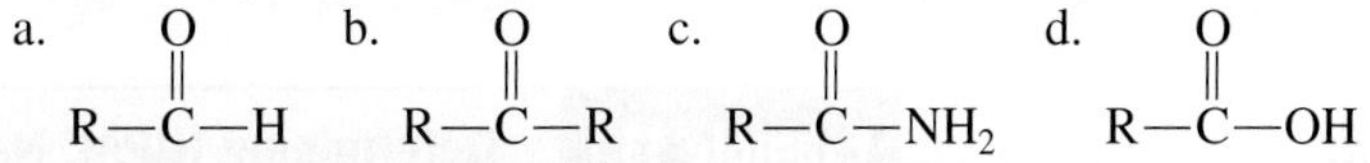

a. $R-\overset{\overset{\displaystyle O}{\|}}{C}-H$ b. $R-\overset{\overset{\displaystyle O}{\|}}{C}-R$ c. $R-\overset{\overset{\displaystyle O}{\|}}{C}-NH_2$ d. $R-\overset{\overset{\displaystyle O}{\|}}{C}-OH$

13.5 Write the structural formula for the simplest member of each of the following types of hydrocarbon derivatives.
a. Aldehyde b. Ester c. Ketone d. Carboxylic acid

13.6 In which of the following types of carbonyl-containing compounds is it possible to have a methyl group attached to the carbonyl carbon atom?
a. Amide b. Carboxylic acid c. Aldehyde d. Ketone

Learning Focus

Know the general structural characteristics of aldehydes; be able to name such compounds given their structural formulas, or vice versa.

▶ The word *aldehyde* is pronounced "AL-da-hide."

13.2 Structural Characteristics and Naming of Aldehydes

An **aldehyde** *is a compound that has at least one hydrogen atom attached to the carbon atom of a carbonyl group.* The remaining group attached to the carbonyl carbon atom can be hydrogen, an alkyl group (R), a cycloalkyl group, or an aryl group (Ar).

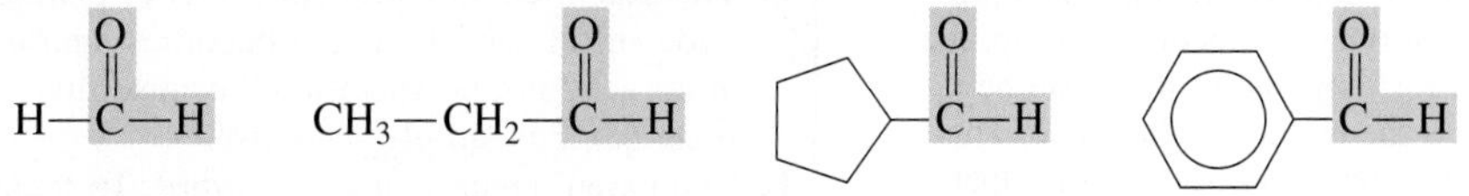

$$H-\overset{\overset{\displaystyle O}{\|}}{C}-H \qquad CH_3-CH_2-\overset{\overset{\displaystyle O}{\|}}{C}-H \qquad C_5H_9-\overset{\overset{\displaystyle O}{\|}}{C}-H \qquad C_6H_5-\overset{\overset{\displaystyle O}{\|}}{C}-H$$

Fourth bond is to a hydrogen atom — Fourth bond is to an alkyl group — Fourth bond is to a cycloalkyl group — Fourth bond is to an aryl group

The aldehyde functional group, the structural feature common to all the preceding compounds, is

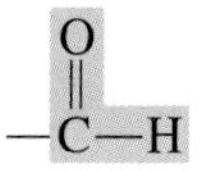

$$-\overset{\overset{\displaystyle O}{\|}}{C}-H$$

Aldehydes occur widely in nature. Several such aldehydes are responsible for the odor and taste of nuts and spices.

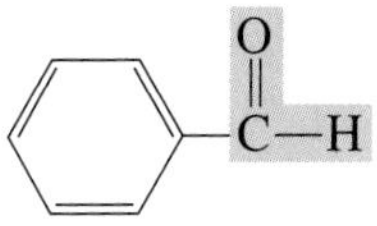

Benzaldehyde (oil of almonds) 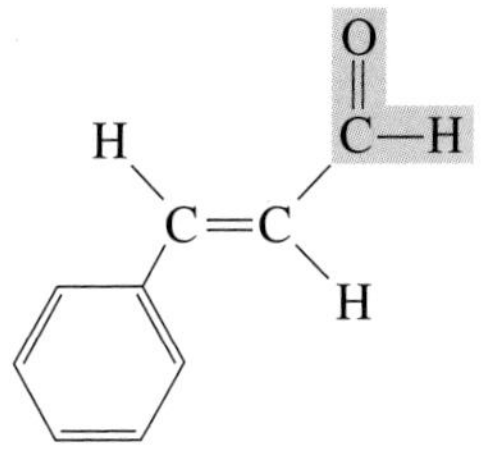 — Cinnamaldehyde (cinnamon) 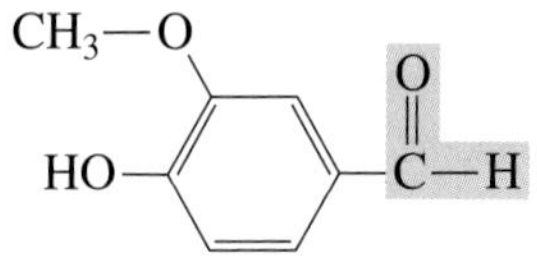 — Vanillin (vanilla bean)

Note the polyfunctional nature of these substances. Benzaldehyde has aldehyde and aromatic functional groups; cinnamaldehyde has aldehyde, alkene, and aromatic functional groups; and vanillin has aldehyde, alcohol, and ether functional groups. Note also that natural products nearly always are known by common names, names that provide no structural information.

Naming of Aldehydes

The IUPAC rules for naming aldehydes are as follows:

1. Select as the parent carbon chain the longest chain that *includes* the carbon atom of the carbonyl group.
2. Name the parent chain by changing the *-e* ending of the corresponding alkane name to *-al*.
3. Number the parent chain by assigning the number 1 to the carbonyl carbon atom of the aldehyde group.
4. Determine the identity and location of any substituents, and append this information to the front of the parent chain name.

▶ The carbonyl carbon atom in an aldehyde cannot have any number but 1, so we do not have to include this number in the aldehyde's IUPAC name.

▶ Be careful about the endings *-al* and *-ol*. They are easily confused. The suffix *-al* (pronounced like the man's name Al) denotes an aldehyde; the suffix *-ol* (pronounced like the *ol* in old) denotes an alcohol.

▶ When a compound contains more than one type of functional group, the suffix for only one of them can be used as the ending of the name. The IUPAC rules establish priorities that specify which suffix is used. For the functional groups we have discussed up to this point in the text, the IUPAC priority system is

Increasing priority (lowest to highest, arrow pointing up):
- aldehyde
- alcohol
- amine
- alkene
- alkyne
- alkoxy, halogen, alkyl — Equal-priority substituents (listed in alphabetical order)

aldehyde–ether

$CH_3-O-CH_2-C(=O)-H$

2-methoxyethanal

aldehyde–alkene

$CH_2=CH-C(=O)-H$

2-propenal

Example 13.1 Determining IUPAC Names for Aldehydes

Assign IUPAC names to the following aldehydes.

a. $CH_3-CH_2-CH_2-CH_2-C(=O)-H$

b. $CH_3-CH(CH_3)-CH_2-C(=O)-H$

c. $CH_3-CH_2-CH_2-CH(CH_2-CH_3)-C(=O)-H$

d. $CH_3-CH_2-CH(OH)-CH_2-C(=O)-H$

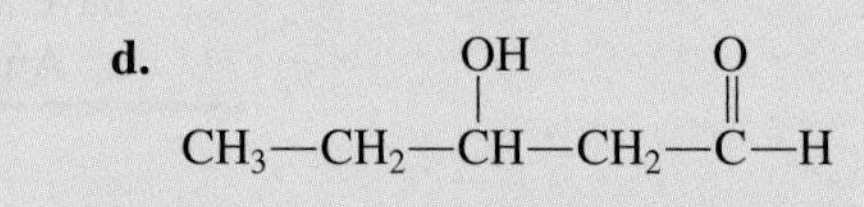

Solution

a. The parent chain name comes from pentane. Remove the *-e* ending and add the aldehyde suffix *-al*. The name becomes *pentanal*. The location of the carbonyl carbon atom need not be specified, because this carbon atom is always number 1. The complete name is simply *pentanal*.

b. The parent chain name is *butanal*. To locate the methyl group, we number the carbon chain beginning with the carbonyl carbon atom. The complete name of the aldehyde is *3-methylbutanal*.

$\overset{4}{C}H_3-\overset{3}{C}H(CH_3)-\overset{2}{C}H_2-\overset{1}{C}(=O)-H$

c. The longest chain containing the carbonyl carbon atom is five carbons long, giving a parent chain name of *pentanal*. An ethyl group is present on carbon 2. Thus the complete name is *2-ethylpentanal*.

$\overset{5}{C}H_3-\overset{4}{C}H_2-\overset{3}{C}H_2-\overset{2}{C}H(CH_2-CH_3)-\overset{1}{C}(=O)-H$

d. This is a hydroxyaldehyde, with the hydroxyl group located on carbon 3.

$\overset{5}{C}H_3-\overset{4}{C}H_2-\overset{3}{C}H(OH)-\overset{2}{C}H_2-\overset{1}{C}(=O)-H$

> The complete name of the compound is *3-hydroxypentanal*. An aldehyde functional group has priority over an alcohol functional group in IUPAC nomenclature. An alcohol group named as a substituent is a *hydroxy* group.

Unbranched aldehydes with four or fewer carbon atoms have common names:

$$H{-}\overset{\overset{\displaystyle O}{\|}}{C}{-}H \qquad CH_3{-}\overset{\overset{\displaystyle O}{\|}}{C}{-}H$$

Formaldehyde   Acetaldehyde

$$CH_3{-}CH_2{-}\overset{\overset{\displaystyle O}{\|}}{C}{-}H \qquad CH_3{-}CH_2{-}CH_2{-}\overset{\overset{\displaystyle O}{\|}}{C}{-}H$$

Propionaldehyde   Butyraldehyde

▶ The common names for simple aldehydes illustrate a second method for counting from one to four: *form-*, *acet-*, *propion-*, and *butyr-*. We will use this method again later in this chapter when we consider the common names for carboxylic acids and esters. (The first method for counting from one to four, with which you are now thoroughly familiar, is *meth-*, *eth-*, *prop-*, and *but-*, as in methane, ethane, propane, and butane.)

Unlike the common names for alcohols and ethers, the common names for aldehydes are *one* word rather than two or three.

In assigning common names to branched aldehyde systems, we name and locate substituents (branches) in the same way as in the IUPAC system.

$$\overset{4}{C}H_3{-}\underset{\underset{\displaystyle CH_3}{|}}{\overset{3}{C}H}{-}\overset{2}{C}H_2{-}\overset{\overset{\displaystyle O}{\|}}{\overset{1}{C}}{-}H \qquad Cl{-}\overset{3}{C}H_2{-}\overset{2}{C}H_2{-}\overset{\overset{\displaystyle O}{\|}}{\overset{1}{C}}{-}H$$

3-Methylbutyraldehyde   3-Chloropropionaldehyde

▶ The contrast between IUPAC names and common names for aldehydes is as follows:

IUPAC (one word)
alkanal
pentanal
Common (one word)
(prefix) aldehyde*
butyraldehyde

*The common-name prefixes are related to natural sources for carboxylic acids with the same number of carbon atoms (see Section 13.6)

Aromatic aldehydes—compounds in which an aldehyde group is attached to a benzene ring—are named as derivatives of benzaldehyde, the parent compound.

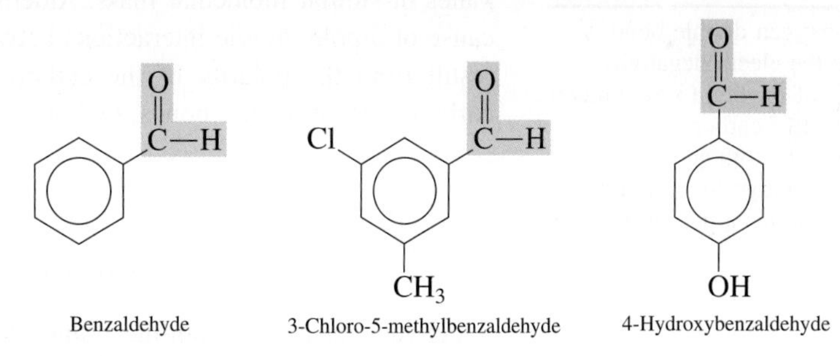

Benzaldehyde   3-Chloro-5-methylbenzaldehyde   4-Hydroxybenzaldehyde

The last of these compounds is named as a benzaldehyde rather than as a phenol because the aldehyde group has priority over the hydroxyl group in the IUPAC naming system.

Practice Problems and Questions

13.7 Indicate whether each of the following structures is an aldehyde.

a. $CH_3{-}CH_2{-}CH_2{-}\overset{\overset{\displaystyle O}{\|}}{C}{-}OH$

b. $CH_3{-}CH_2{-}CH_2{-}CH_2{-}\overset{\overset{\displaystyle O}{\|}}{C}{-}H$

c. $CH_3{-}\overset{\overset{\displaystyle O}{\|}}{C}{-}CH_3$

d. $CH_3{-}O{-}CH_2{-}CH_3$

13.8 Assign an IUPAC name to each of the following aldehydes.

a. $CH_3{-}CH_2{-}CH_2{-}\overset{\overset{\displaystyle O}{\|}}{C}{-}H$

b. $CH_3{-}CH_2{-}\overset{\overset{\displaystyle CH_3}{|}}{C}H{-}\overset{\overset{\displaystyle O}{\|}}{C}{-}H$

c. $C_6H_5{-}\overset{\overset{\displaystyle O}{\|}}{C}{-}H$ (benzene ring attached to C—H)

d. $C_6H_5{-}CH_2{-}CH_2{-}\overset{\overset{\displaystyle O}{\|}}{C}{-}H$ (benzene ring attached to CH₂)

13.9 Draw a structural formula for each of the following aldehydes.
a. 3-Methylpentanal b. 3,4-Dimethylheptanal
c. 2-Ethylhexanal d. 2,2-Dichloropropanal

13.10 What is the common name for each of the following unsubstituted aldehydes?
a. One-carbon aldehyde b. Two-carbon aldehyde
c. Three-carbon aldehyde d. Four-carbon aldehyde

13.11 What is the common name for each of the following aldehydes?

a. $CH_3—CH_2—\overset{\overset{\large O}{\|}}{C}—H$

b. $CH_3—\underset{\underset{\large CH_3}{|}}{CH}—\overset{\overset{\large O}{\|}}{C}—H$

c. $Cl—\underset{\underset{\large Cl}{|}}{CH}—\overset{\overset{\large O}{\|}}{C}—H$

d.
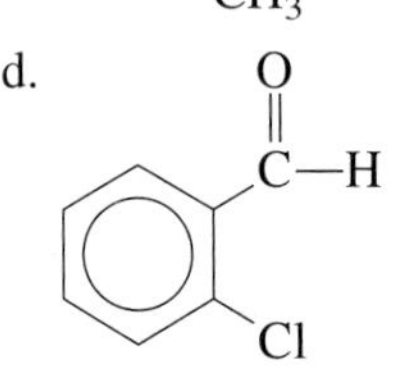

13.3 Properties and Reactions of Aldehydes

Learning Focus
List general physical properties of aldehydes; discuss products from the reduction and oxidation of aldehydes and the reaction of aldehydes with alcohols.

Methanal and ethanal, the two simplest aldehydes, are gases at room temperature. The C_3 through C_{11} straight-chain saturated aldehydes are liquids, and still longer-chain aldehydes are solids. The simplest aromatic aldehyde, benzaldehyde, is a liquid. Benzaldehyde derivatives tend to be solids.

The boiling points of aldehydes are intermediate between those of alcohols and alkanes of similar molecular mass. Aldehydes have higher boiling points than alkanes because of dipole–dipole interactions between molecules. These dipole–dipole interactions result from the polarity of the carbon–oxygen double bond (carbonyl group). Unlike carbon–carbon double bonds, carbon–oxygen double bonds are polar.

▶ A carbon–oxygen double bond is polar because the electronegativity (Section 4.15) of oxygen (3.5) is much greater than that of carbon (2.5). Hence the oxygen atom acquires a partial negative charge (δ^-), and the carbon atom acquires a partial positive charge (δ^+).

$$\overset{\delta^+}{C}=\overset{\delta^-}{O}$$

Polar nature of carbon–oxygen double bond

Aldehydes have lower boiling points than corresponding alcohols because no hydrogen bonding occurs as it does with alcohols. Dipole–dipole interactions are weaker forces than hydrogen bonds (Section 6.13).

Methanal (formaldehyde) and ethanal (acetaldehyde) are very soluble in water because of carbonyl group polarity. C_3 through C_5 aldehydes have limited solubility in water. Hexanal and longer-chain aldehydes are insoluble, as are most aromatic aldehydes.

Formaldehyde is generally available as *formalin,* an aqueous solution containing 37% formaldehyde gas by mass or 40% by volume. This represents the solubility limit of formaldehyde gas in water. Very little free formaldehyde is actually present in formalin; most of it reacts with water, converting it into methylene glycol:

$$\underset{\text{Formaldehyde}}{H—\overset{\overset{O}{\|}}{C}—H} + \underset{\text{Water}}{H—O—H} \longrightarrow \underset{\text{Methylene glycol}}{HO—CH_2—OH}$$

Figure 13.1 Formalin, a 37% by mass solution of formaldehyde in water, is used to preserve biological specimens. The formaldehyde reacts with proteins to change their structure, causing them to resist decay. This coelacanth, a "prehistoric" fish, is preserved in formaldehyde.

Formalin, which is used for preserving biological specimens (see Figure 13.1), is also the most widely used preservative chemical in embalming fluids used by morticians. Its mode of action involves reaction with protein molecules in a manner that links the protein molecules together; the result is a "hardening" of the protein.

Low-molecular-mass aldehydes have pungent, penetrating, unpleasant odors. Higher-molecular-mass aldehydes (above C_8) are more fragrant, especially benzaldehyde derivatives.

Reactions of Aldehydes

We will consider three reactions of aldehydes: (1) reduction, (2) oxidation, and (3) hemiacetal and acetal formation. All three of these types of reactions have biochemical significance; they all occur within the human body in reactions associated with carbohydrates.

Reduction

Aldehydes are easily reduced to primary alcohols by a variety of reducing agents, with appropriate catalysts. Such reductions involve the addition of H atoms to the carbon–oxygen double bond, a process that is similar to the addition of H atoms to a carbon–carbon double bond.

$$\mathrm{>C{=}C< + H_2 \xrightarrow{\text{Catalyst}} >\underset{\mid}{\overset{}{C}}(H){-}\underset{\mid}{C}(H)<}$$

$$\mathrm{>C{=}O + H_2 \xrightarrow{\text{Catalyst}} >C(H){-}O(H)}$$

A specific example of aldehyde reduction is the reduction of ethanal using H_2 gas with Ni as a catalyst; the product is ethanol.

$$\underset{\text{Ethanal}}{\mathrm{CH_3{-}\overset{\overset{\displaystyle O}{\|}}{C}{-}H}} + \mathrm{H_2} \xrightarrow{\text{Ni}} \underset{\text{Ethanol}}{\mathrm{CH_3{-}\overset{\overset{\displaystyle OH}{|}}{\underset{\underset{\displaystyle H}{|}}{C}}{-}H}}$$

The reduction of aldehydes to primary alcohols is the opposite process to the oxidation of primary alcohols to produce aldehydes, a reaction of primary alcohols considered in Section 12.4.

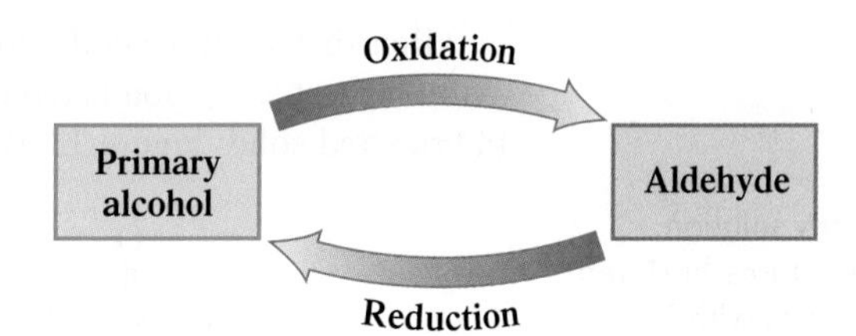

Oxidation

Aldehydes readily undergo oxidation to carboxylic acids (Section 13.1). A general equation for such an oxidation is

$$\underset{\text{Aldehyde}}{\mathrm{R{-}\overset{\overset{\displaystyle O}{\|}}{C}{-}H}} \xrightarrow[\text{oxidizing agent}]{\text{Mild}} \underset{\text{Carboxylic acid}}{\mathrm{R{-}\overset{\overset{\displaystyle O}{\|}}{C}{-}OH}}$$

In this reaction the aldehyde gains an oxygen atom (supplied by the oxidizing agent). Gain of oxygen is one of the definitions for the process of oxidation (Section 8.2); another definition for oxidation is loss of hydrogen (Section 8.2).

(a)

(b)

(c)

Figure 13.2 A positive Tollens test for aldehydes involves the formation of a silver mirror. (a) An aqueous solution of ethanal is added to a solution of silver nitrate in aqueous ammonia and stirred. (b) The solution darkens as ethanal is oxidized to ethanoic acid, and Ag^+ ion is reduced to silver. (c) The inside of the beaker becomes coated with metallic silver.

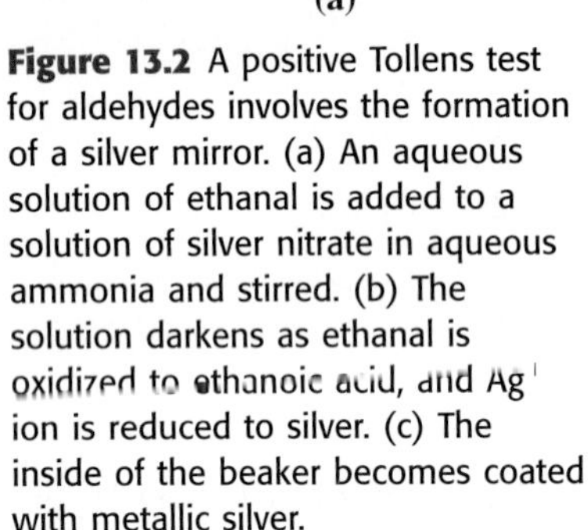

Among the mild oxidizing agents that convert aldehydes to carboxylic acids is oxygen in air. Thus aldehydes must be protected from air. When an aldehyde is prepared from oxidation of a primary alcohol (Section 12.4), it is usually immediately removed from the reaction mixture to prevent it from being further oxidized to a carboxylic acid; aldehydes are oxidized by the same oxidizing agents that oxidize primary alcohols.

Several tests, based on the ease with which aldehydes are oxidized, have been developed for detecting the presence of aldehyde groups in sugars (carbohydrates) or to measure the amounts of these compounds present in solution. The most widely used of these tests are the Tollens test and Benedict's test.

The Tollens test, also called the silver mirror test, involves a solution that contains silver nitrate ($AgNO_3$) and ammonia (NH_3) in water. When Tollens solution is added to an aldehyde, Ag^+ ion (the oxidizing agent) is reduced to silver metal, which deposits on the inside of the test tube, forming a silver mirror. The appearance of this silver mirror (see Figure 13.2) is a positive test for the presence of the aldehyde group.

$$\underset{\text{Aldehyde}}{\mathrm{R{-}\overset{\overset{\displaystyle O}{\|}}{C}{-}H}} + Ag^+ \xrightarrow[\text{heat}]{NH_3,\ H_2O} \underset{\text{Carboxylic acid}}{\mathrm{R{-}\overset{\overset{\displaystyle O}{\|}}{C}{-}OH}} + \underset{\text{Silver metal}}{Ag}$$

Benedict's test is similar to the Tollens test in that a metal ion is the oxidizing agent. With this test, Cu^{2+} ion is reduced to Cu^+ ion, which precipitates from solution as Cu_2O (a brick red solid; Figure 13.3).

Figure 13.3 Benedict's solution, which is blue in color, turns brick red when an aldehyde reacts with it.

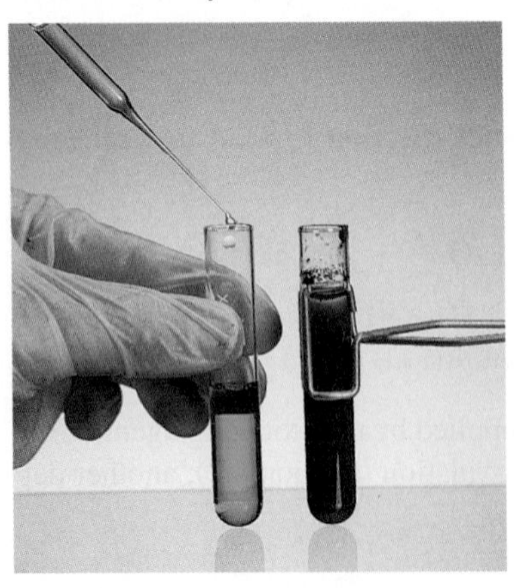

$$\underset{\text{Aldehyde}}{\mathrm{R{-}\overset{\overset{\displaystyle O}{\|}}{C}{-}H}} + Cu^{2+} \longrightarrow \underset{\text{Carboxylic acid}}{\mathrm{R{-}\overset{\overset{\displaystyle O}{\|}}{C}{-}OH}} + \underset{\text{Brick red solid}}{Cu_2O}$$

Benedict's solution is made by dissolving copper sulfate, sodium citrate, and sodium carbonate in water.

Hemiacetal and Acetal Formation

Many substances, besides hydrogen, can add to a carbon–oxygen double bond. In fact, because of its polarity, a carbon–oxygen double bond is even more susceptible to addition reactions than a carbon–carbon double bond.

Addition of an alcohol (R—OH) to a carbon–oxygen double bond produces a *hemiacetal.* In such addition, the H atom from the alcohol adds to the carbonyl oxygen atom, and the R—O portion of the alcohol (an alkoxy group, Section 12.5) adds to the carbonyl carbon atom.

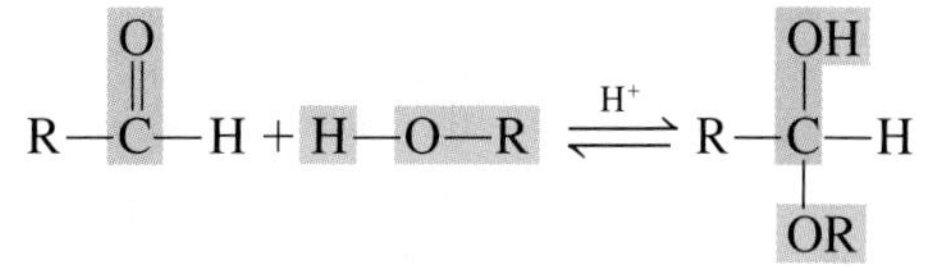

$$\begin{array}{c} \text{O} \\ \| \\ \text{R—C—H} \end{array} + \text{H—O—R} \overset{H^+}{\rightleftharpoons} \begin{array}{c} \text{OH} \\ | \\ \text{R—C—H} \\ | \\ \text{OR} \end{array}$$

Aldehyde Alcohol Hemiacetal

The reaction of ethanal with methanol provides a specific example of hemiacetal formation.

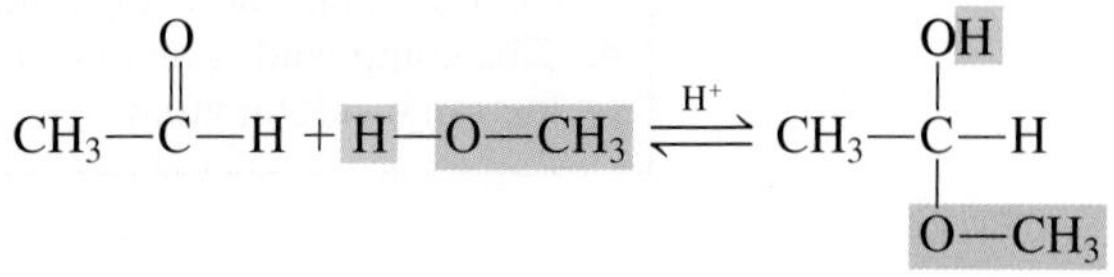

$$\begin{array}{c} \text{O} \\ \| \\ \text{CH}_3\text{—C—H} \end{array} + \text{H—O—CH}_3 \overset{H^+}{\rightleftharpoons} \begin{array}{c} \text{OH} \\ | \\ \text{CH}_3\text{—C—H} \\ | \\ \text{O—CH}_3 \end{array}$$

Ethanol Methanol

Formally defined, a **hemiacetal** *is a compound that has a carbon atom to which a hydroxyl group (—OH), an alkoxy group (—OR), and a hydrogen atom (—H) are attached.*

Reaction mixtures containing hemiacetals are always in equilibrium with the alcohol and aldehyde from which they are made, and the equilibrium lies to the aldehyde side of the reaction (Section 8.7).

Hemiacetals will react with an additional alcohol molecule to produce an acetal. In this process, a molecule of H_2O is also formed.

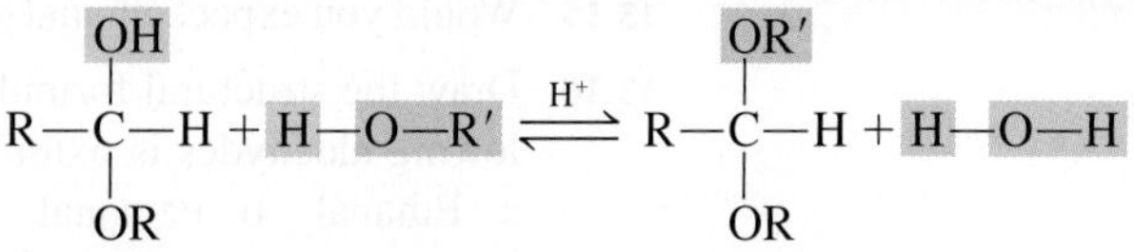

$$\begin{array}{c} \text{OH} \\ | \\ \text{R—C—H} \\ | \\ \text{OR} \end{array} + \text{H—O—R}' \overset{H^+}{\rightleftharpoons} \begin{array}{c} \text{OR}' \\ | \\ \text{R—C—H} \\ | \\ \text{OR} \end{array} + \text{H—O—H}$$

Hemiacetal Alcohol Acetal

Acetal formation, unlike hemiacetal formation, does not involve addition to a double bond. It is a substitution reaction; the —OR′ group of the alcohol replaces the —OH group or the hemiacetal.

An **acetal** *is a compound that has a carbon atom to which two alkoxy groups (—OR) and a hydrogen atom (—H) are attached.* Acetals, unlike hemiacetals, are easily isolated from reaction mixtures. Hemiacetal and acetal formation are important in carbohydrate chemistry, which is the topic of the next chapter.

Example 13.2 Recognizing Hemiacetal and Acetal Structures

Identify each of the following compounds as a hemiacetal, an acetal, or neither of these two types of compounds.

a. 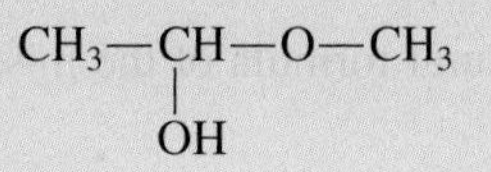 $\begin{array}{l} \text{CH}_3\text{—CH—O—CH}_3 \\ \quad\quad\ \ | \\ \quad\quad\ \text{OH} \end{array}$

b. 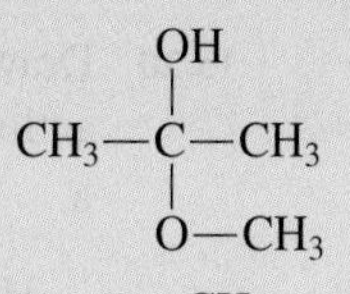 $\begin{array}{c} \text{OH} \\ | \\ \text{CH}_3\text{—C—CH}_3 \\ | \\ \text{O—CH}_3 \end{array}$

c. $\begin{array}{c} \quad\quad\ \text{CH}_3 \\ \quad\quad\ | \\ \text{CH}_3\text{—CH—CH—O—CH}_3 \\ | \quad\quad\quad\quad\quad \\ \text{OH} \quad\quad\quad\quad \end{array}$

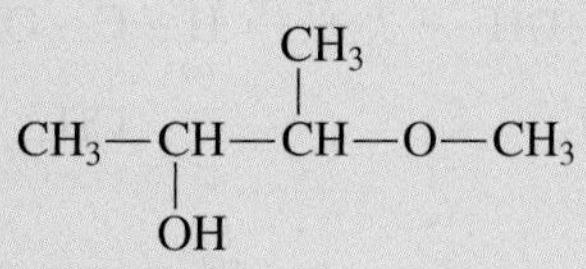

d. 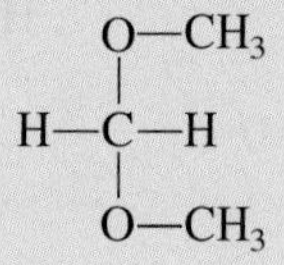 $\begin{array}{c} \text{O—CH}_3 \\ | \\ \text{H—C—H} \\ | \\ \text{O—CH}_3 \end{array}$

Solution

In each part, we will be looking for the following structural features:

1. For a hemiacetal, the presence of an —OH group, an —OR group, and a hydrogen atom, all attached to the same carbon atom.

2. For an acetal, the presence of two —OR groups and a hydrogen atom, all attached to the same carbon atom.

a. We have an —OH group, an —OR group, and a hydrogen atom attached to the same carbon atom. The compound is a *hemiacetal.*
b. We have an —OH group and an —OR group on the same carbon atom, but there is no hydrogen atom. The compound is *neither a hemiacetal nor an acetal.*
c. We have an —OH group and an —OR group, but they are not attached to the same carbon atom. The compound is *neither a hemiacetal nor an acetal.*
d. This compound is an *acetal.* Two —OR groups and a hydrogen atom are attached to the same carbon atom.

Practice Problems and Questions

13.12 Aldehydes have higher boiling points than alkanes of similar molecular mass. Explain why.

13.13 Aldehydes have lower boiling points than alcohols of similar molecular mass. Explain why.

13.14 Would you expect ethanal or octanal to be more soluble in water? Explain your answer.

13.15 Would you expect ethanal or octanal to have the fragrant odor? Explain your answer.

13.16 Draw the structural formula of the carboxylic acid produced when each of the following aldehydes is oxidized using a mild oxidizing agent.
a. Ethanal b. Pentanal c. Formaldehyde d. 3,4-Dichlorohexanal

13.17 Draw the structural formula of the primary alcohol produced when each of the aldehydes in Problem 13.16 is reduced using molecular H_2 and a nickel catalyst.

13.18 Identify the oxidizing agent in
a. Tollens solution b. Benedict's solution

13.19 Identify each of the following compounds as a hemiacetal, an acetal, or neither a hemiacetal nor an acetal.

a. $CH_3-CH_2-CH_2-CH(O-CH_3)(O-CH_3)$ (the CH carbon bears an $O-CH_3$ group above and an $O-CH_3$ group below)

b. $CH_3-CH(O-CH_3)-OH$ (the CH carbon bears an CH_3-O group above)

c. $CH_3-O-CH_2-CH_2-OH$

d. $CH_3-C(O-H)(O-CH_3)-CH_3$ (the central C bears an $O-H$ group above and an $O-CH_3$ group below)

13.20 Draw the structural formula of the missing compound in each of the following reactions.

a. $H-C(O-CH_3)(CH_3)-OH + ? \xrightarrow{H^+} H-C(O-CH_3)(CH_3)-O-CH_3 + H_2O$ (in each, the central C bears $O-CH_3$ above and CH_3 below)

b. $? + CH_3-CH_2-OH \xrightarrow{H^+} CH_3-C(H)(O-CH_2-CH_3)-O-CH_3 + H_2O$ (the central C bears H above and $O-CH_2-CH_3$ below)

c. $CH_3-(CH_2)_2-\overset{\displaystyle O}{\overset{\|}{C}}-H + CH_3-CH_2-OH \underset{}{\overset{H^+}{\rightleftharpoons}} ?$

d.

$$? + CH_3{-}OH \underset{}{\overset{H^+}{\rightleftharpoons}} CH_3{-}CH_2{-}\underset{\displaystyle O{-}CH_3}{\overset{\displaystyle OH}{CH}}$$

Learning Focus

Know the general structural characteristics of ketones; be able to name such compounds given their structural formulas, or vice versa.

▶ The word *ketone* is pronounced "KEY-tone."

▶ In an aldehyde, the carbonyl group is always located at the end of a hydrocarbon chain.

$$CH_3{-}CH_2{-}CH_2{-}CH_2{-}\overset{O}{\overset{\|}{C}}{-}H$$

In a ketone, the carbonyl group is always at a nonterminal (interior) position on the hydrocarbon chain.

$$CH_3{-}CH_2{-}CH_2{-}\overset{O}{\overset{\|}{C}}{-}CH_2{-}CH_3$$

13.4 Structural Characteristics and Naming of Ketones

A **ketone** *is a compound that has two carbon atoms attached to the carbon atom of a carbonyl group.* The groups containing these bonded carbon atoms may be alkyl, cycloalkyl, or aryl.

$$CH_3{-}\overset{O}{\overset{\|}{C}}{-}CH_2{-}CH_3$$

Ketone with two alkyl groups

Ketone with an alkyl group and a cycloalkyl group

Ketone with two aryl groups

The ketone functional group, the structural feature common to all the preceding compounds, is

Unlike aldehydes, ketones can form cyclic structures, such as

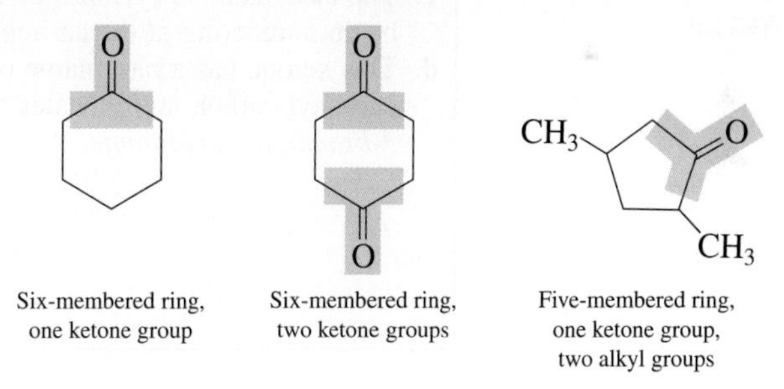

Cyclic ketones are *not* heterocyclic ring systems.

Naming of Ketones

Assigning IUPAC names to ketones is similar to naming aldehydes except that the ending *-one* is used instead of *-al*. The rules for IUPAC ketone nomenclature follow.

▶ Propanone is the simplest possible ketone. One- and two-carbon ketones cannot exist. A minimum of three carbon atoms is required for a ketone: one C atom for the carbonyl group and one C atom for each of the groups attached to the carbonyl carbon atom.

1. Select as the parent carbon chain the longest carbon chain that *includes* the carbon atom of the carbonyl group.
2. Name the parent chain by changing the *-e* ending of the corresponding alkane name to *-one*.
3. Number the carbon chain such that the carbonyl carbon atom receives the lowest possible number. The position of the carbonyl carbon atom is noted by placing a number immediately before the parent chain name.
4. Determine the identity and location of any substituents, and append this information to the front of the parent chain name.
5. Cyclic ketones are named by assigning the number 1 to the carbon atom of the carbonyl group. The ring is then numbered to give the lowest number(s) to the atom(s) bearing substituents.

Example 13.3 Determining IUPAC Names for Ketones

Assign IUPAC names to the following ketones.

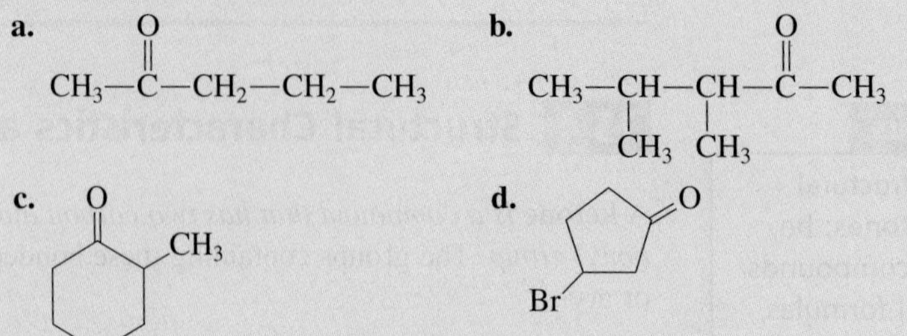

a. $CH_3{-}\overset{\overset{\displaystyle O}{\|}}{C}{-}CH_2{-}CH_2{-}CH_3$

b. $CH_3{-}\underset{\underset{\displaystyle CH_3}{|}}{CH}{-}\underset{\underset{\displaystyle CH_3}{|}}{CH}{-}\overset{\overset{\displaystyle O}{\|}}{C}{-}CH_3$

c. (cyclohexanone ring with O and CH_3)

d. (cyclopentanone ring with O and Br)

Solution

a. The parent chain name is *pentanone*. We number the chain from the end closest to the carbonyl carbon atom. Locating the carbonyl carbon at carbon 2 completes the name: *2-pentanone*.

b. The longest carbon chain of which the carbonyl carbon is a member is five carbons long. The parent chain name is again *pentanone*. There are two methyl groups attached, and the numbering system is from right to left.

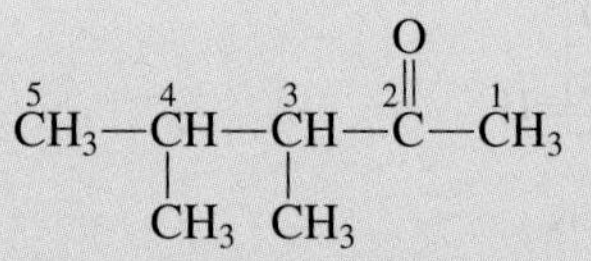

$$\overset{5}{C}H_3{-}\overset{4}{\underset{\underset{\displaystyle CH_3}{|}}{C}}H{-}\overset{3}{\underset{\underset{\displaystyle CH_3}{|}}{C}}H{-}\overset{\overset{\displaystyle O}{\|}}{\overset{2}{C}}{-}\overset{1}{C}H_3$$

The complete name for the compound is *3,4-dimethyl-2-pentanone*.

c. The base name is *cyclohexanone*. The methyl group is bonded to carbon 2 because we begin numbering at the carbonyl carbon: the name is *2-methylcyclohexanone*.

d. This ketone has a base name of *cyclopentanone*. Numbering clockwise from the carbonyl carbon atom locates the bromo group on carbon 3. The complete name is *3-bromocyclopentanone*.

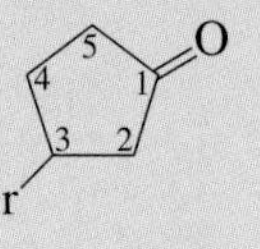

(cyclopentanone ring numbered 1–5, with O on carbon 1 and Br on carbon 3)

▶ In IUPAC nomenclature, the ketone functional group has precedence over all groups we have discussed so far except the aldehyde group. When both aldehyde and ketone groups are present in the same molecule, the ketone group is named as a substituent (the *oxo-* group).

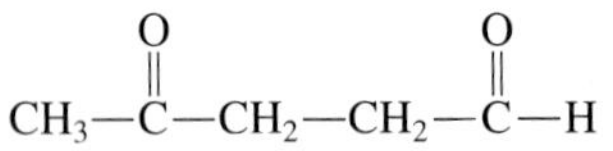

$$CH_3{-}\overset{\overset{\displaystyle O}{\|}}{C}{-}CH_2{-}CH_2{-}\overset{\overset{\displaystyle O}{\|}}{C}{-}H$$

4-Oxopentanal

Common names for ketones are constructed by giving, in alphabetical order, the names of the alkyl or aryl groups attached to the carbonyl functional group and then adding the word *ketone*. Unlike aldehyde common names, which are one word, those for ketones are two or three words.

▶ The procedure for coining common names for ketones is the same as that used for ether common names (Section 12.5).

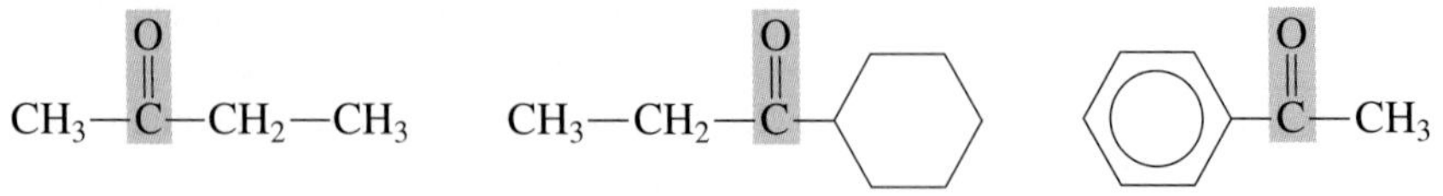

$CH_3{-}\overset{\overset{\displaystyle O}{\|}}{C}{-}CH_2{-}CH_3$

Ethyl methyl ketone

$CH_3{-}CH_2{-}\overset{\overset{\displaystyle O}{\|}}{C}{-}C_6H_{11}$ (cyclohexyl ring)

Cyclohexyl ethyl ketone

$C_6H_5{-}\overset{\overset{\displaystyle O}{\|}}{C}{-}CH_3$ (benzene ring)

Methyl phenyl ketone

▶ The contrast between IUPAC names and common names for ketones is as follows:

IUPAC (one word)
alkanone
2-butanone
Common (three or two words)
alkyl alkyl ketone
ethyl methyl ketone
or
dialkyl ketone
dipropyl ketone

The simplest ketone, dimethyl ketone, has a second common name with which you need to be familiar. It is *acetone,* which is the most often used name for this compound.

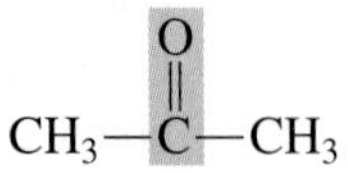

$$CH_3{-}\overset{\overset{\displaystyle O}{\|}}{C}{-}CH_3$$

Acetone
(dimethyl ketone)

Ketones, like aldehydes, occur widely in nature. For example, the major components of both clove and spearmint flavorings are ketones.

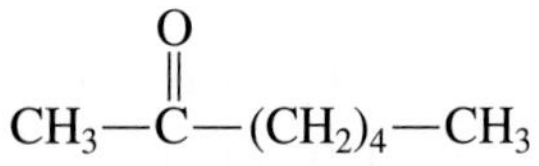

$$CH_3-\overset{\overset{\displaystyle O}{\|}}{C}-(CH_2)_4-CH_3$$

2-Heptanone
(clove flavoring)

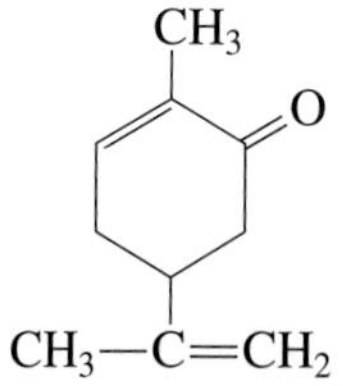

Carvone
(spearmint flavoring)

A lachrymator (pronounced "lack-ra-mater) is a compound that causes the production of tears. Several halogenated ketones have lachrymatic properties. 2-Chloro-1-phenylethanone is a component of the tear gas used by police and the military; it is also the active ingredient in MACE canisters now marketed for use by individuals to protect themselves from attackers. The compound 1-bromopropanone has been used as a military tear gas.

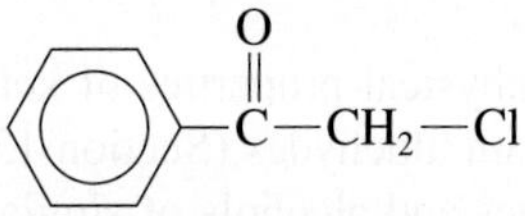

$$C_6H_5-\overset{\overset{\displaystyle O}{\|}}{C}-CH_2-Cl \qquad CH_3-\overset{\overset{\displaystyle O}{\|}}{C}-CH_2-Br$$

2-Chloro-1-phenylethanone 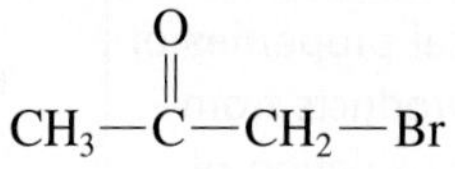 1-Bromopropanone

Practice Problems and Questions

13.21 Indicate whether each of the following structures is a ketone.

a. 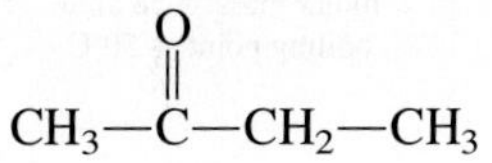 $CH_3-\overset{\overset{\displaystyle O}{\|}}{C}-CH_2-CH_3$

b. $CH_3-CH_2-\overset{\overset{\displaystyle CH_3}{|}}{C}=O$

c. cyclohexyl$-\overset{\overset{\displaystyle O}{\|}}{C}-H$

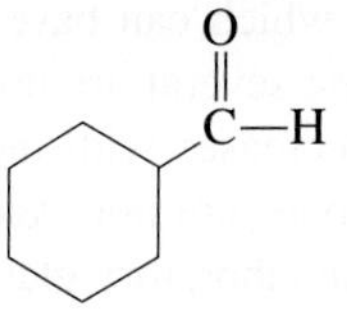

d. 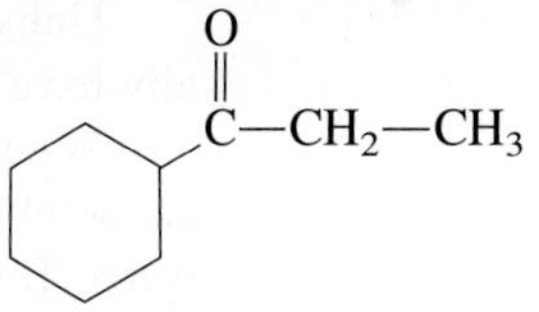 cyclohexyl$-\overset{\overset{\displaystyle O}{\|}}{C}-CH_2-CH_3$

13.22 Assign an IUPAC name to each of the following ketones.

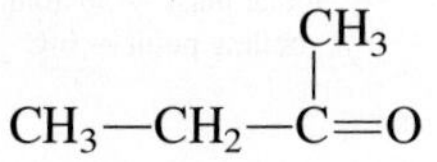

a. 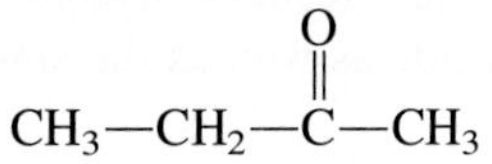 $CH_3-CH_2-\overset{\overset{\displaystyle O}{\|}}{C}-CH_3$

b. $CH_3-CH_2-\overset{\overset{\displaystyle CH_3}{|}}{CH}-\overset{\overset{\displaystyle O}{\|}}{C}-\overset{\overset{\displaystyle CH_3}{|}}{CH}-CH_3$

c. $CH_3-CH_2-\overset{\overset{\displaystyle O}{\|}}{C}-CH_2-\underset{\underset{\displaystyle CH_3}{\underset{|}{\underset{\displaystyle CH_2}{|}}}}{CH}-CH_3$

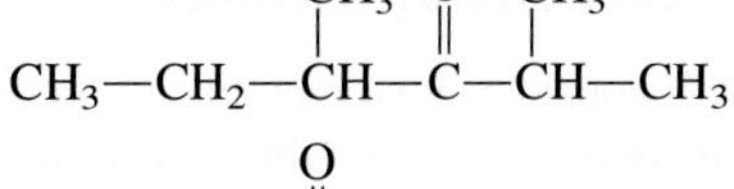

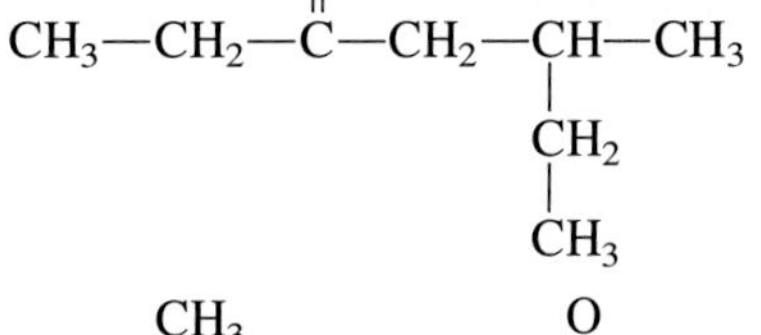

d. 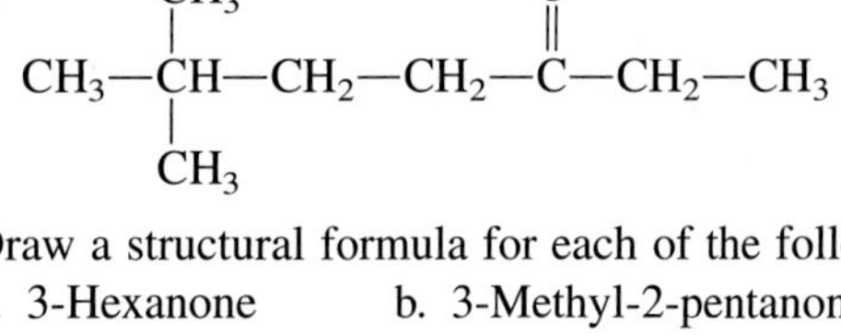 $CH_3-\underset{\underset{\displaystyle CH_3}{|}}{\overset{\overset{\displaystyle CH_3}{|}}{CH}}-CH_2-CH_2-\overset{\overset{\displaystyle O}{\|}}{C}-CH_2-CH_3$

13.23 Draw a structural formula for each of the following ketones.

a. 3-Hexanone b. 3-Methyl-2-pentanone
c. Cyclobutanone d. 1,3-Dichloropropanone

13.24 The simplest ketone has two common names in addition to its IUPAC name. What are these three names?

13.25 What is the common name for each of the following ketones?

a. $CH_3—CH_2—CH_2—\overset{\overset{\displaystyle O}{\|}}{C}—CH_2—CH_2—CH_3$

b. $CH_3—CH_2—CH_2—\overset{\overset{\displaystyle O}{\|}}{C}—CH_2—CH_3$

c. $CH_3—\underset{\underset{\displaystyle CH_3}{|}}{CH}—\underset{\underset{\displaystyle O}{\|}}{C}—CH_3$

d. $CH_3—CH_2—\overset{\overset{\displaystyle O}{\|}}{C}—$(cyclohexyl)

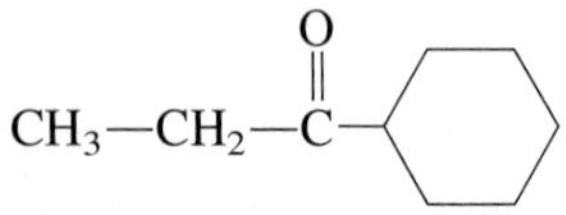

Learning Focus

List general physical properties of ketones; discuss products from the reduction and oxidation of ketones and from the reaction of ketones with alcohols.

13.5 Properties and Reactions of Ketones

The physical properties of ketones parallel closely the physical properties of aldehydes. As with aldehydes (Section 13.3), their boiling points are intermediate between those of alkanes and alcohols of similar molecular mass.

$CH_3—CH_2—CH_2—CH_3$	$CH_3—CH_2—\overset{\overset{\displaystyle O}{\|}}{C}—H$	$CH_3—\overset{\overset{\displaystyle O}{\|}}{C}—CH_3$	$CH_3—CH_2—CH_2—OH$
Butane	Propanal	Propanone	1-Propanol
molar mass = 58 amu	molar mass = 58 amu	molar mass = 58 amu	molar mass = 60 amu
boiling point = 0°C	boiling point = 50°C	boiling point = 56°C	boiling point = 97°C

Unlike aldehydes, which can have either pungent or fragrant odors, ketones generally have pleasant odors; several are used in perfumes.

Water solubility decreases with increasing carbon–chain length; smaller molecules are soluble and larger ones are not. Acetone, the simplest ketone, is infinitely soluble in water. Traces of water in laboratory glassware (test tubes, etc.) can be removed by "swishing" with a small amount of acetone and then pouring out the acetone. The acetone takes up the water, and any residual acetone rapidly vaporizes. Chemical Portraits 21 profiles the acetone molecule as well as the simple aldehydes formaldehyde and acrolein.

Reactions of Ketones

We will consider three reactions of ketones: (1) reduction, (2) oxidation, and (3) reaction with alcohols. These are the same three types of reactions that were considered with aldehydes (Section 13.3).

Reduction

Ketones are easily reduced to secondary alcohols by a variety of reducing agents, with appropriate catalysts. This contrasts with aldehyde reduction, wherein the product is a primary rather than a secondary alcohol. A specific example of ketone reduction is the reduction of propanone using H_2 gas with Ni as a catalyst; the product is 2-propanol (isopropyl alcohol).

$$\underset{\text{Propanone}}{CH_3—\overset{\overset{\displaystyle O}{\|}}{C}—CH_3} + H_2 \xrightarrow{Ni} \underset{\text{2-Propanol}}{CH_3—\overset{\overset{\displaystyle OH}{|}}{\underset{\underset{\displaystyle H}{|}}{C}}—CH_3}$$

The reduction of ketones to secondary alcohols is the opposite process to the oxidation of secondary alcohols to produce ketones, a reaction of secondary alcohols that we considered in Section 12.4.

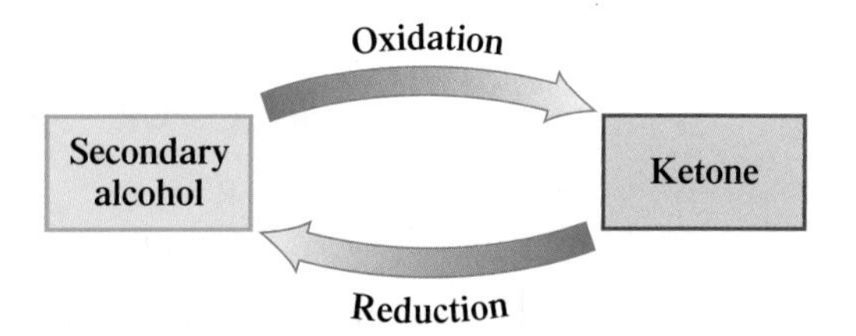

Oxidation

Aldehydes readily undergo oxidation with mild oxidizing agents to carboxylic acids (Section 13.3). Ketones are not oxidized by mild oxidizing agents. Because both aldehydes and ketones contain carbonyl groups, we might expect similar reactions for the two types of compounds. Oxidation of an aldehyde involves breaking a carbon–hydrogen bond, and oxidation of a ketone involves breaking a carbon–carbon bond. The former is much easier to accomplish than the latter. For ketones to be oxidized, strenuous reaction conditions must be employed.

Reaction with Alcohols

Aldehydes can be converted into hemiacetals and then into acetals through reaction with alcohols (Section 13.3). A similar reaction system exists with ketones. Here, however, the products are *hemiketals* and *ketals* rather than *hemiacetals* and *acetals*. Structurally, the general difference between a hemiketal and a hemiacetal and that between a ketal and an acetal are the same as the difference between a ketone and an

Chemical Portraits 21 Simple Carbonyl-Containing Compounds

Formaldehyde

Profile: Formaldehyde, the simplest aldehyde, is a flammable, colorless gas with an irritating odor. It is readily soluble in water and is generally available in aqueous solution form (*formalin*). Its major use is as a starting material for the production of polymers. Phenol-formaldehyde polymers, called phenolics, are used as adhesives in the production of plywood and particle board.

Biochemical considerations: Formaldehyde is a component of wood smoke and is partially responsible for the preservation action of the smoking of foods; the formaldehyde reacts with the protein present in the bacteria; such protein is "embalmed."

What are the sources for formaldehyde as an *indoor* air pollutant?

Acetone

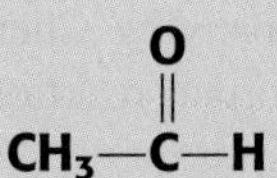

Profile: Acetone, the simplest ketone, is a flammable, colorless, volatile liquid with a pleasant but mildly pungent odor. It is an important industrial solvent, with uses in paints, varnishes, resins, coatings, and fingernail polish.

Biochemical considerations: Small amounts of acetone are produced in the human body in reactions relating to obtaining energy from fats. Normally, such acetone is degraded to CO_2 and H_2O. Diabetic people produce larger amounts of acetone, all of which cannot be degraded. In severe diabetes, the odor of acetone can be detected in the person's breath.

What are the reactants for the industrial production of acetone?

Acrolein (Propenal)

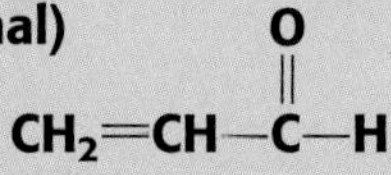

Profile: A colorless, flammable, volatile liquid in the pure state, acrolein is the simplest molecule that contains both a carbon–carbon and carbon–oxygen double bond. It is both an alkene and an aldehyde. It is used in making plastics and aquatic pesticides.

Biochemical considerations: Acrolein is produced when meat is barbecued. The fats in meat break down under heat stress forming glycerin which further breaks down to produce acrolein. Part of the "pleasant" smell associated with a barbecue is due to acrolein. It is a major contributor to the irritating quality of wood smoke, cigarette smoke, and photochemical smog.

What are some of the initial symptoms of overexposure to acrolein?

See the text web site at **www.cengage.com/chemistry/stoker** for answers to the above questions and for further information.

aldehyde. That difference is "R" versus "H." The hemiketal–hemiacetal and ketal–acetal structural contrasts are as follows:

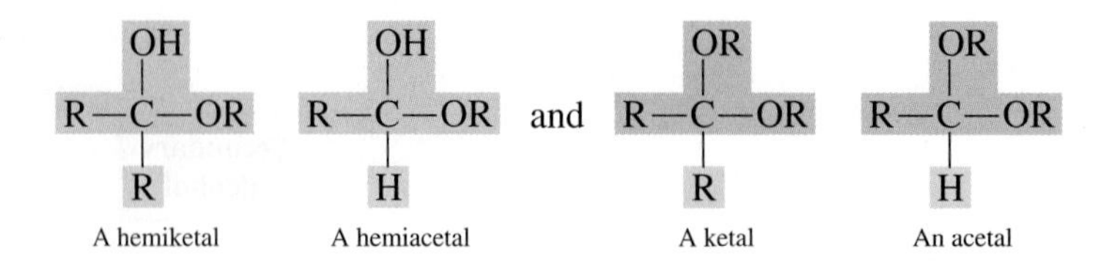

A **hemiketal** *is a compound that has a carbon atom to which a hydroxyl group (—OH) and an alkoxy group (—OR), but no hydrogen atom (—H), are attached.* A hemiketal is changed to a ketal by replacing the hydroxyl group with another alkoxy group. A **ketal** *is a compound that has a carbon atom to which two alkoxy groups (—OR), but no hydrogen atom (—H), are attached.*

A specific example of hemiketal and then ketal formation from a ketone follows.

$$\underset{\text{Ketone}}{CH_3-\overset{\overset{\large O}{\|}}{C}-CH_3} + \underset{\text{Alcohol}}{H-O-CH_3} \overset{H^+}{\rightleftharpoons} \underset{\text{Hemiketal}}{CH_3-\overset{OH}{\overset{|}{\underset{\underset{O-CH_3}{|}}{C}}}-CH_3}$$

$$\underset{\text{Hemiketal}}{CH_3-\overset{OH}{\overset{|}{\underset{\underset{O-CH_3}{|}}{C}}}-CH_3} + \underset{\text{Alcohol}}{H-O-CH_3} \rightleftharpoons \underset{\text{Ketal}}{CH_3-\overset{O-CH_3}{\overset{|}{\underset{\underset{O-CH_3}{|}}{C}}}-CH_3} + \underset{\text{Water}}{H-O-H}$$

The accompanying Chemistry at a Glance summarizes and also contrasts the biochemically important reaction chemistries of aldehydes and ketones.

Practice Problems and Questions

13.26 Draw the structure of the secondary alcohol produced when each of the following ketones is reduced using molecular H_2 and a nickel catalyst.

a. $CH_3-CH_2-\overset{\overset{O}{\|}}{C}-CH_3$

b. $CH_3-\overset{\overset{O}{\|}}{C}-\underset{\underset{CH_3}{|}}{CH}-CH_3$

c. $CH_3-\underset{\underset{CH_3}{|}}{CH}-\underset{\underset{CH_3}{|}}{CH}-\overset{\overset{O}{\|}}{C}-CH_3$

d. $CH_3-\overset{\overset{CH_3}{|}}{\underset{\underset{CH_3}{|}}{C}}-CH_2-\overset{\overset{O}{\|}}{C}-CH_2-CH_3$

13.27 Mild oxidizing agents readily oxidize aldehydes but not ketones. Explain why.

13.28 What is the formula of the ketone needed to produce each of the following alcohols through a reduction reaction?

a. $CH_3-\underset{\underset{CH_3}{|}}{CH}-OH$

b. $CH_3-\underset{\underset{OH}{|}}{CH}-CH_3$

c. $CH_3-\underset{\underset{CH_3}{|}}{CH}-CH_2-\underset{\underset{CH_3}{|}}{CH}-OH$

d. $CH_3-CH_2-CH_2-\underset{\underset{OH}{|}}{CH}-CH_3$

Reactions Involving Aldehydes and Ketones

Primary alcohol → (Oxidizing agent) → **Oxidation** → **ALDEHYDE**

Secondary alcohol → (Oxidizing agent) → **Oxidation** → **KETONE**

ALDEHYDE / KETONE:

- Oxidizing agent → **Oxidation**
  - If aldehyde → Carboxylic acid
  - If ketone → No reaction
- H_2, Ni catalyst → **Reduction**
  - If aldehyde → Primary alcohol
  - If ketone → Secondary alcohol
- Alcohol → **Hemi Formation**
  - If aldehyde → Hemiacetal → (Alcohol) → Acetal
  - If ketone → Hemiketal → (Alcohol) → Ketal

13.29 Identify each of the following compounds as a hemiketal, a ketal, or neither a hemiketal nor a ketal.

a. $CH_3-CH_2-\underset{OH}{\overset{OH}{C}}-CH_3$

b. $CH_3-CH_2-\underset{O-CH_3}{\overset{OH}{C}}-CH_3$

c. $CH_3-CH_2-\underset{O-CH_3}{\overset{O-CH_3}{C}}-CH_3$

d. $CH_3-\underset{O-CH_3}{\overset{H}{C}}-CH_3$

13.30 Draw the structure of the missing compound(s) in each of the following reactions.

a. $CH_3-CH_2-CH_2-\overset{O}{\overset{\|}{C}}-CH_3 + CH_3-CH_2-OH \overset{H^+}{\rightleftharpoons} ?$

b. $? + CH_3-OH \overset{H^+}{\rightleftharpoons} CH_3-CH_2-\underset{O-CH_3}{\overset{OH}{C}}-CH_3$

c. $CH_3-CH_2-\underset{CH_3}{\overset{OH}{C}}-O-CH_3 + CH_3-\underset{CH_3}{CH}-OH \xrightarrow{H^+} ? + H_2O$

d. $\underset{\text{Hemiketal}}{?} + \underset{\text{Alcohol}}{?} \xrightarrow{H^+} CH_3-CH_2-\underset{O-CH_3}{\overset{CH_3}{C}}-O-CH_3 + H_2O$

Learning Focus

Know the general structural characteristics of carboxylic acids; be able to name such compounds given their structural formulas, or vice versa.

13.6 Structural Characteristics and Naming of Carboxylic Acids

A **carboxylic acid** *is a compound whose characteristic functional group is the carboxyl group.* Because of their wide distribution and abundance in natural products, these compounds were some of the first organic compounds studied in detail. A **carboxyl group** *is a carbonyl group (C═O) with a hydroxyl group (—OH) bonded to the carbonyl carbon atom.* Two different structural notations are used to denote a carboxyl group:

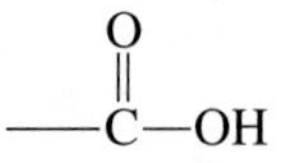

$$-\overset{\overset{\displaystyle O}{\|}}{C}-OH \quad \text{and} \quad -COOH$$

Although we see within a carboxyl group both a carbonyl group (C═O) and a hydroxyl group (—OH), the carboxyl group does not show characteristic behavior of either an alcohol or a carbonyl compound (aldehyde or ketone). Rather, it is a unique functional group with a set of characteristics different from those of its component parts.

The simplest carboxylic acid has a hydrogen atom attached to the carboxyl group carbon atom.

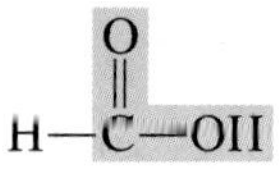

$$H-\overset{\overset{\displaystyle O}{\|}}{C}-OH$$

Structures for the next two simplest carboxylic acids, those with methyl and ethyl alkyl groups, are

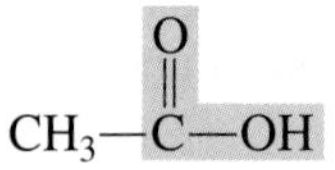

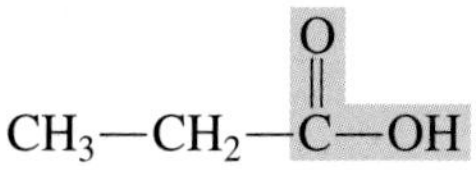

$$CH_3-\overset{\overset{\displaystyle O}{\|}}{C}-OH \qquad CH_3-CH_2-\overset{\overset{\displaystyle O}{\|}}{C}-OH$$

The structure of the simplest aromatic carboxylic acid involves a benzene ring to which a carboxyl group is attached.

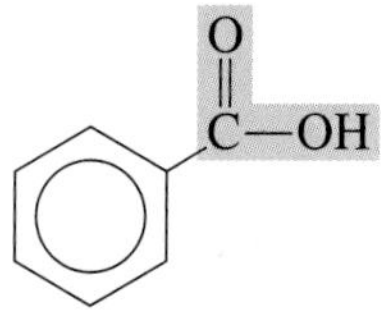

$$C_6H_5-\overset{\overset{\displaystyle O}{\|}}{C}-OH$$

Naming of Carboxylic Acids

IUPAC rules for naming carboxylic acids resemble those for naming aldehydes (Section 13.2).

1. Select as the parent carbon chain the longest carbon chain that *includes* the carbon atom of the carboxyl group.
2. Name the parent chain by changing the *-e* ending of the corresponding alkane to *-oic acid.*
3. Number the parent chain by assigning the number 1 to the carboxyl carbon atom.
4. Determine the identity and location of any substituents in the usual manner, and append this information to the front of the parent chain name.

▶ A carboxyl group must occupy a terminal (end) position in a carbon chain because there can be only one other bond to it.

Example 13.4 Determining IUPAC Names for Carboxylic Acids

Assign IUPAC names to the following carboxylic acids.

a.

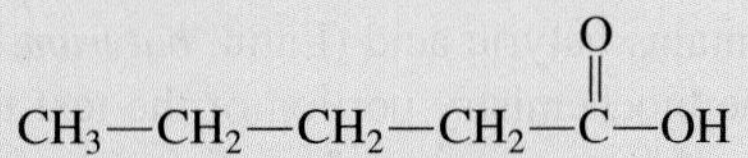

b.

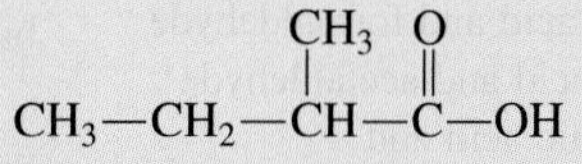

c.

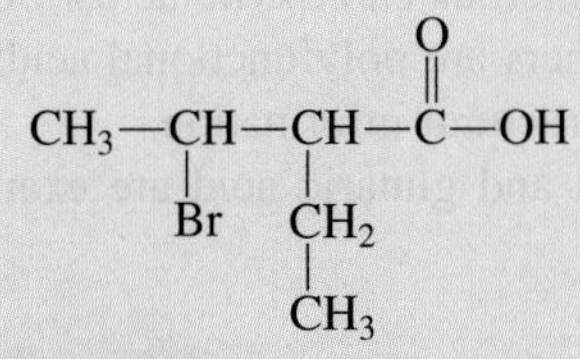

Solution

a. The parent chain name is based on pentane. Removing the *-e* ending from pentane and replacing it with the ending *-oic acid* gives *pentanoic acid.* The location of the carboxyl group need not be specified, because by definition the carboxyl carbon atom is always carbon 1.

b. The parent chain name is *butanoic acid.* To locate the methyl group substituent, we number the carbon chain beginning with the carboxyl carbon atom. The complete name of the acid is *2-methylbutanoic acid.*

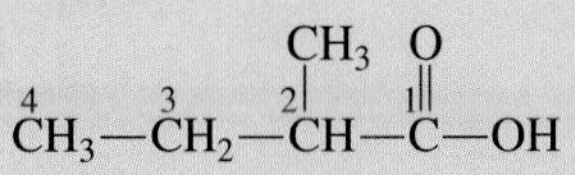

c. The longest carboxyl-carbon-containing chain has four carbon atoms. The parent chain name is thus *butanoic acid.* There are two substituents present, an ethyl group on carbon 2 and a bromo group on carbon 3. The complete name is *3-bromo-2-ethylbutanoic acid.*

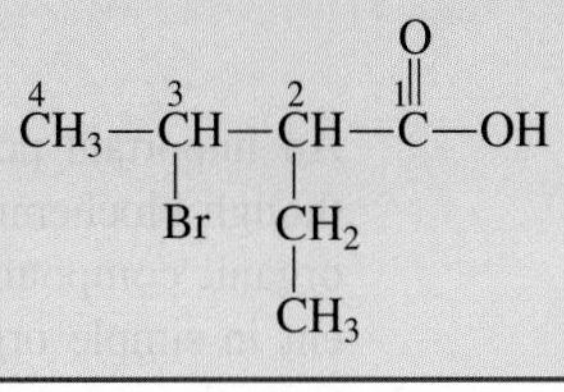

▶ The carboxyl functional group has the highest priority in the IUPAC naming system of all functional groups considered so far. When both a carboxyl group and a carbonyl group (aldehyde, ketone) are present in the same molecule, the prefix *oxo-* is used to denote the carbonyl group.

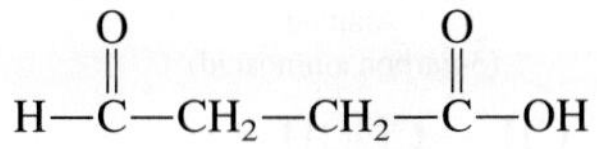

4-Oxobutanoic acid

The simplest aromatic carboxylic acid is called benzoic acid. Other aromatic carboxylic acids are named as derivatives of benzoic acid using the same substituent location rules as for other types of aromatic compounds.

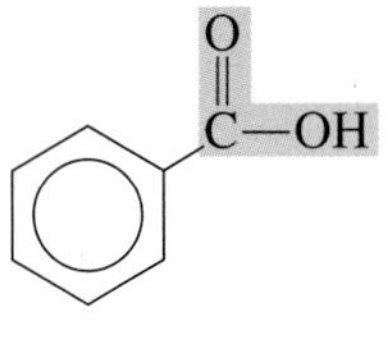

Benzoic acid

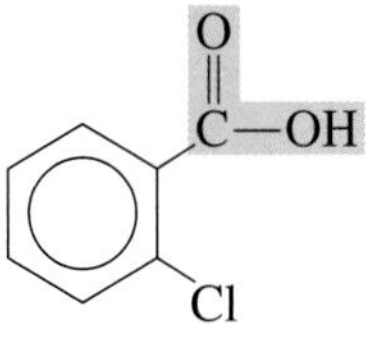

2-Chlorobenzoic acid

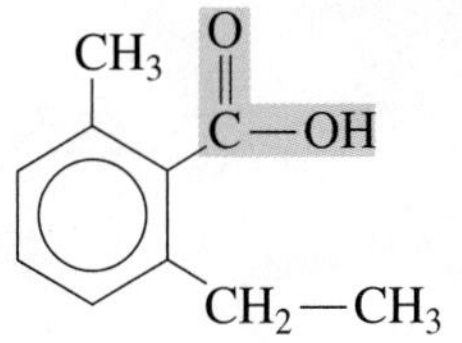

2-Ethyl-6-methylbenzoic acid

The use of common names is more prevalent for carboxylic acids than for any other family of organic compounds. Often these common names indicate the natural sources from which the acids were first isolated. The common names of the first four straight-chain carboxylic acids are

H—COOH	CH_3—COOH	CH_3—CH_2—COOH	CH_3—CH_2—CH_2—COOH
Formic acid	Acetic acid	Propionic acid	Butyric acid

▶ The common names of straight-chain carboxylic acids are the basis for aldehyde common names (Section 13.2).

C_1: formic acid and formaldehyde
C_2: acetic acid and acetaldehyde
C_3: propionic acid and propionaldehyde
C_4: butyric acid and butyraldehyde

▶ The IUPAC name for succinic acid is *butanedioic acid,* and that for glutaric acid is *pentanedioic acid.*

The stinging sensation associated with red ant bites is due in part to formic acid (Latin, *formica,* "ant") (see Figure 13.4). Acetic acid gives vinegar its tartness (sour taste); vinegar is a 4%–8% (v/v) acetic acid solution (Latin, *acetum,* "sour"). Propionic acid is the smallest acid that can be obtained from fats (Greek, *protos,* "first," and *pion,* "fat"). Rancid butter contains butyric acid (Latin, *butyrum,* "butter").

In the biochemistry portion of the text that deals with metabolism (Chapter 18), we will encounter numerous carboxylic acids. Some are simple dicarboxylic acids (two carboxyl groups); others are polyfunctional acids. Nearly all of them go by common names that provide no structural information.

Succinic acid and glutaric acid are examples of dicarboxylic acids encountered in biochemistry.

$$\mathrm{HOOC{-}CH_2{-}CH_2{-}COOH}$$
Succinic acid
(4-carbon diacid)

$$\mathrm{HOOC{-}CH_2{-}CH_2{-}CH_2{-}COOH}$$
Glutaric acid
(5-carbon diacid)

Polyfunctional acids (two or more different functional groups) are common in biochemical systems. Five such molecules of biological importance are

$$\mathrm{CH_3{-}CH(OH){-}COOH}$$
Lactic acid
(3-carbon hydroxyacid)

$$\mathrm{CH_3{-}C({=}O){-}COOH}$$
Pyruvic acid
(3-carbon ketoacid)

$$\mathrm{CH_3{-}CH(NH_2){-}COOH}$$
Alanine
(3-carbon aminoacid)

$$\mathrm{HOOC{-}C({=}O){-}CH_2{-}COOH}$$
Oxaloacetic acid
(4-carbon ketodiacid)

$$\mathrm{HO{-}C(COOH)(CH_2{-}COOH)_2}$$
Citric acid
(6-carbon hydroxytriacid)

An important point to be derived from "surveying" the preceding structures is that although biochemical molecules, in general, have more "complex" structures than simple organic compounds, the functional groups present in them are the same ones that are present in simple organic compounds. These biochemical molecules will undergo the reactions that are characteristic of the functional groups that are present within them. Terms such as *three-carbon ketoacid* and *four-carbon diacid* are terms that can be understood at this point in our study of organic chemistry.

The two commonly used nonprescription pain relievers ibuprofen (Advil) and naproxen (Aleve) are derivatives of propanoic acid, the three-carbon monocarboxylic acid.

$$\mathrm{CH_3{-}CH(CH_3){-}CH_2{-}C_6H_4{-}CH(CH_3){-}COOH}$$
Ibuprofen

$$\mathrm{CH_3{-}O{-}C_{10}H_6{-}CH(CH_3){-}COOH}$$
Naproxen

In ibuprofen, carbon 2 of propanoic acid carries an aromatic ring system that bears an alkyl group. In naproxen, carbon 2 of propanoic acid carries a fused aromatic ring system that bears an alkoxy group. The latter compound is thus polyfunctional, being both an acid and an ether.

Chemical Portraits 22 profiles acetic acid (a monocarboxylic acid), lactic acid (a hydroxy monocarboxylic acid) and citric acid (a hydroxy tricarboxylic acid)

Figure 13.4 Red ants in a defensive position. The irritant in the "bite" of a red ant is methanoic acid (formic acid), the simplest carboxylic acid. The Latin word for "ant" is *formica.*

Chemical Portraits 22 Commonly Encountered Carboxylic Acids

Acetic Acid (CH_3-COOH)

Profile: Pure acetic acid, also known as *glacial* acetic acid, freezes on a moderately cool day (17°C) producing icy-looking crystals. Aqueous solution is the form in which acetic acid is normally encountered. Acetic acid solutions exceeding 50%(m/v) concentration can damage human skin; no immediate pain occurs but painful blisters begin to form in approximately 30 minutes.

Biochemical considerations: Vinegar is a 4–8%(v/v) solution of acetic acid, with other flavoring agents present, produced by fermentation of grapes, apples, and other fruits in the presence of oxygen. Fermentation produces ethyl alcohol first, then acetaldehyde, and finally acetic acid. It is the acetic acid present in vinegar that gives vinegar its tartness (sour taste).

What is the origin of the name *acetic acid*?

Lactic Acid

$$CH_3-\overset{\overset{\displaystyle OH}{|}}{CH}-COOH$$

Profile: Lactic acid, a white solid at room temperature, is encountered naturally in numerous environments. Lactic acid, secreted by bacteria as they metabolize sugars, causes milk to sour. This process, applied in a controlled manner, yields yogurt. The sharp taste of sauerkraut results from lactic acid produced by bacteria present in the brine in which fresh cabbage is soaked.

Biochemical considerations: The buildup of lactic acid in muscle tissue during strenuous exercise is what causes them to feel "weak" and cramp; it is also the cause of soreness the day after the exercise. Such lactic acid production in the body occurs when the body operates under the "oxygen deficient" conditions associated with strenuous exercise.

What is the fate of the lactic acid produced in the body as the result of strenuous exercise?

Citric Acid

$$HO-\underset{\underset{\displaystyle CH_2-COOH}{|}}{\overset{\overset{\displaystyle CH_2-COOH}{|}}{C}}-COOH$$

Profile: Pure citric acid is a white crystalline solid that readily absorbs moisture from the air. It is the acid that gives citrus fruits their tart taste; lemon juice contains 4–8% citric acid, and orange juice is about 1% citric acid. In fresh salads, citric acid prevents enzymatic browning reactions, and in frozen fruits, it prevents deterioration of color and flavor. In jams, jellies, and preserves, it produces tartness and pH adjustment to optimize conditions for gelation.

Biochemical considerations: Citric acid is an important substance in the processes by which energy is obtained from food in the human body. Indeed, one of the main sets of chemical reactions that occurs in the body is called the *citric acid cycle.*

What is the main function of the human body's *citric acid cycle*?

See the text web site at **www.cengage.com/chemistry/stoker** for answers to the above questions and for further information.

Practice Problems and Questions

13.31 In which of the following compounds is a carboxyl group present?

a. $CH_3-CH_2-\overset{\overset{\displaystyle O}{\|}}{C}-OH$

b. $CH_3-CH_2-CH_2-\overset{\overset{\displaystyle O}{\|}}{C}-CH_3$

c. $CH_3-\overset{\overset{\displaystyle OH}{|}}{CH}-\overset{\overset{\displaystyle O}{\|}}{C}-CH_3$

d. CH_3-CH_2-COOH

13.32 Give the IUPAC name for each of the following carboxylic acids.

a. $CH_3-CH_2-CH_2-\overset{\overset{\displaystyle O}{\|}}{C}-OH$

b. $CH_3-CH_2-\overset{\overset{\displaystyle CH_3}{|}}{CH}-\overset{\overset{\displaystyle CH_3}{|}}{CH}-\overset{\overset{\displaystyle O}{\|}}{C}-OH$

c. $\overset{\overset{\displaystyle CH_3}{|}}{CH_2}-\overset{\overset{\displaystyle CH_3}{|}}{CH}-\overset{\overset{\displaystyle O}{\|}}{C}-OH$

d. $CH_3-\overset{\overset{\displaystyle Br}{|}}{CH}-CH_2-CH_2-\overset{\overset{\displaystyle O}{\|}}{C}-OH$

13.33 Draw a condensed structural formula for each of the following carboxylic acids.

a. 2-Ethylbutanoic acid
b. 2,3-Dimethylbutanoic acid
c. Methylpropanoic acid
d. Dichloroethanoic acid

Figure 13.5 "Drug-sniffing" dogs used by narcotics agents can find hidden heroin by detecting the odor of acetic acid (vinegar odor). Acetic acid is a by-product of the final step in illicit heroin production, and trace amounts remain in the heroin.

13.34 Give the IUPAC name for each of the following aromatic carboxylic acids.

a. COOH (on benzene ring)

b. COOH, $CH_2—CH_3$ (on benzene ring)

c. COOH, Cl (on benzene ring)

d. COOH, Cl, Br (on benzene ring)

13.35 Give the common name for each of the following carboxylic acids.
a. Methanoic acid b. Ethanoic acid c. Propanoic acid d. Butanoic acid

13.36 Draw condensed structural formulas that fit each of the following "acid descriptions."
a. Two-carbon diacid
b. Three-carbon hydroxyacid [two different structures (isomers) are possible]
c. Four-carbon hydroxydiacid [two different structures (isomers) are possible]
d. Four-carbon ketodiacid

> **Learning Focus**
> List the general properties of carboxylic acids; understand why carboxylic acids exhibit acidic properties; write equations for the formation of acid salts and the reaction of carboxylic acids with alcohols.

13.7 Properties and Reactions of Carboxylic Acids

Unsubstituted saturated monocarboxylic acids containing up to nine carbon atoms are liquids that have strong, sharp odors (see Figure 13.5). Acids with 10 or more carbon atoms in an unbranched chain are waxy solids that are odorless (because of low volatility). Aromatic carboxylic acids, as well as dicarboxylic acids, are also odorless solids.

Carboxylic acids have the highest boiling points of all organic compounds we have considered so far (Figure 13.6), because hydrogen-bonding opportunities are more extensive for carboxylic acids. A given carboxylic acid molecule can form two hydrogen bonds with another carboxylic acid molecule (Figure 13.7).

Carboxylic acids readily hydrogen-bond to water molecules. Such hydrogen bonding contributes to water solubility for short-chain carboxylic acids. The unsubstituted C_1 to C_4 monocarboxylic acids are completely miscible with water. Solubility then rapidly decreases with carbon number, as shown in Figure 13.8. Short-chain dicarboxylic acids are also water-soluble. In general, aromatic acids are not water-soluble.

Figure 13.6 The boiling points of monocarboxylic acids compared to those of other types of compounds. All compounds in the comparison have unbranched carbon chains.

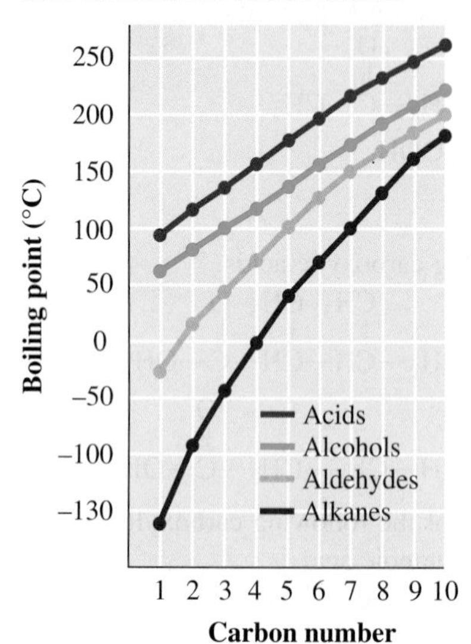

Acidity of Carboxylic Acids

Carboxylic acids, as the name implies, are *acidic*. When a carboxylic acid is placed in water, hydrogen ion transfer (proton transfer; Section 9.2) occurs to produce hydronium ion (the acidic species in water; Section 9.2) and carboxylate ion.

$$R—COOH + H_2O \longrightarrow \underset{\text{Hydronium ion}}{H_3O^+} + \underset{\text{Carboxylate ion}}{R—COO^-}$$

A **carboxylate ion** *is the negative ion produced when a carboxylic acid loses one or more acidic hydrogen atoms.*

Carboxylate ions formed from monocarboxylic acids always carry a −1 charge; only one acidic hydrogen atom is present in such molecules. Dicarboxylic acids, which possess two acidic hydrogen atoms (one in each carboxyl group), can produce carboxylate ions bearing a −2 charge.

Carboxylate ions are named by dropping the *-ic acid* ending from the name of the parent acid and replacing it with *-ate*.

$$\underset{\text{Acetic acid (ethanoic acid)}}{CH_3—\overset{O}{\overset{\|}{C}}—OH} + H_2O \longrightarrow H_3O^+ + \underset{\text{Acetate ion (ethanoate ion)}}{CH_3—\overset{O}{\overset{\|}{C}}—O^-}$$

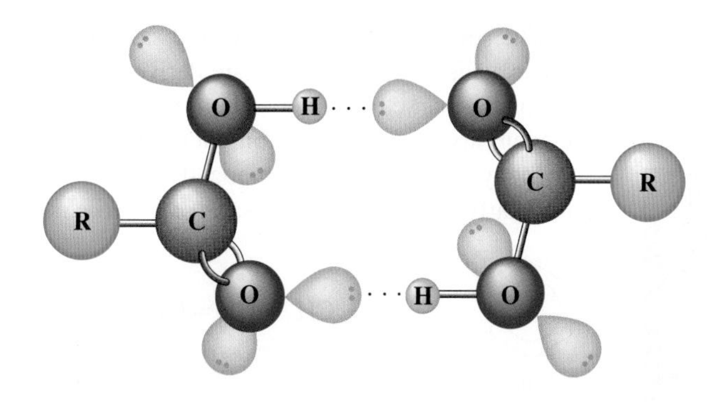

Figure 13.7 A given carboxylic acid molecule can form two hydrogen bonds to another carboxylic acid molecule, producing a "dimer." Dimers have twice the mass of a single molecule, and a higher temperature is needed to boil carboxylic acids than would be needed if no dimers were present.

$$\underset{\substack{\text{Oxalic acid}\\\text{(ethanedioic acid)}}}{HO-\overset{\overset{\displaystyle O}{\|}}{C}-\overset{\overset{\displaystyle O}{\|}}{C}-OH} + 2H_2O \longrightarrow 2H_3O^+ + \underset{\substack{\text{Oxalate ion}\\\text{(ethanedioate ion)}}}{{}^-O-\overset{\overset{\displaystyle O}{\|}}{C}-\overset{\overset{\displaystyle O}{\|}}{C}-O^-}$$

Carboxylic acids are weak acids (Section 9.4). The extent of proton transfer is usually less than 5%; that is, an equilibrium situation exists in which the equilibrium lies far to the left.

$$\underset{\substack{\text{More than 95\%}\\\text{of molecules in this form}}}{R-COOH + H_2O} \rightleftharpoons \underset{\substack{\text{Less than 5\%}\\\text{of molecules in this form}}}{H_3O^+ + R-COO^-}$$

▶ Carboxylic acid salt formation involves an acid–base neutralization reaction (Section 9.6).

Formation of Carboxylic Acid Salts

In a manner similar to that of inorganic acids (Section 9.6), carboxylic acids react completely with strong bases to produce water and a carboxylic acid salt.

$$\underset{\text{Carboxylic acid}}{CH_3-\overset{\overset{\displaystyle O}{\|}}{C}-OH} + \underset{\text{Strong base}}{NaOH} \longrightarrow \underset{\substack{\text{Carboxylic}\\\text{acid salt}}}{CH_3-\overset{\overset{\displaystyle O}{\|}}{C}-O^-\,Na^+} + \underset{\text{Water}}{H_2O}$$

A **carboxylic acid salt** *is an ionic compound in which the negative ion is a carboxylate ion.*

Carboxylic acid salts are named similarly to other ionic compounds (Section 4.19): *The positive ion is named first, followed by a separate word giving the name of the negative ion.* The salt formed in the preceding reaction contains sodium ions and acetate ions (from acetic acid); hence the salt's name is sodium acetate.

Converting a carboxylic acid salt back to a carboxylic acid is very simple. React the salt with a solution of a strong acid such as hydrochloric acid (HCl) or sulfuric acid (H_2SO_4).

$$\underset{\text{Sodium acetate}}{CH_3-\overset{\overset{\displaystyle O}{\|}}{C}-O^-\,Na^+} + \underset{\substack{\text{Hydrochloric}\\\text{acid}}}{HCl} \longrightarrow \underset{\text{Acetic acid}}{CH_3-\overset{\overset{\displaystyle O}{\|}}{C}-OH} + \underset{\substack{\text{Sodium}\\\text{chloride}}}{NaCl}$$

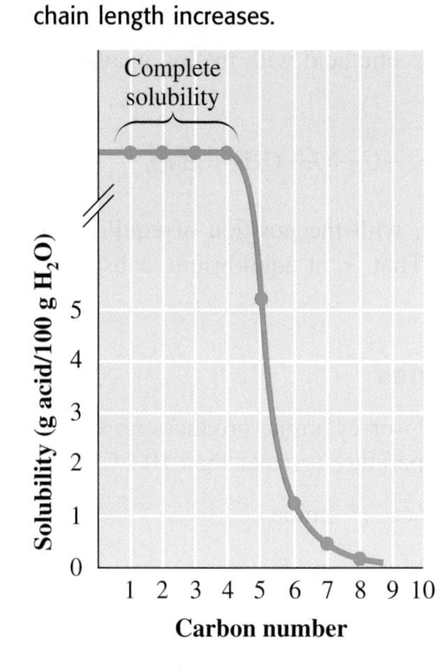

Figure 13.8 The solubility of saturated unbranched-chain carboxylic acids decreases as carbon-chain length increases.

The interconversion reactions between carboxylic acid salts and their "parent" carboxylic acids are so easy to carry out that organic chemists consider these two types of compounds interchangeable.

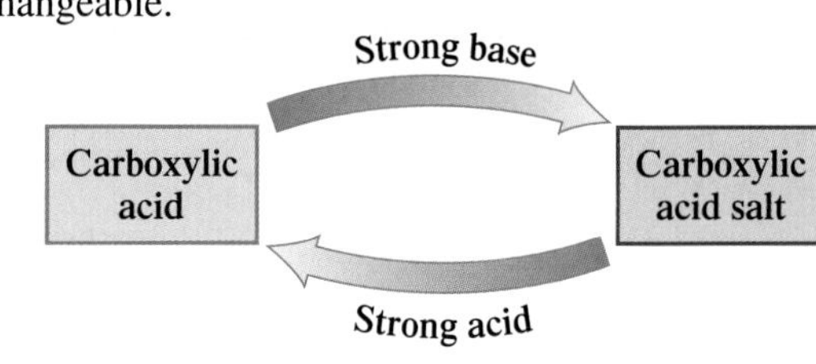

Figure 13.9 Propionates (salts of propionic acid) extend the shelf life of bread by preventing the formation of mold.

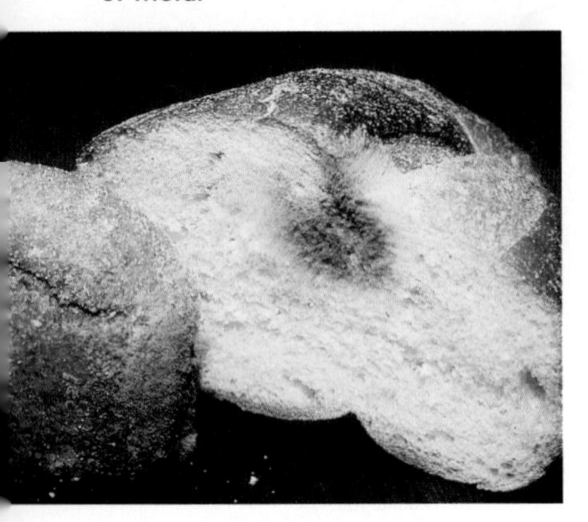

The solubility of carboxylic acid salts in water is much greater than that of the carboxylic acids from which they are derived. Drugs and medicines that contain acid groups are usually marketed as the sodium or potassium salt of the acid. This greatly enhances the solubility of the medication, increasing the ease of its absorption by the body.

Many *antimicrobials,* compounds used as food preservatives, are carboxylic acid salts. Particularly important are the salts of benzoic, sorbic, and propionic acids.

C_6H_5—COOH (Benzoic acid)

$CH_3—CH{=}CH—CH{=}CH—COOH$ (Sorbic acid (2,4-hexadienoic acid))

$CH_3—CH_2—COOH$ (Propionic acid)

Calcium and sodium propionates are used in baked products and also in cheese foods and spreads (see Figure 13.9). Benzoates and sorbates cannot be used in yeast-leavened baked goods, because they affect the activity of the yeast; they are used in beverages, jams, and jellies.

$$(CH_3—CH_2—\overset{O}{\overset{\|}{C}}—O^-)_2\ Ca^{2+} \qquad CH_3—CH_2—\overset{O}{\overset{\|}{C}}—O^-\ Na^+$$

Calcium propionate — Sodium propionate

Reaction of Carboxylic Acids with Alcohols

The reaction of a carboxylic acid with an alcohol, using a strong-acid catalyst (generally H_2SO_4), produces an ester, a type of compound that is discussed in detail in the next section.

$$\underset{\text{Carboxylic acid}}{R—\overset{O}{\overset{\|}{C}}—O—H} + \underset{\text{Alcohol}}{H—O—R'} \overset{H^+}{\rightleftharpoons} \underset{\text{Ester}}{R—\overset{O}{\overset{\|}{C}}—O—R'} + \underset{\text{Water}}{H_2O}$$

In this reaction, called *esterification,* a —OH group is lost from the carboxylic acid, a —H atom is lost from the alcohol, and water is formed as a by-product. The net effect of this reaction is substitution of the —OR group of the alcohol for the —OH group of the acid.

$$R—\overset{O}{\overset{\|}{C}}—(O—H) + H—(O—R') \overset{H^+}{\rightleftharpoons} R—\overset{O}{\overset{\|}{C}}—O—R' + H_2O$$

A specific example of esterification is the reaction of acetic acid with methyl alcohol.

▶ Studies show that in ester formation, the hydroxyl group of the acid (not of the alcohol) becomes part of the water molecule.

$$CH_3—\overset{O}{\overset{\|}{C}}—O—H + H—O—CH_3 \overset{H^+}{\rightleftharpoons} CH_3—\overset{O}{\overset{\|}{C}}—O—CH_3 + H_2O$$

Esterification reactions are equilibrium processes, with the position of equilibrium (Section 8.8) usually only slightly favoring products. That is, at equilibrium, substantial amounts of both reactants and products are present.

Reaction of Carboxylic Acids with Amines

The reaction of a carboxylic acid with ammonia or a 1° or 2° amine produces an amide, provided the reaction is carried out at an elevated temperature (greater than 100°C).

$$\text{carboxylic acid} + \text{ammonia} \xrightarrow{\text{high T}} 1° \text{ amide}$$

$$\text{carboxylic acid} + 1° \text{ amine} \xrightarrow{\text{high T}} 2° \text{ amide}$$

$$\text{carboxylic acid} + 2° \text{ amine} \xrightarrow{\text{high T}} 3° \text{ amide}$$

In such reactions, which are called *amidification* reactions, a —OH group is lost from the carboxylic acid, and a —H atom is lost from the ammonia or amine (to form water) and the organic residues remaining join to form the amide. The general equation for the reaction of ammonia with a carboxylic acid is

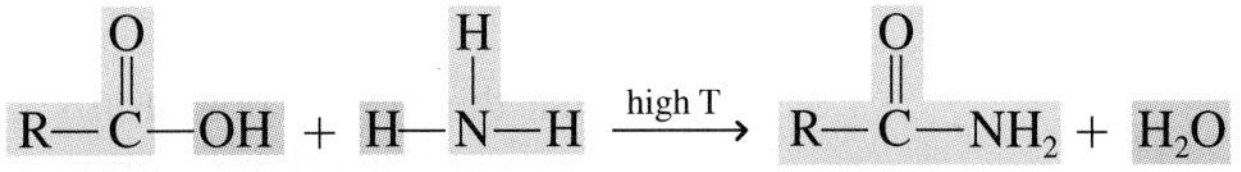

$$\underset{\text{Carboxylic acid}}{R-\overset{\overset{\displaystyle O}{\|}}{C}-OH} + \underset{\text{Ammonia}}{H-\overset{\overset{\displaystyle H}{|}}{N}-H} \xrightarrow{\text{high T}} \underset{1^\circ\text{ Amide}}{R-\overset{\overset{\displaystyle O}{\|}}{C}-NH_2} + \underset{\text{Water}}{H_2O}$$

Amides will be discussed in detail in Section 13.10 and 13.11.

Practice Problems and Exercises

13.37 Carboxylic acids have higher boiling points than alcohols or aldehydes of similar molecular mass. Explain why.

13.38 Which of the hydrogen atoms in a carboxylic acid is involved in the proton transfer reaction that occurs when a carboxylic acid is dissolved in water?

13.39 What is a *carboxylate ion*?

13.40 What is the name of the carboxylate ion that forms when each of the following carboxylic acids ionizes in water?
a. Formic acid b. Lactic acid c. Citric acid d. Ethanoic acid

13.41 What is the complete name of the sodium salt of each of the following carboxylic acids?
a. Acetic acid b. Butanoic acid c. Oxalic acid d. Succinic acid

13.42 Give the IUPAC name for each of the following carboxylic acid salts.

a. 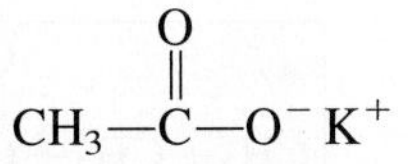 $CH_3-\overset{\overset{\displaystyle O}{\|}}{C}-O^-\ K^+$

b. $CH_3-CH_2-\overset{\overset{\displaystyle O}{\|}}{C}-O^-\ Na^+$

c. $CH_3-CH_2-CH_2-\overset{\overset{\displaystyle O}{\|}}{C}-O^-\ K^+$

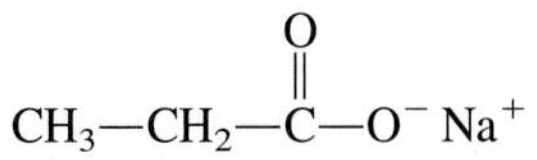

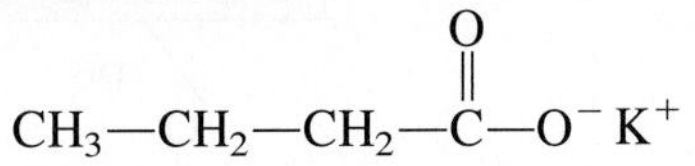

d. 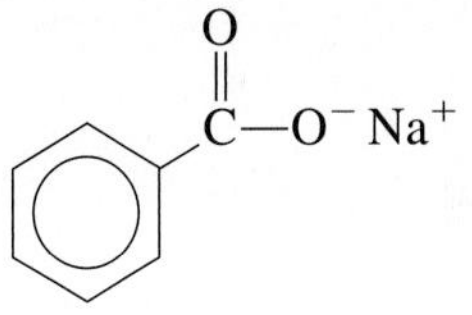 $C_6H_5-\overset{\overset{\displaystyle O}{\|}}{C}-O^-\ Na^+$ (benzene ring)

13.43 Write equations for the reaction of each of the following carboxylic acids with sodium hydroxide (NaOH).
a. Methanoic acid b. Formic acid c. Ethanoic acid d. Acetic acid

13.44 How do the generalized structures of a carboxylic acid and of an ester differ?

13.45 Draw the structure of the ester produced when each of the following pairs of carboxylic acid and alcohol react.
a. Propanoic acid and methanol b. Acetic acid and 1-propanol
c. 2-Methylbutanoic acid and 2-propanol d. Pentanoic acid and 2-butanol

Learning Focus

Know the general structural characteristics of esters; be able to name such compounds given their structural formulas, or vice versa.

13.8 Structural Characteristics and Naming of Esters

Esters are carboxylic acid derivatives in which the —OH group of the carboxylic acid has been replaced with an —OR group.

$$\underset{\text{Carboxylic acid}}{R-\overset{\overset{\displaystyle O}{\|}}{C}-(O-H)} \qquad \underset{\text{Ester}}{R-\overset{\overset{\displaystyle O}{\|}}{C}-(O-R)}$$

An **ester** *is an organic compound whose characteristic functional group is* $-\overset{\overset{\displaystyle O}{\|}}{C}-O-R$.

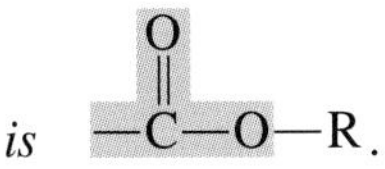

The simplest ester, which has two carbon atoms, has a hydrogen atom attached to the ester functional group.

$$\mathrm{H{-}\overset{\overset{\displaystyle O}{\|}}{C}{-}O{-}CH_3}$$

Note that the two carbon atoms in this ester are not bonded to each other.

There are two three-carbon esters.

$$\mathrm{H{-}\overset{\overset{\displaystyle O}{\|}}{C}{-}O{-}CH_2{-}CH_3} \quad \text{and} \quad \mathrm{CH_3{-}\overset{\overset{\displaystyle O}{\|}}{C}{-}O{-}CH_3}$$

In the previous section we learned that the reaction between a carboxylic acid and an alcohol produces an ester. It is often useful to think of the structure of an ester in terms of its "parent" alcohol and acid molecules; the ester has an acid part and an alcohol part.

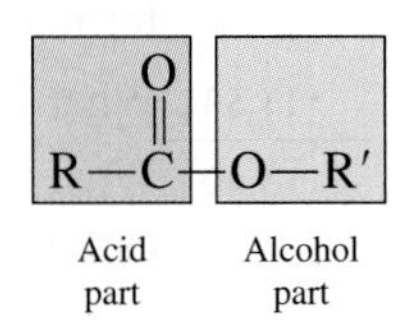

In this context, it is easy to identify the acid and alcohol from which a given ester can be produced; just add a —OH group to the acid part of the ester and a —H atom to the alcohol part to generate the parent molecules.

$$\mathrm{CH_3{-}CH_2{-}\overset{\overset{\displaystyle O}{\|}}{C}{-}O{-}CH_2{-}CH_2{-}CH_3}$$

$$\text{+OH} \swarrow \qquad \searrow \text{+H}$$

$$\underset{\text{"Parent" acid}}{\mathrm{CH_3{-}CH_2{-}\overset{\overset{\displaystyle O}{\|}}{C}{-}OH}} \qquad \underset{\text{"Parent" alcohol}}{\mathrm{H{-}O{-}CH_2{-}CH_2{-}CH_3}}$$

Looking again at the three ester structures given at the beginning of this section, we now note their "parentage." The two-carbon ester was formed from the one-carbon carboxylic acid and the one-carbon alcohol. The first of the three-carbon esters involves the one-carbon carboxylic acid and the two-carbon alcohol; a reverse situation exists for the second three-carbon ester, where the two-carbon carboxylic acid and the one-carbon alcohol are involved.

▶ Salts of carboxylic acids and esters are named in the same way. The name of the positive ion (in the case of a salt) or the name of the organic group attached to the oxygen of the carbonyl group (in the case of an ester) precedes the name of the acid. The *-ic acid* part of the name of the acid is converted to *-ate*.

$$\mathrm{CH_3{-}CH_2{-}CH_2{-}\overset{\overset{\displaystyle O}{\|}}{C}{-}O^-\ Na^+}$$

IUPAC: Sodium butanoate
Common: Sodium butyrate

$$\mathrm{CH_3{-}CH_2{-}CH_2{-}\overset{\overset{\displaystyle O}{\|}}{C}{-}O{-}CH_3}$$

IUPAC: Methyl butanoate
Common: Methyl butyrate

Naming of Esters

Visualizing esters as having an "alcohol part" and an "acid part" is the key to naming them in both the common and the IUPAC systems of nomenclature. The name of the alcohol part of the ester appears first and is followed by a separate word giving the name for the acid part of the ester. The name for the alcohol part of the ester is simply the name of the R group (alkyl, cycloalkyl, or aryl) present in the —OR portion of the ester. The name for the acid part of the ester is obtained by dropping the *-ic acid* ending from the acid's name and adding the suffix *-ate*.

Consider the ester derived from ethanoic acid (acetic acid) and methanol (methyl alcohol). Its name will be *methyl ethanoate* (IUPAC) or *methyl acetate* (common).

$$\mathrm{CH_3{-}\overset{\overset{\displaystyle O}{\|}}{C}{-}OH + HO{-}CH_3 \longrightarrow CH_3{-}\overset{\overset{\displaystyle O}{\|}}{C}{-}O{-}CH_3 + H_2O}$$

IUPAC:	Ethanoic acid	Methanol	Methyl ethanoate
Common:	Acetic acid	Methyl alcohol	Methyl acetate

Dicarboxylic acids can form diesters, with each of the carboxyl groups undergoing esterification. An example of such a molecule and how it is named is

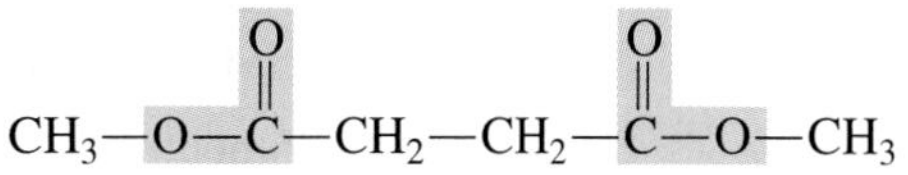

$$CH_3-O-\overset{O}{\overset{\|}{C}}-CH_2-CH_2-\overset{O}{\overset{\|}{C}}-O-CH_3$$

IUPAC: Dimethyl butanedioate
Common: Dimethyl succinate

Example 13.5 Determining IUPAC and Common Names for Esters

Assign both IUPAC and common names to the following esters.

a. $CH_3-CH_2-\overset{O}{\overset{\|}{C}}-O-CH_2-CH_3$

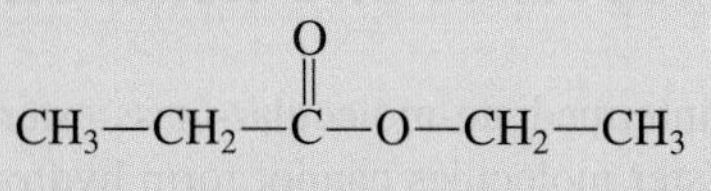

b. $CH_3-\underset{CH_3}{\underset{|}{CH}}-CH_2-\overset{O}{\overset{\|}{C}}-O-CH_3$

c. $C_6H_5-\overset{O}{\overset{\|}{C}}-O-CH_2-CH_2-CH_3$ (benzene ring attached to the carbonyl carbon)

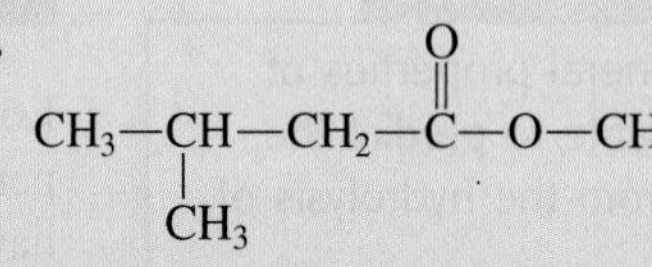

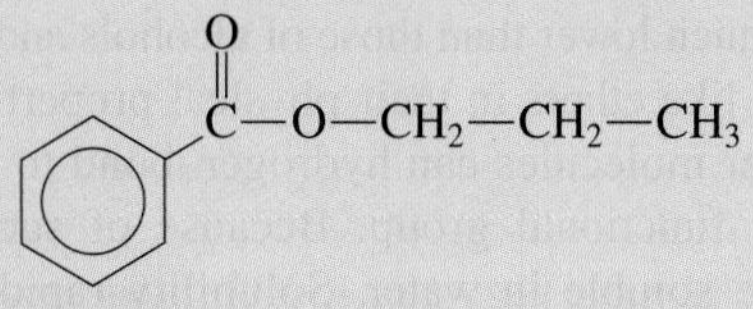

Solution

a. The name *ethyl* characterizes the alcohol part of the molecule. The name of the acid is propanoic acid (IUPAC) or propionic acid (common). Deleting the *-ic acid* ending and adding *-ate* give the name *ethyl propanoate* (IUPAC) or *ethyl propionate* (common).

b. The name of the alcohol part of the molecule is methyl (from methanol or methyl alcohol). The name of the five-carbon acid is 3-methylbutanoic acid or 3-methylbutyric acid. Hence the ester name is *methyl 3-methylbutanoate* (IUPAC) or *methyl 3-methylbutyrate* (common).

c. The name *propyl* characterizes the alcohol part of the molecule. The acid part of the molecule is derived from benzoic acid (both IUPAC and common name). Hence the ester name in both systems is *propyl benzoate.*

Practice Problems and Questions

13.46 Determine whether each of the following structures represents an ester.

a. $CH_3-CH_2-CH_2-\overset{O}{\overset{\|}{C}}-O-CH_3$

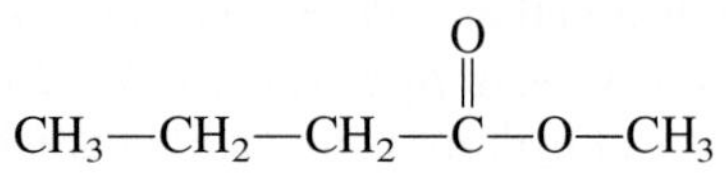

b. $CH_3-O-\overset{O}{\overset{\|}{C}}-CH_3$

c. $CH_3-O-CH_2-\overset{O}{\overset{\|}{C}}-CH_3$

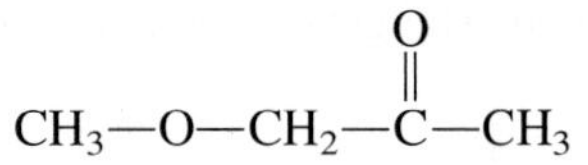

d. $CH_3-\overset{O}{\overset{\|}{C}}-O-CH_2-CH_3$

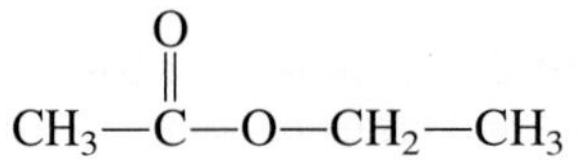

13.47 For each of the following esters, draw the structural formula of the "parent" acid and the "parent" alcohol.

a. $CH_3-CH_2-\overset{O}{\overset{\|}{C}}-O-CH_2-CH_3$

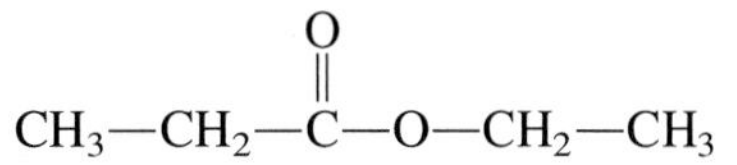

b. $CH_3-CH_2-CH_2-\overset{O}{\overset{\|}{C}}-O-CH_3$

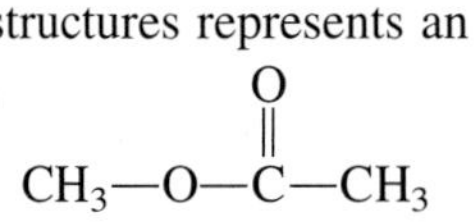

c. $CH_3-\overset{O}{\overset{\|}{C}}-O-C_6H_5$ (benzene ring attached to the oxygen)

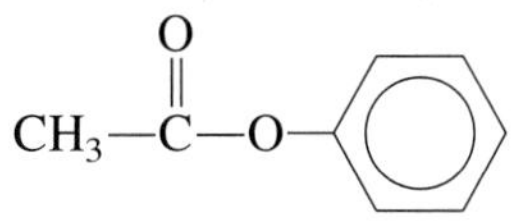

d. $C_6H_5-\overset{O}{\overset{\|}{C}}-O-CH_3$ (benzene ring attached to the carbonyl carbon)

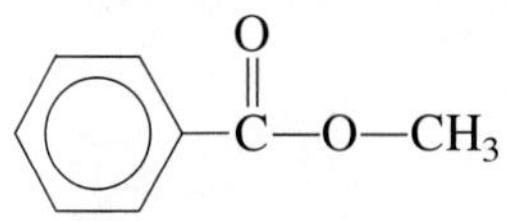

13.48 Assign an IUPAC name to each of the following esters.

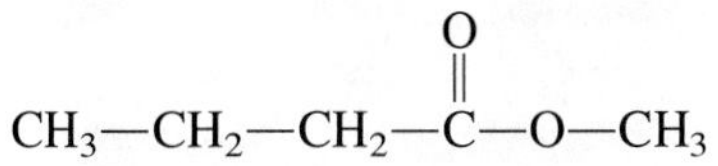

a. $CH_3-CH_2-\overset{O}{\overset{\|}{C}}-O-CH_3$

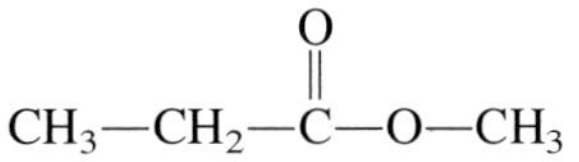

b. $H-\overset{O}{\overset{\|}{C}}-O-CH_3$

c. $CH_3-\overset{O}{\overset{\|}{C}}-O-CH_3$

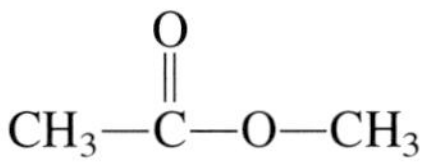

d. $CH_3-CH_2-CH_2-O-\overset{O}{\overset{\|}{C}}-CH_3$

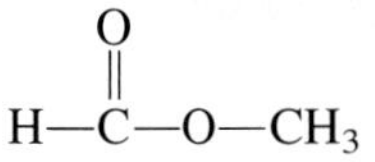

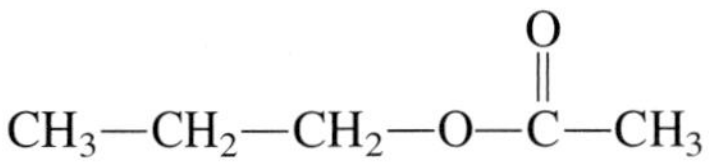

13.49 Assign a common name to each of the esters in Problem 13.48.

13.50 Draw the structural formula of each of the following esters.
a. Methyl formate b. Propyl acetate
c. Methyl methanoate d. Propyl ethanoate

13.51 Without actually drawing the structure, determine how many carbon atoms are in the "parent" alcohol and the "parent" acid for each of the following esters.
a. Octyl decanoate b. Isopropyl acetate
c. Butyl ethanoate d. Phenyl benzoate

Learning Focus

List the general properties of esters; be able to predict the products from the hydrolysis of an ester.

13.9 Properties and Reactions of Esters

Low- and intermediate-molecular-mass esters are usually colorless liquids at room temperature. Ester molecules cannot form hydrogen bonds to each other because they do not have a hydrogen atom bonded to an oxygen atom. Consequently, the boiling points of esters are much lower than those of alcohols and acids of comparable molecular mass. Esters are more like ethers in their physical properties.

Water molecules can hydrogen-bond to esters through the oxygen atoms present in the ester functional group. Because of such hydrogen bonding, low-molecular-mass esters are soluble in water. Solubility rapidly decreases with increasing carbon chain length; borderline solubility situations are reached when three to five carbon atoms are in a chain.

Most esters have pleasant odors. Esters are largely responsible for the flavor and fragrance of fruits and flowers (see Figure 13.10). Generally, a natural flavor or odor is caused by a mixture of esters, with one particular compound being dominant. The synthetic production of these "dominant" compounds is the basis for the flavoring agents used in ice cream, gelatins, soft drinks, and so on. Table 13.1 gives the structures of selected esters used as flavoring agents. What is surprising about the structures in Table 13.1 is how closely some of them resemble each other. For example, the apple and pineapple flavoring agents differ by one carbon atom (methyl versus ethyl); a five-carbon chain versus an eight-carbon chain makes the difference between banana and orange flavor.

Ester Hydrolysis

The most important reaction of esters, from a biochemical viewpoint, is *hydrolysis*. A **hydrolysis reaction** *is the reaction of a compound with water, in which the compound splits into two or more fragments as the elements of water (—H and —OH) are added.* The products of hydrolysis for an ester are the carboxylic acid and alcohol from which the ester was formed. The general equation for ester hydrolysis is

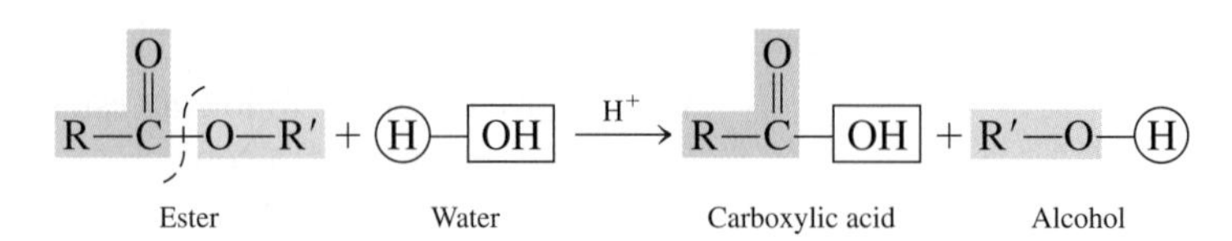

The specific equation for the hydrolysis of methyl acetate is

$$CH_3—\overset{O}{\overset{\|}{C}}—O—CH_3 + H—OH \xrightarrow{H^+} CH_3—\overset{O}{\overset{\|}{C}}—OH + CH_3—O—H$$

Methyl acetate Water Acetic acid Methyl alcohol

Ester hydrolysis requires the presence of a strong-acid catalyst or enzymes.

Ester hydrolysis is the reverse of esterification (Section 13.7), the formation of an ester from a carboxylic acid and an alcohol.

Figure 13.10 Esters are responsible for the odors and tastes of many fruits.

Table 13.1
Selected Esters That Are Used as Flavoring Agents

IUPAC name	Structural formula	Characteristic flavor and odor
isobutyl methanoate	$H-C(=O)-O-CH_2-CH(CH_3)-CH_3$	raspberry
propyl ethanoate	$CH_3-C(=O)-O-(CH_2)_2-CH_3$	pear
pentyl ethanoate	$CH_3-C(=O)-O-(CH_2)_4-CH_3$	banana
octyl ethanoate	$CH_3-C(=O)-O-(CH_2)_7-CH_3$	orange
pentyl propanoate	$CH_3-CH_2-C(=O)-O-(CH_2)_4-CH_3$	apricot
methyl butanoate	$CH_3-(CH_2)_2-C(=O)-O-CH_3$	apple
ethyl butanoate	$CH_3-(CH_2)_2-C(=O)-O-CH_2-CH_3$	pineapple

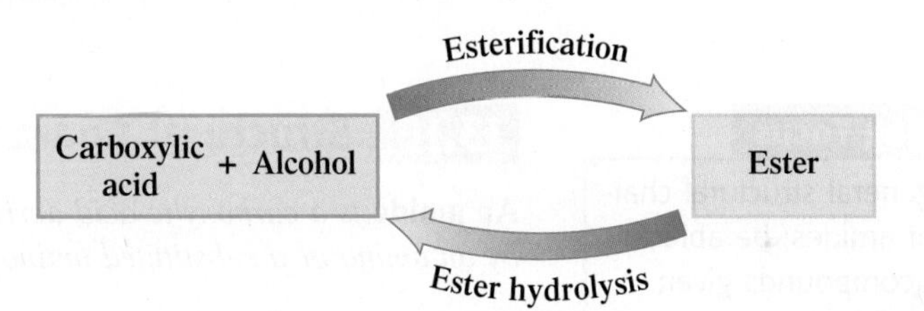

Aspirin, one of the most widely and frequently used medications of all time, is an ester whose mode of action involves ester hydrolysis. It is made from acetic acid (which functions as the carboxylic acid in esterification) and salicylic acid (a hydroxyacid that functions as the alcohol).

$$\underset{\text{Salicylic acid}}{C_6H_4(COOH)(O-H)} + \underset{\text{Acetic acid}}{HO-C(=O)-CH_3} \xrightarrow[\text{Heat}]{H^+} \underset{\text{Aspirin}}{C_6H_4(COOH)(O-C(=O)-CH_3)} + H_2O$$

After ingestion, aspirin undergoes hydrolysis to regenerate salicylic acid and acetic acid. Salicylic acid is the active ingredient of aspirin—the substance that decreases pain, lowers body temperature, and reduces inflammation.

$$\underset{\text{Aspirin}}{C_6H_4(COOH)(O-C(=O)-CH_3)} + H_2O \xrightarrow{H^+} \underset{\text{Salicylic acid}}{C_6H_4(COOH)(OH)} + \underset{\text{Acetic acid}}{CH_3-C(=O)-OH}$$

The reason why the ester of salicylic acid, rather than salicylic acid itself, is used as the medication is that the ester is more palatable; salicylic acid causes mouth and throat irritation.

Practice Problems and Questions

13.52 Ester boiling points are lower than those of corresponding alcohols and carboxylic acids. Explain why.

13.53 With the help of Table 13.1, determine which flavor you would detect if you smelled or tasted each of the following esters.
a. Octyl ethanoate b. Methyl butanoate
c. Propyl ethanoate d. Ethyl butanoate

13.54 Give the name of the carboxylic acid produced when each of the following esters undergoes hydrolysis.
a. Methyl propanoate b. Ethyl acetate
c. Methyl formate d. Propyl propanoate

13.55 Give the name of the alcohol produced when each of the following esters undergoes hydrolysis.
a. Ethyl butanoate b. Ethyl acetate
c. Methyl methanoate d. Isopropyl ethanoate

13.56 Write the structural formulas for the reaction products when each of the following esters undergoes hydrolysis.

a. $CH_3—CH_2—\overset{\overset{\Large O}{||}}{C}—O—CH_2—CH_3$

b. $CH_3—\overset{\overset{\Large O}{||}}{C}—O—CH_2—CH_3$

c.

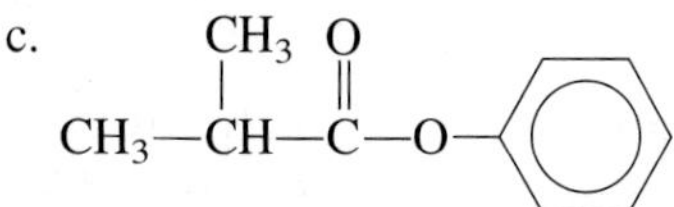

d.

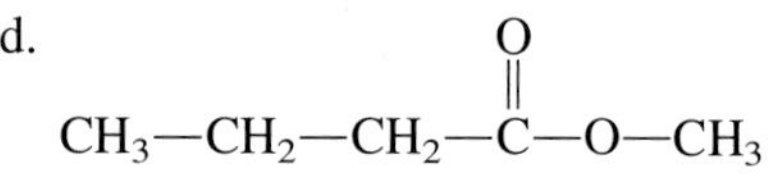

13.57 What alcohol and carboxylic acid are used to produce aspirin?

13.58 What is the "active ingredient" in aspirin and how is it generated?

Learning Focus

Know the general structural characteristics of amides; be able to name such compounds given their structural formulas, or vice versa.

13.10 Structural Characteristics and Naming of Amides

An **amide** *is a carboxylic acid derivative in which the carboxyl —OH group is replaced by an amino or a substituted amino group.* The amide functional group is thus

$$—\overset{\overset{\Large O}{||}}{C}—NH_2 \quad \text{or} \quad —\overset{\overset{\Large O}{||}}{C}—NH—R \quad \text{or} \quad —\overset{\overset{\Large O}{||}}{C}—\underset{\underset{\Large R'}{|}}{N}—R$$

depending on the degree of substitution on the nitrogen atom. Amides, like amines, can be classified as primary, secondary, or tertiary, depending on how many carbon atoms are attached to the nitrogen atom.

$$R—\overset{\overset{\Large O}{||}}{C}—NH_2 \qquad R—\overset{\overset{\Large O}{||}}{C}—NH—R' \qquad R—\overset{\overset{\Large O}{||}}{C}—\underset{\underset{\Large R''}{|}}{N}—R'$$

Primary amide Secondary amide Tertiary amide

Note that the difference between a primary amide and a secondary amide is "H" versus "R" and that the difference between a secondary amide and a tertiary amide is again "H" versus "R." We have encountered this relationship numerous times in our study of hydrocarbon derivatives in this chapter and the previous one. The accompanying Chemistry at a Glance summarizes all of the "H" versus "R" relationships that we have encountered in our study of hydrocarbon derivatives.

The simplest amide has a hydrogen atom attached to an unsubstituted amide functional group.

$$H—\overset{\overset{\Large O}{||}}{C}—NH_2$$

Summary of Structural Relationships for Hydrocarbon Derivatives: "H" versus "R"

Alcohol / Ether	Aldehyde / Ketone	Carboxylic acid / Ester
R—O—H	R—C(=O)—H	R—C(=O)—O—H
R—O—R	R—C(=O)—R	R—C(=O)—O—R

Hemiacetal / Hemiketal	Acetal / Ketal
R—C(OH)(OR)—H	R—C(OR)(OR)—H
R—C(OH)(OR)—R	R—C(OR)(OR)—R

Primary amine / Secondary amine	Secondary amine / Tertiary amine	Primary amide / Secondary amide	Secondary amide / Tertiary amide
R—N(H)—H	R—N(R)—H	R—C(=O)—N(H)—H	R—C(=O)—N(R)—H
R—N(H)—R	R—N(R)—R	R—C(=O)—N(H)—R	R—C(=O)—N(R)—R

Next in complexity are amides in which a methyl group is present. There are two of them, one with the methyl group attached to the carbon atom and the other with the methyl group attached to the nitrogen atom.

$$CH_3—\overset{\overset{\displaystyle O}{\|}}{C}—NH_2 \quad \text{and} \quad H—\overset{\overset{\displaystyle O}{\|}}{C}—NH—CH_3$$

The first of these structures is a 1° amide, and the second structure is that of a 2° amide. The structure of the simplest aromatic amide involves a benzene ring to which an unsubstituted amide functional group is attached.

$$C_6H_5—\overset{\overset{\displaystyle O}{\|}}{C}—NH_2$$

Naming of Amides

The "base" for an amide name is the name of the parent acid from which the amide can be considered to be derived (either the common or the IUPAC name).

Primary amides, amides with unsubstituted —NH_2 groups, are named by replacing the *-oic acid* (IUPAC name) or *-ic acid* (common name) of the corresponding carboxylic acid name with *-amide*. Selected primary amide IUPAC names (with the common name in parentheses) are

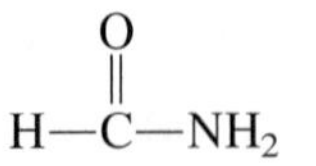

$$H-\overset{O}{\overset{\|}{C}}-NH_2 \qquad CH_3-\overset{O}{\overset{\|}{C}}-NH_2 \qquad CH_3-CH_2-\overset{O}{\overset{\|}{C}}-NH_2$$

Methanamide (formamide) — Ethanamide (acetamide) — Propanamide (propionamide)

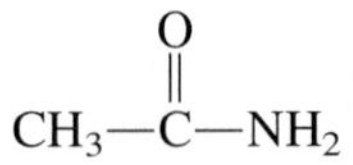

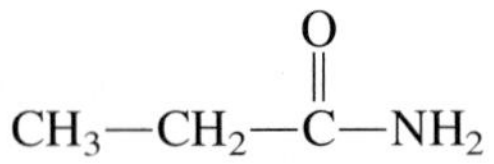

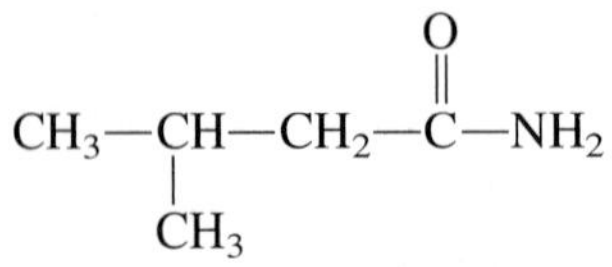

$$CH_3-\underset{CH_3}{\underset{|}{CH}}-CH_2-\overset{O}{\overset{\|}{C}}-NH_2 \qquad CH_3-\overset{CH_3}{\overset{|}{\underset{Cl}{\underset{|}{C}}}}-\overset{O}{\overset{\|}{C}}-NH_2$$

3-Methylbutanamide (3-methylbutyramide) — 2-Chloro-2-methylpropanamide (2-cholro-2-methylpropionamide)

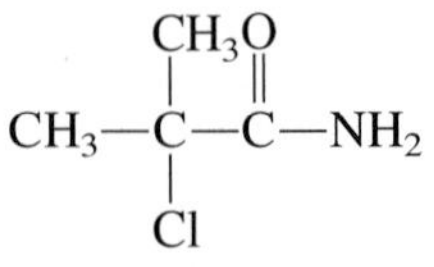

Secondary and tertiary amides, amides with substituted amino groups, have names in which the prefix *N*- is used for each substituent on the nitrogen atom, a practice we previously encountered with amine nomenclature (Section 12.7).

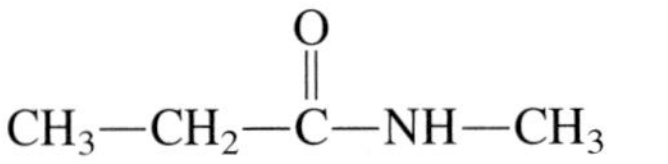

$$CH_3-CH_2-\overset{O}{\overset{\|}{C}}-NH-CH_3 \qquad CH_3-\overset{O}{\overset{\|}{C}}-\overset{CH_3}{\overset{|}{N}}-CH_3$$

N-Methylpropanamide (*N*-methylpropionamide) — *N,N*-Dimethylethanamide (*N,N*-dimethylacetamide)

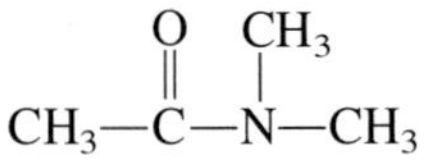

The simplest aromatic amide, a benzene ring bearing an amide functional group, is called *benzamide*. Other simple aromatic amines are named as derivatives of benzamide.

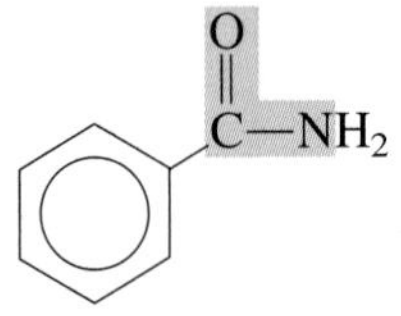

$$C_6H_5-\overset{O}{\overset{\|}{C}}-NH_2 \qquad (2-CH_3)C_6H_4-\overset{O}{\overset{\|}{C}}-NH_2 \qquad C_6H_5-\overset{O}{\overset{\|}{C}}-NH-CH_3$$

Benzamide 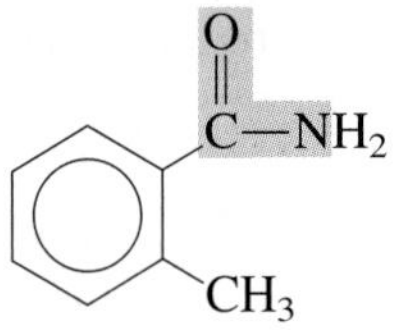 — 2-Methylbenzamide 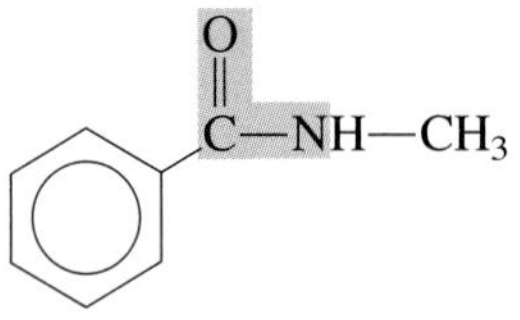 — *N*-Methylbenzamide

▶ The contrast between IUPAC names and common names for unbranched unsubstituted amides is as follows:

IUPAC (one word)
alkanamide
ethanamide

Common (one word)
(prefix)amide*
acetamide

*The common-name prefixes are related to natural sources for the acids.

Example 13.6 Determining IUPAC and Common Names for Amides

Assign both common and IUPAC names to each of the following amides.

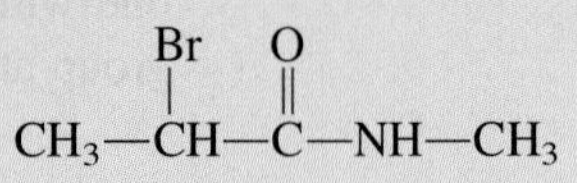

a. $CH_3-CH_2-CH_2-\overset{O}{\overset{\|}{C}}-NH_2$

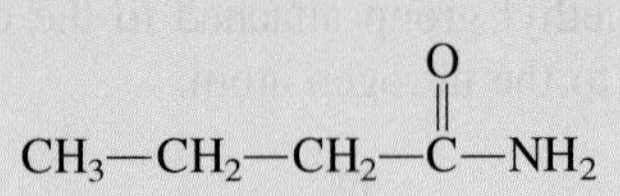

b. $CH_3-\overset{Br}{\overset{|}{CH}}-\overset{O}{\overset{\|}{C}}-NH-CH_3$

c. $C_6H_5-\overset{O}{\overset{\|}{C}}-N(C_6H_5)-C_6H_5$

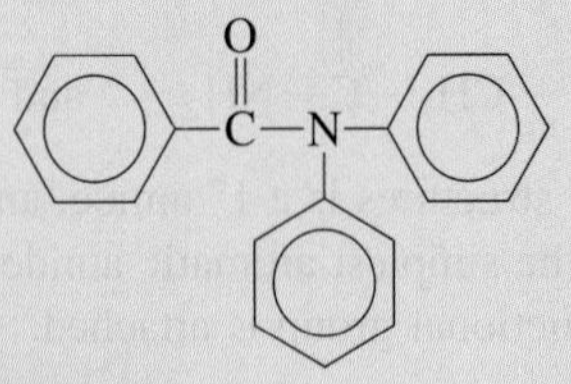

Solution

a. The parent acid for this amide is butyric acid (common) or butanoic acid (IUPAC). The common name for this amide is *butyramide,* and the IUPAC name is *butanamide*.

b. The common and IUPAC names of the acid are very similar; they are propionic acid and propanoic acid, respectively. The common name is *2-bromo-N-methylpropionamide,* and the IUPAC name is *2-bromo-N-methylpropanamide*. The prefix *N*- must be used with the methyl group to indicate that it is attached to the nitrogen atom.

c. In both the common and IUPAC systems of nomenclature, the name of the parent acid is the same: benzoic acid. The name of the amide is *N,N-diphenylbenzamide.*

The simplest naturally occurring amide is urea, a water-soluble white solid produced in the human body from carbon dioxide and ammonia through a complex series of metabolic reactions.

$$CO_2 + 2NH_3 \longrightarrow (H_2N)_2CO + H_2O$$

Urea is a one-carbon diamide. Its molecular structure is

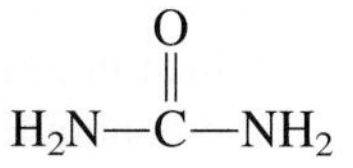

$$H_2N-\overset{\overset{\large O}{\|}}{C}-NH_2$$

Urea

Urea formation is the human body's primary method for eliminating "waste" nitrogen. The kidneys remove urea from the blood and provide for its excretion in urine. With malfunctioning kidneys, urea concentrations in the body can build to toxic levels—a condition called *uremia*.

A number of synthetic amides exhibit physiological activity and are used as drugs in the human body. Foremost among them, in terms of use, is acetaminophen, which in 1992 replaced aspirin as the top-selling over-the-counter pain reliever. Acetaminophen is a derivative of acetamide.

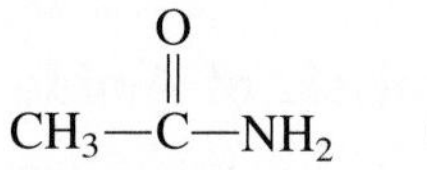

$$CH_3-\overset{\overset{\large O}{\|}}{C}-NH_2 \qquad CH_3-\overset{\overset{\large O}{\|}}{C}-NH-C_6H_4-OH$$

Acetamide 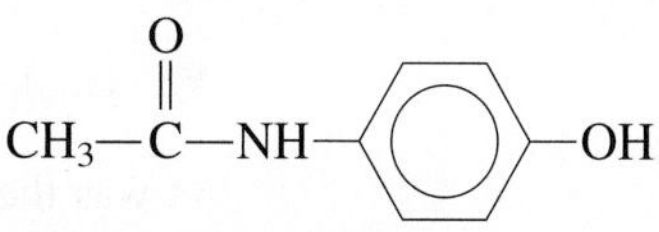 Acetaminophen

The common name *acetaminophen* has built into it reference to acetamide and to the phenol molecule that has been attached to the nitrogen atom. Acetaminophen is the active ingredient in all Tylenol products as well as in Datril, Tempra, and Anacin-3.

Practice Problems and Questions

13.59 Determine whether each of the following structures represents an amide.

a. 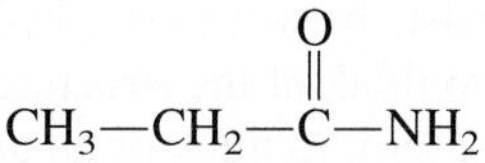 $CH_3-CH_2-\overset{\overset{\large O}{\|}}{C}-NH_2$

b. $C_6H_5-\overset{\overset{\large O}{\|}}{C}-\underset{\underset{\large CH_3}{|}}{N}-CH_2-CH_3$

c. $CH_3-\overset{\overset{\large O}{\|}}{C}-CH_2-CH_2-NH_2$

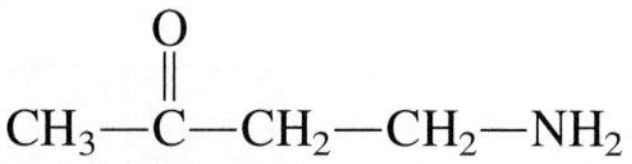

d. 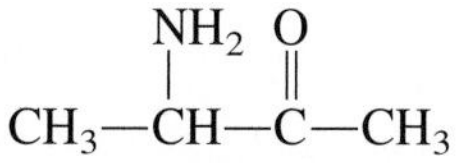 $CH_3-\overset{\overset{\large NH_2}{|}}{CH}-\overset{\overset{\large O}{\|}}{C}-CH_3$

13.60 Assign an IUPAC name to each of the following amides.

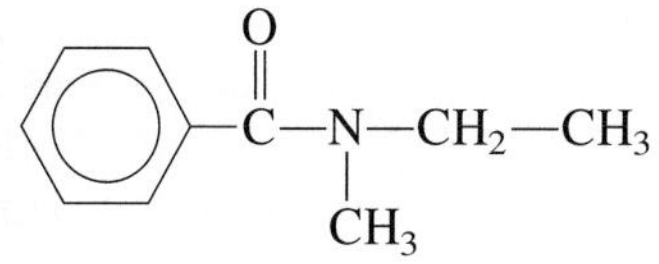

a. 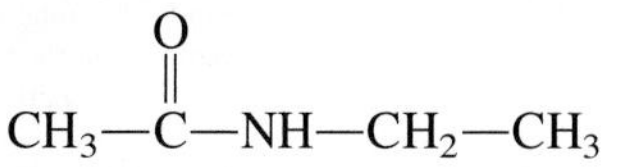 $CH_3-\overset{\overset{\large O}{\|}}{C}-NH-CH_2-CH_3$

b. $CH_3-CH_2-\overset{\overset{\large O}{\|}}{C}-\overset{\overset{\large CH_3}{|}}{N}-CH_3$

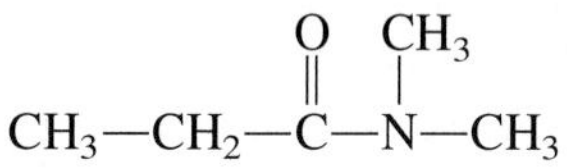

c. $H_2N-\overset{\overset{\large O}{\|}}{C}-CH_2-CH_2-CH_3$

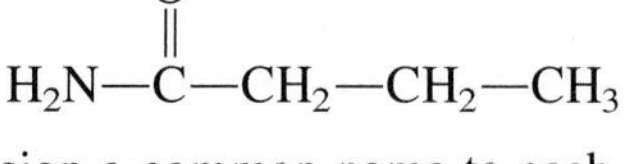

d. 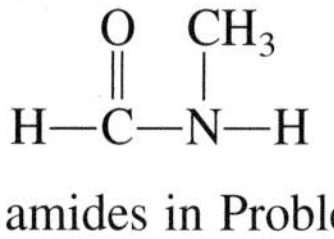 $H-\overset{\overset{\large O}{\|}}{C}-\overset{\overset{\large CH_3}{|}}{N}-H$

13.61 Assign a common name to each of the amides in Problem 13.60.

13.62 Classify each of the amides in Problem 13.60 as a primary, a secondary, or a tertiary amide.

13.63 Classify each of the amides in Problem 13.60 as an unsubstituted, a monosubstituted, or a disubstituted amide.

13.64 Draw the structural formula of each of the following amides.

a. *N,N*-Dimethylacetamide b. 2-Methylbutyramide

c. 3,*N*-Dimethylbutanamide d. *N*-Phenylbenzamide

Learning Focus

List the general properties of amides; be able to predict the products of the hydrolysis of an amide.

13.11 Properties and Reactions of Amides

Amides do not exhibit basic properties in solution as amines do (Section 12.8). Although the nitrogen atom present in amides has a nonbonding pair of electrons, as in amines, these electrons are not available for bonding to a H^+ ion. The oxygen atom of the carbonyl group pulls electron density from the carbonyl carbon (an electronegativity effect), which in turn pulls electron density from the nitrogen atom. The net effect of this is that the nonbinding pair of electrons on the N atom does not exhibit basic behavior.

$$-\overset{\overset{\delta-}{O}}{\overset{\|}{C}}-\overset{\delta+}{\ddot{N}}-$$

With the exception of formamide, which is a liquid, all unsubstituted amides are solids at room temperature. Numerous intermolecular hydrogen-bonding possibilities exist, between amide H atoms and carbonyl O atoms, in unsubstituted amides.

Amides of low molecular mass, up to five or six carbon atoms, are soluble in water. Again, numerous hydrogen-bonding possibilities exist between water and the amide.

Hydrolysis of Amides

As was the case with esters (Section 13.9), the most important reaction of amides is hydrolysis. In amide hydrolysis, the bond between the carbonyl carbon atom and the nitrogen is broken, and free acid and free amine (or ammonia) are produced. Amide hydrolysis is catalyzed by acids, bases, or certain enzymes; sustained heating is also often required.

The type of amide (1°, 2°, or 3°) that is hydrolyzed determines the specific hydrolysis products.

$$\text{Primary amide} \xrightarrow{\text{Hydrolysis}} \text{carboxylic acid} + \text{ammonia}$$

$$\text{Secondary amide} \xrightarrow{\text{Hydrolysis}} \text{carboxylic acid} + \text{primary amine}$$

$$\text{Tertiary amide} \xrightarrow{\text{Hydrolysis}} \text{carboxylic acid} + \text{secondary amine}$$

A parallel exists between ester hydrolysis (Section 13.9) and amide hydrolysis. Just as it was useful to think of the structure of an ester in terms of an "acid part" and an "alcohol part," it is useful to think of an amide in terms of an "acid part" and an "amine (or ammonia) part."

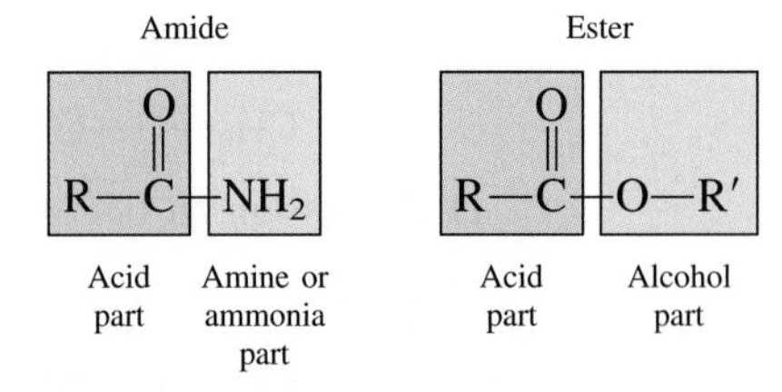

In this context, it is easy to identify the "parent" acid and the "parent" amine (or ammonia) produced from amide hydrolysis; to generate the parent molecules, just add an —OH group to the acid part of the amide and an H atom to the amine part.

The general structural equation for the hydrolysis of a primary amide is

$$\underset{\text{Amide}}{R-\overset{\overset{O}{\|}}{C}-NH_2} + \underset{\text{Water}}{H-OH} \xrightarrow{\text{Heat}} \underset{\text{Carboxylic acid}}{R-\overset{\overset{O}{\|}}{C}-OH} + \underset{\text{Ammonia}}{NH_2-H}$$

The specific equation for the hydrolysis of methyl acetamide (a 2° amide) is

$$\underset{N\text{-Methylacetamide}}{CH_3-\overset{\overset{O}{\|}}{C}-NH-CH_3} + H_2O \xrightarrow{\text{Heat}} \underset{\text{Acetic acid}}{CH_3-\overset{\overset{O}{\|}}{C}-OH} + \underset{\text{Methylamine}}{CH_3-\overset{\overset{H}{|}}{N}H}$$

An amide has a "parent" carboxylic acid and "parent" amine (or ammonia) from which we can consider it to be made. However, the amide *cannot* be made by *directly* reacting the "parent" acid with the "parent" amine (or ammonia). When mixed together, these two types of reactants undergo an acid–base reaction to form an amine salt (Section 12.8); the amine (or ammonia) is a base, and bases react with acids in an acid–base neutralization reaction (Section 9.6). A standard procedure for preparing amides, which avoids the "neutralization problem," is to convert the carboxylic acid into an acid chloride and then react the acid chloride with the amine (or ammonia).

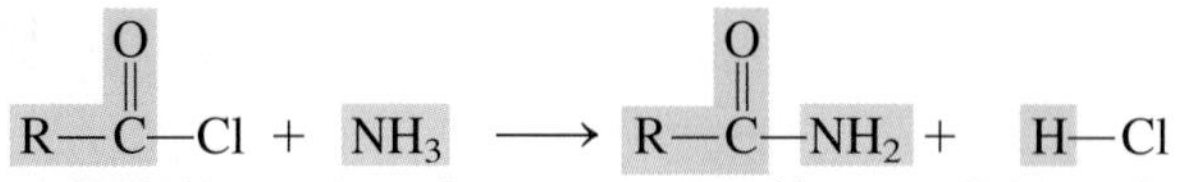

$$\underset{\text{Acid chloride}}{R{-}\overset{\overset{\displaystyle O}{\|}}{C}{-}Cl} + \underset{\text{Ammonia}}{NH_3} \longrightarrow \underset{\text{Amide}}{R{-}\overset{\overset{\displaystyle O}{\|}}{C}{-}NH_2} + \underset{\text{Hydrogen chloride}}{H{-}Cl}$$

Practice Problems and Questions

13.65 Although amides contain a nitrogen atom, they are not bases as amines are. Explain why.

13.66 What are the two "parent" molecules for an amide?

13.67 An amide cannot be made by the direct reaction of its two "parent" molecules. Explain why.

13.68 Give the name of the carboxylic acid produced when each of the following amides undergoes hydrolysis.
a. *N,N*-Dimethylacetamide b. 2-Methylbutyramide
c. Methanamide d. *N*-Methylbenzamide

13.69 Give the name of the nitrogen-containing compound produced when each of the following amides undergoes hydrolysis.
a. Propanamide b. *N*-Methylacetamide
c. *N,N*-Dimethylmethanamide d. 3,3,*N*-Trimethylbutyramide

13.70 Write the structural formulas of the reaction products when each of the following amides undergoes hydrolysis.

a. 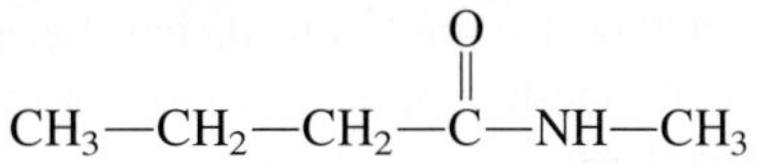 $CH_3{-}CH_2{-}CH_2{-}\overset{\overset{\displaystyle O}{\|}}{C}{-}NH{-}CH_3$

b. $CH_3{-}CH_2{-}CH_2{-}\overset{\overset{\displaystyle O}{\|}}{C}{-}NH_2$

c. $CH_3{-}CH_2{-}CH_2{-}\overset{\overset{\displaystyle O}{\|}}{C}{-}\overset{\overset{\displaystyle CH_3}{|}}{N}{-}CH_3$

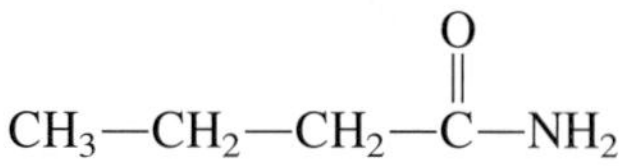

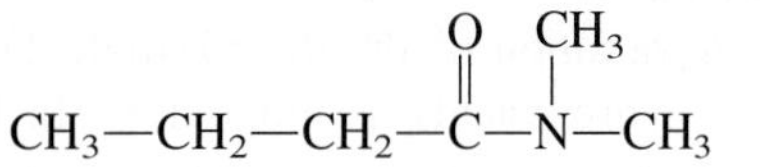

d. 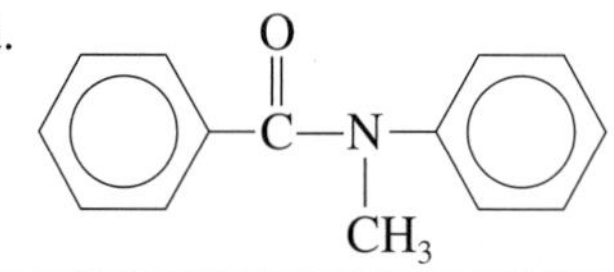 $C_6H_5{-}\overset{\overset{\displaystyle O}{\|}}{C}{-}\underset{\underset{\displaystyle CH_3}{|}}{N}{-}C_6H_5$

Learning Focus

Give a general description for a condensation polymer; discuss the formation and uses of polyesters and polyamides.

13.12 Condensation Polymers: Polyesters and Polyamides

A **condensation polymer** *is a polymer formed by reacting difunctional monomers to give a polymer and some small molecule.* Condensation polymers differ from addition polymers such as polyethylene (Section 11.6), where the polymer is the *only* product. Two of the most common types of condensation polymers are *polyesters* and *polyamides.*

A **polyester** *is a condensation polymer in which the monomers are joined through ester linkages.* Dicarboxylic acids and dialcohols are the monomers used to form polyesters.

The best known of the many polyesters now marketed is *poly(ethylene terephthalate),* which is also known by the acronym *PET.* The monomers used to produce PET are terephthalic acid (a diacid) and ethylene glycol (a dialcohol).

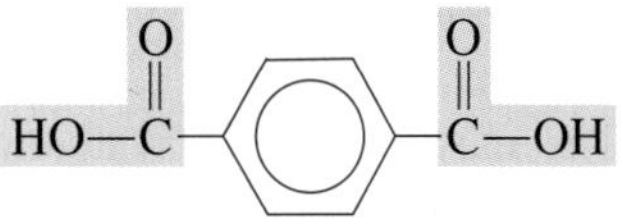

$$\underset{\text{Terephthalic acid}}{HO{-}\overset{\overset{\displaystyle O}{\|}}{C}{-}C_6H_4{-}\overset{\overset{\displaystyle O}{\|}}{C}{-}OH} \qquad \underset{\text{Ethylene glycol}}{HO{-}CH_2{-}CH_2{-}OH}$$

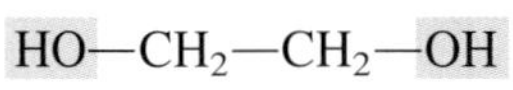

The reaction of one acid group of the diacid with one alcohol group of the dialcohol initially produces an ester molecule, with an acid group left over on one end and an alcohol group left over on the other end.

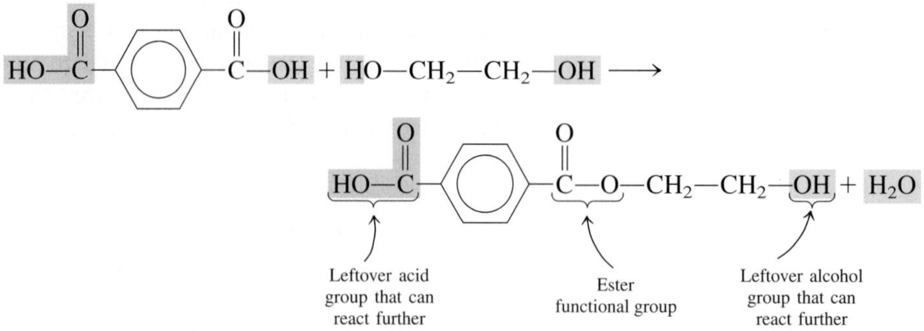

$$HO-\overset{O}{\overset{\|}{C}}-C_6H_4-\overset{O}{\overset{\|}{C}}-OH + HO-CH_2-CH_2-OH \longrightarrow HO-\overset{O}{\overset{\|}{C}}-C_6H_4-\overset{O}{\overset{\|}{C}}-O-CH_2-CH_2-OH + H_2O$$

Leftover acid group that can react further

Ester functional group

Leftover alcohol group that can react further

This species can react further. The remaining acid group can react with an alcohol group from another monomer, and the alcohol group can react with an acid group from another monomer. This process continues until an extremely long polymer molecule called a *polyester* is produced.

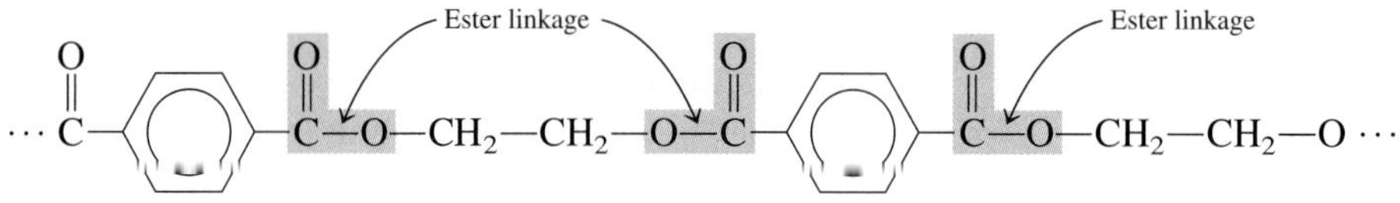

$$\cdots\overset{O}{\overset{\|}{C}}-C_6H_4-\overset{O}{\overset{\|}{C}}-O-CH_2-CH_2-O-\overset{O}{\overset{\|}{C}}-C_6H_4-\overset{O}{\overset{\|}{C}}-O-CH_2-CH_2-O\cdots$$

Ester linkage

Ester linkage

Poly(ethylene terephthalate), a polyester

Producing PET in the form of a filament generates the world's leading synthetic clothing fiber, *Dacron*. When PET is formed as a film, rather than a fiber, it is called *Mylar,* which is used as the plastic backing for audio and video tapes and computer diskettes. Its chemical name PET is applied when this polyester is used in clear, flexible soft-drink bottles and as the wrapping material for frozen foods and boil-in bags for foods.

PET is also used in medicine. Because it is physiologically inert, PET is used in the form of a mesh to replace diseased sections of arteries. It has also been used in synthetic heart valves.

A variation of the diacid–dialcohol monomer formulation for polyesters involves using hydroxyacids as monomers. In this situation, both of the functional groups required are present in the same molecule. A polyester in which lactic acid and glycolic acid (both hydroxyacids, Section 13.6) are monomers produces a biodegradable material (trade name *Lactomer*) that is used as surgical staples in several types of surgery. Traditional suture materials must be removed later on, after they have served their purpose. Lactomer staples start to dissolve (hydrolyze) after a period of several weeks. The hydrolysis products are the starting monomers, lactic acid and glycolic acid, both of which are normally present in the human body. By the time tissue has fully healed, the staples have fully degraded.

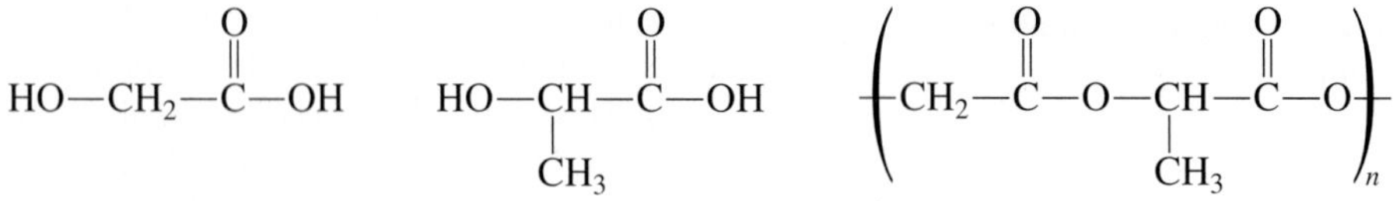

$$HO-CH_2-\overset{O}{\overset{\|}{C}}-OH \qquad HO-\underset{CH_3}{\underset{|}{CH}}-\overset{O}{\overset{\|}{C}}-OH \qquad \left(CH_2-\overset{O}{\overset{\|}{C}}-O-\underset{CH_3}{\underset{|}{CH}}-\overset{O}{\overset{\|}{C}}-O\right)_n$$

Glycolic acid monomer

Lactic acid monomer

Lactomer (repeating unit in polymer)

A **polyamide** *is a condensation polymer in which the monomers are joined through amide linkages.* Dicarboxylic acids and diamines are the monomers used to form polyamides.

The most important synthetic polyamide is *nylon*. Nylon is used in clothing and hosiery, as well as in carpets, tire cord, rope, and parachutes. It also has nonfiber uses;

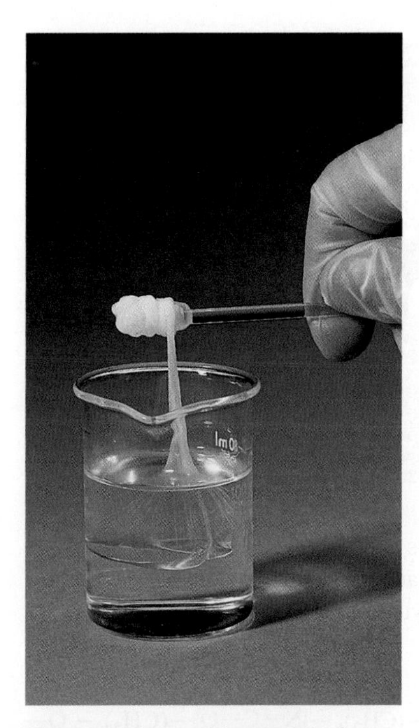

Figure 13.11 A white strand of a nylon polymer forms between two layers of a solution containing a diacid (bottom layer) and a diamine (top layer).

for example, it is used in paint brushes, electrical parts, valves, and fasteners. It is a tough, strong, nontoxic, nonflammable material that is resistant to chemicals. Surgical suture is made of nylon because it is such a strong fiber.

There are actually many different types of nylon, all of which are based on diamine and diacid monomers (Figure 13.11). The most important nylon is nylon 66, which is made by using 1,6-hexanediamine and hexanedioic acid as monomers.

$$\underset{\text{1,6-Hexanediamine}}{H-\overset{\overset{\displaystyle H}{|}}{N}-(CH_2)_6-\overset{\overset{\displaystyle H}{|}}{N}-H} \qquad \underset{\text{Hexanedioic acid}}{HO-\overset{\overset{\displaystyle O}{\|}}{C}-(CH_2)_4-\overset{\overset{\displaystyle O}{\|}}{C}-OH}$$

The reaction of one acid group of the diacid with one amine group of the diamine initially produces an amide molecule; an acid group is left over on one end, and an amine group is left over on the other end.

$$HO-\overset{\overset{\displaystyle O}{\|}}{C}-(CH_2)_4-\overset{\overset{\displaystyle O}{\|}}{C}-OH + H-\overset{\overset{\displaystyle H}{|}}{N}-(CH_2)_6-\overset{\overset{\displaystyle H}{|}}{N}-H \longrightarrow$$

$$HO-\overset{\overset{\displaystyle O}{\|}}{C}-(CH_2)_4-\overset{\overset{\displaystyle O}{\|}}{C}-\overset{\overset{\displaystyle H}{|}}{N}-(CH_2)_6-\overset{\overset{\displaystyle H}{|}}{N}-H + H_2O$$

Leftover acid group that can react further — Amide linkage — Leftover amine group that can react further

This species then reacts further, and the process continues until a long polymeric molecule, nylon, has been produced.

$$\cdots\overset{\overset{\displaystyle O}{\|}}{C}-(CH_2)_4-\overset{\overset{\displaystyle O}{\|}}{C}-\overset{\overset{\displaystyle H}{|}}{N}-(CH_2)_6-\overset{\overset{\displaystyle H}{|}}{N}-\overset{\overset{\displaystyle O}{\|}}{C}-(CH_2)_4-\overset{\overset{\displaystyle O}{\|}}{C}-\overset{\overset{\displaystyle H}{|}}{N}-(CH_2)_6-\overset{\overset{\displaystyle H}{|}}{N}-\overset{\overset{\displaystyle O}{\|}}{C}-(CH_2)_4-\overset{\overset{\displaystyle O}{\|}}{C}\cdots$$

Amide linkages — Amide linkages

A portion of the polyamide nylon 66

Additional stiffness and toughness are imparted to polyamides if aromatic rings are present in the polymer "backbone." The polyamide Kevlar is now used in place of steel in bullet-resistant vests. The polymeric repeating unit in Kevlar is

$$\left[\overset{\overset{\displaystyle H}{|}}{N}-C_6H_4-\overset{\overset{\displaystyle H}{|}}{N}-\overset{\overset{\displaystyle O}{\|}}{C}-C_6H_4-\overset{\overset{\displaystyle O}{\|}}{C}\right]_n$$

Kevlar

Figure 13.12 Firefighters with flame-resistant clothing containing Nomex.

Nomex is a polymer whose structure is a variation of that of Kevlar. With Nomex, the monomers are *meta* isomers rather than *para* isomers. Nomex is used in flame-resistant clothing for firefighters and race car drivers (see Figure 13.12).

Silk and wool are examples of *naturally occurring* polyamide polymers. Silk and wool are proteins, and protein are polyamide polymers. Because much of the human body is protein material, much of the human body is polyamide polymer material. The monomers for proteins are amino acids, difunctional molecules containing both amino and carboxyl groups. Representative structures for amino acids, of which there are many, include

$$H_2N-CH_2-COOH \qquad H_2N-\underset{\underset{\displaystyle CH_3}{|}}{CH}-COOH \qquad H_2N-\underset{\underset{\underset{\underset{\displaystyle CH_3}{|}}{CH-CH_3}}{|}}{CH}-COOH$$

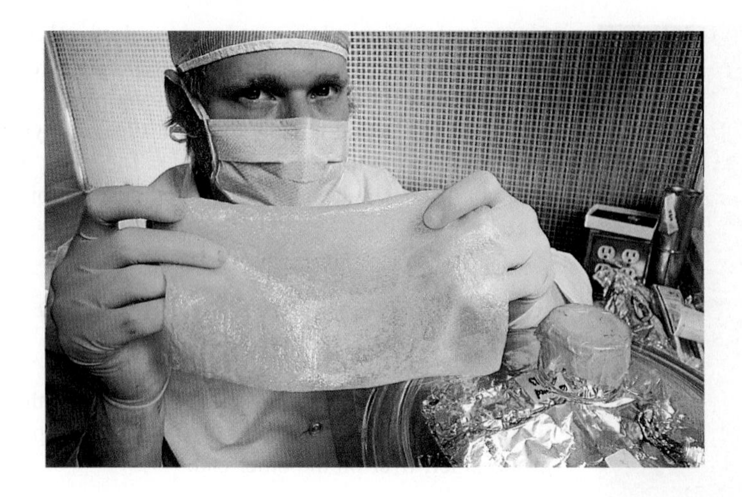

Figure 13.13 Polyurethanes have medical applications. For example, polyurethane membranes are used as skin substitutes for severe burn victims. Because they allow only oxygen and water to pass through, these membranes help patients recover more rapidly.

Polyurethanes are polymers related to polyesters and polyamides. The backbone of a polyurethane polymer contains aspects of both ester and amide functional groups. The following is a portion of the structure of a typical polyurethane polymer.

$$\text{---}\overset{O}{\overset{\|}{C}}-\overset{H}{\overset{|}{N}}-(CH_2)_6-\overset{H}{\overset{|}{N}}-\overset{O}{\overset{\|}{C}}-O-(CH_2)_4-O-\overset{O}{\overset{\|}{C}}-\overset{H}{\overset{|}{N}}-(CH_2)_6-\overset{H}{\overset{|}{N}}-\overset{O}{\overset{\|}{C}}-O-(CH_2)_4-O\text{---}$$

Foam rubber in furniture upholstery, packaging materials, life preservers, elastic fibers, and many other products contain polyurethane polymers (Figure 13.13).

Practice Problems and Questions

13.71 Contrast a condensation polymer with an addition polymer in terms of number of products produced.

13.72 Indicate, in generalized terms, what types of monomers are used to produce
a. a polyester b. a polyamide

13.73 What do the letters in the acronym PET stand for?

13.74 Indicate whether each of the following polymers is a polyester or a polyamide.
a. Dacron b. PET c. Nylon d. Silk

13.75 Characterize, in terms of functional groups present, the monomers that are the basis for the following types of polymers.
a. Proteins b. PET c. Nylon d. Kevlar

13.76 What is the difference, in terms of linkages between monomers, between a polyamide and a polyurethane?

CONCEPTS TO REMEMBER

The carbonyl group. A carbonyl group consists of a carbon atom bonded to an oxygen atom through a double bond. Aldehydes and ketones are compounds that contain a carbonyl functional group. The carbonyl carbon in an aldehyde has at least one hydrogen attached to it, and the carbonyl carbon in a ketone has no hydrogens.

Nomenclature of aldehydes and ketones. The IUPAC names of aldehydes and ketones are based on the longest carbon chain that contains the carbonyl group. The chain numbering is done from the end that results in the lowest number for the carbonyl group. The names of aldehydes end in *-al,* those of ketones in *-one.*

Physical properties of aldehydes and ketones. The boiling points of aldehydes and ketones are intermediate between those of alcohols and alkanes. The polarity of the carbonyl groups enables aldehyde and ketone molecules to interact with each other through dipole–dipole interactions. They cannot, however, hydrogen-bond to each other. Lower-molecular-mass aldehydes and ketones are soluble in water.

Reactions of aldehydes and ketones. Oxidation of primary and secondary alcohols, using mild oxidizing agents, produces aldehydes and ketones, respectively. Aldehydes are easily oxidized to carboxylic acids; ketones do not readily undergo oxidation. Reduction of aldehydes and ketones produces primary and secondary alcohols, respectively.

Hemiacetals, hemiketals, acetals, and ketals. A characteristic reaction of aldehydes and ketones is the addition of an alcohol across the carbonyl double bond to produce hemiacetals and hemiketals.

The reaction of a second alcohol molecule with a hemiacetal or hemiketal produces an acetal or a ketal.

The carboxyl group. The functional group present in carboxylic acids is the carboxyl group. A carboxyl group is composed of a hydroxyl group bonded to a carbonyl carbon atom. It thus contains two oxygen atoms directly bonded to the same carbon atom.

Nomenclature of carboxylic acids. The IUPAC name for a monocarboxylic acid is formed by replacing the final *-e* of the hydrocarbon parent name with *-oic acid*. The longest carbon chain containing the functional group is identified, and it is numbered starting with the carboxyl carbon atom. Common-name usage is more prevalent for carboxylic acids than for any other type of organic compound.

Physical properties of carboxylic acids. Low-molecular-mass carboxylic acids are liquids at room temperature and have sharp or unpleasant odors. Long-chain acids are wax-like solids. The carboxyl group is polar and forms hydrogen bonds to other carboxyl groups or other molecules. Thus carboxylic acids have relatively high boiling points, and those with lower molecular masses are soluble in water.

Acidity of carboxylic acids. Soluble carboxylic acids behave as weak acids, donating protons to water molecules. The portion of the acid molecule left after proton loss is called a carboxylate ion.

Reactions of carboxylic acids. Carboxylic acids are neutralized by bases to produce carboxylic acid salts. Such salts are usually more soluble in water than are the acids themselves. Carboxylic acid salts are named by changing the *-ic* ending of the acid to *-ate*. Carboxylic acids react with alcohols to produce esters.

Esters. Esters are formed by the reaction of an acid with an alcohol. In such reactions, the —OR group from the alcohol replaces the —OH group in the carboxylic acid. Esters are polar compounds, but they cannot form hydrogen bonds to each other. Therefore, their boiling points are lower than those of alcohols and acids of similar molecular mass.

Nomenclature of esters. An ester is named as an alkyl (from the name of the alcohol reactant) carboxylate (from the name of the acid reactant).

Hydrolysis of esters. In ester hydrolysis, the bond between the carbonyl carbon atom and the oxygen is broken, and free acid and free alcohol are produced.

Classification of amides. An amide is derived from a carboxylic acid by replacing the hydroxyl group with an amino or a substituted amino group. Amides, like amines, can be classified as primary, secondary, or tertiary, depending on how many nonhydrogen atoms are attached to the nitrogen atom.

Amides. The nomenclature for amides is derived from that for carboxylic acids by changing the *-oic acid* ending to *-amide*. Groups attached to the nitrogen atom of the amide are included as prefixes, using a capital *N*- with each group to indicate location.

Properties of amides. Amides do not exhibit basic properties in solution. The electronegative oxygen atom in the carbonyl group draws electron density away from the nitrogen, leaving very little electron density on the nitrogen to bond to an incoming proton. Most unbranched amides are solids at room temperature and have correspondingly high boiling points because of strong hydrogen bonds between molecules.

Hydrolysis of amides. In amide hydrolysis, the bond between the carbonyl carbon atom and the nitrogen is broken, and free acid and free amine (or ammonia) are produced.

Condensation polymers. A condensation polymer is formed by the reaction of difunctional monomers to give a polymer and some small molecule. Polyesters, made using diacid and dialcohol monomers, contain ester linkages between monomers. Polyamides, made using diacid and diamine monomers, contain amide linkages between monomers.

KEY REACTIONS AND EQUATIONS

1. Reduction of an aldehyde to give a primary alcohol (Section 13.3)

$$\text{R—C(=O)—H} + H_2 \xrightarrow{\text{Catalyst}} \text{R—CH(OH)—H}$$

2. Oxidation of an aldehyde to give a carboxylic acid (Section 13.3)

$$\text{R—C(=O)—H} \xrightarrow{[O]} \text{R—C(=O)—OH}$$

3. Addition of an alcohol to an aldehyde to form a hemiacetal and then an acetal (Section 13.3)

$$\underset{\text{Aldehyde}}{R_1\text{—C(=O)—H}} + R_2\text{—O—H} \overset{H^+}{\rightleftharpoons} \underset{\text{Hemiacetal}}{R_1\text{—C(OH)(H)—}OR_2}$$

$$\underset{\text{Hemiacetal}}{R_1\text{—C(OH)(H)—}OR_2} + R_3\text{—OH} \overset{H^+}{\rightleftharpoons} \underset{\text{Acetal}}{R_1\text{—C(}OR_3\text{)(H)—}OR_2} + H_2O$$

4. Reduction of a ketone to give a secondary alcohol (Section 13.5)

$$\text{R—C(=O)—R}' + H_2 \xrightarrow{\text{Catalyst}} \text{R—C(OH)(H)—R}'$$

5. Attempted oxidation of a ketone (Section 13.5)

$$\text{R—C(=O)—R}' \xrightarrow{[O]} \text{no reaction}$$

6. Addition of an alcohol to a ketone to form a hemiketal and a ketal (Section 13.5)

$$\underset{\text{Ketone}}{R_1\text{—C(=O)—}R_2} + R_3\text{—O—H} \overset{H^+}{\rightleftharpoons} \underset{\text{Hemiketal}}{R_1\text{—C(OH)(}R_2\text{)—}OR_3}$$

$$\underset{\text{Hemiketal}}{R_1\text{—C(OH)(}R_2\text{)—}OR_3} + R_4\text{—OH} \overset{H^+}{\rightleftharpoons} \underset{\text{Ketal}}{R_1\text{—C(}OR_4\text{)(}R_2\text{)—}OR_3} + H_2O$$

7. Ionization of a carboxylic acid to give a carboxylate ion and a hydronium ion (Section 13.7)

$$R{-}\overset{O}{\overset{\|}{C}}{-}OH + H_2O \rightleftharpoons R{-}\overset{O}{\overset{\|}{C}}{-}O^- + H_3O^+$$

8. Reaction of a carboxylic acid with a base to produce a carboxylic acid salt plus water (Section 13.7)

$$R{-}\overset{O}{\overset{\|}{C}}{-}OH + NaOH \longrightarrow R{-}\overset{O}{\overset{\|}{C}}{-}O^-Na^+ + H_2O$$

9. Preparation of an ester from an acid and an alcohol (Section 13.7)

$$R{-}\overset{O}{\overset{\|}{C}}{-}OH + R'{-}OH \overset{H^+}{\rightleftharpoons} R{-}\overset{O}{\overset{\|}{C}}{-}O{-}R' + H_2O$$

10. Ester hydrolysis to produce a carboxylic acid and an alcohol (Section 13.9)

$$R{-}\overset{O}{\overset{\|}{C}}{-}O{-}R' + H{-}OH \overset{H^+}{\rightleftharpoons} R{-}\overset{O}{\overset{\|}{C}}{-}OH + R'{-}OH$$

11. Amide hydrolysis to produce a carboxylic acid and an amine (Section 13.11)

$$R{-}\overset{O}{\overset{\|}{C}}{-}NH{-}R' + H{-}OH \longrightarrow R{-}\overset{O}{\overset{\|}{C}}{-}OH + R'{-}NH_2$$

KEY TERMS

Acetal (13.3)
Aldehyde (13.2)
Amide (13.10)
Carbonyl group (13.1)
Carboxyl group (13.6)
Carboxylate ion (13.7)
Carboxylic acid (13.6)
Carboxylic acid salt (13.7)
Condensation polymer (13.12)
Ester (13.8)
Hemiacetal (13.3)
Hemiketal (13.5)
Hydrolysis reaction (13.9)
Ketal (13.5)
Ketone (13.4)
Polyamide (13.12)
Polyester (13.12)

ADDITIONAL PROBLEMS

13.77 What is the generalized formula for the functional group present in each of the following types of hydrocarbon derivatives?
a. Ester b. Amide c. Carboxylic acid d. Aldehyde

13.78 How many carbon–oxygen bonds (single or double) are present in each of the following types of hydrocarbon derivatives?
a. Ketone b. Aldehyde c. Amide d. Ester

13.79 What is the molecular formula for each of the following compounds?
a. Methanal b. Methanoic acid
c. Methyl methanoate d. Methanamide

13.80 Draw the structures of the following aromatic compounds.
a. Benzaldehyde b. Benzoic acid
c. Methyl benzoate d. Benzamide

13.81 Draw the structure of the simplest compounds (fewest carbon atoms) in each of the following types of hydrocarbon derivatives.
a. Aldehyde b. Ketone c. Ester d. Amide

13.82 What type of hydrocarbon derivative is produced from each of the following types of reactions?
a. Oxidation of an aldehyde
b. Reduction of a ketone
c. Reaction of a carboxylic acid and an alcohol
d. Reaction of an aldehyde and an alcohol

13.83 Name the two types of products produced in each of the following types of reactions.
a. Hydrolysis of a secondary amide
b. Hydrolysis of an ester
c. Reaction of an alcohol and a hemiacetal
d. Reaction of a carboxylic acid and an alcohol

13.84 How many different molecules exist that fit each of the following descriptions? Assume that no carbon–carbon multiple bonds are present.
a. Three-carbon carboxylic acid b. Three-carbon ester
c. Three-carbon amide d. Three-carbon ketone

13.85 How many different molecules exist that fit each of the following descriptions? Assume that no carbon–carbon multiple bonds are present.
a. Three-carbon diacid b. Three-carbon amino acid
c. Three-carbon ketoacid d. Three-carbon hydroxyester

13.86 Indicate whether the members of each of the following pairs of compounds have the same molecular formula.
a. Methanal and methanoic acid
b. Propanone and propanal
c. Butanoic acid and methyl propanoate
d. 2-Methylbutanamide and *N*-methylbutanamide

PRACTICE TEST ▸ True/False

13.87 In a ketone, both of the additional bonds that the carbonyl carbon atom forms must be to other carbon atoms.

13.88 Both esters and amides must contain the elements oxygen and nitrogen.

13.89 In naming aldehydes, the carbon chain is always numbered such that the carbonyl carbon atom is carbon 1.

13.90 Aldehydes are easily reduced to primary alcohols and easily oxidized to ketones.

13.91 The reactants needed to produce a hemiacetal are an aldehyde and an alcohol.

13.92 *Dimethyl ketone* and *acetone* are two names for the same compound.

13.93 The notation —COOH represents a carboxyl group.

13.94 When a monocarboxylic acid loses its acidic hydrogen atom, the resulting carboxylate ion carries a charge of $+1$.

13.95 Carboxylic acid salts contain carboxylate ions.

13.96 The reactants needed to produce an ester are a carboxylic acid and an alcohol.

13.97 Two oxygen atoms must be present in all unsubstituted esters.

13.98 Ethanoic acid and formic acid are two names for the same compound.

13.99 Esters are largely responsible for the odors and flavors of flowers and fruits.

13.100 Both ester hydrolysis and amide hydrolysis produce a carboxylic acid as one of the products.

13.101 Primary amides contain one nitrogen atom, and secondary amides contain two nitrogen atoms.

13.102 PET is a polyester, and nylon is a polyamide.

PRACTICE TEST ▸ Multiple Choice

13.103 Which of the following structural features is possessed by aldehydes but not by ketones?
a. At least one hydrogen atom is bonded to the carbonyl carbon atom.
b. At least one hydroxyl group is bonded to the carbonyl carbon atom.
c. The carbonyl carbon atom is bonded to two other carbon atoms.
d. The carbonyl carbon atom is part of a ring structure.

13.104 The simplest aldehyde and the simplest ketone contain, respectively, how many carbon atoms?
a. one and one b. one and three
c. two and two d. two and three

13.105 Which of the following statements about the oxidation of aldehydes and ketones is *correct*?
a. Aldehydes readily undergo oxidation, and ketones are resistant to oxidation.
b. Ketones readily undergo oxidation, and aldehydes are resistant to oxidation.
c. Both aldehydes and ketones readily undergo oxidation.
d. Both aldehydes and ketones are resistant to oxidation.

13.106 The structural difference between a hemiacetal and an acetal is the replacement of a
a. —OH group with an —OR group
b. —OR group with an —OH group
c. —H atom with an —OR group
d. —H atom with a —OH group

13.107 Which of the following statements concerning the acid strength of carboxylic acids is *correct*?
a. All are weak acids.
b. All are strong acids.
c. Nonaromatic acids are weak, and aromatic acids are strong.
d. Unbranched acids are weak, and branched acids are strong.

13.108 Carboxylic acid salts may be converted back into carboxylic acids by reacting them with
a. a strong base b. a strong acid
c. an alcohol d. an aldehyde

13.109 The ester obtained by reacting ethanol and butanoic acid is called
a. ethyl butanoate b. butyl ethanoate
c. ethyl esterate d. butyl esterate

13.110 The number of oxygen atoms present in an ester and in a monocarboxylic acid are, respectively,
a. one and one b. two and two
c. two and three d. three and two

13.111 A monosubstituted amide is also called a
a. primary amide b. secondary amide
c. tertiary amide d. amide salt

13.112 Which of the following sets of monomers would produce a polyamide?
a. Dicarboxylic acid and dialcohol
b. Dicarboxylic acid and diamine
c. Dicarboxylic acid and dialdehyde
d. Diamine and dialcohol

CHAPTER OUTLINE

CHAPTER FOURTEEN

Carbohydrates

Carbohydrates in the form of cotton and linen may be woven into clothing materials.

Beginning with this chapter, we will focus almost exclusively on biochemistry, the chemistry of living systems. Like organic chemistry, biochemistry is a vast subject, and we can discuss only a few of its facets. Our approach to biochemistry will be similar to our approach to organic chemistry. We will devote individual chapters to each of the major classes of biochemical compounds, which are carbohydrates, lipids, proteins, and nucleic acids. Then we will consider the topic of metabolism in living organisms. In this first "biochapter," carbohydrates are considered.

Learning Focus

Distinguish between bioorganic substances and bioinorganic substances; know the major classes of bioorganic substances.

14.1 Biochemistry—An Overview

Biochemistry *is the study of the chemical substances found in living systems and the chemical interactions of these substances with each other*. Biochemistry is a field in which new discoveries are made almost daily about how cells manufacture the molecules needed for life and how the chemical reactions by which life is maintained occur. The knowledge explosion that occurred in the field of biochemistry during the last decades of the twentieth century is truly phenomenal.

A **biochemical substance** *is a chemical substance found within a living organism*. Biochemical substances are divided into two groups: bioinorganic substances and bioorganic substances. *Bioinorganic substances* include water and inorganic salts. *Bioorganic substances* include carbohydrates, lipids, proteins, and nucleic acids. Figure 14.1 gives an approximate mass composition for the human body in terms of types of biochemical substances present.

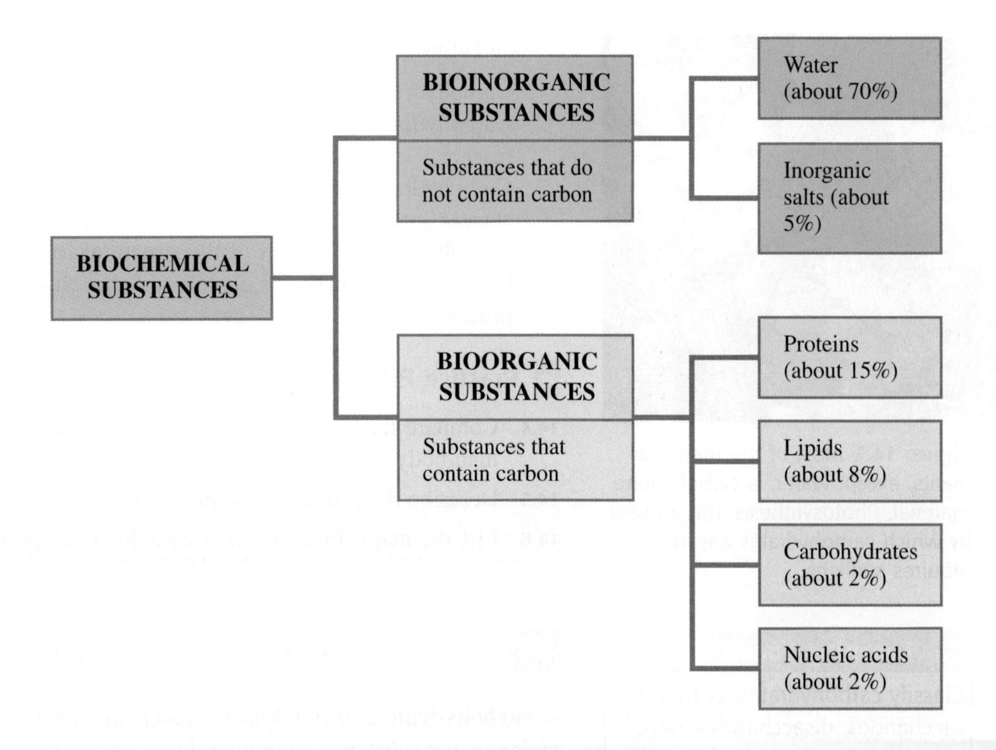

Figure 14.1 Mass composition data for the human body in terms of major types of biochemical substances.

▶ The functional groups present in bioorganic substances are the same ones found in the organic compounds studied in previous chapters. The major difference between simple organic and bioorganic compounds is structural complexity; bioorganic substances generally have several different functional groups present.

▶ As isolated compounds, bioinorganic and bioorganic substances have no life in and of themselves. Yet when these substances are gathered together in a cell, their chemical interactions are able to sustain life.

Although we tend to think of the human body as made up of organic substances, bioorganic molecules make up only about one-fourth of body mass. The bioinorganic substance water constitutes over two-thirds of the mass of the human body, and another 4%–5% of body mass comes from inorganic salts (Section 9.5).

Practice Problems and Questions

14.1 Contrast the relative amounts, by mass, of bioorganic and bioinorganic substances present in the human body.

14.2 What are the four major types of bioorganic substances present in the human body?

14.3 Indicate which member of each of the following pairs of bioorganic substances is the more abundant in the human body.

a. Proteins and nucleic acids b. Proteins and carbohydrates
c. Lipids and carbohydrates d. Lipids and nucleic acids

Learning Focus

Know the major source of and major functions of carbohydrates in the human body.

14.2 Occurrence and Functions of Carbohydrates

Carbohydrates are the most abundant class of bioorganic molecules on planet Earth. Although their abundance in the human body is relatively low (Section 14.1), carbohydrates constitute about 75% by mass of dry plant materials.

Green (chlorophyll-containing) plants produce carbohydrates via *photosynthesis*. In this process, carbon dioxide from the air and water from the soil are the reactants, and sunlight absorbed by chlorophyll is the energy source (see Figure 14.2).

$$CO_2 + H_2O + \text{solar energy} \xrightarrow[\text{Plant enzymes}]{\text{Chlorophyll}} \text{carbohydrates} + O_2$$

▶ It is estimated that more than half of all organic carbon atoms are found in the carbohydrate materials of plants.

Plants have two main uses for the carbohydrates they produce. In the form of *cellulose*, carbohydrates serve as structural elements, and in the form of *starch*, they provide energy reserves for the plants.

▶ Human uses for carbohydrates of the plant kingdom extend beyond food. Carbohydrates in the form of cotton and linen are used as clothing. Carbohydrates in the form of wood are used for shelter and heating and in making paper.

Dietary intake of plant materials is the major carbohydrate source for humans and animals. The average human diet should ideally be about two-thirds carbohydrate by mass.

Figure 14.2 Most of the matter in plants, except water, is carbohydrate material. Photosynthesis, the process by which carbohydrates are made, requires sunlight.

Carbohydrates have the following functions in humans:

1. Carbohydrate oxidation provides energy.
2. Carbohydrate storage, in the form of glycogen, provides a short-term energy reserve.
3. Carbohydrates supply carbon atoms for the synthesis of other biochemical substances (proteins, lipids, and nucleic acids).
4. Carbohydrates form part of the structural framework of DNA and RNA molecules.
5. Carbohydrate "markers" on cell surfaces play key roles in cell–cell recognition processes.

Practice Problems and Questions

14.4 Compare the amount of carbohydrates present in plant materials to that in the human body.

14.5 Describe the process by which carbohydrates are produced by plants.

14.6 List the major functions that carbohydrates perform in the human body.

Learning Focus

Classify carbohydrates as monosaccharides, disaccharides, oligosaccharides, or polysaccharides.

14.3 General Types of Carbohydrates

A **carbohydrate** *is a polyhydroxy aldehyde, a polyhydroxy ketone, or a compound that yields such a substance upon hydrolysis.* The carbohydrate glucose is a polyhydroxy aldehyde, and the carbohydrate fructose is a polyhydroxy ketone.

► In the previous chapter, —COOH was used as a linear designation for the acid functional group. In a like manner, the linear designation for an aldehyde functional group is —CHO. Note that the designation is —CHO rather than —COH, which is the designation for an alcohol.

```
Aldehyde ──→ CHO                    CH2OH
group         |                      |
           H—C—OH                    C=O  ←── Ketone
              |                      |        group
          HO—C—H                 HO—C—H
              |                      |
           H—C—OH                 H—C—OH
              |                      |
           H—C—OH                 H—C—OH
              |                      |
             CH2OH                  CH2OH
           Glucose                 Fructose
  (a polyhydroxy aldehyde)   (a polyhydroxy ketone)
```

A striking structural feature of carbohydrates is the large number of functional groups present. In glucose and fructose there is a functional group attached to each carbon atom.

Carbohydrates are classified on the basis of molecular size as monosaccharides, oligosaccharides, and polysaccharides.

► The term *monosaccharide* is pronounced "mon-oh-SACK-uh-ride."

A **monosaccharide** *is a carbohydrate that contains a single polyhydroxy aldehyde or polyhydroxy ketone unit.* Monosaccharides cannot be broken down into simpler units by hydrolysis reactions. Both glucose and fructose are monosaccharides. Naturally occurring monosaccharides have from three to seven carbon atoms; five- and six-carbon species are especially common. Pure monosaccharides are water-soluble, white, crystalline solids.

► The *oligo* in the term *oligosaccharides* comes from the Greek *oligos,* which means "small" or "few." The term *oligosaccharide* is pronounced "OL-ee-go-SACK-uh-ride."

An **oligosaccharide** *is a carbohydrate that contains from two to ten monosaccharide units covalently bonded to each other.* Disaccharides are the most common type of oligosaccharide. A **disaccharide** *is a carbohydrate that contains two monosaccharide units covalently bonded to each other.* Like monosaccharides, disaccharides are crystalline, water-soluble substances. Sucrose (table sugar) and lactose (milk sugar) are disaccharides.

Within the human body, oligosaccharides are often found associated with proteins and lipids in complexes that have both structural and regulatory functions. Free oligosaccharides, other than disaccharides, are seldom encountered in biological systems.

Complete hydrolysis of an oligosaccharide produces monosaccharides. Upon hydrolysis, a disaccharide produces two monosaccharides, a trisaccharide three monosaccharides, a hexasaccharide six monosaccharides, and so on.

▶ Types of carbohydrates are related to each other through hydrolysis.

Polysaccharides
↓ Hydrolysis
Oligosaccharides
↓ Hydrolysis
Monosaccharides

A **polysaccharide** *is a carbohydrate that contains many monosaccharide units covalently bonded to each other.* Polysaccharides, which are polymers, often consist of tens of thousands of monosaccharide units. Both cellulose and starch are polysaccharides. We encounter these two substances everywhere. The paper on which this book is printed is mainly cellulose, as are the cotton in our clothes and the wood in our houses. Starch is a component of many types of foods, including bread, pasta, potatoes, rice, corn, beans, and peas.

Practice Problems and Questions

14.7 What functional group is present in all carbohydrates?

14.8 What two types of carbonyl-containing functional groups are found in carbohydrates?

14.9 Indicate how many monosaccharide units are present in each of the following?
a. Disaccharide b. Tetrasaccharide
c. Oligosaccharide d. Polysaccharide

14.10 Identify, in general terms, the product produced from the complete hydrolysis of each of the following types of carbohydrates.
a. Disaccharide b. Tetrasaccharide
c. Oligosaccharide d. Polysaccharide

Learning Focus

Classify a monosaccharide by the number of carbons present and by the type of carbonyl group present.

14.4 Classification of Monosaccharides

Monosaccharides are classified by (1) the number of carbon atoms present and (2) by the type of carbonyl-containing functional group present.

Although there is no limit to the number of carbon atoms that can be present in a monosaccharide, only monosaccharides with three to seven carbon atoms are commonly found in nature. A three-carbon monosaccharide is called a *triose,* and those that contain four, five, and six carbon atoms are called *tetroses, pentoses,* and *hexoses,* respectively.

Monosaccharides are classified as *aldoses* or *ketoses* on the basis of type of carbonyl group (Section 13.1) present. An **aldose** *is a monosaccharide that contains an aldehyde functional group.* A **ketose** *is a monosaccharide that contains a ketone functional group.*

Monosaccharides are often classified by both their number of carbon atoms and their functional group. A six-carbon monosaccharide with an aldehyde functional group is an *aldohexose;* a five-carbon monosaccharide with a ketone functional group is a *ketopentose.*

▶ The Latin word for "sugar" is *saccharum,* from whence comes the term *saccharide.*

Monosaccharides are also often called sugars. Hexoses are six-carbon sugars, pentoses five-carbon sugars, and so on. The word *sugar* is associated with "sweetness," and most (but not all) monosaccharides have a sweet taste. The designation *sugar* is also applied to disaccharides, many of which also have a sweet taste. Thus **sugar** *is a general designation for either a monosaccharide or a disaccharide.*

Example 14.1 Classifying Monosaccharides on the Basis of Their Structural Characteristics

Classify each of the following monosaccharides according to both the number of carbon atoms and the type of carbonyl group present.

a.
CHO
HO—H
H—OH
H—OH
CH_2OH

b.
CH_2OH
C=O
H—OH
HO—H
HO—H
CH_2OH

c.
CHO
HO—H
H—OH
HO—H
HO—H
CH_2OH

d.
CH_2OH
C=O
H—OH
H—OH
CH_2OH

Solution

a. An aldehyde functional group is present as well as five carbon atoms. This monosaccharide is thus an *aldopentose*.
b. This monosaccharide contains a ketone group and six carbon atoms, so it is a *ketohexose*.
c. Six carbon atoms and an aldehyde group in a monosaccharide are characteristic of an *aldohexose*.
d. This monosaccharide is a *ketopentose*.

Practice Problems and Questions

14.11 What is the difference between a pentose and a hexose?

14.12 What is the difference between an aldose and a ketose?

14.13 What are the functional groups present and the number of carbons in a ketohexose?

14.14 Classify the following monosaccharides on the basis of (1) the number of carbon atoms present, (2) the type of carbonyl-containing functional group present, and (3) both the number of carbon atoms and the type of carbonyl-containing functional group.

a.
CH_2OH
$C{=}O$
HO—|—H
H—|—OH
H—|—OH
CH_2OH

b.
CHO
H—|—OH
HO—|—H
H—|—OH
H—|—OH
CH_2OH

Learning Focus

Understand the concept of handedness as it applies to monosaccharides; be able to recognize the D form and the L form of a monosaccharide given a structural representation that shows chiral centers.

14.5 Handedness in Monosaccharides

Most monosaccharides exist in two forms—a "left-handed" form and a "right-handed" form. These two forms are related to each other in the same way your left hand and right hand are related to each other: as *mirror images*. Figure 14.3 shows this mirror-image relationship of human hands.

The concept of *mirror images* is the key to understanding molecular handedness. All objects, including all molecules, have mirror images. A **mirror image** *of an object is the object's reflection in a mirror.* Objects can be divided into two classes on the basis of their mirror images: (1) object with *superimposable* mirror images, and (2) objects with *nonsuperimposable* mirror images. Superimposable mirror images have parts that coincide exactly at all points when the objects are laid upon each other. A dinner plate with

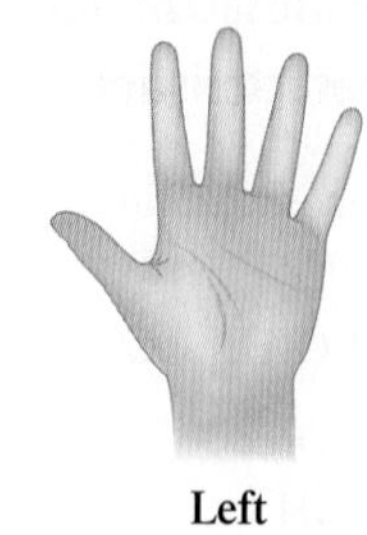

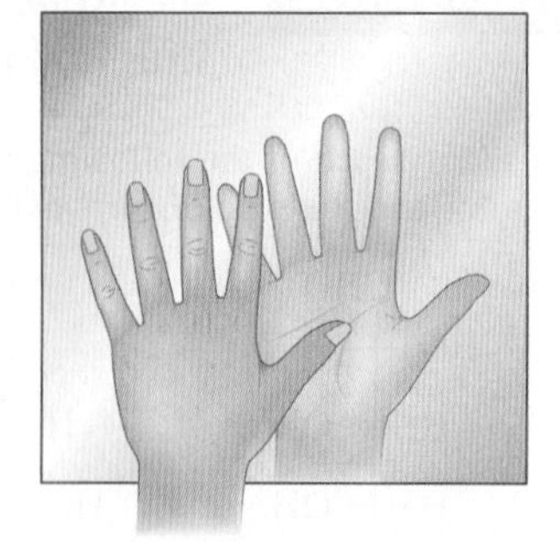

Mirror image of left hand
is in the back of the mirror

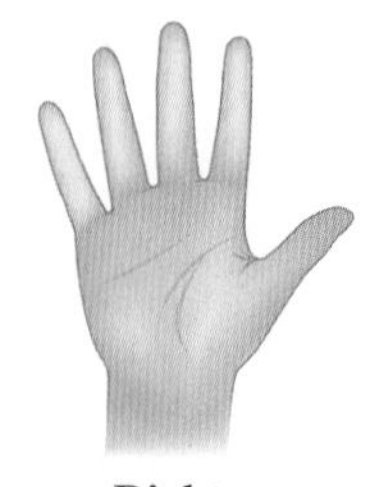

Figure 14.3 The mirror image of the right hand is the left hand. Conversely, the mirror image of the left hand is the right hand.

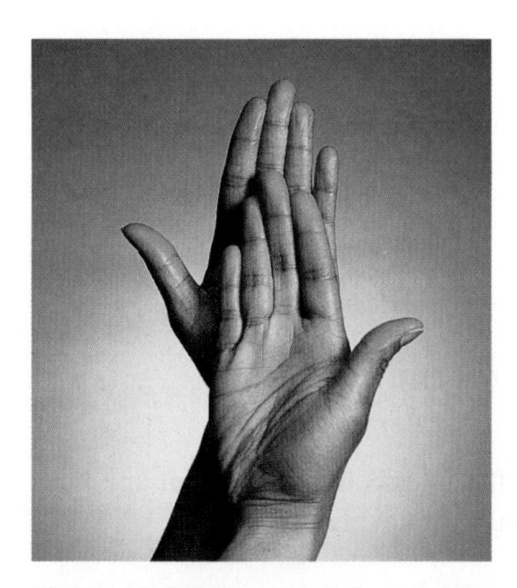

Figure 14.4 A person's left and right hands are not superimposable upon each other.

no design features is an example of an object that has superimposable mirror images. For nonsuperimposable mirror images, not all points coincide when the objects are laid upon each other. Human hands are nonsuperimposable mirror images, as is shown in Figure 14.4; note that the two thumbs point in opposite directions and that the fingers do not align correctly. All objects with nonsuperimposable mirror images exist in left- and right-handed forms.

One of the simplest examples of a molecule that possesses handedness—that is, a molecule that has nonsuperimposable mirror images, is the trisubstituted methane molecule bromochloroiodomethane.

$$\begin{array}{c} \text{H} \\ | \\ \text{Br—C—Cl} \\ | \\ \text{I} \end{array}$$

Figure 14.5a shows the nonsuperimposability of the mirror-image forms of this molecule. The simplest example of a monosaccharide with a nonsuperimposable mirror image is the three-carbon molecule glyceraldehyde.

$$\begin{array}{c} \text{CHO} \\ | \\ \text{H—C—OH} \\ | \\ \text{CH}_2\text{OH} \end{array}$$

The nonsuperimposability of the mirror-image forms of glyceraldehyde is shown in Figure 14.5b.

Not all molecules possess handedness; that is, not all molecules have left- and right-handed forms. What is the prerequisite for handedness? Any organic molecule that contains a carbon atom with four *different* groups attached to it possesses handedness. Such a carbon atom is called a *chiral center*. A **chiral center** *is an atom in a molecule that has four different groups tetrahedrally bonded to it.* Note that both of the molecules in Figure 14.5 has a chiral center.

A molecule that contains a chiral center is said to be *chiral*. A **chiral molecule** *is a molecule whose mirror image is not superimposable upon it.* Chiral molecules have handedness. An **achiral molecule** *is a molecule whose mirror image is superimposable upon it.* Achiral molecules do not possess handedness.

Figure 14.5 Examples of simple molecules that have nonsuperimposable mirror images. (a) The molecule bromochloroiodomethane. (b) The molecule glyceraldehyde.

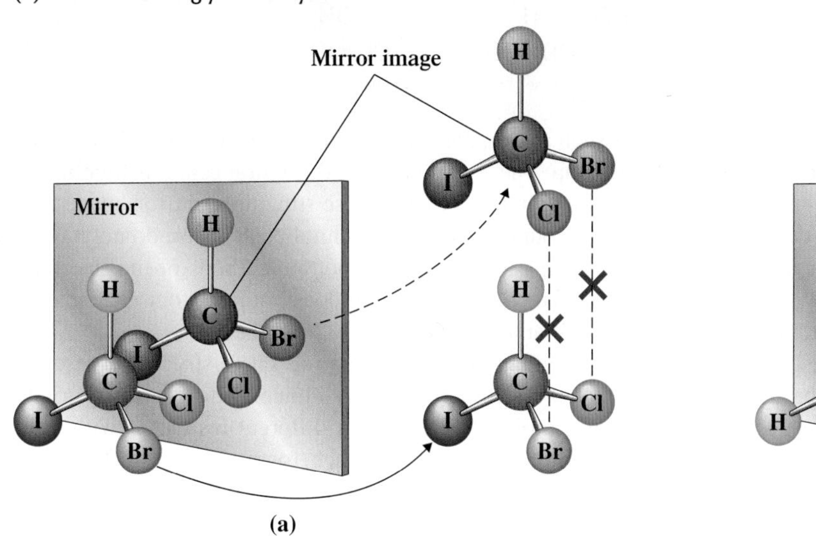

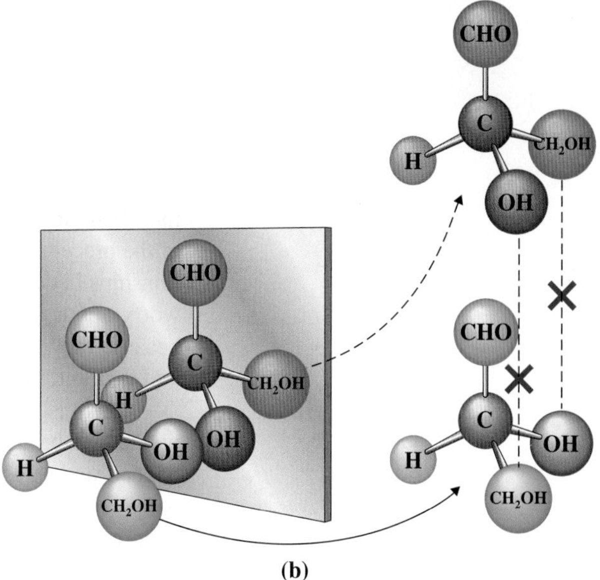

Designating Molecular Handedness

The aldotriose glyceraldehyde is the simplest monosaccharide that possesses a chiral center. Notation for its two nonsuperimposable mirror-image forms is

$$\begin{array}{ccc} & CHO & \\ HO & - & H \\ & CH_2OH & \end{array} \quad \vdots \quad \begin{array}{ccc} & CHO & \\ H & - & OH \\ & CH_2OH & \end{array}$$

Mirror

In this notation, the chiral center is represented as the intersection of vertical and horizontal lines. The atom at the center, which is carbon, is not explicitly shown.

Chiral center

By convention, when the carbon chain is drawn vertically with the carbonyl group at the top, the right-handed form of the molecule has the chiral-center —OH group on the right, and the left-handed form has the chiral-center —OH group on the left. The designation D- (from the Latin *dextro* for "right") is used for the right-handed molecule, and the designation L- (from the Latin *levo* for "left") is used for the left-handed molecule.

$$\begin{array}{ccc} & CHO & \\ HO & - & H \\ & CH_2OH & \end{array} \quad \vdots \quad \begin{array}{ccc} & CHO & \\ H & - & OH \\ & CH_2OH & \end{array}$$

L-Glyceraldehyde D-Glyceraldehyde

As the number of carbon atoms in a monosaccharide increases, the number of chiral centers present in the molecule increases. For example, the aldohexose glucose has four chiral centers. Structural notation for the D and L forms of this monosaccharide are

$$\begin{array}{ccc} & CHO & \\ HO & - & H \\ H & - & OH \\ HO & - & H \\ HO & - & H \\ & CH_2OH & \end{array} \quad \vdots \quad \begin{array}{ccc} & CHO & \\ H & - & OH \\ HO & - & H \\ H & - & OH \\ H & - & OH \\ & CH_2OH & \end{array}$$

L-Glucose D-Glucose

In multi-chiral-center situations such as this, handedness is determined by the chiral center most distant from the carbonyl group. The —OH group at this chiral center is highlighted in the preceding structures. Also note, in these two glucose structures, mirror-image relationships between —H and —OH groups at *each* of the chiral centers. We do not have to worry about mirror-image relationships for the top and bottom carbons of the structures because they are not chiral centers. The top carbon atom does not have four attachments, and the bottom carbon atom has two attachments that are the same.

Obviously, handedness is important in monosaccharide chemistry; otherwise we would not be discussing it. What is its importance? In human body chemistry, right-handed and left-handed forms of a molecule often elicit different responses within the body. Sometimes both forms are biologically active, each form giving a different response; sometimes both give the same response, but one form's response is many times greater than that of the other; and sometimes only one of the two forms is biologically active. For example, studies show that the body's response to the D form of the hormone epinephrine (Section 12.7) is 20 times greater than its response to the L form.

Naturally occurring carbohydrate molecules are always made up of right-handed (D-form) monosaccharides. Plants produce only right-handed monosaccharide molecules. Interestingly, when we consider protein chemistry (Chapter 16), we will find that amino acids, the building blocks for proteins, are always left-handed molecules.

Practice Problems and Questions

14.15 What is the difference between a chiral object and an achiral object?

14.16 Indicate whether each of the following objects is chiral or achiral. Assume each object to be without markings or lettering.
a. Nail b. Hammer c. Baseball cap d. Baseball glove

14.17 Indicate whether each of the following words is chiral or achiral.
a. MOM b. DAD c. TOT d. TOOT

14.18 Indicate whether the circled carbon atom in each of the following molecules is a chiral center.

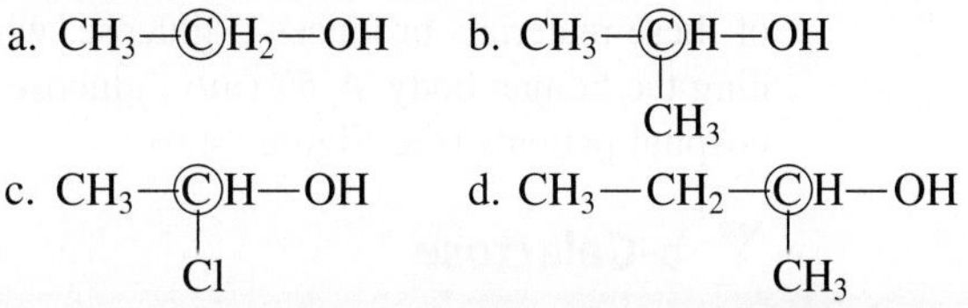

a. $CH_3-\text{Ⓒ}H_2-OH$

b. $CH_3-\underset{\displaystyle CH_3}{\underset{|}{\text{Ⓒ}}}H-OH$

c. $CH_3-\underset{\displaystyle Cl}{\underset{|}{\text{Ⓒ}}}H-OH$

d. $CH_3-CH_2-\underset{\displaystyle CH_3}{\underset{|}{\text{Ⓒ}}}H-OH$

14.19 Classify each of the following monosaccharide molecules as a right-handed molecule or as a left-handed molecule.

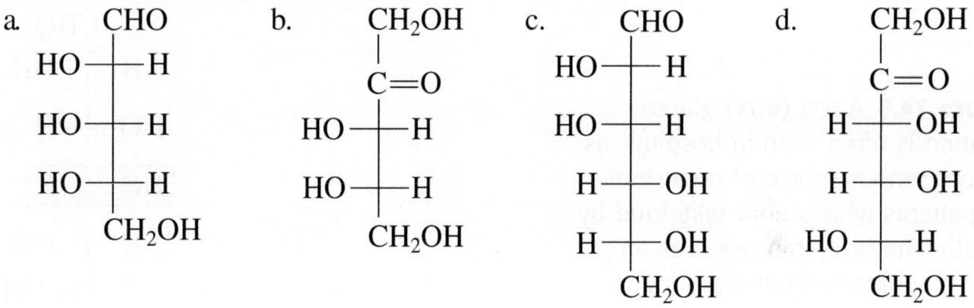

a.
$$\begin{array}{c} CHO \\ HO-\!\!\!+\!\!\!-H \\ HO-\!\!\!+\!\!\!-H \\ HO-\!\!\!+\!\!\!-H \\ CH_2OH \end{array}$$

b.
$$\begin{array}{c} CH_2OH \\ | \\ C{=}O \\ HO-\!\!\!+\!\!\!-H \\ HO-\!\!\!+\!\!\!-H \\ CH_2OH \end{array}$$

c.
$$\begin{array}{c} CHO \\ HO-\!\!\!+\!\!\!-H \\ HO-\!\!\!+\!\!\!-H \\ H-\!\!\!+\!\!\!-OH \\ H-\!\!\!+\!\!\!-OH \\ CH_2OH \end{array}$$

d.
$$\begin{array}{c} CH_2OH \\ | \\ C{=}O \\ H-\!\!\!+\!\!\!-OH \\ H-\!\!\!+\!\!\!-OH \\ HO-\!\!\!+\!\!\!-H \\ CH_2OH \end{array}$$

14.20 Classify each of the monosaccharide molecules in Problem 14.19 as the D form of the molecule or as the L form of the molecule.

14.21 How many chiral centers are present in each of the monosaccharide molecules in Problem 14.19?

Learning Focus

Be able to draw the structures of the D forms of the monosaccharides glyceraldehyde, dihydroxyacetone, glucose, galactose, fructose, and ribose.

14.6 Names and Structures of Biochemically Important Monosaccharides

All monosaccharides go by common names. The two simplest monosaccharides, both trioses, have the names glyceraldehyde (an aldotriose) and dihydroxyacetone (a ketotriose).

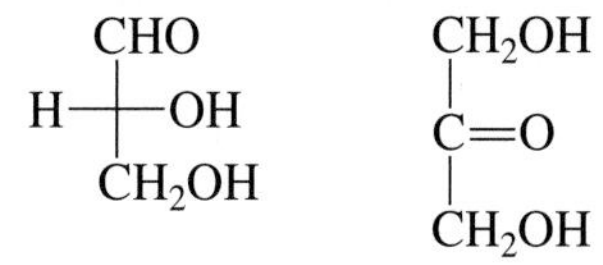

$$\begin{array}{c} CHO \\ H-\!\!\!+\!\!\!-OH \\ CH_2OH \end{array} \qquad \begin{array}{c} CH_2OH \\ | \\ C{=}O \\ | \\ CH_2OH \end{array}$$

D-Glyceraldehyde Dihydroxyacetone

Beginning with tetroses, monosaccharides have names with the characteristic ending *-ose*.

The most important monosaccharides, in terms of human body chemistry, are the D forms of glucose, galactose, fructose, and ribose. Glucose and galactose are aldohexoses, fructose is a ketohexose, and ribose is an aldopentose. All four of these monosaccharides are water-soluble, white, crystalline solids.

▶ You should memorize the structures of the monosaccharides considered in this section.

D-Glucose

▶ D-Glucose tastes sweet, is nutritious, and is an important component of the human diet. L-Glucose, on the other hand, is tasteless, and the body cannot use it.

Of all monosaccharides, D-glucose is the most abundant in nature and the most important from a nutritional standpoint. Its structure is

CHO
H—OH
HO—H
H—OH
H—OH
CH_2OH
D-Glucose

Note that four chiral centers are present and that the hydroxyl group is positioned to the "right" at the first, third, and fourth chiral centers and to the "left" at the second chiral center.

Most dietary carbohydrates are polysaccharides that are glucose polymers. Digestion of these materials produces D-glucose, which becomes the major energy source for running the human body. A 5%(m/v) glucose solution is often administered intravenously to hospital patients (see Figure 14.6).

D-Galactose

A comparison of the structures for D-galactose and D-glucose shows that these two compounds differ only in the configuration of the —OH group and —H group on carbon-4.

D-Galactose	D-Glucose
CHO	CHO
H—OH	H—OH
HO—H	HO—H
HO—H	H—OH
H—OH	H—OH
CH_2OH	CH_2OH

D-galactose and D-glucose are isomers. These two aldohexoses both have the molecular formula $C_6H_{12}O_6$. Both of these monosaccharides have four chiral centers, the only difference being the —OH group orientation at carbon-4 (the third chiral center). This "orientation difference" is sufficient, however, to give the two compounds different biochemical properties.

Figure 14.6 A 5% (m/v) glucose solution is often used in hospitals as an intravenous source of nourishment for patients who cannot take food by mouth. The body can use it as an energy source without digesting it.

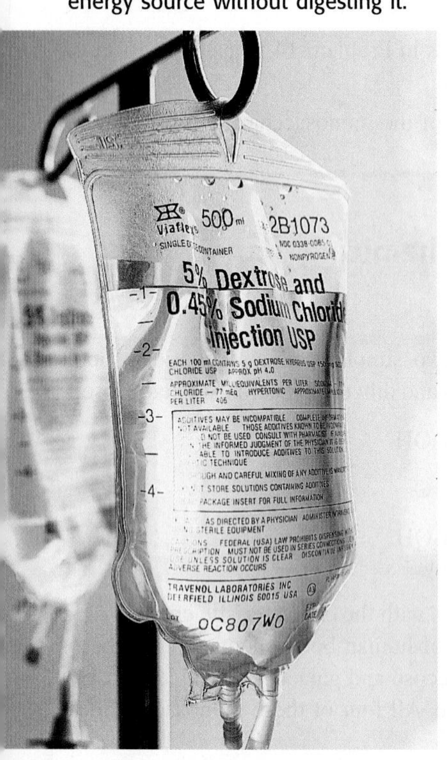

D-Fructose

D-fructose, like D-glucose and D-galactose, is a hexose. However, unlike these two other monosaccharides, D-fructose is a ketose rather than an aldose. However, its structure can still be easily compared to that of D-glucose. From the third to the sixth carbon, the structure of D-fructose is identical to that of D-glucose. Differences at carbons 1 and 2 are related to the presence of a ketone group in fructose and an aldehyde group in glucose.

D-Fructose	D-Glucose
CH_2OH	CHO
C=O	H—OH
HO—H	HO—H
H—OH	H—OH
H—OH	H—OH
CH_2OH	CH_2OH

Same structure

Despite the functional group difference, D-fructose is still isomeric with D-glucose and D-galactose. All three of these compounds have the molecular formula $C_6H_{12}O_6$. Chemical Portraits 23 gives some nonstructural information about these three isomeric hexoses.

D-Ribose

The monosaccharides glucose, galactose, and fructose are hexoses. D-Ribose is a pentose. If carbon-3 and its accompanying —H and —OH groups were eliminated from the structure of D-glucose, the remaining structure would be that of D-ribose.

D-Glucose	D-Ribose
CHO	CHO
H—OH	H—OH
HO—H	
H—OH	H—OH
H—OH	H—OH
CH_2OH	CH_2OH

D-Ribose is a component of a variety of complex molecules, including ribonucleic acids (RNAs) and energy-rich compounds such as adenosine triphosphate (ATP). The compound 2-deoxy-D-ribose is also important in nucleic acid chemistry. This monosaccharide

Chemical Portraits 23 Three Important Isomeric Monosaccharides

D-Glucose (an aldohexose)

Profile: D-glucose is the most abundant monosaccharide in nature and the most important from a human nutritional standpoint. Cells of the brain and nervous system depend *primarily* on glucose for energy. Another name for glucose is *blood sugar*: blood contains 80 to 120 mg/dL dissolved glucose. Blood glucose levels are tightly controlled by hormones.

Biochemical considerations: Humans seldom consume glucose directly; its source is usually dietary intake of disaccharides and polysaccharides which, when digested, release glucose. Ripe fruits, particularly grapes, have some free glucose. Glucose is sometimes called *grape sugar*; ripe grapes contain 20 to 30 mass percent glucose.

How does diabetes affect blood glucose levels?

D-Galactose (an aldohexose)

Profile: D-galactose is rarely found in nature as a free monosaccharide. It is a component, however, of several important biochemical substances including lactose (a glucose-galactose combination present in milk) and compounds present in brain and nerve tissue. It is sometimes called *brain sugar*. It is also present in the chemical markers that distinguish various types of blood—A, B, AB, and O. Free galactose is used as an ultrasound contrast agent in diagnostic medicine.

Biochemical considerations: In the human body, galactose is synthesized from glucose in the mammary glands for use in producing lactose, the main sugar in mammalian milk. Lactose is the main carbohydrate energy source for infants.

In terms of galactose, how do the chemical markers for the various blood types differ?

D-Fructose (a ketohexose)

Profile: Fructose, also known as *fruit sugar*, occurs naturally in fruits and honey. With glucose, it is a component of table sugar (sucrose). The major source of free fructose as a food ingredient is HFCS (high-fructose corn syrup); HFCS is widely used in baked goods and canned fruit.

Biochemical considerations: Fructose is the sweetest tasting of all sugars. It is sometimes used as a dietary sugar, not because it has fewer calories per gram than other sugars but because less is needed for the same amount of sweetness. On a scale where table sugar has a sweetness rating of 100, the rating for fructose is 173. Fructose is preferred over table sugar in soft drinks because it dissolves more readily.

How is high-fructose corn syrup produced?

See the text web site at **www.cengage.com/chemistry/stoker** for answers to the above questions and for further information.

is a component of DNA molecules. The prefix *deoxy-* means "minus an oxygen"; the structures of ribose and 2-deoxyribose differ in that the latter compound lacks an oxygen atom at carbon-2.

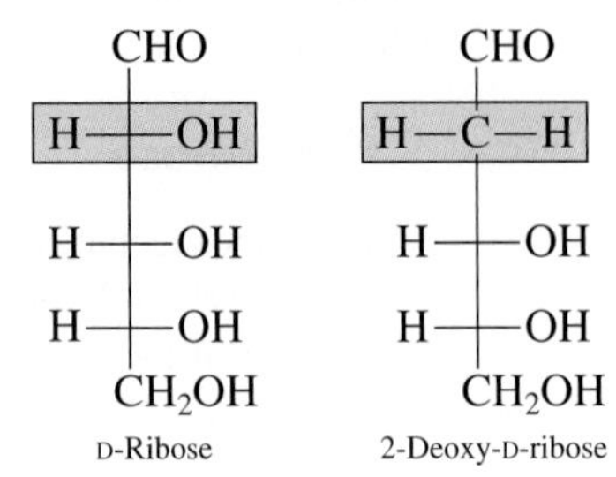

Practice Problems and Questions

14.22 List the similarities and the differences between the members of each of the following pairs of monosaccharide structures.
a. D-Glucose and D-galactose b. D-Glucose and D-fructose
c. D-Glucose and D-ribose d. D-Ribose and 2-deoxy-D-ribose

14.23 Draw a structural representation for each of the following monosaccharides.
a. D-Glucose b. D-Galactose c. D-Fructose d. D-Ribose

14.24 Which of the monosaccharides in Problem 14.23 fits each of the following descriptions?
a. Aldose b. Ketose c. Hexose d. Aldohexose

14.25 Which of the monosaccharides in Problem 14.23 fits each of the following descriptions?
a. Blood sugar b. Grape sugar c. Brain sugar d. Fruit sugar

Learning Focus

Draw and identify the cyclic structures for common monosaccharides.

14.7 Cyclic Forms of Monosaccharides

So far in this chapter, the structures of monosaccharides have been depicted as open-chain polyhydroxy aldehydes or ketones. Experimental studies indicate, however, that for monosaccharides containing five or more carbon atoms, the open-chain form is readily converted into cyclic structures and that the cyclic structures are the dominant forms for such monosaccharides.

The cyclic forms of monosaccharides result from the ability of their carbonyl group to react intramolecularly with a hydroxyl group at the other "end" of the molecule. The result is a cyclic hemiacetal (for aldoses) or a cyclic hemiketal (for ketoses). In Sections 13.3 and 13.5, we learned that hemiacetal formation and hemiketal formation involve the reaction between an alcohol and a carbonyl group (aldehyde or ketone). Monosaccharides are polyfunctional, having both an alcohol and an aldehyde or ketone group. The alcohol and carbonyl "ends" of the molecule react with each other; the result is cyclization of the structure.

In aldohexoses (glucose, galactose), cyclization involves the reaction of the hydroxyl group on carbon 5 with the aldehyde group at the other end of the molecule. The result is a six-membered heterocyclic ring in which one oxygen atom is present. Figure 14.7 shows diagrammatically the interactions that occur during the cyclization process in the case of D-glucose. The structure at the left in Figure 14.7 is the open-chain form of D-glucose with the two functional groups that are going to react with each other highlighted. In the second structure, the open-chain form has been folded such that the carbon atoms have locations similar to those found for carbon atoms in a six-membered ring. The third structure diagrams the functional group reaction, the addition of the —OH group on carbon 5 across the carbon–oxygen double bond. The cyclic structure that results has two forms; the newly formed hydroxyl group on carbon 1 can be up or down relative to other substituents on the ring. When this hydroxyl group is in a "down position," we have the alpha form (α form) of cyclic D-glucose, and when the hydroxyl group is in the "up position," we have the beta form (β form) of cyclic D-glucose (see Figure 14.7). (The im-

Figure 14.7 The cyclic hemiacetal forms of D-glucose result from the intramolecular reaction between the carbonyl group and the hydroxyl group on carbon 5.

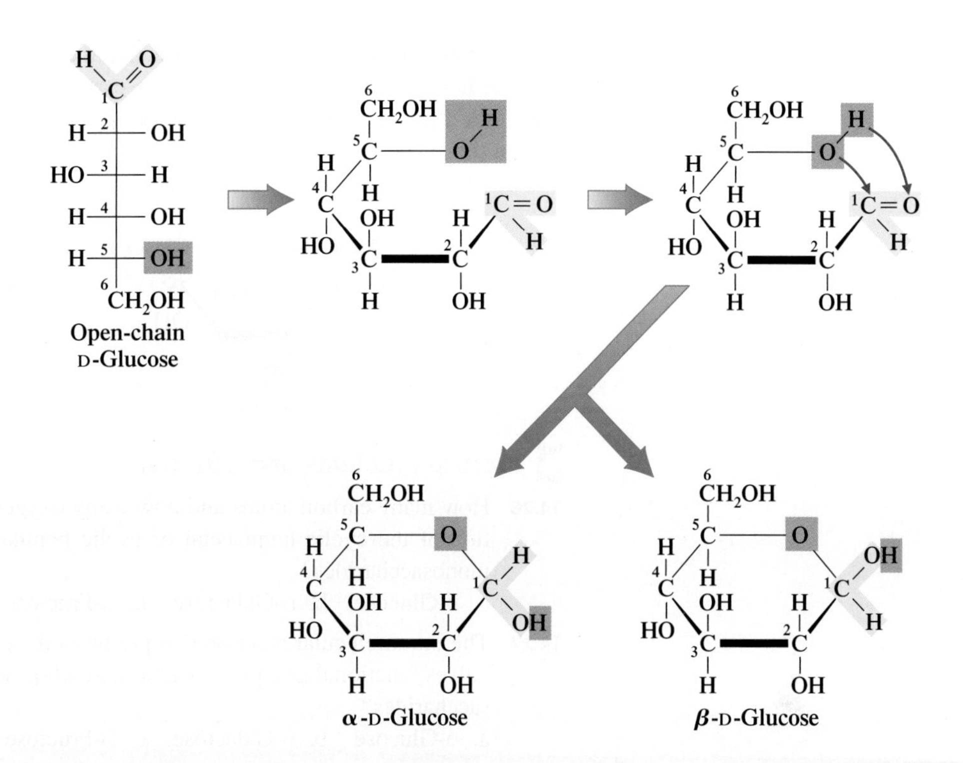

portance of the fact that cyclic monosaccharide structures have two forms will emerge later in this chapter when we consider the formation and reactions of larger carbohydrates.)

Further aspects of the relationship between the open-chain form of D-glucose and its cyclic forms are as follows:

1. Chiral-center —OH groups that point to the right in the open-chain form point down in the cyclic forms. This relationship pertains to carbons 3, 4, and 5.
2. Positioning of the —OH group on carbon 5 of the open-chain form is not a concern because it no longer exists; it is the —OH that added to the carbon–oxygen double bond.
3. Carbon 6 of the open-chain form does not become part of the ring. The position of the —OH group on carbon 6 does not matter because carbon 6 is not a chiral center.

In an aqueous solution of D-glucose, a dynamic equilibrium exists among the α, β, and open-chain forms, and there is continual interconversion among them. For example, a freshly mixed solution of pure α-D-glucose slowly converts to a mixture of both α- and β-D-glucose by an opening and a closing of the cyclic structure. When equilibrium is established, 63% of the molecules are β-D-glucose, 37% are α-D-glucose, and less than 0.01% are in the open-chain form.

Our preceding discussion has focused on D-glucose. Everything we have said about it also applies to D-galactose, because both are aldohexoses. The only difference between these two monosaccharides is the orientation of the carbon-4 —OH group. In the following comparison of the structures of α-D-glucose and α-D-galactose, note how this difference at carbon 4 "shows up."

α-D-Glucose α-D-Galactose

Both D-fructose and D-ribose form five-membered rings. For D-fructose, a ketohexose, the hemiacetal ring closure involves carbons 2 and 5; carbon 2 is the carbonyl carbon atom. Both carbons 1 and 6 of the open-chain D-fructose structure end up outside the ring in the cyclic structure. For D-ribose, an aldopentose, ring closure involves carbons 1 and 4, with carbon 5 ending up outside the ring.

$HOCH_2$ (6), O, CH_2OH (1), HO, OH, OH — α-D-Fructose

$HOCH_2$ (5), O, OH, OH, OH — α-D-Ribose

Practice Problems and Questions

14.26 How many carbon atoms and how many oxygen atoms are present in the ring portion of the cyclic hemiacetal or cyclic hemiketal form of each of the following monosaccharides?
a. D-Glucose b. D-Galactose c. D-Fructose d. D-Ribose

14.27 The intramolecular reaction that produces the cyclic forms of monosaccharides involves functional groups on which two carbon atoms in each of the following monosaccharides?
a. D-Glucose b. D-Galactose c. D-Fructose d. D-Ribose

14.28 What is the structural difference between the alpha and beta forms of D-glucose?

14.29 What is the structural difference between the alpha forms of D-glucose and D-galactose?

14.30 D-Glucose and D-fructose are both hexoses. D-Glucose forms six-membered ring cyclic structures, and D-fructose cyclic structures involve five-membered rings. Explain why the ring sizes are different.

14.31 Identify each of the following structures as an α-D-monosaccharide or as a β-D-monosaccharide.

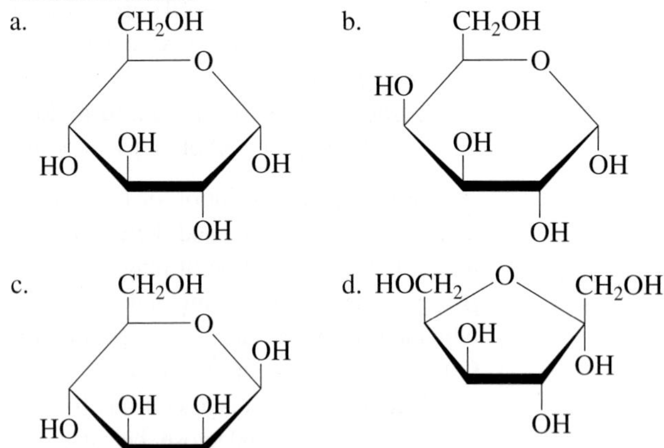

14.32 Which of the structures in Problem 14.31 is a representative of α-D-glucose?

14.33 Draw the open-chain form of the monosaccharide represented by the second cyclic structure in Problem 14.31.

Learning Focus

Know the characteristics of the following reactions that monosaccharides undergo: oxidation and reduction, glycoside formation, and phosphate ester formation.

14.8 Reactions of Monosaccharides

Four important reactions of monosaccharides are oxidation, reduction, glycoside formation, and phosphate ester formation. In considering these reactions, we will use glucose as the monosaccharide reactant. Remember, however, that other aldoses as well as ketoses undergo similar reactions.

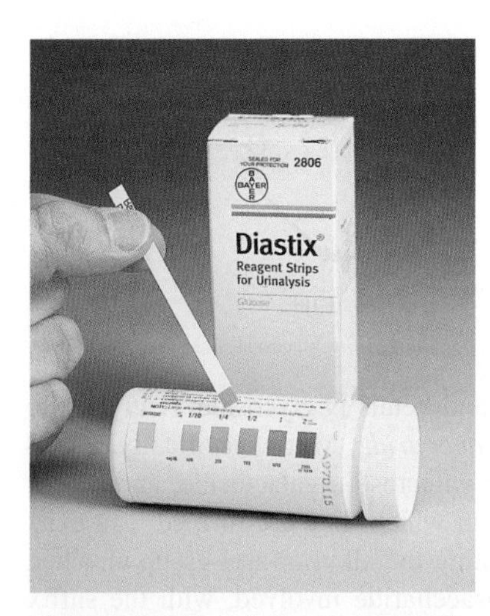

Figure 14.8 The glucose content of urine can be determined by dipping a plastic strip treated with oxidizing agents into the urine sample and comparing the color change of the strip to a color chart that indicates glucose concentration.

Oxidation and Reduction

The aldehyde group of aldoses is readily oxidized to a carboxyl group. Oxidation of the aldehyde end of glucose produces gluconic acid.

$$\begin{array}{c} \text{CHO} \\ \text{H}-\!\!\!-\text{OH} \\ \text{HO}-\!\!\!-\text{H} \\ \text{H}-\!\!\!-\text{OH} \\ \text{H}-\!\!\!-\text{OH} \\ \text{CH}_2\text{OH} \\ \text{D-Glucose} \end{array} \xrightarrow[\text{oxidizing agent}]{\text{Weak}} \begin{array}{c} \text{COOH} \\ \text{H}-\!\!\!-\text{OH} \\ \text{HO}-\!\!\!-\text{H} \\ \text{H}-\!\!\!-\text{OH} \\ \text{H}-\!\!\!-\text{OH} \\ \text{CH}_2\text{OH} \\ \text{D-Gluconic acid} \end{array}$$

A substance that is oxidized in a reaction is also the reducing agent in the reaction (Section 8.3). Because monosaccharides act as reducing agents in such reactions, they are called *reducing sugars*. With Tollens solution, glucose reduces Ag^+ ion to Ag, and with Benedict's solution, glucose reduces Cu^{2+} ion to Cu^+ ion (see Section 13.3). A **reducing sugar** *is a carbohydrate that gives a positive test with Tollens and Benedict's solutions.* All monosaccharides are reducing sugars.

Tollens and Benedict's solutions can be used to test for glucose in urine, a symptom of diabetes. For example, using Benedict's solution, we observe that if no glucose is present in the urine (a normal condition), the Benedict's solution remains blue. The presence of glucose is indicated by the formation of a red precipitate. Testing for the presence of glucose in urine is such a common laboratory procedure that much effort has been put into the development of easy-to-use test methods (Figure 14.8).

The carbonyl group present in a monosaccharide (either an aldose or a ketose) can be reduced to a hydroxyl group, using hydrogen as the reducing agent. For aldoses and ketoses, the product of the reduction is the corresponding polyhydroxy alcohol, which is sometimes called a *sugar alcohol*. For example, the reduction of D-glucose gives D-glucitol.

$$\begin{array}{c} \text{CHO} \\ \text{H—C—OH} \\ \text{HO—C—H} \\ \text{H—C—OH} \\ \text{H—C—OH} \\ \text{CH}_2\text{OH} \\ \text{D-Glucose} \end{array} \xrightarrow[\text{catalyst}]{\text{H}_2} \begin{array}{c} \text{CH}_2\text{OH} \\ \text{H—C—OH} \\ \text{HO—C—H} \\ \text{H—C—OH} \\ \text{H—C—OH} \\ \text{CH}_2\text{OH} \\ \text{D-Glucitol} \end{array}$$

▶ D-Sorbitol accumulation in the eye is a major factor in the formation of cataracts due to diabetes.

D-Glucitol is also known by the common name D-sorbitol. Hexahydroxy alcohols such as D-sorbitol have properties similar to those of the trihydroxy alcohol *glycerol* (Section 12.3). These alcohols are used as moisturizing agents in foods and cosmetics because of their affinity for water. D-Sorbitol is also used as a sweetening agent in chewing gum; bacteria that cause tooth decay cannot use polyalcohols as food sources, as they can glucose and many other monosaccharides.

Glycoside Formation

▶ Remember, from Sections 13.3 and 13.5, that acetals and ketals have two —OR groups attached to the same carbon atom.

In Sections 13.3 and 13.5 we learned that hemiacetals and hemiketals can react with alcohols in acid solution to produce acetals and ketals. Because the cyclic forms of monosaccharides are hemiacetals and hemiketals, they react with alcohols to form acetals and ketals, as is illustrated for the reaction of β-D-glucose with methyl alcohol.

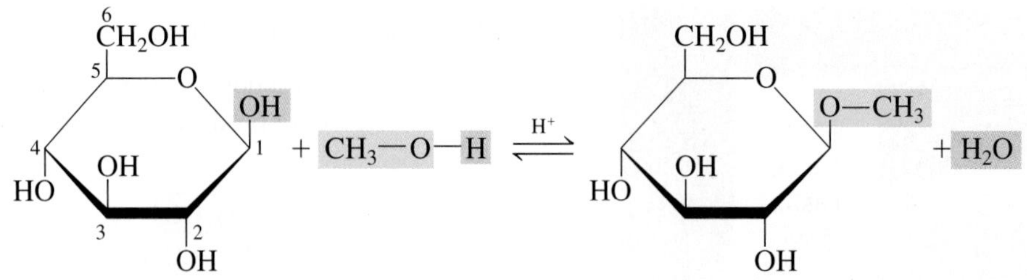

β-D-Glucose + $CH_3—O—H$ $\xrightleftharpoons{H^+}$ Methyl-β-D-glucoside + H_2O

The general name for monosaccharide acetals and ketals is glycoside. A **glycoside** *is an acetal or a ketal formed from a cyclic monosaccharide.* More specifically, a glycoside produced from glucose is a glucoside, from galactose a galactoside, and so on. Glycosides, like the hemiacetals and hemiketals from which they are formed, can exist in both α and β forms. Glycosides are named by listing the alkyl or aryl group attached to the oxygen, followed by the name of the monosaccharide involved, with the suffix *-ide* appended to it.

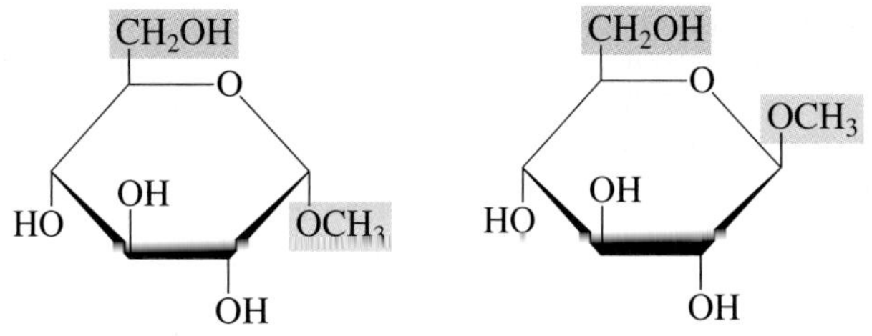

Methyl-α-D-glucoside

Methyl-β-D-glucoside

Phosphate Ester Formation

The hydroxyl groups of a monosaccharide can react with carboxylic acids to form carboxylic esters (Section 13.8). Monosaccharide hydroxyl groups will also undergo ester formation with inorganic acids. Phosphate esters, formed from phosphoric acid (H_3PO_4) and various monosaccharides, are commonly encountered in biological systems. For example, specific enzymes in the human body catalyze the esterification of the carbonyl group (carbon 1) and the primary alcohol group (carbon 6) in glucose to produce the compounds glucose 1-phosphate and glucose 6-phosphate, respectively.

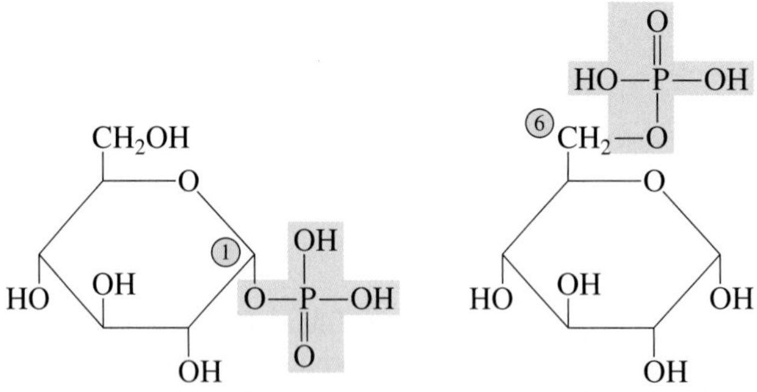

α-D-Glucose 1-phosphate

α-D-Glucose 6-phosphate

These phosphate esters of glucose are stable in aqueous solution and play important roles in the metabolism of carbohydrates.

Practice Problems and Questions

14.34 Indicate whether each of the following reactions involves oxidation or reduction.

a. Production of D-gluconic acid from D-glucose

b. Production of D-galactonic acid from D-galactose

c. Production of D-glucitol from D-glucose
d. Production of D-sorbitol from D-glucose

14.35 Which of the following monosaccharides is a *reducing sugar*?
a. D-Glucose b. D-Galactose c. D-Fructose d. D-Ribose

14.36 Which of the following monosaccharides will give a positive test with Benedict's solution?
a. D-Glucose b. D-Galactose c. D-Fructose d. D-Ribose

14.37 Indicate whether each of the following structures is that of a glycoside.

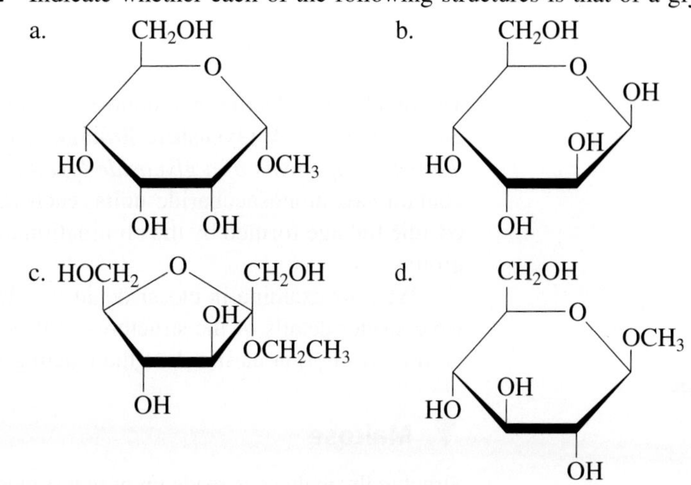

14.38 Which of the structures in Problem 14.37 is that of a β-glycoside?

14.39 Which of the structures in Problem 14.37 is a methyl-D-glucoside?

14.40 Draw the structure of α-D-galactose-1-phosphate.

Learning Focus

Specify the monosaccharides present and the type of linkage between monosaccharides for the disaccharides maltose, lactose, and sucrose.

14.9 Disaccharides

As the name implies, *disaccharides* (Section 14.3) contain two monosaccharide units bonded together. For the three most common disaccharides—maltose, lactose, and sucrose—the monosaccharides present are as follows:

Maltose:	D-glucose	D-glucose
Lactose:	D-galactose	D-glucose
Sucrose:	D-glucose	D-fructose

To understand disaccharide chemistry, we need more information than simply the identity of the monosaccharides present. Understanding how the monosaccharides in a disaccharide are covalently bonded to each other is the key to understanding disaccharide chemistry.

In the previous section we saw that a cyclic monosaccharide (a hemiacetal or a hemiketal) can react with an alcohol to form a glycoside (an acetal or a ketal). The same type of reaction, between two monosaccharides, occurs in disaccharide formation. One of the monosaccharides functions as the hemiacetal or hemiketal, and the other functions as the alcohol.

$$\underset{\left(\substack{\text{Functioning as a}\\\text{hemiacetal or}\\\text{as a hemiketal}}\right)}{\text{Monosaccharide}} + \underset{\left(\substack{\text{Functioning as}\\\text{an alcohol}}\right)}{\text{monosaccharide}} \longrightarrow \underset{\text{(Glycoside)}}{\text{disaccharide}} + H_2O$$

Specifically, with D-glucose molecules as the monosaccharides, the reaction is

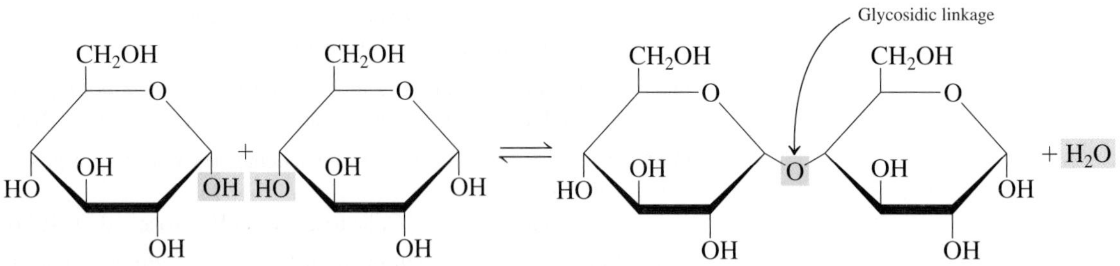

The bond that links the two monosaccharides of a disaccharide together is called a glycosidic linkage. A **glycosidic linkage** *is the carbon–oxygen–carbon bond that joins the two components of a glycoside together*. Thus disaccharides are carbohydrates that contain two monosaccharide units, each in its cyclic form, joined together by a glycosidic linkage formed by the elimination of a molecule of water between two hydroxyl groups.

We now examine in closer detail the structures of maltose, lactose, and sucrose. As we consider details of the structures of these disaccharides, we will find that the configuration (α or β) at carbon 1 of the reacting monosaccharides is of prime importance.

Maltose

Structurally, maltose is made up of two D-glucose units, one of which must be α-D-glucose. The formation of maltose from two glucose molecules is as follows:

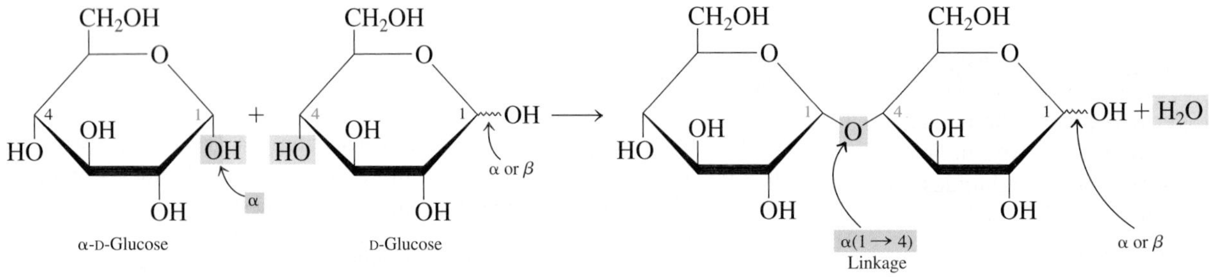

The glycosidic linkage between the two glucose units is called an $\alpha(1 \rightarrow 4)$ linkage. The two —OH groups that form the linkage are attached, respectively, to carbon 1 of the first glucose unit (in an α configuration) and to carbon 4 of the second.

Maltose is a reducing sugar (Section 14.8), because the glucose unit on the right has a hemiacetal carbon atom (C-1). Thus this glucose unit can open and close; it is in equilibrium with its open-chain aldehyde form (Section 14.7). This means there are actually three forms of the maltose molecule: α-maltose, β-maltose, and the open-chain form. Structures for these three maltose forms are shown in Figure 14.9. In the solid state, the β form is dominant.

The most important chemical reaction of maltose is that of hydrolysis. Hydrolysis of D-maltose, whether in a laboratory flask or in a living organism, produces two molecules of D-glucose. Acidic conditions or the enzyme *maltase* are needed for the hydrolysis to occur.

$$\text{D-Maltose} + H_2O \xrightarrow{H^+ \text{ or maltase}} 2\ \text{D-Glucose}$$

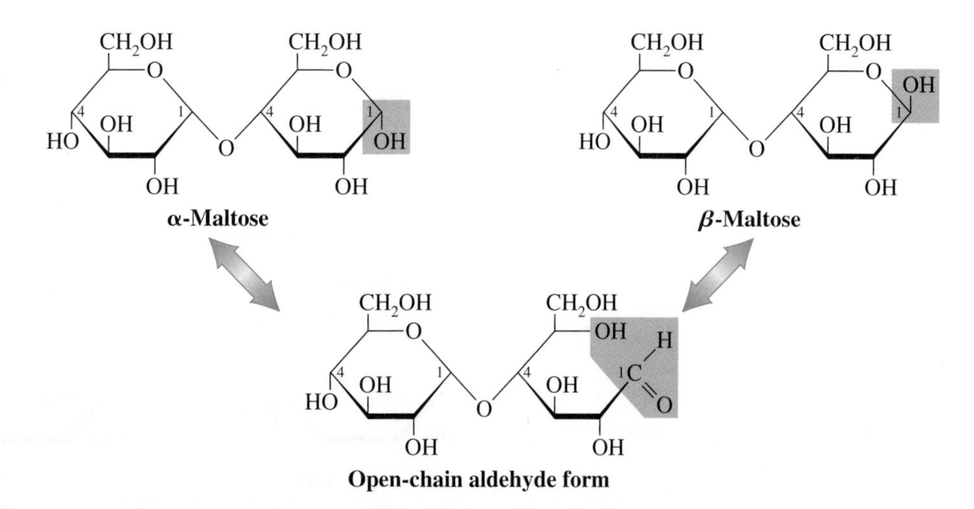

Figure 14.9 The three forms of maltose present in aqueous solution.

Lactose

In maltose the two monosaccharide units present are identical; both are glucose units. However, the two monosaccharide units in a disaccharide need not be identical. Lactose is made up of a β-D-galactose unit and a D-glucose unit joined by a $\beta(1 \rightarrow 4)$ glycosidic linkage.

CH_2OH ... OH + HO ... $\xrightarrow{}$ ... OH + H_2O

β-D-Glucose D-Glucose β-D-Galactose β(1→ 4) Linkage D-Glucose Lactose α or β

In a $\beta(1 \rightarrow 4)$ linkage, the oxygen of the glycosidic linkage is in the beta position, and the linkage involves carbon 1 of galactose and carbon 4 of glucose. Because carbon 1 of glucose is unaffected when galactose bonds to glucose in the formation of lactose, lactose is a reducing sugar (the glucose ring can open to give an aldehyde).

It is important to distinguish between the structural notation used for a $\alpha(1 \rightarrow 4)$ glycosidic linkage and that used for a $\beta(1 \rightarrow 4)$ glycosidic linkage.

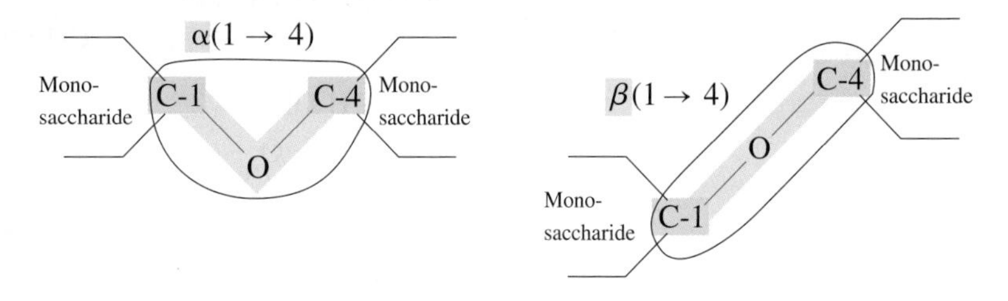

Lactose can be hydrolyzed by acid or by the enzyme *lactase*, forming an equimolar mixture of galactose and glucose.

$$\text{D-Lactose} + H_2O \xrightarrow{H^+ \text{ or lactase}} \text{D-galactose} + \text{D-glucose}$$

Sucrose

The two monosaccharide units present in a D-sucrose molecule are α-D-glucose and β-D-fructose. The glycosidic linkage is not a $(1 \rightarrow 4)$ linkage, as was the case for maltose and lactose. It is instead an $\alpha,\beta(1 \rightarrow 2)$ glycosidic linkage. The —OH group on carbon 2 of D-fructose (the hemiketal carbon) reacts with the —OH group on carbon 1 of D-glucose (the hemiacetal carbon).

▶ The glycosidic linkage in sucrose is very different from that in maltose and lactose. The linkages in the latter two compounds can be characterized as "head-to-tail" linkages—that is, the front end (carbon 1) of one monosaccharide is linked to the back end (carbon 4) of the other monosaccharide. Sucrose has a "head-to-head" glycosidic linkage; the front ends of the two monosaccharides (carbon 1 for glucose and carbon 2 for fructose) are linked.

Sucrose, unlike maltose and lactose, is a *nonreducing sugar*. No hemiacetal or hemiketal center is present in the molecule, because the glycosidic linkage involves the reducing ends of both monosaccharides. Sucrose, in the solid state and in solution, exists in only one form—there are no α and β isomers, and an open-chain form is not possible.

Sucrase, the enzyme needed to break the $\alpha,\beta(1 \rightarrow 2)$ linkage in sucrose, is present in the human body. Hence sucrose is an easily digested substance. Sucrose hydrolysis (digestion) produces an equimolar mixture of glucose and fructose.

$$\text{D-Sucrose} + H_2O \xrightarrow{H^+ \text{ or sucrase}} \text{D-glucose} + \text{D-fructose}$$

Chemical Portraits 24 gives nonstructural details about the three disaccharides that we have just considered—maltose, lactose, and sucrose. Interestingly, these three very different disaccharides all have the same molecular formula, $C_{12}H_{22}O_{11}$; that is, they are isomers.

Practice Problems and Questions

14.41 Identify the type of glycosidic linkage [such as $\alpha(1 \rightarrow 4)$] present in each of the following disaccharides.

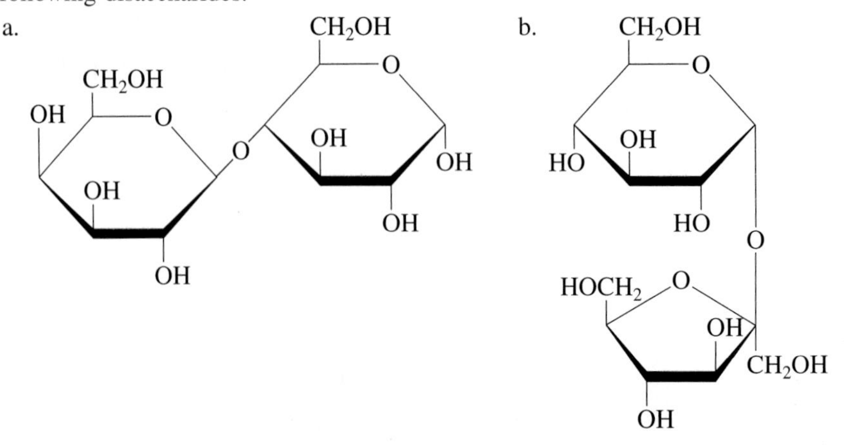

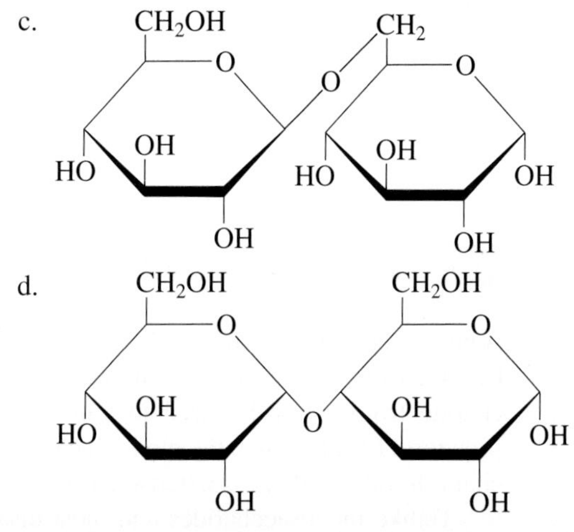

14.42 For each of the disaccharides in Problem 14.41, specify whether the disaccharide is in an α configuration, a β configuration, or neither.

14.43 Indicate whether each of the disaccharides in Problem 14.41 is a reducing sugar or a nonreducing sugar.

14.44 Indicate whether glucose is a component of each of the disaccharides in Problem 14.41.

14.45 Identify the two monosaccharides produced from the hydrolysis of each of the following disaccharides.
a. Lactose b. Sucrose c. Maltose

Chemical Portraits 24 Biochemically Important Disaccharides

Maltose (Glucose-Glucose)

Profile: Maltose is produced whenever starch (polymeric glucose) breaks down, as happens when seeds germinate and in human beings during starch digestion. Maltose's sweetness is about 30% that of table sugar. Maltose is responsible for the slightly sweet taste experienced when bread is held in the mouth for a few minutes; saliva begins the starch digestion process. Maltose is a common ingredient in baby foods because glucose is the only digestion product.

Uses: Malt (germinated barley that has been baked and ground) has a high maltose content; hence the name *malt sugar* for maltose. Malt is used to make beer. An enzyme present in yeast converts the maltose present in malt to glucose and then another yeast enzyme facilitates the conversion of the glucose to ethanol (a fermentation process).

What are the physical characteristics of a pure maltose sample?

Lactose (Galactose-Glucose)

Profile: Lactose, also known as *milk sugar*, is the major sugar found in milk. Its sweetness is 16% that of sucrose. The lactose content of mother's milk obtained by nursing infants [7–8%(m/m)] is almost twice that in cow's milk [4–5%(m/m)]. Lactose accounts for about 40% of the energy in human milk.

Biochemical considerations: Many adults lack the enzyme *lactase*, necessary for the digestion of lactose; this causes *lactose intolerance*. The level of lactase in humans varies with age. Most people have sufficient lactase during childhood years when milk is a needed calcium source. In adulthood, lactase levels decrease and lactose intolerance develops. This explains changes in milk-drinking habits of many adults. About one-third of Americans suffer from lactose intolerance.

What are the symptoms of lactose intolerance, and how is this intolerance treated?

Sucrose (Glucose-Fructose)

Profile: Sucrose, common white *table sugar*, is the most abundant of all disaccharides; it occurs in many fruits, in the nectar of flowers, and in the juices of many plants. Commercial sources for sucrose are sugar cane and sugar beets, which are respectively, 20% and 17% by mass sucrose.

Uses: Sucrose is one of the most heavily used food additives in the United States. Annual per capita consumption of sucrose is approximately 100 pounds, two-thirds of which is "food additive sugar" and one-third of which is "sugar naturally present" in food. Mass percent sucrose is 10% for a cola drink, 12% for fruit juice, 11% for Kool-Aid, and 44% for milk chocolate. Sucrose is the only sweetener that can be called "sugar" in the ingredient list on food labels in the United States.

What is the process for obtaining "table sugar" from sugar beets and sugar cane?

See the text web site at **www.cengage.com/chemistry/stoker** for answers to the above questions and for further information.

14.46 Identify the type of glycosidic linkage [such as $\alpha(1 \rightarrow 4)$] present in each of the following disaccharides.
a. Sucrose b. Maltose c. Lactose

14.47 Identify the disaccharide that fits each of the following descriptions.
a. Table sugar b. Milk sugar c. Malt sugar

Learning Focus

List properties of and structures for the polysaccharides starch, cellulose, glycogen, and chitin; describe the biological significance of α and β glycosidic linkages within polysaccharides.

14.10 Polysaccharides

A polysaccharide (Section 14.3) contains many monosaccharide units bonded to each other by glycosidic linkages. The number of monosaccharide units varies with the polysaccharide from a few hundred to hundreds of thousands. Polysaccharides are polymers (Section 11.6). In some, the monosaccharides are bonded together in a linear (unbranched) chain. In others, there is extensive branching of the chains (Figure 14.10).

Unlike monosaccharides and most disaccharides, polysaccharides are not sweet and do not test positive in Tollens and Benedict's solutions. They have limited water solubility because of their size. However, the —OH groups present can individually become hydrated by water molecules. The result is usually a thick colloidal suspension of the polysaccharide in water. Polysaccharides, such as flour and cornstarch, are often used as thickening agents in sauces, desserts, and gravy.

▶ In nutrition discussions, monosaccharides and disaccharides are called *simple carbohydrates,* and polysaccharides are called *complex carbohydrates.*

Although there are many naturally occurring polysaccharides, in this section we will focus on only four of them: cellulose, starch, glycogen, and chitin. All play vital roles in living systems—cellulose and starch in plants, glycogen in humans and other animals, and chitin in arthropods.

Cellulose

Cellulose is the most abundant polysaccharide. It is the structural component of the cell walls of plants. Approximately half of all the carbon atoms in the plant kingdom are contained in cellulose molecules. Structurally, cellulose is a linear (unbranched) D-glucose polymer in which the glucose units are linked by $\beta(1 \rightarrow 4)$ glycosidic bonds.

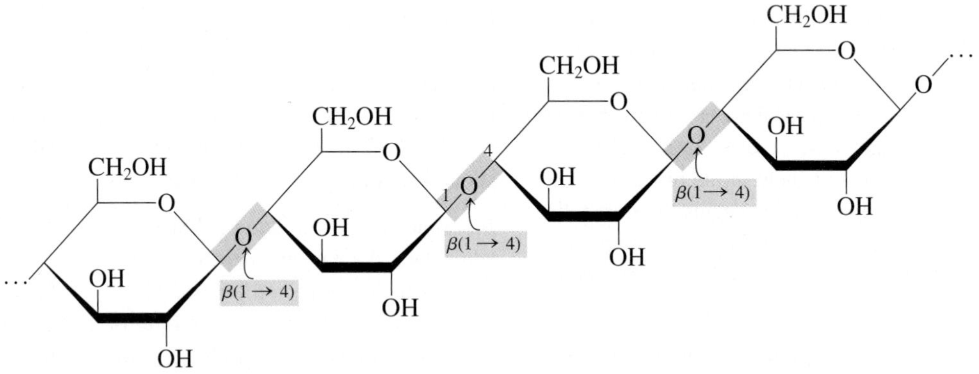

Typically, cellulose chains contain about 5000 glucose units, which gives macromolecules with molecular masses of about 900,000 amu. Cotton is almost pure cellulose (95%), and wood is about 50% cellulose.

Even though it is a glucose polymer, cellulose is not a source of nutrition for human beings. Humans lack the enzymes capable of catalyzing the hydrolysis of $\beta(1 \rightarrow 4)$ linkages in cellulose. Even grazing animals lack the enzymes necessary for cellulose digestion. However, the intestinal tracts of animals such as horses, cows, and sheep contain bacteria that produce *cellulase,* an enzyme that can hydrolyze $\beta(1 \rightarrow 4)$ linkages and produce free glucose from cellulose. Thus grasses and other plant materials are a source of nutrition for grazing animals. The intestinal tracts of termites contain the same microorganisms, which enable termites to use wood as their source of food. Microorganisms in the soil can also metabolize cellulose, which makes possible the biodegradation of dead plants.

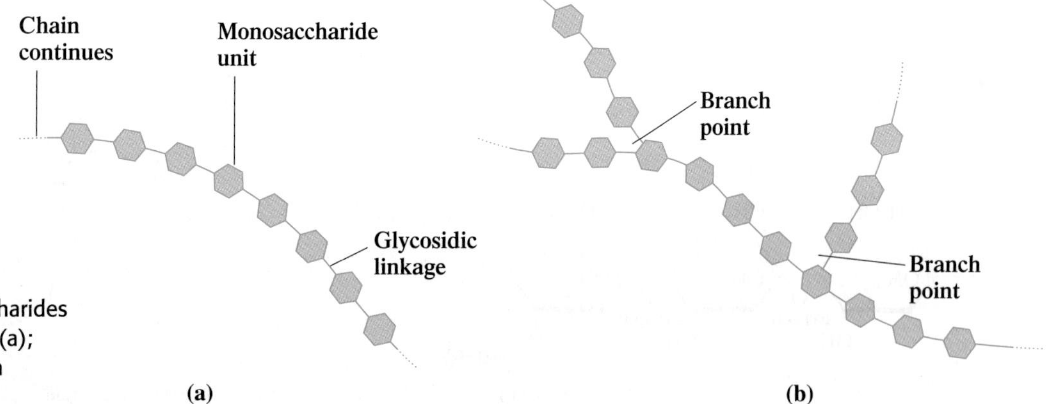

Figure 14.10 Some polysaccharides have a linear chain structure (a); others have a branched chain structure (b).

Despite its nondigestibility, cellulose is still an important component of a balanced diet. It serves as dietary fiber. Dietary fiber provides the digestive tract with "bulk" that helps move food through the intestinal tract and facilitates the excretion of solid wastes. Cellulose readily absorbs water, leading to softer stools and frequent bowel action. Links have been found between the length of time stools spend in the colon and possible colon cancer.

Starch

Starch, like cellulose, is a polysaccharide containing only glucose units. It is the storage polysaccharide in plants. If excess glucose enters a plant cell, it is converted to starch and stored for later use. When the cell cannot get enough glucose from outside the cell, it hydrolyzes starch to release glucose.

▶ Amylose and cellulose are both linear chains of D-glucose molecules. They differ in the configuration at carbon 1 of each D-glucose unit. In amylose, α-D-glucose is present; in cellulose, β-D-glucose.

Two different polyglucose polysaccharides can be isolated from most starches: amylose and amylopectin. *Amylose,* a straight-chain glucose polymer, usually accounts for 15%–20% of the starch; *amylopectin,* a highly branched glucose polymer, accounts for the remaining 80%–85% of the starch.

In amylose's structure, the glucose units are connected by $\alpha(1 \rightarrow 4)$ glycosidic linkages.

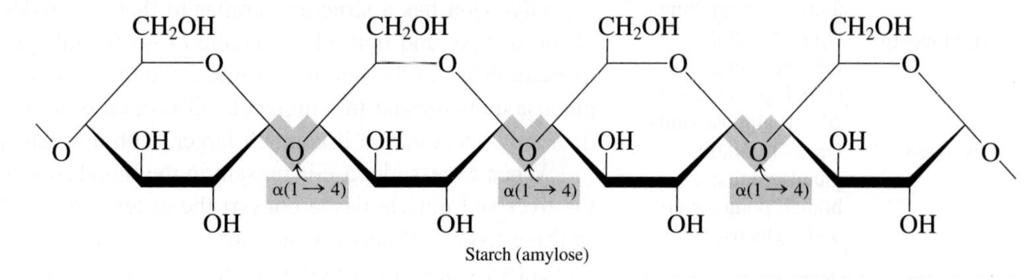

The number of glucose units present in an amylose chain depends on the source of the starch; 300–500 monomer units are usually present.

Amylopectin, the other polysaccharide in starch, is similar to amylose in that all linkages are α linkages. It is different in that there is a high degree of branching in the polymer. A branch occurs about once every 25–30 glucose units. The branch points involve $\alpha(1 \rightarrow 6)$ linkages (Figure 14.11). Because of the branching, amylopectin has a larger average molecular mass than the linear amylose. The average molecular mass of amylose is 50,000 amu or more; it is 300,000 or more for amylopectin.

Note that all of the glycosidic linkages in starch (both amylose and amylopectin) are of the α type. In amylose, they are all $(1 \rightarrow 4)$; in amylopectin, both $(1 \rightarrow 4)$ and $(1 \rightarrow 6)$ linkages are present. Because α linkages can be broken through hydrolysis within the human digestive tract (with the help of the enzyme *amylase*), starch has nutritional value for humans. The starches present in potatoes and cereal grains (wheat, rice, corn, etc.) account for approximately two-thirds of the world's food consumption.

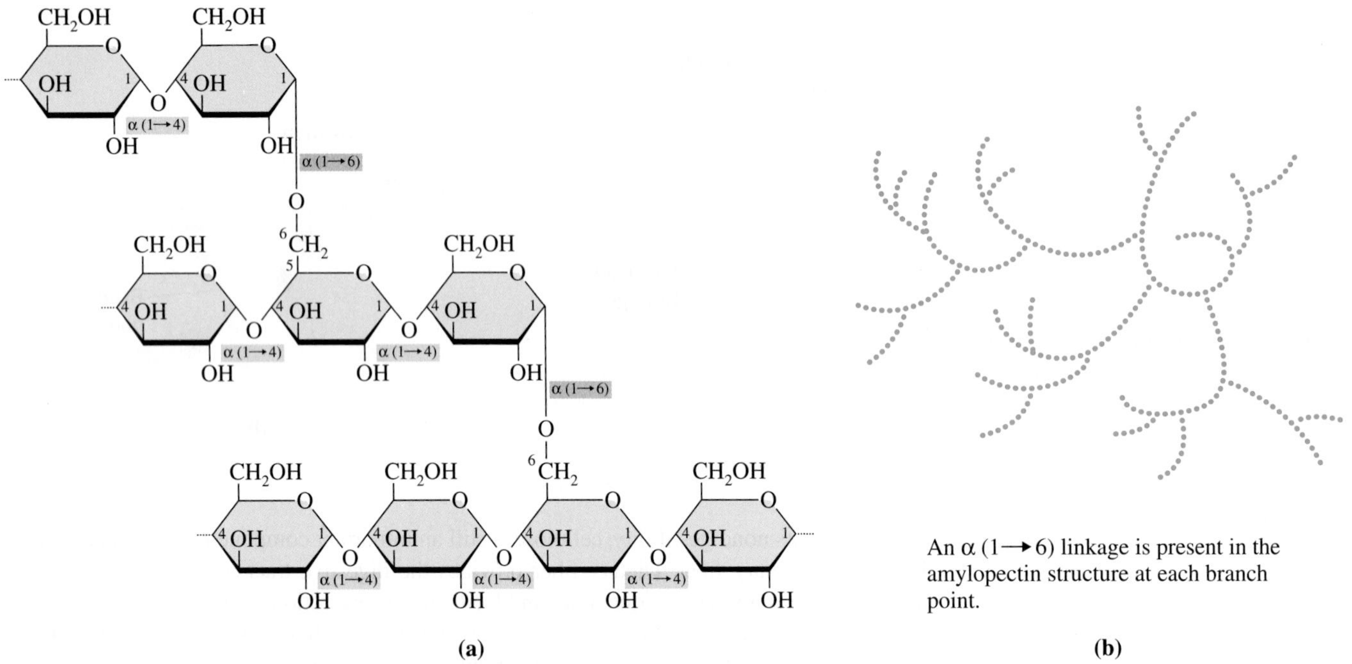

An α (1→6) linkage is present in the amylopectin structure at each branch point.

(a) (b)

Figure 14.11 Two perspectives on the structure of the polysaccharide amylopectin. (a) Molecular structure of amylopectin. (b) An overview of the branching that occurs in the amylopectin structure. Each circle is a glucose unit.

▶ The glucose polymers amylose, amylopectin, and glycogen compare as follows in molecular size and degree of branching.

Amylose:	Up to 1000 glucose units; no branching
Amylopectin:	Up to 100,000 glucose units; branch points every 24–30 glucose units
Glycogen:	Up to 1,000,000 glucose units; branch points every 8–12 glucose units

▶ The amount of stored glycogen in the human body is relatively small. Muscle tissue is approximately 1% glycogen, and liver tissue 2%–3%. However, this amount is sufficient to take care of normal-activity glucose demands for about 15 hours. During strenuous exercise, glycogen supplies can be exhausted rapidly. At this point, the body begins to oxidize fat as a source of energy.

Many marathon runners eat large quantities of starch foods the day before a race. This practice, called *carbohydrate loading,* maximizes body glycogen reserves.

Glycogen

Glycogen, like cellulose and starch, is a polysaccharide containing only glucose units. It is the glucose storage polysaccharide in humans and animals. Its function is thus similar to that of starch in plants, and it is sometimes referred to as *animal starch.* Liver cells and muscle cells are the storage sites for glycogen in humans.

Glycogen has a structure similar to that of amylopectin; all glycosidic linkages are of the α type, and both $(1 \rightarrow 4)$ and $(1 \rightarrow 6)$ linkages are present. Glycogen and amylopectin differ in the number of glucose units between branches and the total number of glucose units present in a molecule. Glycogen is about three times more highly branched than amylopectin, and it is much larger, with a molar mass of up to 3,000,000 amu.

When excess glucose is present in the blood (normally from eating too much starch), the liver and muscle tissue convert the excess glucose to glycogen, which is then stored in these tissues. Whenever the glucose blood level drops (from exercise, fasting, or normal activities), some stored glycogen is hydrolyzed back to glucose. These two opposing processes are called *glycogenesis* and *glycogenolysis*, the formation and decomposition of glycogen, respectively.

$$\text{Glucose} \underset{\text{Glycogenolysis}}{\overset{\text{Glycogenesis}}{\rightleftarrows}} \text{glycogen}$$

Glycogen is an ideal storage form for glucose. The large size of these macromolecules prevents them from diffusing out of cells. Also, conversion of glucose to glycogen reduces osmotic pressure (Section 7.8). Cells would burst because of increased osmotic pressure if all of the glucose in glycogen were present in cells in free form. High concentrations of glycogen in a cell sometimes precipitate or crystallize into *glycogen granules*. These granules are discernible in photographs of cells under electron microscope magnification (Figure 14.12).

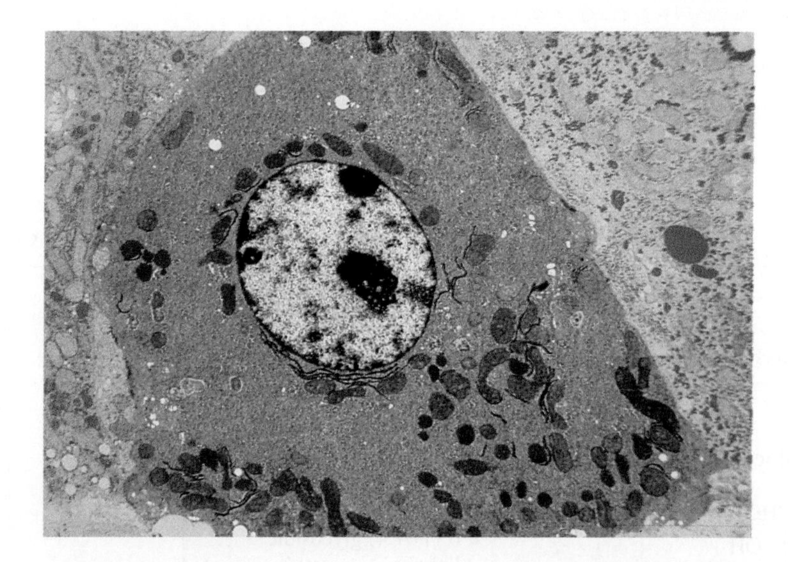

Figure 14.12 The small, dense particles within this electron micrograph of a liver cell are glycogen granules.

Chitin

▶ The word *chitin* is pronounced "kye-ten"; it rhymes with *Titan*.

Chitin is a polysaccharide that is similar to cellulose in both function and structure. Its function is to give rigidity to the exoskeletons of crabs, lobsters, shrimp, insects, and other arthropods (see Figure 14.13). It also occurs in the cell walls of fungi.

Structurally, chitin is a linear polymer (no branching) with all $\beta(1 \rightarrow 4)$ glycosidic linkages, as is cellulose. Chitin differs from cellulose in that the monosaccharide present is an *N*-acetyl amino derivative of D-glucose. Figure 14.14 contrasts the structures of chitin and cellulose.

The accompanying Chemistry at a Glance contrasts the various carbohydrates that have been considered in this chapter in terms of structural characteristics.

Figure 14.13 Chitin, a linear $\beta(1 \rightarrow 4)$ polysaccharide produces the rigidity in the exoskeletons of crabs and other arthropods.

Practice Problems and Questions

14.48 Match each of the following structural characteristics to the polysaccharides amylopectin, amylose, glycogen, cellulose, and chitin. (More than one substance may be correct in a given situation.)

a. Contains both $\alpha(1 \rightarrow 4)$ and $\alpha(1 \rightarrow 6)$ glycosidic linkages
b. Contains only D-glucose monosaccharide units
c. Composed of unbranched monosaccharide chains
d. Contains only $\beta(1 \rightarrow 4)$ glycosidic linkages

14.49 What is the difference between "plant starch" and "animal starch"?

14.50 Both cellulose and starch are D-glucose polymers. Why can humans digest starch

Figure 14.14 The structures of cellulose (a) and chitin (b). In both substances, all glycosidic linkages are of the $\beta(1 \rightarrow 4)$ type.

β (1→4) Glycosidic linkage

(a) (b)

Chemistry *at a Glance*

Biochemically Important Carbohydrates

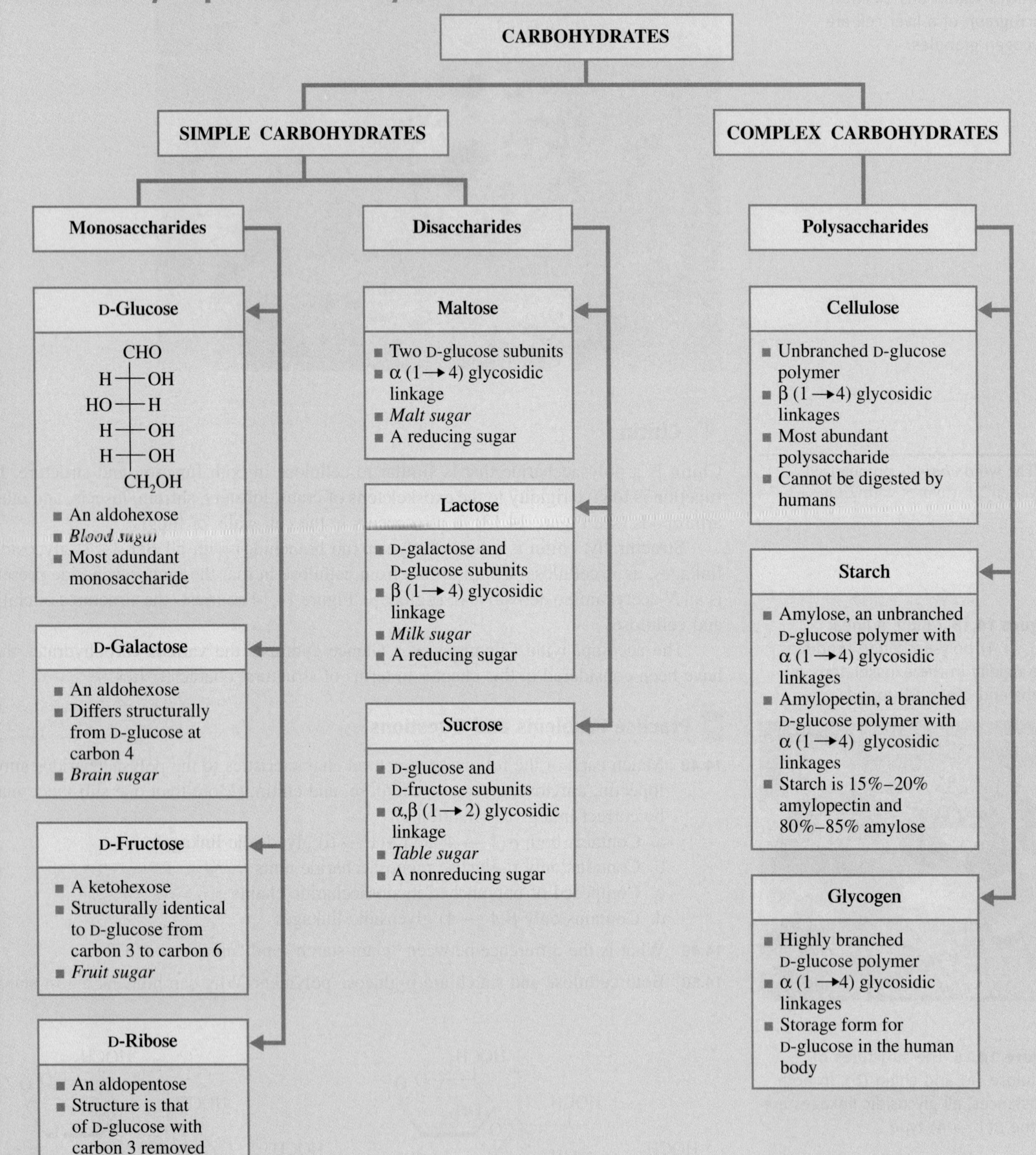

CARBOHYDRATES
SIMPLE CARBOHYDRATES
COMPLEX CARBOHYDRATES
Monosaccharides
Disaccharides
Polysaccharides
D-Glucose
CHO
H—OH
HO—H
H—OH
H—OH
CH2OH
An aldohexose
Blood sugar
Most adundant monosaccharide
D-Galactose
An aldohexose
Differs structurally from D-glucose at carbon 4
Brain sugar
D-Fructose
A ketohexose
Structurally identical to D-glucose from carbon 3 to carbon 6
Fruit sugar
D-Ribose
An aldopentose
Structure is that of D-glucose with carbon 3 removed
Maltose
Two D-glucose subunits
α (1→4) glycosidic linkage
Malt sugar
A reducing sugar
Lactose
D-galactose and D-glucose subunits
β (1→4) glycosidic linkage
Milk sugar
A reducing sugar
Sucrose
D-glucose and D-fructose subunits
α,β (1→2) glycosidic linkage
Table sugar
A nonreducing sugar
Cellulose
Unbranched D-glucose polymer
β (1→4) glycosidic linkages
Most abundant polysaccharide
Cannot be digested by humans
Starch
Amylose, an unbranched D-glucose polymer with α (1→4) glycosidic linkages
Amylopectin, a branched D-glucose polymer with α (1→4) glycosidic linkages
Starch is 15%–20% amylopectin and 80%–85% amylose
Glycogen
Highly branched D-glucose polymer
α (1→4) glycosidic linkages
Storage form for D-glucose in the human body

14.51 Why can cattle digest cellulose whereas humans cannot?

14.52 Humans can digest both amylose and amylopectin. Describe the structural differences between these two types of polysaccharides.

14.53 Describe the structural differences between cellulose and chitin.

CONCEPTS TO REMEMBER

Biochemistry. Biochemistry is the study of the chemical substances found in living systems and the chemical interactions of these substances with each other.

Carbohydrates. Carbohydrates are polyhydroxy aldehydes, polyhydroxy ketones, or compounds that yield such substances upon hydrolysis. Plants contain large quantities of carbohydrates produced via photosynthesis.

Carbohydrate classification. Carbohydrates are classified into three groups: monosaccharides, oligosaccharides, and polysaccharides.

Classification of monosaccharides. Monosaccharides are classified as aldoses or ketoses on the basis of the type of carbonyl group present. They are further classified as trioses, tetroses, pentoses, etc. on the basis of the number of carbon atoms present.

Handedness in monosaccharides. Most monosaccharides exhibit handedness—that is, they exist in a "left-handed" form (L form) and a "right-handed" form (D form). Handedness requires a chiral center, an atom that has four different groups tetrahedrally bonded to it. Naturally occurring monosaccharides are "right-handed" molecules.

Important monosaccharides. Important monosaccharides include glucose, galactose, fructose, and ribose. Glucose and galactose are aldohexoses, fructose is a ketohexose, and ribose is an aldopentose.

Cyclic monosaccharides. Cyclic monosaccharides form through an intramolecular reaction between the carbonyl group and an alcohol group of an open-chain monosaccharide. These cyclic forms predominate in solution.

Reactions of monosaccharides. Four important reactions of monosaccharides are (1) oxidation to a polyhydroxy acid, (2) reduction to a polyhydroxy alcohol, (3) glycoside formation, and (4) phosphate ester formation.

Disaccharides. Disaccharides are glycosides formed from the linking together of two monosaccharides. The most important disaccharides are maltose, lactose, and sucrose. Each of these has at least one glucose unit in its structure.

Polysaccharides. Polysaccharides are polymers of monosaccharides. Cellulose and starch are polymers of glucose, but the linkages between glucose units are different. The α linkages in starch are easily broken by enzymatic hydrolysis in humans, thus providing great sources of dietary energy. However, the β linkages in cellulose prevent enzymatic hydrolysis in humans, so cellulose provides no usable energy in our diets. Glycogen is a highly branched glucose polymer utilized for energy storage within the body.

KEY REACTIONS AND EQUATIONS

1. Monosaccharide oxidation (Section 14.8)

   Aldose + weak oxidizing agent $\longrightarrow$ polyhydroxy acid

2. Monosaccharide reduction (Section 14.8)

   Aldose or ketose + H_2 $\xrightarrow{\text{Catalyst}}$ polyhydroxy alcohol

3. Glycoside (acetal or ketal) formation (Section 14.8)

   Cyclic monosaccharide + alcohol $\longrightarrow$ glycoside (acetal or ketal) + H_2O

4. Monosaccharide ester formation (Section 14.8)

   Monosaccharide + oxyacid $\longrightarrow$ ester + H_2O

5. Hydrolysis of disaccharide (Section 14.9)

   Disaccharide + H_2O $\xrightarrow{\text{Catalyst}}$ two monosaccharides

6. Hydrolysis of maltose (Section 14.9)

   D-Maltose + H_2O $\xrightarrow{H^+ \text{ or maltase}}$ 2 D-glucose

7. Hydrolysis of lactose (Section 14.9)

   D-Lactose + H_2O $\xrightarrow{H^+ \text{ or lactose}}$ D-galactose + D-glucose

8. Hydrolysis of sucrose (Section 14.9)

   D-Sucrose + H_2O $\xrightarrow{H^+ \text{ or sucrase}}$ D-fructose + D-glucose

9. Complete hydrolysis of polysaccharide (Section 14.10)

   Polysaccharide + H_2O $\xrightarrow{H^+ \text{ or enzymes}}$ many monosaccharides

10. Complete hydrolysis of starch (Section 14.10)

    Starch + H_2O $\xrightarrow{H^+ \text{ or enzymes}}$ many D-glucose

11. Complete hydrolysis of glycogen (Section 14.10)

    Glycogen + H_2O $\xrightarrow{H^+ \text{ or enzymes}}$ many D-glucose

KEY TERMS

Achiral molecule (14.5)
Aldose (14.4)
Biochemical substance (14.1)
Biochemistry (14.1)
Carbohydrate (14.3)
Chiral center (14.5)
Chiral molecule (14.5)
Disaccharide (14.3)
Glycoside (14.8)
Glycosidic linkage (14.9)
Ketose (14.4)
Mirror image (14.5)
Monosaccharide (14.3)
Oligosaccharide (14.3)
Polysaccharide (14.3)
Reducing sugar (14.8)
Sugar (14.4)

ADDITIONAL PROBLEMS

14.54 Classify each of the following substances as a *monosaccharide,* a *disaccharide,* or a *polysaccharide.*
a. Sucrose b. Starch
c. Fructose d. Lactose

14.55 Indicate the number of D-glucose subunits present in each of the following *disaccharides* or *polysaccharides.* Use "many" for the answer in the case of glucose-containing polysaccharides.
a. Lactose b. Maltose
c. Glycogen d. Amylose

14.56 How many glycosidic linkages are present in each of the following carbohydrates?
a. Fructose b. Sucrose
c. Lactose d. Ribose

14.57 In which of the following pairs of monosaccharides or disaccharides do both members of the pair contain the same number of carbon atoms?
a. Sucrose and lactose
b. Glucose and fructose
c. Maltose and lactose
d. Glucose and ribose

14.58 What monosaccharide(s) is(are) obtained from the hydrolysis of each of the following carbohydrates?
a. Glycogen b. Sucrose
c. Amylopectin d. Maltose

14.59 Which of the following carbohydrates is classified as a reducing sugar?
a. Glucose b. Sucrose
c. Starch d. Ribose

14.60 One of the products of the partial hydrolysis of cellulose is *cellobiose,* a disaccharide.
a. What monosaccharide(s) is(are) present in cellobiose?
b. What type of glycosidic linkage is present in cellobiose?

14.61 If α-D-galactose is dissolved in water, β-D-galactose is eventually present. Explain what is happening?

PRACTICE TEST ▸ True/False

14.62 Complete hydrolysis of an oligosaccharide produces monosaccharides.

14.63 The designation *sugar* applies to aldoses but not to ketoses.

14.64 Monosaccharide molecules that exhibit handedness have mirror images that are superimposable upon each other.

14.65 Naturally occurring carbohydrate molecules are nearly always "right-handed" molecules.

14.66 All monosaccharides have names that end in the suffix *-ose.*

14.67 Glucose and galactose are aldohexoses, and fructose and ribose are ketohexoses.

14.68 The structure of D-galactose differs from that of D-glucose only in the configuration of the —OH group and —H group on carbon 4.

14.69 The cyclic forms of monosaccharides result from the ability of their carbonyl group to react intramolecularly with a hydroxy group at the other "end" of the molecule.

14.70 The cyclic forms of D-glucose involve a six-membered ring, and those for D-fructose involve a five-membered ring.

14.71 All monosaccharides are reducing sugars.

14.72 The general name for monosaccharide cyclic hemiacetals and cyclic hemiketals is *glycoside.*

14.73 The disaccharides maltose, galactose, and glucose all contain at least one D-glucose monosaccharide unit.

14.74 The two monosaccharide building blocks of a disaccharide cannot be the same monosaccharide.

14.75 Both cellulose and glycogen are unbranched glucose polymers.

14.76 Complete hydrolysis of amylose and complete hydrolysis of amylopectin (the two forms of starch) yield the same product.

PRACTICE TEST ▸ Multiple Choice

14.77 Which of the following compounds is both an aldose and a hexose?

a.
CHO
H—OH
H—OH
CH_2OH

b.
CHO
H—OH
H—OH
H—OH
CH_2OH

c.
CHO
H—OH
HO—H
HO—H
H—OH
CH_2OH

d.
CH_2OH
C=O
HO—H
H—OH
H—OH
CH_2OH

14.78 Which of the following structures is that of an L-monosaccharide?

a.
CHO
HO—H
H—OH
H—OH
CH_2OH

b.
CHO
H—OH
H—OH
H—OH
CH_2OH

c.
CHO
H—OH
H—OH
H—OH
HO—H
CH_2OH

d.
CHO
HO—H
H—OH
H—OH
H—OH
CH_2OH

14.79 Which of the following structures is that of D-glucose?

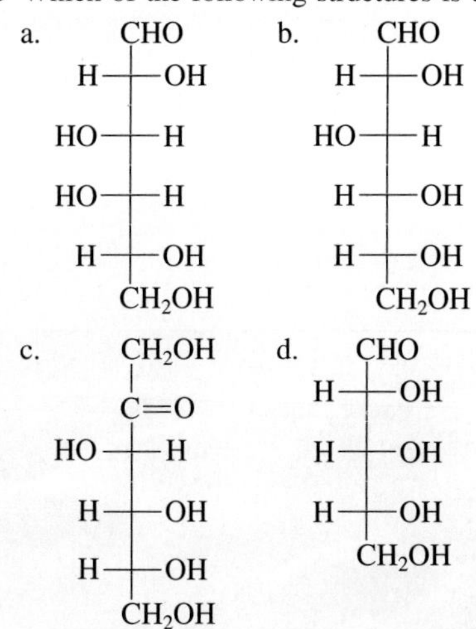

14.80 In which of the following pairs of molecules do the members of the pair contain different numbers of carbon atoms?
a. Fructose and ribose
b. Ribose and deoxyribose
c. Glucose and fructose
d. Glyceraldehyde and dihydoxyacetone

14.81 Which of the following structures represent a beta-monosaccharide?

a.
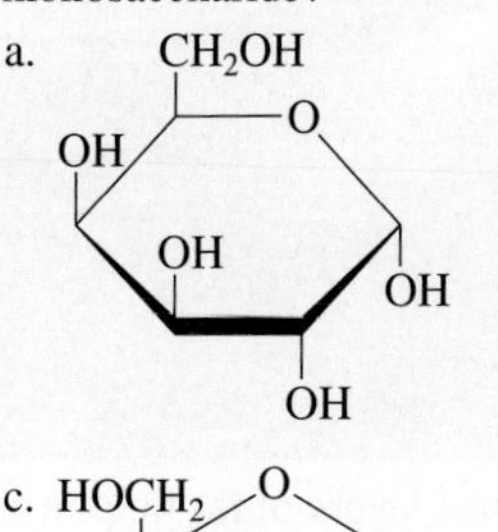

b.
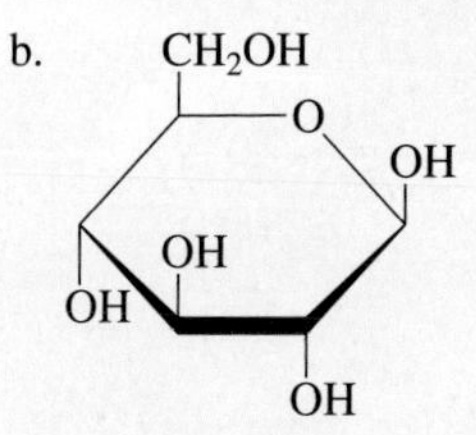

c.
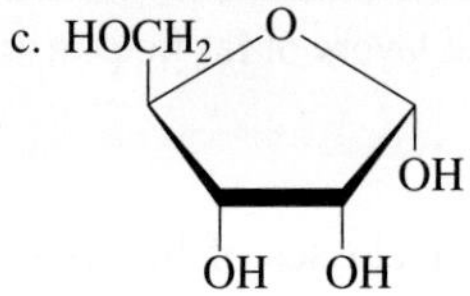

d.
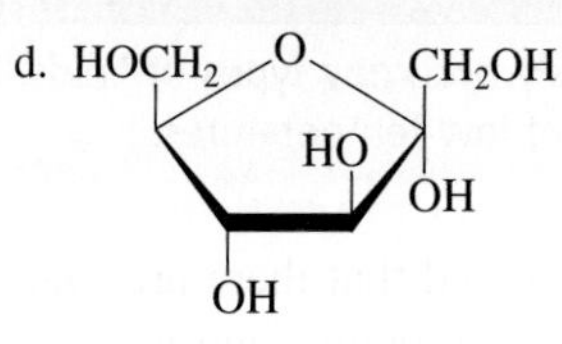

14.82 In which of the following pairs of carbohydrates are both members of the pair disaccharides?
a. Glucose and fructose
b. Lactose and amylose
c. Sucrose and maltose
d. Cellulose and starch

14.83 In which of the following carbohydrates is the bonding between monosaccharide units "head-to-head" rather than "head-to-tail?"
a. Amylose b. Maltose
c. Lactose d. Sucrose

14.84 The major structural difference between cellulose and starch is found in the
a. identity of the monosaccharide units present
b. ring size of the monosaccharide units present
c. linkages between the monosaccharide units present
d. handedness of the monosaccharide units present

14.85 In which of the following pairs of polysaccharides are both members of the pair branched polysaccharides?
a. Amylopectin and glycogen
b. Cellulose and glycogen
c. Cellulose and amylose
d. Amylose and amylopectin

14.86 Which of the following sugars will *not* give a positive Benedict's test?
a. Galactose b. Lactose
c. Sucrose d. Fructose

CHAPTER OUTLINE

CHAPTER FIFTEEN

Lipids

Fats and oils are the most widely occurring types of lipids. Thick layers of fat help insulate polar bears against the effects of low temperatures.

In the previous chapter, we noted that there are four major classes of bioorganic substances: carbohydrates, lipids, proteins, and nucleic acids. We now turn our attention to the second of the bioorganic classes, the compounds we call lipids. Lipids are a structurally heterogeneous class of bioorganic compounds and include such diverse types of molecules as fats and oils, cholesterol, bile salts, and sex hormones. What these substances have in common is limited solubility in water.

Learning Focus

Know how lipids are distinguished from other biochemical substances and explain the structural classification scheme for lipids.

15.1 Characteristics of Lipids

Lipids form a large class of relatively water-insoluble bioorganic compounds. In humans and many animals, excess carbohydrates and other energy-yielding foods are converted to, and stored in the body in the form of, lipids called fats. These fat reservoirs constitute a major way of storing chemical energy and carbon atoms in the body. Fats and other lipids also surround and insulate vital body organs, providing protection from mechanical shock and helping to maintain correct body temperature. Lipids function as coverings for nerve fibers and as the basic structural components of all cell membranes. Many chemical messengers in the human body, substances called hormones, are lipids.

Lipids, unlike carbohydrates and most other classes of compounds, cannot be defined from a structural viewpoint. A variety of functional groups and structural features are found in molecules classified as lipids. What lipids share are their solubility properties.

▶ The term *lipid* comes from the Greek *lipos,* which means "fat" or "lard."

A **lipid** *is an organic molecule found in living systems that is insoluble (or only sparingly soluble) in water but soluble in nonpolar organic solvents.* When biological material (animal or plant tissue) is homogenized in a blender and mixed with a nonpolar organic solvent, the substances that dissolve in the solvent are the lipids.

The lipids we consider in this chapter can be divided into three major categories on the basis of polarity and type of structural subunits present. The three types of lipids are

1. Nonpolar fatty-acid-containing lipids
2. Polar fatty-acid-containing lipids
3. Non-fatty-acid-containing lipids

As is obvious from this list, fatty acids are an important consideration when discussing lipids. We will begin our discussion of lipids with an in-depth look at fatty acids.

Practice Problems and Questions

15.1 What characteristic do all lipids have in common?

15.2 What structural feature, if any, do all lipids have in common? Explain your answer.

15.3 List the three major types of lipids to be discussed in this chapter.

15.2 Structural Characteristics of Fatty Acids

Learning Focus

Recognize the general structural features of fatty acids.

▶ Fatty acids were first isolated from naturally occurring fats; hence the designation *fatty acids.*

A **fatty acid** *is a naturally occurring carboxylic acid with an unbranched carbon chain and an even number of carbon atoms.* Fatty acids are rarely found free in nature but rather occur mostly in esterified form in the structures of many lipids. Because of the pathway by which fatty acids are biosynthesized they almost always contain an even number of carbon atoms. *Long-chain* fatty acids (12 to 26 carbon atoms) are found in meats and fish; *medium-chain* fatty acids (6 to 10 carbon atoms) and *short-chain* fatty acids (fewer than 6 carbon atoms) occur primarily in dairy products.

Saturated and Unsaturated Fatty Acids

Fatty acids are further classified as saturated, monounsaturated, or polyunsaturated. A **saturated fatty acid** *has a carbon chain in which all carbon–carbon bonds are single bonds.* An example is hexadecanoic acid, a 16-carbon acid whose common name is palmitic acid.

```
    H  H  H  H  H  H  H  H  H  H  H  H  H  H  H  O
    |  |  |  |  |  |  |  |  |  |  |  |  |  |  |  ||
H—C—C—C—C—C—C—C—C—C—C—C—C—C—C—C—C—OH
    |  |  |  |  |  |  |  |  |  |  |  |  |  |  |
    H  H  H  H  H  H  H  H  H  H  H  H  H  H  H
```

Hexadecanoic acid (palmitic acid)

The structural formulas of fatty acids are usually written in a more condensed form than the preceding structural formula. An alternative notation for palmitic acid's structure is

$$CH_3—(CH_2)_{14}—COOH$$

A **monounsaturated fatty acid** *has a carbon chain in which one carbon–carbon double bond is present.* In naturally occurring and biochemically important monounsaturated fatty acids, the configuration about the double bond is almost always *cis* (Section 11.4). Different ways of depicting the structure of a monounsaturated fatty acid are as follows; the molecule shown is the 18-carbon acid with one double bond (*cis*-9-octadecenoic acid, or oleic acid).

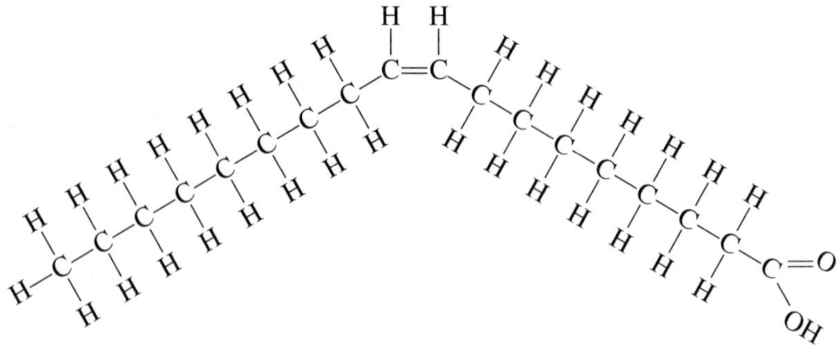

$$CH_3-(CH_2)_7-\overset{H}{\overset{|}{C}}=\overset{H}{\overset{|}{C}}-(CH_2)_7-COOH$$

The first of these structures correctly emphasizes that the presence of a *cis* double bond in the carbon chain puts a rigid 30° bend in the chain. Such a bend affects the physical properties of a fatty acid, as we will see in Section 15.3. In the second structure, the *cis* nature of the double bond is indicated by having both hydrogens associated with the double bond point in the same direction.

A **polyunsaturated fatty acid** *has a carbon chain in which two or more carbon–carbon double bonds are present.* Up to six double bonds are found in biochemically important unsaturated fatty acids.

Table 15.1 gives structures and names for the naturally occurring fatty acids most often found in lipid structures. Fatty acids are almost always referred to by their common names. IUPAC names, although easily understandable, are usually quite long. Selected IUPAC names are given as a footnote to Table 15.1. A convenient shorthand notation for fatty acids, introduced in Table 15.1, involves two numbers separated by a

▶ More than 500 different fatty acids have been isolated from the lipids of microorganisms, plants, animals, and humans. These fatty acids differ from one another in the length of their carbon chains, their degree of unsaturation (number of double bonds), and the positions of the double bonds in the chains.

Table 15.1 Selected Fatty Acids of Biological Importance

Structure notation	Common name	Structure
Saturated Fatty Acids		
12:0	lauric acid	$CH_3-(CH_2)_{10}-COOH$
14:0	myristic acid	$CH_3-(CH_2)_{12}-COOH$
16:0	palmitic acid	$CH_3-(CH_2)_{14}-COOH$
18:0	stearic acid	$CH_3-(CH_2)_{16}-COOH$
Monounsaturated Fatty Acids		
16:1	palmitoleic acid	$CH_3-(CH_2)_5-\overset{H}{\overset{\mid}{C}}=\overset{H}{\overset{\mid}{C}}-(CH_2)_7-COOH$
18:1	oleic acid	$CH_3-(CH_2)_7-\overset{H}{\overset{\mid}{C}}=\overset{H}{\overset{\mid}{C}}-(CH_2)_7-COOH$
Polyunsaturated Fatty Acids		
18:2	linoleic acid	$CH_3-(CH_2)_4-\overset{H}{\overset{\mid}{C}}=\overset{H}{\overset{\mid}{C}}-CH_2-\overset{H}{\overset{\mid}{C}}=\overset{H}{\overset{\mid}{C}}-(CH_2)_7-COOH$
18:3	linolenic acid	$CH_3-CH_2-\overset{H}{\overset{\mid}{C}}=\overset{H}{\overset{\mid}{C}}-CH_2-\overset{H}{\overset{\mid}{C}}=\overset{H}{\overset{\mid}{C}}-CH_2-\overset{H}{\overset{\mid}{C}}=\overset{H}{\overset{\mid}{C}}-(CH_2)_7-COOH$

Common names are used instead of IUPAC names in most situations because of the length of the IUPAC names. Selected examples of IUPAC names follow.
16:1 acid *cis*-9-hexadecenoic acid
18:2 acid *cis-cis*-9,12-octadecadienoic acid

colon. The first number indicates how many carbon atoms are present in the fatty acid, and the second number specifies the number of carbon–carbon double bonds present.

Omega-3 and Omega-6 Fatty Acids

Not only do the fatty acids come in three categories (saturated, monounsaturated, and polyunsaturated), but the polyunsaturates also belong to families, two of which are particularly important in human body chemistry. These important families are the omega-6 and the omega-3 fatty acids.

The basis for the omega classification system involves the following considerations. A fatty acid has two ends, designated as the methyl (CH_3) end and the carboxyl (COOH) end.

$$\underbrace{CH_3{-}(CH_2)_4}_{\text{Methyl end}}{-}(CH{=}CH{-}CH_2)_4{-}\underbrace{(CH_2)_2{-}COOH}_{\text{Carboxyl end}}$$

In the omega classification system, the carbon chain is numbered beginning at the methyl end, which is the reverse of the usual way carbon chains are numbered in the naming of simple acids. The "reverse" numbering system is used because of the mechanism by which fatty acid carbon chains are lengthened during biotransformations within the body. Lengthening involves adding carbon atoms, two at a time, at the carboxyl end of the chain. Thus, by numbering polyunsaturated fatty acid chains from the methyl end, which does not change, chemists ease the task of keeping track of fatty acid identities. When an omega-3 acid is lengthened, the new acid is still an omega-3 acid.

An **omega-3 fatty acid** *is a polyunsaturated fatty acid with its endmost double bond three carbons away from its methyl end.* An **omega-6 fatty acid** *is a polyunsaturated fatty acid with its endmost double bond six carbons away from its methyl end.*

Essential Fatty Acids

An **essential fatty acid** *is a fatty acid that is needed by the human body and that must be obtained from dietary sources because it cannot be synthesized within the body from other substances.* There are two essential fatty acids: *linoleic acid and linolenic acid.* Both are 18-carbon polyunsaturated fatty acids. Chemical Portraits 25 profiles these two essential fatty acids as well as arachidonic acid, an acid the body synthesizes from linoleic acid.

Example 15.1 Classifying Fatty Acids in Various Ways

A fatty acid has the structure

$$CH_3{-}(CH_2)_4{-}\overset{\displaystyle H}{\overset{|}{C}}{=}\overset{\displaystyle H}{\overset{|}{C}}{-}(CH_2)_{10}{-}COOH$$

Classify this fatty acid as

a. A saturated, a monounsaturated, or a polyunsaturated fatty acid
b. An omega-3, an omega-6, or neither an omega-3 nor an omega-6 fatty acid
c. An 18:0, an 18:1, a 20:0, or a 20:1 fatty acid
d. An essential or a nonessential fatty acid

Solution

a. monounsaturated; one carbon–carbon double bond is present
b. omega-6; counting carbon atoms from the methyl end, the double bond begins with carbon 6
c. 18:1; there are 18 carbons present and one carbon–carbon double bond present
d. nonessential; there are only two essential fatty acids, and both are polyunsaturated

Chemical Portraits 25 Essential Fatty Acids

Linolenic Acid (an 18:3 fatty acid)

Profile: Linolenic acid is the primary member of the omega-3 family of fatty acids. From dietary linolenic acid, the body can make the 20- and 22-carbon members of the omega-3 fatty acid series, fatty acids essential for normal growth and development. Omega-3 fatty acids are important for the structure and function of cell membranes, particularly in the retina of the eye and the central nervous system.

Biochemical considerations: Optimal amounts of linolenic acid are often lacking in the typical American diet. Because fish are a good omega-3 fatty acid source, nutritionists recommend adding more fish—several servings each week—to the diet. Cold-water fish tend to have higher omega-3 fatty acid concentrations than do warm-water fish.

What common species of fish are the best sources of omega-3 fatty acids?

Linoleic Acid (an 18:2 fatty acid)

Profile: Linoleic acid is the primary member of the omega-6 family of fatty acids. Given dietary linoleic acid, the body can make necessary longer carbon-chain members of the omega-6 fatty acid series such as arachidonic acid (20:4). Normally, vegetable oils and meats supply enough linoleic acid to meet the body's needs for it.

Biochemical considerations: Omega-6 fatty acids are important for growth, skin integrity, fertility, and maintaining red blood cell structure. Lack of linoleic acid causes the skin to redden and become irritated. Infants have especial need of linoleic acid for their growth. Human breast milk has a much higher percentage of it than cow's milk.

How do formula-fed infants obtain adequate amounts of linoleic acid?

Arachidonic Acid (a 20:4 fatty acid)

Profile: Arachidonic acid is an omega-6 fatty acid biosynthesized from linoleic acid. In the body, it serves as the percursor for a family of molecules called eicosanoids, which are oxygenated derivatives of this acid. Eicosanoids are present in all cells except red blood cells.

Biochemical considerations: Eicosanoids regulate a wide range of body functions including blood pressure, blood clotting, blood lipid levels, the sleep/wake cycle, and the inflammation response to injury and infection. Eicosanoids are hormone-like molecules rather than hormones; they are not transported in the bloodstream to their site of action but rather exert their effects in the tissues where they are synthesized.

In what way is the action of aspirin related to arachidonic acid?

See the text web site at **www.cengage.com/chemistry/stoker** for answers to the above questions and for further information.

Practice Problems and Questions

15.4 Classify each of the following fatty acids as long-chain, medium-chain, or short-chain.
a. Myristic (14:0) b. Caproic (6:0) c. Arachidic (20:0) d. Capric (10:0)

15.5 Classify each the following fatty acids as saturated, monounsaturated, or polyunsaturated.
a. Stearic (18:0) b. Linolenic (18:3) c. Oleic (18:1) d. Lauric (12:0)

15.6 With the help of Table 15.1, classify each of the following acids as an omega-3 acid, an omega-6 acid, or neither an omega-3 nor an omega-6 acid.
a. Stearic (18:0) b. Oleic (18:1) c. Linoleic (18:2) d. Linolenic (18:3)

15.7 What effect does a *cis* double bond have on the shape of a fatty acid molecule?

15.3 Physical Properties of Fatty Acids

Learning Focus
Understand the relationship between structure and melting point for fatty acids.

The melting points of fatty acids depend on both the length of their hydrocarbon chains and their degree of unsaturation (number of double bonds per molecule). The graph in Figure 15.1 shows melting-point variation as a function of both of these variables. A trend of particular significance is that saturated acids have higher melting points than unsaturated acids with the same number of carbon atoms. The greater the degree of unsaturation, the

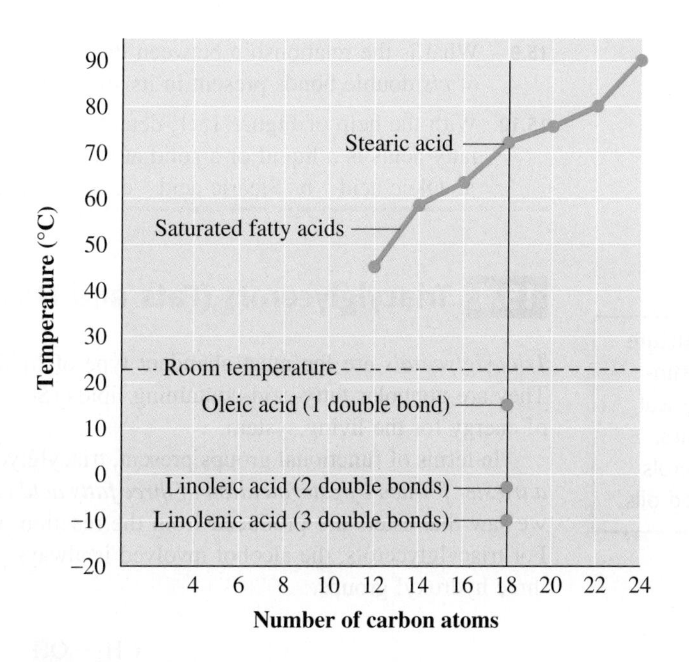

Figure 15.1 The melting point of a fatty acid depends on the length of the carbon chain and on the number of double bonds present in the carbon chain.

greater the reduction in melting point. Figure 15.1 shows this effect for the 18-carbon acids with zero, one, two, and three double bonds.

The decreasing melting point associated with increasing degree of unsaturation in fatty acids is explained by decreased molecular attractions between carbon chains. The double bonds in unsaturated fatty acids, which generally have the *cis* configuration, produce "bends" in the carbon chains of these molecules (see Figure 15.2). These "bends" prevent unsaturated fatty acids from packing together as tightly as fully saturated fatty acids. The greater the number of double bonds, the less efficient the packing. As a result, unsaturated fatty acids always have fewer intermolecular attractions and therefore lower melting points than their saturated counterparts.

▶ Fatty acids have low water solubilities, which decrease with increasing carbon chain length; at 30°C, lauric acid (12:0) has a water solubility of 0.063 g/L and stearic acid (18:0) a solubility of 0.0034 g/L. Contrast this with glucose's solubility in water at the same temperature, 1100 g/L.

Practice Problems and Questions

15.8 In each of the following pairs, select the fatty acid that has the lower melting point.

a. 18:0 acid and 18:1 acid
b. 18:2 acid and 18:3 acid
c. 14:0 acid and 16:0 acid
d. 18:1 acid and 20:0 acid

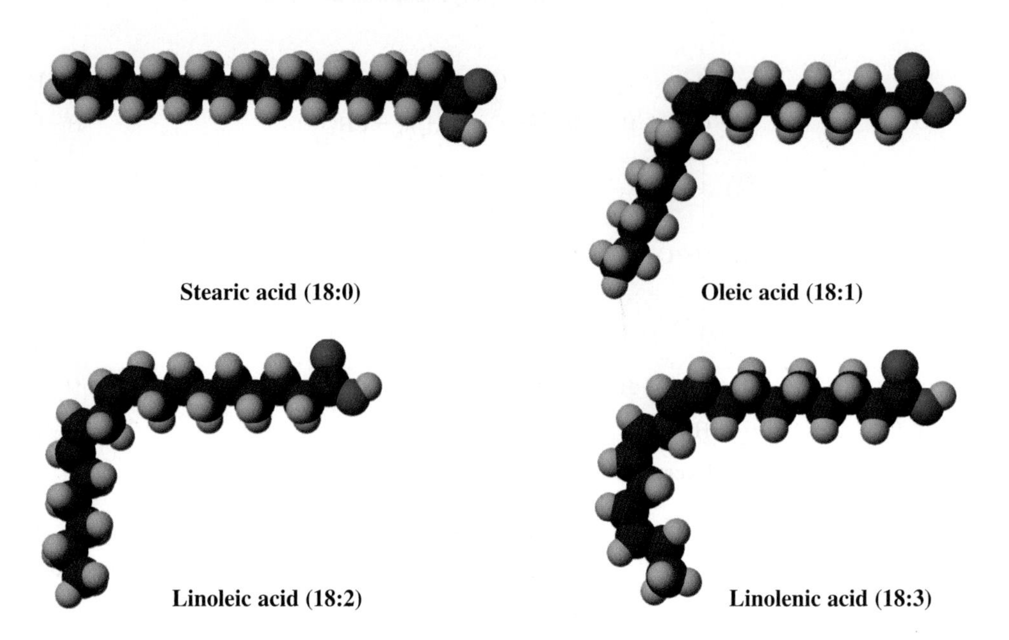

Figure 15.2 Space-filling models of four 18-carbon fatty acids, which differ in the number of double bonds present. Note how the presence of double bonds changes the shape of the molecule.

15.9 What is the relationship between the melting point of a fatty acid and the number of *cis* double bonds present in its structure?

15.10 With the help of Figure 15.1, determine whether each of the following 18-carbon fatty acids is a liquid or a solid at room temperature.
a. Oleic acid b. Stearic acid c. Linoleic acid d. Linolenic acid

Learning Focus

Recognize the general structure for triacylglycerols, and distinguish, on the basis of physical and structural characteristics, between those triacylglycerols called fats and those called oils.

15.4 Triacylglycerols (Fats and Oils)

Triacylglycerols are the most abundant type of lipid present in humans and in animals. They are nonpolar fatty-acid-containing lipids (Section 15.1) that represent stored forms of energy for the living system.

In terms of functional groups present, triacylglycerols are esters. A **triacylglycerol** *is a triester formed by esterification of three fatty acid molecules to glycerol.* In Section 13.8, we saw that esters are produced from the reaction of an alcohol with a carboxylic acid. For triacylglycerols, the alcohol involved is always glycerol, a three-carbon alcohol with three hydroxyl groups.

$CH_2—OH$
$CH—OH$
$CH_2—OH$
Glycerol

Fatty acids are the carboxylic acids present in a triacylglycerol. In triacylglycerol formation, each of the three hydroxyl groups of glycerol is esterified with a fatty acid. Figure 15.3 shows the triple esterification reaction that occurs between glycerol and three stearic acid (18:0) molecules.

The triacylglycerol produced from glycerol and three molecules of stearic acid (Figure 15.3) is an example of a simple triacylglycerol. A **simple triacylglycerol** *is a triester formed from the reaction of glycerol with three identical fatty acid molecules.* If the reacting fatty acid molecules are not all identical, then the result is a mixed triacylglycerol. A **mixed triacylglycerol** *is a triester formed from the reaction of glycerol with more than one kind of fatty acid molecule.* Figure 15.4 shows the structure of a mixed triacylglycerol in which one fatty acid is saturated, another monounsaturated, and the third polyunsaturated.

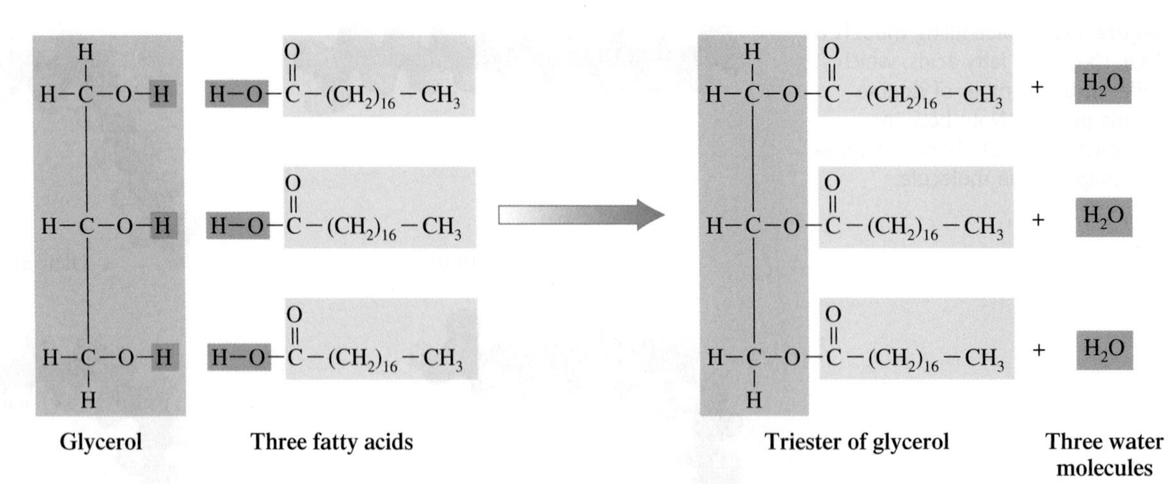

Figure 15.3 Triple esterification reaction between glycerol and three molecules of stearic acid (18:0 acid). Three molecules of water are a by-product of this reaction.

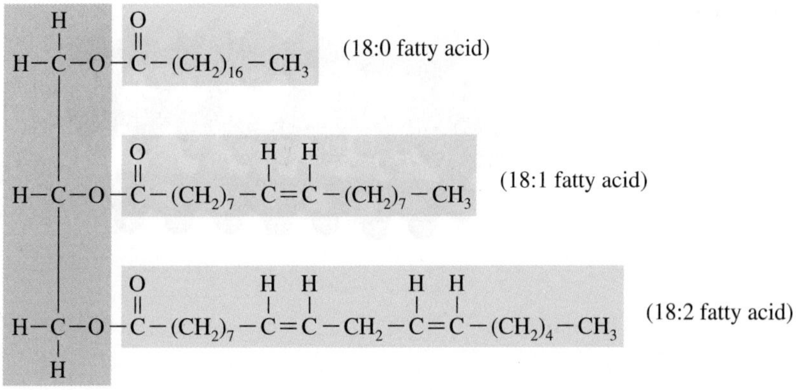

Figure 15.4 Structure of a mixed triacylglycerol in which three different fatty acid residues are present.

$H-C(H)-O-C(=O)-(CH_2)_{16}-CH_3$ (18:0 fatty acid)

$H-C-O-C(=O)-(CH_2)_7-CH=CH-(CH_2)_7-CH_3$ (18:1 fatty acid)

$H-C(H)-O-C(=O)-(CH_2)_7-CH=CH-CH_2-CH=CH-(CH_2)_4-CH_3$ (18:2 fatty acid)

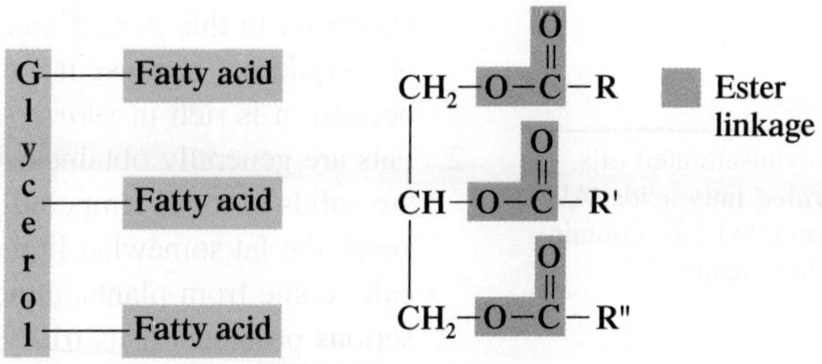

▶ Triacylglycerols do not actually contain glycerol and three fatty acids, as the block diagram for a triacylglycerol implies. They actually contain a glycerol *residue* and three fatty acid *residues*. In the formation of the triacylglycerol, three molecules of water have been removed from the structural components of the fat or oil, leaving residues of the reacting molecules.

▶ An *acyl group* is the group that remains after the —OH group is removed from a fatty acid. Thus, as the name implies, triacylglycerols contain three fatty acid residues (acyl groups) esterified to glycerol.

▶ *Petroleum oils* (Section 10.12) are structurally different from *lipid oils*. The former are mixtures of alkanes and cycloalkanes. The latter are triesters of glycerol.

Two general representations for the structure of triacylglycerols are

Glycerol—Fatty acid
Glycerol—Fatty acid
Glycerol—Fatty acid

$CH_2-O-C(=O)-R$
$CH-O-C(=O)-R'$
$CH_2-O-C(=O)-R''$

Ester linkage

The first representation, a block diagram, shows that four structural components are present in the structure: glycerol and three fatty acids. The second representation, a general structural formula, shows the triester nature of triacylglycerols. Each of the fatty acids is attached to glycerol through an ester linkage.

Note that all triacylglycerols contain glycerol but that the fatty acids present vary from molecule to molecule. *Triacylglycerol* is thus the name for a whole family of similarly structured molecules rather than the name for a unique molecule. In nature, simple triacylglycerols are rare. Most naturally occurring triacylglycerols are mixed triacylglycerols. A naturally occurring triacylglycerol sample would be a complex mixture of many mixed triacylglycerols.

Triacylglycerols are known by several other names. *Triglycerides* is an older term that still finds extensive use in the fields of nutrition and clinical chemistry. Blood chemistry reports often list *triglyceride* levels. A **fat** *is a triglyceride mixture that is a solid or semi-solid at room temperature (25°C).* Beef tallow and pork lard are fats. An **oil** *is a triglyceride mixture that is a liquid at room temperature (25°C).* We speak of soybean oil, canola oil, and peanut oil; all are liquid-state triacylglycerol mixtures.

Some generalizations about distinguishing between triacylglycerols on the basis of physical state (fats and oils) are:

1. The melting points of triacylglycerols are directly related to the identity of the fatty acids present within them. Fats are composed largely of triacylglycerols in which saturated fatty acids predominate, although some unsaturated fatty acids are present. Such triacylglycerols can pack closely together because of the "linearity" of their fatty acid chains (Figure 15.5a), causing the higher melting points associated with fats (Section 15.3). Oils contain triacylglycerols with larger amounts of mono- and polyunsaturated fatty acids than those in fats. Such triacylglycerols cannot pack as tightly together because of the "bends" in their fatty acid chains (Figure 15.5b). The result is lower melting points.

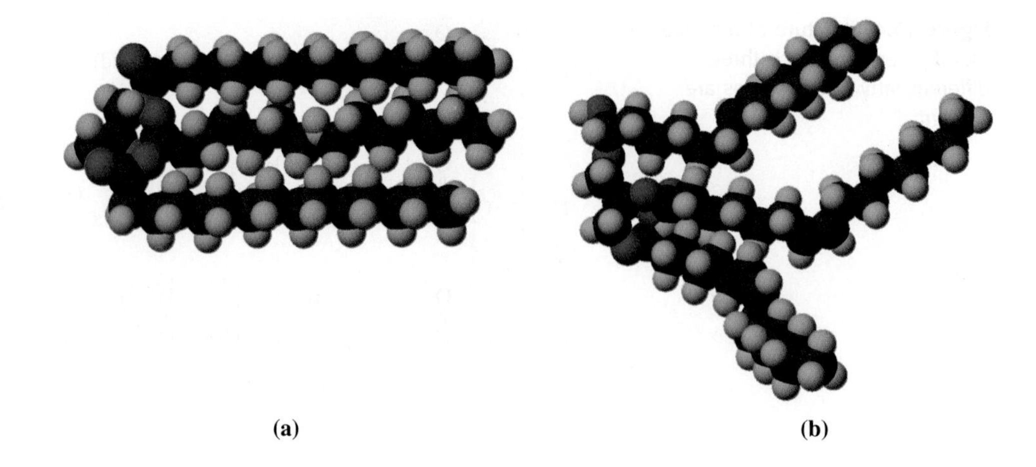

Figure 15.5 Representative triacylglycerols from (a) a fat and (b) an oil.

Figure 15.6 gives the percentages of saturated, monounsaturated, and polyunsaturated fatty acids found in common dietary oils and fats. In general, a higher degree of fatty acid unsaturation is associated with oils than with fats. A notable exception to this generalization is coconut oil, which is highly saturated. This oil is a liquid not because it contains many double bonds within the fatty acids but because it is rich in *shorter-chain* fatty acids, particularly lauric acid (12:0).

2. Fats are generally obtained from animals; hence the term *animal fat*. Although fats are solids at room temperature, the warmer body temperature of the living animal keeps the fat somewhat liquid (semi-solid) and thus allows for movement. Oils typically come from plants, although there are also fish oils. A fish would have some serious problems if its triacylglycerols "solidified" when it encountered cold waters.

▶ All oils, even polyunsaturated oils, contain some *saturated* fatty acids. All fats, even highly saturated fats, contain some *unsaturated* fatty acids.

When a person consumes too much food, much of the excess energy (calories) is stored as fat. This fat is concentrated in special cells (adipocytes) that are nearly filled with large fat droplets. Adipose tissue containing these cells is found in various parts of the body—under the skin, in the abdominal cavity, in the mammary glands, and around various organs (Figure 15.7).

In the past decade, considerable attention has been paid to the role of dietary factors as a cause of disease (obesity, diabetes, cancer, hypertension, and atherosclerosis). Several organizations have proposed new dietary guidelines on the basis of these studies. The guidelines strongly suggest that total fat intake should be reduced, with particular emphasis on reduction of saturated fatty acid intake.

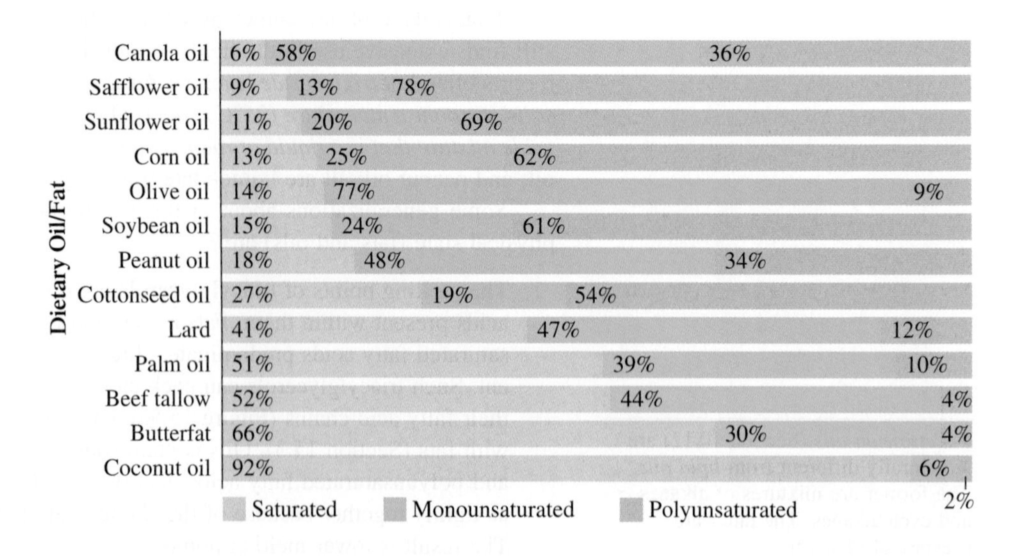

Figure 15.6 Percentages of saturated, monounsaturated, and polyunsaturated fatty acids in the triacylglycerols of various dietary fats and oils.

Figure 15.7 An electron micrograph of fat cells. Note their bulging spherical shape.

Practice Questions and Problems

15.11 What is the functional group present in a triacylglycerol and how many of this functional group are present?

15.12 What are the four structural subunits that make up the structure of a triacylglycerol?

15.13 Draw the generalized block diagram structure for a triacylglycerol.

15.14 What is the difference between a simple triacylglycerol and a mixed triacylglycerol?

15.15 Draw block diagram structures for the four different triacylglycerols that can be produced from glycerol, stearic acid, and linolenic acid.

15.16 Identify, with the help of Table 15.1, the fatty acids that are present in the following triacylglycerols.

a.

$$\begin{array}{l} CH_2-O-\overset{\overset{\displaystyle O}{\|}}{C}-(CH_2)_{14}-CH_3 \\ | \\ CH-O-\overset{\overset{\displaystyle O}{\|}}{C}-(CH_2)_{12}-CH_3 \\ | \\ CH_2-O-\overset{\overset{\displaystyle O}{\|}}{C}-(CH_2)_7-CH{=}CH-(CH_2)_7-CH_3 \end{array}$$

b.

$$\begin{array}{l} CH_2-O-\overset{\overset{\displaystyle O}{\|}}{C}-(CH_2)_{16}-CH_3 \\ | \\ CH-O-\overset{\overset{\displaystyle O}{\|}}{C}-(CH_2)_7-CH{=}CH-(CH_2)_7-CH_3 \\ | \\ CH_2-O-\overset{\overset{\displaystyle O}{\|}}{C}-(CH_2)_{12}-CH_3 \end{array}$$

15.17 What is the difference in meaning, if any, between the members of each of the following pairs of terms?
a. Triacylglycerol and triglyceride
b. Triacylglycerol and fat
c. Triacylglycerol and mixed triacylglycerol
d. Fat and oil

15.18 Structurally, what features do fats and oils share?

15.19 Structurally, how do fats and oils differ?

15.20 With the help of Figure 15.6, indicate which member of each of the following pairs of triacylglycerols has the higher percent of unsaturated fatty acids.
a. Beef tallow and lard
b. Cottonseed oil and canola oil
c. Peanut oil and olive oil
d. Coconut oil and lard

Learning Focus

Understand four types of reactions that are characteristic of triacylglycerols: hydrolysis, saponification, hydrogenation, and oxidation.

15.5 Chemical Reactions of Triacylglycerols

Hydrolysis

Hydrolysis of a triacylglycerol is the reverse of the esterification reaction by which it was formed. *Complete* hydrolysis of a triacylglycerol molecule always gives one glycerol molecule and three fatty acid molecules as products.

Triacylglycerol hydrolysis within the human body requires the help of enzymes (protein catalysts; Section 16.13) produced by the pancreas. These enzymes cause the triacylglycerol to be hydrolyzed in a *stepwise* fashion. First one of the outer fatty acids is

▶ Naturally occurring mono- and diacylglycerols are seldom encountered. *Synthetic* mono- and diacylglycerols are used as emulsifiers in many food products. Emulsifiers prevent particles suspended in solutions from coalescing and settling. Emulsifiers are usually present in so-called fat-free cakes and other fat-free products.

removed, then the other outer one, leaving a monoacylglycerol. In most cases this is the end product of the initial hydrolysis (digestion) of the triacylglycerol. Sometimes, enzymes remove all three fatty acids, leaving a free molecule of glycerol.

Saponification

Saponification is a hydrolysis reaction carried out in an alkaline (basic) solution. For fats and oils, the products of saponification are glycerol and fatty acid *salts*.

The overall reaction of triacylglycerol saponification can be thought of as occurring in two steps. The first step is the hydrolysis of the ester linkages to produce glycerol and three fatty acid molecules:

$$\text{Fat or oil} + 3H_2O \longrightarrow 3 \text{ fatty acids} + \text{glycerol}$$

▶ Recall, from Section 13.7 the structural difference between a carboxylic acid and a carboxylic acid salt.

$$\underset{\text{Carboxylic acid}}{R-\overset{\overset{\displaystyle O}{\|}}{C}-OH} \qquad \underset{\text{Carboxylic acid salt}}{R-\overset{\overset{\displaystyle O}{\|}}{C}-O^- \ Na^+}$$

The second step involves a reaction between the acid molecules and the base (usually NaOH) in the alkaline solution. This is an acid–base reaction that produces water plus salts:

$$3 \text{ Fatty acids} + 3NaOH \longrightarrow 3 \text{ fatty acid salts} + 3H_2O$$

Saponification of animal fat is the process by which soap was made in pioneer times. Soap making involved heating lard (fat) with lye (ashes of wood, an impure form of KOH). Today most soap is prepared by hydrolyzing fats and oils (animal fat and coconut oil) under high pressure and high temperature. Sodium carbonate is used as the base.

Hydrogenation

Hydrogenation is a reaction we encountered in Section 11.5. It involves hydrogen addition across carbon–carbon multiple bonds, which increases the degree of saturation as some double bonds are converted to single bonds. With this change, there is a corresponding increase in the melting point of the substance.

The carbon–carbon double bonds in vegetable oils are partially hydrogenated (some, but not all, of the double bonds are converted to single bonds) to produce semi-solid rather than liquid products. Many food products are produced in this way. The peanut oil in many popular brands of peanut butter has been partially hydrogenated to convert the oil into a solid that does not separate out of the mixture. Hydrogenation is used to produce solid cooking shortenings or margarines from liquid vegetable oils.

Soft-spread margarines are partially hydrogenated oils. The extent of hydrogenation is carefully controlled to make the margarine soft at refrigerated temperatures (4°C).

Oxidation

Carbon–carbon double bonds in fatty acid carbon chains are subject to oxidation with molecular oxygen (from air) as the oxidizing agent. Such oxidation breaks the carbon–carbon double bonds, producing both aldehyde and carboxylic acid products.

$$\underset{\text{Unsaturated fatty acids}}{-CH{=}CH-} \xrightarrow{\text{Oxidation}} \underset{\text{Short-chain aldehydes}}{-\overset{\overset{\displaystyle O}{\|}}{C}-H + H-\overset{\overset{\displaystyle O}{\|}}{C}-} \xrightarrow{\text{Oxidation}} \underset{\text{Short-chain carboxylic acids}}{-\overset{\overset{\displaystyle O}{\|}}{C}-OH + HO-\overset{\overset{\displaystyle O}{\|}}{C}-}$$

The short-chain aldehydes or carboxylic acids so produced often have objectionable odors, and fats and oils containing them are said to have become *rancid*. To avoid this unwanted oxidation process, commercially prepared foods containing fats and oils nearly always contain *antioxidants*—substances that are more easily oxidized than the food.

Perspiration generated by strenuous exercise contains numerous oils. Rapid oxidation of these oils, promoted by microorganisms on the skin, generates the body odor that may also occur (see Figure 15.8).

Figure 15.8 The oils present in skin perspiration rapidly undergo oxidation. The oxidation products, aldehydes and carboxylic acid, often have strong odors.

Example 15.2 Recognizing Various Characteristics of Triacylglycerols

A triacylglycerol has the structure

$$\begin{array}{l} \overset{\displaystyle O}{\|} \\ CH_2-O-C-(CH_2)_{14}-COOH \\ |\overset{\displaystyle O}{\|} \\ CH-O-C-(CH_2)_{14}-COOH \\ |\overset{\displaystyle O}{\|} \\ CH_2-O-C-(CH_2)_{14}-COOH \end{array}$$

a. Is this a simple or a mixed triacylglycerol?
b. Will this triacylglycerol undergo hydrogenation, and if so, what are the products?
c. Will this triacylglycerol undergo hydrolysis, and if so, what are the products?
d. Will this triacylglycerol readily undergo oxidation with molecular oxygen, and if so, what are the products?

Solution

a. This is a **simple** triacylglycerol because all three fatty acid subunits are identical.
b. This triacylglycerol **will not** undergo hydrogenation; there are no carbon–carbon double bonds present in the fatty acid subunits.
c. This triacylglycerol **will** undergo hydrolysis, as do all triacylglycerols; the products are glycerol and three fatty acid molecules.
d. This triacylglycerol **will not** readily undergo oxidation with molecular oxygen; there are no carbon–carbon double bonds present in the fatty acid subunits for the oxygen to "attack."

Practice Questions and Problems

15.21 Name, in general terms, the products of each of the following reactions.
a. The complete hydrolysis of a triacylglycerol
b. The complete hydrogenation of a triacylglycerol

15.22 An oil that has been stored for a long period of time is found to be rancid. Explain the chemical changes that have occurred.

15.23 What is the relationship between rancidity and antioxidants?

15.24 What chemical changes occur when an oil is partially hydrogenated?

15.25 Draw the condensed structural formulas for all products that would be obtained from the complete hydrolysis of the following triacylglycerol.

$$\begin{array}{l} \overset{\displaystyle O}{\|} \\ CH_2-O-C-(CH_2)_{14}-CH_3 \\ |\overset{\displaystyle O}{\|} \\ CH-O-C-(CH_2)_{12}-CH_3 \\ |\overset{\displaystyle O}{\|} \\ CH_2-O-C-(CH_2)_7-CH=CH-(CH_2)_7-CH_3 \end{array}$$

Learning Focus

List the general function of and the general structure for phosphoacylglycerols.

15.6 Phosphoacylglycerols

Triacylglycerols (fats and oils) represent the most abundant type of *nonpolar* fatty-acid-containing lipids (Sections 15.4 and 15.5). We now turn our attention to *polar* fatty-acid-containing lipids, the second of the three categories of lipids mentioned in Section 15.1. Two important families of lipids that fall within this category are the phosphoacylglycerols (discussed in this section) and the sphingolipids (to be discussed in the next section). In general, *polar* fatty-acid-containing lipids (phosphoacylglycerols and sphingolipids) function as components of cell membranes. This contrasts with *nonpolar* fatty-acid-containing lipids (triacylglycerols), which serve as energy storage molecules.

A **phosphoacylglycerol** *is a triester of glycerol in which two —OH groups are esterified with fatty acids and the third is esterified with phosphoric acid, which in turn is esterified to an alcohol.* The block diagram for a phosphoacylglycerol has the following general structure.

▶ Phosphoacylglycerols differ from fats and oils in that one of the fatty acids has been replaced with phosphoric acid (a triprotic acid), which is further esterified with an alcohol.

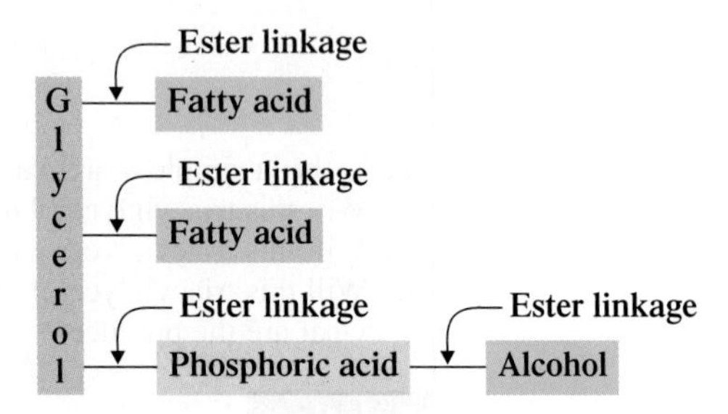

▶ An amino alcohol molecule contains both a hydroxyl group, —OH, and an amino group, $—NH_2$.

The most abundant phosphoacylglycerols have an amino alcohol (choline, ethanolamine, or serine) attached to the phosphate group. The structures of these three amino alcohols, given in terms of the charged forms that they adopt in neutral solution, are

$$HO—CH_2—CH_2—\overset{+}{N}(CH_3)_3 \qquad HO—CH_2—CH_2—\overset{+}{N}H_3 \qquad HO—CH_2—\underset{\displaystyle COO^-}{\underset{|}{CH}}—\overset{+}{N}H_3$$

Choline Ethanolamine Serine

Phosphoacylglycerols containing these three amino alcohols are respectively known as phosphatidylcholines, phosphatidylethanolamines, and phosphatidylserines. The fatty acid, glycerol, and phosphoric acid portions of a phosphoacylglycerol structure constitute a *phosphatidyl* group.

Phosphoacylglycerols are membrane lipids—that is, they serve as components of cell membranes. Further consideration of general phosphoacylglycerol structure reveals an additional structural characteristic associated with membrane lipids. To illustrate this characteristic, let us specifically consider a phosphatidylcholine containing stearic and oleic acids. The chemical structure of this molecule is shown in Figure 15.9a.

A molecular model for this compound, which gives the orientation of groups in space, is illustrated in Figure 15.9b. There are two important things to notice about this model: (1) There is a "head" part, the choline and phosphate, and (2) there are two "tails," the two fatty acid carbon chains. The head part is polar. The two tails, the carbon chains, are nonpolar.

All phosphoacylglycerols have structures similar to that shown in Figure 15.9. All have a "head" and two "tails." A simplified representation for this structure uses a circle to represent the polar head and two wavy lines to represent the nonpolar tails.

▶ Phosphoacylglycerols have a *hydrophobic* ("water-hating") portion, the nonpolar fatty acid groups, and a *hydrophilic* ("water-loving") portion, the polar phosphate-amino alcohol–ester group.

Polar head group Nonpolar chains

The polar head group of a phosphoacylglycerol is soluble in water. The nonpolar tail chains are insoluble in water but soluble in nonpolar substances.

Phosphatidylcholines

▶ The amino alcohol in phosphatidylcholines (pronounced fahs-fuh-TIDE-ul-KOH-leen) is choline.

Phosphatidylcholines are also known as lecithins. There are a number of different phosphatidylcholines because different fatty acids may be bonded to the glycerol portion of the phosphatidylcholine structure. In general, phosphatidylcholines are waxy solids that form colloidal suspensions in water. Egg yolks and soybeans are good dietary sources of these lipids. Within the body, phosphatidylcholines are prevalent in cell membranes.

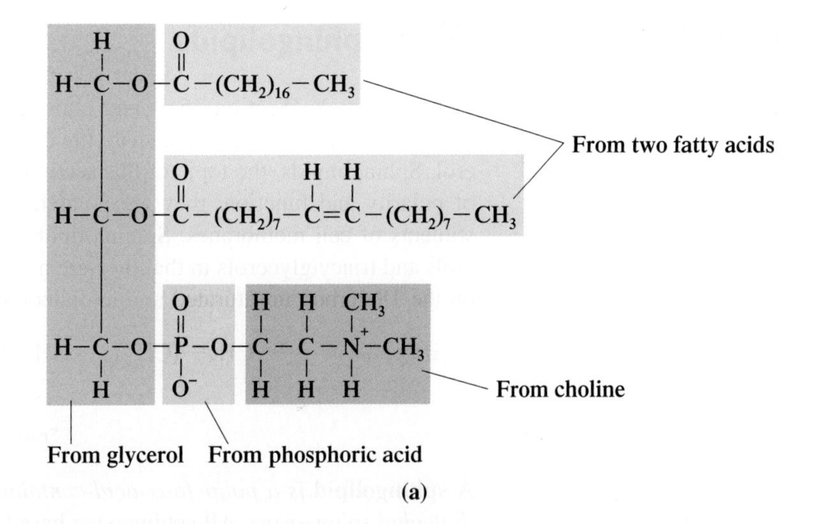

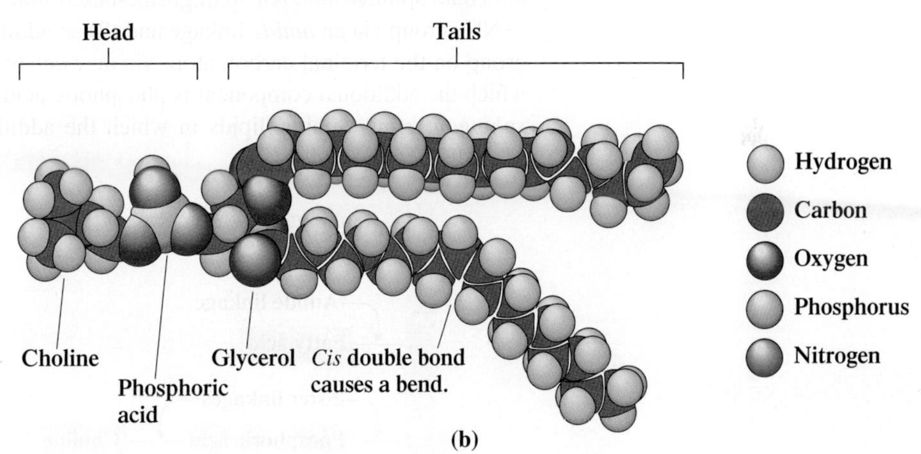

Figure 15.9 (a) Structural formula and (b) molecular model showing the "head and two tails" structure of a phosphatidylcholine molecule containing stearic acid (18:0) and oleic acid (18:1).

Periodically, claims arise that phosphatidylcholine should be taken as a nutritive supplement; some claims indicate it will improve memory. There is no evidence that these supplements are useful. The enzyme lecithinase in the intestine hydrolyzes most of the phosphatidylcholine taken orally before it passes into body fluids, so it does not reach body tissues. The phosphatidylcholine present in cell membranes is made by the liver; thus phosphatidylcholines are not essential nutrients.

Phosphatidylethanolamines and Phosphatidylserines

Phosphatidylethanolamines and phosphatidylserines are also known as cephalins. These compounds are found in heart and liver tissue and in high concentrations in the brain. They are important in blood clotting. Much is yet to be learned about how these compounds function within the human body.

Practice Problems and Questions

15.26 Draw the general block diagram structure for a phosphoacylglycerol.

15.27 How do phosphoacylglycerols differ structurally from triacylglycerols?

15.28 Draw the structures of the three amino alcohols commonly esterified to the phosphate group in phosphoacylglycerols.

15.29 How do a lecithin and a cephalin differ in structure?

15.30 How many ester linkages are present in a phosphoacylglycerol?

15.31 Phosphoacylglycerols have a "head and two tails" structure. What is the chemical identity of the head and of each of the two tails?

Learning Focus

List the general function of and the general structure for sphingolipids.

15.7 Sphingolipids

Triacylglycerols (Section 15.4) and phosphoacylglycerols (Section 15.6), although they differ in polarity and function, share the common characteristic of being triesters of glycerol. Sphingolipids, the topic of this section, are similar to phosphoacylglycerols in terms of polarity and function; they are polar fatty-acid-containing lipids that are major constituents of cell membranes. Sphingolipids differ, however, from both phosphoacylglycerols and triacylglycerols in that they are not glycerol-based lipids; instead, they are based on the 18-carbon unsaturated amino dialcohol *sphingosine.*

$$CH_3-(CH_2)_{12}-CH=CH-\underset{OH}{\underset{|}{CH}}-\underset{NH_2}{\underset{|}{CH}}-\underset{OH}{\underset{|}{CH_2}}$$

Sphingosine

A **sphingolipid** *is a polar fatty-acid-containing lipid derived from the 18-carbon amino dialcohol sphingosine.* All sphingosine-based lipids have (1) a fatty acid connected to the $-NH_2$ group via an *amide* linkage and (2) an additional component attached to the $-OH$ group on the terminal carbon atom via an *ester* or a *glycosidic* linkage. Sphingolipids in which the additional component is phosphoric acid to which choline is attached are called *sphingomyelins*. Sphingolipids in which the additional component is a monosaccharide are called *cerebrosides.*

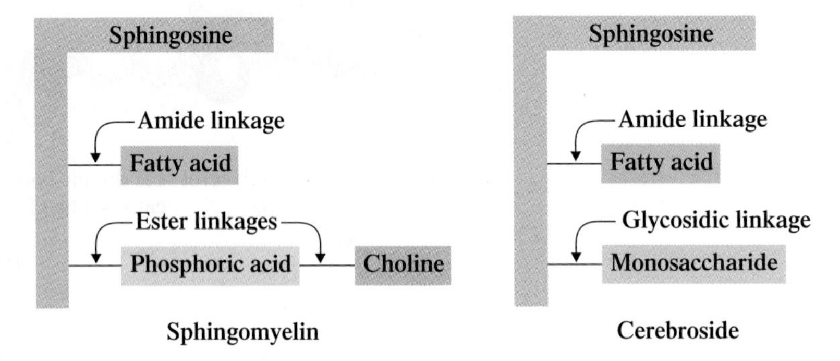

Sphingomyelins are found in all cell membranes and are important structural components of the myelin sheath—the protective and insulating coating that surrounds nerves. Cerebrosides, which usually have glucose or galactose as the monosaccharide unit, occur primarily in the brain (7 percent of dry mass) and in the myelin sheath of nerves. The specific structure of the cerebroside in which stearic acid (18:0) is the fatty acid and galactose is the monosaccharide is

Sphingosine; $HO-CH-CH=CH-(CH_2)_{12}-CH_3$; $CH-NH-C(=O)-(CH_2)_{16}-CH_3$; $O-CH_2$; Galactose (CH_2OH, HO, OH, OH ring); Amide linkage; Stearic acid; Glycosidic linkage

In Section 15.6 we noted that phosphoacylglycerols have a "head and two tail" structure and that this structural characteristic is found in most membrane lipids. Sphingolipids also have a "head and two tails" structure, as is shown in Figure 15.10, which is consis-

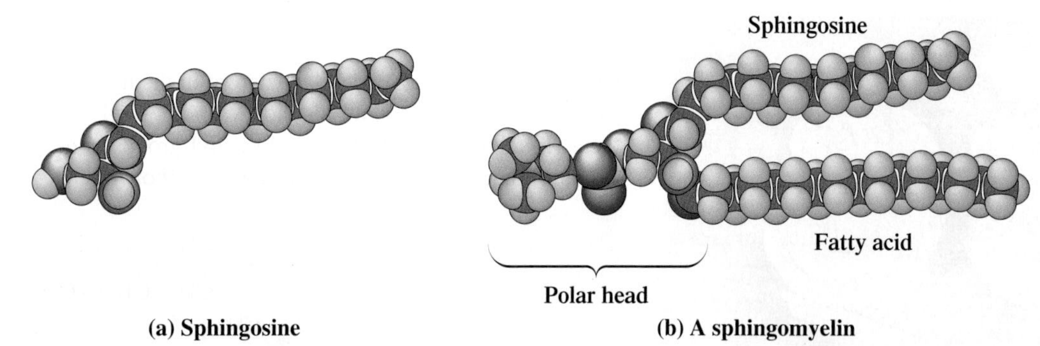

Figure 15.10 Molecular models showing (a) sphingosine and (b) a sphingomyelin. Note the "head and two tails" structure of the sphingomyelin, a characteristic associated with most membrane lipids.

tent with their membrane function. For sphingolipids, the fatty acid is one of the tails, and the long carbon chain of sphingosine itself is the other tail. The "additional component," which is a phosphoric-acid–choline group in the model shown, is the head.

Practice Problems and Questions

15.32 Draw the general block diagram structure for a sphingolipid.

15.33 Describe the location of each of the following in a sphingolipid.
a. The amide linkage b. The ester or glycosidic linkage

15.34 Sphingolipids have a "head and two tails" structure. What is the chemical identity of the head and of each of the two tails?

15.35 Are phosphoric acid and fatty acid components present in all sphingolipid structures? Explain.

15.36 What is the major structural difference between a sphingomyelin and a cerebroside?

Learning Focus

Recognize the fused hydrocarbon ring system that is characteristic of steroids, and understand the function and general structure of various steroids.

15.8 Steroids

In Section 15.1 we noted three major structural types of lipids to be considered in this chapter: (1) nonpolar fatty-acid-containing lipids, (2) polar fatty-acid-containing lipids, and (3) non-fatty-acid-containing lipids. Triglycerides (Section 15.4) represent the most abundant family of polar fatty-acid-containing lipids. Phosphoacylglycerols (Section 15.6) and sphingolipids (Section 15.7) are examples of polar fatty-acid-containing lipids. In this section we focus on *steroids,* the most abundant type of non-fatty-acid-containing lipid.

A **steroid** *is a lipid with a structure that is based on a fused-ring system involving three 6-membered rings and one 5-membered ring.* The fused-ring system of steroids, which is called a steroid nucleus, has the following structure:

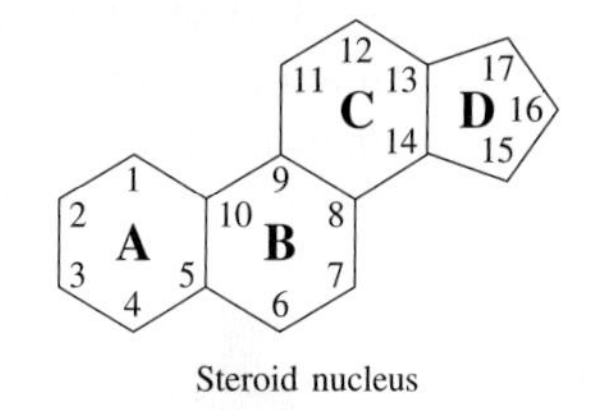

Note that the rings are customarily labeled with letters and that each carbon atom is labeled with a number.

Numerous steroids have been isolated from plants, animals, and human beings. The location of double bonds within the fused-ring system and the nature and location of substituents distinguish one steroid from another. Most steroids have an oxygen functional group (=O or —OH) at carbon 3 and some kind of side chain at carbon 17. Many also have a double bond from carbon 5 to either carbon 4 or carbon 6.

The steroids discussed in this section are cholesterol, bile salts, and steroid hormones.

Figure 15.11 Cholesterol contributes to many gallbladder attacks caused by gallstones. A large percentage of gallstones are almost pure crystallized cholesterol that has precipitated from bile solution.

Cholesterol

Cholesterol is the most abundant steroid in the human body. The name *cholesterol* has the *-ol* ending because it is an alcohol, with an —OH group on carbon 3 of the steroid nucleus. In addition, cholesterol has methyl groups bonded to carbon atoms 10 and 13 and a small branched hydrocarbon chain on carbon 17.

CH_3 CH_3
$CH—CH_2—CH_2—CH_2—CH—CH_3$
CH_3
13 17
CH_3
10
3
HO

Cholesterol

Within the human body, cholesterol is found in cell membranes (up to 25% by mass), nerve tissue, and brain tissue (about 10% by dry mass), and it is the main component of gallstones (see Figure 15.11). Human blood plasma contains about 50 mg of free cholesterol per 100 mL and about 170 mg of cholesterol esterified with various fatty acids.

A space-filling model of the cholesterol molecule (Figure 15.12) shows the rather compact nature of this molecule. The "head and two tails" arrangement found in many lipids is not present. The lack of a large polar head causes cholesterol to have limited water solubility. The —OH group on carbon 3 is considered the head of the molecule.

Cholesterol plays a vital biochemical role in chemical synthesis within the human body. It is the starting material for the synthesis of numerous steroid hormones, vitamin D, and bile salts. Its presence in the body is essential to life.

The human body, mainly within the liver, synthesizes about 1 gram of cholesterol each day, an amount sufficient to meet the body's biosynthetic needs. Therefore, cholesterol is not necessary in the diet. When we ingest cholesterol, the amount synthesized by the body is reduced. However, the reduction is less than the amount ingested. Therefore, total body cholesterol level increases with dietary cholesterol level.

Medical science now considers high blood cholesterol, along with high blood pressure and smoking, as the major risk factors for cardiovascular disease (CVD). High blood cholesterol contributes to atherosclerosis, the main form of CVD, which is characterized by the buildup of plaque along the inner walls of the arteries. Plaque is a mound of lipid material mixed with smooth muscle cells and calcium. Much of the lipid material in plaque is cholesterol.

People who want to reduce their level of dietary cholesterol should reduce the amount of animal products they eat (meat, dairy products, etc.) and eat more fruit and vegetables. Plant foods contain no cholesterol; it is found only in foods of animal origin. Table 15.2 lists the cholesterol content of selected meats and dairy products.

Table 15.2
The Cholesterol Content of Selected Meats and Dairy Products

Food	Cholesterol (mg)
liver (3 oz)	410
egg (1 large)	213
shrimp (3 oz)	166
pork chop (3 oz)	83
chicken (3 oz)	75
beef steak (3 oz)	70
fish fillet (3 oz)	54
whole milk (1 cup)	33
cheddar cheese (1 oz)	30
Swiss cheese (1 oz)	26
low-fat milk (1 cup)	22

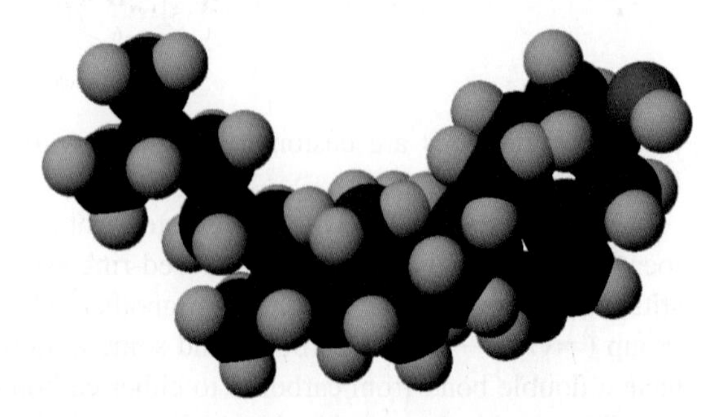

Figure 15.12 A molecular model showing the compact nature of the cholesterol molecule.

▶ The source for bile salts is *bile,* a fluid secreted by the liver, stored in the gallbladder, and released into the small intestine during digestion. Besides bile salts, bile also contains bile pigments (breakdown products of hemoglobin) and electrolytes such as bicarbonate ion.

Bile Salts

A **bile salt** *is an emulsifying agent that makes dietary lipids soluble in the aqueous environment of the digestive tract.* During digestion, bile salts are released into the intestine from the gallbladder, where they help digestion by emulsifying (solubilizing) fats and oils. Their mode of action is much like that of soap during washing.

Bile salts are cholesterol oxidation products. They are trihydroxy cholesterol derivatives in which the carbon-17 side chain has been oxidized to a carboxylic acid. This acid side chain is then bonded to an amino acid through an amide linkage. The two principal bile salts are sodium glycocholate (glycine is the amino acid) and sodium taurocholate (taurine is the amino acid).

CH_3, OH, CH_3, CH_3, HO, OH; side chain: $CH—CH_2—CH_2—C(=O)—NH—CH_2—COO^-\ Na^+$ (Glycine portion)

Sodium glycocholate

CH_3, OH, CH_3, CH_3, HO, OH; side chain: $CH—CH_2—CH_2—C(=O)—NH—CH_2—CH_2—SO_3^-\ Na^+$ (Taurine portion)

Sodium taurocholate

Steroid Hormones

A **hormone** *is a chemical messenger produced by a ductless gland.* Hormones serve as a means of communication between various tissues. Many, but not all, hormones in the human body are steroids. Cholesterol is the ultimate starting material for the production of all steroid hormones, so they contain its characteristic system of four fused rings. Steroid hormone synthesis is always a multistep process.

There are two major classes of steroid hormones: the *sex hormones,* which control reproduction and secondary sex characteristics, and the *adrenocortical hormones,* which regulate numerous biochemical processes in the body.

The sex hormones can be classified into three major groups:

1. Estrogens—the female sex hormones
2. Androgens—the male sex hormones
3. Progestins—the pregnancy hormones

▶ Estrogens are a class of molecules rather than a single molecule. Statements like "the estrogen level is high" should be rephrased as "a high level of estrogens."

Estrogens are synthesized in the ovaries and adrenal cortex and are responsible for the development of female secondary sex characteristics at the onset of puberty and for regulation of the menstrual cycle. They also stimulate the development of the mammary glands during pregnancy and induce estrus (heat) in animals.

Androgens are synthesized in the testes and adrenal cortex and promote the development of secondary male characteristics. They also promote muscle growth.

Progestins are synthesized in the ovaries and the placenta and prepare the lining of the uterus for implantation of fertilized ovum. They also suppress ovulation.

The upper part of Figure 15.13 gives the structure of the primary hormone in each of the three subclasses of sex hormones. Other members of these hormone families are metabolized forms of the primary hormone.

Note, in Figure 15.13a, how similar the structures are for these principal hormones, and yet how different their functions. The fact that seemingly minor changes in structure effect great changes in biofunction points out, again, the extreme specificity of the enzymes that control biochemical reactions.

Increased knowledge of the structures and functions of sex hormones has led to the development of a number of *synthetic* steroids whose actions often mimic those of the natural hormones. Among the best known of the synthetic steroids are oral contraceptives and anabolic agents.

▶ The C≡C functional group, which occurs in both norethynodrel (Enovid) and RU-486, is rarely found in biomolecules.

Oral contraceptives are used to suppress ovulation as a method of birth control. Generally, a mixture of a synthetic estrogen and a synthetic progestin is used. The synthetic estrogen regulates the menstrual cycle, and the synthetic progestin prevents ovulation, thus creating a false state of pregnancy. The structure of norethynodrel (Enovid), a synthetic progestin, is given in the lower half of Figure 15.13. Compare its structure to that of progesterone (the real hormone); the structures are very similar.

Interestingly, the controversial "morning after" pill developed in France and known as RU-486 is also similar in structure to progesterone. RU-486 interferes with gestation of a fertilized egg and terminates a pregnancy within the first 9 weeks of gestation more effectively and safely than surgical methods. The structure of RU-486 appears next to that of norethynodrel in Figure 15.13.

Figure 15.13 Structures of selected sex hormones and synthetic compounds that have similar actions.

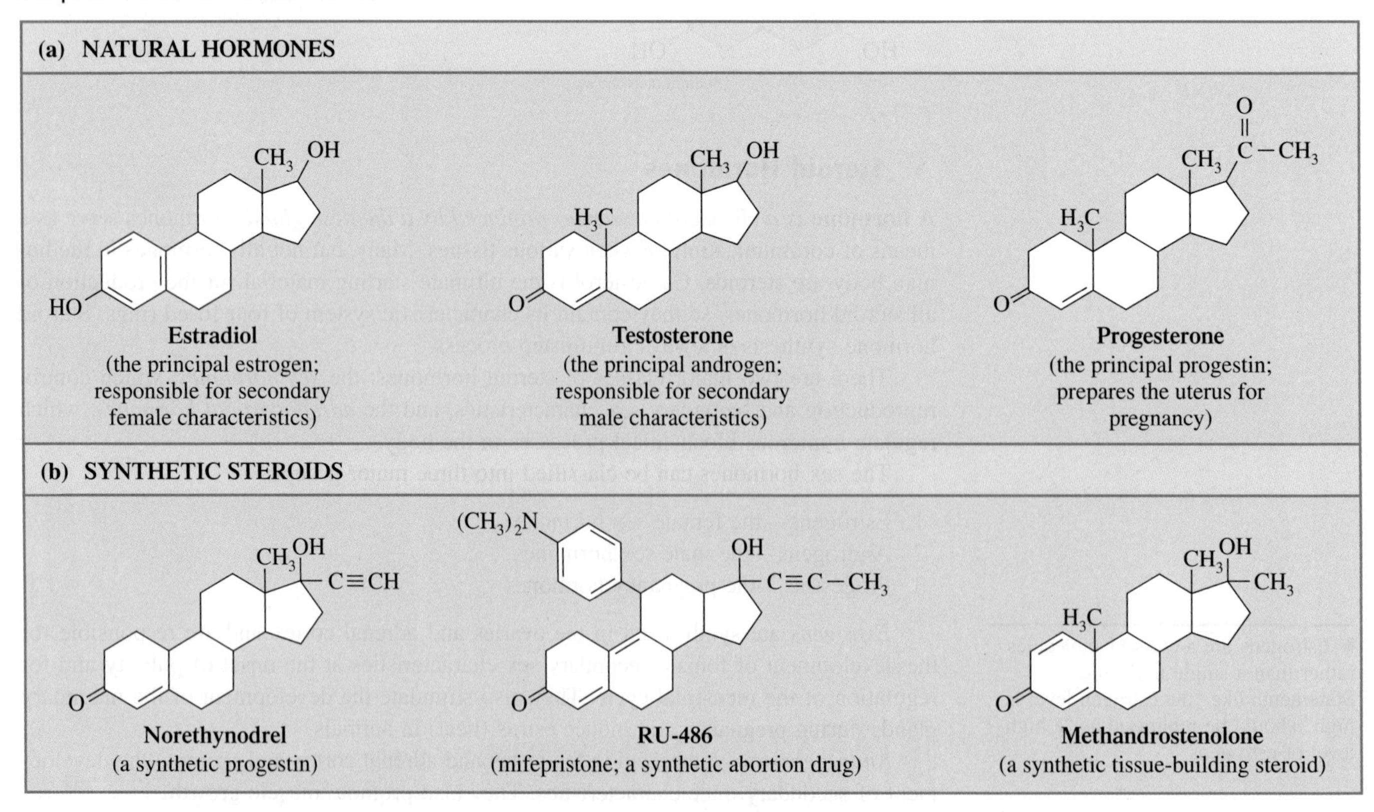

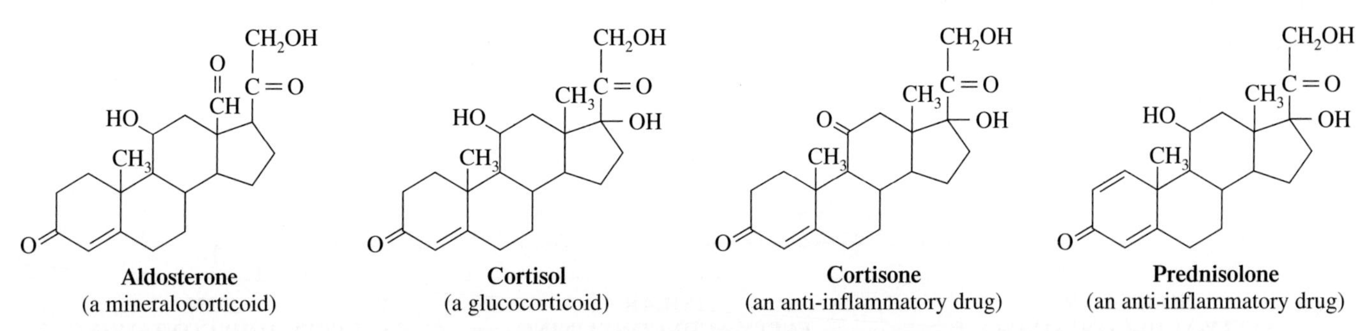

Figure 15.14 Structures of selected adrenocortical hormones and related synthetic compounds.

Anabolic agents include the illegal steroid drugs used by some athletes to build up muscle strength and enhance endurance. Anabolic agents are now known to have serious side effects on the user. The structure of one of the more commonly used anabolic agents, methandrostenolone, is given in the lower half of Figure 15.13. Note the similarities between its structure and that of testosterone.

The second major group of steroid hormones consists of the adrenocortical hormones. Produced by the adrenal glands, small organs located on top of each kidney, at least 28 different hormones have been isolated from the adrenal cortex (the outer part of the glands).

There are two types of adrenocortical hormones.

1. *Mineralocorticoids* control the balance of Na^+ and K^+ ions in cells.
2. *Glucocorticoids* control glucose metabolism and counteract inflammation.

The major mineralocorticoid is aldosterone, and the major glucocorticoid is cortisol (hydrocortisone). Cortisol is the hormone synthesized in the largest amount by the adrenal glands. Cortisol and its synthetic ketone derivative cortisone exert powerful anti-inflammatory effects in the body. Both cortisone and prednisolone, a similar synthetic derivative, are used as prescription drugs to control inflammatory diseases such as rheumatoid arthritis. Figure 15.14 gives the structures of aldosterone and cortisol as well as those of the synthetic steroids cortisone and prednisolone.

The accompanying Chemistry at a Glance summarizes the features of the major types of lipids that we have considered in this chapter.

Practice Problems and Questions

15.37 Draw and number the fused hydrocarbon ring system that is characteristic of all steroids.

15.38 What positions in the steroid nucleus are particularly likely to bear substituents?

15.39 Describe the structure of cholesterol in terms of substituents attached to the steroid nucleus.

15.40 Explain the fallacy in the statement that a cholesterol-free diet would eventually reduce body cholesterol levels to near zero.

15.41 What is the role of bile salts in the human digestive process?

15.42 List the general structural differences between cholesterol and bile salts.

15.43 What are the two major classes of steroid hormones?

15.44 Describe the general function of each of the following types of steroid hormones.

a. Estrogens b. Androgens
c. Progestins d. Mineralocorticoids

Chemistry *at a Glance*

Major Types of Lipids

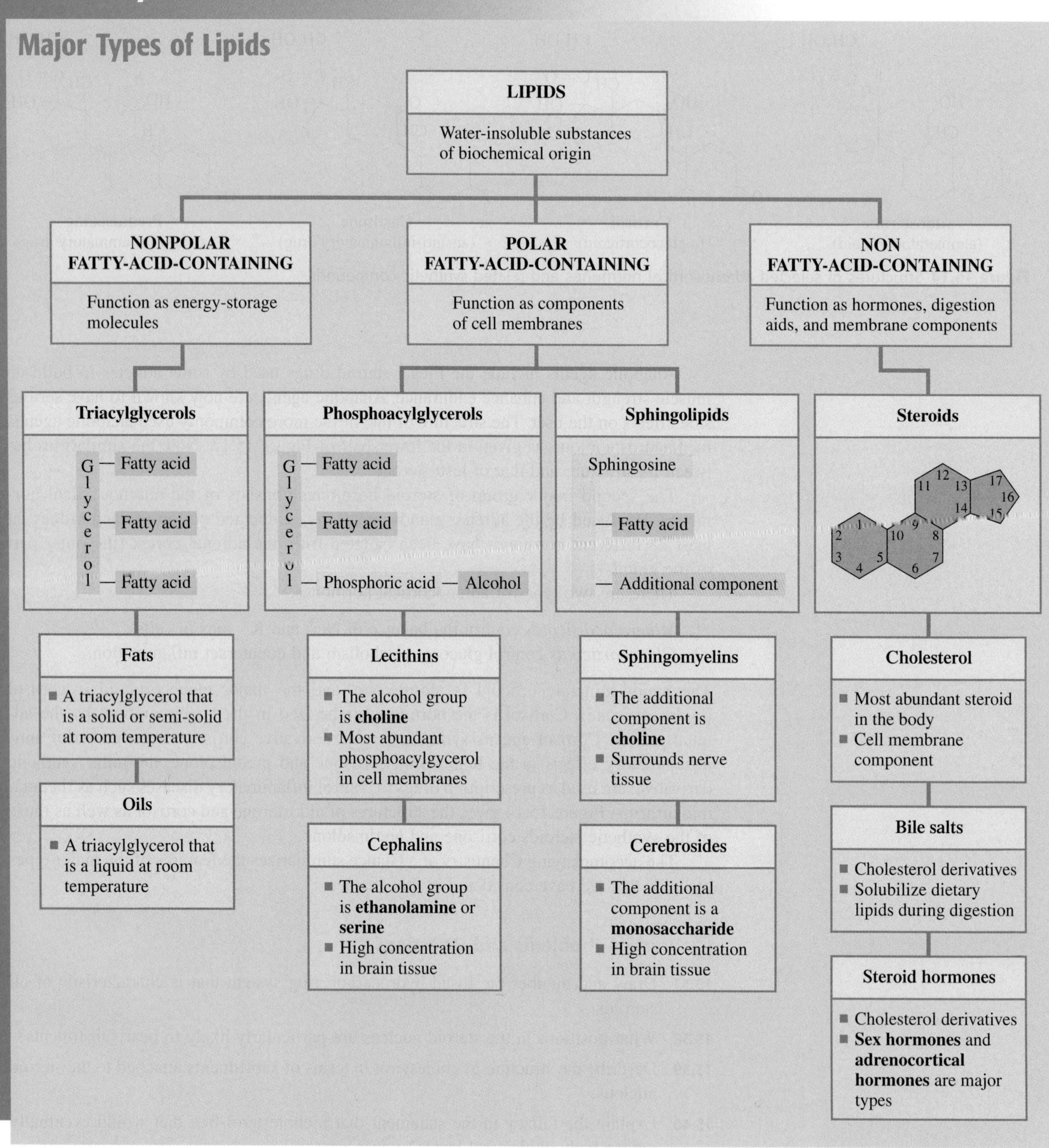

Learning Focus

Describe cell membrane structure in terms of the structural arrangement of lipid molecules present.

15.9 Structure of Cell Membranes

Living cells contain approximately 10,000 kinds of molecules in an aqueous environment confined by a *cell membrane*. A **cell membrane** *is a structure that separates a cell's aqueous-based contents from the aqueous environment surrounding the cell.* Besides its "separation" function, a cell membrane also controls the movement of substances into and out of the cell. Up to 80% of the mass of a cell membrane is lipid material; hence our consideration of cell membranes in this chapter.

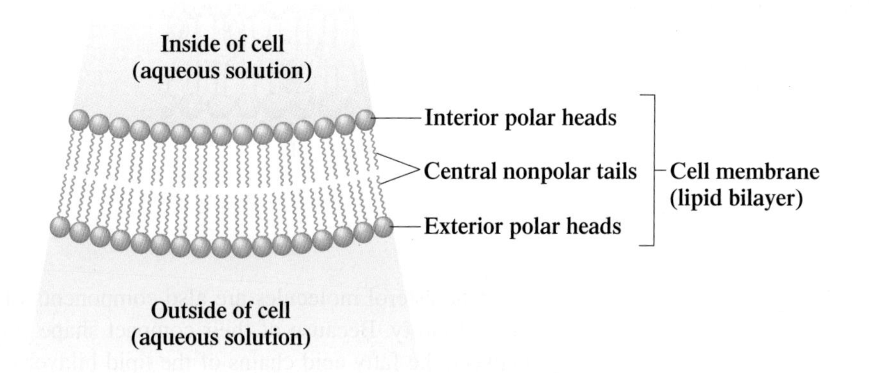

Figure 15.15 Cross section of a lipid bilayer. The circles represent the polar heads of the lipid components, and the wavy lines represent the nonpolar tails of the lipid components. The "heads" occupy surface positions, and the "tails" occupy internal positions.

Cell membranes are remarkable bits of molecular architecture containing various phosphoacylglycerols and sphingolipids. The "head and two tail" structure of these types of lipids (Sections 15.6 and 15.7) is of key importance to an understanding of a basic structural feature of cell membranes, a feature called a *lipid bilayer*. A **lipid bilayer** *is a two-layer-thick structure of lipid molecules aligned so that the nonpolar tails of the lipids form the interior of the structure and the polar heads form the outside surfaces.* Figure 15.15 is a cross section of part of a lipid bilayer.

A lipid bilayer is 6 to 9 billionths of a meter thick—that is, 6 to 9 nanometers thick. There are three distinct parts to such a layer: the exterior polar "heads," the interior polar "heads," and the central nonpolar "tails." These three distinct regions are labeled in Figure 15.15.

A lipid bilayer is held together by dipole–dipole interactions, not by covalent bonds. This means each phospholipid or sphingolipid is free to diffuse laterally within the lipid bilayer. Most lipid molecules in the bilayer contain at least one unsaturated fatty acid. The presence of such acids, with the kinks in their carbon chains (Section 15.3), prevent tight packing of fatty acid chains (Figure 15.16). The open packing imparts a liquid-like character to the membrane—a necessity because numerous types of biochemicals must pass into and out of a cell.

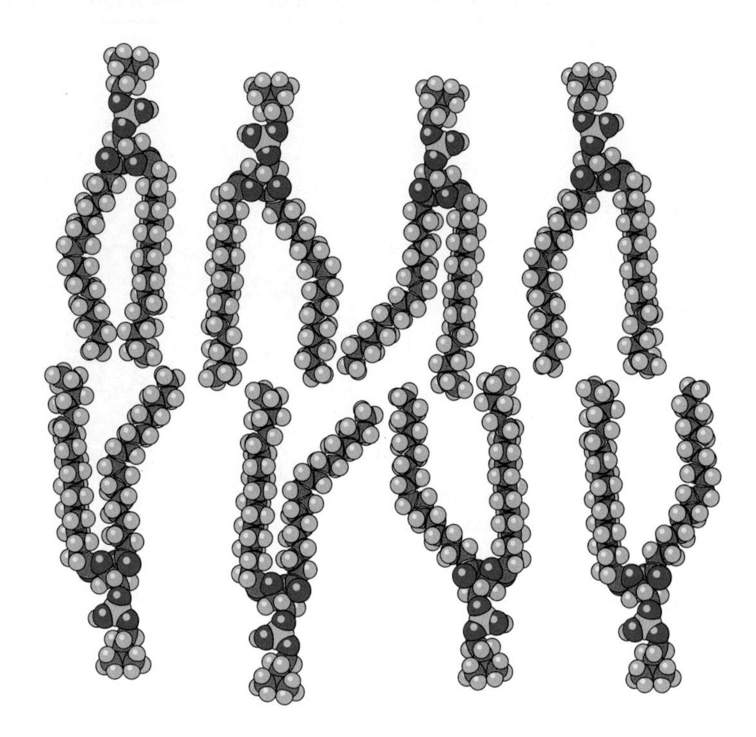

Figure 15.16 The kinks associated with *cis* double bonds in fatty acid chains prevent tight packing of the lipid molecules in a lipid bilayer.

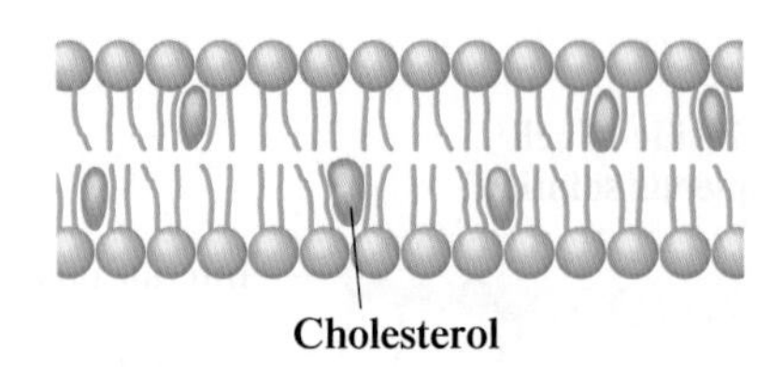

Figure 15.17 Cholesterol molecules fit between fatty acid chains in a lipid bilayer. They regulate the rigidity of the structure.

▶ The percentage of lipid and protein components in a cell membrane is related to the function of the cell. The lipid/protein ratio ranges from about 80% lipid/20% protein by mass in the myelin sheath of nerve cells to the unique 20% lipid/80% protein ratio for the inner mitochondrial membrane. Red blood cell membranes contain approximately equal amounts of lipid and protein. A typical membrane also has a carbohydrate content that varies between 2% and 10% by mass.

Cholesterol molecules are also components of cell membranes. They regulate membrane fluidity. Because of their compact shape (Figure 15.12), cholesterol molecules fit between the fatty acid chains of the lipid bilayer (Figure 15.17), restricting movement of the fatty acid chains and making the bilayer more rigid.

Proteins are also components of lipid bilayers. The proteins are responsible for moving substances such as nutrients and electrolytes across the membrane, and they also act as receptors that bind hormones and neurotransmitters (Figure 15.18).

Small carbohydrate molecules are also components of cell membranes. They are found on the *outer* membrane surface covalently bonded to protein molecules (a glycoprotein) or lipid molecules (a glycolipid). The carbohydrate portions of glycoproteins and glycolipids function as *markers,* substances that play key roles in the process by which different cells recognize each other.

Practice Problems and Questions

15.45 What are the two major types of lipids that are present in a cell membrane?

15.46 What is a lipid bilayer?

15.47 Draw a short section of a lipid bilayer and label the parts of the molecules with respect to polarity.

15.48 What is the function of unsaturation in the hydrocarbon tails of membrane lipids?

15.49 What is the function of cholesterol present in cell membranes?

15.50 What is the function of proteins present in cell membranes?

15.51 What is the function of carbohydrates present in cell membranes?

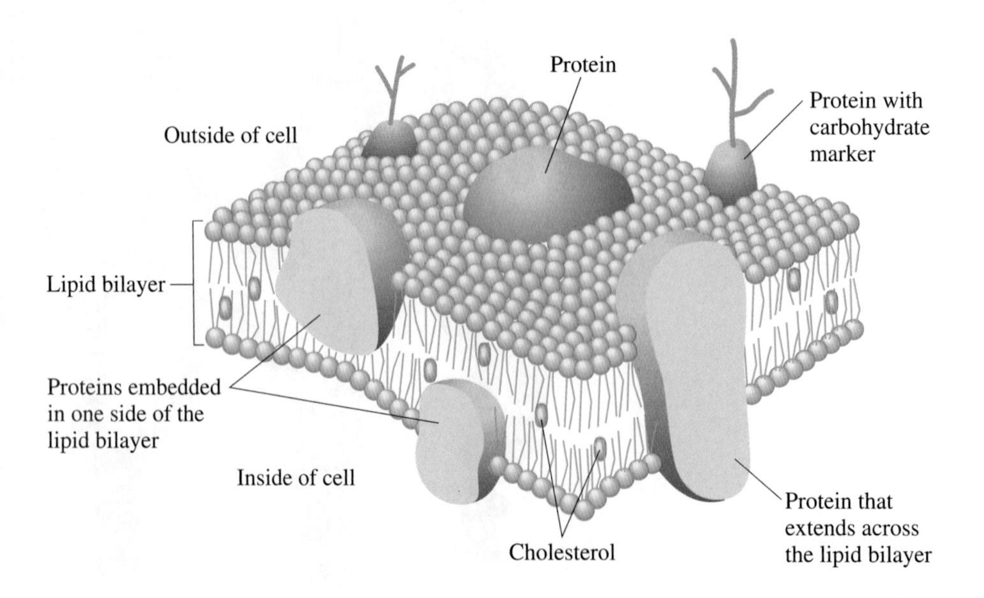

Figure 15.18 Proteins are important structural components of cell membranes.

CONCEPTS TO REMEMBER

Lipids. Lipids are a structurally heterogeneous group of compounds that are soluble in nonpolar organic solvents and insoluble in water. Three major types of lipids are (1) nonpolar fatty-acid-containing lipids, (2) polar fatty-acid-containing lipids, and (3) non-fatty-acid-containing lipids.

Fatty acids. Fatty acids are long-chain unbranched monocarboxylic acids. The carbon chain may be saturated, monounsaturated, or polyunsaturated. Length of carbon chain, degree of unsaturation, and location of unsaturation influence the properties of fatty acids.

Triacylglycerols. Triacylglycerols are triesters formed by esterification of three fatty acid molecules to glycerol. Fats are triacylglycerols that are solids at room temperature, and oils are triacylglycerols that are liquids at room temperature. Chemical reactions of triacylglycerols include hydrolysis, saponification, hydrogenation, and oxidation. Triacylglycerols represent stored forms of energy for living systems.

Phosphoacylglycerols. Phosphoacylglycerols are triesters of glycerol in which two —OH groups are esterified with fatty acids and one —OH group is esterified with phosphoric acid, which in turn is esterified to an alcohol. Phosphoacylglycerols have a "head and two tails" structure and are major components of cell membranes. Lecithins and cephalins are types of phosphoacylglycerols.

Sphingolipids. Sphingolipids have structures based on the long-chain amino dialcohol sphingosine rather than on glycerol. A fatty acid is bonded to the sphingosine via an amide linkage, and an additional group is bonded to the terminal hydroxyl group of sphingosine through an ester linkage. Sphingolipids have a "head and two tails" structure and are components of cell membranes. Sphingolmyelins and cerebrosides are types of sphingolipids.

Steroids. Steroids are non-fatty-acid-containing lipids that have structures involving a fused-ring system containing three 6-membered rings and one 5-membered ring. Cholesterol is the most abundant steroid in the human body. Bile salts and steroid hormones are cholesterol derivatives.

Lipid bilayer. A lipid bilayer is the fundamental structural feature of a cell membrane. It is a two-layer-thick structure of lipid molecules (phosphoacylglycerols and sphingolipids) aligned so that the nonpolar tails of the lipids form the interior of the structure and the polar heads form the outside surfaces. Cholesterol, proteins, and carbohydrates can also be present in cell membranes.

KEY REACTIONS AND EQUATIONS

1. Formation of a triacylglycerol (Section 15.4)

Glycerol + 3 fatty acids $\xrightarrow{\text{Enzymes}}$ Glycerol—Fatty acid, —Fatty acid, —Fatty acid + 3 H_2O

2. Hydrolysis of a triacylglycerol to produce glycerol + 3 fatty acids (Section 15.5)

Glycerol—Fatty acid, —Fatty acid, —Fatty acid + 3 H_2O $\xrightarrow[\text{enzymes}]{H^+ \text{ or}}$ Glycerol + 3 fatty acids

3. Saponification of a triacylglycerol to produce glycerol + 3 fatty acid salts (Section 15.5)

Glycerol—Fatty acid, —Fatty acid, —Fatty acid + 3 H_2O $\xrightarrow{OH^-}$ Glycerol + 3 fatty acid salts

4. Hydrogenation of a triacylglycerol to reduce the unsaturation of its fatty acid components (Section 15.5)

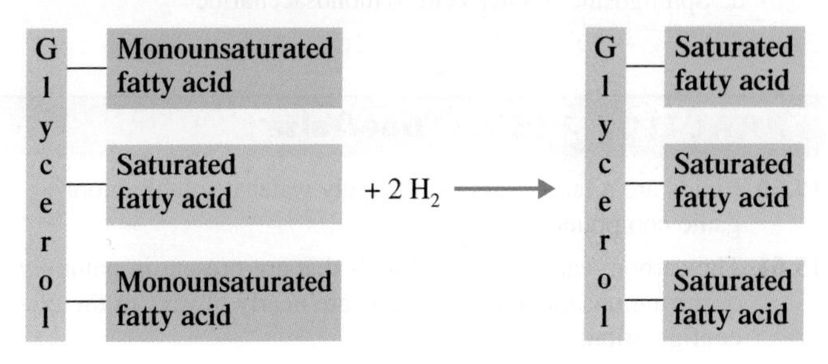

5. Block diagram for a phosphoacylglycerol (Section 15.6)

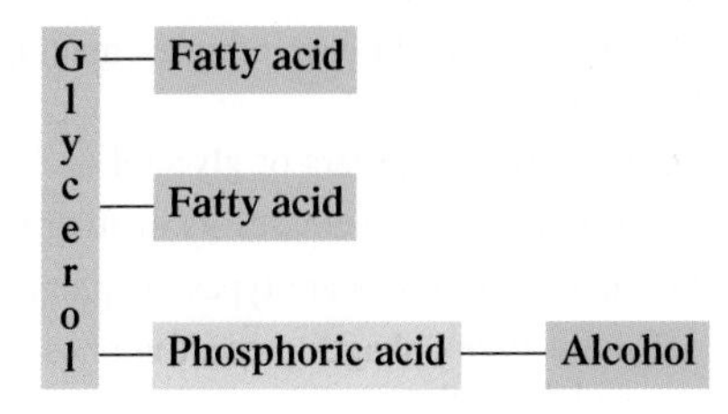

6. Block diagram for a sphingolipid (Section 15.7)

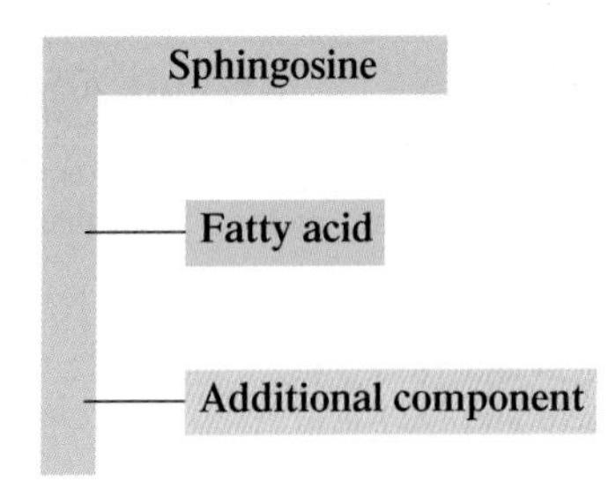

KEY TERMS

Bile salt (15.8)
Cell membrane (15.9)
Essential fatty acid (15.2)
Fat (15.4)
Fatty acid (15.2)
Hormone (15.8)
Lipid (15.1)
Lipid bilayer (15.9)
Mixed triacylglycerol (15.4)
Monounsaturated fatty acid (15.2)
Oil (15.4)
Omega-3 fatty acid (15.2)
Omega-6 fatty acid (15.2)
Phosphoacylglycerol (15.6)
Polyunsaturated fatty acid (15.2)
Saturated fatty acid (15.2)
Simple triacylglycerol (15.4)
Sphingolipid (15.7)
Steroid (15.8)
Triacylglycerol (15.4)

ADDITIONAL PROBLEMS

15.52 Classify each of the following types of lipids as (1) nonpolar fatty-acid-containing, (2) polar fatty-acid-containing, or (3) non-fatty-acid-containing.
a. Triacylglycerols b. Phosphoacylglycerols
c. Fats d. Steroids

15.53 Classify each of the following types of lipids as (1) glycerol-based, (2) sphingosine-based, or (3) neither glycerol-based nor sphingosine-based.
a. Fats b. Oils c. Cholesterol d. Cerebrosides

15.54 Indicate whether each of the lipids in Problem 15.53 has a "head and two tails" structure.

15.55 Identify the type of lipid that contains each of the following sets of structural subunits.
a. Glycerol + three fatty acids
b. Sphingosine + fatty acid + phosphoric acid + choline
c. Glycerol + two fatty acids + phosphoric acid + choline
d. Sphingosine + fatty acid + monosaccharide

15.56 Classify each of the following types of lipids as (1) an energy-storage lipid, (2) a membrane lipid, (3) a polar lipid, or (4) nonpolar lipid. More than one classification may apply to a given type of lipid.
a. Fats b. Sphingomyelins c. Lecithins d. Cephalins

15.57 Specify the number of ester linkages, amide linkages, and glycosidic linkages present in each of the lipids in Problem 15.56.

15.58 Indicate whether each of the following types of lipids contains a "steroid nucleus" as part of its structure.
a. Estrogens b. Cephalins c. Bile salts d. Cerebrosides

15.59 Classify the fatty acid content of each of the following types of lipids as (1) saturated only, (2) unsaturated only, or (3) both saturated and unsaturated.
a. Fats b. Oils
c. Mixed triacylglycerols d. Phosphoacylglycerols

PRACTICE TEST ▶ True/False

15.60 Lipids are a large class of relatively water-insoluble bioorganic compounds.

15.61 The carbon–carbon double bonds that are present in naturally occurring unsaturated fatty acids are nearly always in the *cis* configuration.

15.62 Omega-6 fatty acids always contain one more double bond than omega-3 fatty acids.

15.63 An 18:1 fatty acid should have a higher melting point than an 18:0 fatty acid.

15.64 Both fats and oils are triesters of glycerol.

15.65 Hydrogenation of an oil decreases its melting point.

15.66 Cephalins and cerebrosides are types of sphingolipids.

15.67 There are more ester linkages present in phosphoacylglycerols than in triacylglycerols.

15.68 Sphingosine is an 18-carbon unsaturated amino dialcohol.

15.69 Both sphingolipids and phosphoacylglycerols have "head and two tails" structures.

15.70 A "steroid nucleus" is a fused-ring system involving two 6-membered rings and two 5-membered rings.

15.71 Cholesterol is the most abundant steroid in the human body.

15.72 Bile salts, estrogens, and lecithins are cholesterol derivatives.

15.73 The interior of a lipid bilayer is polar and its exterior is nonpolar.

15.74 Both cholesterol and protein molecules can be components of cell membranes.

PRACTICE TEST ▶ Multiple Choice

15.75 In which of the following pairs of fatty acids does the first-listed acid have a higher melting point than the second-listed acid?
a. 16:1 acid and 16:0 acid
b. 20:0 acid and 18:0 acid
c. 18:3 acid and 18:1 acid
d. 18:2 acid and 20:0 acid

15.76 Which of the following fatty acids is both polyunsaturated and an omega-6 fatty acid?
a. $CH_3—(CH_2)_{18}—COOH$
b. $CH_3—(CH_2)_5—CH{=}CH—(CH_2)_7—COOH$
c. $CH_3—(CH_2)_4—CH{=}CH—CH_2—CH{=}CH—(CH_2)_7—COOH$
d. $CH_3—CH_2—CH{=}CH—CH_2—CH{=}CH—(CH_2)_{10}—COOH$

15.77 Unsaturated fatty acids are structural components of
a. both fats and oils b. neither fats nor oils
c. fats but not oils d. oils but not fats

15.78 In the oxidation of fats and oils, which part of the molecule is attacked by the oxidizing agent?
a. Carbon–carbon double bonds b. Ester linkages
c. Hydroxyl groups d. Carboxyl groups

15.79 The products of the complete hydrolysis of a triacylglycerol include
a. fatty acids b. fatty acid salts
c. acyl groups d. an amino alcohol

15.80 In a phosphoacylglycerol, glycerin's three —OH groups are esterified, respectively, with
a. one fatty acid and two phosphoric acid molecules
b. two fatty acid and one phosphoric acid molecules
c. three phosphoric acid molecules
d. one fatty acid, one phosphoric acid, and one amino alcohol molecule

15.81 Which of the following types of lipids does *not* have a "head and two tails" structure?
a. Sphingolipids
b. Triacylglycerols
c. Phosphoacylglycerols
d. Lecithins

15.82 All lipids that contain sphingosine also have
a. two fatty acids participating in ester linkages
b. two fatty acids participating in amide linkages
c. one fatty acid participating in an ester linkage
d. one fatty acid participating in an amide linkage

15.83 The "steroid nucleus" common to all steroid structures involves a fused-ring system that has
a. two rings b. three rings c. four rings d. five rings

15.84 Which of the following polarity-based descriptions is correct for a lipid bilayer?
a. Both the outer and the inner surfaces contain polar "heads."
b. Both the outer and the inner surfaces contain nonpolar "heads."
c. Both the outer and the inner surfaces contain nonpolar "tails."
d. The outer surface contains polar "heads" and the inner surface contains nonpolar "tails."

CHAPTER OUTLINE

CHAPTER SIXTEEN

Proteins

The protein made by spiders to produce a web is a form of silk that can be exceptionally strong.

In this chapter we consider the third of the bioorganic classes of molecules (Section 14.1), the compounds called proteins. An extraordinary number of different proteins, each with a different function, exist in the human body. A typical human cell contains about 9000 different kinds of proteins, and the human body contains about 100,000 different proteins. Proteins are needed for the synthesis of enzymes, certain hormones, and some blood components; for the maintenance and repair of existing tissues; for the synthesis of new tissue; and sometimes for energy.

16.1 Characteristics of Proteins

Learning Focus

Know the general characteristics of proteins.

Next to water, proteins are the most abundant substances in most cells; they account for 10% to 20% of a cell's overall mass (Section 14.1) and almost half of a cell's dry mass. All proteins contain the elements carbon, hydrogen, oxygen, and nitrogen; most also contain sulfur. The presence of nitrogen in proteins sets them apart from carbohydrates and lipids, which generally do not contain nitrogen. The average nitrogen content of proteins is 15% by mass. Other elements, such as phosphorus and iron, are essential constituents of certain specialized proteins. Casein, the main protein of milk, contains phosphorus, an element very important in the diet of infants and children. Hemoglobin, the oxygen-transporting protein of blood, contains iron.

▶ The word *protein* comes from the Greek *proteios,* which means "of first importance." This derivation alludes to the key role that proteins play in life processes.

A **protein** *is a polymer in which the monomer units are amino acids.* Thus the starting point for a discussion of proteins is an understanding of the structures and chemical properties of amino acids.

Practice Problems and Questions

16.1 What element is *always* present in proteins that is seldom present in carbohydrates and lipids?

16.2 How does the amount of protein present in a cell, on the basis of dry mass, compare with that of other biochemical substances?

16.3 What is the general name for the building blocks (monomers) from which a protein is made?

Learning Focus

Recognize the distinguishing aspects of the standard amino acids present in proteins.

16.2 Amino Acids: The Building Blocks for Proteins

An **amino acid** *is an organic compound that contains both an amino* (—NH_2) *group and a carboxyl* (—COOH) *group.* The amino acids found in proteins are always α-amino acids. An **α-amino acid** *is an amino acid in which the amino group and the carboxyl group are attached to the same carbon atom.* The carbon atom to which these two groups are attached is called the *α-carbon atom.* The general structural formula for an α-amino acid is

$$H_2N-\underset{\displaystyle R}{\overset{\displaystyle H}{\underset{|}{\overset{|}{C}}}}-COOH$$

α-Carbon atom; Amino group; Carboxyl group; Side chain

▶ The nature of the side chain (R group) distinguishes α-amino acids from each other, both physically and chemically.

The R group present in an α-amino acid is called the amino acid *side chain.* The nature of this side chain distinguishes α-amino acids from each other. Side chains vary in size, shape, charge, acidity, functional groups present, hydrogen-bonding ability, and chemical reactivity.

Over 700 different naturally occurring amino acids are known, but only 20 of them, called standard amino acids, are normally present in proteins. A **standard amino acid** *is one of the 20 α-amino acids normally found in proteins.* The structures of the 20 standard amino acids are given in Table 16.1. Within Table 16.1, amino acids are grouped according to side-chain polarity. In this system there are four categories: (1) nonpolar amino acids, (2) polar neutral amino acids, (3) polar acidic amino acids, and (4) polar basic amino acids. This classification system gives insights into how various types of amino acid side chains help determine the properties of proteins (Section 16.9).

▶ The nonpolar amino acid *proline* has a structural feature not found in any other standard amino acid. Its side chain, a propyl group, is bonded to both the α-carbon atom and the amino nitrogen atom, giving a cyclic side chain.

(Structure: HN—C(H)—COOH ring with H_2C, CH_2, CH_2)

Proline

A nonpolar amino acid *contains one amino group, one carboxyl group, and a nonpolar side chain.* When incorporated into a protein, such amino acids are *hydrophobic* ("water-fearing"); that is, they are not attracted to water molecules. They are generally found in the interior of proteins, where there is limited contact with water. There are eight nonpolar amino acids.

The three types of polar amino acids have varying degrees of affinity for water. Within a protein, such amino acids are said to be *hydrophilic* ("water-loving"). Hydrophilic amino acids are often found on the surfaces of proteins.

A polar neutral amino acid *contains one amino group, one carboxyl group, and a side chain that is polar but neutral.* The side chain is neutral in that it is neither acidic nor basic in solution at physiological pH. There are seven polar neutral amino acids.

A polar acidic amino acid *contains one amino group and two carboxyl groups, the second carboxyl group being part of the side chain.* In solution at physiological pH, the side chain of a polar acidic amino acid bears a negative charge; the side-chain carboxyl group has lost its acidic hydrogen atom. There are two polar acidic amino acids: aspartic acid and glutamic acid.

A polar basic amino acid *contains two amino groups and one carboxyl group, the second amino group being part of the side chain.* In solution at physiological pH, the side chain of a polar basic amino acid bears a positive charge; the nitrogen atom of the amino group has accepted a proton (basic behavior; Section 12.8). There are three polar basic amino acids: lysine, arginine, and histidine.

Table 16.1 The 20 Standard Amino Acids, Grouped According to Side-Chain Polarity. Below each amino acid's structure are its name (with pronunciation), its three-letter abbreviation, and its one-letter abbreviation.

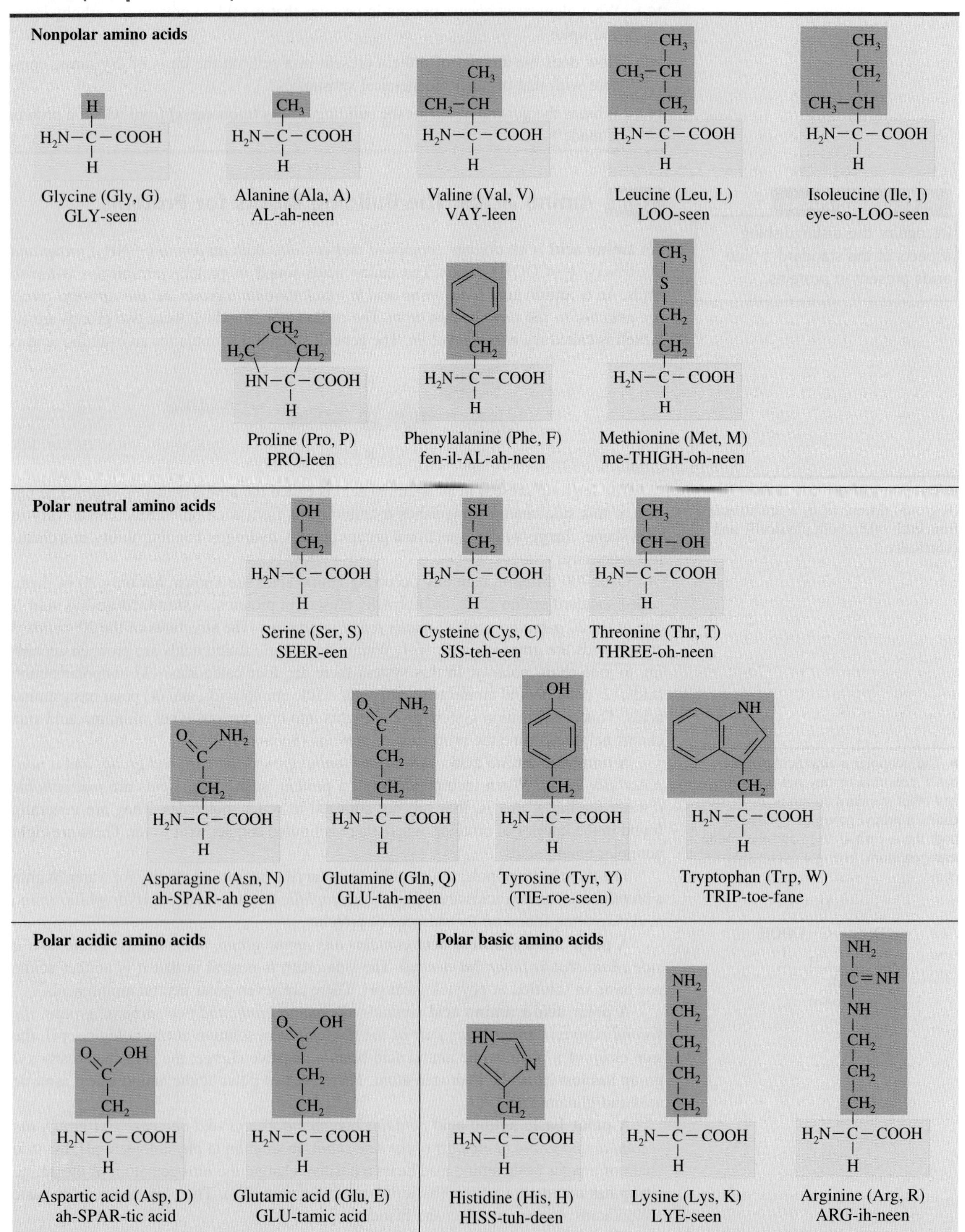

Table 16.2
The Essential Amino Acids for Humans

Arginine[a]	Methionine
Histidine	Phenylalanine
Isoleucine	Threonine
Leucine	Tryptophan
Lysine	Valine

[a]Arginine is required for growth in children but is not required by adults.

The names of the standard amino acids are often abbreviated using three-letter codes. Except in four cases, these abbreviations are the first three letters of the amino acid's name. In addition, a new one-letter code for amino acid names is currently gaining popularity (particularly in computer applications). These abbreviations are used extensively in describing peptides and proteins, which contain tens and hundreds of amino acid units. Both types of abbreviations are given in Table 16.1.

Essential Amino Acids

All of the amino acids in Table 16.1 are necessary constituents of human protein. Adequate amounts of 11 of the 20 amino acids can be synthesized from carbohydrates and lipids in the body if a source of nitrogen is also available. Because the human body is incapable of producing 9 of these 20 acids fast enough or in sufficient quantities to sustain normal growth, these 9 amino acids, which are called *essential* amino acids, must be obtained from food. An **essential amino acid** *is an amino acid that must be obtained from food.* Table 16.2 lists the essential amino acids.

Most animal proteins, including casein from milk and proteins found in meat, fish, and eggs, contain all of the essential amino acids, although gelatin is an exception (it lacks tryptophan). Proteins from plants (vegetables, grains, legumes) have quite diverse amino acid patterns, and some tend to be lacking in one or more of the essential amino acids. Thus vegetarians must eat a variety of plant foods to obtain all of the essential amino acids in appropriate quantities.

Practice Problems and Questions

16.4 What is the major structural difference among the standard amino acids?

16.5 On the basis of polarity, what are the four types of side chains found in the standard amino acids?

16.6 Indicate the distinguishing characteristic of a
a. polar basic amino acid b. polar acidic amino acid

16.7 In what way is the structure of the amino acid proline different from that of the other 19 standard amino acids?

16.8 With the help of Table 16.1, determine what amino acids the following abbreviations stand for.
a. Ala b. Leu c. Met d. Trp

16.9 What four standard amino acids have three-letter abbreviations that are not the first three letters of their common names?

16.10 What is an essential amino acid?

16.11 With the help of Table 16.2, determine which of the amino acids listed in Problem 16.8 are essential amino acids?

Learning Focus

Recognize how handedness is designated in amino acid structures.

16.3 Handedness in Amino Acids

Four different groups are attached to the α-carbon atom in all of the standard amino acids except glycine, where the R group is a hydrogen atom.

$$\begin{array}{c} (H) \\ | \\ (H_2N)\!-\!C\!-\!(COOH) \\ | \\ (R) \end{array}$$

▶ Glycine, the simplest of the standard amino acids, is achiral. All of the other standard amino acids are chiral.

This means that the structures of 19 of the 20 standard amino acids possess a chiral center (Section 14.5) and thus exhibit handedness (left-handed and right-handed forms).

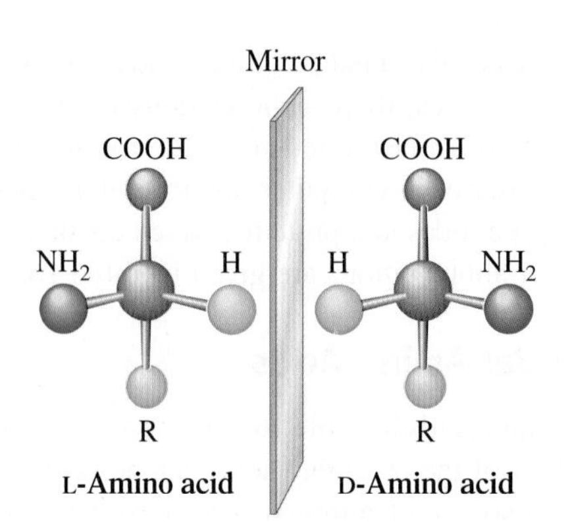

Figure 16.1 Designation of handedness in standard amino acid structures involves aligning the carbon chain vertically and looking at the position of the horizonally aligned —NH_2 group. The L form has the —NH_2 group on the left and the D form has the —NH_2 group on the right.

▶ Because only L amino acids are constituents of proteins, the enantiomer designation of L or D will be omitted in subsequent amino acid and protein discussions. It is understood that it is the L isomer that is always present.

With few exceptions (in some bacteria), the amino acids found in nature and in proteins are L isomers. Thus, as is the case with monosaccharides (Section 14.5), nature favors one mirror-image form over the other. Interestingly, for amino acids the L isomer is the preferred form, whereas for monosaccharides the D isomer is preferred.

The rules for designating handedness in amino acids when drawing their structures are

1. The —COOH group is put at the top of the structure, the R group at the bottom. This positions the carbon chain vertically.
2. The —NH_2 group is in a horizontal position. Positioning it on the left denotes the L isomer, and positioning it on the right denotes the D isomer.

Figure 16.1 illustrates the use of these rules.

Practice Problems and Questions

16.12 To which of the handedness families do nearly all naturally occurring amino acids belong?

16.13 In terms of handedness, how is the structure of glycine, the simplest amino acid, different from that of all the other standard amino acids?

16.14 Draw a structural formula for each of the following amino acids.
a. L-Serine b. D-Serine c. D-Alanine d. L-Leucine

16.4 Acid–Base Properties of Amino Acids

Learning Focus

Understand the *internal* acid–base reaction that occurs within amino acids; be able to draw the structure of the zwitterion form, the positive-ion form, and the negative-ion form of an amino acid.

In pure form, amino acids are white crystalline solids with relatively high decomposition points. (Most amino acids decompose before they melt.) Also most amino acids are *not* very soluble in water because of strong intermolecular forces within their crystal structures. Such properties are those often exhibited by compounds in which charged species are present. Studies of amino acids confirm that they are charged species both in the solid state and in solution. Why is this so?

Both an acidic group (—COOH) and a basic group (—NH_2) are present on the same carbon in an α-amino acid.

▶ In drawing amino acid structures, where handedness designation is not required, the placement of the four groups about the α-carbon atom is arbitrary. From this point on in the text, we will be consistent in drawing the —COOH group on the left, the —NH_2 group on the right, the R group pointing down, and the H atom pointing up. Drawing amino acids in this "arrangement" makes it easier to draw structures where amino acids are linked together to form longer amino acid chains.

Basic group → H_2N—C(H)(R)—COOH ← Acidic group

In Section 13.7, we learned that in neutral solution, carboxyl groups have a tendency to lose protons (H^+), producing a negatively charged species:

$$\text{—COOH} \longrightarrow \text{—COO}^- + H^+$$

In Section 12.8, we learned that in neutral solution, amino groups have a tendency to accept protons (H^+), producing a positively charged species:

$$—NH_2 + H^+ \longrightarrow —\overset{+}{N}H_3$$

As is consistent with the behavior of these groups, in neutral solution, the —COOH group of an amino acid donates a proton to the $—NH_2$ of the same amino acid. We can characterize this behavior as an *internal* acid–base reaction. The net result is that in neutral solution, amino acid molecules have the structure

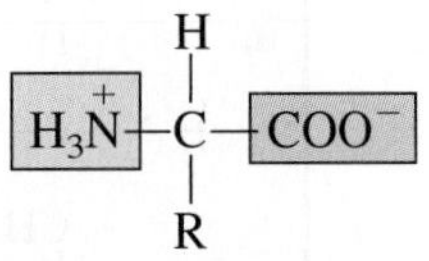

$$H_3\overset{+}{N}—\underset{R}{\overset{H}{C}}—COO^-$$

▶ Strong intermolecular forces between the positive and negative centers of zwitterions are the cause of the high melting points of amino acids.

Such a molecule is known as a zwitterion, from the German term meaning "double ion." A **zwitterion** *is a molecule that has a positive charge on one atom and a negative charge on another atom.* Note that the net charge on a zwitterion is zero even though parts of the molecule carry charges. In solution and also in the solid state, α-amino acids are zwitterions.

Zwitterion structure changes when the pH of a solution containing an amino acid is changed from neutral either to acidic (low pH) by adding an acid such as HCl or to basic (high pH) by adding a base such as NaOH. In an acidic solution, the zwitterion accepts a proton (H^+) to form a positively charged ion.

▶ From this point on in the text, the structures of amino acids will be drawn in their zwitterion form unless information given about the pH of the solution indicates otherwise.

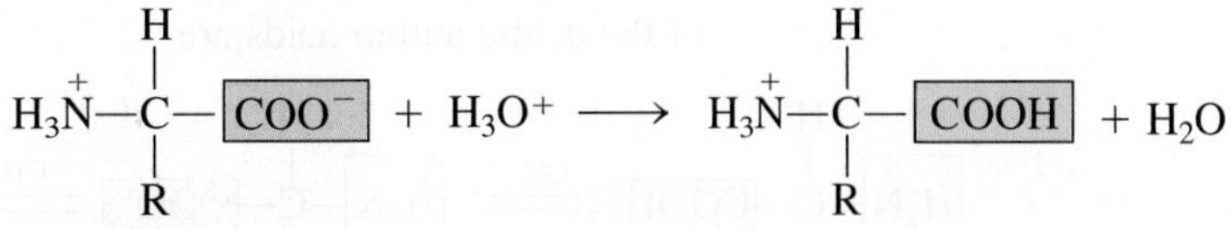

$$H_3\overset{+}{N}—\underset{R}{\overset{H}{C}}—COO^- + H_3O^+ \longrightarrow H_3\overset{+}{N}—\underset{R}{\overset{H}{C}}—COOH + H_2O$$

Zwitterion (no net charge) — Positively charged ion

In basic solution, the $—\overset{+}{N}H_3$ of the zwitterion loses a proton, and a negatively charged species is formed.

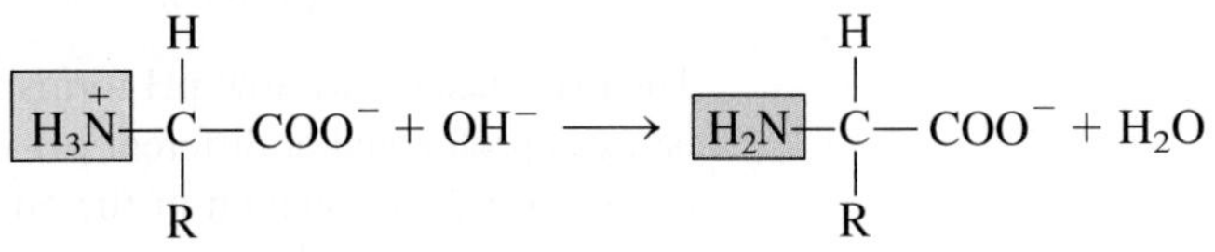

$$H_3\overset{+}{N}—\underset{R}{\overset{H}{C}}—COO^- + OH^- \longrightarrow H_2N—\underset{R}{\overset{H}{C}}—COO^- + H_2O$$

Zwitterion (no net charge) — Negatively charged ion

Thus, in solution, three different amino acid forms can exist (zwitterion, negative ion, and positive ion). The three species are actually in equilibrium with each other, and the equilibrium shifts with pH change. The overall equilibrium process can be represented as follows:

$$H_3\overset{+}{N}—\underset{R}{\overset{H}{C}}—COOH \underset{H_3O^+}{\overset{OH^-}{\rightleftharpoons}} H_3\overset{+}{N}—\underset{R}{\overset{H}{C}}—COO^- \underset{H_3O^+}{\overset{OH^-}{\rightleftharpoons}} H_2N—\underset{R}{\overset{H}{C}}—COO^-$$

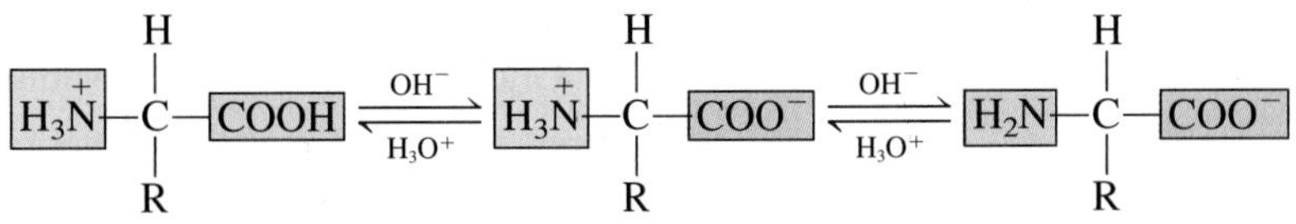

Acidic solution (low pH) — Neutral solution (pH = 7.0) — Basic solution (high pH)

▶ Guidelines for amino acid form as a function of solution pH are

Low pH: All acid groups are protonated (—COOH). All amino groups are protonated ($—\overset{+}{N}H_3$).

High pH: All acid groups are deprotonated ($—COO^-$). All amino groups are deprotonated ($—NH_2$).

Neutral pH: All acid groups are deprotonated ($—COO^-$). All amino groups are protonated ($—\overset{+}{N}H_3$).

In acidic solution, the positively charged species on the left predominates; nearly neutral solutions have the middle species (the zwitterion) as the dominant species; in basic solution, the negatively charged species on the right predominates.

Example 16.1 Determining Amino Acid Form in Solutions of Various pH

Draw an appropriate structural form for the amino acid alanine that predominates in solution at each of the following pH values.

a. pH = 1.0 **b.** pH = 7.0 **c.** pH = 11.0

Solution

At low pH, both amino and carboxyl groups are protonated. At high pH, both groups have lost their protons. At neutral pH, the zwitterion is present.

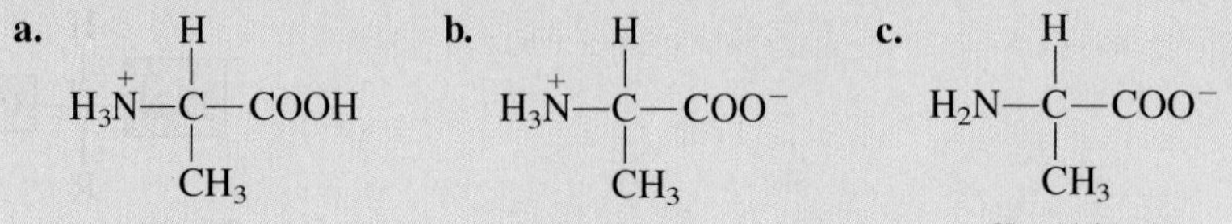

a. $H_3\overset{+}{N}—C(H)(CH_3)—COOH$ pH = 1.0 (net charge of +1)

b. $H_3\overset{+}{N}—C(H)(CH_3)—COO^-$ pH = 7.0 (no net charge)

c. $H_2N—C(H)(CH_3)—COO^-$ pH = 11.0 (net charge of −1)

▶ The term *protonated* denotes gain of a H^+ ion, and the term *deprotonated* denotes loss of a H^+ ion.

The previous discussion assumed that the side chain (R group) of an amino acid remains unchanged in solution as the pH is varied. This is the case for neutral amino acids but not for acidic or basic ones. For these latter compounds, the side chain can also acquire a charge, because it contains an amino or a carboxyl group that can, respectively, gain or lose a proton.

Because of the extra site that can be protonated or deprotonated, acidic and basic amino acids have four charged forms in solution. These four forms for aspartic acid, one of the acidic amino acids are

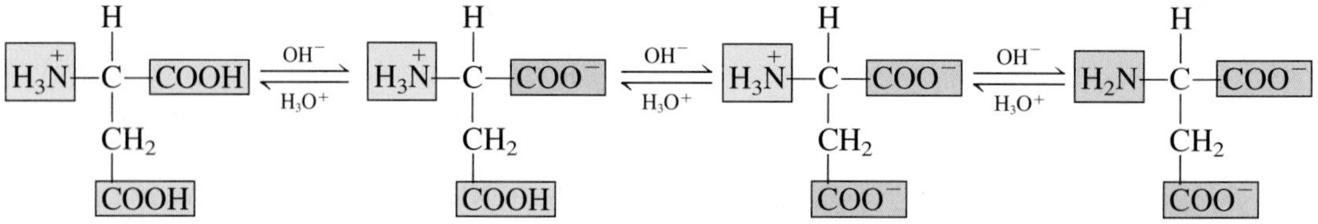

$H_3\overset{+}{N}—C(H)(CH_2—COOH)—COOH \underset{H_3O^+}{\overset{OH^-}{\rightleftharpoons}} H_3\overset{+}{N}—C(H)(CH_2—COOH)—COO^- \underset{H_3O^+}{\overset{OH^-}{\rightleftharpoons}} H_3\overset{+}{N}—C(H)(CH_2—COO^-)—COO^- \underset{H_3O^+}{\overset{OH^-}{\rightleftharpoons}} H_2N—C(H)(CH_2—COO^-)—COO^-$

Low-pH form (+1 charge); Moderately low-pH form (no net charge) (zwitterion); Neutral-pH form (−1 net charge); High-pH form (−2 net charge)

The existence of two low-pH forms for aspartic acid results from the two carboxyl groups being deprotonated at different pH values. For basic amino acids, two high-pH forms exist because deprotonation of the amino groups does not occur simultaneously. The side-chain amino group deprotonates before the α-amino group.

Practice Problems and Questions

16.15 Amino acids are both acids and bases. Explain why this is so.

16.16 Describe the net effect of the *internal* acid–base reaction that occurs in amino acid molecules.

16.17 Draw the structures of the following amino acids in both the un-ionized and the zwitterion forms.
a. Glycine b. Alanine c. Valine d. Threonine

16.18 Draw the structure of the zwitterion form of the following amino acids.
a. Serine b. Cysteine c. Leucine d. Isoleucine

16.19 Draw the structure of the ionized low-pH form of each of the amino acids in Problem 16.18.

16.20 Draw the structure of the ionized high-pH form of each of the amino acids in Problem 16.18.

16.21 Glutamic acid, a polar acidic amino acid, exists in two low-pH forms instead of the usual one. Explain why.

> **Learning Focus**
> Name and describe the bond that links amino acids together in peptides; draw complete structural formulas for simple peptides.

16.5 Peptide Formation

In Section 13.11, we learned that a carboxylic acid and an amine can be considered the "parent" molecules for an amide. Their combination to produce the amide can be diagramed as follows:

$$\underset{\text{Acid}}{R-\overset{O}{\overset{\|}{C}}-OH} + \underset{\text{Amine}}{H-\overset{H}{\overset{|}{N}}-R} \longrightarrow \underset{\text{Amide}}{R-\overset{O}{\overset{\|}{C}}-\overset{H}{\overset{|}{N}}-R} + H_2O$$

Two amino acids can combine in a similar way—the carboxyl group of one amino acid interacts with the amino group of the other amino acid. The products are a molecule of water and a molecule containing the two amino acids linked by an amide bond.

$$H_3\overset{+}{N}-\underset{R_1}{\overset{H}{\overset{|}{\underset{|}{C}}}}-COO^- + H_3\overset{+}{N}-\underset{R_2}{\overset{H}{\overset{|}{\underset{|}{C}}}}-COO^- \longrightarrow H_3\overset{+}{N}-\underset{R_1}{\overset{H}{\overset{|}{\underset{|}{C}}}}-\underbrace{\overset{O}{\overset{\|}{C}}-\overset{H}{\overset{|}{N}}}_{\text{Amide bond}}-\underset{R_2}{\overset{H}{\overset{|}{\underset{|}{C}}}}-COO^- + H_2O$$

Removal of the elements of water from the reacting carboxyl and amino groups and the ensuing formation of the amide bond are better visualized when expanded structural formulas for the reacting groups are used.

$$\underset{\substack{\text{Carboxyl}\\ \text{group}\\ (-COO^-)}}{-\overset{O}{\overset{\|}{C}}-O} + \underset{\substack{\text{Amino}\\ \text{group}\\ (H_3\overset{+}{N}-)}}{H-\overset{H}{\overset{|}{\underset{\underset{H}{|}}{N}}}-} \longrightarrow \underset{\substack{\text{Amide}\\ \text{bond}}}{-\overset{O}{\overset{\|}{C}}-\overset{H}{\overset{|}{N}}-} + H_2O$$

In amino acid chemistry, amide bonds that link amino acids together are given the specific name of peptide bond. A **peptide bond** *is a bond between the carboxyl group of one amino acid and the amino group of another amino acid.*

Under proper conditions, many amino acids can bond together to give chains of amino acids containing numerous peptide bonds. For example, four peptide bonds are present in a chain of five amino acids.

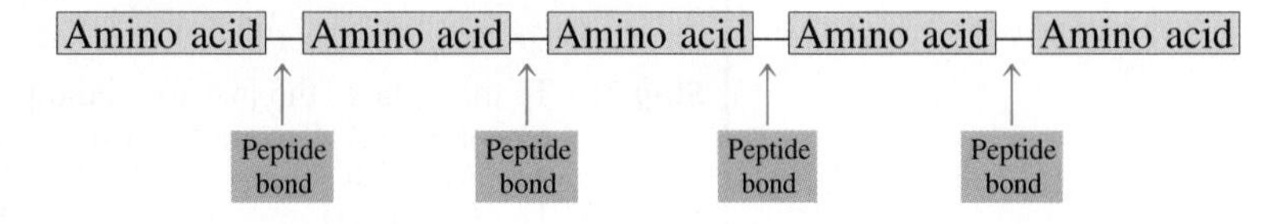

Chains of amino acids are known as peptides. A **peptide** *is a sequence of amino acids, in which the amino acids are joined together through amide (peptide) bonds.* A compound containing two amino acids joined by a peptide bond is specifically called a *dipeptide;* three amino acids in a chain constitute a *tripeptide;* and so on. The name *oligopeptide* is loosely used to refer to peptides with 10 to 20 amino acid residues and *polypeptide* to larger peptides.

In all peptides, the amino acid at one end of the amino acid sequence has a free $H_3\overset{+}{N}$ group, and the amino acid at the other end of the sequence has a free COO^- group. The end with the free $H_3\overset{+}{N}$ group is called the *N-terminal end,* and the end with the free COO^- group is called the *C-terminal end.* By convention, the sequence of amino acids in a peptide is written with the N-terminal end amino acid at the left. The individual amino acids within a peptide chain are called *amino acid residues.*

► A peptide chain has *directionality* because its two ends are different. There is an N-terminal end and a C-terminal end. By convention, the direction of the peptide chain is always.

N-terminal end ⟶ C-terminal end

The N-terminal end is always on the left, and the C-terminal end is always on the right.

The structural formula for a polypeptide may be written out in full, or the sequence of amino acids present may be indicated by using the standard three-letter amino acid abbreviations. The abbreviated formula for the tripeptide

$$\text{N-terminal end} \longrightarrow \overset{+}{H_3N}-\underset{H}{\overset{H}{C}}-\overset{O}{\overset{\|}{C}}-\overset{H}{N}-\underset{CH_3}{\overset{H}{C}}-\overset{O}{\overset{\|}{C}}-\overset{H}{N}-\underset{\underset{OH}{CH_2}}{\overset{H}{C}}-COO^- \longleftarrow \text{C-terminal end}$$

Glycine Alanine Serine

which contains the amino acids glycine, alanine, and serine, is Gly–Ala–Ser. When we use this abbreviated notation, by convention, the amino acid at the N-terminal end of the peptide is always written on the left.

Example 16.2 Converting an Abbreviated Peptide Formula to a Structural Peptide Formula

Draw the structural formula for the tripeptide Ala–Gly–Val.

Solution

Step 1: The N-terminal end of the peptide involves alanine. Its structure is written first.

$$\overset{+}{H_3N}-\underset{CH_3}{\overset{H}{C}}-COO^-$$

Step 2: The structure of glycine is written to the right of the alanine structure and a peptide bond is formed between the two amino acids by removing the elements of H_2O and bonding the N of glycine to the carboxyl C of alanine.

$$\overset{+}{H_3N}-\underset{CH_3}{\overset{H}{C}}-COO^- + \overset{+}{H_3N}-\underset{H}{\overset{H}{C}}-COO^- \longrightarrow$$

$$\overset{+}{H_3N}-\underset{CH_3}{\overset{H}{C}}-\overset{O}{\overset{\|}{C}}-\overset{H}{N}-\underset{H}{\overset{H}{C}}-COO^- + H_2O$$

Step 3: To the right of the just-formed dipeptide, draw the structure of valine. Then repeat Step 2 to form the desired tripeptide.

$$\overset{+}{H_3N}-\underset{CH_3}{\overset{H}{C}}-\overset{O}{\overset{\|}{C}}-\overset{H}{N}-\underset{H}{\overset{H}{C}}-COO^- + \overset{+}{H_3N}-\underset{\underset{CH_3}{CH-CH_3}}{\overset{H}{C}}-COO^- \longrightarrow$$

$$\overset{+}{H_3N}-\underset{CH_3}{\overset{H}{C}}-\overset{O}{\overset{\|}{C}}-\overset{H}{N}-\underset{H}{\overset{H}{C}}-\overset{O}{\overset{\|}{C}}-\overset{H}{N}-\underset{\underset{CH_3}{CH-CH_3}}{\overset{H}{C}}-COO^- + H_2O$$

The repeating chain of peptide bonds and α-carbon atoms in a peptide is referred to as the *backbone* of the peptide.

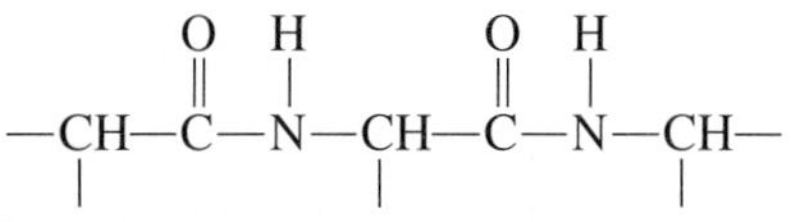

$$-\underset{|}{CH}-\overset{O}{\overset{\|}{C}}-\overset{H}{\overset{|}{N}}-\underset{|}{CH}-\overset{O}{\overset{\|}{C}}-\overset{H}{\overset{|}{N}}-\underset{|}{CH}-$$

The backbones of all peptides are the same except for length. It is the R groups that are attached to the backbone that distinguish one peptide from another.

Peptides that contain the same amino acids but in different order are different molecules (structural isomers) with different properties. For example, two different dipeptides can be formed from one molecule of alanine and one molecule of glycine.

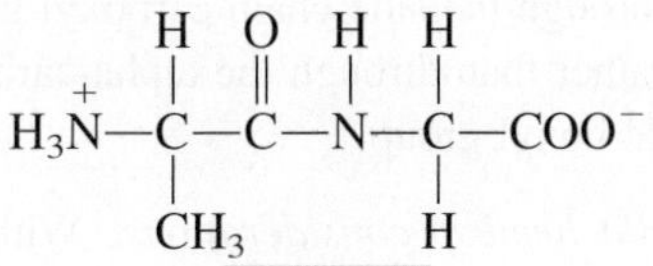

$$H_3\overset{+}{N}-\overset{H}{\overset{|}{\underset{\underset{CH_3}{|}}{C}}}-\overset{O}{\overset{\|}{C}}-\overset{H}{\overset{|}{N}}-\overset{H}{\overset{|}{\underset{\underset{H}{|}}{C}}}-COO^- \qquad H_3\overset{+}{N}-\overset{H}{\overset{|}{\underset{\underset{H}{|}}{C}}}-\overset{O}{\overset{\|}{C}}-\overset{H}{\overset{|}{N}}-\overset{H}{\overset{|}{\underset{\underset{CH_3}{|}}{C}}}-COO^-$$

Ala–Gly 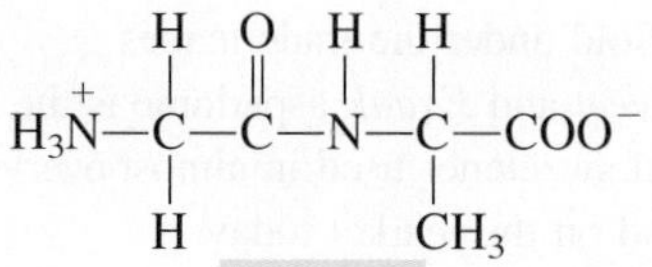   Gly–Ala

In the first dipeptide, the alanine is the N-terminal residue, and in the second molecule, it is the C-terminal residue. These two compounds are isomers with different chemical and physical properties.

▶ Amino acid sequence in a peptide has biochemical importance. Isomeric peptides give different biochemical responses; that is, they have different biochemical specificities.

The number of isomeric peptides possible increases rapidly as the length of the peptide chain increases. Let us consider the tripeptide Ala–Ser–Cys as another example. In addition to this sequence, five other arrangements of these three components are possible, each representing another isomeric tripeptide: Ala–Cys–Ser, Ser–Ala–Cys, Ser–Cys–Ala, Cys–Ala–Ser, and Cys–Ser–Ala. For a pentapeptide containing 5 different amino acids, 120 isomers are possible.

More than two hundred peptides have been isolated and identified as essential to the proper functioning of the human body. In general, these substances serve as hormones or neurotransmitters. Their functions range from controlling pain to controlling muscle contraction or kidney fluid excretion. Chemical Portraits 26 considers three biochemically important small peptides.

Practice Problems and Questions

16.22 What is a peptide?

16.23 What is a peptide bond?

16.24 What two functional groups are involved in the formation of a peptide bond?

16.25 What is meant by the N-terminal end and the C-terminal end of a peptide?

16.26 Write out the full structure of the tripeptide Val–Phe–Cys.

16.27 Explain why the notations Ser–Cys and Cys–Ser represent two different dipeptides rather than the same dipeptide.

16.28 There are a total of six different sequences for a tripeptide containing one molecule each of serine, valine, and glycine. Using three-letter abbreviations for the amino acids, draw the six possible sequences of amino acids.

16.29 Identify the amino acids present in each of the following tripeptides.

a.
$$H_3\overset{+}{N}-\underset{\underset{\underset{OH}{|}}{\underset{CH_2}{|}}}{CH}-\overset{O}{\overset{\|}{C}}-\overset{H}{\overset{|}{N}}-\underset{\underset{CH_3}{|}}{CH}-\overset{O}{\overset{\|}{C}}-\overset{H}{\overset{|}{N}}-\underset{\underset{\underset{SH}{|}}{\underset{CH_2}{|}}}{CH}-COO^-$$

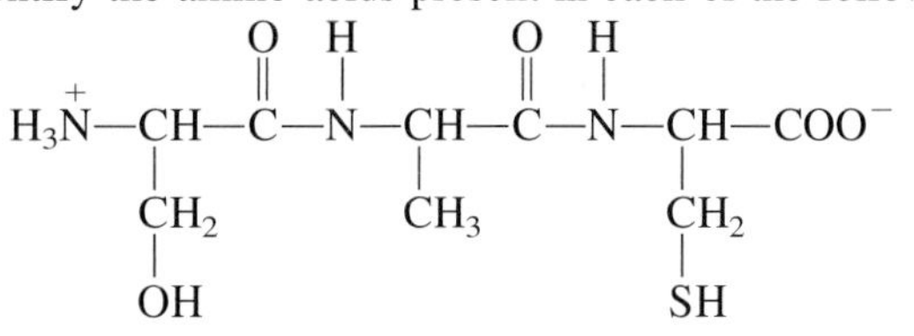

Chemical Portraits 26 Biochemically Important Small Peptides

Aspartame (Asp-Phe)

Profile: The dipeptide aspartame has a sweet taste. Both amino acids present in aspartame must be in the L-form for the sweet taste to occur; the L-D, D-L, and D-D forms have a bitter taste. Aspartame does not have a "pure" dipeptide structure; the carboxyl group of Phe has been esterified with a methyl group.

Uses: Sold under the trade names *Nutrasweet* and *Equal,* aspartame is the artificial sweetener used in almost every diet food on the market today. Aspartame's sweetness is 180 times that of sucrose; its caloric content is the same as sucrose. But because so little is used, its caloric contribution is negligible. It has no bitter aftertaste as do some artificial sweeteners.

Within the human body, how is aspartame processed?

Glutathione (Glu-Cys-Gly)

Profile: The tripeptide glutathione, produced by the body itself, is present in significant concentrations in most cells and is of considerable physiological importance as a regulator of oxidation-reduction reactions. An unusual feature in glutathione's structure is the manner in which Glu (an acidic amino acid) is bonded to Cys. Glu is bonded to Cys through the side-chain carboxyl group rather than through the alpha-carbon carboxyl group.

Biochemical considerations: Within cells glutathione functions as an antioxidant, protecting cellular contents from oxidizing agents such as peroxides and superoxides (highly reactive forms of oxygen often generated within a cell).

Within cells, what is the source of oxidizing agents such as peroxides and superoxides?

Enkephalins (Tyr-Gly-Gly-Phe-Leu) (Tyr-Gly-Gly-Phe-Met)

Profile: Enkephalins, of which there are two, are pentapeptides produced by the brain itself that bind at receptor sites in the brain to reduce pain. The two enkephalins differ only in the amino acid at the carboxyl end of the peptide chain.

Biochemical considerations: The pain-reducing effects of enkephalin action play a role in the "high" reported by long-distance runners, the competitive athlete who manages to finish the game despite being injured, and in the pain-relieving effects of acupuncture. The action of the prescription painkillers morphine and codeine is based on their binding at the same receptor sites in the brain as the naturally occurring enkephalins.

How do morphine and codeine compare to enkephalins in terms of structure?

See the text web site at **www.cengage.com/chemistry/stoker** for answers to the above questions and for further information.

b.
```
          O  H      O  H
          ‖  |      ‖  |
  +
H3N—CH—C—N—CH—C—N—CH—COO⁻
     |         |         |
    CH2      CH—OH      CH2
     |         |         |
    COO⁻      CH3       C—NH2
                         ‖
                         O
```

16.30 How many peptide linkages are present in each of the tripeptides in Problem 16.29?

16.31 How many carbon atoms are present in the "backbone" of each of the tripeptides in Problem 16.29?

16.6 Levels of Protein Structure

Learning Focus

Distinguish between polypeptides and proteins; list the four levels of substructure associated with a protein.

A **protein** *is a polypeptide that contains more than 50 amino acid residues.* The dividing line (50) between the use of the terms polypeptide and protein is actually somewhat arbitrary. The important point is that proteins are polymers that contain a large number of amino acid units linked by peptide bonds. Polypeptides are shorter chains of amino acids. Some proteins have molecular masses in the millions. Some proteins also contain more than one polypeptide chain.

▶ Proteins are the second type of biochemical polymer we have encountered; the other was polysaccharides (Section 14.10). Protein monomers are amino acids, whereas polysaccharide monomers are monosaccharides.

To aid us in describing protein structure, we will consider four levels of substructure: primary, secondary, tertiary, and quaternary. Even though we consider these structure levels one by one, remember that it is the combination of all four levels of structure that controls protein function.

Practice Problems and Questions

16.32 Explain the statement "All proteins are polypeptides, but not all polypeptides are proteins."

16.33 What are the four levels of protein substructure?

Learning Focus

Understand the concept of primary structure of a protein.

▶ The primary structure of a protein is the *sequence* of amino acids in a protein chain—that is, the order in which the amino acids are connected to each other.

16.7 Primary Structure of Proteins

The **primary structure of a protein** *is the order in which the amino acids are linked together in the protein.* Every protein has its own *unique* amino acid sequence.

A segment of a protein with the primary structure

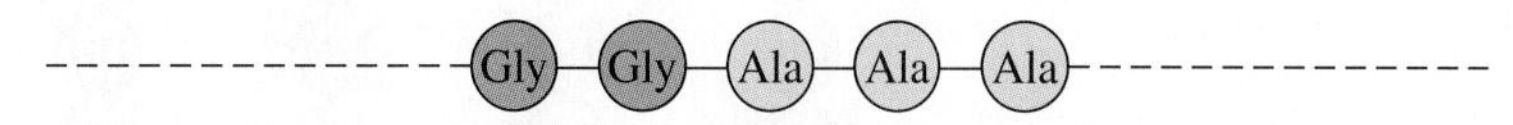

has a different primary structure than a protein segment with the primary structure

even though both primary structures involve three alanine and two glycine amino acid residues. Primary protein structure involves more than just the numbers and kinds of amino acids present; it also involves the *order of attachment* of the amino acids.

Insulin, the hormone that regulates blood-glucose level, was the first protein for which primary structure was determined; the "sequencing" of its 51 amino acids was completed in 1953, after 8 years of work. Today, primary structures are known for thousands of proteins, and the sequencing procedures involve automated methods that require relatively short periods of time. Figure 16.2 shows the primary structure of myoglobin, a protein involved in oxygen transport in muscles; it contains 153 amino acids assembled in the particular, definite order shown.

The primary structure of a specific protein is always the same, regardless of where the protein is found within an organism. The structures of certain proteins are even similar among different species of animals. For example, the primary structures of insulin in cows, pigs, sheep, and horses are very similar both to each other and to human insulin.

An analogy is often drawn between the primary structure of proteins and words. Words, which convey information, are formed when the 26 letters of the English alphabet are properly sequenced. Proteins are formed from proper sequences of the 20 standard amino acids. The proper sequence of letters in a word is necessary for it to make sense, just as the proper sequence of amino acids is necessary to make biologically active protein. Furthermore, the letters that form a word are written from left to right, as are amino acids in protein formulas. As any dictionary of the English language will document, a tremendous variety of words can be formed by different letter sequences. Imagine the number of amino acid sequences possible for a large protein. There are 1.55×10^{66} sequences possible for the 51 amino acids found in insulin! From these possibilities, the body reliably produces only *one,* illustrating the remarkable precision of life processes. From the simplest bacterium to the human brain cell, only those amino acid sequences needed by the cell are produced. The fascinating process of protein biosynthesis and the way in which genes in DNA direct this process will be discussed in Chapter 17.

Practice Problems and Questions

16.34 What is meant by the *primary structure* of a protein?

16.35 How can two proteins with the same numbers and types of amino acids have *different* primary structures?

16.36 What types of bonds are responsible for the primary structure of a protein?

16.37 A segment of a protein contains two alanine and two glycine amino acid residues. How many different primary structures are possible for this segment of the protein?

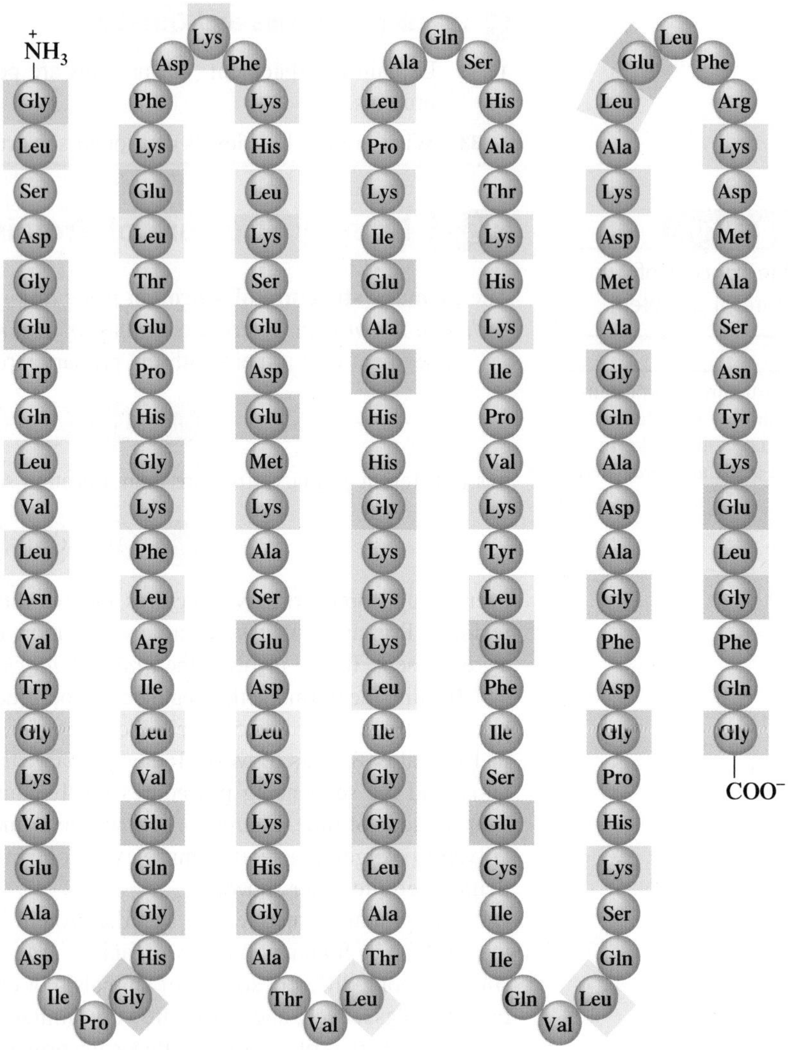

$\overset{+}{NH_3}$ – Gly – Leu – Ser – Asp – Gly – Glu – Trp – Gln – Leu – Val – Leu – Asn – Val – Trp – Gly – Lys – Val – Glu – Ala – Asp – Ile – Pro – Gly – His – Gly – Gln – Glu – Val – Leu – Ile – Arg – Leu – Phe – Lys – Gly – His – Pro – Glu – Thr – Leu – Glu – Lys – Phe – Asp – Lys – Phe – Lys – His – Leu – Lys – Ser – Glu – Asp – Glu – Met – Lys – Ala – Ser – Glu – Asp – Leu – Lys – Lys – His – Gly – Ala – Thr – Val – Leu – Thr – Ala – Leu – Gly – Gly – Ile – Leu – Lys – Lys – Lys – Gly – His – His – Glu – Ala – Glu – Ile – Lys – Pro – Leu – Ala – Gln – Ser – His – Ala – Thr – Lys – His – Lys – Ile – Pro – Val – Lys – Tyr – Leu – Glu – Phe – Ile – Ser – Glu – Cys – Ile – Ile – Gln – Val – Leu – Gln – Ser – Lys – His – Pro – Gly – Asp – Phe – Gly – Ala – Asp – Ala – Gln – Gly – Ala – Met – Asp – Lys – Ala – Leu – Glu – Leu – Phe – Arg – Lys – Asp – Met – Ala – Ser – Asn – Tyr – Lys – Glu – Leu – Gly – Phe – Gln – Gly – COO^-

Figure 16.2 The primary structure of human myoglobin. This diagram gives only the sequence of the amino acids present and conveys no information about the actual three-dimensional shape of the protein. The "wavy" pattern for the amino acid sequence was chosen to minimize the space used to present the needed information. The actual shape of the protein is determined by secondary and tertiary levels of protein structure, levels yet to be discussed.

16.8 Secondary Structure of Proteins

Learning Focus

Understand the type of attractive interaction that contributes to protein secondary structure, and describe the two common types of secondary structure.

The **secondary structure of a protein** *is the arrangement in space adopted by the backbone portion of the protein.* There are two common types of protein secondary structure: (1) α helix and (2) β pleated sheet. The major force responsible for both types of secondary structure is hydrogen bonding (Section 6.13) between a carbonyl oxygen atom of a peptide linkage and the hydrogen atom of an amino group (—NH) of another peptide linkage farther along the backbone. This hydrogen-bonding interaction may be diagrammed as follows:

$$>N{-}H\cdots\cdots O{=}C<$$

The Alpha Helix

The *alpha helix* (α helix) structure resembles a coiled helical spring, with the coil configuration maintained by hydrogen bonds between $>N{-}H$ and $>C{=}O$ groups of every fourth amino acid, as is shown diagrammatically in Figure 16.3.

Proteins have varying amounts of α-helical secondary structure, ranging from a few percent to nearly 100%. In an α helix, all of the amino acid side chains (R groups) lie outside the helix; there is not enough room for them in the interior. Figure 16.3d illustrates this situation. This structural feature of the α helix is the basis for protein tertiary structure (Section 16.9).

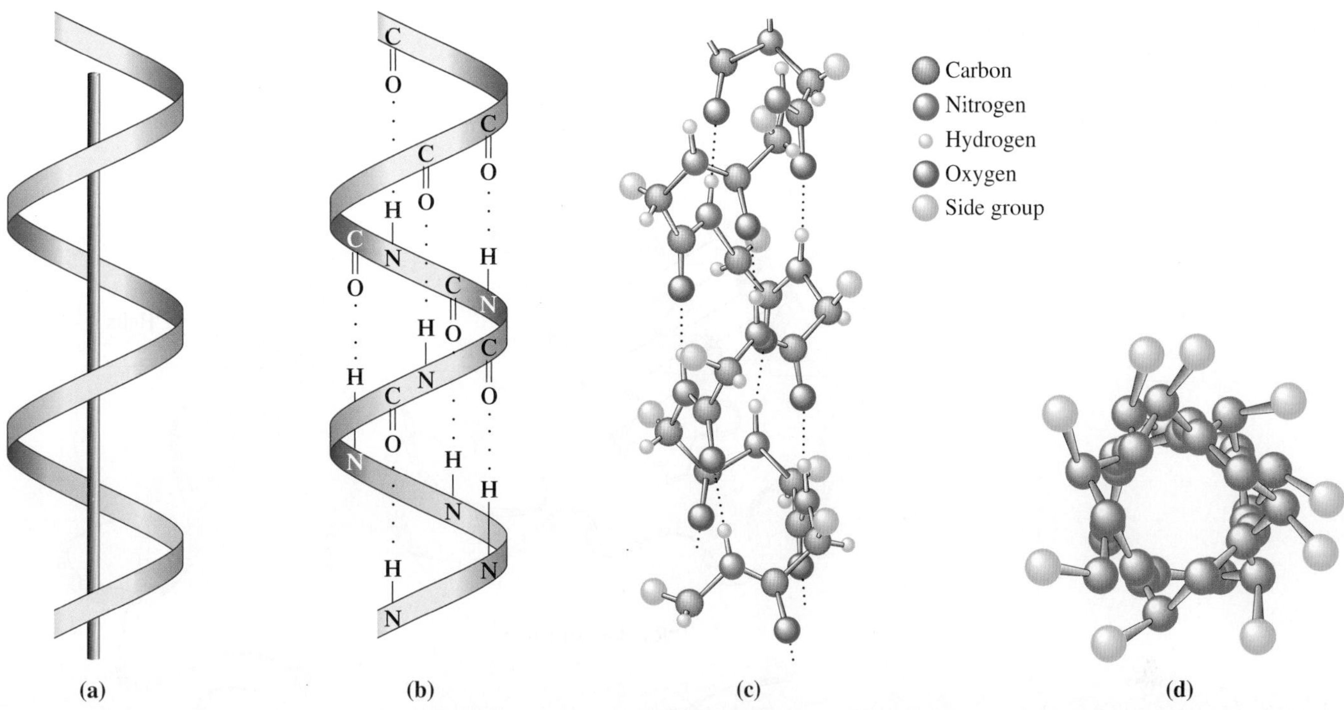

Figure 16.3 Four representations of the α helix protein structure. (a) Arrangement of protein backbone with no structural detail shown. (b) Backbone arrangement with hydrogen-bonding interactions shown. (c) Backbone atomic detail shown, as well as hydrogen-bonding. (d) Top view of an α-helix showing that amino acid side chains (R groups) point away from the long axis of the helix.

▶ The hydrogen bonding present in an α helix is *intra*molecular. In a β pleated sheet, the hydrogen bonding can be *inter*molecular (between two different chains) or *intra*molecular (a single chain folding back on itself).

The Beta Pleated Sheet

The *beta pleated sheet* (β pleated sheet) secondary structure involves amino acid chains (backbones) that are almost completely extended. Hydrogen bonds form between atoms that are either in different parts of a single chain that folds back on itself (intrachain bonds) or in different polypeptide chains in those proteins that contain more than one chain (interchain bonds). (Many proteins exist in which several polypeptide chains are present.)

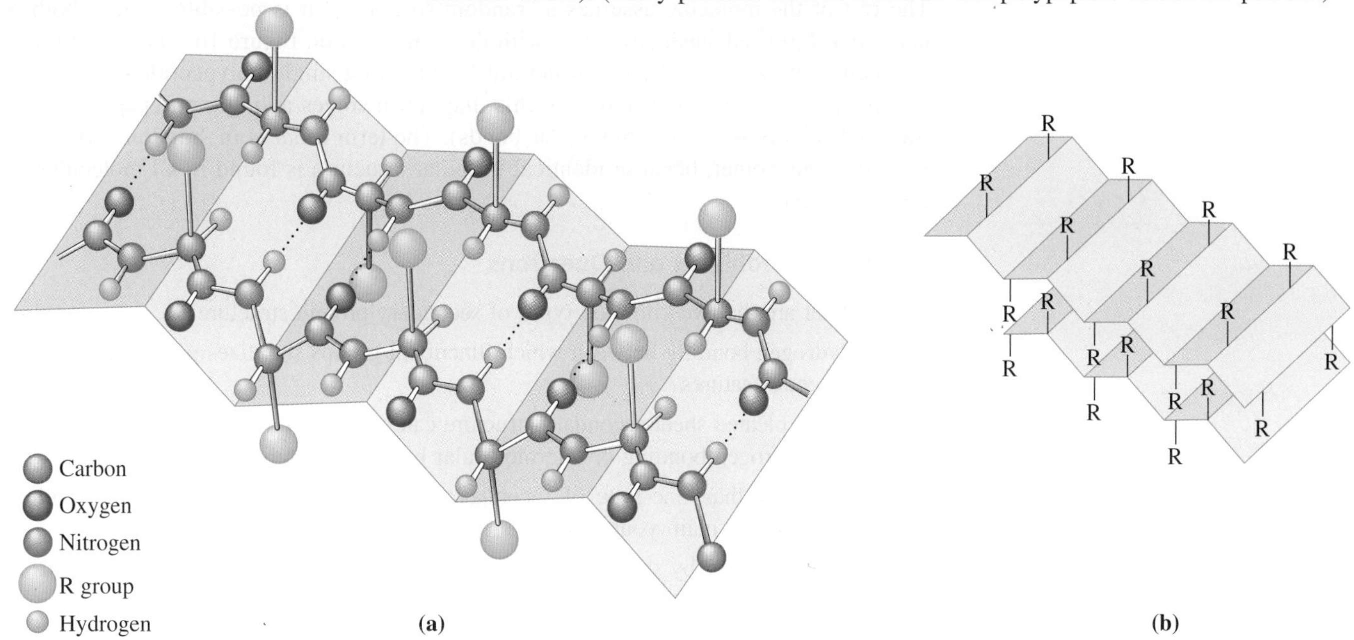

Figure 16.4 Two representations of the β pleated sheet protein structure. (a) A representation emphasizing the hydrogen bonds between protein chains. (b) A representation emphasizing the pleats and the location of the R groups.

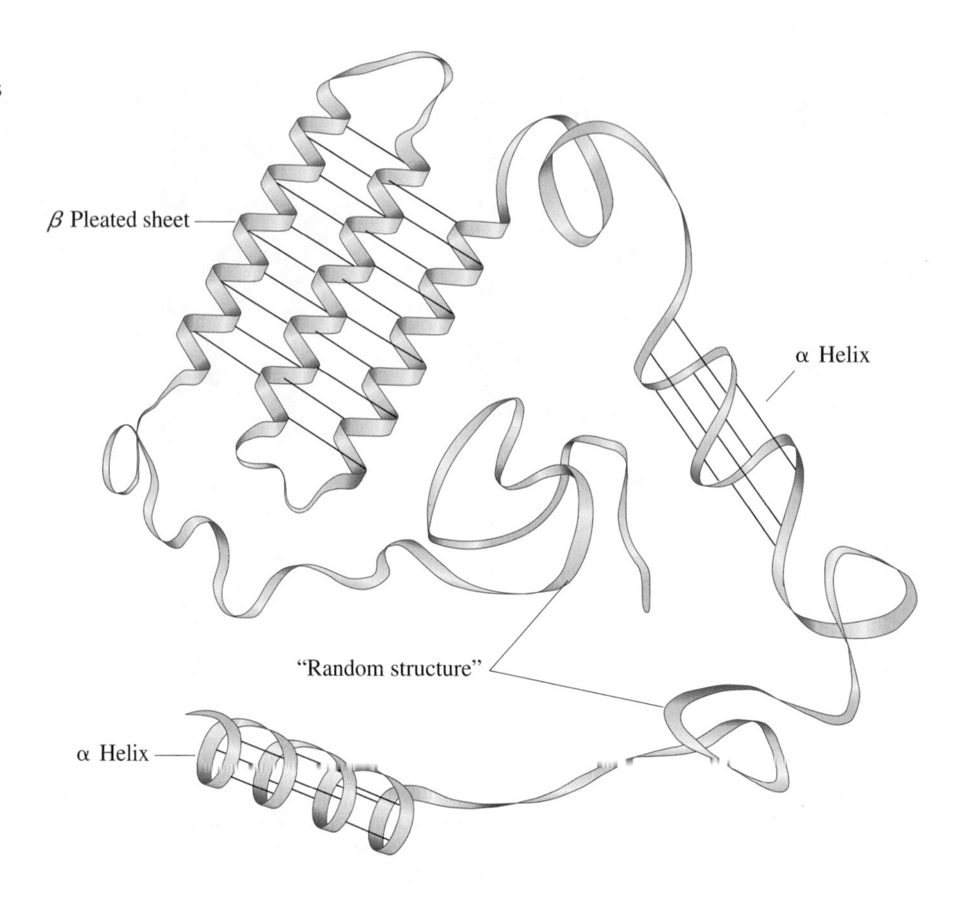

Figure 16.5 The secondary structure of a single protein often shows areas of α helix and β pleated sheet configurations, as well as areas of random coiling.

Figure 16.4a shows a representation of the β pleated sheet structure in the situation where portions of two different polypeptide chains are aligned parallel to each other. The term *pleated sheet* arises from the repeated zigzag pattern in the structure (Figure 16.4b). Amino acid side chains are located above and below the plane of the sheet.

Very few proteins have 100% α helix or β pleated sheet structures. Instead, most proteins have only certain portions of their amino acid residues in these conformations. The rest of the molecule assumes a "random structure." It is possible to have both α helix and β pleated sheet structures with the same protein. Figure 16.5 is a diagram of a protein that has both of these structural features in a single polypeptide chain. Note that the β pleated sheet structure in this diagram involves a single polypeptide chain folding back on itself (intramolecular bonds). The term *random* in "random structure" is really a misnomer, because identical irregular structure is found in all molecules of a given protein.

Practice Problems and Questions

16.38 What are the two common types of secondary protein structures?

16.39 Hydrogen bonding between which functional groups stabilizes protein secondary-structure features?

16.40 The β pleated sheet secondary structure can be formed through either intramolecular hydrogen bonding or intermolecular hydrogen bonding. Explain why.

16.41 Can more than one type of secondary structure be present in the same protein molecule? Explain your answer.

16.42 What is meant by the statement that a section of a protein has a "random structure" arrangement?

16.43 What happens to the primary structure of a protein when a protein adopts secondary-structure features?

Learning Focus

Recognize the four types of attractive interactions that contribute to the tertiary structure of a protein.

16.9 Tertiary Structure of Proteins

The **tertiary structure of a protein** *is the overall three-dimensional shape that results from the attractive forces between amino acid side chains (R groups) that are widely separated from each other within the chain.*

A good analogy for the relationships among the primary, secondary, and tertiary structures of a protein is that of a telephone cord (Figure 16.6). The primary structure is the long, straight cord. The coiling of the cord into a helical arrangement gives the secondary structure. The supercoiling arrangement the cord adopts after you hang up the receiver is the tertiary structure.

Interactions Responsible for Tertiary Structure

Four types of attractive interactions contribute to the tertiary structure of a protein: (1) covalent disulfide bonds, (2) electrostatic attractions (salt bridges), (3) hydrogen bonds, and (4) hydrophobic attractions. All four of these interactions are interactions between amino acid R groups. This is a major distinction between tertiary-structure interactions and secondary-structure interactions. Tertiary-structure interactions involve the R groups of amino acids; secondary-structure interactions involve the peptide linkages between amino acid units.

Disulfide bonds, the strongest of the tertiary-structure interactions, result from the —SH groups of two cysteine molecules reacting with each other to form a covalent disulfide. This type of interaction is the only one of the four tertiary-structure interactions that involves a covalent bond. Recall, from Section 12.6, that —SH groups are readily oxidized to give a disulfide bond, —S—S—.

$$\underset{\text{Cysteine}}{-CH_2-SH} \quad \underset{\text{Cysteine}}{HS-CH_2-} \xrightarrow{\text{Oxidation}} \underset{\text{Cystine}}{-CH_2-S-S-CH_2-} + 2H$$

▶ Cysteine is the only α-amino acid that contains a sulfhydryl group (—SH).

Disulfide bonds may involve two cysteine units in the same chain or in different chains. Insulin, a 51-amino-acid-residue two-chain protein (Figure 16.7) contains both types of disulfide bonds.

Electrostatic interactions, also called *salt bridges,* always involve amino acids with charged side chains. These amino acids are the acidic and basic amino acids. The two R groups, one acidic and one basic, interact through ion–ion attractions. Figure 16.8a shows an electrostatic interaction.

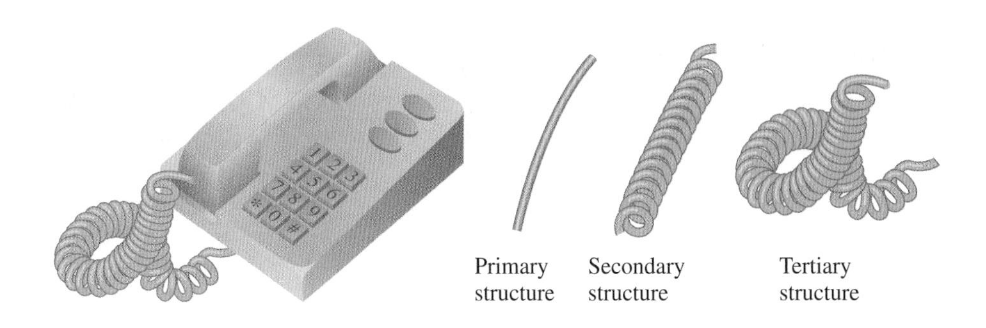

Figure 16.6 A telephone cord has three levels of structure. These structural levels are a good analogy for the first three levels of protein structure.

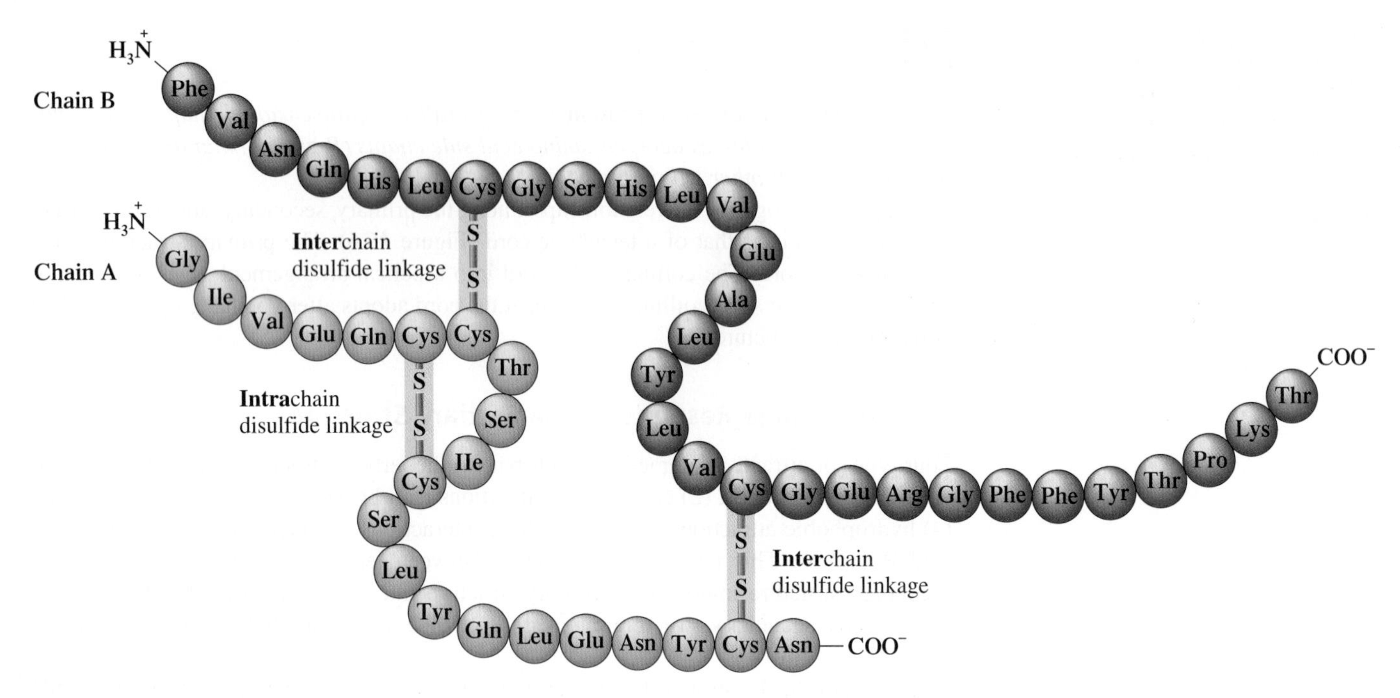

Figure 16.7 Human insulin, a small two-chain protein, has both intrachain and interchain disulfide linkages as part of its tertiary structure.

Hydrogen bonds can occur between amino acids with polar R groups. A variety of polar side chains can be involved, especially those that possess the following functional groups:

$$-OH \qquad -NH_2 \qquad -\overset{\overset{\displaystyle O}{\|}}{C}-OH \qquad -\overset{\overset{\displaystyle O}{\|}}{C}-NH_2$$

Hydrogen bonds are relatively weak and are easily disrupted by changes in pH and temperature. Figure 16.8b shows the hydrogen-bonding interactions between the R groups of glutamine and serine.

Hydrophobic interactions result when two nonpolar side chains are close to each other. In aqueous solution, many proteins have their polar R groups outward, toward the aqueous solvent (which is also polar), and their nonpolar R groups inward (away from the polar water molecules). The nonpolar R groups then interact with each other. Hydrophobic interactions are common between phenyl rings and alkyl side chains. Although hydrophobic interactions are weaker than hydrogen bonds or electrostatic interactions, they are a significant force in some proteins because there are so many of them; their cumulative effect can be greater in magnitude than the effects of hydrogen bonding. Figure 16.8c shows the hydrophobic interactions between the R groups of phenylalanine and leucine.

(a) Electrostatic interaction — Asp, Lys

(b) Hydrogen bond — Gln, Ser

(c) Hydrophobic interaction — Phe, Leu

Figure 16.8 Noncovalent R group interactions that contribute to the tertiary structure of a protein: (a) electrostatic interaction, (b) hydrogen bond, and (c) hydrophobic interaction.

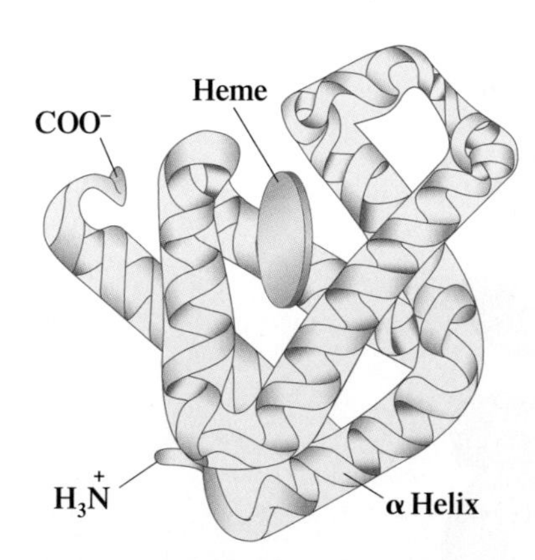

Figure 16.9 A schematic diagram showing the tertiary structure of the single-chain protein myoglobin.

In 1959, a protein tertiary structure was determined for the first time, that of the oxygen-storage protein myoglobin. Figure 16.9 shows myoglobin's tertiary structure. Besides amino acids, the structure also contains a heme group, an iron-containing group with the ability to bind oxygen (see Section 16.11). The primary structure of myoglobin appears in Figure 16.2. In comparing the two structures, note how different the perspectives of primary structure and tertiary structure are.

Practice Problems and Questions

16.44 Give the type of amino acid R group that is involved in each of the following interactions, which contribute to tertiary protein structure.
a. Hydrophobic interaction b. Hydrogen bond
c. Disulfide bond d. Electrostatic interaction

16.45 Classify each of the interactions listed in Problem 16.44 as (1) a covalent bond or (2) a noncovalent interaction.

16.46 What is the difference between the types of hydrogen bonds that occur in secondary and tertiary structures?

16.47 With the help of Table 16.1, specify the type of tertiary interaction (hydrophobic interaction, hydrogen bond, disulfide bond, or electrostatic interaction) that occurs when the R groups of the following pairs of amino acid residues interact.
a. Phenylalanine and leucine b. Arginine and glutamic acid
c. Two cysteines d. Serine and tyrosine

Learning Focus

Recognize the four types of attractive interactions that contribute to the tertiary structure of a protein.

16.10 Quaternary Structure of Proteins

Quaternary structure is the highest level of protein organization. It is found only in proteins that have structures involving two or more polypeptide chains that are independent of each other—that is, are not covalently bonded to each other. These multichain proteins are often called *oligomeric proteins*. The **quaternary structure of a protein** *involves the associations among the separate chains in an oligomeric protein.*

Most oligomeric proteins contain an even number of subunits (two subunits = a dimer, four subunits = a tetramer, and so on). The subunits are held together mainly by noncovalent interactions between amino acid R groups. The noncovalent interactions that contribute to tertiary structure (electrostatic interactions, hydrogen bonds, and hydrophobic interactions) are also responsible for the maintenance of quaternary structure.

The noncovalent interactions maintaining quaternary structure are more easily interrupted than those for tertiary structure. For example, only small changes in cellular conditions can cause a tetrameric protein to fall apart, dissociating into dimers or perhaps four separate subunits, with a resulting temporary loss of protein activity. As conditions change back, the oligomer automatically re-forms, and normal protein function is restored.

An example of a protein with quaternary structure is hemoglobin, the oxygen-carrying protein in blood (Figure 16.10). It is a tetramer in which there are two identical α chains and two identical β chains. Each chain enfolds a heme group, the site where oxygen binds to the protein.

Practice Problems and Questions

16.48 What is an *oligomeric* protein?

16.49 What is the quaternary structure of a protein?

16.50 What types of R group interactions contribute to quaternary structure?

The Chemistry at a Glance, on p. 453, reviews what we have said about levels of protein structure in this and the three preceding sections.

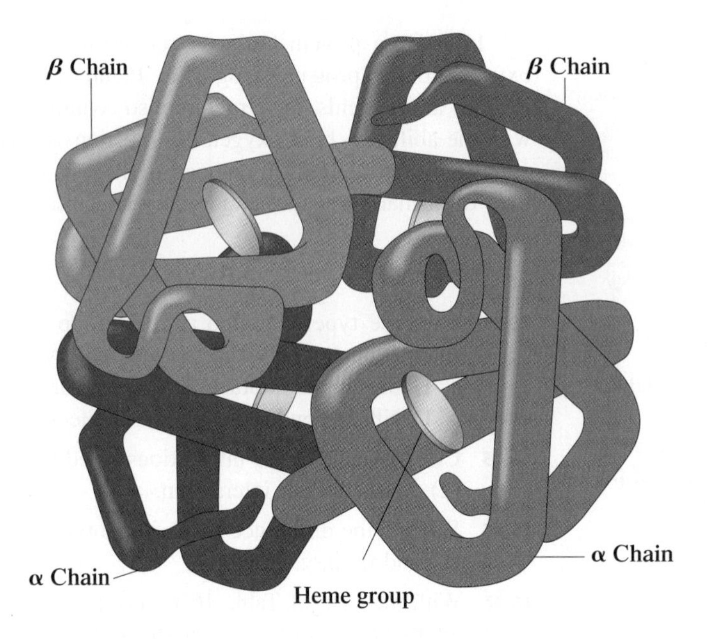

Figure 16.10 A schematic diagram showing the tertiary and quaternary structure of the oxygen-carrying protein hemoglobin.

Learning Focus

Classify proteins in terms of components present and in terms of general shape.

16.11 Protein Classifications and Functions

Proteins as classified as either simple proteins or conjugated proteins on the basis of the components present. A **simple protein** *contains only amino acid residues*. More than one polypeptide chain may be present, but all chains contain only amino acids. A **conjugated protein** *has other chemical components in addition to amino acids*. These additional components, which may be organic or inorganic, are called prosthetic groups. A **prosthetic group** *is a non-amino acid unit permanently associated with a protein*. Myoglobin and hemoglobin, whose tertiary structures were previously given, are both conjugated proteins whose prosthetic group is heme.

Conjugated proteins may be further classified according to the nature of the prosthetic group. For example, proteins containing lipids, those containing carbohydrates, and those containing metal ions are called lipoproteins, glycoproteins, and metalloproteins, respectively. Table 16.3 gives further examples of the types of conjugated proteins.

Table 16.3 Types of Conjugated Proteins

Class	Prosthetic group	Specific example	Function of example
hemoproteins	heme unit	hemoglobin	carrier of O_2 in blood
		myoglobin	oxygen binder in muscles
lipoproteins	lipid	low-density lipoprotein (LDL)	lipid carrier
		high-density lipoprotein (HDL)	lipid carrier
glycoproteins	carbohydrate	gamma globulin	antibody
		mucin	lubricant in mucous secretions
		interferon	antiviral protection
phosphoproteins	phosphate group	glycogen phosphorylase	enzyme in glycogen phosphorylation
nucleoproteins	nucleic acid	ribosomes	site for protein synthesis in cells
		viruses	self-replicating, infectious complex
metalloproteins	metal ion	iron–ferritin	storage complex for iron
		zinc–alcohol dehydrogenase	enzyme in alcohol oxidation

Protein Structure

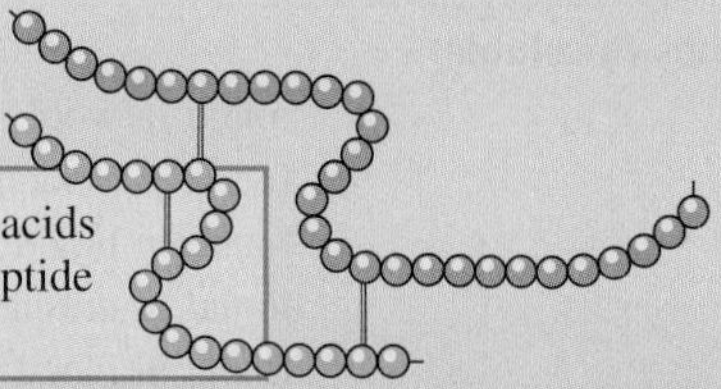

PRIMARY STRUCTURE	The sequence of amino acids present in a protein's peptide chain or chains		
SECONDARY STRUCTURE	The regularly repeating ordered spatial arrangements of amino acids near each other in the protein chain, that result from hydrogen bonds between carbonyl oxygen atoms and amino hydrogen atoms	**Alpha Helix**	Hydrogen bonds between every fourth amino acid
		Beta Pleated Sheet	Hydrogen bonds between two side-by-side chains, or a single chain that is folded back on itself
TERTIARY STRUCTURE	The overall three-dimensional shape that results from the attractive forces between amino acid side chains (R groups) that are not near each other in the protein chain		■ Disulfide Bonds ■ Electrostatic Interactions ■ Hydrogen Bonds ■ Hydrophobic Interactions
QUATERNARY STRUCTURE	The three-dimensional shape of a protein consisting of two or more independent polypeptide chains, that results from hydrophobic interactions between R groups		■ Electrostatic Interactions ■ Hydrogen Bonds ■ Hydrophobic Interactions

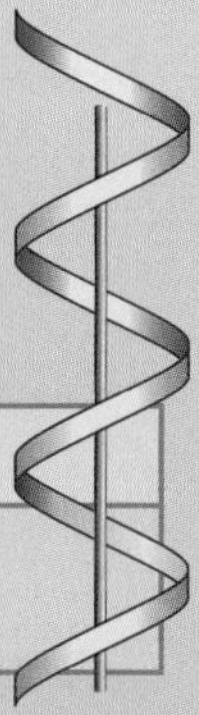

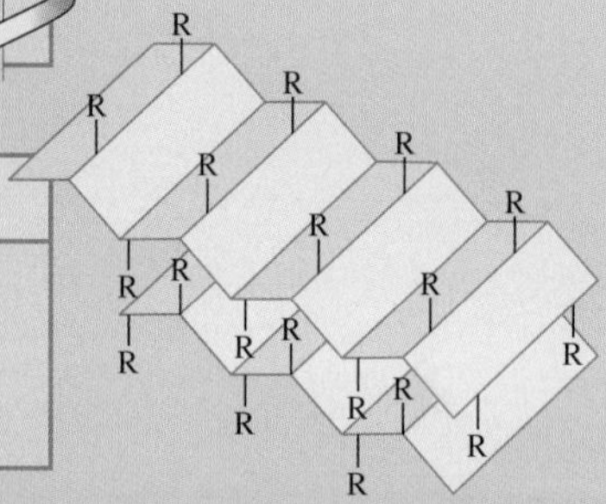

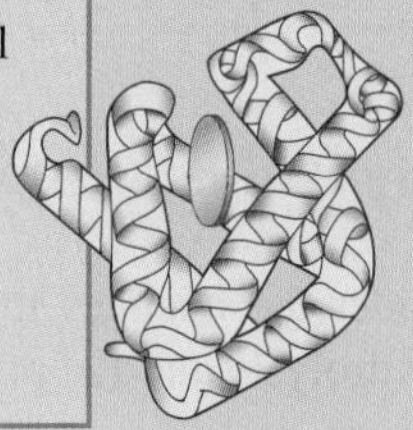

▶ Secondary structure is primarily responsible for the nature of fibrous proteins. By contrast, tertiary structure is of prime importance in determining the overall shapes of globular proteins.

On the basis of structural shape, proteins can be classified into two major types: fibrous proteins and globular proteins. A **fibrous protein** *is a protein in which polypeptide chains are arranged in long strands or sheets.* Such proteins have long rod-shaped or string-like molecules that can intertwine with one another and form strong fibers. They are water-insoluble and generally have structural functions within the human body. A **globular protein** *is a protein in which polypeptide chains are folded into spherical or globular shapes.* Globular proteins either dissolve in water or form stable suspensions in water, which allows them to travel through the blood and other body fluids to sites where their activity is needed. Table 16.4 gives examples of selected common fibrous and globular proteins.

Table 16.4
Some Common Fibrous and Globular Proteins

Name	Occurrence and function
Fibrous proteins (insoluble)	
keratins	found in wool, feathers, hooves, silk, and fingernails
collagens	found in tendons, bone, and other connective tissue
elastins	found in blood vessels and ligaments
myosins	found in muscle tissue
fibrin	found in blood clots
Globular proteins (soluble)	
insulin	regulatory hormone for controlling glucose metabolism
myoglobin	protein involved in oxygen transport in muscles
hemoglobin	protein involved in oxygen transport in blood
transferrin	protein involved in iron transport in blood
immunoglobulins	proteins involved in immune response

Practice Problems and Questions

16.51 What is the major difference between a simple protein and a conjugated protein?

16.52 What is a *prosthetic* group?

16.53 What prosthetic group is present in each of the following types of proteins?
a. Lipoproteins b. Glycoproteins c. Phosphoproteins d. Metalloproteins

16.54 Contrast the properties of fibrous and globular proteins in terms of
a. general structural shape b. solubility characteristics in water

Learning Focus

Be able to describe the structural changes that occur in a protein when it undergoes hydrolysis and when it is denatured.

16.12 Protein Hydrolysis and Denaturation

In this section we consider the changes that occur in proteins when they undergo two processes: (1) protein hydrolysis and (2) protein denaturation.

Protein Hydrolysis

When a protein or polypeptide in a solution of strong acid or strong base is heated, the peptide bonds of the amino acid chain are hydrolyzed and free amino acids are produced. The hydrolysis reaction is the reverse of the formation reaction for a peptide bond. Amine and carboxylic acid functional groups are regenerated.

Let us consider the hydrolysis of the tripeptide Ala–Gly–Cys under acidic conditions. Complete hydrolysis produces one unit each of the amino acids alanine, glycine, and cysteine. The equation for the hydrolysis is

$$\underset{\text{Ala–Gly–Cys}}{H_3\overset{+}{N}-\underset{CH_3}{\overset{H}{C}}-\overset{O}{\overset{\|}{C}}-\overset{H}{N}-\underset{H}{\overset{H}{C}}-\overset{O}{\overset{\|}{C}}-\overset{H}{N}-\underset{CH_2-SH}{\overset{H}{C}}-COOH} \xrightarrow[\text{heat}]{H_2O,\ H^+} \underset{\text{Ala}}{H_3\overset{+}{N}-\underset{CH_3}{\overset{H}{C}}-COOH} + \underset{\text{Gly}}{H_3\overset{+}{N}-\underset{H}{\overset{H}{C}}-COOH} + \underset{\text{Cys}}{H_3\overset{+}{N}-\underset{CH_2-SH}{\overset{H}{C}}-COOH}$$

Note that the product amino acids in this reaction are written in positive ion form because of the acidic reaction conditions.

▶ Protein hydrolysis produces free amino acids. This process is the reverse of protein synthesis, where free amino acids are combined.

Protein digestion is simply enzyme-catalyzed hydrolysis of ingested protein. The free amino acids produced from this process are absorbed through the intestinal wall into the bloodstream and transported to the liver. Here they become the raw materials for the synthesis of new protein tissue. Also, the hydrolysis of cellular proteins to amino acids is an ongoing process, as the body resynthesizes needed molecules and tissue.

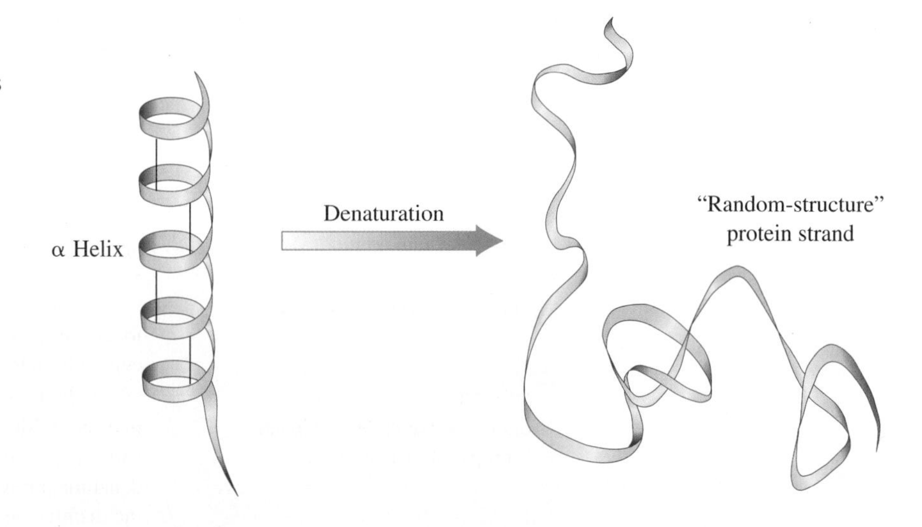

Figure 16.11 Protein denaturation involves loss of the protein's three-dimensional structure. Complete loss of such structure produces a random-coil protein strand.

Protein Denaturation

► A consequence of protein denaturation, the partial or complete loss of a protein's three-dimensional structure, is loss of biochemical activity for the protein.

Protein denaturation *is the partial or complete disorganization of a protein's characteristic three-dimensional shape as a result of disruption of its secondary, tertiary, and quaternary structural interactions.* Because the biochemical function of a protein depends on its three-dimensional shape, the result of denaturation is loss of biochemical activity. Protein denaturation does not affect the primary structure of a protein.

Although some proteins lose all of their three-dimensional structural characteristics upon denaturation (Figure 16.11), most proteins maintain some three-dimensional structure. For a few small proteins, it is possible to find conditions under which the effects of denaturation can be reversed; this restoration process in which the protein is "refolded" is called *renaturation*. Denaturation is irreversible, however, for most proteins.

Loss of water solubility is a frequent physical consequence of protein denaturation. The precipitation out of biological solution of denatured protein is called *coagulation*.

A most dramatic example of protein denaturation occurs when egg white (a concentrated solution of the protein albumin) is poured onto a hot surface. The clear albumin solution immediately changes into a white solid with a jelly-like consistency (see Figure 16.12). A similar process occurs when hamburger juices encounter a hot surface. A brown, jelly-like solid forms.

When protein-containing foods are cooked, protein denaturation occurs. Such "cooked" protein is more easily digested because it is easier for digestive enzymes to "work on" denatured (unraveled) protein. Cooking foods also kills microorganisms through protein denaturation. For example, ham and bacon can harbor parasites that cause trichinosis. Cooking the ham or bacon denatures parasite protein.

Figure 16.12 Heat denatures the protein in egg white, producing a white, jelly-like solid. The primary structure of the protein remains intact, but all higher levels of protein structure are disrupted.

In surgery, heat is often used to seal small blood vessels. This process is called *cauterization*. Small wounds can also be sealed by cauterization. Heat-induced denaturation is used in sterilizing surgical instruments and in canning foods; bacteria are destroyed when the heat denatures their protein.

The body temperature of a patient with fever may rise to 102°F, 103°F, or even 104°F without serious consequences. A temperature above 106°F (41°C) is extremely dangerous, for at this level, the enzymes of the body begin to be inactivated. Enzymes, which function as catalysts for almost all body reactions, are protein. Inactivation of enzymes, through denaturation, can have lethal effects on body chemistry.

The effect of ultraviolet radiation from the sun is similar to that of heat. Denatured skin proteins cause most of the problems associated with sunburn.

A curdy precipitate of casein, the principal protein in milk, is formed in the stomach when the hydrochloric acid of gastric juice denatures the casein. The curdling of milk that takes place when milk sours or cheese is made results from the presence of lactic acid, a

Table 16.5
Selected Physical and Chemical Denaturing Agents

Denaturing agent	Mode of action
heat	disrupts hydrogen bonds by making molecules vibrate too violently; produces coagulation, as in the frying of an egg
microwave radiation	causes violent vibrations of molecules that disrupt hydrogen bonds
ultraviolet radiation	operates very similarly to the action of heat (e.g., sunburning)
violent whipping or shaking	causes molecules in globular shapes to extend to longer lengths, which then entangle (e.g., beating egg white into meringue)
detergent	affects hydrogen bonds and salt linkages
organic solvents (e.g., ethanol, 2-propanol, acetone)	interfere with hydrogen bonds, because these solvents also can form hydrogen bonds; quickly denatures proteins in bacteria, killing them (e.g., the disinfectant action of 70% ethanol)
strong acids and bases	disrupt hydrogen bonds and salt bridges; prolonged action leads to actual hydrolysis of peptide bonds
salts of heavy metals (e.g., salts of Hg^{2+}, Ag^{+}, Pb^{2+})	metal ions combine with —SH groups and form poisonous salts
reducing agents	oxidize disulfide linkages to produce —SH groups

by-product of bacterial growth. Yogurt is prepared by growing lactic-acid-producing bacteria in skim milk. The coagulated denatured protein gives yogurt its semi-solid consistency.

Serious eye damage can result from eye tissue contact with acids or bases, when irreversibly denatured and coagulated protein causes a clouded cornea. This reaction is part of the basis for the rule that students wear protective eyewear in the chemistry laboratory.

Alcohols are an important type of denaturing agent. Denaturation of bacterial protein takes place when isopropyl or ethyl alcohol is used as a disinfectant. This accounts for the common practice of swabbing the skin with alcohol before giving an injection. Interestingly, pure isopropyl or ethyl alcohol is less effective than the commonly used 70% alcohol solution. Pure alcohol quickly denatures and coagulates the bacterial surface, thereby forming an effective barrier to further penetration by the alcohol. The 70% solution denatures more slowly and allows complete penetration to be achieved before coagulation of the surface proteins takes place.

Table 16.5 is a list of selected physical and chemical agents that cause denaturation. The effectiveness of a given denaturing agent depends on the type of protein upon which it is acting.

Practice Problems and Questions

16.55 Which structural levels of a protein are affected by hydrolysis?

16.56 What products would result from the complete hydrolysis of Val–Ser–Gly?

16.57 Will hydrolysis of Ala–Val and Val–Ala yield the same products? Explain your answer.

16.58 What dipeptides would be produced from the *partial* hydrolysis of Val–Gly–Gly–Val?

16.59 Which structural levels of a protein are affected by denaturation?

16.60 Suppose a sample of protein is completely hydrolyzed and another sample of the same protein is denatured. Compare the final products of these two processes.

16.61 In what way is the protein in a cooked egg the same as that in a raw egg?

Learning Focus

Understand the general characteristics of enzymes and their importance in biochemical systems.

16.13 General Characteristics of Enzymes

An **Enzyme** *is a catalyst for biochemical reaction*. Each cell in the human body contains thousands of different enzymes, because almost every reaction in a cell requires its own specific enzyme. Enzymes cause cellular reactions to occur millions of times faster than corresponding uncatalyzed reactions. As catalysts (Section 8.6), enzymes are not consumed during the reaction but merely help the reaction occur more rapidly.

Most enzymes are globular proteins (Section 16.11). Some are simple proteins, consisting entirely of amino acid chains. Others are conjugated proteins, containing additional chemical components (Section 16.14). Until the 1980s, it was thought that *all* enzymes were proteins. A few enzymes are now known that are made of ribonucleic acids (RNA; Section 17.7) and catalyze cellular reactions involving nucleic acids. In this chapter, we will consider only enzymes that are proteins.

Enzymes undergo all the reactions of proteins, including *denaturation* (Section 16.12). Slight alterations in pH, temperature, or other protein denaturants affect enzyme activity dramatically. Good cooks realize that overheating yeast kills the action of the yeast. A person suffering from a high fever (greater than 106°F) runs the risk of denaturing certain enzymes. The biochemist must exercise extreme caution in handling enzymes to avoid the loss of their activity. Even vigorous shaking of an enzyme solution can destroy enzyme activity.

Enzymes differ from laboratory catalysts in that their activity is usually regulated by other substances present in the cell in which they are found. Most laboratory catalysts need to be removed from the reaction to stop their catalytic action; not so with enzymes. In some cases, if a certain chemical is needed in the cell, then the enzyme responsible for its production is activated by other cellular components. When a sufficient quantity has been produced, the enzyme is then deactivated. In other situations, the cell may produce more or less enzyme as required. Because different enzymes catalyze nearly all reactions in the cell, certain necessary reactions can be accelerated without affecting the rest of the cellular chemistry.

The word *enzyme* comes from the Greek words *en,* which means "in," and *zyme,* which means "yeast." Long before their chemical nature was understood, yeast enzymes were used in the production of bread and alcoholic beverages. The action of yeast on sugars produces the carbon dioxide gas that causes bread to rise (Figure 16.13). Fermentation of sugars in fruit juices with the same yeast enzymes produces alcoholic beverages.

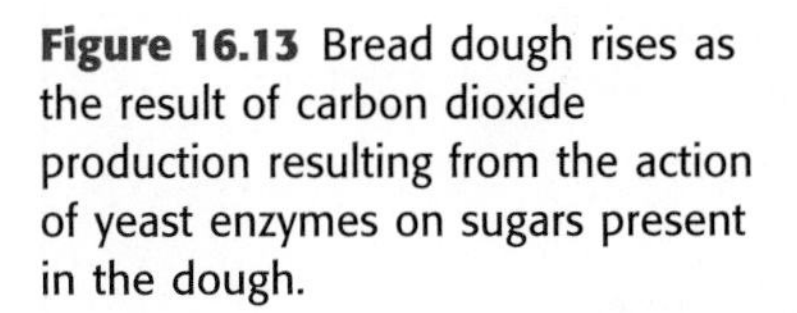

Figure 16.13 Bread dough rises as the result of carbon dioxide production resulting from the action of yeast enzymes on sugars present in the dough.

Practice Problems and Questions

16.62 What is the general role of enzymes in the human body?

16.63 Why does the body need so many different enzymes?

16.64 Most enzymes are what type of compound?

16.65 Enzymes are very sensitive to heat. Explain why.

Learning Focus

Understand the system for naming enzymes; know the structural difference between simple enzymes and conjugated enzymes.

▶ Enzymes, the most efficient catalysts known, increase the rates of biochemical reactions by factors of up to 10^{20} over uncatalyzed reactions. Nonenzymatic catalysts, on the other hand, typically enhance the rate of a reaction by factors of 10^2 to 10^4.

16.14 Nomenclature and Classification of Enzymes

Enzymes are most commonly named by using a system that attempts to provide information about the *function* (rather than the structure) of the enzyme. Type of reaction catalyzed and *substrate* identity are focal points for the nomenclature. A **substrate** *is the reactant in an enzyme-catalyzed reaction.* The substrate is the substance upon which the enzyme "acts."

Three important aspects of the naming process are the following:

1. The suffix *-ase* identifies a substance as an enzyme. Thus ure*ase,* sucr*ase,* and lip*ase* are all *enzyme designations.* The suffix *-in* is still found in the names of some of the first enzymes studied, many of which are digestive enzymes. Such names include *trypsin, chymotrypsin,* and *pepsin.*
2. The type of reaction catalyzed by an enzyme is often noted with a prefix. An *oxidase* enzyme catalyzes an oxidation reaction, and a *hydrolase* enzyme catalyzes a hydrolysis reaction.
3. The identity of the substrate is often noted in addition to the type of reaction. Enzyme names of this type include *glucose oxidase, pyruvate carboxylase,* and *succinate dehydrogenase.* Infrequently, the substrate but not the reaction type is given, as in the names *urease* and *lactase.* In such names, the reaction involved is hydrolysis; *urease* catalyzes the hydrolysis of urea, *lactase* the hydrolysis of lactose.

Example 16.3 **Predicting Enzyme Function from an Enzyme's Name**

Predict the function of the following enzymes.

a. Cellulase **b.** Sucrase
c. L-Amino acid oxidase **d.** Aspartate aminotransferase

Solution

a. Cellulase catalyzes the hydrolysis of cellulose.
b. Sucrase catalyzes the hydrolysis of the disaccharide sucrose.
c. L-Amino acid oxidase catalyzes the oxidation of L-amino acids.
d. Aspartate aminotransferase catalyzes the transfer of an amino group from aspartate to a different molecule.

Classification of Enzymes

▶ *Holoenzyme* is an alternative name for a conjugated enzyme. A holoenzyme, a biochemically active entity, results from the combination of an apoenzyme and a cofactor.

Apoenzyme + cofactor = holoenzyme

Enzymes can be divided into two general structural classes: simple enzymes and conjugated enzymes. A **simple enzyme** *is composed only of protein (amino acid chains).* A **conjugated enzyme** *has a nonprotein portion in addition to a protein portion.* By itself, neither the protein part nor the nonprotein portion of a conjugated enzyme has catalytic properties. An **apoenzyme** *is the protein portion of a conjugated enzyme.* A **cofactor** *is the nonprotein portion of a conjugated enzyme.* Only the combination of apoenzyme and cofactor—the conjugated enzyme—shows biochemical activity.

Why do many enzymes need cofactors? Cofactors provide additional chemically reactive functional groups besides those present in the amino acid side chains of apoenzymes.

A cofactor is generally either a small organic molecule or an inorganic ion (usually a metal ion). A **coenzyme** *is a small organic molecule that serves as a cofactor in a conjugated enzyme.* Many vitamins have coenzyme functions in the human body.

Typical inorganic ion cofactors include Zn^{2+}, Mg^{2+}, Mn^{2+}, and Fe^{2+}. The nonmetallic Cl^- ion occasionally acts as a cofactor. Dietary minerals are an important source of inorganic ion cofactors.

Practice Problems and Questions

16.66 Which of the following substances are enzymes?
a. Sucrase b. Galactose c. Pepsin d. Glutamine synthetase

16.67 Predict the function of each of the following enzymes.
a. Cytochrome oxidase b. Alcohol dehydrogenase
c. L-Amino acid reductase d. Lactase

16.68 Suggest a name for an enzyme that catalyzes each of the following reactions.
a. Hydrolysis of sucrose b. Isomerization of glucose
c. Removal of hydrogen from lactate d. Reduction of oxalate

16.69 Give the name of the substrate on which each of the following enzymes acts.
a. Pyruvate carboxylase b. Galactase
c. Succinate dehydrogenase d. L-Amino acid reductase

16.70 Indicate whether each of the following phrases describes a simple or a conjugated enzyme.
a. An enzyme that contains a carbohydrate portion
b. An enzyme that contains only protein
c. An enzyme that requires Mg^{2+} for activity
d. An enzyme that has a vitamin as part of its structure

16.71 What is the difference between an apoenzyme and a holoenzyme?

16.72 All coenzymes are cofactors, but not all cofactors are coenzymes. Explain this statement.

16.73 Why are cofactors present in most enzymes?

Learning Focus

Describe the two common models used to explain enzyme action.

16.15 Models of Enzyme Action

Explanations of *how* enzymes function as catalysts in biochemical systems are based on the concepts of an enzyme active site and enzyme–substrate complex formation.

Enzyme Active Site

Studies show that only a small portion of an enzyme molecule, called the active site, participates in the interaction with a substrate or substrates during a reaction. An **active site** *is the relatively small part of an enzyme that is actually involved in catalysis.*

▶ It is possible for an enzyme to have more than one active site.

The active site in an enzyme is a three-dimensional entity formed by groups that come from different parts of the protein chain(s); these groups are brought together by the folding and bending (secondary and tertiary structure; Sections 16.8 and 16.9) of the protein. The active site is usually a "crevice-like" location in the enzyme (see Figure 16.14).

Enzyme–Substrate Complex

Catalysts offer an alternative pathway with lower activation energy through which a reaction can occur (Section 8.6). In enzyme-controlled reactions, this alternative pathway involves the formation of an enzyme–substrate complex as an intermediate species in the reaction. An **enzyme–substrate complex** *is the intermediate reaction species that is*

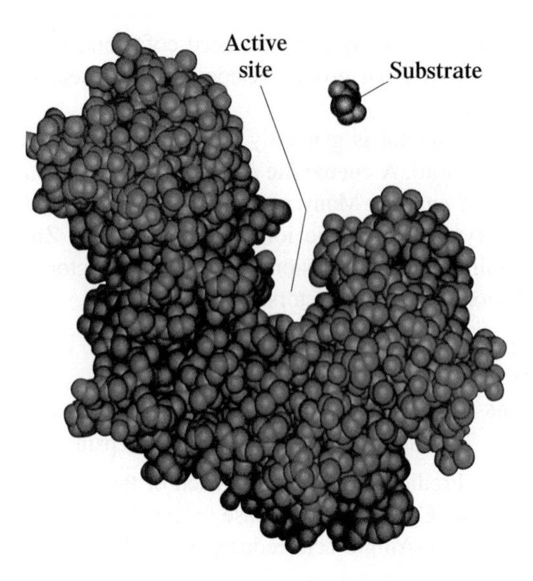

Figure 16.14 The active site of an enzyme is usually a crevice-like region formed as the result of the protein's secondary and tertiary structural characteristics.

formed when a substrate binds to the active site of an enzyme. Within the enzyme–substrate complex, proximity effects and orientation effects create more favorable reaction conditions than if substrates were free. The result is faster formation of products.

Lock-and-Key Model

▶ The lock-and-key model is more than just a "shape fit." In addition, there are weak binding forces (R group interactions) between parts.

To account for the highly specific way an enzyme selects a substrate and binds it to the active site, researchers have proposed several models. The simplest of these models is the lock-and-key model.

In the lock-and-key model, the active site in the enzyme has a fixed, rigid geometrical conformation. Only substrates with a complementary geometry can be accommodated at such a site, much as a lock accepts only certain keys. Figure 16.15 illustrates the lock-and-key concept of substrate–enzyme interaction.

Induced-Fit Model

The lock-and-key model explains the action of many enzymes. It is, however, too restrictive for the action of other enzymes. Experimental evidence indicates that many enzymes have flexibility in their shapes. They are not rigid and static; there is constant change in their shape. The induced-fit model is used for this type of situation.

The induced-fit model allows for small changes in the shape or geometry of the active site of an enzyme in order to accommodate a substrate. An analogy would be the changes that occur in the shape of a glove when a hand is inserted into it. The induced fit is a result of the enzyme's flexibility; it adapts to accept the incoming substrate. This model, illustrated in Figure 16.16, is a more thorough explanation for the active-site properties of an enzyme because it includes the specificity of the lock-and-key model coupled with the flexibility of the enzyme protein.

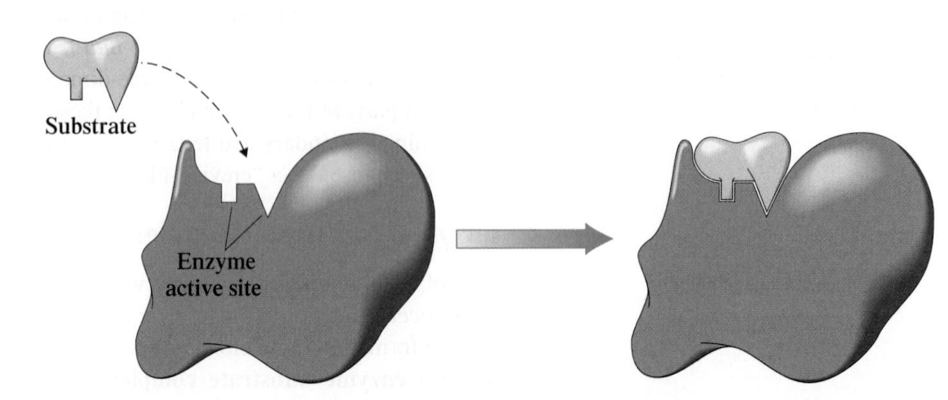

Figure 16.15 The lock-and-key model for enzyme activity. Only a substrate whose shape and chemical nature are complementary to those of the active site can interact with the enzyme.

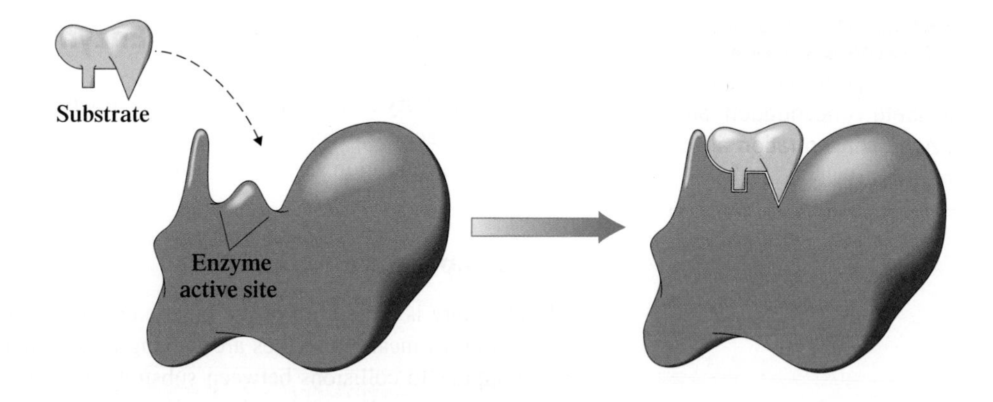

Figure 16.16 The induced-fit model for enzyme activity. The enzyme active site, although not exactly complementary in shape to that of the substrate, is flexible enough that it can adapt to the shape of the substrate.

The forces that draw the substrate into the active site are many of the same forces that maintain tertiary structure in the folding of polypeptide chains. Electrostatic interactions, hydrogen bonds, and hydrophobic interactions all help attract and bind substrate molecules. For example, a protonated (positively charged) amino group in a substrate could be attracted and held at the active site by a negatively charged aspartate or glutamate residue. Alternatively, cofactors such as positively charged metal ions often help bind substrate molecules. Figure 16.17 is a schematic representation of the amino acid R group interactions that bind a substrate to an enzyme active site.

Practice Problems and Questions

16.74 What is the function of an enzyme active site?

16.75 What is an enzyme–substrate complex?

16.76 A substrate is complementary to the active site of an enzyme. Explain this statement.

16.77 How does the lock-and-key model of enzyme action explain the fact that some enzymes act on just one substrate?

16.78 How does the induced-fit model of enzyme action explain the fact that some enzymes act on several related substrates?

16.79 Describe the attractive forces that hold a substrate at an enzyme active site.

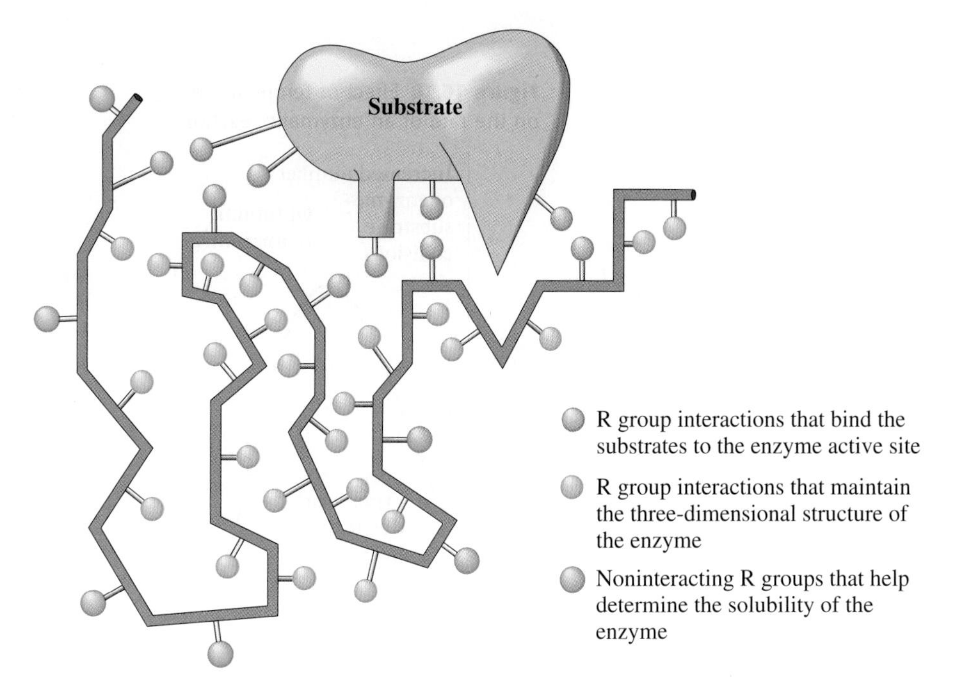

Figure 16.17 A schematic diagram representing amino acid R group interactions that bind a substrate to an enzyme active site. The R group interactions that maintain the three-dimensional structure of the enzyme (secondary and tertiary structure) are also shown.

Learning Focus

Explain how temperature, pH, substrate concentration, and enzyme concentration affect enzyme activity.

16.16 Factors That Affect Enzyme Activity

Enzyme activity *is a measure of the rate at which an enzyme converts substrate to products.* Four factors affect enzyme activity: temperature, pH, substrate concentration, and enzyme concentration.

 Temperature

Temperature is a measure of the kinetic energy (energy of motion) of molecules. Higher temperatures mean molecules are moving faster and colliding more frequently. This concept applies to collisions between substrate molecules and enzymes. As the temperature of an enzymatically catalyzed reaction increases, so does the rate (velocity) of the reaction.

However, when the temperature increases beyond a certain point, the increased energy begins to cause disruptions in the tertiary structure of the enzyme; denaturation is occuring. Tertiary-structure change at the active site impedes catalytic action, and the enzyme activity quickly decreases as the temperature climbs past this point (Figure 16.18). The temperature that produces maximum activity for an enzyme is known as the optimum temperature for that enzyme. **Optimum temperature** *is the temperature at which the rate of an enzyme-catalyzed reaction is the greatest.* For human enzymes, the optimum temperature is often 37°C, normal body temperature.

pH

The pH of an enzyme's environment can affect its activity. This is not surprising, because the *charge* on acidic and basic amino acids (Section 16.4) located at the active site depends on pH. Small changes in pH (less than one unit) can result in enzyme denaturation (Section 16.12) and subsequent loss of catalytic activity.

Most enzymes exhibit maximum activity over a very narrow pH range. Only within this narrow pH range do the enzyme's amino acids exist in properly charged forms (Section 16.4). **Optimum pH** *is the pH at which an enzyme has maximum activity.* Figure 16.19 shows the effect of pH on an enzyme's activity. Biochemical buffers (Section 9.9) help maintain the optimum pH for an enzyme.

Figure 16.18 Effect of temperature on the rate of an enzymatic reaction.

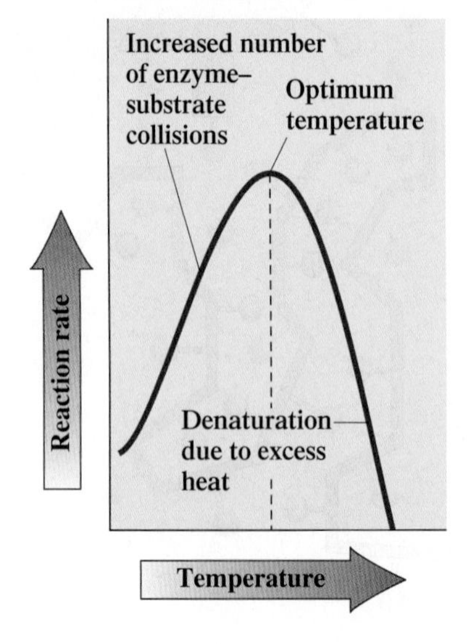

Figure 16.19 Effect of pH on an enzyme's activity.

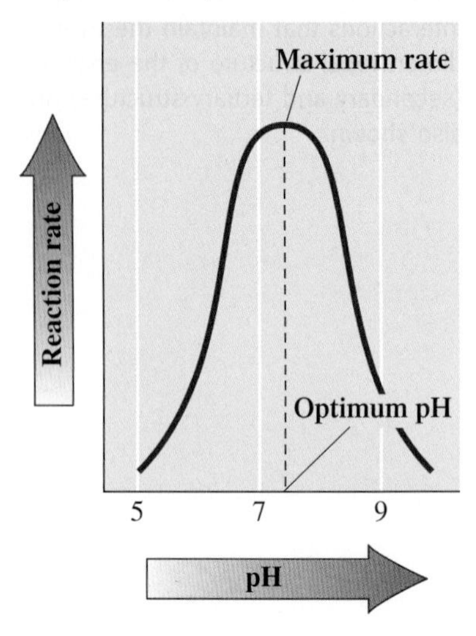

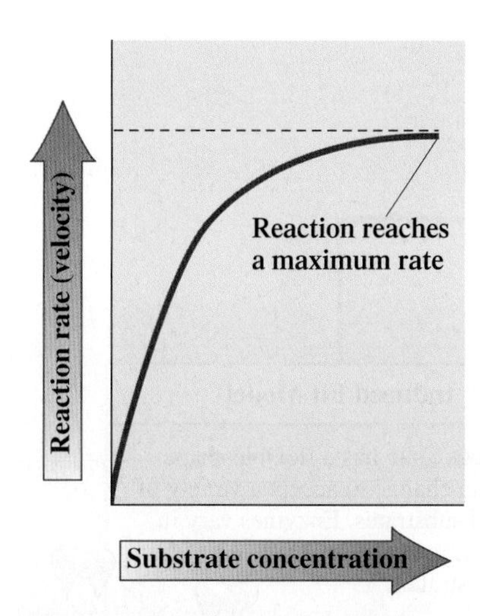

Figure 16.20 A graph showing the change in enzyme activity with a change in substrate concentration at constant temperature, pH, and enzyme concentration. Enzyme activity remains constant after a certain substrate concentration is reached.

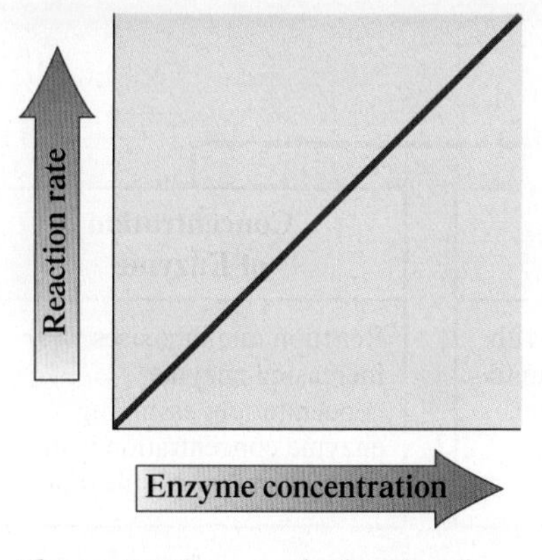

Figure 16.21 A graph showing the change in reaction rate with a change in enzyme concentration for an enzymatic reaction. Temperature, pH, and substrate concentration are constant. The substrate concentration is high relative to enzyme concentration.

Each enzyme has a characteristic optimum pH, which usually falls within the physiological pH range of 7.0–7.5. Notable exceptions to this generalization are the digestive enzymes pepsin and trypsin. Pepsin, which is active in the stomach, functions best at a pH of 2.0. On the other hand, trypsin, which operates in the small intestine, functions best at a pH of 8.0.

A variation from normal pH can also affect substrates, causing either protonation or deprotonation of groups on the substrate. The interaction between the altered substrate and the enzyme active site may be less efficient than normal—or even impossible.

Substrate Concentration

When the concentration of an enzyme is kept constant and the concentration of substrate is increased, the enzyme activity pattern shown in Figure 16.20 is obtained. This activity pattern is called a *saturation curve*. Enzyme activity increases up to a certain substrate concentration and thereafter remains constant.

What limits enzymatic activity to a certain maximum value? As substrate concentration increases, the point is eventually reached where enzyme capabilities are used to their maximum extent. The rate remains constant from this point on (Figure 16.20). Each substrate must occupy an enzyme active site for a finite amount of time, and the products must leave the site before the cycle can be repeated. When each enzyme molecule is working at full capacity, the incoming substrate molecules must "wait their turn" for an empty active site. At this point, the enzyme is said to be under saturating conditions.

The rate at which an enzyme accepts and releases substrate molecules at substrate saturation is given by its turnover number. An enzyme's **turnover number** *is equal to the number of substrate molecules transformed per second by one molecule of enzyme under optimum conditions of temperature, pH, and saturation.* Table 16.6 gives turnover numbers for selected enzymes. Some enzymes have a much faster mode of operation than others.

Enzyme Concentration

Because enzymes are not consumed in the reactions they catalyze, the cell usually keeps the number of enzymes low compared with the number of substrate molecules. This is efficient; the cell avoids paying the energy costs of synthesizing and maintaining a large work force of enzyme molecules. Thus, in general, the concentration of substrate in a reaction is much higher than that of the enzyme.

If the amount of substrate present is kept constant and the enzyme concentration is increased, the reaction rate increases because more substrate molecules can be accommodated in a given amount of time. A plot of enzyme activity versus enzyme concentration, at a constant substrate concentration that is high relative to enzyme concentration, is shown in Figure 16.21. The greater the enzyme concentration, the greater the reaction rate.

The accompanying Chemistry at a Glance reviews what we have said about enzyme action and enzyme activity in this section and the preceding section.

Table 16.6
Turnover Numbers for Selected Enzymes

Enzyme	Turnover number (molecules of substrate per second per enzyme molecule)	Turnover time (seconds per molecule of product)
carbonic anhydrase	600,000	2×10^{-6}
α-glucosidase	17,000	6×10^{-5}
glutamate dehydrogenase	500	2×10^{-3}
phosphoglucomutase	21	5×10^{-2}
chymotrypsin	2	5×10^{-1}

Chemistry at a Glance

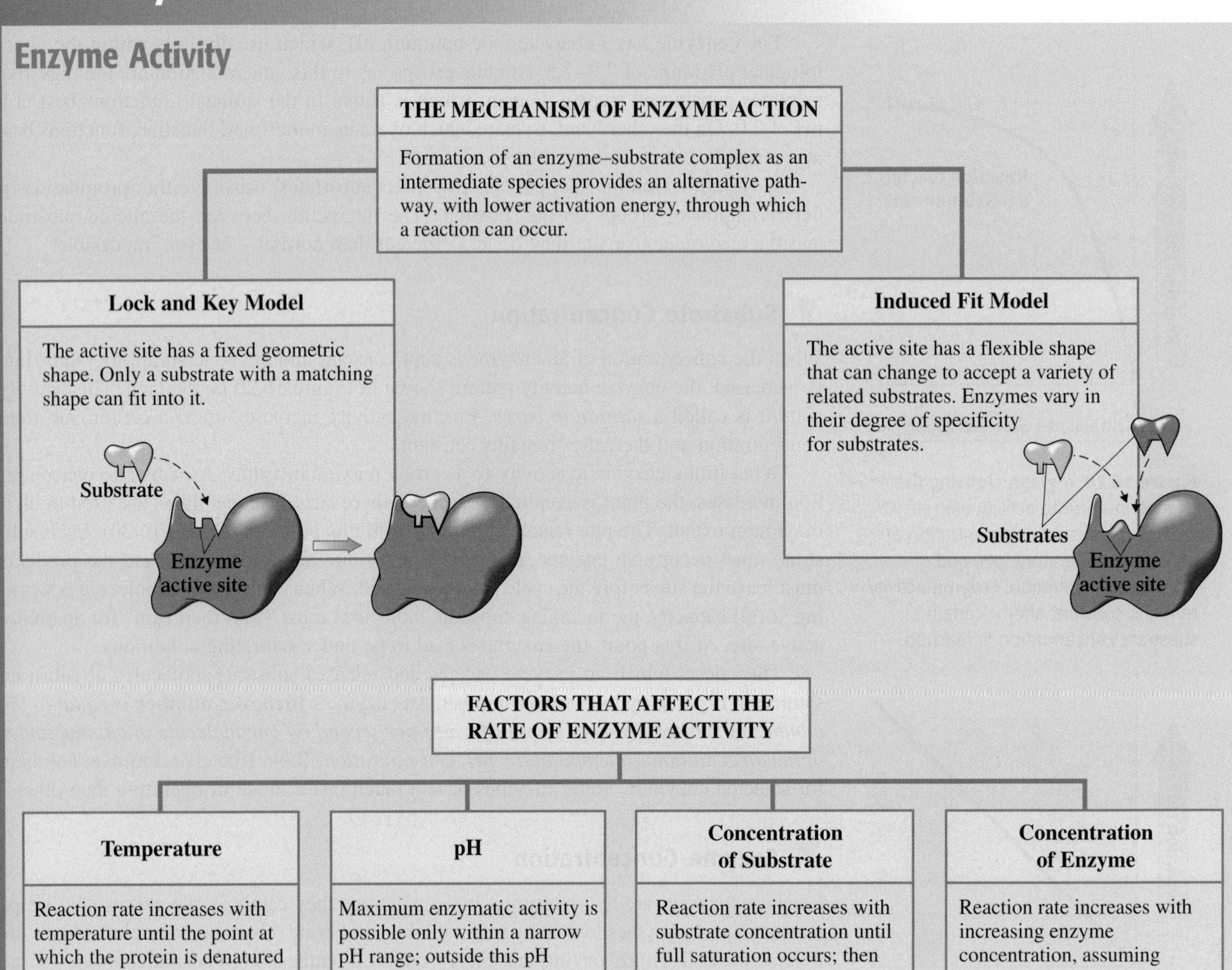

Enzyme Activity
THE MECHANISM OF ENZYME ACTION
Formation of an enzyme–substrate complex as an intermediate species provides an alternative pathway, with lower activation energy, through which a reaction can occur.
Lock and Key Model
The active site has a fixed geometric shape. Only a substrate with a matching shape can fit into it.
Substrate
Enzyme active site
Induced Fit Model
The active site has a flexible shape that can change to accept a variety of related substrates. Enzymes vary in their degree of specificity for substrates.
Substrates
Enzyme active site
FACTORS THAT AFFECT THE RATE OF ENZYME ACTIVITY
Temperature
Reaction rate increases with temperature until the point at which the protein is denatured and activity drops sharply.
pH
Maximum enzymatic activity is possible only within a narrow pH range; outside this pH range, the protein is denatured and activity drops sharply.
Concentration of Substrate
Reaction rate increases with substrate concentration until full saturation occurs; then the rate levels off.
Concentration of Enzyme
Reaction rate increases with increasing enzyme concentration, assuming enzyme concentration is much lower than that of substrate.

Practice Problems and Questions

16.80 Temperature affects enzymatic reaction rates in two ways. An increase in temperature can accelerate the reaction rate, or it can stop the reaction. Explain each of these effects.

16.81 Why does an enzyme lose activity when the pH is drastically changed from the optimum pH?

16.82 In an enzyme-catalyzed reaction, all of the enzyme active sites are saturated by substrate molecules at a certain substrate concentration. What happens to the rate of the reaction when the substrate concentration is doubled?

16.83 What is an enzyme turnover number?

CONCEPTS TO REMEMBER

Protein. A protein is a polymer in which the monomer units are amino acids.

α-Amino acid. An α-amino acid is an amino acid in which the amino group and carboxyl group are attached to the same carbon atom.

Standard amino acid. A standard amino acid is one of the 20 α-amino acids that are normally present in protein.

Amino acid classifications. Amino acids are classified as nonpolar, polar neutral, polar basic, or polar acidic, depending on the nature of the side chain (R group) present.

Chirality of amino acids. Amino acids found in proteins are always left-handed (L isomer).

Zwitterion. A zwitterion is a molecule that has a positive charge on one atom and a negative charge on another atom. α-Amino acids exist as zwitterions because of an internal acid–base reaction between the amino and carboxyl groups present.

Peptides and peptide bonds. A peptide is a sequence of amino acids in which the amino acids are joined together through peptide bonds. A peptide bond is the bond between the carboxyl group of one amino acid and the amino group of another amino acid.

Protein primary structure. The primary structure of a protein is the sequence of amino acids present in the peptide chain or chains of the protein.

Protein secondary structure. The secondary structure of a protein is the arrangement in space of the atoms in the backbone portion of the protein. The two major types of protein secondary structure are the alpha helix and the beta pleated sheet.

Protein tertiary structure. The tertiary structure of a protein is the overall three-dimensional shape that results from the attractive forces among amino acid side chains (R groups).

Protein quaternary structure. The quaternary structure of a protein involves the associations among various polypeptide chains present in the protein.

Protein classifications. Proteins as classified as *simple* or *conjugated* on the basis of whether a non-amino-acid component is present. Proteins are classified as *fibrous* or *globular* on the basis of structural shape.

Protein hydrolysis. Protein hydrolysis is a chemical reaction in which peptide bonds within a protein are broken through reaction with water. Complete hydrolysis produces free amino acids.

Protein denaturation. Protein denaturation is the partial or complete disorganization of a protein's characteristic three-dimensional shape as a result of disruption of its secondary, tertiary, and quaternary structural interactions.

Enzyme. Enzymes are highly specialized protein molecules that act as biochemical catalysts. Enzymes have common names that provide information about their function rather than their structure. The suffix *-ase* is characteristic of most enzyme names.

Enzyme structure. Simple enzymes are composed only of protein (amino acids). Conjugated enzymes have a nonprotein portion (cofactor) in addition to a protein portion (apoenzyme). Cofactors may be small organic molecules (coenzymes) or inorganic ions.

Enzyme active site. An enzyme active site is the relatively small part of the enzyme that is actually involved in catalysis. It is where substrate binds to the enzyme.

Lock-and-key model of enzyme activity. The active site in an enzyme has a fixed, rigid geometrical conformation. Only substrates with a complementary geometry can be accommodated at the active site.

Induced-fit model of enzyme activity. The active site in an enzyme can undergo small changes in shape or geometry in order to accommodate a series of related substrates.

Enzyme activity. Enzyme activity is a measure of the rate at which an enzyme converts substrate to products. Four factors that affect enzyme activity are temperature, pH, substrate concentration, and enzyme concentration.

KEY REACTIONS AND EQUATIONS

1. Formation of a zwitterion at pH 7.0 (Section 16.4)

$$H_2N{-}\underset{\large R}{\underset{|}{CH}}{-}COOH \longrightarrow \overset{+}{H_3N}{-}\underset{\large R}{\underset{|}{CH}}{-}COO^-$$

2. Conversion of a zwitterion to a positive ion in acidic solution (Section 16.4)

$$\overset{+}{H_3N}{-}\underset{\large R}{\underset{|}{CH}}{-}COO^- + H_3O^+ \longrightarrow \overset{+}{H_3N}{-}\underset{\large R}{\underset{|}{CH}}{-}COOH + H_2O$$

3. Conversion of a zwitterion to a negative ion in basic solution (Section 16.4)

$$\overset{+}{H_3N}{-}\underset{\large R}{\underset{|}{CH}}{-}COO^- + OH^- \longrightarrow H_2N{-}\underset{\large R}{\underset{|}{CH}}{-}COO^- + H_2O$$

4. Formation of a peptide bond (Section 16.5)

$$\overset{+}{H_3N}{-}\underset{\large R}{\underset{|}{CH}}{-}COO^- + \overset{+}{H_3N}{-}\underset{\large R'}{\underset{|}{CH}}{-}COO^- \longrightarrow$$

$$\overset{+}{H_3N}{-}\underset{\large R}{\underset{|}{CH}}{-}\overset{\overset{\large O}{\|}}{C}{-}\overset{\overset{\large H}{|}}{N}{-}\underset{\large R'}{\underset{|}{CH}}{-}COO^- + H_2O$$

5. Hydrolysis of a protein in acidic solution (Section 16.12)

$$\text{Protein} + H_2O \xrightarrow{H^+} \text{smaller peptides} \xrightarrow{H^+} \text{amino acids}$$

6. Denaturation of a protein (Section 16.12)

$$\begin{array}{c}\text{Protein with}\\ 1°, 2°, \text{and}\\ 3° \text{ structure}\end{array} \xrightarrow[\text{agent}]{\text{Denaturing}} \begin{array}{c}\text{Protein with } 1°\\ \text{structure only}\end{array}$$

7. Conversion of an apoenzyme to an active enzyme (Section 16.14)

$$\text{Apoenzyme} + \text{cofactor} \longrightarrow \text{active enzyme (holoenzyme)}$$

8. Mechanism of enzyme action (Section 16.15)

$$\text{Enzyme} + \text{substrate} \longrightarrow \text{enzyme–substrate complex}$$

$$\text{Enzyme–substrate complex} \longrightarrow \text{enzyme} + \text{product}$$

KEY TERMS

Active site (16.15)
α-Amino acid (16.2)
Amino acid (16.2)
Apoenzyme (16.14)
Coenzyme (16.14)
Cofactor (16.14)
Conjugated enzyme (16.14)
Conjugated protein (16.11)
Enzyme (16.13)
Enzyme activity (16.16)
Enzyme–substrate complex (16.15)
Essential amino acid (16.2)
Fibrous protein (16.11)
Globular protein (16.11)
Nonpolar amino acid (16.2)
Optimum pH (16.16)
Optimum temperature (16.16)
Peptide (16.5)
Peptide bond (16.5)
Polar acidic amino acid (16.2)
Polar basic amino acid (16.2)
Polar neutral amino acid (16.2)
Primary protein structure (16.7)
Prosthetic group (16.11)
Protein (16.1, 16.6)
Protein denaturation (16.12)
Quaternary protein structure (16.10)
Secondary protein structure (16.8)
Simple enzyme (16.14)
Simple protein (16.11)
Standard amino acid (16.2)
Substrate (16.14)
Tertiary protein structure (16.9)
Turnover number (16.16)
Zwitterion (16.4)

ADDITIONAL PROBLEMS

16.84 Classify each of the following amino acids as (1) a standard amino acid, (2) an essential amino acid, (3) an α-amino acid, or (4) a nonpolar amino acid. More than one answer may apply in a given case.
a. Alanine b. Tyrosine c. Proline d. Valine

16.85 State whether each of the following statements applies to primary, secondary, tertiary, or quaternary protein structure.
a. A disulfide bond forms between two cysteine residues in different protein chains.
b. A salt bridge forms between amino acids with acidic and basic side chains.
c. Hydrogen bonding between carbonyl oxygen atoms and amino-group hydrogen atoms causes a peptide to coil into a helix.
d. Peptide linkages hold amino acids together in a polypeptide change.

16.86 What is the net charge at a pH of 1.0 for each of the following peptides?
a. Val–Ala–Leu b. Tyr–Trp–Thr
c. Asp–Asp–Glu–Gly d. His–Arg–Ser–Ser

16.87 What is the net charge at a pH of 13.0 for each of the peptides in Problem 16.86?

16.88 Indicate how many structurally isomeric tetrapeptides are possible for a tetrapeptide in which
a. four different amino acids are present
b. three different amino acids are present
c. two different amino acids are present

16.89 Explain what is meant by the equation

$$E + S \rightleftharpoons ES \longrightarrow E + P$$

given that ES stands for enzyme–substrate complex.

16.90 Alcohol dehydrogenase catalyzes the conversion of ethanol to acetaldehyde. This enzyme, in its active state, consists of a protein molecule and a zinc ion. On the basis of this information, identify each of the following for this chemical system.
a. Substrate b. Cofactor c. Apoenzyme d. Holoenzyme

PRACTICE TEST True/False

16.91 In α-amino acids, the amino group and the carboxyl group are attached to the same carbon atom.

16.92 All of the standard amino acids are essential amino acids.

16.93 Naturally occurring amino acids are nearly always right-handed molecules.

16.94 Only amino acids with polar side chains form zwitterions.

16.95 Both nitrogen and oxygen atoms are present in the "backbone" of a protein.

16.96 A β pleated sheet arrangement is a characteristic associated with protein tertiary structure.

16.97 A disulfide bond is a characteristic associated with protein secondary structure.

16.98 Some proteins have more than one polypeptide chain in their structure.

16.99 Both protein hydrolysis and protein denaturation disrupt protein secondary structure.

16.100 All conjugated proteins contain prosthetic groups.

16.101 Most enzymes have names that end in the suffix *-ase*.

16.102 Coenzymes and metal ions are two types of cofactors.

16.103 Enzyme function depends on a substrate occupying the active site of the enzyme.

16.104 Enzyme activity always increases with increasing temperature.

16.105 Enzymes with fast modes of operation have high turnover numbers.

PRACTICE TEST ▶ Multiple Choice

16.106 The 20 standard amino acids differ from each other in the
a. location of the amino group
b. location of the carboxyl group
c. number of carbon atoms between the amino group and the carboxyl group
d. identity of the R group (side chain)

16.107 Which of the following forms of the amino acid alanine (Ala) is the zwitterion form?

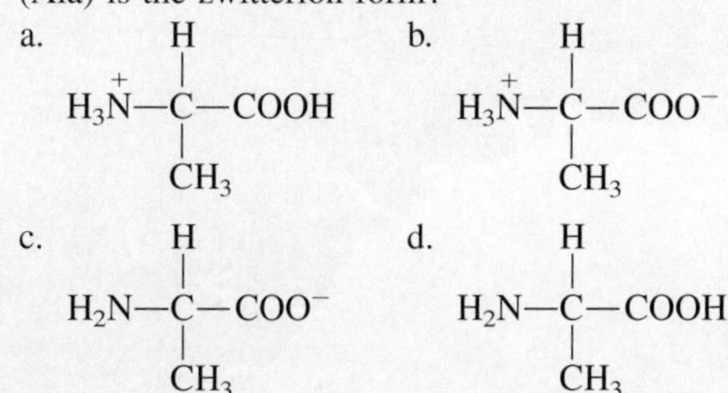

16.108 The joining together of two amino acids to form a dipeptide involves the reaction between
a. two amino groups
b. two carboxyl groups
c. an amino group and a carboxyl group
d. a carboxyl group and a hydroxyl group

16.109 Which of the following would be a correct representation of a portion of the "backbone" of a protein?

a. —C(=O)—N(H)—CH—C(=O)—N(H)—

b. —C(=O)—CH—N(H)—CH—C(=O)—CH—N(H)—

c. —C(=O)—N(H)—C(=O)—N(H)—C(=O)—N(H)—

d. —C(=O)—N(H)—N(H)—C(=O)—C(=O)—N(H)—N(H)—C(=O)—

16.110 In the tetrapeptide Ala–Cys–Val–Leu, the C-terminal amino acid is
a. Ala b. Cys c. Val d. Leu

16.111 Interactions between amino acid R groups are responsible for which of the following levels of protein structure?
a. Primary b. Secondary
c. Tertiary d. Both secondary and tertiary

16.112 The *complete* hydrolysis of a protein produces a mixture of
a. polypeptides
b. free amino acids
c. polypeptides and free amino acids
d. dipeptides and free amino acids

16.113 Which of the following is always present in both conjugated enzymes and simple enzymes?
a. Protein b. Substrate c. A cofactor d. A coenzyme

16.114 An enzyme active site is the location in an enzyme where
a. interaction with a substrate occurs
b. interaction with a cofactor occurs
c. substrate molecules are generated
d. cofactor molecules are generated

16.115 A plot of enzyme activity (y axis) versus temperature (x axis) with other variables constant is a
a. straight line with an upward slope
b. straight horizontal line
c. line with an upward slope and a long, flat top
d. line with an upward slope followed by a downward slope

CHAPTER OUTLINE

CHAPTER SEVENTEEN

Nucleic Acids

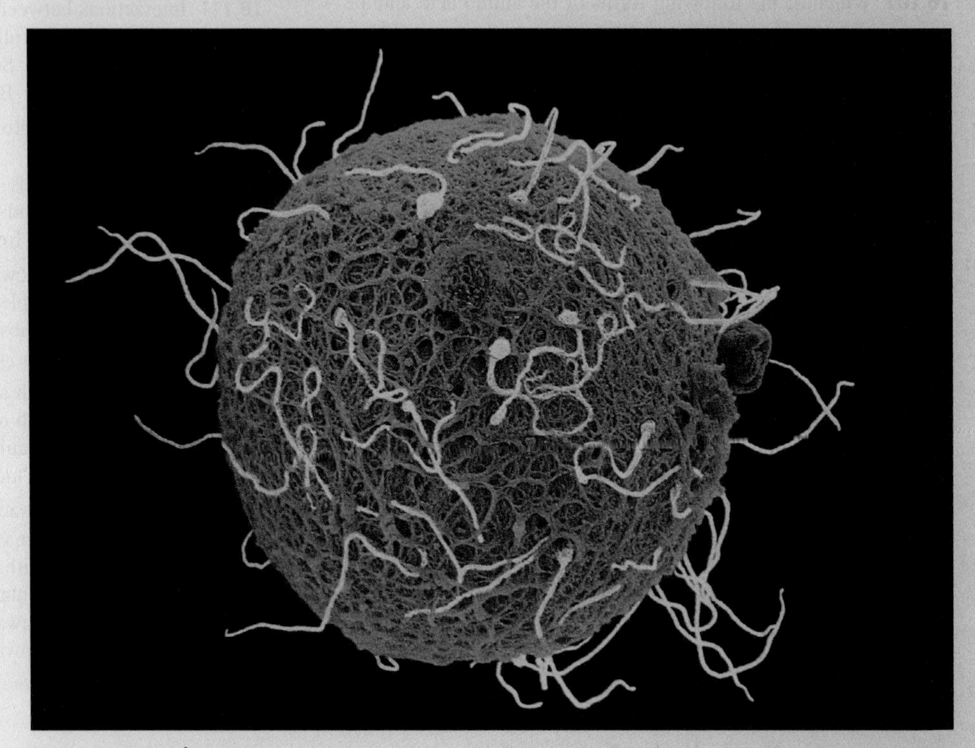

Human egg and sperms.

A most remarkable property of living cells is their ability to produce exact replicas of themselves. Furthermore, cells contain all the instructions needed for making the complete organism of which they are a part. The molecules within a cell that are responsible for these amazing capabilities are nucleic acids.

The Swiss physiologist Friedrich Miescher (1844–1895) discovered nucleic acids in 1869 while studying the nuclei of white blood cells. The fact that they were initially found in cell nuclei and are acidic accounts for the name *nucleic acid*. Although we now know that nucleic acids are found throughout a cell, not just in the nucleus, the name is still used for such materials.

17.1 Characteristics of Nucleic Acids

Learning Focus

Know the general characteristics of nucleic acids.

Two types of nucleic acids are found within cells of higher organisms: *deoxyribonucleic acid* (DNA) and *ribonucleic acid* (RNA). Nearly all the DNA is found within the cell nucleus. Its primary function is the storage and transfer of genetic information. This information is used (indirectly) to control many functions of a living cell. In addition, DNA is passed from existing cells to new cells during cell division. RNA occurs in all parts of a cell. It functions primarily in synthesis of proteins, the molecules that carry out essential cellular functions. The structural distinctions between DNA and RNA molecules are considered in Section 17.2.

All nucleic acid molecules are polymers. A **nucleic acid** *is a polymer in which the monomer units are nucleotides*. Thus the starting point for a discussion of nucleic acids is an understanding of the structures and chemical properties of nucleotides.

▶ It was not until 1944, 75 years after the discovery of nucleic acids, that scientists obtained the first evidence that these molecules are responsible for the storage and transfer of genetic information.

Practice Problems and Questions

17.1 What do DNA and RNA stand for?

17.2 In what parts of a cell are DNA and RNA found?

17.3 Identify the primary function for
a. DNA b. RNA

17.4 What is the general name for the building blocks (monomers) from which a nucleic acid is made?

Learning Focus

Know the general structural aspects of and nomenclature for nucleotides.

17.2 Nucleotides: Building Blocks for Nucleic Acids

A **nucleotide** *is a three-subunit molecule in which a pentose sugar is bonded to both a phosphate group and a nitrogen-containing heterocyclic base.* With a three-subunit structure, nucleotides are more complex monomers than the monosaccharides of polysaccharides and the amino acids of proteins. A block structural diagram for a nucleotide is

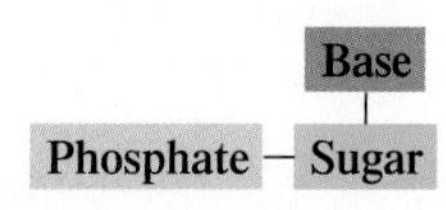

Pentose Sugars

The sugar unit of a nucleotide is either the pentose *ribose* or the pentose *2-deoxyribose.*

▶ The systems for numbering the atoms in the pentose and nitrogen-containing base subunits of a nucleotide are important and will be used extensively in later sections of this chapter. The convention is that
1. Pentose ring atoms are designated with *primed* numbers.
2. Nitrogen-containing base ring atoms are designated with *unprimed* numbers.

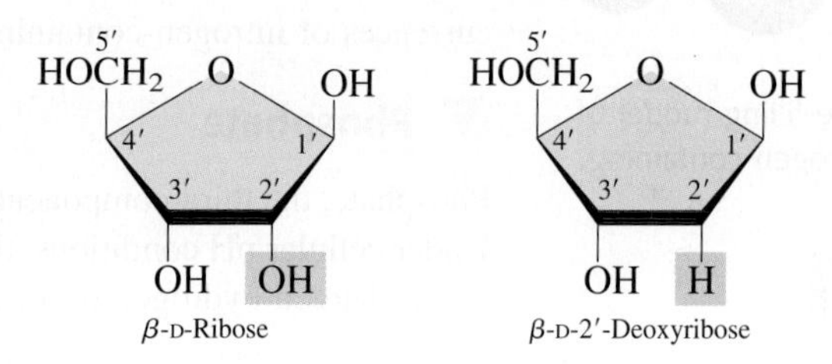

Structurally, the only difference between these two sugars occurs at carbon 2′. The —OH group present on this carbon in ribose becomes a —H atom in 2-deoxyribose. (The prefix *deoxy-* means "without oxygen.")

RNA and DNA differ in the identity of the sugar unit in their nucleotides. In RNA the sugar unit is *r*ibose—hence the *R* in RNA. In DNA the sugar unit is 2-*d*eoxyribose—hence the *D* in DNA.

Nitrogen-Containing Heterocyclic Bases

Five nitrogen-containing heterocyclic bases are nucleotide components. Three of them are derivatives of pyrimidine, a monocyclic base with a six-membered ring, and two are derivatives of purine, a bicyclic base with fused five- and six-membered rings.

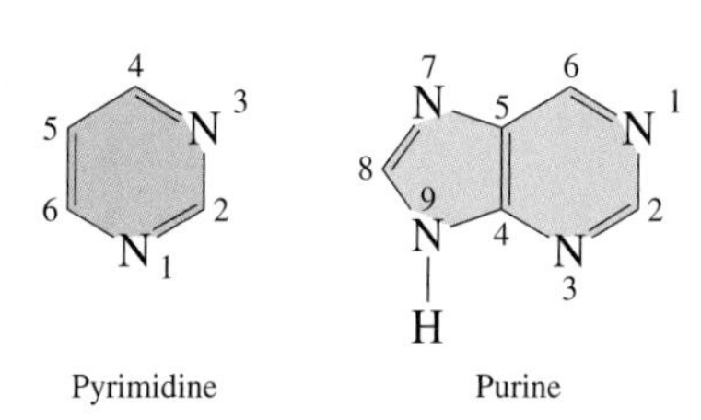

Both of these heterocyclic compounds are bases because they contain amine functional groups (secondary or tertiary), and amine functional groups exhibit basic behavior (proton acceptors; Section 12.8).

The three pyrimidine derivatives found in nucleotides are thymine (T), cytosine (C), and uracil (U).

Thymine (T) Cytosine (C) Uracil (U)

Thymine is the 5-methyl-2,4-dioxo derivative, cytosine the 4-amino-2-oxo derivative, and uracil the 2,4-dioxo derivative of pyrimidine.

The two purine derivatives found in nucleotides are adenine (A) and guanine (G).

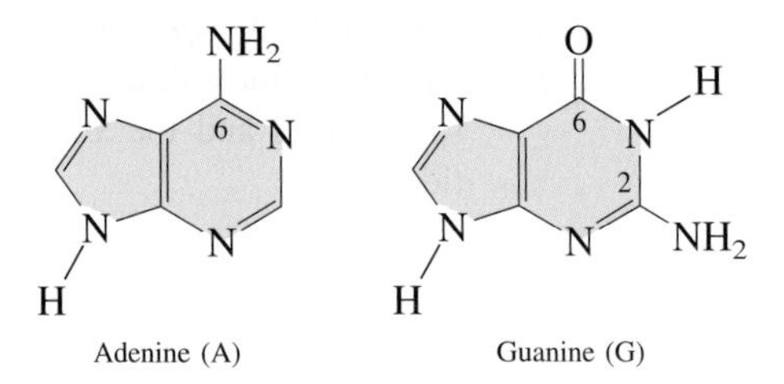

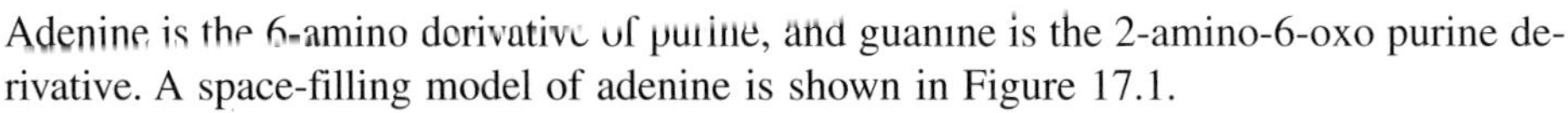

Adenine is the 6-amino derivative of purine, and guanine is the 2-amino-6-oxo purine derivative. A space-filling model of adenine is shown in Figure 17.1.

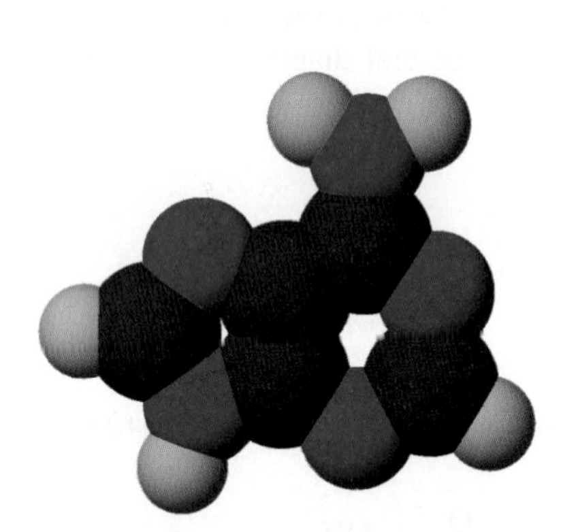

Figure 17.1 A space-filling model of the heterocyclic nitrogen-containing base adenine.

Adenine, guanine, and cytosine are found in both DNA and RNA. Uracil is found only in RNA, and thymine usually occurs only in DNA. Figure 17.2 summarizes the occurrences of nitrogen-containing bases in nucleic acids.

Phosphate

Phosphate, the third component of a nucleotide, is derived from phosphoric acid (H_3PO_4). Under cellular pH conditions, the phosphoric acid loses two of its hydrogen atoms to give a −2-charged hydrogen phosphate ion.

Phosphoric acid   Hydrogen phosphate ion under cellular pH conditions

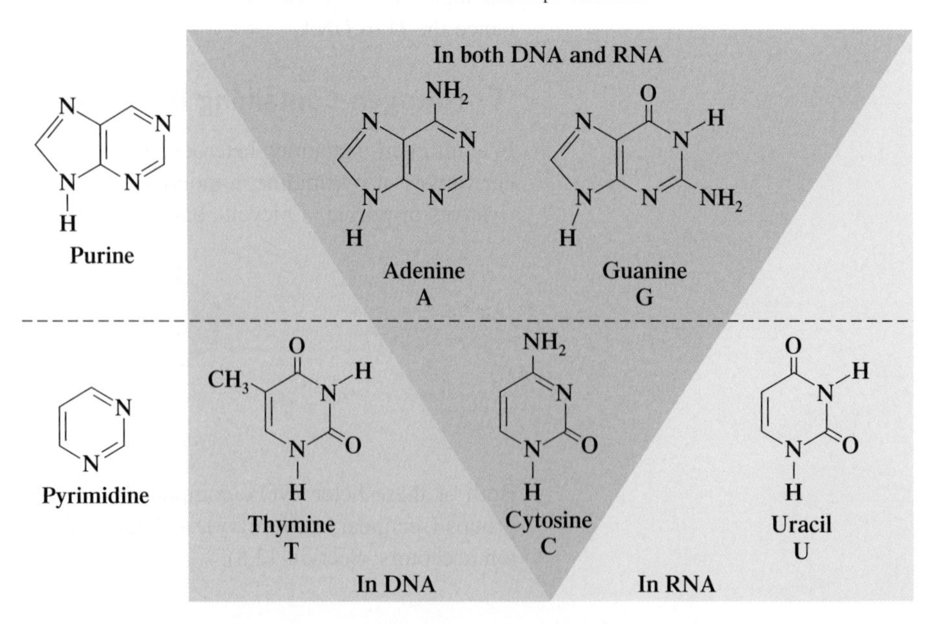

Figure 17.2 Two purine bases and three pyrimidine bases are found in the nucleotides present in nucleic acids.

▶ To remember which two of the five nucleotide bases are the purine derivatives (fused rings), use the phrase "pure silver" and the chemical symbol for silver, which is Ag.

pure Ag
purine A and G

Nucleotide Formation

The formation of a nucleotide from sugar, base, and phosphate can be visualized as occurring in the following manner:

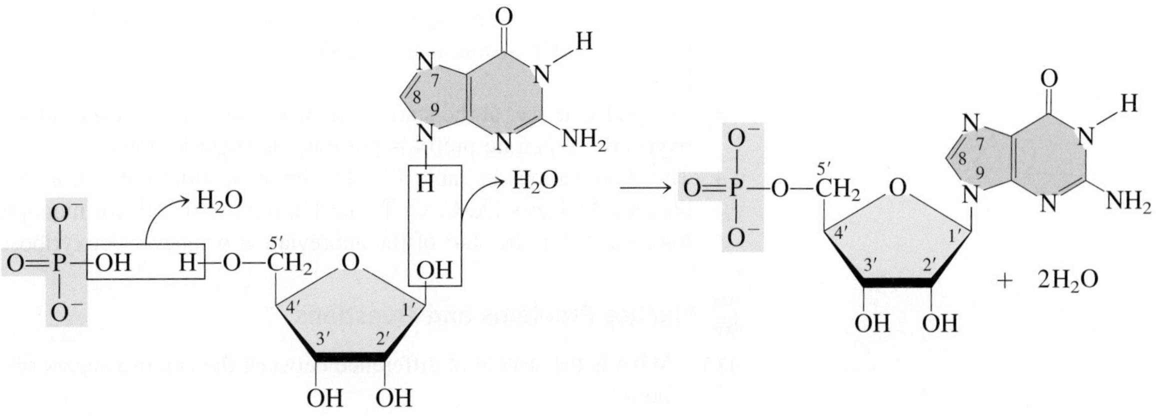

Phosphate Sugar Nucleotide

Important characteristics of this combining of three molecules into one molecule (the nucleotide) are that

1. Condensation, with formation of a water molecule, occurs at two locations: between sugar and base, and between sugar and phosphate.
2. The base is always attached at the C-1′ position of the sugar. For purine bases, attachment is through N-9; for pyrimidine bases, N-1 is involved. The C-1′ carbon atom of the ribose unit is always in a β configuration (Section 14.7).
3. The phosphate group is attached to the sugar at the C-5′ position through a phosphate–ester linkage.

There are four possible RNA nucleotides, differing in the base present (A, C, G, or U) and four possible DNA nucleotides, differing in the base present (A, C, G, or T).

Nucleotide Nomenclature

The eight nucleotides of DNA and RNA molecules have common names and abbreviations (see Table 17.1). It is important to be familiar with them, because they are frequently encountered in biochemistry.

Table 17.1 The Names of the Eight Nucleotides Found in DNA and RNA

Base	Sugar	Nucleotide name	Nucleotide abbreviation
DNA Nucleotides			
adenine	deoxyribose	deoxyadenosine 5′-monophosphate	dAMP
guanine	deoxyribose	deoxyguanosine 5′-monophosphate	dGMP
cytosine	deoxyribose	deoxycytidine 5′-monophosphate	dCMP
thymine	deoxyribose	deoxythymidine 5′-monophosphate	dTMP
RNA Nucleotides			
adenine	ribose	adenosine 5′-monophosphate	AMP
guanine	ribose	guanosine 5′-monophosphate	GMP
cytosine	ribose	cytidine 5′-monophosphate	CMP
uracil	ribose	uridine 5′-monophosphate	UMP

We can make several generalizations about the nomenclature given in Table 17.1.

1. All of the names end in 5′-monophosphate, which signifies the presence of a phosphate group attached to the 5′ carbon atom of ribose or deoxyribose. (In Chapter 18 we will encounter nucleotides that contain two or three phosphate groups—diphosphates and triphosphates.)
2. Preceding the monophosphate ending is the name of the base present in a modified form. The suffix *-osine* is used with purine bases, the suffix *-idine* with pyrimidine bases.
3. The prefix *deoxy-* at the start of the name signifies that the sugar present is deoxyribose. When no prefix is present, the sugar is ribose.
4. The abbreviations in Table 17.1 for the nucleotides come from the one-letter symbols for the bases (A, C, G, T, and U), the use of MP for monophosphate, and a lower-case *d* at the start of the abbreviation whenever deoxyribose is the sugar.

Practice Problems and Questions

17.5 What is the structural difference between the pentose sugars ribose and 2-deoxyribose?

17.6 What are the names of the pentose sugars present, respectively, in DNA and RNA molecules?

17.7 Indicate whether each of the following nitrogen-containing heterocyclic bases is a purine derivative or a pyrimidine derivative.
a. Thymine b. Cytosine c. Adenine d. Guanine

17.8 Characterize each of the following nitrogen-containing bases as a component of (1) both DNA and RNA nucleotides, (2) DNA but not RNA nucleotides, or (3) RNA but not DNA nucleotides.
a. Adenine b. Thymine c. Uracil d. Cytosine

17.9 How many different choices are there for each of the following subunits in the specified type of nucleotide?
a. Pentose sugar subunit in DNA nucleotides
b. Nitrogen-containing base subunit in RNA nucleotides
c. Phosphate subunit in DNA nucleotides
d. Phosphate subunit in RNA nucleotides

17.10 Which nitrogen-containing base is present in each of the following nucleotides?
a. AMP b. dGMP c. dTMP d. UMP

17.11 Which pentose sugar is present in each of the nucleotides in Problem 17.10?

17.12 Characterize as true or false each of the following statements about the given nucleotide.

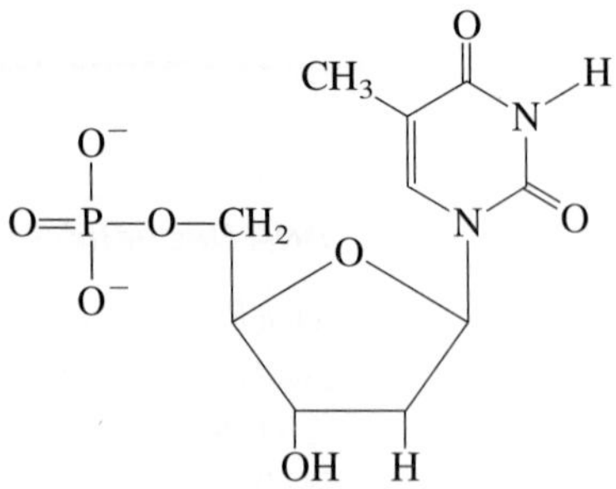

a. The nitrogen-containing base is a purine derivative.
b. The phosphate group is attached to the sugar unit at carbon 3′.
c. The sugar unit is ribose.
d. The nucleotide could be a component of both DNA and RNA.

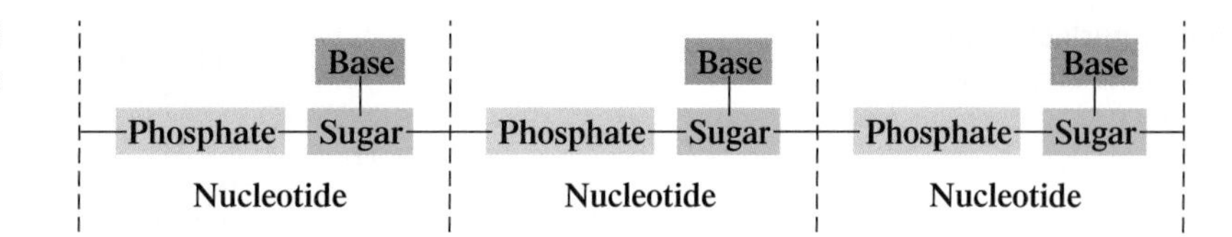

Figure 17.3 The general structure of a nucleic acid in terms of nucleotide subunits.

Learning Focus

Describe the primary structures of DNA and RNA.

▶ Nucleotides are related to nucleic acids in the same way that amino acids are related to proteins.

▶ The backbone of a nucleic acid structure is always an alternating sequence of phosphate and sugar groups. The sugar is ribose in RNA and deoxyribose in DNA.

17.3 Primary Structure of Nucleic Acids

Nucleic acids are polymers in which the repeating units, the monomers, are nucleotides (Section 17.2). The nucleotide units within a nucleic acid molecule are linked to each other through sugar–phosphate bonds. The resulting molecular structure (Figure 17.3) involves a chain of alternating sugar and phosphate groups with a base group protruding from the chain at regular intervals.

The alternating sugar–phosphate chain in a nucleic acid structure is often called the *nucleic acid backbone*. This backbone is constant throughout the entire nucleic acid structure. For DNA molecules, the backbone consists of alternating phosphate and *deoxyribose* sugar units; for RNA molecules, the backbone consists of alternating phosphate and *ribose* sugar units (see Figure 17.4).

The variable portion of nucleic acid structure is the sequence of bases attached to the sugar units of the backbone. The sequence of these base side chains distinguishes various DNAs from each other and various RNAs from each other. Only four types of bases are found in any given nucleic acid structure. This situation is much simpler than that for proteins, where 20 side-chain entities (amino acids) are available (Section 16.2). In both RNA and DNA, adenine, guanine, and cytosine are encountered as side-chain components; thymine is found mainly in DNA, and uracil only in RNA (Figure 17.2).

The **primary structure of a nucleic acid** *is the order in which the nucleotides are linked together in the nucleic acid.* Because the sugar–phosphate backbone of a given nucleic acid does not vary, the primary structure of the nucleic acid depends only on the sequence of bases present. Further information about nucleic acid structure can be obtained by considering the detailed four-nucleotide segment of a DNA molecule shown in Figure 17.5.

The following list describes some important points of DNA structure that are illustrated in Figure 17.5.

1. Each nonterminal phosphate group of the sugar–phosphate backbone is bonded to two sugar molecules through a *3′,5′-phosphodiester linkage*. There is a phosphoester bond to the 5′ carbon of one sugar unit and a phosphoester bond to the 3′ carbon of the other sugar.
2. A nucleotide chain has *directionality*. One end of the nucleotide chain, the *5′ end*, normally carries a free phosphate group attached to the 5′ carbon atom. The other end of the nucleotide chain, the *3′ end*, normally has a free hydroxyl group attached to the 3′ carbon atom. By convention, the sequence of bases of a nucleic acid strand is read from the 5′ end to the 3′ end.
3. Each nonterminal phosphate group in the backbone of a nucleic acid carries a −1 charge. The parent phosphoric acid molecule from which the phosphate was derived originally had three —OH groups (Section 17.2). Two of these become

Figure 17.4 The sugar present in the nucleic acid backbone differs for DNAs and RNAs.

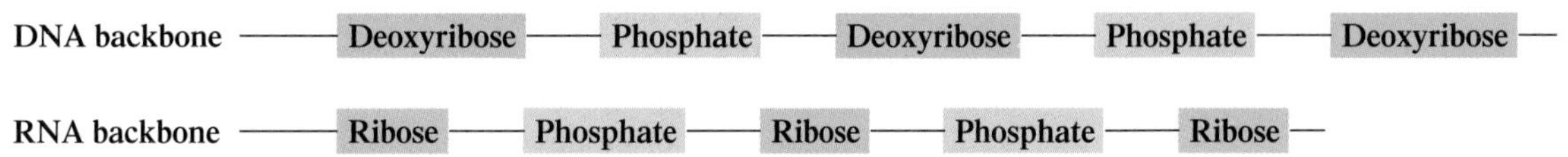

Figure 17.5 A four-nucleotide-long segment of DNA. (The choice of bases was arbitrary.)

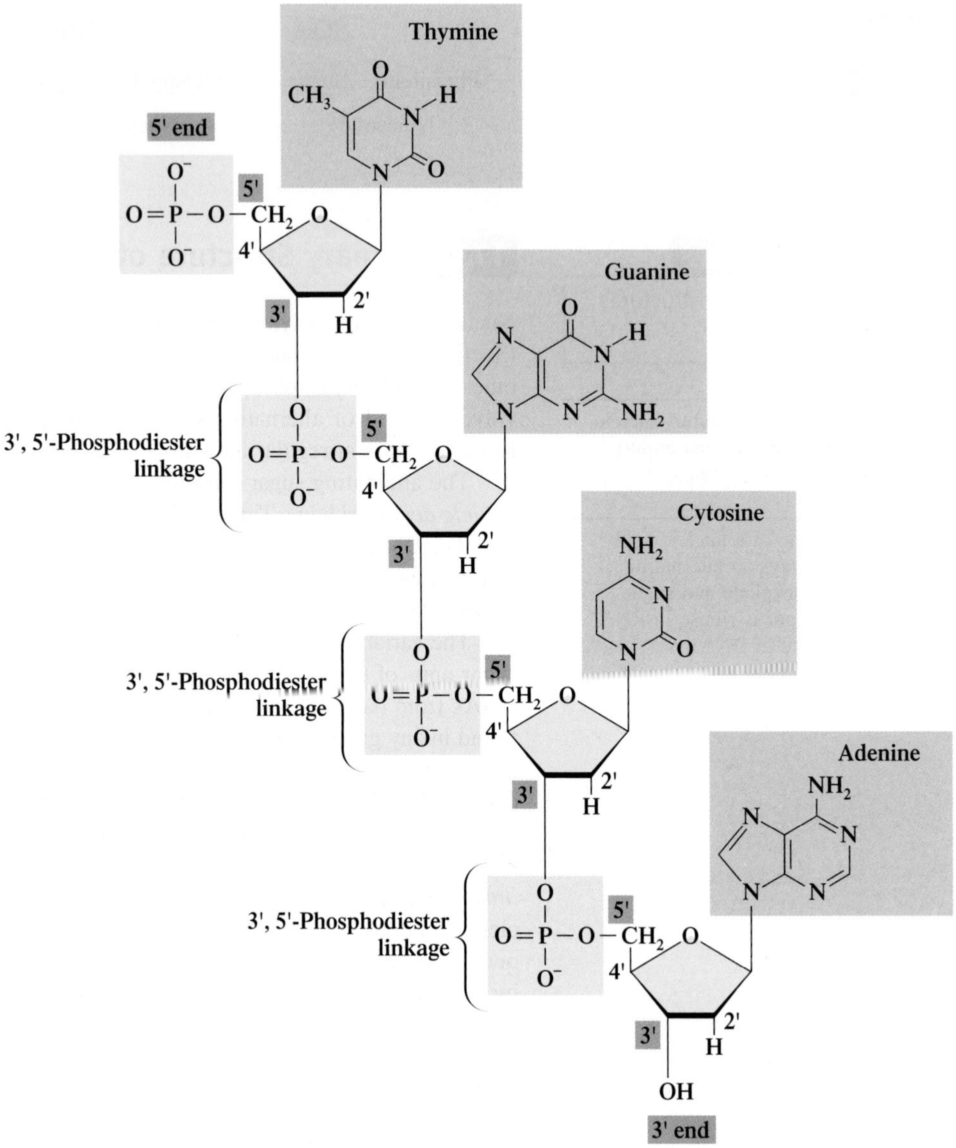

involved in the 3′,5′-phosphodiester linkage. The remaining —OH group is free to exhibit acidic behavior—that is, to produce an H^+ ion.

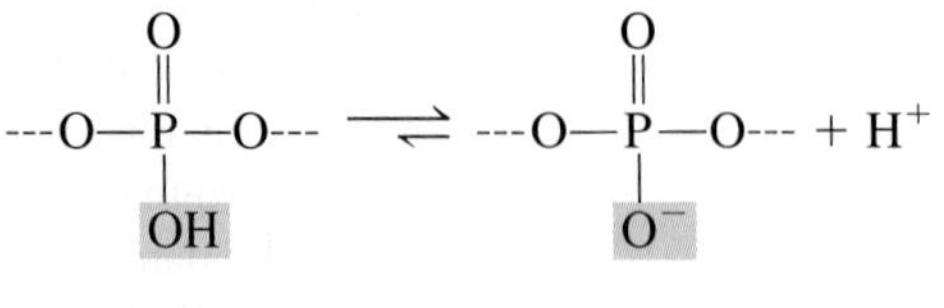

$$\text{---O—}\underset{\text{OH}}{\overset{\text{O}}{\overset{\|}{\text{P}}}}\text{—O---} \rightleftharpoons \text{---O—}\underset{\text{O}^-}{\overset{\text{O}}{\overset{\|}{\text{P}}}}\text{—O---} + H^+$$

This behavior by the many phosphate groups in a nucleic acid backbone gives nucleic acids their acidic properties.

Three parallels between primary nucleic acid structure and primary protein structure are worth noting.

1. Both nucleic acids and proteins have backbones that do not vary in structure (see Figure 17.6).
2. The differences among various nucleic acids and among various proteins are related to the order of groups attached to the backbone: bases in nucleic acids and amino acid R groups in proteins.

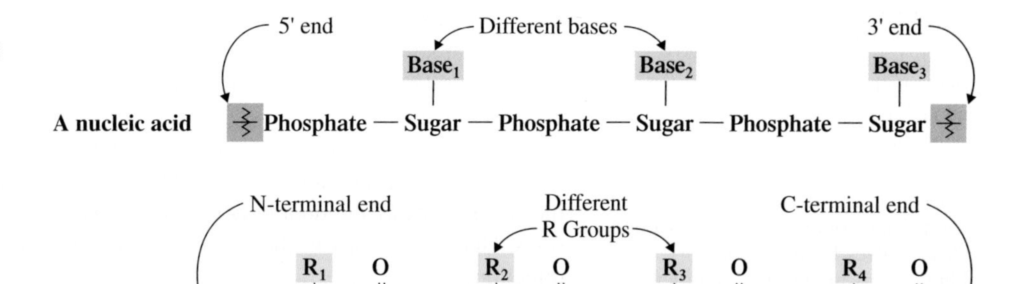

Figure 17.6 A comparison of the general primary structures of nucleic acids and proteins.

3. Both nucleic acids and protein polymer chains have directionality; for nucleic acids there are a 5′ end and a 3′ end, and for proteins there are an N-terminal end and a C-terminal end.

Practice Problems and Questions

17.13 What is meant by the phrase *backbone of a nucleic acid*?

17.14 How does the "backbone" for a DNA molecule differ from that for an RNA molecule?

17.15 To what parts of a nucleic acid "backbone" are the nitrogen-containing bases attached?

17.16 What distinguishes various DNA molecules (or RNA molecules) from each other?

17.17 What is the difference between a nucleic acid's 3′ end and its 5′ end?

17.18 In the lengthening of a polynucleotide chain, which type of nucleotide subunit would bond to the 3′ end of the polynucleotide chain?

17.19 What are the nucleotide subunits that participate in a 3′,5′-phosphodiester linkage?

17.20 How many 3′,5′-phosphodiester linkages are present in a tetranucleotide segment of a nucleic acid?

Learning Focus

Know general structural features for the DNA double helix; discuss the significance of A = T and G = C as related to the DNA double helix.

17.4 The DNA Double Helix

Like proteins, nucleic acids have secondary, or three-dimensional, structure as well as primary structure. The secondary structures of DNAs and RNAs differ, and we will discuss them separately.

The amounts of the bases A, T, G, and C present in DNA molecules were the key to determination of the general three-dimensional structure of DNA molecules. Base composition data for DNA molecules from many different organisms revealed a definite pattern of base occurrence. The amounts of A and T were always equal, and the amounts of C and G were always equal, as were the amounts of total purines and total pyrimidines.

The relative amounts of these base pairs in DNA vary depending on the life form from which the DNA is obtained. (Each animal or plant has a unique base composition.) However, the relationships

$$\%A = \%T \quad \text{and} \quad \%C = \%G$$

always hold true. For example, human DNA contains 30% adenine, 30% thymine, 20% guanine, and 20% cytosine.

In 1953, an explanation for the base composition patterns associated with DNA molecules was proposed by the American microbiologist James Watson and the English biophysicist Francis Crick. Their model, which has now been validated in numerous ways, involves a double-helix structure that accounts for the equality of bases present, as well

▶ The *antiparallel* nature of the two polynucleotide chains in the DNA double helix means that there is a 5′ end and a 3′ end at both ends of the double helix.

▶ A mnemonic device for recalling base-pairing combinations in DNA and RNA involves listing the base abbreviations in alphabetical order. Then the first and last bases pair, and so do the middle two bases.

DNA: A C G T

RNA: A C G U

▶ Hydrogen bonding is responsible for the secondary structure (double helix) of DNA. Hydrogen bonding is also responsible for secondary structure in proteins (Section 16.8).

▶ The two strands of DNA in a double helix are complementary. This means that if you know the order of bases in one strand, you can predict the order of bases in the other strand.

Figure 17.7 A schematic drawing of the DNA double helix that emphasizes the hydrogen bonding between bases on the two chains.

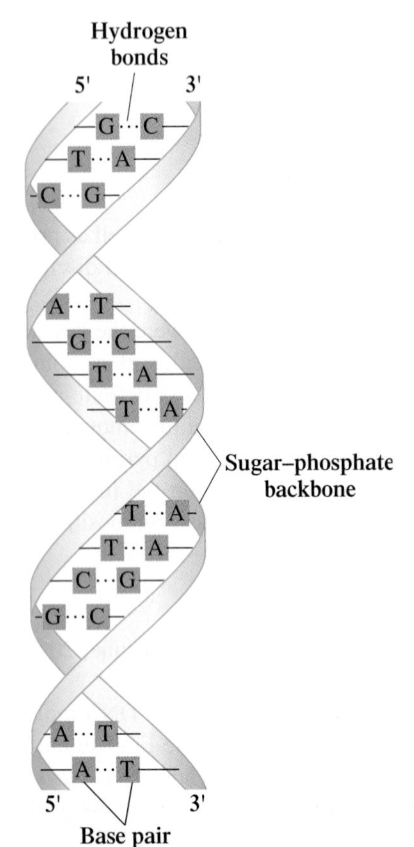

The DNA double helix involves two polynucleotide strands coiled around each other in a manner somewhat like a spiral staircase. The sugar–phosphate backbones of the two polynucleotide strands can be thought of as being the outside banisters of the spiral staircase (Figure 17.7).

The bases (side chains) of each backbone extend inward toward the bases of the other strand. The two strands are connected by *hydrogen bonds* between their bases. Additionally, the two strands of the double helix are *antiparallel*—that is, they run in opposite directions. One strand runs in the 5′-to-3′ direction, and the other is oriented in the 3′-to-5′ direction.

Base Pairing

A physical restriction, the size of the interior of the DNA double helix, limits the base pairs that can hydrogen-bond to one another. Only pairs involving one small base (a pyrimidine) and one large base (a purine) correctly "fit" within the helix interior. There is not enough room for two large purine bases to fit opposite each other (they overlap), and two small pyrimidine bases are too far apart to hydrogen-bond to one another effectively. Of the four possible purine–pyrimidine combinations (A–T, A–C, G–T, and G–C), hydrogen-bonding possibilities are *most favorable* for the A–T and G–C pairings, and these two combinations are the *only two* that normally occur in DNA. Figure 17.8 shows the specific hydrogen-bonding interactions for the two purine–pyrimidine base-pairing combinations that do occur.

The pairing of A with T and of G with C is said to be *complementary*. A and T are complementary bases, as are G and C. **Complementary bases** *are specific pairs of bases in nucleic acid structures that hydrogen-bond to each other.* The fact that complementary base pairing occurs in DNA molecules explains, very simply, why the amounts of the bases A and T present are always equal, as are the amounts of G and C.

Hydrogen bonding holds the two strands of the DNA double helix together. Although hydrogen bonds are relatively weak forces, each DNA molecule contains so many base pairs that the hydrogen-bonding attractions are sufficient in magnitude, collectively, to prevent the two entwined DNA strands from separating spontaneously under normal physiological conditions.

The two strands of DNA in a double helix are *not identical*—they are complementary. Wherever G occurs in one strand, there is a C in the other strand; wherever T occurs in one strand, there is an A in the other strand. An important ramification of this complementary relationship is that knowing the base sequence of one strand of DNA enables us to predict the base sequence of the complementary strand.

In specifying the base sequence of a segment of a strand of DNA (or RNA), we list the bases in sequential order (using their one-letter abbreviations) in the direction from the 5′ end to the 3′ end of the segment.

5′ A–A–G–C–T–A–G–C–T–T–A–C–T 3′

Example 17.1 Predicting Base Sequence in a Complementary DNA Strand

Predict the sequence of bases in the DNA strand that is complementary to the single DNA strand shown.

5′ C–G–A–A–T–C–C–T–A 3′

Solution

Because only A forms a complementary base pair with T, and only G with C, the complementary strand is as follows:

Given:	5′ C–G–A–A–T–C–C–T–T–A 3′
Complementary strand:	3′ G–C–T–T–A–G–G–A–T 5′

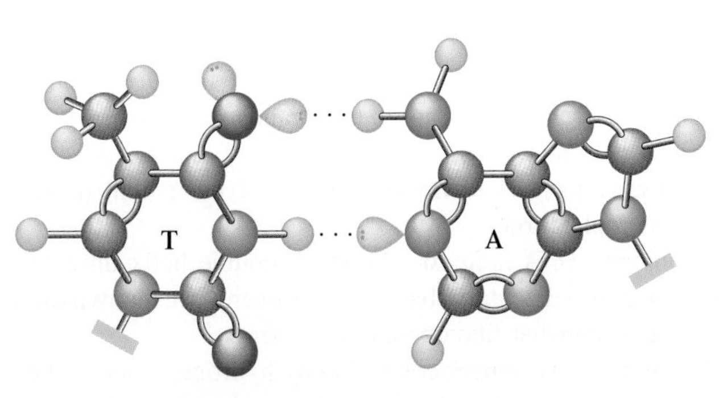

(a) Thymine–Adenine Base Pairing
(two hydrogen bonds form)

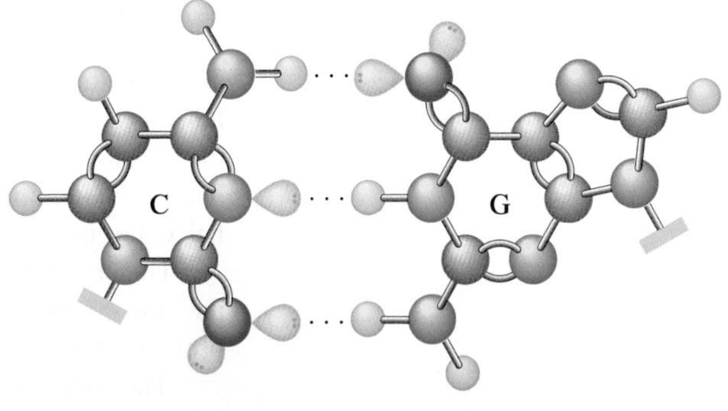

(b) Cytosine–Guanine Base Pairing
(three hydrogen bonds form)

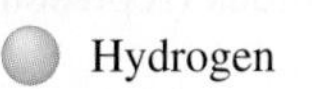

Figure 17.8 (a) Thymine–adenine base pairing involves two hydrogen bonds. (b) Cytosine–guanine base pairing involves three hydrogen bonds.

> Note the reversal of the numbering of the ends of the complementary strand compared to the given strand. This is due to the antiparallel nature of the two strands in a DNA double helix.

Practice Problems and Questions

17.21 Describe the DNA double helix in terms of
a. general shape
b. what is on the outside of the helix and what is within the interior of the helix

17.22 The base content of a particular DNA molecule is 36% thymine. What is the percentage of each of the following bases in the molecule?
a. Adenine b. Guanine c. Cytosine

17.23 What base sequence is implied by the designation T–T–A–G–C?

17.24 Identify the 3′ end and the 5′ end for the base sequence in Problem 17.23.

17.25 The two-base DNA sequences TA and AT represent different dinucleotides. Explain why this is so.

17.26 Indicate whether each of the following base pairs could be part of the structure of a DNA double helix.
a. A–A b. A–T c. G–C d. G–A

17.27 The two strands of DNA in a DNA double helix are *antiparallel*. Explain what is meant by this statement.

17.28 Using the concept of complementary base pairing, write the complementary DNA strands, with their 5′ and 3′ ends labeled, for each of the following DNA base sequences.
a. 5′ A–C–G–T–A–T 3′ b. 5′ T–T–A–C–C–G 3′
c. 3′ T–T–T–A–G–A 5′ d. C–A–T–T–A–C

Learning Focus

Explain the major steps in the process of DNA replication.

17.5 Replication of DNA Molecules

DNA molecules are the carriers of the genetic information within a cell; that is, they are the molecules of heredity. Each time a cell divides, an exact copy of the DNA of the parent cell is needed for the new daughter cell. The process by which new DNA molecules are generated is DNA replication. **DNA replication** *is the process by which DNA molecules produce exact duplicates of themselves.*

DNA Replication Overview

To understand DNA replication, we must regard the two strands of the DNA double helix as a pair of *templates,* or patterns. During replication, the strands separate. Each can then act as a template for the synthesis of a new, complementary strand. The result is two daughter DNA molecules with base sequences identical to those of the parent double helix. Let us consider details of this replication.

Under the influence of the enzyme *DNA helicase,* the DNA double helix unwinds, and the hydrogen bonds between complementary bases are broken. This unwinding process, as shown in Figure 17.9, is somewhat like opening a zipper.

The bases of the separated strands are no longer connected by hydrogen bonds. They can pair with *free* individual nucleotides present in the cell's nucleus. As shown in Figure 17.9, the base pairing always involves C pairing with G and A pairing with T. After the free nucleotides have formed hydrogen bonds with the old strand (the template), the enzyme *DNA polymerase* verifies that the base pairing is correct and catalyzes the formation of new phosphodiester linkages between nucleotides (represented by colored ribbons in Figure 17.9).

Each of the two daughter molecules of double-stranded DNA formed in the DNA replication process contains one strand from the original parent molecule and one newly formed strand.

The Replication Process in Finer Detail

Though simple in principle, the DNA replication process has many intricacies.

1. The enzyme *DNA polymerase* can operate on a forming DNA daughter strand only in the $5'$-to-$3'$ direction. Because the two strands of parent DNA run in opposite directions (one is $5'$ to $3'$ and the other $3'$ to $5'$, Section 17.4), only one strand can grow continuously in the $5'$-to-$3'$ direction. The other strand must be formed in short segments, called *Okazaki fragments* (after their discoverer, Reiji Okazaki), as the DNA unwinds (see Figure 17.10). The breaks or gaps in this daughter strand are called *nicks.* To complete the formation of this strand, the Okazaki fragments are connected by action of the enzyme *DNA ligase.*
2. The process of DNA unwinding does not have to begin at an end of the DNA molecule. It may occur at any location within the molecule. Indeed, studies show that unwinding usually occurs at several interior locations simultaneously and that DNA replication is bidirectional for these locations; that is, it proceeds in both directions from the unwinding sites. As shown in Figure 17.11, the result of this multiple-site

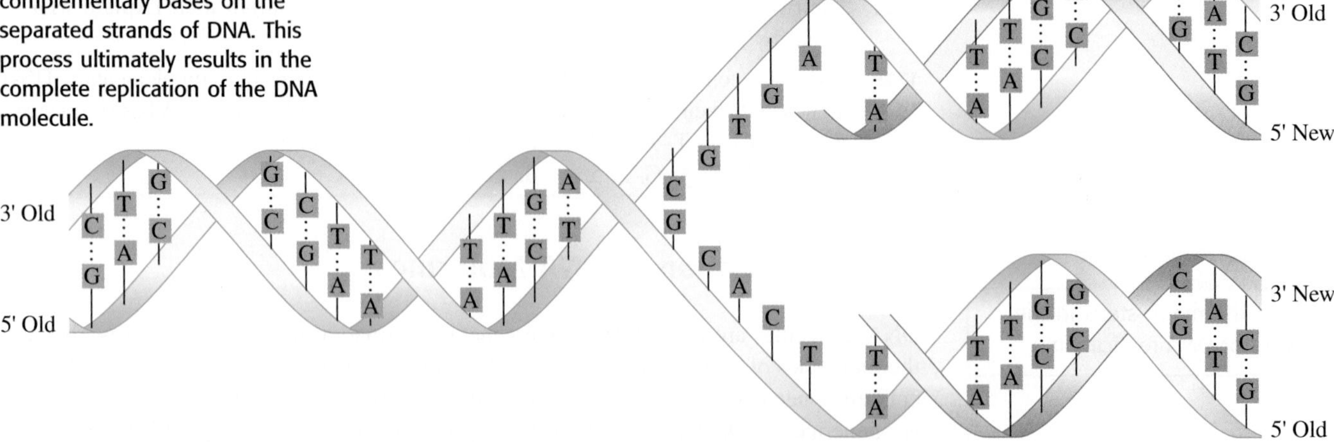

Figure 17.9 In DNA replication, the two strands of the DNA double helix unwind, the separated strands serving as templates for the formation of new DNA strands. Free nucleotides pair with the complementary bases on the separated strands of DNA. This process ultimately results in the complete replication of the DNA molecule.

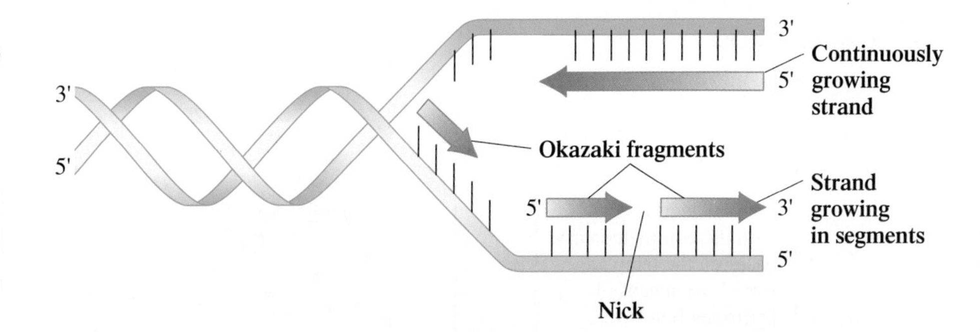

Figure 17.10 Because the enzyme DNA polymerase can act only in the 5′-to-3′ direction, one strand (top) grows continuously in the direction of the unwinding, and the other strand grows in segments in the opposite direction. The segments in this latter chain are then connected by a different enzyme, DNA ligase.

replication process is formation of "bubbles" of newly synthesized DNA. The bubbles grow larger and eventually coalesce, giving rise to two complete daughter DNAs. Multiple-site replication enables large DNA molecules to be replicated rapidly.

Chromosomes

Once the DNA within a cell has been replicated, it interacts with specific proteins in the cell to form structural units called chromosomes. A **chromosome** *is an individual DNA molecule bound to a group of proteins.* Typically, a chromosome is about 15% by mass DNA and 85% by mass protein.

Cells from different kinds of organisms have different numbers of chromosomes. A human has 46 chromosomes per cell, a mosquito 6, a frog 26, a dog 78, and a turkey 82.

Chromosomes occur in matched (*homologous*) pairs. The 46 chromosomes of a human cell constitute 23 homologous pairs. One member of each homologous pair is derived from a chromosome inherited from the father, and the other is a copy of one of the chromosomes inherited from the mother. Homologous chromosomes have similar, but not identical, DNA base sequences; both code for the same traits but for different forms of the trait (for example, blue eyes versus brown eyes).

▶ Chromosomes are *nucleoproteins*. They are a combination of nucleic acid (DNA) and various proteins.

The accompanying Chemistry at a Glance summarizes the steps in DNA replication.

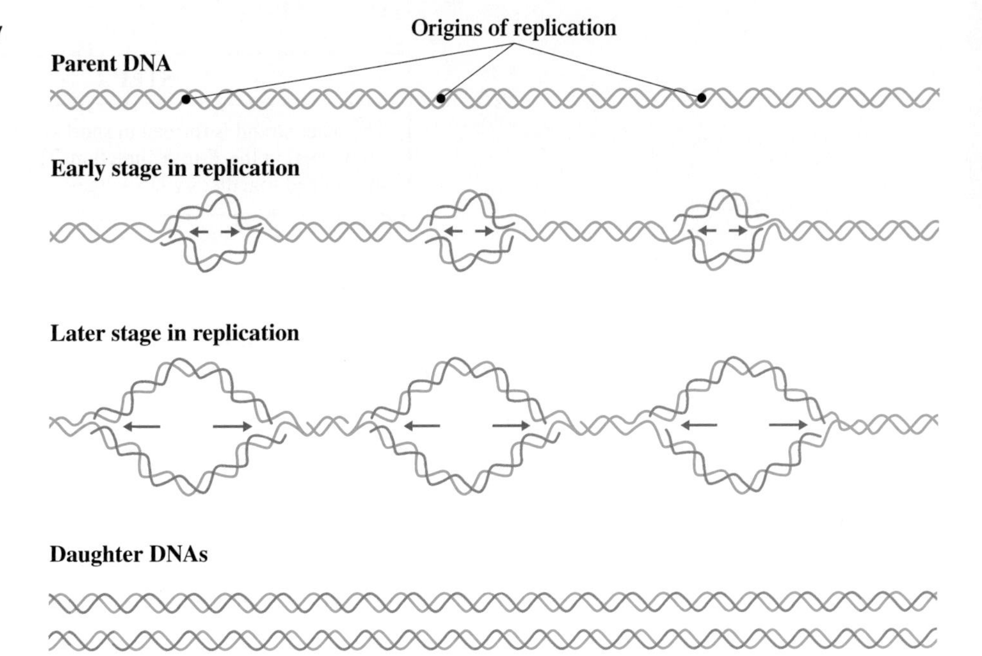

Figure 17.11 DNA replication usually occurs at multiple sites within a molecule, and the replication is bidirectional from these sites.

DNA Replication

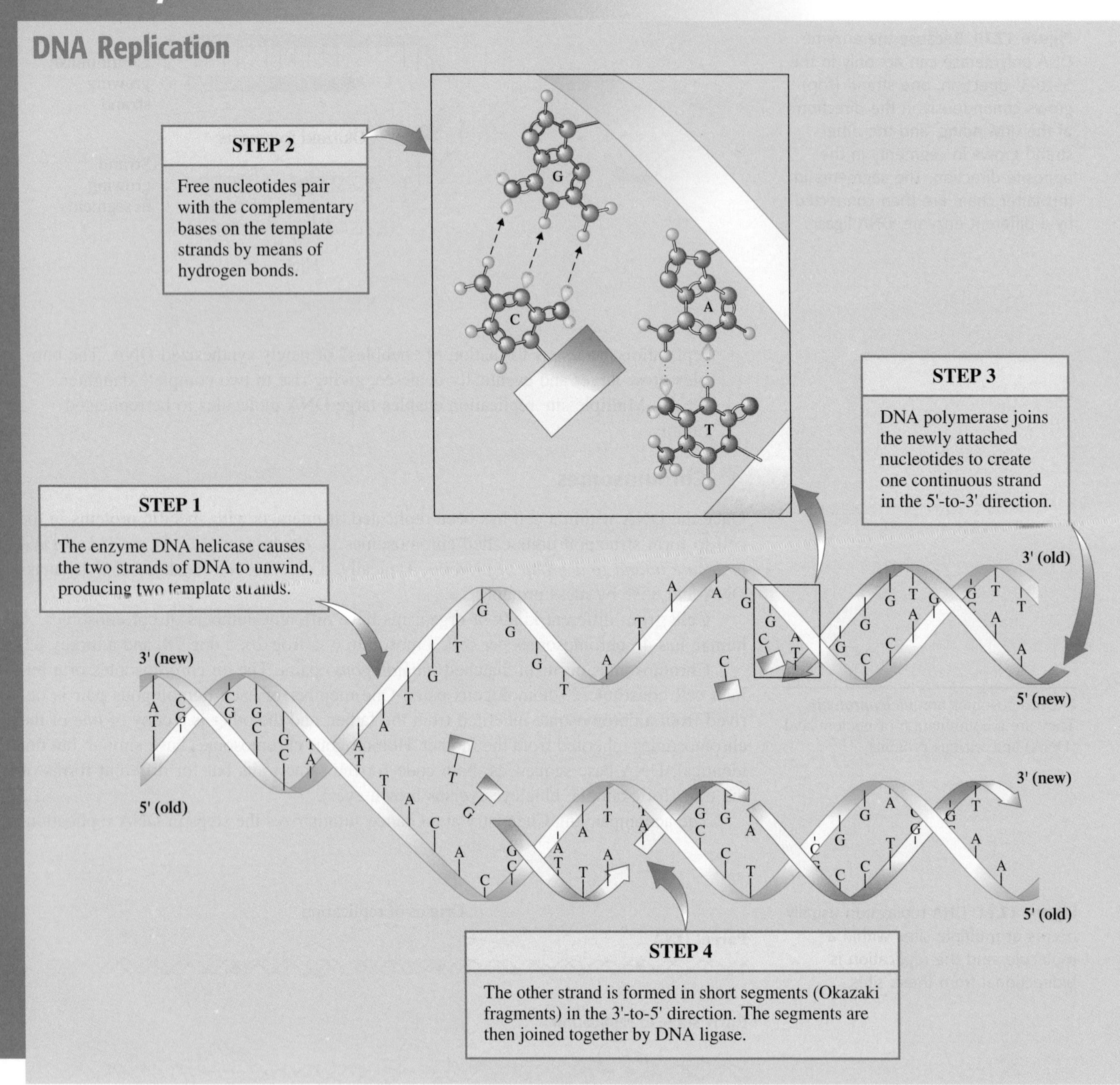

STEP 1
The enzyme DNA helicase causes the two strands of DNA to unwind, producing two template strands.
STEP 2
Free nucleotides pair with the complementary bases on the template strands by means of hydrogen bonds.
STEP 3
DNA polymerase joins the newly attached nucleotides to create one continuous strand in the 5'-to-3' direction.
STEP 4
The other strand is formed in short segments (Okazaki fragments) in the 3'-to-5' direction. The segments are then joined together by DNA ligase.
3' (new)
5' (old)
3' (old)
5' (new)
3' (new)
5' (old)

Practice Problems and Questions

17.29 What is the function of the enzyme *DNA helicase* in the DNA replication process?

17.30 What are two functions of the enzyme *DNA polymerase* in the DNA replication process?

17.31 What is the fate of each strand of the parent DNA molecule in the DNA replication process whereby two daughter molecules are produced?

17.32 In the replication of a DNA molecule, two daughter molecules, Q and R, are formed. The following base sequence is part of the newly formed strand in daughter molecule Q.

5′ A–C–T–T–A–G 3′

Indicate the corresponding base sequence in
a. the newly formed strand in daughter molecule R
b. the "parent" strand in daughter molecule Q
c. the "parent" strand in daughter molecule R

17.33 During DNA replication, one of the newly formed strands grows continuously, whereas the other grows in segments that are later connected together. Explain why this is so.

17.34 What is a chromosome?

Learning Focus

Describe the two major phases of protein synthesis.

17.6 Overview of Protein Synthesis

We saw in the previous section how the replication of DNA makes it possible for a new cell to contain the same genetic information as its parent cell. We will now consider how the genetic information contained in a cell is expressed in cell operation. This brings us to the topic of protein synthesis. The synthesis of proteins (skin, hair, enzymes, hormones, and so on) is under the direction of DNA molecules. It is this role of DNA that establishes the similarities between parent and offspring that we regard as hereditary characteristics (see Figure 17.12).

We can divide the overall process of protein synthesis into two steps. The first step is called transcription and the second translation. **Transcription** *is the process by which DNA directs the synthesis of RNA molecules that carry the coded information needed for protein synthesis.* **Translation** *is the process by which the codes within RNA molecules are deciphered and a particular protein molecule is formed.* The following diagram summarizes the relationship between transcription and translation.

$$\text{DNA} \xrightarrow{\text{Transcription}} \text{RNA} \xrightarrow{\text{Translation}} \text{protein}$$

Before discussing the details of transcription and translation, we need to learn more about RNA molecules. We will be particularly concerned with differences between RNA and DNA and various types of RNA molecules.

Practice Problems and Questions

17.35 In the process of protein synthesis, does the transcription step or the translation step occur first?

17.36 Describe, in general terms, the *transcription* phase of protein synthesis.

17.37 Describe, in general terms, the *translation* phase of protein synthesis.

Figure 17.12 The identical physical characteristics of identical twins result from identical DNA molecules.

Learning Focus

Be familiar with the characteristics and function of the four major types of RNA molecules.

17.7 Ribonucleic Acids

Four major differences exist between RNA molecules and DNA molecules.

1. The sugar unit in the backbone of RNA is ribose; it is deoxyribose in DNA.
2. The base thymine found in DNA is replaced by uracil in RNA (Figure 17.2). Uracil, instead of thymine, pairs with (forms hydrogen bonds with) adenine in RNA.
3. RNA is a single-stranded molecule; DNA is double-stranded (double helix). Thus RNA, unlike DNA, does not contain equal amounts of specific bases.
4. RNA molecules are much smaller than DNA molecules, ranging from as few as 75 nucleotides to a few thousand nucleotides.

▶ The bases thymine (T) and uracil (U) have similar structures. Thymine is a methyluracil (Section 17.2). The hydrogen-bonding patterns (Figure 17.8) for the A–U base pair (RNA) and the A–T base pair (DNA) are identical.

We should note that the single-stranded nature of RNA does not prevent *portions* of an RNA molecule from folding back upon itself and forming double-helical regions. If the base sequences along two portions of an RNA strand are complementary, a structure with a hairpin loop results, as shown in Figure 17.13. The amount of double-helical structure present in a RNA varies with RNA type, but a value of 50% is not atypical.

Types of RNA Molecules

Through transcription, DNA produces four types of RNA, distinguished by their function. The four types are ribosomal RNA (rRNA), messenger RNA (mRNA), primary transcript RNA (ptRNA), and transfer RNA (tRNA).

▶ Primary transcript RNA was the last of the four types of RNA to be characterized. There is some disagreement on what it should be called. Some biochemistry reference sources call it *pre-RNA,* and others call it *heterogeneous nuclear RNA* (hnRNA).

Ribosomal RNA *combines with a series of proteins to form complex structures, called ribosomes, that serve as the physical sites for protein synthesis.* Ribosomes have molecular masses on the order of 3 million. The rRNA present in ribosomes has no informational function.

Messenger RNA *carries genetic information (instructions for protein synthesis) from DNA to the ribosomes.* The size (molecular mass) of mRNA varies with the length of the protein whose synthesis it will direct. Each kind of protein in the body has its own mRNA.

Primary transcript RNA *is the material from which messenger RNA is made.*

Transfer RNA *delivers specific individual amino acids to the ribosomes, the sites of protein synthesis.* These RNAs are the smallest of the RNAs, possessing only 75–90 nucleotide units.

▶ The most abundant type of RNA in a cell is ribosomal RNA (75% to 80% by mass). Transfer RNA constitutes 10%–15% of cellular RNA; messenger RNA and its precursor, primary transcript RNA, make up the remaining 5%–10% of RNA material in the cell.

Practice Problems and Questions

17.38 List four major differences between RNA molecules and DNA molecules.

17.39 Give the names and abbreviations for the four major types of RNA molecules.

17.40 State whether each of the following phrases applies to ptRNA, mRNA, tRNA, or rRNA.
a. Material from which messenger RNA is made
b. Delivers amino acids to protein synthesis sites
c. Smallest of the RNAs in terms of nucleotide units present
d. Most abundant type of RNA in a cell

Figure 17.13 A hairpin loop is produced when single-stranded RNA doubles back on itself and complementary base pairing occurs.

Learning Focus

Describe the transcription process and understand the relationship among DNA, ptRNA, mRNA, and genes.

17.8 Transcription: RNA Synthesis

In the process of transcription (Section 17.6) a ptRNA molecule is made using a segment of a DNA molecule as a template. The ptRNA so produced is then "edited" to produce the other types of RNA. Our discussion of transcription focuses on the production of mRNA molecules, the molecules that *direct* the synthesis of the many different proteins needed for cellular function.

Within a strand of a DNA molecule are instructions for the synthesis of numerous mRNA molecules. During transcription, a DNA molecule unwinds at the particular spot where the appropriate base sequence is found for the mRNA of concern, and the "exposed" base sequence is transcribed (copied). A short segment of DNA of this type, containing instructions for the formation of a particular mRNA, is called a *gene*. A **gene** *is a segment of a DNA molecule that contains the base sequence for the production of specific RNA molecules, which in turn guide the production of specific protein molecules.*

In humans, most genes are composed of 1000–3500 nucleotide units. Hundreds of genes can exist along a DNA molecule strand. Information concerning the total number of genes and the total number of nucleotide base pairs present in human DNA was "updated" in 2001. In that year, results of the Human Genome Project, a decade-long internationally based research project to determine the location and base sequence of each of the genes in the human genome, were announced. A **genome** *is all of the genetic material in the chromosomes of an organism.*

The DNA found in a human cell contains an estimated 2.9 billion nucleotide base pairs, making up approximately 40,000 genes. The estimate of 40,000 genes in the human genome is significantly lower than the previous estimate of 100,000 genes that had been made before the results of the Human Genome Project were in. These results also indicated that the base pairs present in these genes constituent only a very small percentage (2%) of the total number of base pairs present in the chromosomes of the human genome.

Steps in the Transcription Process

The mechanics of transcription are in many ways similar to those of DNA replication. Four steps are involved.

1. A *portion* of the DNA double helix unwinds, exposing some bases (a gene). The unwinding process is governed by the enzyme *RNA polymerase* rather than by *DNA helicase* (replication enzyme).
2. Free *ribo*nucleotides align along *one* of the exposed strands of DNA bases, forming new base pairs. In this process, U rather than T aligns with A in the base-pairing process. Because ribonucleotides rather than deoxyribonucleotides are involved in the base pairing, ribose, rather than deoxyribose, becomes incorporated into the new nucleic acid backbone.
3. RNA polymerase links the aligned ribonucleotides.
4. Transcription ends when the RNA polymerase enzyme encounters a sequence of bases that is "read" as a stop signal. The newly formed ptRNA molecule and the RNA polymerase enzyme are released, and the DNA then rewinds to reform the original double helix.

▶ In DNA–RNA base pairing, the complementary base pairs are

DNA		RNA
A	—	U
G	—	C
C	—	G
T	—	A

RNA molecules contain the base U instead of the base T.

Figure 17.14 shows the overall process of transcription of DNA to form RNA. All types of RNA molecules (ptRNA/mRNA, rRNA, and tRNA) are synthesized in the nucleus of a cell via this general transcription process.

Example 17.2 Base Pairing Associated with the Transcription Process

Determine the RNA base sequence that is complementary to the following DNA "template,"

5′ A–T–G–C–C–C–G–A–G–T–T 3′

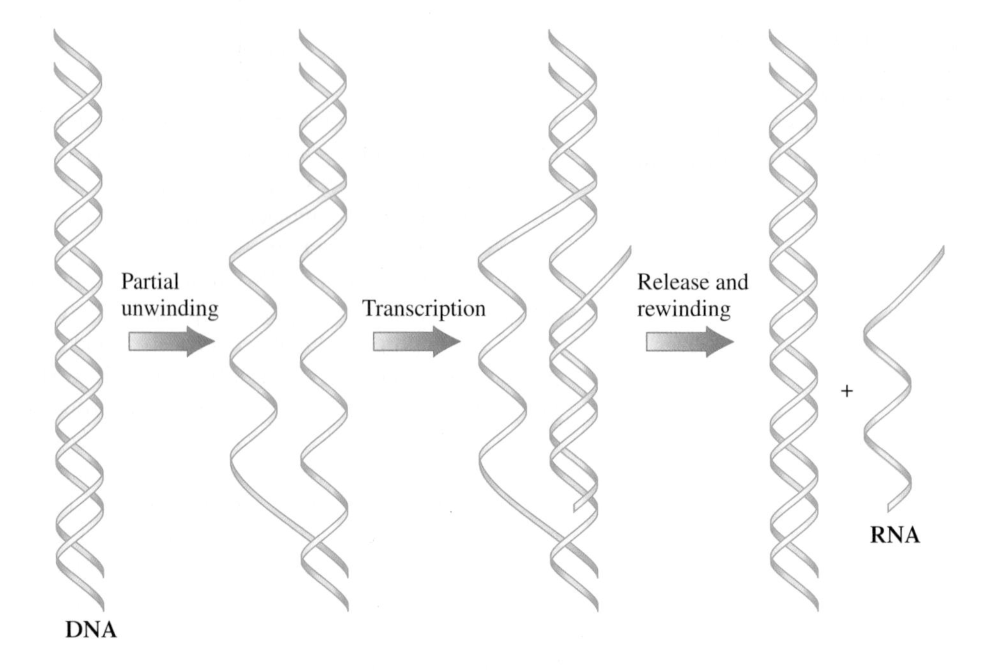

Figure 17.14 The transcription of DNA to form RNA involves an unwinding of a portion of the DNA double helix. Only one strand of the DNA is copied during transcription.

Solution

An RNA molecule cannot contain the base T. The base U is present instead. Therefore, U pairs with DNA A units instead of T. The transcription process will therefore produce the RNA sequence

$$3' \text{ U–A–C–G–G–G–C–U–C–A–A } 5'$$

Note that the DNA and RNA strands are antiparallel.

Post-Transcription Processing: Formation of mRNA

The RNA initially produced through transcription is ptRNA. The conversion of ptRNA into mRNA involves *post-transcription processing* of the ptRNA. In this processing, certain parts of the ptRNA are deleted and the retained parts are then spliced together. This leads us to the concepts of *exons* and *introns*.

It is now known that not all bases in a gene convey genetic information. Instead, a gene is *segmented* in that it has portions called *exons* that contain genetic information and portions called *introns* that do not convey genetic information.

An **exon** *is a DNA segment that conveys (codes for) genetic information. E*xons are DNA segments that help *ex*press a genetic message. An **intron** *is a DNA segment that does not convey (code for) genetic information. Int*rons are DNA segments that *int*errupt a genetic message. A gene consists of alternating exon and intron segments (Figure 17.15). At present, it is not known why introns occur in genes, but determining their function is an active area of biochemical research.

Both the exons and the introns of a gene are transcribed during production of *primary transcript RNA* (ptRNA). The ptRNA is then "edited," under the direction of enzymes, to remove the introns. The remaining exons are joined together to form a shortened RNA strand that carries the genetic information of the transcribed gene. This "edited" RNA is the messenger RNA (mRNA) that serves as a blueprint for protein assembly.

Current research suggests that several different mRNA molecules are obtainable from a single type of ptRNA. This involves the concept of *alternative splicing,* whereby some but not all of the "exon information" in a ptRNA is spliced together.

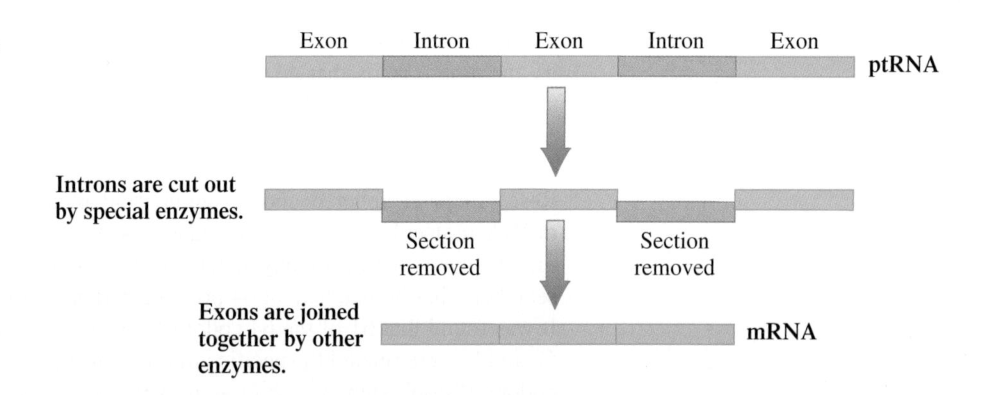

Figure 17.15 Primary transcript RNA contains both exons and introns. Messenger RNA is primary transcript RNA from which the introns have been excised.

Human Genome Project results indicate that the number of known proteins significantly exceeds the estimated number of genes. Thus the "one-protein-one-gene" dogma, long a part of biochemical theory, no longer seems plausible. A process such as alternative splicing is necessary to account for the excess of proteins over genes. Much research is still needed in this relatively unexplored area.

Practice Problems and Questions

17.41 What is the function of the enzyme *RNA polymerase* in the process of transcription?

17.42 What serves as a template in the process of transcription?

17.43 What type of RNA is the initial product of the transcription process?

17.44 What are the complementary base pairs in DNA–RNA interactions?

17.45 Write the base sequence of the ptRNA formed by transcription of the following DNA base sequence.

5′ A–T–G–C–T–T–A 3′

17.46 From what DNA base sequence was the following ptRNA sequence transcribed?

5′ U–U–C–G–C–A–G 3′

17.47 Explain the relationship between a gene and

a. an exon b. an intron

17.48 What mRNA base sequence would be obtained from the following portion of a gene?

Exon | Intron | Exon

5′ TCAG–TAGC–TTCA 3′

Learning Focus

Understand the relationships among nucleotide bases, codons, and the genetic code.

17.9 The Genetic Code

The nucleotide (base) sequence of an mRNA molecule is the informational part of such a molecule. This base sequence in a given mRNA determines the amino acid sequence for the protein synthesized under that mRNA's direction.

How can the base sequence of an mRNA molecule (which involves only *4* different bases—A, C, G, and U) encode enough information to direct proper sequencing of *20* amino acids in proteins? If each base encoded for a particular standard amino acid, then only 4 amino acids would be specified out of the 20 needed for protein synthesis, a clearly inadequate number. If two-base sequences were used to code amino acids, then there would be $4^2 = 16$ possible combinations, so 16 amino acids could be represented uniquely. This is still an inadequate number. If three-base sequences were used to code for amino acids, there would be $4^3 = 64$ possible combinations, which is more than enough combinations for uniquely specifying each of the 20 standard amino acids found in proteins.

Research has verified that sequences of three nucleotides in mRNA molecules specify the amino acids that go into synthesis of a protein. Such three-nucleotide sequences are called codons. A **codon** *is a sequence of three nucleotides in an mRNA molecule that codes for a specific amino acid.*

Which amino acid is specified by which codon? (We have 64 codons to choose from.) Researchers deciphered codon–amino acid relationships by adding different *synthetic* mRNA molecules (whose base sequences were known) to cell extracts and then determining the structure of any newly formed protein. After many such experiments, researchers finally matched all 64 possible codons with their functions in protein synthesis. It was found that 61 of the 64 codons formed by various combinations of the bases A, C, G, and U were related to specific amino acids; the other 3 combinations were termination codons ("stop" signals) for protein synthesis. Collectively, these relationships between three-nucleotide sequences in mRNA and amino acid identities are known as the genetic code. The **genetic code** *gives the assignment of the 64 mRNA codons to specific amino acids (or stop signals).* The determination of this code is one of the most remarkable of twentieth-century scientific achievements. The 1968 Noble Prize in chemistry was awarded to Marshall Nirenberg and Gobind Khovana for their work in illuminating this essential code.

The complete genetic code is given in Table 17.2. Examination of this table indicates that the genetic code has several remarkable features.

▶ There is a rough correlation between the number of codons for a particular amino acid and that amino acid's frequency of occurrence in proteins. For example, the two amino acids that have a single codon, Met and Trp, are two of the least common amino acids in proteins.

1. *The genetic code is highly degenerate; that is, many amino acids are designated by more than one codon.* Three amino acids (Arg, Leu, and Ser) are represented by six codons. Two or more codons exist for all other amino acids except Met and Trp, which have only a single codon. Codons that specify the same amino acid are called *synonyms.*
2. *There is a pattern to the arrangement of synonyms in the genetic code table.* All synonyms for an amino acid fall within a single box in Table 17.2, unless there are more than four synonyms, where two boxes are needed. The significance of the "single box" pattern is that with synonyms, the first two bases of the codon are the same—they differ only in the third base. For example, the four synonyms for the amino acid Pro are CCU, CCC, CCA, and CCG.

Table 17.2
The Universal Genetic Code.
The code is composed of 64 three-nucleotide sequences (codons), which can be read from the table. The left-hand column indicates the nucleotide base found in the first (5′) position of the codon. The nucleotides in the second (middle) position of the codon are in the middle columns. The right-hand column indicates the nucleotide found in the third (3′) position. Thus the codon ACG encodes for the amino acid Thr, and the codon GGG encodes for the amino acid Gly.

First position (5′ end)	Second Position				Third position (3′ end)
	U	C	A	G	
U	Phe	Ser	Tyr	Cys	U
	Phe	Ser	Tyr	Cys	C
	Leu	Ser	Stop	Stop	A
	Leu	Ser	Stop	Trp	G
C	Leu	Pro	His	Arg	U
	Leu	Pro	His	Arg	C
	Leu	Pro	Gln	Arg	A
	Leu	Pro	Gln	Arg	G
A	Ile	Thr	Asn	Ser	U
	Ile	Thr	Asn	Ser	C
	Ile	Thr	Lys	Arg	A
	Met	Thr	Lys	Arg	G
G	Val	Ala	Asp	Gly	U
	Val	Ala	Asp	Gly	C
	Val	Ala	Glu	Gly	A
	Val	Ala	Glu	Gly	G

3. *The genetic code is almost universal.* Although Table 17.2 does not show this feature, studies of many organisms indicate that with minor exceptions, the code is the same in all of them. The same codon specifies the same amino acid whether the cell is a bacterial cell, a corn plant cell, or a human cell.
4. *An initiation codon exists.* The existence of "stop" codons (UAG, UAA, and UGA) suggests the existence of "start" codons. There is one initiation codon. Besides coding for the amino acid methionine, the codon AUG functions as an initiator of protein synthesis when it occurs as the first codon in an amino acid sequence.

Example 17.3 Using the Genetic Code and mRNA Codons to Predict Amino Acid Sequences

Using the genetic code in Table 17.2, determine the sequence of amino acids encoded by the mRNA codon sequence

5′ GCC–AUG–GUA–AAA–UGC–GAC–CCA 3′

Solution

Matching the codons with the right amino acids using Table 17.2 yields

mRNA:	5′ GCC–AUG–GUA–AAA–UGC–GAC–CCA 3′
Peptide:	Ala- Met- Val- Lys- Cys- Asp- Pro-

Practice Problems and Questions

17.49 What is a codon?

17.50 On what type of RNA molecule are codons found?

17.51 Why can't two-base codons instead of three-base codons be used in the genetic code to code for the amino acids?

17.52 Using the information in Table 17.2, determine what amino acid is coded for by each of the following codons.
a. CUU b. AAU c. AGU d. GGG

17.53 Using the information in Table 17.2, determine the synonyms, if any, of each of the codons in Problem 17.52.

17.54 Explain why the base sequence ATC could not be a codon.

17.55 Predict the sequence of amino acids coded for by the mRNA sequence

5′ AUG–AAA–GAA–GAC–CUA 3′

17.10 Anticodons and tRNA Molecules

Learning Focus

Be able to write the anticodon for a given codon, and be able to determine the amino acid sequence of a peptide on the basis of codon or anticodon information.

The amino acids used in protein synthesis do not directly interact with the codons of a mRNA molecule. Instead, tRNA molecules function as intermediaries that deliver amino acids to the mRNA. At least one type of tRNA molecule exists for each of the 20 amino acids found in proteins.

All tRNA molecules have the same general shape, and this shape is crucial to how they function. Figure 17.16 shows the general *two-dimensional* "cloverleaf" shape of a tRNA molecule, a shape produced by the molecule's folding and twisting into regions of parallel strands and regions of hairpin loops. (The actual three-dimensional shape of a tRNA molecule involves considerable additional twisting of the "cloverleaf" shape.)

Two features of the tRNA structure are of particular importance.

1. The 3′ end of the open part of the cloverleaf structure is where an amino acid becomes *covalently* bonded to the tRNA molecule through an ester bond. Each of the

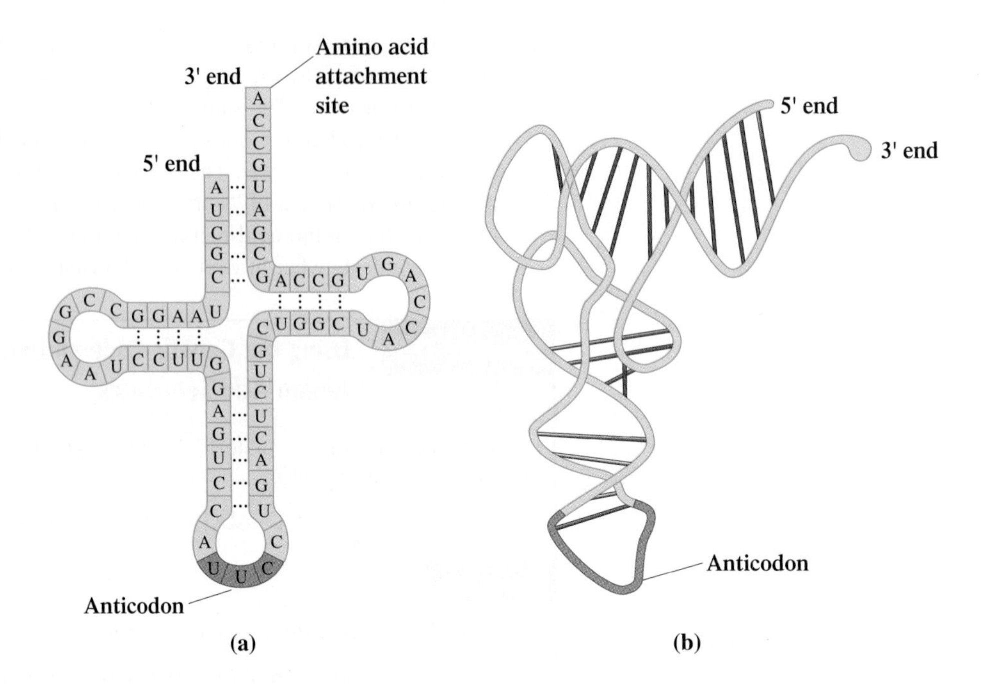

Figure 17.16 A tRNA molecule. The amino acid attachment site is at the open end of the cloverleaf (the 3′ end), and the anticodon is located in the loop opposite the open end.

different tRNA molecules is specifically recognized by an *aminoacyl synthetase* enzyme. These enzymes also recognize the one kind of amino acid that "belongs" with the particular tRNA and facilitates its bonding to the tRNA (see Figure 17.17).

2. The loop *opposite* the open end of the cloverleaf is the site for a sequence of three bases called an anticodon. An **anticodon** *is a three-nucleotide sequence in tRNA that is complementary to the mRNA codon for the amino acid that bonds to the tRNA.*

The interaction between the anticodon of the tRNA and the codon of the mRNA leads to the proper placement of an amino acid into a growing peptide chain during protein synthesis. This interaction, which involves complementary base pairing, is shown in Figure 17.18.

Practice Problems and Questions

17.56 Describe the general structure of a tRNA molecule.

17.57 Where is the anticodon site on a tRNA molecule?

17.58 By what type of bond is an amino acid attached to a tRNA molecule?

17.59 What principle governs the codon–anticodon interaction that leads to proper placement of amino acids in proteins?

17.60 What is the anticodon that would interact with each of the following codons?
a. AGA b. CGU c. UUU d. CAA

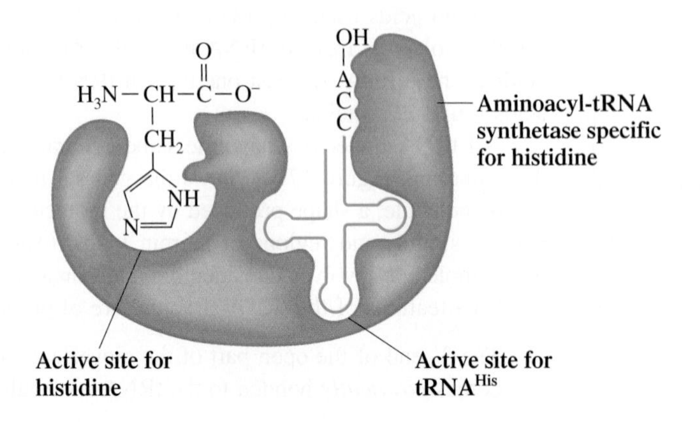

Figure 17.17 An aminoacyl–tRNA synthetase has an active site for tRNA and a binding site for the particular amino acid that is to be attached to that tRNA.

17.61 What is the codon that would interact with each of the following anticodons?
a. AGA b. CGU c. UUU d. CAA

17.62 Identify the amino acid that is associated with each of the codons in Problem 17.60.

17.63 Identify the amino acid that is associated with each of the anticodons in Problem 17.61.

Learning Focus

Describe the steps involved in the translation phase of protein synthesis.

17.11 Translation: Protein Synthesis

The substances needed for the translation phase of protein synthesis are mRNA molecules, tRNA molecules, amino acids, ribosomes, and a number of different enzymes. Ribosomes, which serve as sites for protein synthesis, have structures involving two subunits—a large subunit and a small subunit (see Figure 17.19). Each subunit has a composition of approximately 65% rRNA and 35% protein. The rRNA helps maintain the structure of the ribosome and also provides sites where mRNA can attach itself to the ribosome.

There are five general steps to the translation process: (1) activation of tRNA, (2) initiation, (3) elongation, (4) termination, and (5) post-translational processing.

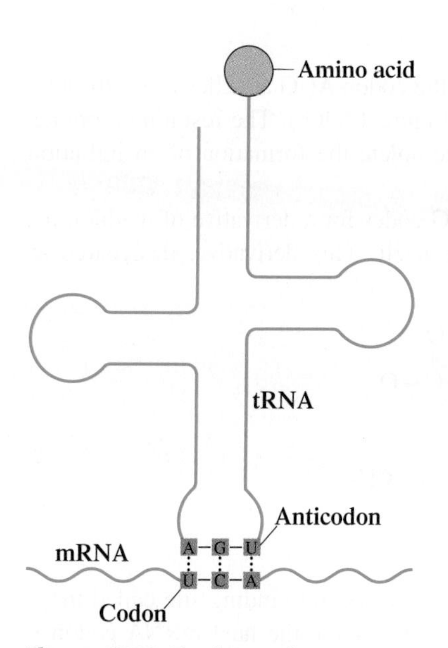

Figure 17.18 The interaction between anticodon (tRNA) and codon (mRNA), which involves complementary base pairing, governs the proper placement of amino acids in a protein.

Activation of tRNA

There are two steps involved in tRNA activation. First, an amino acid interacts with an activator molecule (ATP; Section 18.3) to form a highly energetic complex. This complex then reacts with the appropriate tRNA molecule to produce an *activated tRNA molecule,* a tRNA molecule that has an amino acid covalently bonded to it at its 3′ end through an ester linkage.

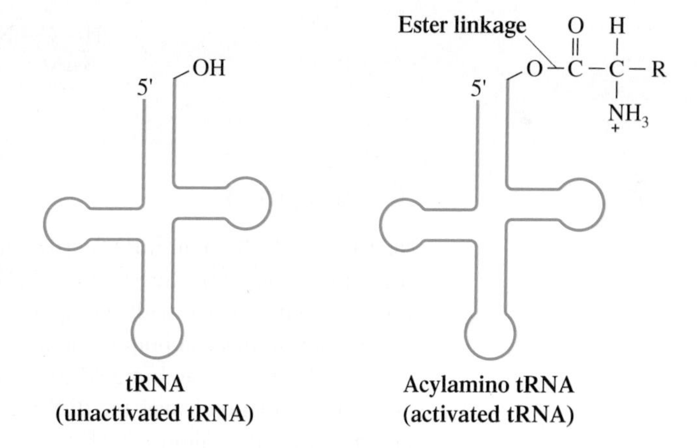

Initiation

The initiation of protein synthesis begins when mRNA attaches itself to the surface of a small ribosomal subunit such that its first codon, which is always the initiating codon AUG, occupies a site called the P site (peptidyl site). (See Figure 17.20a.) An activated

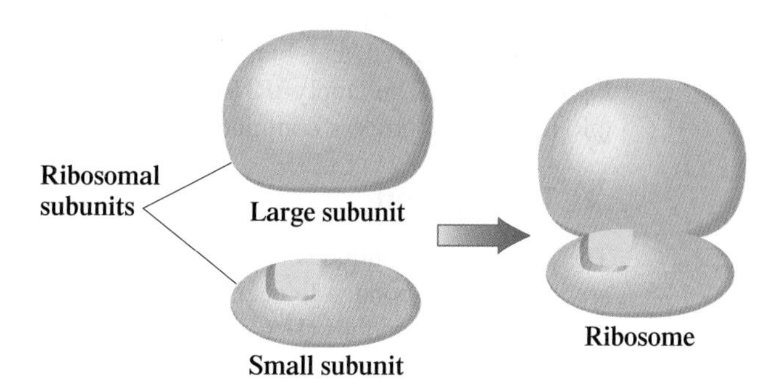

Figure 17.19 Ribosomes, which contain both rRNA and protein, have structures that contain two subunits. One subunit is much larger than the other.

Figure 17.20 Initiation of protein synthesis begins with the formation of an initiation complex.

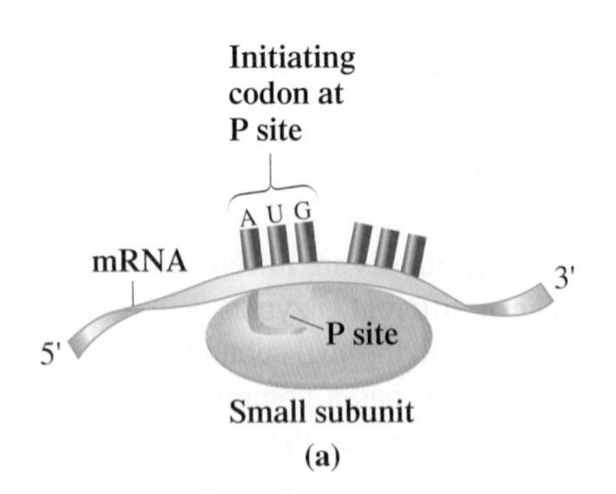

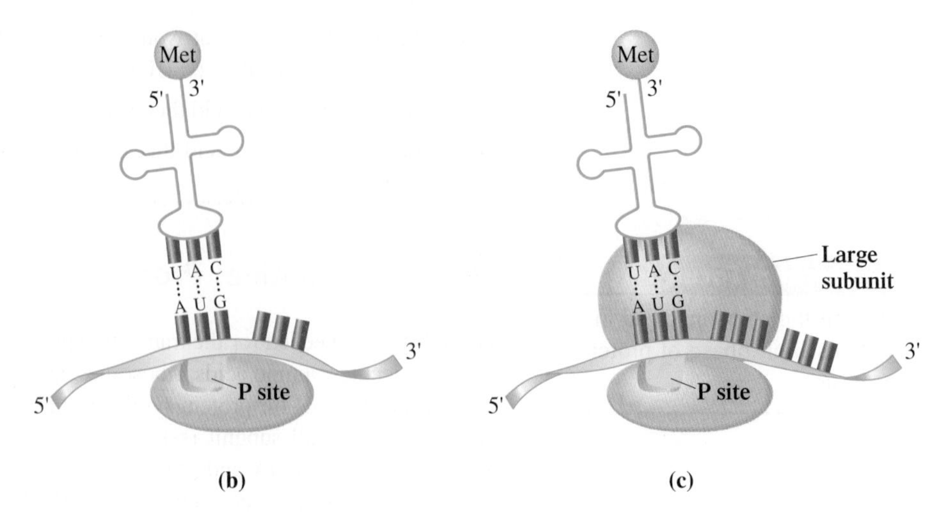

tRNA molecule with anticodon complementary to the codon AUG attaches itself, through complementary base pairing, to the AUG codon (Figure 17.20b). The resulting complex then interacts with a large ribosomal subunit to complete the formation of an initiation complex (Figure 17.20c).

When functioning as the initiating codon, AUG codes for a derivative of methionine, *N*-formyl methionine, rather than for methionine itself. This derivative, designated as f-Met, has the structure

$$\underbrace{H-\overset{\overset{\displaystyle O}{\|}}{C}}_{\text{Formyl group}}-NH-\underset{\underset{\displaystyle CH_2-S-CH_3}{|}}{\underset{\displaystyle CH_2}{\underset{|}{CH}}}-\overset{\overset{\displaystyle O}{\|}}{C}-O^-$$

Elongation

Next to the P site in an mRNA–ribosome complex is a second binding site called the A site (aminoacyl site). (See Figure 17.21a). At this second site the next mRNA codon is exposed, and a tRNA with the appropriate anticodon binds to it (Figure 17.21b). With amino acids in place at both P and the A sites, the enzyme *peptidyl transferase* effects the linking of the P site amino acid to the A site amino acid to form a dipeptide. Such peptide bond formation leaves the tRNA at the P site empty and the tRNA at the A site bearing the dipeptide (Figure 17.21c).

The empty tRNA at the P site now leaves that site and is free to pick up another molecule of its specific amino acid. Simultaneously with the release of tRNA from the P site, the ribosome shifts along the mRNA. This shift puts the newly formed dipeptide at the P site, and the third codon of mRNA is now available, at site A, to accept a tRNA molecule whose anticodon complements this codon (see Figure 17.21d).

▶ In elongation, the polypeptide chain grows one amino acid at a time.

Now a repetitious process begins. The third codon, now at the A site, accepts an incoming tRNA with its accompanying amino acid; and then the entire dipeptide at the P site is transferred and bonded to the A site amino acid to give a tripeptide (see Figure 17.21e). The empty tRNA at the P site is released, the ribosome shifts along the mRNA, and the process continues.

Termination

The polypeptide continues to grow by way of translation until all necessary amino acids are in place and bonded to each other. Appearance in the mRNA codon sequence of one of the three stop codons (UAA, UAG, or UGA) terminates the process. No tRNA has an anticodon that can base-pair with these stop codons. The polypeptide is then cleaved from the tRNA through hydrolysis.

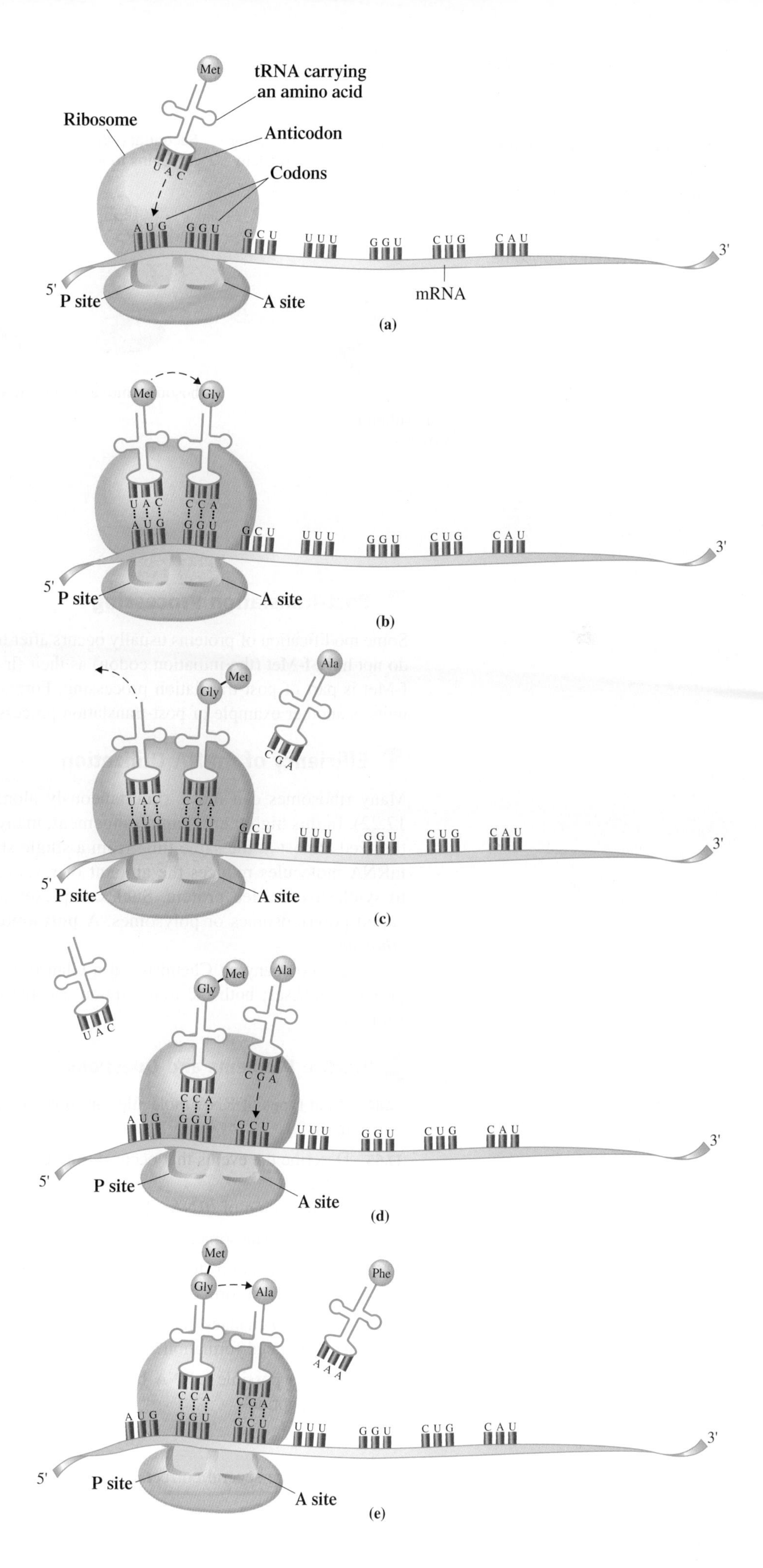

Figure 17.21 The process of translation that occurs during protein synthesis. The anticodons of tRNA molecules are paired with the codons of an mRNA molecule to bring the appropriate amino acids into sequence for protein formation.

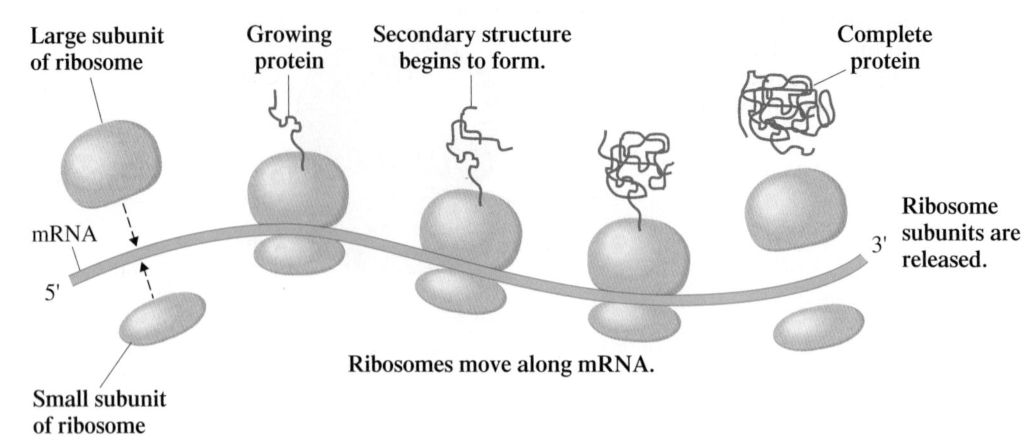

Figure 17.22 Several ribosomes can simultaneously proceed along a single strand of mRNA one after another. Such a complex of mRNA and ribosomes is called a polysome.

Post-Translation Processing

Some modification of proteins usually occurs after translation. For example, most proteins do not have f-Met (the initiation codon) as their first amino acid. Cleavage of N-terminal f-Met is part of post-translation processing. Formation of S—S bonds between cysteine units is another example of post-translation processing.

Efficiency of mRNA Utilization

Many ribosomes can move simultaneously along a single mRNA molecule (Figure 17.22). In this highly efficient arrangement, many identical protein chains can be synthesized almost at the same time from a single strand of mRNA. This multiple use of mRNA molecules reduces the amount of resources and energy expended by the cell to synthesize needed protein. Such complexes of several ribosomes and mRNA are called polyribosomes or polysomes. A **polysome** *is a complex of mRNA and several ribosomes.*

The accompanying Chemistry at a Glance is an overall summary of the process of protein syntheses; both the transcription and the translation phases are included in the summary.

Practice Problems and Questions

17.64 What types of RNA molecules are involved in the translation phase of protein synthesis?

17.65 Describe the events that occur during each of the following steps of the translation process.
a. Activation of tRNA b. Initiation c. Elongation d. Termination

17.66 What is a ribosome, and what role do ribosomes play in protein synthesis?

17.67 In the elongation step of protein synthesis, at which site in the ribosome does new peptide bond formation actually take place?

17.68 What two changes occur at a ribosome during protein synthesis immediately after peptide bond formation?

17.69 Write a possible mRNA base sequence that would lead to the production of the pentapeptide

Gly–Ala–Cys–Val–Tyr

(There are several correct answers.)

Protein Synthesis

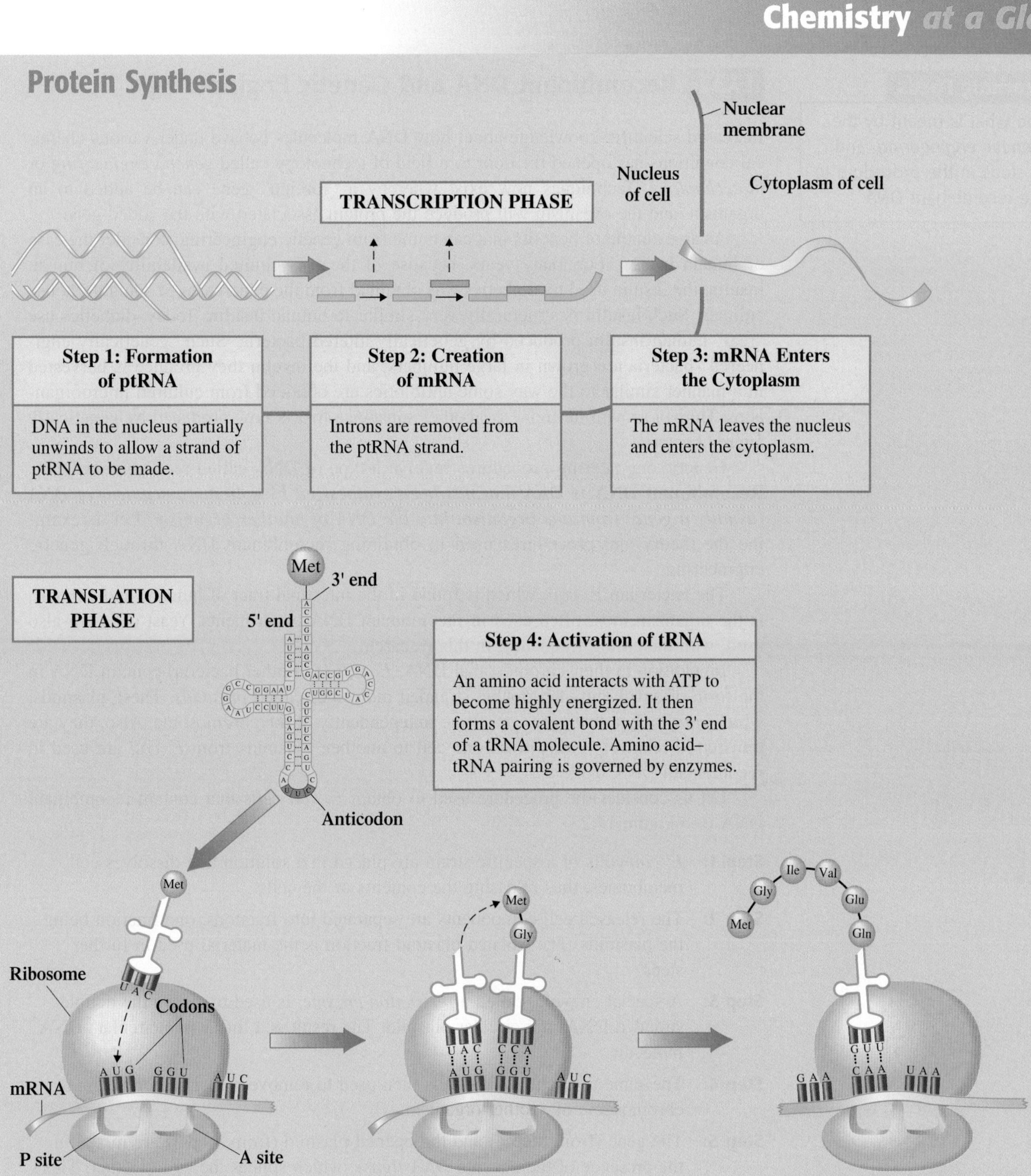

TRANSCRIPTION PHASE
Nuclear membrane
Nucleus of cell
Cytoplasm of cell
Step 1: Formation of ptRNA
DNA in the nucleus partially unwinds to allow a strand of ptRNA to be made.
Step 2: Creation of mRNA
Introns are removed from the ptRNA strand.
Step 3: mRNA Enters the Cytoplasm
The mRNA leaves the nucleus and enters the cytoplasm.
TRANSLATION PHASE
Met
3' end
5' end
Anticodon
Step 4: Activation of tRNA
An amino acid interacts with ATP to become highly energized. It then forms a covalent bond with the 3' end of a tRNA molecule. Amino acid–tRNA pairing is governed by enzymes.
Ribosome
Codons
mRNA
P site
A site
Gly
Ile
Val
Glu
Gln

Step 5: Initiaton

The mRNA attaches to a ribosome so that the first codon (AUG) is at the P site. A tRNA carrying N-formyl methionine, an amino acid derivitave, attaches to the first codon.

Step 6: Elongation

Another tRNA with the second amino acid binds at the A site. The methionine derivative transfers from the P site to the A site. The ribosome shifts to the next codon, making its A site available for the tRNA carrying the third amino acid.

Steps 7 and 8: Termination and Post-Translation Processing

The polypeptide chain continues to lengthen until a stop codon appears on the mRNA. The new protein is cleaved from the last tRNA.

During post-translation processing, cleavage of f-Met (the initiation codon) usually occurs. S—S bonds between Cys units also can form.

Learning Focus

Describe what is meant by the term *genetic engineering,* and list the steps in the procedure to produce recombinant DNA.

17.12 Recombinant DNA and Genetic Engineering

Increased scientific knowledge about how DNA molecules behave under various chemical conditions has opened the door to a field of technology called *genetic engineering* or *biotechnology.* Techniques now exist whereby a "foreign" gene can be added to an organism, and the organism will produce the protein associated with the added gene.

As an example of benefits that can come from genetic engineering, consider the case of human insulin. For many years, because of the very limited availability of human insulin, the insulin used by diabetics was obtained from the pancreases of slaughterhouse animals. Such insulin is structurally very similar to human insulin. Today, diabetics use "real" human insulin produced by genetically altered bacteria. Such "genetically engineered" bacteria are grown in large numbers, and the insulin they produce is harvested in a manner similar to the way some antibiotics are obtained from cultured microorganisms. Human growth hormone is another substance that is now produced by genetically altered bacteria.

Genetic engineering procedures involve a type of DNA called recombinant DNA. **Recombinant DNA** *is DNA that has been synthesized by splicing a segment of DNA (usually a gene) from one organism into the DNA of another organism.* Let us examine the theory and procedures used in obtaining recombinant DNA through genetic engineering.

The bacterium *E. coli,* which is found in the intestinal tract of humans and animals, is the organism most often used in recombinant DNA experiments. Yeast cells are also used, with increasing frequency, in this research.

In addition to their chromosomal DNA, *E. coli* (and other bacteria) contain DNA in the form of small, circular, double-stranded molecules called *plasmids.* These plasmids, which carry only a few genes, replicate independently of the chromosome. Also, they are transferred relatively easily from one cell to another. Plasmids from *E. coli* are used in recombinant DNA work.

Let us consider the procedure used to obtain *E. coli* cells that contain recombinant DNA (see Figure 17.23).

Step 1: *E. coli* cells of a specific strain are placed in a solution that dissolves cell membranes, thus releasing the contents of the cells.

Step 2: The released cell components are separated into fractions, one fraction being the plasmids. The isolated plasmid fraction is the material used in further steps.

Step 3: A special enzyme, called a *restriction enzyme,* is used to cleave the double-stranded DNA of a circular plasmid. The result is a linear (noncircular) DNA molecule.

Step 4: The same restriction enzyme is then used to remove a desired gene from a chromosome of another organism.

Step 5: The gene (from Step 4) and the opened plasmid (from Step 3) are mixed in the presence of the enzyme *DNA ligase,* which splices the two together. This splicing, which attaches one end of the gene to one end of the opened plasmid and attaches the other end of the gene to the other end of the plasmid, results in an altered circular plasmid (the recombinant DNA).

Step 6: The altered plasmids (recombinant DNA) are placed in a live *E. coli* culture, where they are taken up by the *E. coli* bacteria. The *E. coli* culture into which the plasmids are placed need not be identical to that from which the plasmids were originally obtained.

In Step 3 we noted that the conversion of a circular plasmid into a linear DNA molecule requires a restriction enzyme. A **restriction enzyme** *is an enzyme that recognize specific base sequences in DNA and cleaves the DNA in a predictable manner at these sequences.* The discovery of restriction enzymes made genetic engineering possible.

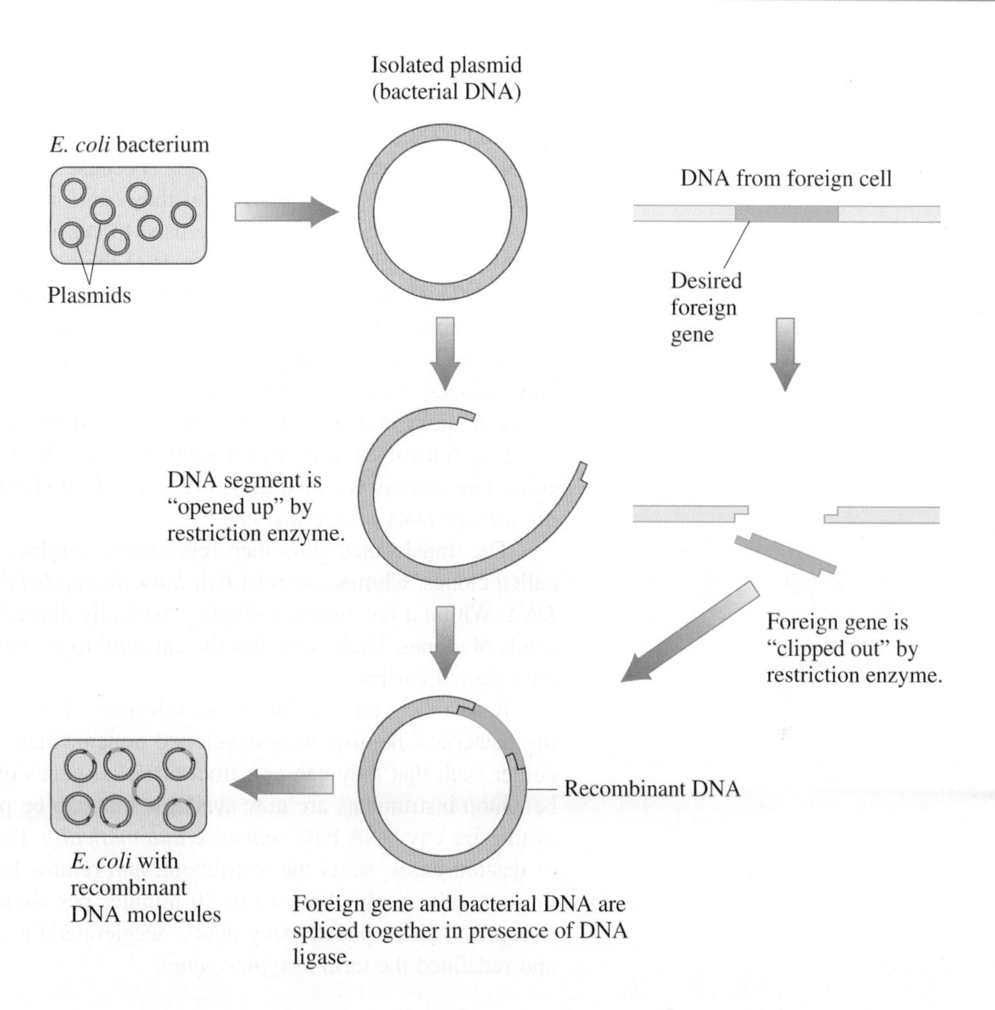

Figure 17.23 Recombinant DNA is made by inserting a gene obtained from a cell of one kind of organism into the DNA of another kind of organism.

Restriction enzymes occur naturally in numerous types of bacterial cells. Their function is to protect the bacteria from invasion by foreign DNA by catalyzing the cleavage of the invading DNA. Their names are derived from their placing a "restriction" on the type of DNA allowed into the bacterial cells.

To understand how a restriction enzyme works, let us consider one that cleaves DNA between G and A bases in the 5′-to-3′ direction in the sequence G–A–A–T–T–C. This enzyme will cleave the double helix structure of a DNA molecule in the manner shown in Figure 17.24.

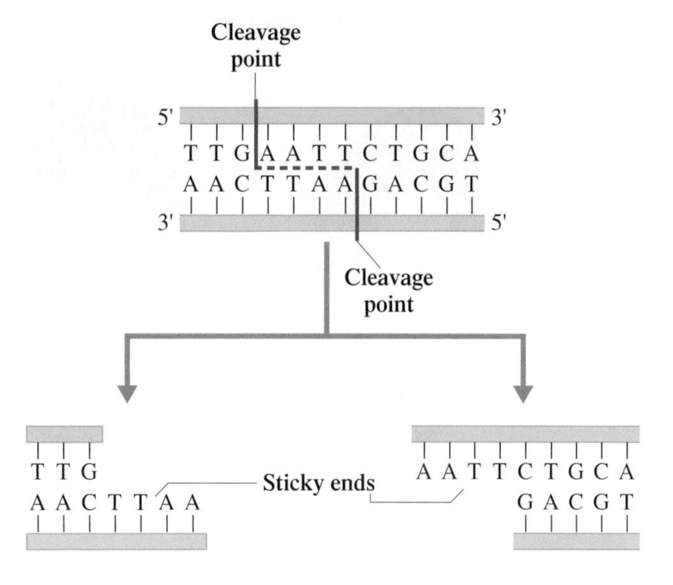

Figure 17.24 Cleavage pattern resulting from the use of a restriction enzyme that cleaves DNA between G and A bases in the 5′-to-3′ direction in the sequence G–A–A–T–T–C. The double helix structure is not cut straight across.

Note that the double helix is not cut straight across, the individual strands are cut at different points, giving a staircase cut. (Both cuts must be between G and A in the 5′-to-3′ direction.) This staircase cut leaves unpaired bases on each cut strand. These ends with unpaired bases are called "sticky ends" because they are ready to "stick to" (pair up with) a complementary section of DNA if they can find one.

If the same restriction enzyme used to cut a plasmid is also used to cut a gene from another DNA molecule, the sticky ends of the gene will be complementary to those of the plasmid. This enables the plasmid and gene to combine readily, forming a new, modified plasmid molecule. This modified plasmid molecule is called recombinant DNA. In addition to the newly spliced gene, the recombinant DNA plasmid contains all of the genes and characteristics of the original plasmid. Figure 17.25 shows diagrammatically the match between sticky ends that occurs when plasmid and gene combine.

Step 6 involves inserting the recombinant DNA (modified plasmids) back into *E. coli* cells. The process is called transformation. **Transformation** *is the process of incorporating foreign DNA into a host cell.*

The transformed cells then reproduce, resulting in large numbers of identical cells called clones. **Clones** *are cells that have descended from a single cell and have identical DNA.* Within a few hours, a single genetically altered bacterial cell can give rise to thousands of clones. Each clone has the capacity to synthesize the protein directed by the foreign gene it carries.

Researchers are not limited to selection of naturally occurring genes for transforming bacteria. Chemists have developed nonenzymatic methods of linking nucleotides together such that they can construct artificial genes of any sequence they desire. In fact, benchtop instruments are now available that can be programmed by a microprocessor to synthesize any DNA base sequence *automatically.* The operator merely enters a sequence of desired bases, starts the instrument, and returns later to obtain the product. This synthesis requires only about 15 to 20 minutes per nucleotide. Such flexibility in manufacturing DNA has opened many doors, accelerated the pace of recombinant DNA research, and redefined the term *designer genes!*

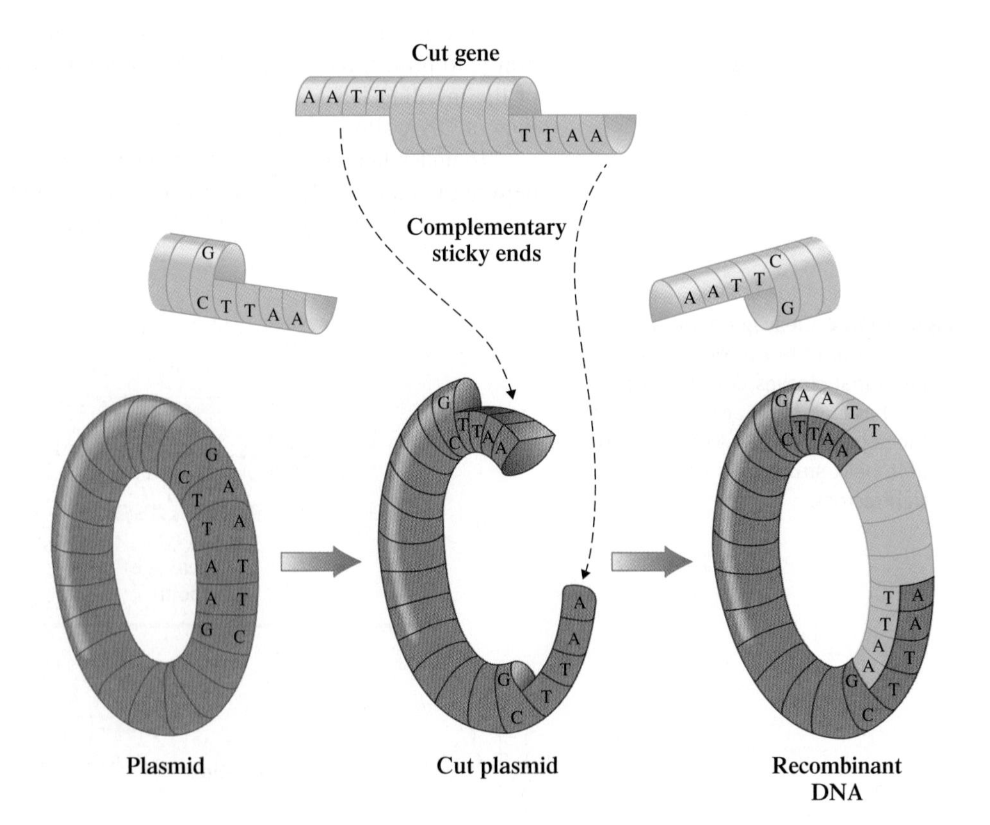

Figure 17.25 The "sticky ends" of the cut plasmid and the cut gene are complementary and combine to form recombinant DNA.

Practice Problems and Questions

17.70 How does recombinant DNA differ from "normal" DNA?

17.71 Give two reasons why bacterial cells are used for recombinant DNA procedures.

17.72 What role do plasmids play in recombinant DNA procedures?

17.73 Describe what happens when a particular restriction enzyme operates on a segment of double-stranded DNA.

17.74 How are plasmids obtained from *E. coli* bacteria?

17.75 Describe the process of transformation in the context of genetic engineering.

17.76 A particular restriction enzyme will cleave DNA between A and A in the sequence A–A–G–C–T–T in the 5′-to-3′ direction. Draw a diagram showing the structural details of the "sticky ends" that result from cleavage of the following DNA segment.

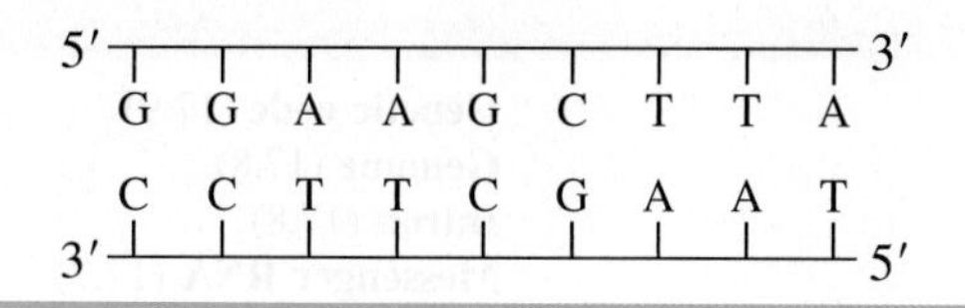

CONCEPTS TO REMEMBER

Nucleic acids. Nucleic acids are polymeric molecules in which the repeating units are nucleotides. Cells contain two kinds of nucleic acids—deoxyribonucleic acids (DNA) and ribonucleic acids (RNA). The major biological functions of DNA and RNA are, respectively, transfer of genetic information and synthesis of proteins.

Nucleotides. Nucleotides, the monomers of nucleic acid polymers, are molecules composed of a pentose sugar bonded to both a phosphate group and a nitrogen-containing heterocyclic base. The pentose sugar must be either ribose or deoxyribose. Five nitrogen-containing bases are found in nucleotides: adenine (A), guanine (G), cytosine (C), thymine (T), and uracil (U).

Primary structure of nucleic acids. The "backbone" of a nucleic acid molecule is a constant alternating sequence of sugar and phosphate groups. Each sugar unit has a nitrogen-containing base attached to it.

Secondary structure of DNA. A DNA molecule exists as two polynucleotide chains coiled around each other in a double helix arrangement. The double helix is held together by hydrogen bonding between complementary pairs of bases. Only two base-pairing combinations occur: A with T, and C with G.

DNA replication. DNA replication occurs when the two strands of a parent DNA double helix separate and act as templates for the synthesis of new chains using the principle of complementary base pairing.

Chromosome. A chromosome is a cell structure that consists of an individual DNA molecule bound to a group of proteins.

RNA molecules. Four important types of RNA molecules, distinguished by their function, are ribosomal RNA (rRNA), messenger RNA (mRNA), primary transcript RNA (ptRNA), and transfer RNA (tRNA).

Transcription. Transcription is the process in which the genetic information encoded in the base sequence of DNA is copied into RNA molecules.

Complementary bases. Complementary bases are specific pairs of bases in nucleic acid structures that hydrogen-bond to each other.

Gene. A gene is a segment of a DNA molecule that contains the base sequence for the production of specific RNA molecules, which in turn guide the production of specific protein molecules. Genes are segmented, with portions called exons that contain genetic information and portions called introns that do not convey genetic information.

Codon. A codon is a three-nucleotide sequence in mRNA that codes for a specific amino acid needed during the process of protein synthesis.

Genetic code. The genetic code consists of all the mRNA codons that specify either a particular amino acid or the termination of protein synthesis.

Anticodon. An anticodon is a three-nucleotide sequence in tRNA that binds to a complementary sequence (a codon) in mRNA.

Translation. Translation is the stage of protein synthesis in which the codons in mRNA are translated into amino acid sequences of new proteins. Translation involves interactions between the codons of mRNA and the anticodons of tRNA.

Recombinant DNA. Recombinant DNA molecules are synthesized by splicing a segment of DNA, usually a gene, from one organism into the DNA of another organism.

KEY REACTIONS AND EQUATIONS

1. Formation of a nucleotide (Section 17.2)

   Pentose sugar (ribose or deoxyribose) + phosphate group + nitrogen-containing heterocyclic base $\longrightarrow$

   Phosphate – Sugar (– Base) + $2H_2O$

2. Formation of a nucleic acid (Section 17.3)

   Many deoxyribose-containing nucleotides $\longrightarrow$ DNA

   Many ribose-containing nucleotides $\longrightarrow$ RNA

3. Protein synthesis (Section 17.6)

   $$\text{DNA} \xrightarrow{\text{Transcription}} \text{RNA} \xrightarrow{\text{Translation}} \text{Protein}$$

KEY TERMS

Anticodon (17.10)
Chromosome (17.5)
Clones (17.12)
Codon (17.9)
Complementary bases (17.4)
DNA replication (17.5)
Exon (17.8)
Gene (17.8)
Genetic code (17.9)
Genome (17.8)
Intron (17.8)
Messenger RNA (17.7)
Nucleic acid (17.1)
Nucleotide (17.2)
Polysome (17.11)
Primary nucleic acid structure (17.3)
Primary transcript RNA (17.7)
Recombinant DNA (17.12)
Restriction enzyme (17.12)
Ribosomal RNA (17.7)
Transcription (17.6)
Transfer RNA (17.7)
Transformation (17.12)
Translation (17.6)

ADDITIONAL PROBLEMS

17.77 With the help of the structures given in Section 17.2, describe the structural differences between the following pairs of nucleotide bases.
a. T and U b. G and C

17.78 The following is a sequence of bases for an exon portion of a strand of a gene.

5′ C–A–T–A–C–A–G–C–C–T–G–G–A–A–G–C–T–A 3′

a. What is the sequence of bases on the strand of DNA complementary to this segment?
b. What is the sequence of bases on the mRNA molecule synthesized from this strand?
c. What codons are present on the mRNA molecule from part **b**?
d. What anticodons will be found on the tRNA molecules that interact with the codons from part **c**?
e. What is the sequence of amino acids in the peptide formed using these protein synthesis instructions?

17.79 Which of the RNA types—mRNA, ptRNA, rRNA, or tRNA—is most closely associated with each of the following terms?
a. Codon b. Anticodon c. Intron d. Amino acid carrier

17.80 Which of the processes—the translation phase of protein synthesis, the transcription phase of protein synthesis, the replication of DNA, or the formation of recombinant DNA—is associated with each of the following events?
a. Complete unwinding of a DNA molecule
b. Partial unwinding of a DNA molecule
c. Formation of an mRNA-ribosome complex
d. Formation of Okazaki fragments

17.81 What is the relationship among DNA, chromosomes, and genes?

17.82 Which of these characterizations—found in DNA but not in RNA, found in RNA but not in DNA, found in both DNA and RNA, or not found in DNA or RNA—fits each of the following mono-, di- or trinucleotides?
a. 5′ dAMP–dAMP 3′ b. 5′ AMP–AMP–CMP 3′
c. 5′ dAMP–CMP 3′ d. 5′ GGA 3′

17.83 Compare the processes of replication and transcription in terms of
a. similarities b. differences

17.84 Which of these base-pairing situations—between two DNA segments, between two RNA segments, between a DNA segment and an RNA segment, or between a codon and an anticodon—fits each of the following base-pairing situations? More than one response may apply to a given base-pairing situation.

a.	b.	c.	d.
A G T	A C T	A G U	C C G
| | |	| | |	| | |	| | |
U C A	T G A	U C A	G G C

PRACTICE TEST True/False

17.85 A nucleotide is a two-subunit molecule in which a pentose sugar is bonded to a nitrogen-containing heterocyclic base.

17.86 Both adenine and guanine are purine derivatives.

17.87 The nucleotide dAMP is a component of ribonucleic acids.

17.88 An alternating sugar–phosphate sequence forms the "backbone" of a nucleic acid.

17.89 In DNA molecules, the percentages of A and C present are always equal.

17.90 A chromosome is a gene that is bound to a DNA molecule.

17.91 RNAs are single-stranded molecules, and DNAs are double-stranded molecules.

17.92 The entirety of a DNA molecule is segmented into genes.

17.93 An exon is a DNA segment that conveys genetic information.

17.94 Codons are found on mRNA molecules, and anticodons are found on rRNA molecules.

17.95 Most amino acids are designated by more than one codon.

17.96 tRNA molecules function as amino acid carriers.

17.97 During translation, many ribosomes can move simultaneously along a single mRNA molecule.

17.98 Recombinant DNA molecules contain genes from more than one type of organism.

17.99 "Sticky ends" are associated with the action of DNA polymerase enzymes.

PRACTICE TEST Multiple Choice

17.100 Any given nucleotide in a nucleic acid contains
a. two bases and a sugar
b. one sugar, two bases, and one phosphate
c. two sugars and one phosphate
d. one sugar, one base, and one phosphate

17.101 The number of different kinds of nucleotides present in any RNA or DNA molecule is
a. four b. five c. six d. eight

17.102 If 35% of the bases in a certain DNA molecule are found to be T, what percent of the bases in this same molecule are G?
a. 15% b. 25% c. 35% d. 65%

17.103 In a DNA double helix, the base pairs are
a. part of the backbone structure
b. located inside the helix
c. located outside the helix
d. covalently bonded to each other

17.104 Replication of DNA produces two daughter DNA molecules in which
a. one daughter molecule contains both parent strands, and one daughter molecule contains both newly synthesized strands
b. each daughter molecule contains one parent strand and one newly synthesized strand
c. each daughter molecule contains two newly synthesized strands
d. each daughter molecule contains a segment of both parent strands

17.105 RNA molecules differ from DNA molecules in that they
a. are single-stranded rather than double-stranded
b. contain five different bases instead of four
c. are much larger
d. contain 2-deoxyglucose instead of 2-deoxyribose

17.106 The region of a DNA strand that carries the information needed for synthesis of a specific protein is called a
a. codon b. chromosome
c. gene d. complementary base pair

17.107 The genetic code is a listing that gives relationships between
a. codons and anticodons
b. codons and amino acids
c. anticodons and amino acids
d. codons and genes

17.108 Which of the following is *not* necessary for protein synthesis at the time and place where synthesis occurs?
a. mRNA b. tRNA c. Ribosomes d. DNA

17.109 The role of *E. coli* plasmids in recombinant DNA work is to
a. splice DNA strands together to form circular DNA molecules
b. cleave double-stranded DNA molecules such that "sticky ends" are produced
c. provide the restriction enzymes needed for repairing defective DNA
d. serve as a host for a "foreign gene"

CHAPTER OUTLINE

CHAPTER EIGHTEEN

Metabolism

The energy consumed by these scarlet ibises in flight is generated by numerous sequences of biochemical reactions that occur within their bodies.

In this final chapter of the text, we consider the general topic of *metabolism,* the study of the chemical reactions that occur in the human body. Obviously, thousands of different reactions occur within the human body, and a one-chapter introduction to the subject can cover only a small part of what occurs. In our survey of the subject, we will be particularly concerned with reactions that produce the energy needed to "run" the body and with how the body stores and uses the energy so produced.

Learning Focus

Distinguish between catabolic and anabolic processes; understand the concept of a metabolic pathway.

18.1 Types of Metabolic Reactions

Metabolism *is the sum total of all the chemical reactions that take place in a living organism.* Human metabolism is quite remarkable. An average human adult whose weight remains the same for 40 years processes about 6 *tons* of solid food and 10,000 gallons of water, during which time the composition of the body is essentially constant. Just as we must put gasoline in a car to make it go or plug in a kitchen appliance to make it run, we also need a source of energy to think, breathe, exercise, or work. As we have seen in previous chapters, even the simplest living cell is continually carrying on energy-demanding processes such as protein synthesis, DNA replication, and RNA transcription.

Metabolic reactions fall into one of two subtypes: catabolism and anabolism. **Catabolism** *includes all metabolic reactions in which large molecules are broken down to smaller ones.* Catabolic reactions usually release energy. The reactions involved in the oxidation of glucose are catabolic. **Anabolism** *includes all metabolic reactions in which*

▶ *Catabolism* is pronounced ca-TAB-o-lism, and *anabolism* is pronounced an-ABB-o-lism. *Catabolic* is pronounced CAT-a-bol-ic, and *anabolic* is pronounced AN-a-bol-ic.

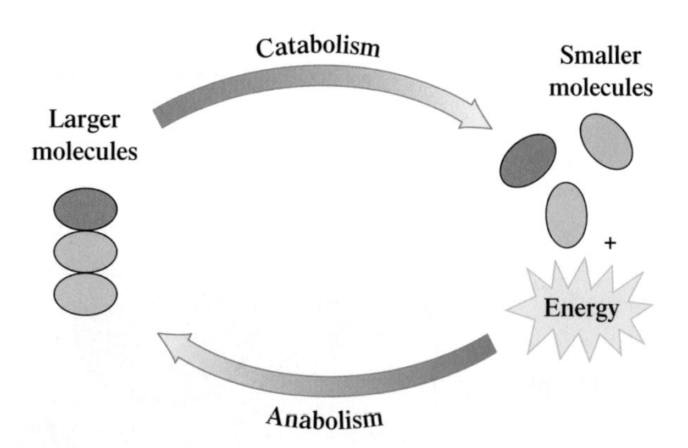

Figure 18.1 The processes of catabolism and anabolism are opposite in nature. The first usually produces energy, and the second usually consumes energy.

small molecules are put together to form larger ones. Anabolic reactions usually require energy in order to proceed. The synthesis of proteins from amino acids is an anabolic process. Figure 18.1 contrasts catabolic and anabolic processes.

The metabolic reactions that occur in a cell are usually organized into sequences called *metabolic pathways.* A **metabolic pathway** *is a series of consecutive biochemical reactions used to convert a starting material into an end product.* Such pathways may be *linear,* in which a series of reactions generates a final product, or *cyclic,* in which a series of reactions regenerates the first reactant.

Linear metabolic pathway: $A \longrightarrow B \longrightarrow C \longrightarrow D$

Cyclic metabolic pathway:

$$\begin{array}{ccc} A & \longrightarrow & B \\ \uparrow & & \downarrow \\ D & \longleftarrow & C \end{array}$$

The major metabolic pathways for all life forms are similar. This enables scientists to study metabolic reactions in simpler life forms and use the results to help understand the corresponding metabolic reactions in more complex organisms, including humans.

Practice Problems and Questions

18.1 Contrast catabolic and anabolic reactions in metabolism in terms of
a. relative sizes of reactants and products
b. energy production or energy consumption

18.2 What is a metabolic pathway?

18.3 What is the difference between a linear metabolic pathway and a cyclic metabolic pathway?

18.2 Metabolism and Cell Structure

Learning Focus

Understand cell structure as it pertains to locations where metabolic reactions occur.

▶ The term *eukaryotic,* pronounced you-KAHR-ee-ah-tic, is from the Greek *eu,* meaning "true," and *karyon,* meaning "nucleus." The term *prokaryotic,* which contains the Greek *pro,* meaning "before," literally means "before the nucleus."

Knowledge of the major structural features of a cell is a prerequisite to understanding *where* metabolic reactions take place.

Cells are of two types: prokaryotic and eukaryotic. *Prokaryotic cells* have no nucleus and are found only in bacteria. The DNA that governs the reproduction of prokaryotic cells is usually a single circular molecule found near the center of the cell in a region called the *nucleoid.* In *eukaryotic cells,* the DNA is found in the membrane-enclosed nucleus. Cells of this type, which are found in all higher organisms, are about 1000 times larger than bacterial cells. Our focus in the remainder of this section will be on eukaryotic cells, the type present in humans.

The *cytoplasm* of a eukaryotic cell is the material that lies between the nucleus and the outer plasma membrane of the cell. Within the water-based cytoplasm are small bodies called *organelles.* An **organelle** *is a minute structure within the cell cytoplasm that*

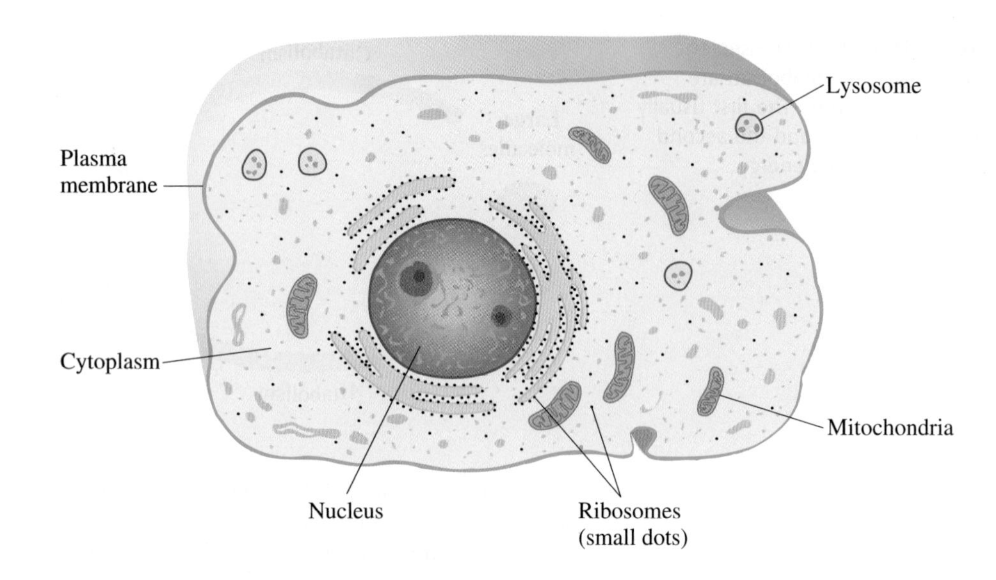

Figure 18.2 A schematic representation of a eukaryotic cell with selected internal components identified.

▶ Eukaryotic and prokaryotic cells differ in that the former contain a well-defined nucleus, set off from the rest of the cell by a membrane.

▶ *Mitochondria,* pronounced my-toe-KON-dree-ah, is plural. The singular form of the term is *mitochondrion.* The thread-like shape of mitochondria is responsible for this organelle's name; *mitos* is Greek for "thread," and *chondrion* is Greek for "granule."

carries out a specific cellular function. Among the organelles are *ribosomes, lysosomes,* and *mitochondria.* Ribosomes are the sites of protein synthesis (Section 17.7). Lysosomes contain hydrolytic enzymes, which are needed for cellular rebuilding, repair, and degradation. Mitochondria house some of the enzymes required for energy generation in the cell. A **mitochondrion** *is an organelle that has a central role in the production of energy.* Figure 18.2 shows the general internal structure of a eukaryotic cell.

Mitochondria are sausage-shaped organelles containing both an *outer membrane* and a *multifolded inner membrane* (see Figure 18.3). The outer membrane, which is about 50% lipid and 50% protein, is freely permeable to small molecules. The inner membrane, which is about 20% lipid and 80% protein, is highly impermeable to most substances. The nonpermeable nature of the inner membrane divides a mitochondrion into two separate compartments—an interior region called the *matrix* and the region between the inner and outer membranes, called the *intermembrane space.* The folds of the inner membrane that protrude into the matrix are called *cristae.*

The invention of high-resolution electron microscopes allowed researchers to see the interior structure of the mitochondrion more clearly and led to the discovery of small spherical knobs attached to the cristae called *ATP synthase complexes.* As their name

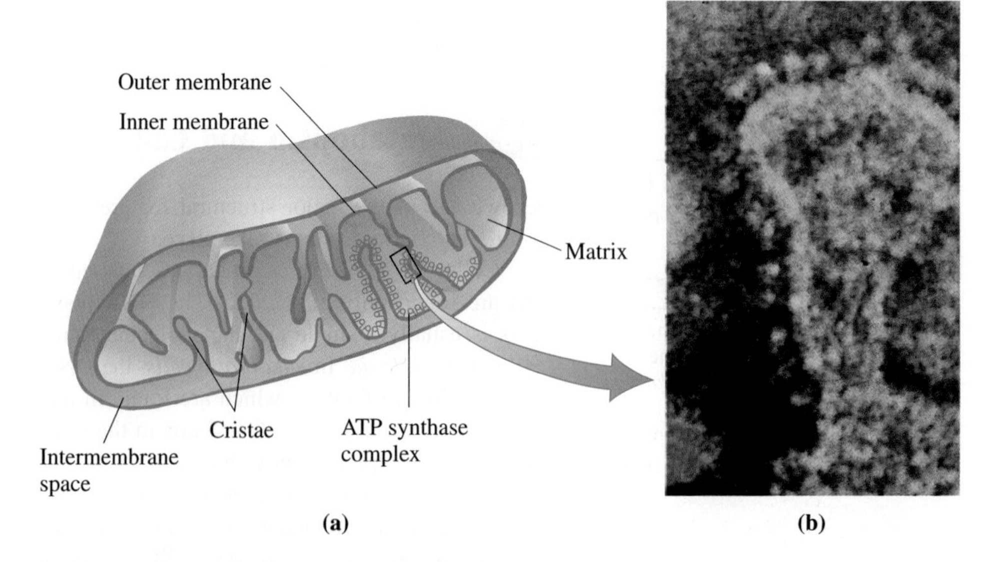

Figure 18.3 (a) A schematic representation of a mitochondrion, showing key features of its internal structure. (b) An electron micrograph of a single mitochondrial crista, showing the ATP synthase knobs extending into the matrix.

implies, these relatively small knobs, which are located on the matrix side of the inner membrane, are responsible for ATP synthesis, and their association with the inner membrane is critically important for this task. More will be said about ATP in the next section.

Practice Problems and Questions

18.4 What kinds of organisms have prokaryotic cells and what kinds have eukaryotic cells?

18.5 What is an organelle?

18.6 What is a mitochondrion?

18.7 Contrast the permeability properties of the outer and inner membranes of mitochondria.

18.8 Contrast the location of the matrix and that of the intermembrane space in a mitochondrion.

18.3 Important Intermediate Compounds in Metabolic Pathways

Learning Focus

Understand what is meant by the notations ATP, ADP, AMP, FAD, $FADH_2$, NAD^+, NADH, and coenzyme A.

▶ Nucleotides are the monomer units from which nucleic acids are made, and they are also present in several *nonpolymeric* molecules that are important in energy production in living things.

As a prelude to an overview presentation (Section 18.4) of the metabolic processes by which our food is converted to energy, we now consider several compounds that repeatedly function as key intermediates in these metabolic pathways. Knowing about these compounds will make it easier to understand the details of metabolic pathways. The compounds to be discussed all have nucleotides (Section 17.2) as part of their structures.

Adenosine Phosphates (ATP, ADP, and AMP)

Several adenosine phosphates exist. Of importance in metabolism are adenosine *mono*phosphate (AMP), adenosine *di*phosphate (ADP), and adenosine *tri*phosphate (ATP). AMP is not a new molecule to us; it is one of the nucleotides present in ribonucleic acid molecules (Section 17.2). ADP and ATP differ structurally from AMP only in the number of phosphate groups present. Block structural diagrams for these three molecules follow.

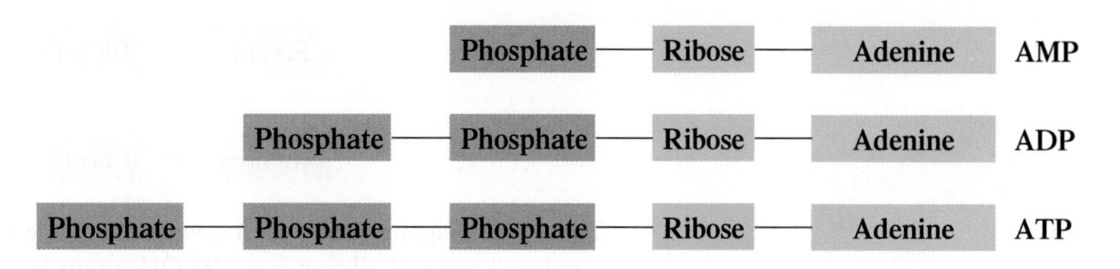

Figure 18.4 shows structural formulas for these three adenosine phosphates.

ATP and ADP molecules readily undergo hydrolysis reactions in which phosphate groups (P_i, inorganic phosphate) are released:

$$\text{ATP} + H_2O \longrightarrow \text{ADP} + P_i$$
$$\text{ADP} + H_2O \longrightarrow \text{AMP} + P_i$$
$$\text{ATP} + 2H_2O \longrightarrow \text{AMP} + 2P_i$$

These hydrolyses are energy-producing reactions that are used to drive cellular processes that require energy input. The phosphate–phosphate bonds in ATP and ADP are *very reactive* bonds that require less energy than normal to break. The presence of such reactive bonds, which are often called *strained bonds,* is the basis for the net energy

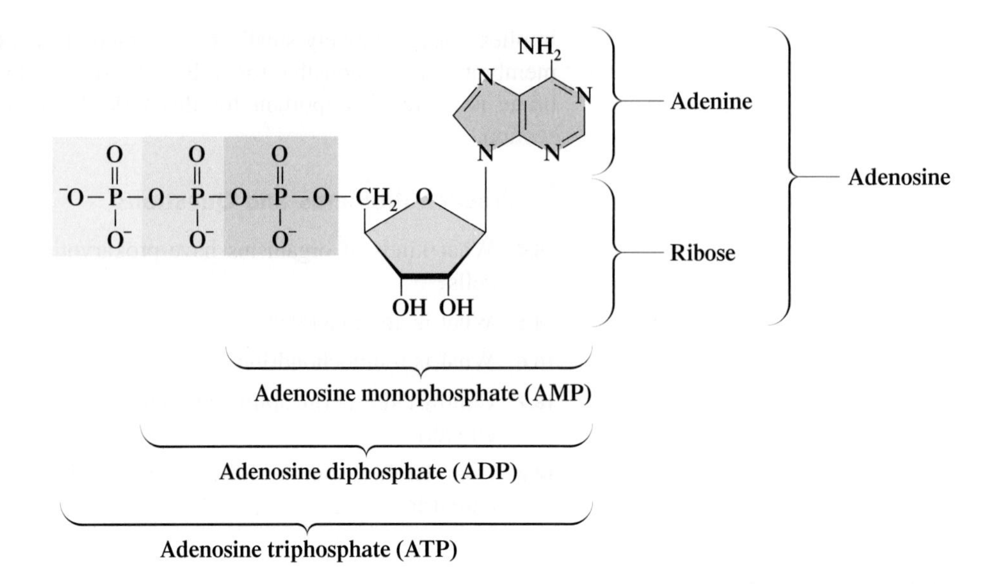

Figure 18.4 Structural relationships among AMP, ADP, and ATP molecules.

production that accompanies hydrolysis. Greater-than-normal electron–electron repulsive forces at specific locations within a molecule are the cause for bond strain; in ATP and ADP, it is the highly electronegative charged oxygen atoms in the additional phosphate groups that cause the increased repulsive strain.

Flavin Adenine Dinucleotide (FAD, $FADH_2$)

Flavin adenine dinucleotide (FAD) is a coenzyme (Section 16.14) required in numerous metabolic redox reactions. Structurally, FAD can be visualized as containing either three subunits or six subunits. A block diagram of FAD from the three-subunit viewpoint is

Flavin — Ribitol — ADP

Flavin and ribitol, the two components attached to the ADP unit, together constitute the B vitamin riboflavin. The block diagram for FAD from the six-subunit viewpoint is

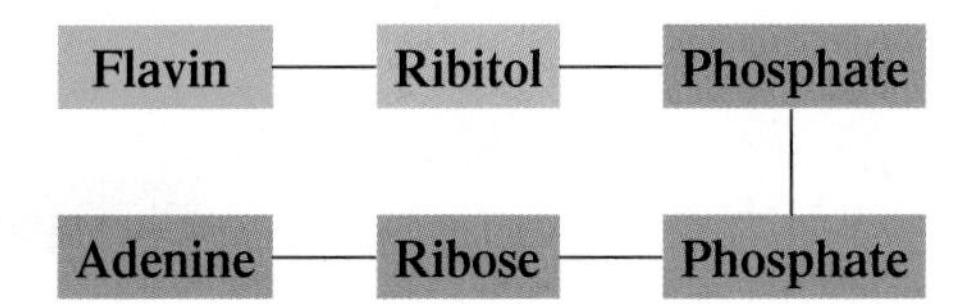

This block diagram shows the basis for the name *f*lavin *a*denine *d*inucleotide. Ribitol is a reduced form of ribose; a $—CH_2OH$ group is present in place of the —CHO group (Section 14.8).

D-Ribose	D-Ribitol
CHO	CH_2OH
H—OH	H—OH
H—OH	H—OH
H—OH	H—OH
CH_2OH	CH_2OH

The complete structural formula of FAD is given in Figure 18.5a.

The active portion of FAD in redox reactions is the flavin subunit of the molecule. The flavin is reduced, converting the FAD to $FADH_2$, a molecule with two additional hydrogen atoms. Thus FAD is the *oxidized* form of the molecule, $FADH_2$ the *reduced* form.

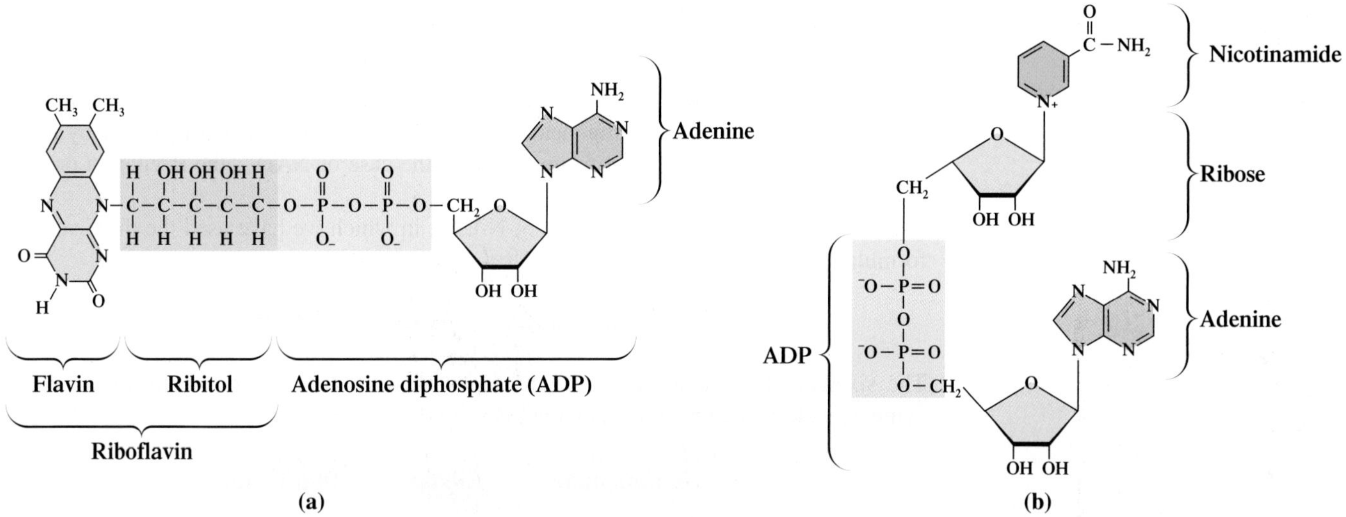

Flavin Ribitol Adenosine diphosphate (ADP)

Riboflavin

(a)

Nicotinamide

Ribose

Adenine

ADP

(b)

Figure 18.5 Structural formulas of the molecules flavin adenine dinucleotide, FAD (a) and nicotinamide adenine dinucleotide, NAD^+ (b).

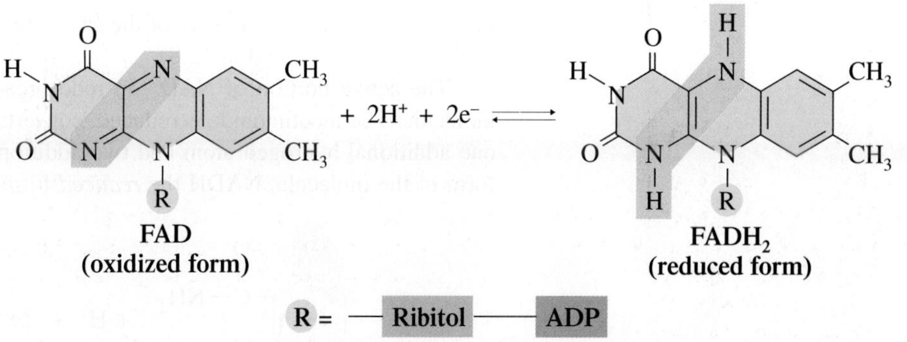

$$\text{FAD (oxidized form)} + 2H^+ + 2e^- \rightleftharpoons \text{FADH}_2 \text{ (reduced form)}$$

R = —Ribitol—ADP

A typical cellular reaction in which FAD serves as the oxidizing agent involves a —CH_2—CH_2— portion of a substrate being oxidized to produce a carbon–carbon double bond.

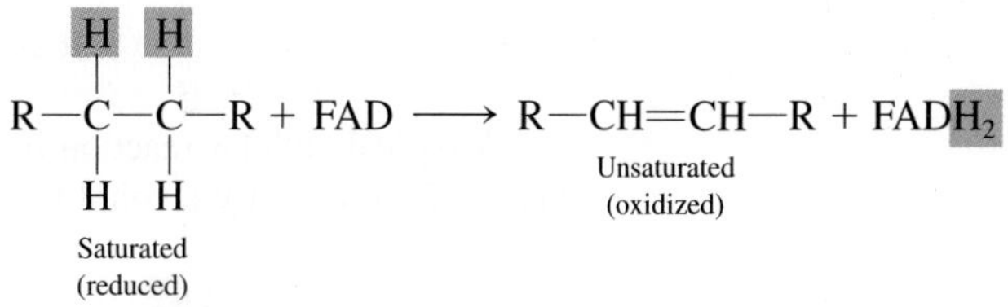

$$\underset{\text{Saturated (reduced)}}{R\text{—CH}_2\text{—CH}_2\text{—R}} + \text{FAD} \longrightarrow \underset{\text{Unsaturated (oxidized)}}{R\text{—CH=CH—R}} + \text{FADH}_2$$

For an enzyme-catalyzed redox reaction involving removal of two hydrogen atoms, such as this, each removed hydrogen atom is equivalent to a hydrogen *ion,* H^+, plus an electron, e^-.

$$2 \text{ H atoms (removed)} \longrightarrow 2H^+ + 2e^-$$

On the basis of this equivalency, the summary equation relating the oxidized and reduced forms of flavin adenine dinucleotide is usually written as

$$\underbrace{2H^+ + 2e^-}_{2 \text{ H atoms}} + \text{FAD} \rightleftharpoons \text{FADH}_2$$

▶ In metabolic pathways in which it is involved, flavin adenine dinucleotide continually changes back and forth between its oxidized and reduced forms.

$$2H^+ + 2e^- + \text{FAD} \rightleftharpoons \text{FADH}_2$$

Nicotinamide Adenine Dinucleotide (NAD^+, NADH)

Several parallels exist between the characteristics of nicotinamide adenine dinucleotide (NAD^+) and FAD. Both have coenzyme functions in metabolic redox pathways, both have a B vitamin as a structural component, and both can be represented structurally by using a three-subunit or a six-subunit formulation. In the case of NAD^+, the B vitamin present is nicotinamide.

A block diagram of the structure of NAD^+, in which we have used the three-subunit formulation, is

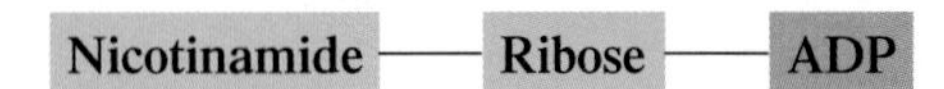

The six-subunit block diagram, which emphasizes the dinucleotide nature of the coenzyme as well as the origin of its name, is

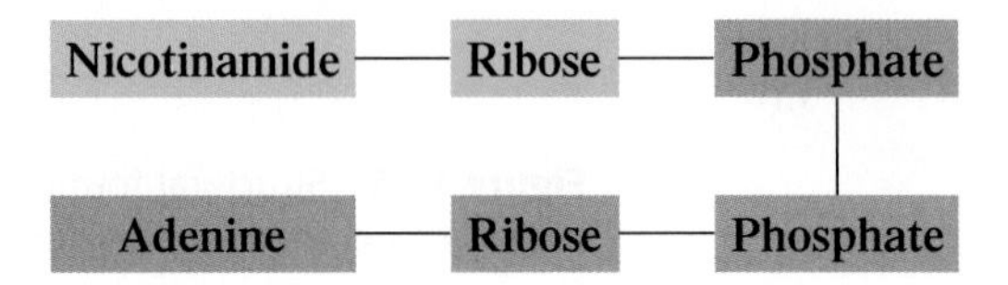

Examination of the detailed structure of NAD^+ (Figure 18.5b) reveals the basis for the positive electrical charge. The + sign refers to the positive charge on the nitrogen atom in the nicotinamide component of the structure; this nitrogen atom has four bonds instead of the usual three (Section 12.8).

The active portion of NAD^+ in redox reactions is the nicotinamide subunit of the molecule. The nicotinamide is reduced, converting the NAD^+ to NADH, a molecule with one additional hydrogen atom and two additional electrons. Thus NAD^+ is the *oxidized* form of the molecule, NADH the *reduced* form.

$$NAD^+ + H^+ + 2e^- \rightleftharpoons NADH$$

NAD⁺ (oxidized form) NADH (reduced form)

R = — Ribose — ADP

A typical cellular reaction in which NAD^+ serves as the oxidizing agent is the oxidation of a secondary alcohol to give a ketone.

$$R-\underset{H}{\overset{OH}{C}}-R + NAD^+ \longrightarrow R-\overset{O}{\overset{\|}{C}}-R + NADH + H^+$$

2° alcohol Ketone

In this reaction, one hydrogen atom of the alcohol substrate is directly transferred to NAD^+, whereas the other appears in solution as H^+ ion. Both electrons lost by the alcohol go to the nicotinamide ring in $NADH^+$. (Two electrons are required, rather than one, because of the original positive charge on NAD^+.) Thus the summary equation relating the oxidized and reduced forms of flavin adenine dinucleotide is written as

$$\underbrace{2H^+ + 2e^-}_{\text{2 H atoms}} + NAD^+ \rightleftharpoons NADH + H^+$$

▶ In metabolic pathways, nicotinamide adenine dinucleotide continually changes back and forth between its oxidized and reduced forms.

$$2H^+ + 2e^- + NAD^+ \rightleftharpoons NADH + H^+$$

Coenzyme A (CoA—SH)

Another important coenzyme in metabolic pathways is coenzyme A, a derivative of the B vitamin pantothenic acid. The three-subunit and six-subunit block diagrams for coenzyme A are

2-Aminoethanethiol — Pantothenic acid — Phosphorylated ADP

and

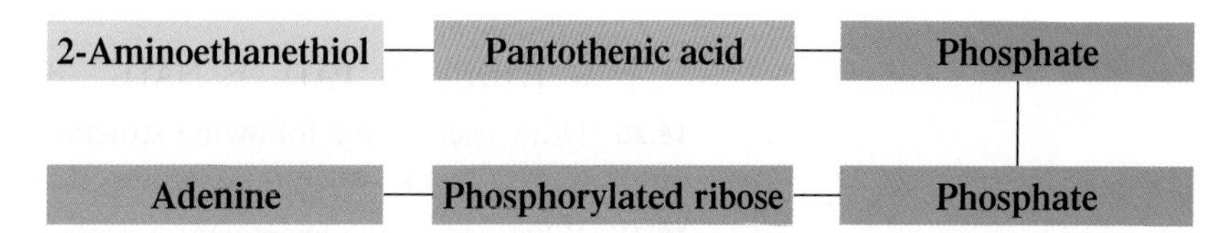

Note, in the three-subunit block diagram, that the ADP subunit present is phosphorylated. As shown in the following complete structural formula for coenzyme A (Figure 18.6), the phosphorylated version of ADP carries an extra phosphate group attached to carbon 3′ of its ribose.

The active portion of coenzyme A is the sulfhydryl group (—SH group; Section 12.6) in the ethanethiol subunit of the coenzyme. For this reason, the abbreviation CoA—SH is used for coenzyme A.

Think of the letter A in the name *coenzyme A* as reflecting a general metabolic function of this substance; it is the transfer of acetyl groups in metabolic pathways. Such groups bond to CoA—SH through a thioester bond to give acetyl CoA.

$$CH_3{-}\overset{\overset{\displaystyle O}{\|}}{C}{-}S{-}CoA$$

Thioester bond

Acetyl CoA

▶ An acetyl group, which can be considered to be derived from acetic acid, has the structure

$$CH_3{-}\overset{\overset{\displaystyle O}{\|}}{C}{-} \qquad CH_3{-}\overset{\overset{\displaystyle O}{\|}}{C}{-}OH$$

Acetyl group Acetic acid

▶ A *thioester* differs from an *ester* in that a S atom has replaced an oxygen atom.

$${-}\overset{\overset{\displaystyle O}{\|}}{C}{-}O{-}R \qquad {-}\overset{\overset{\displaystyle O}{\|}}{C}{-}S{-}R$$

Ester functional group Thioester functional group

Practice Problems and Questions

18.9 What does each letter stand for in each of the following?
a. ATP b. ADP c. AMP d. GTP

18.10 Draw block structural diagrams for each of the compounds in Problem 18.9.

18.11 In terms of hydrolysis, what is the relationship between the members of each of the following pairs of substances?
a. ATP and ADP b. ADP and AMP c. ATP and AMP

18.12 What does each letter in FAD stand for?

18.13 What are the letter designations for the oxidized and reduced forms of the substance in Problem 18.12?

Figure 18.6 Structural formula for coenzyme A.

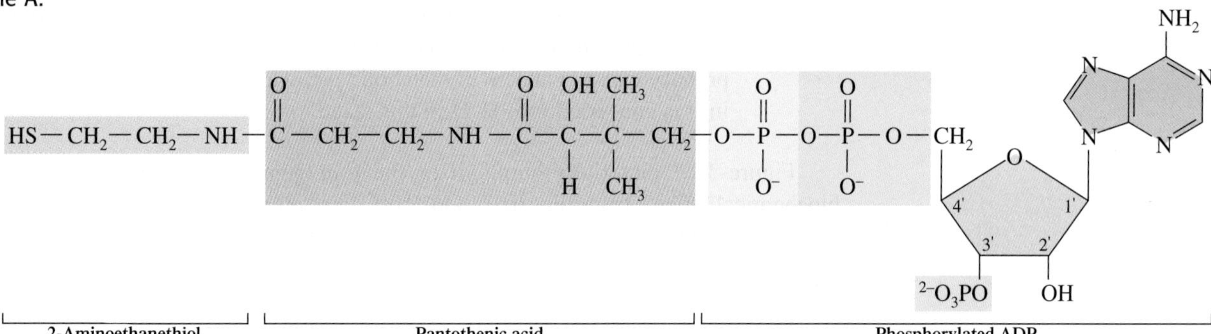

18.14 Draw each of the following structural representations for FAD.
a. A 3-block structural diagram b. A 6-block structural diagram

18.15 What part of the FAD molecule is the active participant in a redox reaction?

18.16 What does each letter in NAD^+ stand for?

18.17 Draw each of the following structural representations for NAD^+.
a. A 3-block structural diagram b. A 6-block structural diagram

18.18 What part of the NAD^+ molecule is the active participant in a redox reaction?

18.19 Indicate whether each of the following is the oxidized or the reduced form of the substance.
a. $FADH_2$ b. FAD c. NAD^+ d. NADH

18.20 Draw each of the following structural representations for coenzyme A.
a. A 3-block structural diagram b. A 6-block structural diagram

18.21 What part of the coenzyme A molecule is the active participant in its reactions?

18.22 Name the B vitamin that is part of the structure of each of the following molecules.
a. FAD b. NAD^+ c. Coenzyme A

Learning Focus

Be able to describe the four general stages in the biochemical energy production process.

▶ The first stage of biochemical energy production, digestion, is not considered part of metabolism because it is extracellular. Metabolic processes are intracellular.

18.4 An Overview of Biochemical Energy Production

The energy needed to run the human body is obtained from ingested food through a multistep process that involves several different catabolic pathways. There are four general stages in the biochemical energy production process, and numerous reactions are associated with each stage.

Stage 1: The first stage, *digestion,* begins in the mouth (saliva contains starch-digesting enzymes), continues in the stomach (gastric juices), and is completed in the small intestine (the majority of digestive enzymes and bile salts). The end products of digestion—glucose and other monosaccharides from carbohydrates, amino acids from proteins, and fatty acids and glycerol from fats and oils—are small enough to pass across intestinal membranes and into the blood, where they are transported to the body's cells.

Stage 2: The second stage, *acetyl group formation,* involves numerous reactions, some of which occur in the cytoplasm of cells and some in cellular mitochondria. The small molecules from digestion are further oxidized during this stage. Primary products include two-carbon acetyl units (which become attached to coenzyme A to give acetyl CoA) and the reduced coenzyme NADH.

Stage 3: The third stage, the *citric acid cycle,* occurs inside mitochondria. Here acetyl groups are oxidized to produce CO_2 (which we exhale during breathing) and energy. Some of the energy released by these reactions is lost as heat, and some is carried by the reduced coenzymes NADH and $FADH_2$ to the fourth stage.

Stage 4: The fourth stage, the *electron transport chain and oxidative phosphorylation,* also occurs inside mitochondria. NADH and $FADH_2$ supply the "fuel" (hydrogen ions and electrons) needed for the production of ATP molecules, the primary energy carriers in metabolic pathways. Molecular O_2 (from breathing) is converted into H_2O in this stage.

Figure 18.7 pictorially summarizes the four general stages in the production of biochemical energy from ingested food. This diagram is a *simplified* version of the "energy generation" process that occurs in the human body, as will be clear in the discussions presented in the later sections of this chapter, which give further details of the process.

The key concepts, associated with Figure 18.7, upon which those "further details" will be based are as follows:

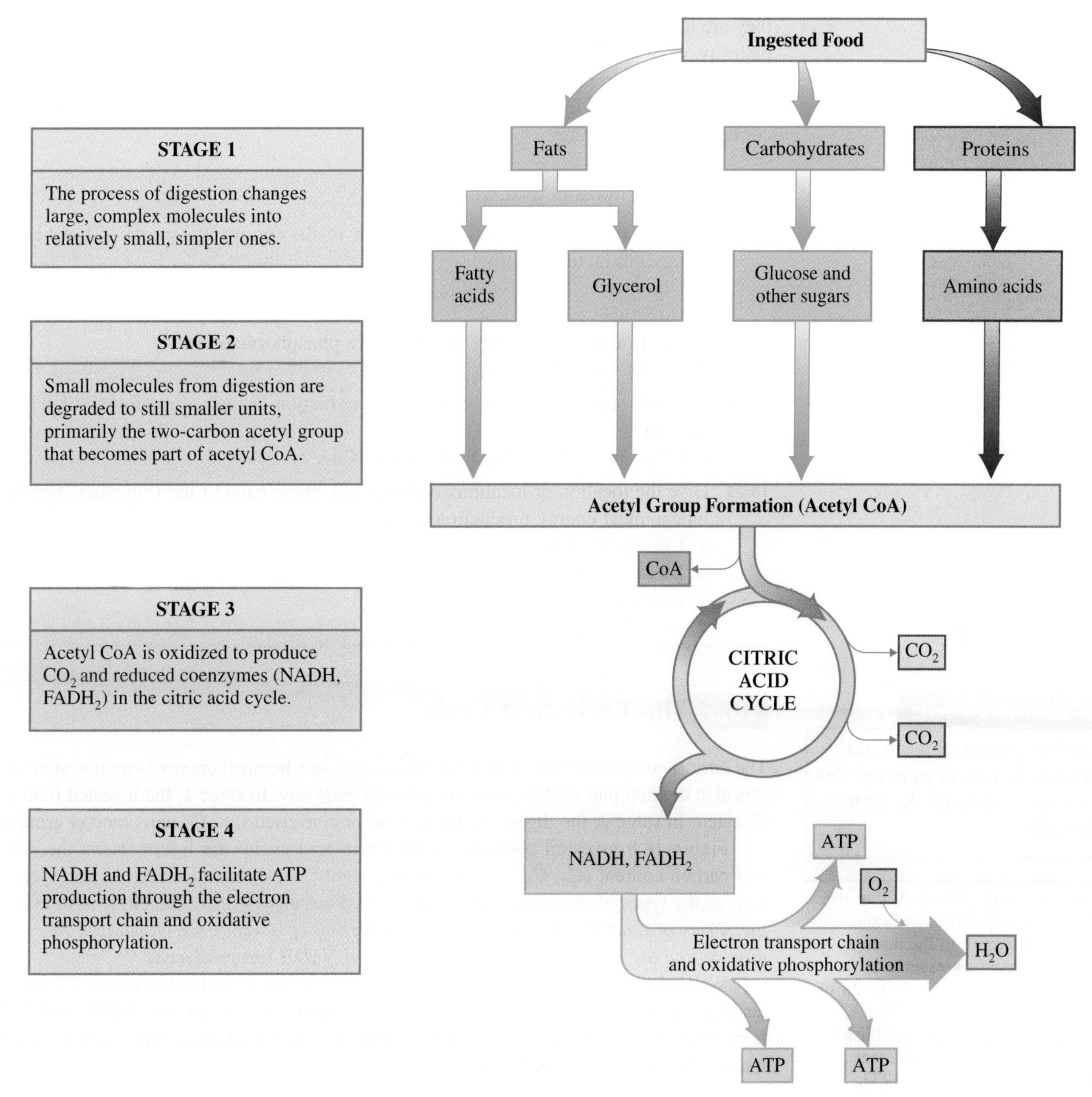

Ingested Food
Fats
Carbohydrates
Proteins
STAGE 1
The process of digestion changes large, complex molecules into relatively small, simpler ones.
Fatty acids
Glycerol
Glucose and other sugars
Amino acids
STAGE 2
Small molecules from digestion are degraded to still smaller units, primarily the two-carbon acetyl group that becomes part of acetyl CoA.
Acetyl Group Formation (Acetyl CoA)
CoA
STAGE 3
Acetyl CoA is oxidized to produce CO_2 and reduced coenzymes (NADH, $FADH_2$) in the citric acid cycle.
CITRIC ACID CYCLE
CO_2
CO_2
STAGE 4
NADH and $FADH_2$ facilitate ATP production through the electron transport chain and oxidative phosphorylation.
NADH, $FADH_2$
ATP
O_2
Electron transport chain and oxidative phosphorylation
H_2O
ATP
ATP

Figure 18.7 The four general stages in the biochemical energy production process in the human body.

1. Relatively simple molecules, all of which have been considered in previous chapters—fatty acids (Chapter 15), glycerol (Chapter 12), sugars (Chapter 14), and amino acids (Chapter 16)—are the products of digestion (stage 1).
2. All four of the major types of digestion products, *through different processes,* are converted into the same thing (stage 2): acetyl units (C_2 units). This greatly simplifies the further processing that occurs in stages 3 and 4, because there is only one starting material for these latter stages.

Our approach to considering the details of biochemical energy production will be first to consider the reactions that occur in stages 3 and 4 of biochemical energy production, the reactions common to the processing of carbohydrates, fats, and proteins for energy content. Collectively, these reactions are known as the *common metabolic pathway*. The **common metabolic pathway** *is the sum of the reactions that occur in (1) the citric acid cycle, (2) the electron transport chain, and (3) oxidative phosphorylation.* The three "parts" of the common metabolic pathway will be considered separately, in the order in which they are listed in the preceding definition. After our discussion of the common metabolic pathway, we will consider how the C_2 units (acetyl groups) that are processed through the common metabolic pathway are themselves formed (stage 2). Here, as representative of what occurs, we will focus on how glucose, the major carbohydrate digestion product, is converted into C_2 groups.

Practice Problems and Questions

18.23 To what stage (1, 2, 3, or 4) does each of the following aspects of biochemical energy production belong?
a. Digestion
b. Citric acid cycle
c. Electron transport chain and oxidative phosphorylation
d. Acetyl group formation

18.24 In which stage of biochemical energy production is each of the following a major product?
a. CO_2 b. ATP c. H_2O d. Acetyl CoA

18.25 Give the location or locations within a cell where each of the following aspects of biochemical energy production occurs.
a. Citric acid cycle
b. Electron transport chain
c. Oxidative phosphorylation
d. Acetyl group formation

Learning Focus

Describe, in general terms, the reactions that occur as acetyl CoA is processed through the citric acid cycle.

18.5 The Citric Acid Cycle

The citric acid cycle is stage 3 in the production of biochemical energy from ingested food; it is also the first part of the *common metabolic pathway*. In stage 1, the ingested food was digested. In stage 2, the digestion products were converted into C_2 units (acetyl groups).

Figure 18.8 gives an overview of the citric acid cycle; the figure shows the names and carbon content (C_2, C_4, C_6, etc.) of the eight compounds produced in the cycle, as well as the types of reactions that are involved. Formally defined, the **citric acid cycle** *is the series of reactions in which the acetyl portion of acetyl CoA is oxidized to carbon dioxide and the reduced coenzymes $FADH_2$ and NADH are produced.*

We shall now consider the individual steps in the cycle in further detail. Of particular importance to us will be the steps where (1) redox reactions occur (NADH or $FADH_2$ is produced), and (2) decarboxylation occurs (the carbon chain is shortened by the removal of a carbon atom as CO_2).

▶ The citric acid cycle (CAC) is also called the *tricarboxylic acid cycle* (TCA), in reference to the three carboxylate groups present in citric acid, and the *Krebs cycle,* in honor of the British biochemist Sir Hans A. Krebs (1900–1981), who established relationships among the different compounds in the cycle in 1937.

Reactions of the Citric Acid Cycle

Step 1: *Formation of Citrate.* Acetyl CoA, the two-carbon degradation product of carbohydrates, fats, and proteins, enters the cycle by combining with the four-carbon keto dicarboxylate species oxaloacetate. This results in the transfer of the acetyl group from coenzyme A to oxaloacetate, producing the C_6 citrate species and free coenzyme A. The enzyme *citrate synthase* is involved, as is a molecule of H_2O.

▶ The formation of citrate is called a condensation reaction because a new carbon–carbon bond is formed.

$$\underset{\text{Oxaloacetate}}{{}^{-}OOC-C(=O)-CH_2-COO^{-}} + \underset{\text{Acetyl CoA}}{CH_3-C(=O)-S-CoA} \xrightarrow[H_2O]{\text{Citrate synthase}} \underset{\text{Citrate}}{HO-C(CH_2-COO^{-})(COO^{-})-CH_2-COO^{-}} + CoA-SH + H^{+}$$

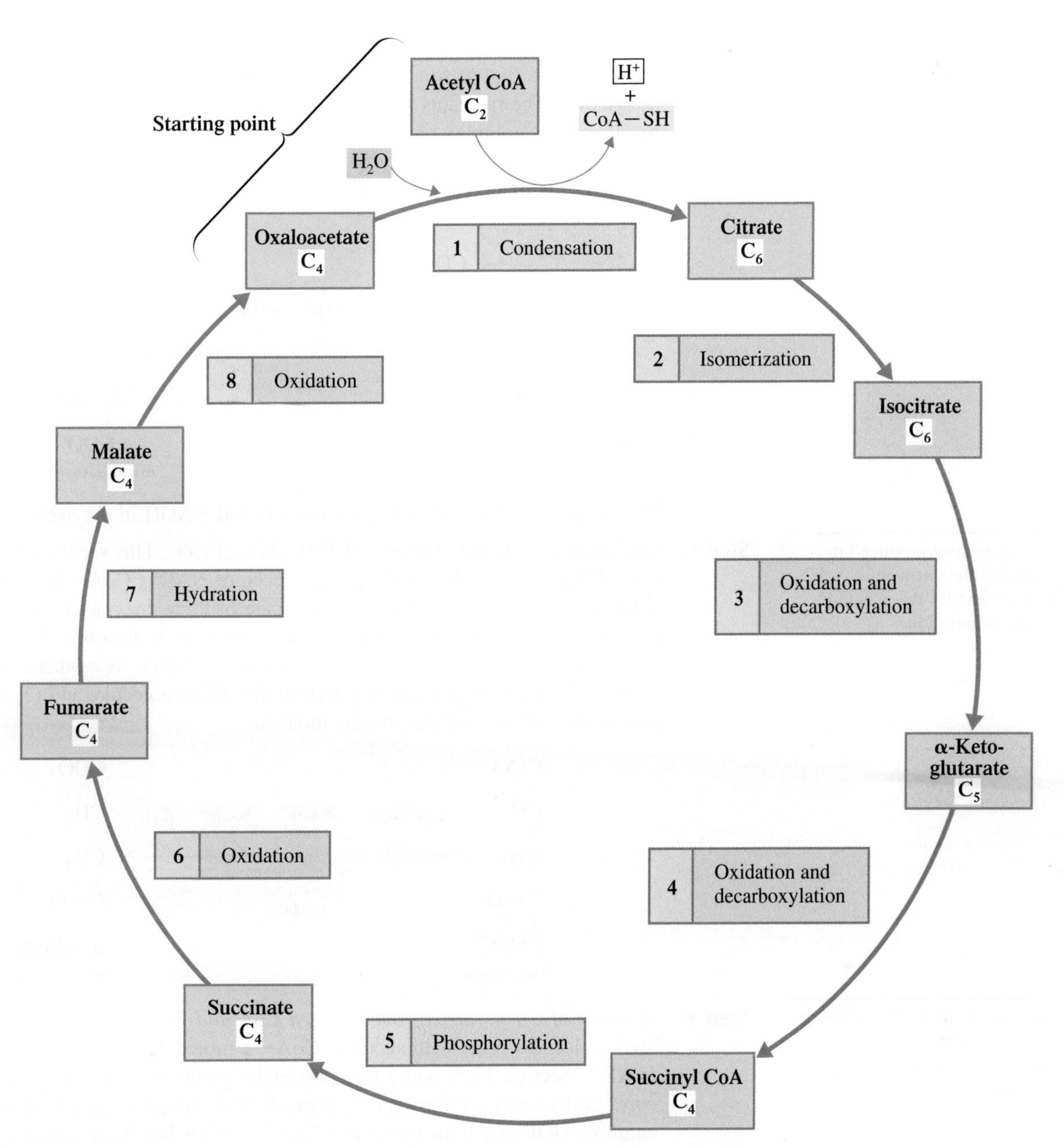

Figure 18.8 The citric acid cycle. Details concerning the numbered steps are given in the text.

Step 2: *Formation of Isocitrate.* Citrate is converted to its less symmetrical isomer isocitrate in an isomerization process catalyzed by the enzyme *aconitase*. The net result of these reactions is that the —OH group from citrate is moved from carbon 3 to carbon 2.

$$
\begin{array}{c}
COO^- \\
| \\
CH_2 \\
| \\
HO\text{—}C\text{—}COO^- \\
| \\
H\text{—}C\text{—}H \\
| \\
COO^- \\
\text{Citrate}
\end{array}
\xrightarrow{\text{Aconitase}}
\begin{array}{c}
COO^- \\
| \\
CH_2 \\
| \\
H\text{—}C\text{—}COO^- \\
| \\
HO\text{—}C\text{—}H \\
| \\
COO^- \\
\text{Isocitrate}
\end{array}
$$

Citrate is a tertiary alcohol and isocitrate a secondary alcohol. Tertiary alcohols are not readily oxidized; secondary alcohols are easier to oxidize (Section 12.4). The next step in the cycle involves oxidation.

▶ All acids found in the citric acid cycle exist as negative ions (carboxylate ions; Section 13.7) at cellular pH.

Step 3: *Oxidation of Isocitrate and Formation of CO_2.* This step involves oxidation-reduction (the first of four redox reactions in the citric acid cycle) and decarboxylation. The reactants are a NAD^+ molecule and isocitrate. The major results of this complex reaction, which is catalyzed by *isocitrate dehydrogenase,* are that (1) the hydroxyl group is oxidized to a ketone group, (2) NAD^+ is oxidized to NADH, and (3) a carboxyl group is removed as a CO_2 molecule.

$$\underset{\text{Isocitrate}}{COO^- - CH_2 - CH(COO^-) - CH(OH) - COO^-} \xrightarrow[\text{Isocitrate dehydrogenase}]{NAD^+ \;\; NADH \;\; CO_2} \underset{\alpha\text{-Ketoglutarate}}{COO^- - CH_2 - CH_2 - C(=O) - COO^-}$$

This step yields the first molecules of CO_2 and NADH in the cycle.

▶ The CO_2 molecules produced in Steps 3 and 4 of the citric acid cycle are the CO_2 molecules we exhale in the process of respiration.

Step 4: *Oxidation of α-Ketoglutarate and Formation of CO_2.* This second redox reaction of the cycle involves one molecule each of NAD^+, CoA—SH, and α-ketoglutarate. The catalyst is an aggregate of three enzymes called the *α-ketoglutarate dehydrogenase complex.* As in Step 3, decarboxylation also occurs. The major results of this step are that (1) NAD^+ is oxidized to NADH, (2) a carboxyl group is removed as a CO_2 molecule, and (3) CoA is part of the structure of the product molecule.

$$\underset{\alpha\text{-Ketoglutarate}}{COO^- - CH_2 - CH_2 - C(=O) - COO^-} \xrightarrow[\alpha\text{-Ketoglutarate dehydrogenase complex}]{CoA{-}SH \;\; NAD^+ \;\; NADH \;\; CO_2} \underset{\text{Succinyl CoA}}{COO^- - CH_2 - CH_2 - C(=O) - S{-}CoA}$$

▶ The function of the GTP produced is similar to that of ATP: to store energy in the form of a high-energy phosphate bond (Section 18.3).

Step 5: *Removal of Coenzyme A from Succinyl CoA and Phosphorylation of GDP.* Two molecules react with succinyl CoA—a molecule of GDP (similar to ADP; Section 18.3) and a free phosphate group (P_i). The enzyme *succinyl CoA synthase* removes coenzyme A. The energy released is used to combine GDP and P_i to form GTP. Succinyl CoA has been converted to succinate.

$$\underset{\text{Succinyl CoA}}{COO^- - CH_2 - CH_2 - C(=O) - S{-}CoA} \xrightarrow[\text{Succinyl CoA synthase}]{GDP \;\; P_i \;\; GTP \;\; CoA{-}SH} \underset{\text{Succinate}}{COO^- - CH_2 - CH_2 - COO^-}$$

Steps 6 through 8 involve a sequence of functional-group changes that we have encountered many times in the organic sections of the text. The reaction sequence is

$$\text{Alkane} \xrightarrow[\text{(dehydrogenation)}]{① \text{ Oxidation}} \text{alkene} \xrightarrow{② \text{ Hydration}} \text{secondary alcohol} \xrightarrow[\text{(dehydrogenation)}]{③ \text{ Oxidation}} \text{ketone}$$

Step 6: *Oxidation of Succinate.* This is the third redox reaction of the cycle. The enzyme involved is *succinate dehydrogenase,* and the oxidizing agent is FAD rather than NAD^+. Two hydrogen atoms are removed from the succinate to produce fumarate, a C_4 species with a *trans* double bond. FAD is reduced to $FADH_2$ in the process.

► Fumarate, with its *trans* double bond, is an essential metabolic intermediate in both plants and animals. Its isomer, with a *cis* double bond, is called maleate, and it is toxic and irritating to tissues. Succinate dehydrogenase produces only the *trans* isomer of this unsaturated diacid.

$$COO^- - CH_2 - CH_2 - COO^- \xrightarrow[\text{Succinate dehydrogenase}]{FAD \rightarrow FADH_2} {}^-OOC(H)C{=}C(H)COO^-$$

Succinate → Fumarate

Step 7: *Hydration of Fumarate.* The enzyme *fumarase* catalyzes the addition of water to the double bond of fumarate. The enzyme is stereospecific, so only the L isomer of the product malate is produced.

$$COO^- - CH{=}HC - COO^- \xrightarrow[\text{Fumarase}]{H_2O} COO^- - C(HO)(H) - C(H)(H) - COO^-$$

Fumarate → L-Malate

Step 8: *Oxidation of L-Malate to Regenerate Oxaloacetate.* In the fourth oxidation–reduction reaction of the cycle, a molecule of NAD^+ reacts with malate, oxidizing the —OH group to a ketone group. The product of this reaction, in which *malate dehydrogenase* is the enzyme, is oxaloacetate, the molecule that started off the cycle in Step 1. The oxaloacetate so formed can combine with another molecule of acetyl CoA (Step 1), and the cycle repeats itself.

$$COO^- - C(HO)(H) - CH_2 - COO^- \xrightarrow[\text{Malate dehydrogenase}]{NAD^+ \rightarrow NADH} COO^- - C{=}O - CH_2 - COO^-$$

L-Malate → Oxaloacetate

Summary of the Citric Acid Cycle

An overall summary equation for the citric acid cycle is obtained by adding together the individual reactions of the cycle:

$$\text{Acetyl CoA} + 3NAD^+ + FAD + GDP + P_i + 2H_2O \longrightarrow 2CO_2 + CoA{-}SH + 3NADH + 2H^+ + FADH_2 + GTP$$

Important features of the cycle include the following:

1. The reactions of the cycle take place in the mitochondrial matrix, except the succinate dehydrogenase reaction that involves FAD. The enzyme that catalyzes this reaction is an integral part of the inner mitochondrial membrane.
2. The "fuel" for the cycle is acetyl CoA, obtained from the breakdown of carbohydrates, fats, and proteins.
3. Four of the cycle reactions involve oxidation and reduction. The oxidizing agent is either NAD^+ (three times) or FAD (once). The operation of the cycle depends on the availability of these oxidizing agents.

4. In redox reactions, NAD^+ is the oxidizing agent when a carbon–oxygen double bond is formed; FAD is the oxidizing agent when a carbon–carbon double bond is formed.
5. The three NADH and one $FADH_2$ that are formed during the cycle carry electrons and H^+ to the electron transport chain (Section 18.6) through which ATP is synthesized.
6. Two carbon atoms enter the cycle as the acetyl unit of acetyl CoA, and two carbon atoms leave the cycle as two molecules of CO_2.
7. Four B vitamins are necessary for the proper functioning of the cycle: riboflavin (in both FAD and the α-ketoglutarate dehydrogenase complex), nicotinamide (in NAD^+), pantothenic acid (in CoA—SH), and thiamin (in the α-ketoglutarate dehydrogenase complex).
8. One high-energy GTP molecule is produced by phosphorylation.

Practice Problems and Questions

18.26 What are two other names for the citric acid cycle (CAC)?

18.27 What is the basis for the name *citric acid cycle*?

18.28 What is the "fuel" for the CAC?

18.29 What is the fate of the CO_2 produced in the CAC?

18.30 What is the fate of the NADH and the $FADH_2$ produced in the CAC?

18.31 Consider the reactions that occur during *one turn* of the CAC in answering the following questions.
a. How many CO_2 molecules are formed?
b. How many molecules of $FADH_2$ are formed?
c. How many molecules of NADH are formed?
d. How many GTP molecules are formed?

18.32 Consider the reactions that occur during *one turn* of the CAC in answering each of the following questions.
a. How many oxidation–reduction reactions occur?
b. How many decarboxylation reactions occur?
c. How many isomerization reactions occur?
d. How many hydration reactions occur?

18.33 There are eight steps in the CAC. List those steps, by number, that involve
a. both oxidation and decarboxylation
b. oxidation but not decarboxylation
c. phosphorylation
d. condensation

18.34 In how many steps in the CAC does each of the following occur?
a. A C_6 compound is converted into another C_6 compound.
b. A C_6 compound is converted into a C_5 compound.
c. A C_5 compound is converted into a C_4 compound.
d. A C_4 compound is converted into another C_4 compound.

18.35 List the two CAC intermediates involved in the reaction governed by each of the following enzymes. List the reactant first.
a. Fumarase
b. Malate dehydrogenase
c. Isocitrate dehydrogenase
d. Aconitase

Learning Focus

Know the purposes of the electron transport chain, and describe, in general terms, the reactions that occur in the electron transport chain.

18.6 The Electron Transport Chain

The NADH and $FADH_2$ produced in the citric acid cycle pass to the electron transport chain, where they serve as the "fuel" for this series of reactions. The **electron transport chain** *is a series of reactions in which electrons and hydrogen ions from NADH and* $FADH_2$ *are passed to intermediate carriers and ultimately react with molecular oxygen to produce water.* The oxidation reactions for NADH and $FADH_2$ are

$$NADH + H^+ \longrightarrow NAD^+ + 2H^+ + 2e^-$$
$$FADH_2 \longrightarrow FAD + 2H^+ + 2e^-$$

▶ The oxygen involved in the water formation associated with the electron transport chain is the oxygen we breathe.

Water is formed when the electrons and hydrogen ions that originate from these reactions react with molecular oxygen.

$$O_2 + 4e^- + 4H^+ \longrightarrow 2H_2O$$

The preceding definition for the electron transport chain indicates that electrons and H^+ ions are passed to intermediate carriers during operation of this chain of reactions. There are eight such intermediate carriers involved in the electron transport chain. Listed in the order in which they are encountered, they are

1. Flavin mononucleotide (FMN)
2. Iron–sulfur protein (FeSP)
3. Coenzyme Q (CoQ)
4. Cytochrome b (cyt b)
5. Cytochrome c_1 (cyt c_1)
6. Cytochrome c (cyt c)
7. Cytochrome a (cyt a)
8. Cytochrome a_3 (cyt a_3)

With the exception of coenzyme Q and cytochrome c, these electron carriers occur as part of enzyme complexes that have *fixed* locations within the inner mitochondrial membrane (Figure 18.9). Coenzyme Q and cytochrome c are *mobile*—free to move between specific fixed enzyme complexes.

Each of these carriers is alternately reduced and oxidized as it first accepts electrons and then passes them on to the next carrier in the electron transport chain. In the final reaction, oxygen, electrons, and H^+ ions interact to form water.

Let us now consider more specific details about the reactions of the electron transport chain. We first note that only three electron carriers (FMN, CoQ, and cyt a_3) have involvement with both electrons and H^+ ions and that the other carriers (all iron-containing substances) have involvement only with electrons.

Step 1: *Flavin Mononucleotide (FMN).* In this initial step, NADH is oxidized to NAD^+, and FMN is reduced to $FMNH_2$ as two H^+ ions and two electrons pass to $FMNH_2$.

$$NADH + H^+ \xrightarrow{\text{Oxidation}} NAD^+ + 2H^+ + 2e^-$$
$$2H^+ + 2e^- + FMN \xrightarrow{\text{Reduction}} FMNH_2$$

NADH supplies both electrons and one of the H^+ ions; the other H^+ ion comes from cellular fluid. The NAD^+ so produced can again participate in citric acid cycle operations, and the $FMNH_2$ produced participates in the second step of the electron transport chain.

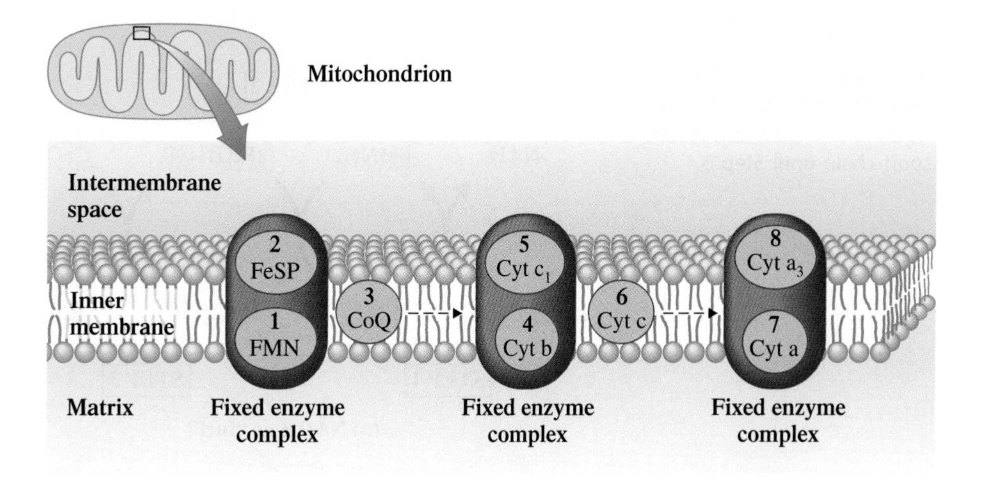

Figure 18.9 Six of the eight electron carriers (enzymes) of the electron transport chain are found at fixed sites within the inner mitochondrial membrane. The other two electron carriers are mobile, moving between fixed sites.

Step 2: *Iron/Sulfur Protein (FeSP).* In this step, $FMNH_2$ is oxidized back to FMN as it transfers the electrons it received to FeSP located in the same fixed-site enzyme complex as it is. The iron present in FeSP is Fe^{3+}, which is reduced to Fe^{2+}. The two H atoms of $FMNH_2$ are released to solution as two H^+ ions. Two FeSP proteins are needed to accommodate the two electrons released by $FMNH_2$ because an Fe^{3+}/Fe^{2+} reduction involves only one electron.

$$FMNH_2 \xrightarrow{\text{Oxidation}} FMN + 2H^+ + 2e^-$$
$$2H^+ + 2e^- + 2Fe(III)SP \xrightarrow{\text{Reduction}} 2Fe(II)SP + 2H^+$$

Step 3: In this step, Fe(II)SP is reconverted into Fe(III)SP as each of two Fe(II)SP passes an electron to the *mobile* coenzyme Q (CoQ). Coenzyme Q also picks up two H^+ ions from solution in addition to the two electrons, forming $CoQH_2$.

$$2Fe(II)SP + 2H^+ \xrightarrow{\text{Oxidation}} 2Fe(III)SP + 2H^+ + 2e^-$$
$$2H^+ + 2e^- + CoQ \xrightarrow{\text{Reduction}} CoQH_2$$

Both the $FADH_2$ and the NADH generated in the citric acid cycle are "fuels" for the electron transport chain. NADH participates in the electron transport chain from the beginning. $FADH_2$ enters the electron transport chain at the CoQ stage. The equations for its interaction with CoQ molecules are

$$FADH_2 \xrightarrow{\text{Oxidation}} FAD + 2H^+ + 2e^-$$
$$2H^+ + 2e^- + CoQ \xrightarrow{\text{Reduction}} CoQH_2$$

The regenerated FAD can then again participate in the citric acid cycle.

Figure 18.10 summarizes the electron transport chain reactions through Step 3. Note the general pattern that is developing for the electron carriers. They are oxidized in one step (accept electrons) and then regenerated (reduced; lose electrons) in the next step so that they can again participate in the electron transport chain.

The electrons that pass through the various steps of the electron transport chain (ETC) lose some energy with each transfer along the chain. Some of this "lost" energy is used to make ATP from ADP (oxidative phosphorylation), as we will see in Section 18.7.

Steps 4–8: *Cytochromes.* The remaining steps in the electron transport chain all involve iron-containing electron carriers called cytochromes. A **cytochrome** *is a heme-containing protein that undergoes reversible oxidation and*

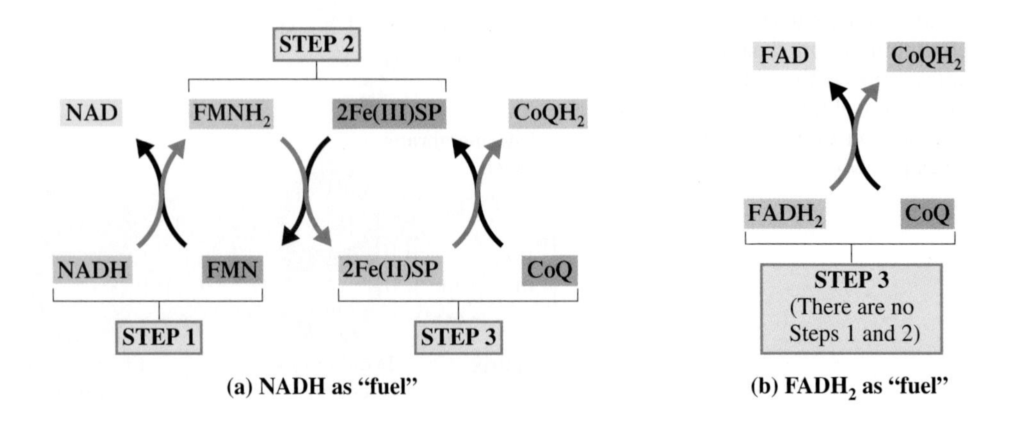

Figure 18.10 $NADH_2$ enters the electron transport chain in Step 1. $FADH_2$ does not enter the electron transport chain until Step 3.

reduction of its Fe atoms. Heme, a compound also present in hemoglobin and myoglobin, has the structure

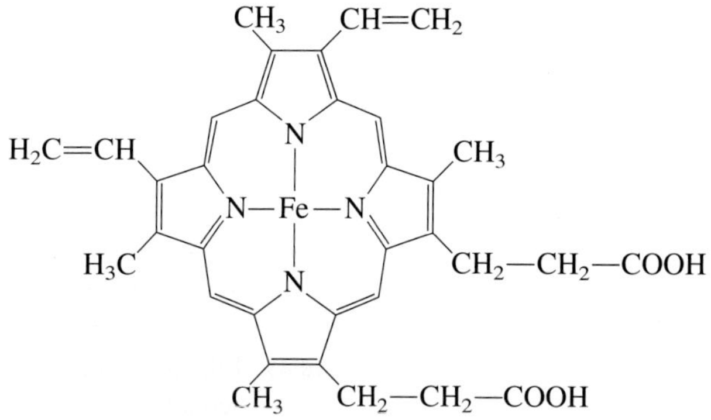

► In cytochromes the iron is involved in redox reactions in which the iron changes back and forth between the +2 and +3 oxidation states.

► Iron/sulfur protein (FeSP) is a *nonheme iron protein.* Most proteins of this type contain sulfur, as is the case with FeSP. Often the iron is bound to the sulfur atom in the amino acid cysteine.

Heme-containing proteins function similarly to FeSP; iron changes back and forth between the +3 and +2 oxidation states.

Various cytochromes, abbreviated cyt a, cyt b, cyt c, and so on, differ from each other in (1) their protein constituents, (2) the manner in which the heme is bound to the protein, and (3) attachments to the heme ring. Again, because the Fe^{3+}/Fe^{2+} system involves only a one-electron change, two cytochrome molecules are needed in these steps of the ETC to move two electrons along the chain.

► A feature all steps in the ETC share is that as each electron carrier passes electrons along the chain, it becomes reoxidized and thus able to accept more electrons.

All the cytochrome steps in the electron transport chain involve just electrons except for the last one (Step 8), where H^+ ions are also involved. The cytochrome in Step 8, cyt a_3, has the ability to bind molecular oxygen to itself and effect its reduction. The reduced oxygen atoms combine with H^+ ions to produce water.

$$O_2 + 4H^+ + 4e^- \longrightarrow 2H_2O$$

It is estimated that 95% of the oxygen used by cells serves as the final electron acceptor for the ETC.

The fixed enzyme complex that contains cytochromes a and a_3 is called *cytochrome oxidase.* Its structure is known to include two different metals, iron and copper. The electrons are transferred from copper to iron to oxygen.

Figure 18.11 summarizes the "electron flow" that occurs in Steps 4–8 of the electron transport chain.

Practice Problems and Questions

18.36 What two molecules are the "fuel" for the electron transport chain?

18.37 Which two substances generated in the citric acid cycle participate in the electron transport chain?

Figure 18.11 The cytochrome steps in the electron transport chain. All steps except the last one involve just electrons; H^+ ions are also involved in the last step.

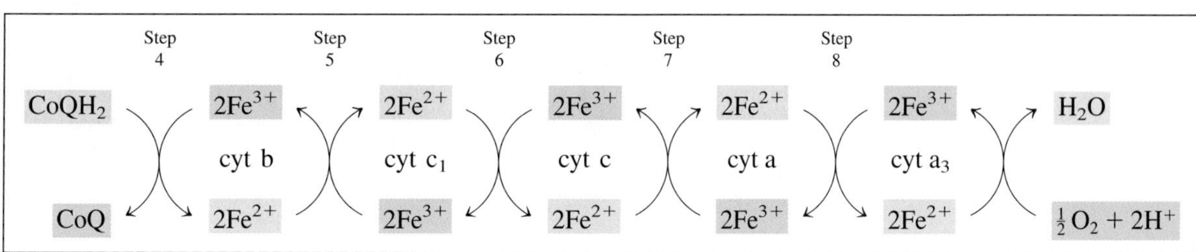

18.38 What do the following abbreviations stand for?
a. FMN b. Cyt c. FeSP d. CoQ

18.39 At what location in a cell do the reactions of the electron transport chain occur?

18.40 Classify each of the following electron carriers as a mobile-site or a fixed-site carrier.
a. cyt c b. cyt b c. FMN d. CoQ

18.41 Put the following substances in the correct order of their participation in the electron transport chain: cyt c_1, cyt a_3, FeSP, and $FADH_2$.

18.42 Indicate whether each of the following changes represents oxidation or reduction.
a. $CoQH_2$ to CoQ b. NAD^+ to NADH
c. cyt c (Fe^{2+}) to cyt c (Fe^{3+}) d. Fe(III)SP to Fe(II)SP

18.43 Fill in the missing substances in the following electron transport chain reaction sequences.
a. FAD, $CoQH_2$, ?, ?, ?, $2Fe^{2+}$
b. ?, 2Fe(III)SP, ?, FMN, ?, CoQ

> **Learning Focus**
> Describe the relationship between the electron transport chain and oxidative phosphorylation, and describe how ATP synthesis occurs during oxidative phosphorylation.

18.7 Oxidative Phosphorylation

Oxidative phosphorylation *is the process by which ATP is synthesized from ADP using energy released in the electron transport chain.* A key aspect of this process is coupled reactions. **Coupled reactions** *are pairs of chemical reactions in which energy released from one reaction changes the equilibrium position of a second reaction.*

The interdependence (coupling) of ATP synthesis with the reactions of the ETC is related to the movement of protons (H^+ ions) across the inner mitochondrial membrane. The three fixed enzyme complexes involved in the ETC chain have a second function besides that of electron transfer down the chain. They also serve as "proton pumps," transferring protons from the matrix side of the inner mitochondrial membrane to the intermembrane space (Figure 18.12).

Some of the H^+ ions crossing the inner mitochondrial membrane come from the reduced electron carriers, and some come from the matrix; the details of how the H^+ ions cross the inner mitochondrial membrane are not fully understood.

For every two electrons passed through the ETC, four protons cross the inner mitochondrial membrane at the first fixed enzyme site, two at the second fixed enzyme site, and four more at the third fixed enzyme site. This proton flow causes a buildup of H^+

▶ *Oxidative phosphorylation* is not the only process by which ATP is produced in cells. A second process, *substrate phosphorylation* (Section 18.9), can also be an ATP source. However, the amount of ATP produced by this second process is much less than that produced by oxidative phosphorylation.

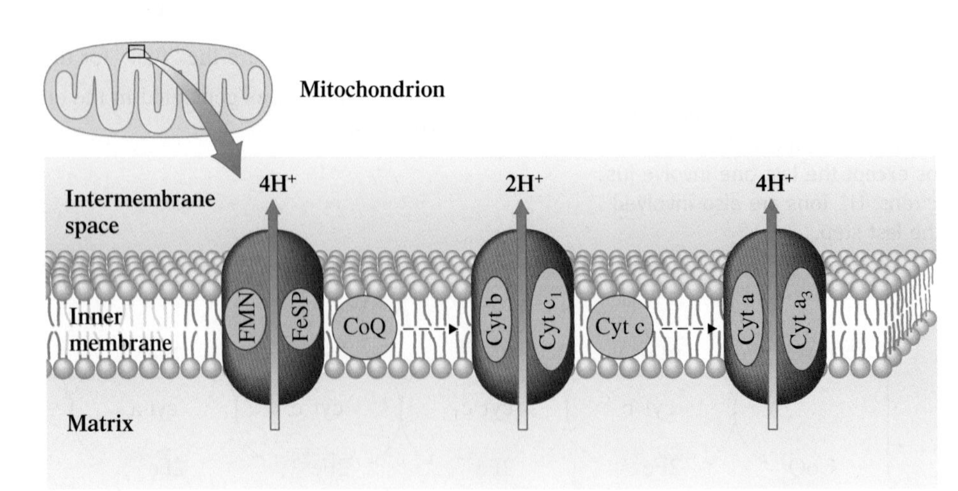

Figure 18.12 A second function for the fixed enzyme sites involved in the electron transport chain is that of proton pumps. For every two electrons passed through the ETC, 10 H^+ ions are transferred from the mitochondrial matrix to the intermembrane space.

ions (protons) in the intermembrane space; this high concentration of protons becomes the basis for ATP synthesis.

The main concepts involved in explaining ATP synthesis, based on "proton flow" across the inner mitochondrial membrane, are as follows:

▶ Some of the energy released at each of the three fixed enzyme complex sites is consumed in the movement of H^+ ions across the inner membrane from the matrix into the intermembrane space. Movement of ions from a region of lower concentration (the matrix) to one of higher concentration (the intermembrane space) requires the expenditure of energy because it opposes the natural tendency, as exhibited in the process of osmosis (Section 7.8), to equalize concentrations.

1. The result of the pumping of protons from the mitochondrial matrix across the inner mitochondrial membrane is a higher concentration of protons in the intermembrane space than in the matrix. This concentration difference constitutes an *electrochemical (proton) gradient.* A chemical gradient exists whenever a substance has a higher concentration in one region than in another. Because the proton has an electrical charge (is an ion), an electrical gradient also exists.
2. A spontaneous flow of protons from the region of high concentration to the region of low concentration occurs because of the electrochemical gradient. This proton flow is not through the membrane itself (it is not permeable to H^+ ions) but rather through enzyme complexes called *ATP synthases* located on the inner mitochondrial membrane (Section 18.2). These enzymes catalyze the conversion of ADP to ATP.

$$\text{ADP} + \text{P}_i \xrightarrow{\text{ATP synthase}} \text{ATP} + \text{H}_2\text{O}$$

3. ATP synthase has two subunits, the F_0 and F_1 subunits (Figure 18.13). The F_0 part of the synthase is the channel for proton flow, whereas the formation of ATP takes place in the F_1 subunit. As protons return to the mitochondrial matrix through the F_0 subunit, the potential energy associated with the electrochemical gradient is released and used in the F_1 subunit for the synthesis of ATP.

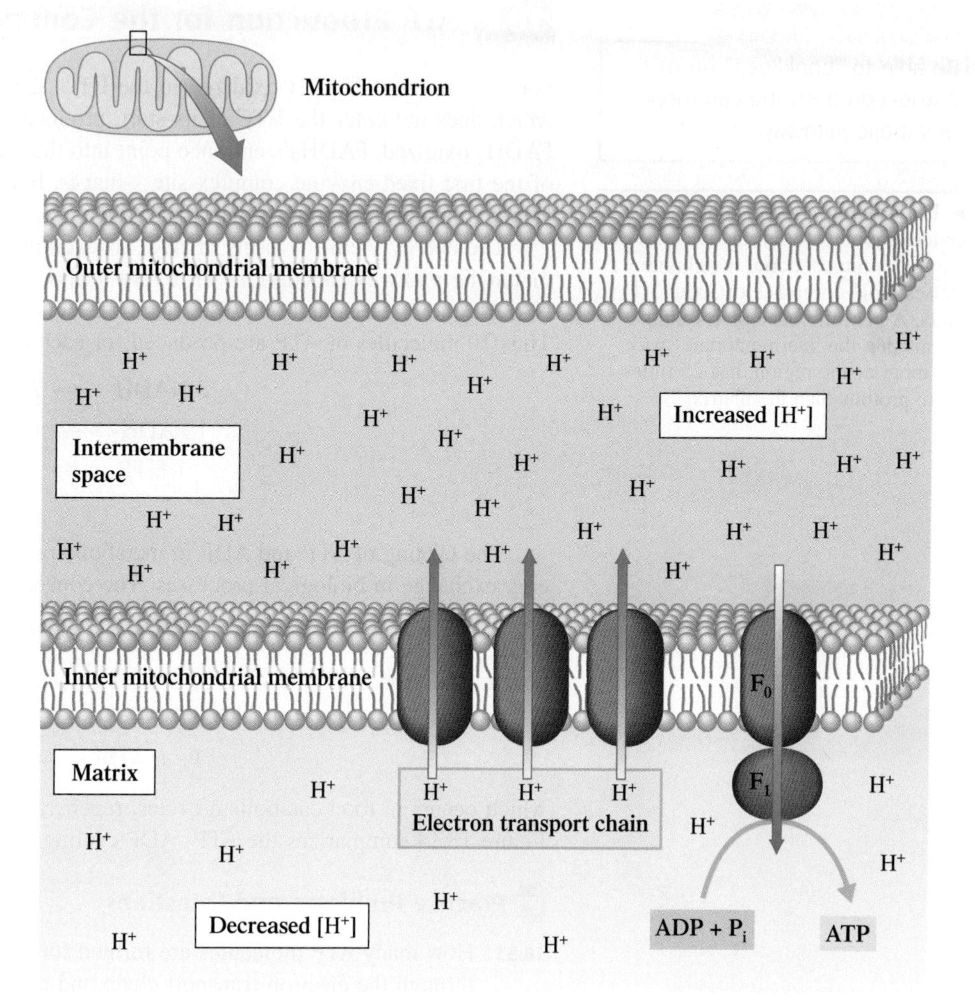

Figure 18.13 Formation of ATP accompanies the flow of protons from the intermembrane space back into the mitochondrial matrix. The proton flow results from an electrochemical gradient across the inner mitochondrial membrane.

Practice Problems and Questions

18.44 The coupling of ATP synthesis with the reactions of the electron transport chain is related to the movement of what chemical species across the inner mitochondrial membrane?

18.45 At what enzyme location(s) in the electron transport chain does proton pumping occur?

18.46 At what mitochondrial location does H^+ ion buildup occur as the result of proton pumping?

18.47 How many protons cross the inner mitochondrial membrane for every 2 electrons that are passed through the electron transport chain?

18.48 The "gradient" of protons across the inner mitochondrial membrane is called both an electrical gradient and a chemical gradient. Explain why this is so.

18.49 What is the name of the enzyme that catalyzes ATP production during oxidative phosphorylation?

18.50 Where is the enzyme located that uses stored energy in a proton gradient to drive the reaction that produces ATP?

18.51 How is the proton gradient that is associated with ATP synthesis dissipated?

18.52 What are the "starting materials" (reactants) from which ATP is synthesized as the proton gradient is dissipated?

18.8 ATP Production for the Common Metabolic Pathway

Learning Focus

Be able to "bookkeep" on ATP production from the common metabolic pathway.

▶ The difference in H^+ ion concentration between the two sides of the inner mitochondrial membrane causes a pH difference of about 1.4 units. A pH difference of 1.4 units means that the intermembrane space, the more acidic region, has 25 times more protons than the matrix.

For each mole of NADH oxidized in the ETC, 2.5 moles of ATP are formed. $FADH_2$, which does not enter the ETC at its start, produces only 1.5 moles of ATP per mole of $FADH_2$ oxidized. $FADH_2$'s entrance point into the chain, coenzyme Q, is beyond the site of the first fixed enzyme complex site—that is, beyond the first "proton-pumping" site. Hence less ATP is produced than for NADH.

The energy yield, in terms of ATP production, can now be totaled for the common metabolic pathway (Section 18.4). Every acetyl CoA entering the CAC produces three NADH, one $FADH_2$, and one GTP (which is equivalent in energy to ATP; Section 18.5). Thus 10 molecules of ATP are produced for each acetyl CoA catabolized.

$$\begin{aligned} 3\ \text{NADH} &\longrightarrow 7.5\ \text{ATP} \\ 1\ \text{FADH}_2 &\longrightarrow 1.5\ \text{ATP} \\ 1\ \text{GTP} &\longrightarrow \underline{\ \ 1\ \text{ATP}} \\ & \qquad 10\ \text{ATP} \end{aligned}$$

The cycling of ATP and ADP in metabolic processes is the principal medium for energy exchange in biological processes. The conversion

$$\text{ATP} \longrightarrow \text{ADP} + \text{P}_i$$

powers life processes (the biosynthesis of essential compounds, muscle contraction, nutrient transport, and so on). The conversion

$$\text{P}_i + \text{ADP} \longrightarrow \text{ATP}$$

which occurs in food catabolism cycles, regenerates the ATP expended in cell operation. Figure 18.14 summarizes the ATP–ADP cycling process.

Practice Problems and Questions

18.53 How many ATP molecules are formed for each NADH molecule that is processed through the electron transport chain and oxidative phosphorylation?

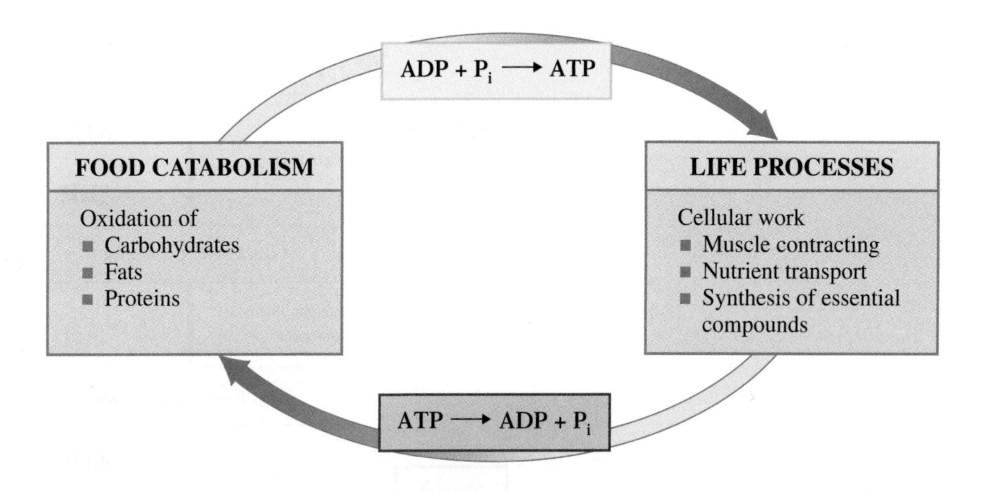

Figure 18.14 The interconversion of ATP and ADP is the principal medium for energy exchange in biological processes.

18.54 How many ATP molecules are formed for each $FADH_2$ molecule that is processed through the electron transport chain and oxidative phosphorylation?

18.55 NADH and FAD_2 molecules do not yield the same number of ATP molecules. Explain why this is so.

18.56 How many ATP molecules are ultimately produced when an acetyl CoA passes through the common metabolic pathway?

Learning Focus

Describe, in general terms, the reactions involved in the conversion of glucose to pyruvate.

18.9 Glycolysis: Oxidation of Glucose

The citric acid cycle (Section 18.5), the electron transport chain (Section 18.6), and oxidative phosphorylation (Section 18.7) together constitute the *common metabolic pathway*. The "fuel" for operation of this pathway is acetyl CoA produced from amino acids (proteins), fatty acids and glycerol (fats), and monosaccharides (carbohydrates), the molecules obtained from food digestion. In this section we consider *how* acetyl CoA is obtained from food digestion products such as those just listed; this is Stage 2 of biochemical energy production; (Section 18.4).

A *common* pathway for acetyl CoA production does not exist. As previously shown in Figure 18.7, protein, fat, and carbohydrate digestion products are processed in different ways to obtain acetyl CoA. In this section, as representative of such processing, we look at details concerning how acetyl CoA is obtained from glucose, the main carbohydrate digestion product. (We will not consider in this text the pathways by which proteins and fats are converted to acetyl CoA.)

Our discussion of acetyl CoA production from glucose is divided into two parts. In the remainder of this section, the processing of C_6 glucose molecules into C_3 pyruvate molecules—a process called glycolysis—is considered. In the following section, we consider the processing of C_3 pyruvate molecules into C_2 acetyl CoA molecules.

▶ The term *glycolysis,* pronounced "gligh-KOLL-ih-sis," comes from the Greek *glyco,* meaning "sweet," and *lysis,* meaning "breakdown."

▶ *Pyruvate,* pronounced "PIE-roo-vate," is the carboxylate ion (Section 13.7) produced when pyruvic acid (a three-carbon keto acid) loses its acidic hydrogen atom.

$$\underset{\text{Pyruvic acid}}{CH_3-C(=O)-COOH} \longrightarrow \underset{\text{Pyruvate ion}}{CH_3-C(=O)-COO^-} + H^+$$

▶ *Anaerobic* is pronounced "AN-air-ROE-bic." *Aerobic* is pronounced "air-ROE-bic."

Glycolysis

Glycolysis *is the metabolic pathway by which glucose (a C_6 molecule) is converted into two molecules of pyruvate (a C_3 molecule).* This metabolic pathway functions in almost all cells.

The conversion of glucose to pyruvate is an oxidation process in which no molecular oxygen is utilized. The oxidizing agent is the coenzyme NAD^+. Metabolic pathways in which molecular oxygen is not a participant are called *anaerobic* pathways. Pathways that require molecular oxygen are called *aerobic* pathways. Glycolysis is an anaerobic pathway.

Glycolysis is a ten-step process (compared to the eight steps of the citric acid cycle; Section 18.5) in which every step is enzyme catalyzed. Figure 18.15 gives an overview

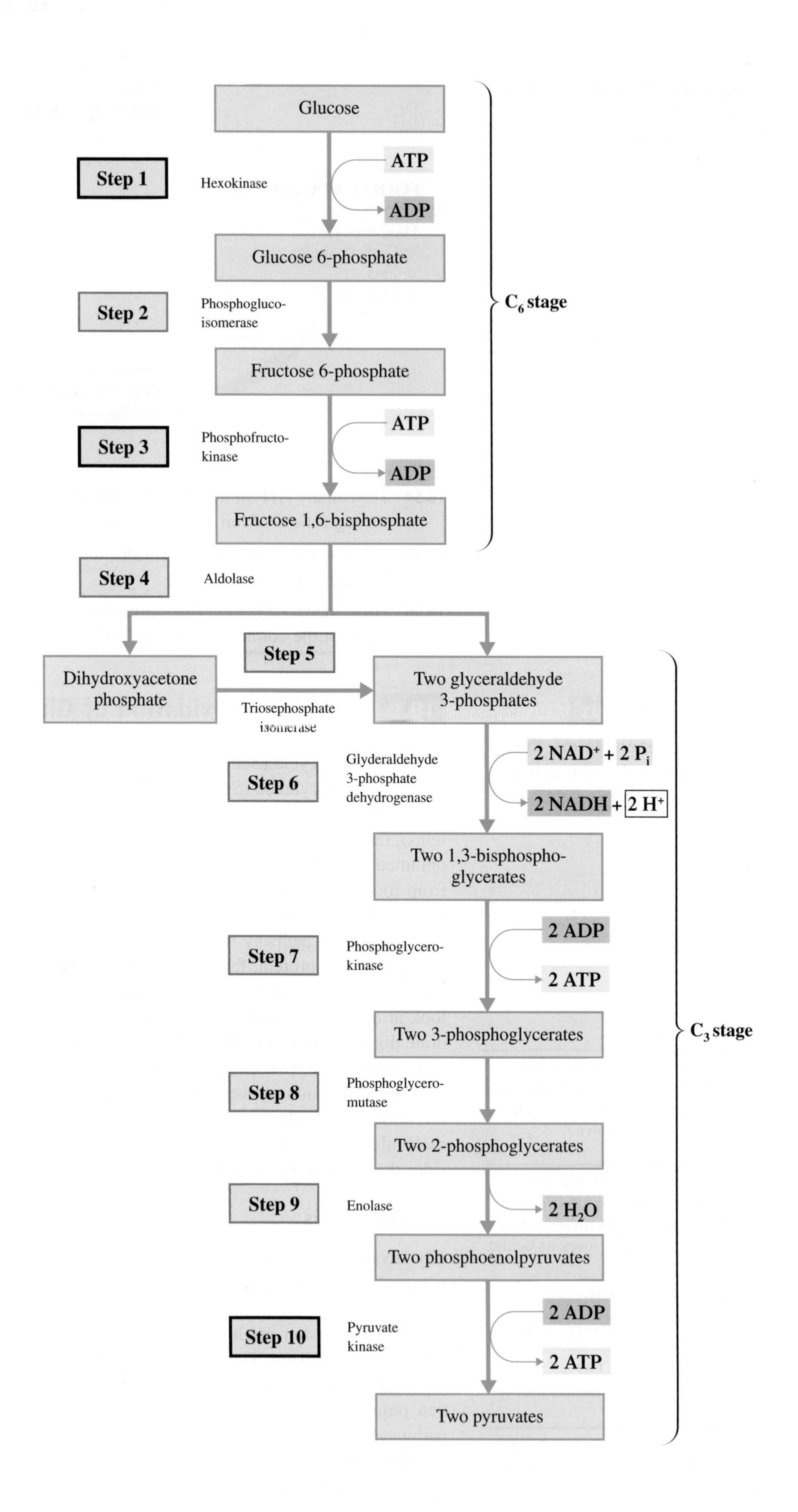

Figure 18.15 An overview of glycolysis.

of glycolysis. There are two stages in the overall process, a *six-carbon stage* (Steps 1–3) and a *three-carbon stage* (Steps 4–10). All of the enzymes needed for glycolysis are present in the cell cytoplasm (Section 18.2), which is where glycolysis takes place. Details of the individual steps within the glycolysis pathway are now considered.

Six-Carbon Stage of Glycolysis (Steps 1–3)

The intermediates of the six-carbon stage of glycolysis are all either *glucose* or *fructose* derivatives in which phosphate groups are present.

▶ A *kinase* is an enzyme that catalyzes the transfer of a phosphoryl group (PO_3^{2-}) from ATP (or some other high-energy phosphate compound) to a substrate.

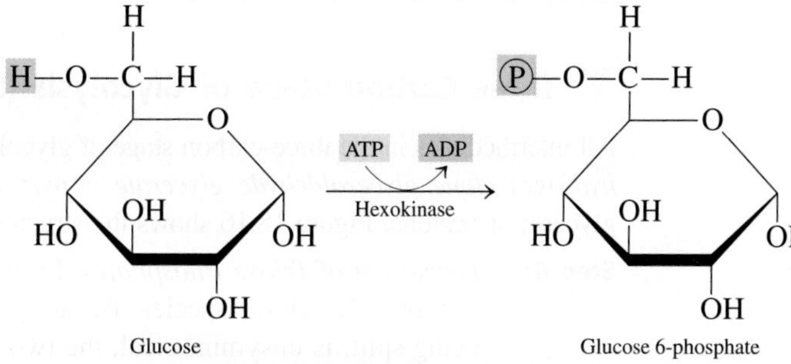

Step 1: *Formation of Glucose 6-phosphate.* Glycolysis begins with the phosphorylation of glucose to yield glucose 6-phosphate, a glucose molecule with a phosphate group attached to the hydroxyl oxygen on carbon 6 (the carbon atom outside the ring). The phosphate group is from an ATP molecule. *Hexokinase,* an enzyme that requires Mg^{2+} ion for its activity, catalyzes the reaction.

Glucose —(ATP → ADP; Hexokinase)→ Glucose 6-phosphate

The symbol Ⓟ is a shorthand notation for a PO_3^{2-} unit.

This reaction requires energy, which is provided by the breakdown of an ATP molecule. This energy expenditure will be recouped later in the cycle. Phosphorylation of glucose provides a way of "trapping" glucose within a cell. Glucose can cross cell membranes, but glucose 6-phosphate cannot.

Step 2: *Formation of Fructose 6-phosphate.* Glucose 6-phosphate is isomerized to fructose 6-phosphate by *phosphoglucoisomerase.*

Glucose 6-phosphate ⇌(Phosphoglucoisomerase) Fructose 6-phosphate

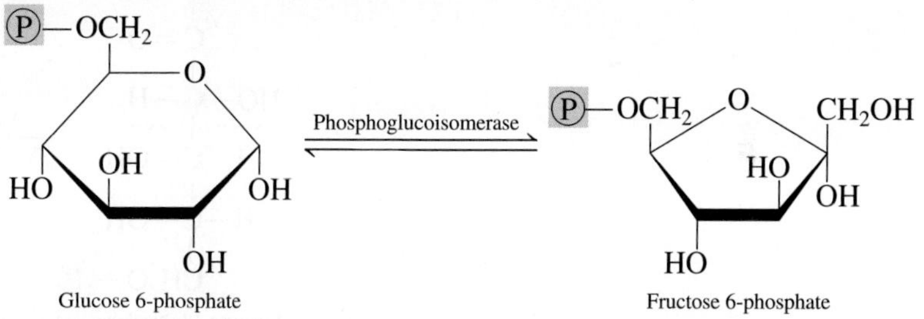

The net result of this change is that carbon 1 of glucose is no longer part of the ring structure. (Glucose, an aldose, forms a six-membered ring, and fructose, a ketose, forms a five-membered ring [Section 14.7]; both sugars, however, contain six carbon atoms.)

▶ Step 3 of glycolysis commits the original glucose molecule to the glycolysis pathway. Glucose 6-phosphate (Step 1) and fructose 6-phosphate (Step 2) can enter other metabolic pathways, but fructose 1,6-bisphosphate can only enter glycolysis.

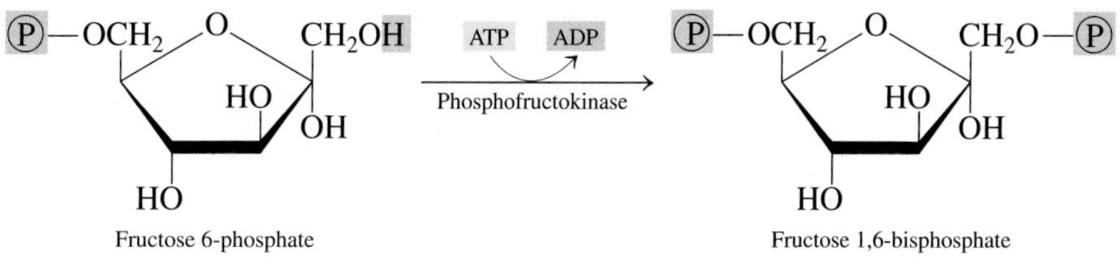

Step 3: *Formation of Fructose 1,6-bisphosphate.* This step, like Step 1, is a phosphorylation reaction and therefore requires the expenditure of energy. ATP is the source of the phosphate and the energy. The enzyme involved, *phosphofructokinase,* is another enzyme that requires Mg^{2+} ion for its activity. The fructose molecule now contains two phosphate groups.

Fructose 6-phosphate —(ATP → ADP; Phosphofructokinase)→ Fructose 1,6-bisphosphate

Figure 18.16 Structural relationships among glycerol and acetone and the C_3 intermediates in the process of glycolysis.

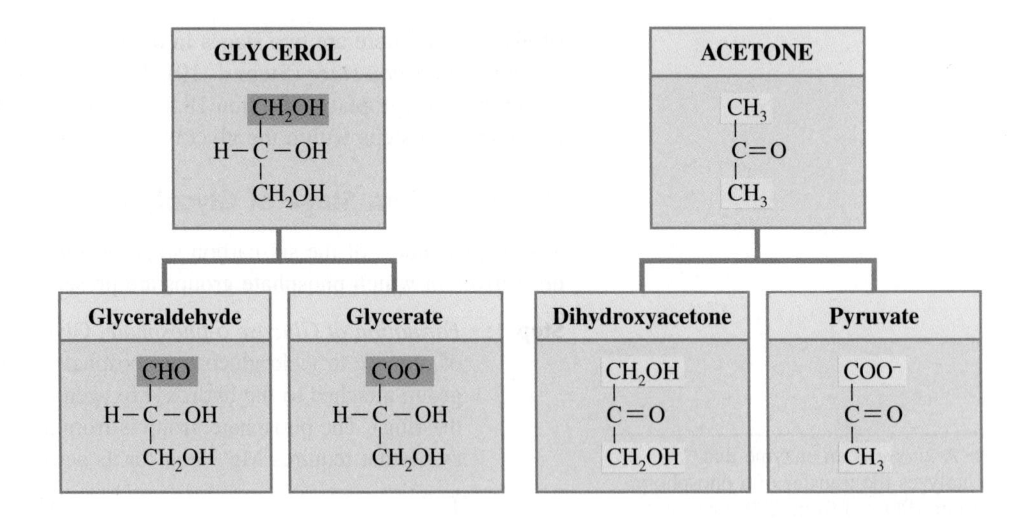

Three-Carbon Stage of Glycolysis (Steps 4–10)

All intermediates in the three-carbon stage of glycolysis are phosphorylated derivatives of *dihydroxyacetone, glyceraldehyde, glycerate,* or *pyruvate,* which in turn are derivatives of either glycerol or acetone. Figure 18.16 shows the structural relationships among these molecules.

Step 4: *Formation of Triose Phosphates.* In this step, the reacting C_6 species is split into two C_3 (triose) species. Because fructose 1,6-bisphosphate, the molecule being split, is unsymmetrical, the two trioses produced are not identical. One product is dihydroxyacetone phosphate, and the other is glyceraldehyde 3-phosphate. *Aldolase* is the enzyme that catalyzes this reaction. A better understanding of the structural relationships between reactant and products is obtained if the fructose 1,6-bisphosphate is written in its open-chain form (Section 14.7) rather than in its cyclic form.

Fructose 1,6-bisphosphate (open-chain form) $\xrightleftharpoons{\text{Aldolase}}$ Dihydroxyacetone phosphate + Glyceraldehyde 3-phosphate

Step 5: *Isomerization of Triose Phosphates.* Only one of the two trioses produced in Step 4, glyceraldehyde 3-phosphate, is a glycolysis intermediate. Dihydroxyacetone phosphate, the other triose, can, however, be readily converted into glyceraldehyde 3-phosphate. Dihydroxyacetone phosphate (a ketose) and glyceraldehyde 3-phosphate (an aldose) are isomers, and the isomerization process from ketose to aldose is catalyzed by the enzyme *triosephosphate isomerase*.

Dihydroxyacetone phosphate $\xrightarrow{\text{Triosephosphate isomerase}}$ Glyceraldehyde 3-phosphate

Step 6: *Formation of 1,3-Bisphosphoglycerate*. In a reaction catalyzed by *glyceraldehyde 3-phosphate dehydrogenase,* a phosphate group is added to glyceraldehyde 3-phosphate to produce 1,3-bisphosphoglycerate. The hydrogen of the aldehyde group becomes part of NADH.

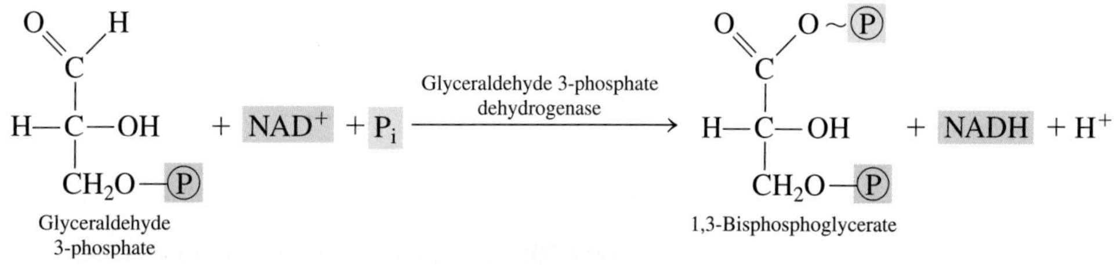

Glyceraldehyde 3-phosphate + NAD^+ + P_i —(Glyceraldehyde 3-phosphate dehydrogenase)→ 1,3-Bisphosphoglycerate + NADH + H^+

▶ Keep in mind that from Step 6 onward, two molecules of each of the C_3 compounds take part in every reaction for each original C_6 glucose molecule.

The newly added phosphate group in 1,3-bisphosphoglycerate is a high-energy phosphate group. A high-energy phosphate group is produced when a phosphate group is attached to a carbon atom that is also participating in a carbon–carbon or carbon–oxygen double bond.

Note that a molecule of the reduced coenzyme NADH is a product of this reaction and also that the source of the added phosphate is inorganic phosphate (P_i).

Step 7: *Formation of 3-Phosphoglycerate*. In this step, the diphosphate species just formed is converted back to a monophosphate species. This is an ATP-producing step in which the C-1 phosphate group of 1,3-bisphosphoglycerate (the high-energy phosphate) is transferred to an ADP molecule to form the ATP. The enzyme involved is *phosphoglycerokinase*.

1,3-Bisphosphoglycerate ⇌ (ADP → ATP; Phosphoglycerokinase) 3-Phosphoglycerate

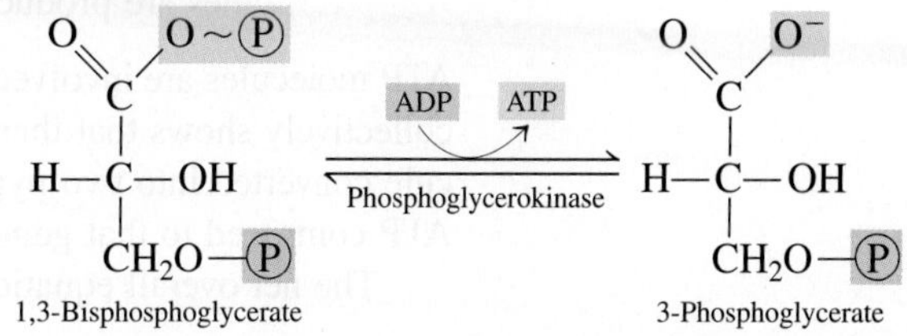

Remember that two ATP molecules are produced for each original glucose molecule, because both C_3 molecules produced from the glucose react.

ATP production in this step involves substrate-level phosphorylation. **Substrate-level phosphorylation** *is the direct transfer of a high-energy phosphate group from an intermediate compound (substrate) to an ADP molecule to produce ATP*. Substrate-level phosphorylation differs from oxidative phosphorylation (Section 18.7) in that the latter process involves the transfer of free phosphate ions in solution (P_i) to ADP molecules to form ATP.

▶ A *mutase* is an enzyme that effects the shift of a phosphoryl group (PO_3^{2-}) from one oxygen atom to another within a molecule.

Step 8: *Formation of 2-Phosphoglycerate*. In this isomerization step, the phosphate group of 3-phosphoglycerate is moved from carbon 3 to carbon 2. The enzyme *phosphoglyceromutase* catalyzes the exchange of the phosphate group between the two carbons.

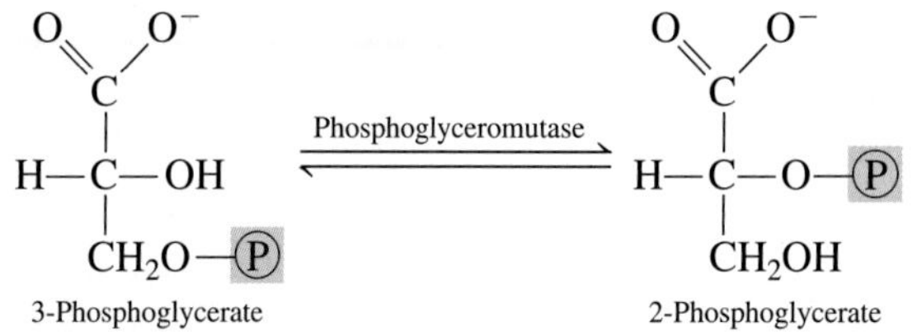

3-Phosphoglycerate ⇌ (Phosphoglyceromutase) 2-Phosphoglycerate

▶ An *enol* (from *ene* + *ol*), as in phospho*enol*pyruvate, is a compound in which an —OH group is attached to a carbon atom involved in a carbon–carbon double bond. Note that in phosphoenolpyruvate the —OH group has been phosphorylated.

Step 9: *Formation of Phosphoenolpyruvate*. This is an alcohol dehydration reaction that proceeds with the enzyme *enolase,* another Mg^{2+}-requiring enzyme. The result is another compound containing a high-energy phosphate group; the phosphate group is attached to a carbon atom that is involved in a carbon–carbon double bond.

$$\text{2-Phosphoglycerate} \xrightleftharpoons{\text{Enolase}} \text{Phosphoenolpyruvate} + \text{HOH}$$

Step 10: *Formation of Pyruvate.* In this step, substrate-level phosphorylation again occurs. Phosphoenolpyruvate transfers its high-energy phosphate group to an ADP molecule to produce ATP and pyruvate.

$$\text{Phosphoenolpyruvate} \xrightarrow[\text{Pyruvate kinase}]{\text{ADP} \quad \text{ATP}} \text{Pyruvate}$$

The enzyme involved, *pyruvate kinase,* requires both Mg^{2+} and K^{+} ions for its activity. Again, because two C_3 molecules are reacting, two ATP molecules are produced.

ATP molecules are involved in Steps 1, 3, 7, and 10 of glycolysis. Considering these steps collectively shows that there is a net gain of two ATP molecules for every glucose molecule converted into two pyruvates (Table 18.1). Though useful, this is a small amount of ATP compared to that generated in oxidative phosphorylation (Section 18.8).

The net overall equation for the process of glycolysis is

$$\text{Glucose} + 2\ \text{NAD}^+ + 2\ \text{ADP} + 2\ \text{P}_i \longrightarrow 2\ \text{pyruvate} + 2\ \text{NADH} + 2\ \text{ATP} + 2\text{H}^+ + 2\text{H}_2\text{O}$$

Practice Problems and Questions

18.57 In the process of glycolysis, indicate the
a. starting material b. end product

18.58 What is the product of the first step in glycolysis, and why is it important in retaining glucose inside the cell?

18.59 In Step 4 of the glycolysis pathway, the C_6 chain is broken into two C_3 fragments.
a. What are the names of these two C_3 molecules?
b. Only one of the C_3 molecules can be further degraded. What happens to the other C_3 molecule?

Table 18.1 ATP Production and Consumption During Glycolysis

Step	Reaction	ATP change per glucose
1	Glucose ⟶ glucose 6-phosphate	−1
3	Fructose 6-phosphate ⟶ fructose 1,6-bisphosphate	−1
7	2(1,3-Bisphosphoglycerate ⟶ 3-phosphoglycerate)	+2
10	2(Phosphoenolpyruvate ⟶ pyruvate)	+2
		Net +2

18.60 Consider the process of glycolysis.
a. How many pyruvate molecules are produced per glucose molecule?
b. How many steps produce ATP?
c. How many steps consume ATP?
d. How many NADH molecules are produced per glucose molecule?

18.61 In the process of glycolysis, indicate how many steps involve
a. a phosphate-containing reactant
b. a reactant containing two phosphate groups
c. phosphorylation (addition of a phosphate group)
d. dephosphorylation (removal of a phosphate group)

18.62 At what location in a cell does glycolysis occur?

18.63 With the help of Figure 18.15, replace the question mark in each of the following word equations with the name of a substance.
a. Glucose + ATP $\xrightarrow{\text{Hexokinase}}$? + ADP + H^+
b. ? $\xrightarrow{\text{Enolase}}$ phosphoenolpyruvate + water
c. 3-Phosphoglycerate $\xrightarrow{?}$ 2-phosphoglycerate
d. 1,3-Bisphosphoglycerate + ? $\xrightarrow[\text{kinase}]{\text{Phosphoglycero-}}$ 3-phosphoglycerate + ATP

18.64 With the help of Figure 18.15, determine the step in glycolysis where each of the following occurs?
a. Second substrate-level phosphorylation reaction
b. First ATP-consuming reaction
c. Third isomerization reaction
d. First ATP-producing reaction

18.65 Write the equation for the net reaction that occurs during glycolysis.

Learning Focus

Give the conditions associated with each of the three fates for pyruvate.

18.10 Fates of Pyruvate

The pyruvate produced by glycolysis has three common fates, which are shown in Figure 18.17. The pyruvate's fate depends on cellular conditions (aerobic or anaerobic) and on the particular organism involved. Some microorganisms process pyruvate in a manner different from that in humans.

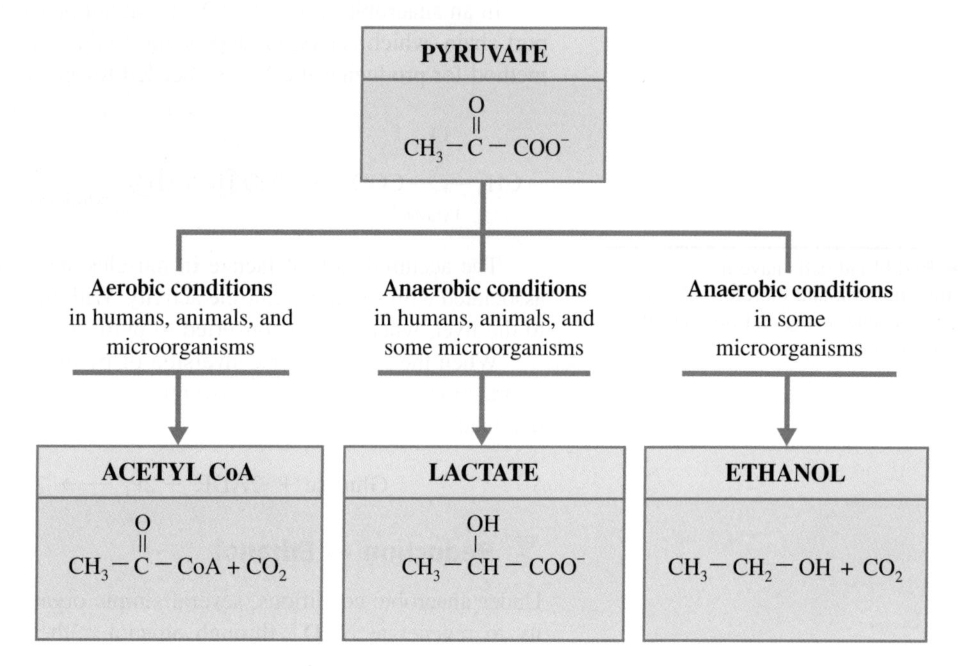

Figure 18.17 The three common fates of pyruvate generated by glycolysis.

A common feature of all three pyruvate metabolic pathways is a provision for regeneration of NAD^+ from NADH; such regenerated NAD^+ is needed for the continuous operation of the glycolysis pathway.

Oxidation to Acetyl CoA

Under *aerobic* (oxygen-rich) conditions, pyruvate is oxidized to acetyl CoA. Pyruvate formed in the cytoplasm through glycolysis crosses the two mitochondrial membranes and enters the matrix, where the oxidation takes place. The overall reaction, in simplified terms, is

$$\underset{\text{Pyruvate}}{CH_3-\overset{\overset{\displaystyle O}{\|}}{C}-COO^-} + CoA-SH + NAD^+ \xrightarrow[\text{complex}]{\text{Pyruvate dehydrogenase}} \underset{\text{Acetyl CoA}}{CH_3-\overset{\overset{\displaystyle O}{\|}}{C}-S-CoA} + NADH + CO_2$$

This reaction, which involves both oxidation and decarboxylation (CO_2 is produced), is far more complex than the simple stoichiometry of the equation suggests. The enzyme complex involved contains three different enzymes, each with numerous subunits. The overall reaction process involves four separate steps and requires NAD^+, CoA—SH, FAD, and two other coenzymes (lipoic acid and thiamin pyrophosphate, the latter derived from the B vitamin thiamin).

▶ Not all acetyl CoA produced from pyruvate enters the citric acid cycle. Particularly when high levels of acetyl CoA are produced (from excess ingestion of dietary carbohydrates), some acetyl CoA is used as the starting material for the production of the fatty acids needed for fat (triacylglycerol) formation.

Most acetyl CoA molecules produced from pyruvate enter the citric acid cycle. Citric acid cycle operations change more NAD^+ to its reduced form, NADH. The NADH from glycolysis, from the conversion of pyruvate to acetyl CoA, and from the citric acid cycle enters the electron transport chain directly or indirectly (Section 18.6). In the ETC, electrons from NADH are transferred to O_2, and the NADH is changed back to NAD^+. The NAD^+ needed for glycolysis, pyruvate–acetyl CoA conversion, and the citric acid cycle has been regenerated.

The net overall equation for processing glucose to two molecules of acetyl CoA is

$$\text{Glucose} + 2\ ADP + 2\ P_i + 4\ NAD^+ + 2\ CoA \longrightarrow 2\ \text{acetyl CoA} + 2CO_2 + 2\ ATP + 4\ NADH + 4H^+ + 2H_2O$$

Reduction to Lactate

Acetyl CoA production is associated with oxygen-rich conditions. In an oxygen-poor environment (anaerobic conditions), pyruvate is reduced to lactate in humans and many other organisms. In humans, oxygen-poor conditions often accompany strenuous muscle activity.

In an anaerobic situation, NADH cannot be oxidized to NAD^+ by the electron transport chain, which is oxygen-dependent. Reduction of pyruvate to lactate is an alternative method for producing the NAD^+ needed for glycolysis.

$$\underset{\text{Pyruvate}}{CH_3-\overset{\overset{\displaystyle O}{\|}}{C}-COO^-} + NAD(H) + (H^+) \xrightarrow[\text{dehydrogenase}]{\text{Lactate}} \underset{\text{Lactate}}{CH_3-\overset{\overset{\displaystyle O(H)}{|}}{C}(H)-COO^-} + NAD^+$$

▶ Red blood cells have no mitochondria and therefore always form lactate as the end product of glycolysis.

The accumulation of lactate in muscles and in the blood is what causes the fatigue associated with strenuous muscle activity. With time, the lactate so produced is transported to the liver, where it is reconverted to pyruvate.

When the reaction for conversion of pyruvate to lactate is added to the net glycolysis reaction (Section 18.9), an overall reaction for the conversion of glucose to lactate is obtained.

$$\text{Glucose} + 2ADP + 2P_i \longrightarrow 2\text{lactate} + 2ATP + 2H_2O$$

Reduction to Ethanol

Under anaerobic conditions, several simple organisms, including yeast, possess the ability to regenerate NAD^+ through ethanol, rather than lactate, production. Such a process is called alcohol fermentation. **Alcohol fermentation** *is the enzymatic conversion of* *pyruvate to ethanol and carbon dioxide.* Fermentation causes bread and related products to rise as a result of CO_2 bubbles being released during baking. Beer, wine, and other alcoholic drinks are produced by fermentation of the sugars in grain and fruit products.

The first step in conversion of pyruvate to ethanol is a decarboxylation reaction to produce acetaldehyde.

$$\underset{\text{Pyruvate}}{CH_3-\overset{\overset{\large O}{\|}}{C}-COO^-} + H^+ \xrightarrow[\text{decarboxylase}]{\text{Pyruvate}} \underset{\text{Acetaldehyde}}{CH_3-\overset{\overset{\large O}{\|}}{C}-H} + CO_2$$

The second step involves acetaldehyde reduction to produce ethanol.

$$\underset{\text{Acetaldehyde}}{CH_3-\overset{\overset{\large O}{\|}}{C}-H} + \boxed{NAD\text{Ⓗ}} + \boxed{\text{Ⓗ}^+} \xrightarrow[\text{dehydrogenase}]{\text{Alcohol}} \underset{\text{Ethanol}}{CH_3-\overset{\overset{\large O\text{Ⓗ}}{|}}{\underset{\underset{\text{Ⓗ}}{|}}{C}}-H} + \boxed{NAD^+}$$

The overall equation for the conversion of pyruvate to ethanol (the sum of the two steps) is

$$\underset{\text{Pyruvate}}{CH_3-\overset{\overset{\large O}{\|}}{C}-COO^-} + 2H^+ + \boxed{NADH} \xrightarrow{\text{Two steps}} \underset{\text{Ethanol}}{CH_3-CH_2-OH} + \boxed{NAD^+} + CO_2$$

An overall reaction for the production of ethanol from glucose is obtained by combining the reaction for the conversion of pyruvate with the net reaction for glycolysis (Section 18.9).

$$\text{Glucose} + \boxed{2ADP} + 2P_i \longrightarrow 2\text{ethanol} + 2CO_2 + \boxed{2ATP} + 2H_2$$

Practice Problems and Questions

18.66 Indicate to what chemical species pyruvate is processed in the human body under
a. aerobic conditions b. anaerobic conditions

18.67 Write the net chemical equation for the conversion of pyruvate into
a. acetyl CoA b. lactate

18.68 Explain how lactate production allows glycolysis to continue under anaerobic conditions.

18.69 How is alcohol fermentation in yeast similar to lactate production in humans?

18.70 In alcohol fermentation, a C_3 pyruvate molecule is changed to a C_2 ethanol molecule. What is the fate of the third pyruvate carbon atom?

18.71 What are the structural differences between pyruvate and lactate?

18.72 Write the net chemical equation for the conversion of one glucose molecule into
a. two lactate molecules b. two ethanol molecules

Learning Focus

Be able to "bookkeep" on the ATP production for the complete oxidation of glucose.

18.11 ATP Production for the Complete Oxidation of Glucose

The final item we consider in this chapter is *total* ATP production for the complete oxidation of glucose. Here we collect together ATP production figures for glycolysis, oxidation of pyruvate to acetyl CoA, the citric acid cycle, the electron transport chain, and oxidative phosphorylation. The result, with one added piece of information, gives the ATP yield for the *complete* oxidation of one molecule of glucose.

The new piece of information involves the NADH produced during Step 6 of glycolysis. This NADH, produced in the cytoplasm, cannot *directly* participate in the electron transport chain, because mitochondria are impermeable to NADH (and NAD^+). A transport system exists that shuttles the electrons from cytoplasmic NADH, but not NADH itself, across the mitochondrial membrane. The net reaction for this shuttle process is

$$\underset{\text{(cytoplasmic)}}{NADH} + H^+ + \underset{\text{(mitochondrial)}}{FAD} \longrightarrow \underset{\text{(cytoplasmic)}}{NAD^+} + \underset{\text{(mitochondrial)}}{FADH_2}$$

The consequence of this reaction is that only 1.5, rather than 2.5, molecules of ATP are formed for each cytoplasmic NADH, because $FADH_2$ yields one less ATP than does NADH in the electron transport chain.

Table 18.2 shows ATP production for the complete oxidation of a molecule of glucose. The final number is 30 ATP, 26 of which come from the oxidative phosphorylation associated with the electron transport chain. This total of 30 ATP for complete oxidation contrasts markedly with a total of 2 ATP for oxidation of glucose to lactate and 2 ATP for oxidation of glucose to ethanol. Neither of these latter processes involves the citric acid cycle or the electron transport chain. Thus the aerobic oxidation of glucose is 15 times more efficient in the production of ATP than the anaerobic lactate and ethanol processes.

The net overall reaction for the *complete* metabolism of a glucose molecule is the simple equation

$$\text{Glucose} + 6O_2 + 30\ \text{ADP} + 30\ P_i \longrightarrow 6CO_2 + 6H_2O + 30\ \text{ATP}$$

Note that substances such as NADH, NAD^+, and FAD are not part of this equation. Why? They cancel out—that is, they are consumed in one step (reactant) and regenerated in another step (product). Note also what the net equation does not acknowledge: the many dozens of reactions that are needed to generate the 30 molecules of ATP. The accompanying Chemistry at a Glance summarizes the glucose metabolism process in terms of stages or phases and also gives the total number of reactions involved in each stage. The processing of 1 glucose molecule involves 157 reactions; it is simply astounding how complex—and yet how simple—human body chemistry can be.

Table 18.2 Yield of ATP from the Complete Oxidation of One Glucose Molecule in a Skeletal Muscle Cell

Reaction	Comments	Yield of ATP
Glycolysis		
glucose ⟶ glucose 6-phosphate	consumes 1 ATP	−1
glucose 6-phosphate ⟶ fructose 1,6-bisphosphate	consumes 1 ATP	−1
2(glyceraldehyde 3-phosphate ⟶ 1,3-bisphosphoglycerate)	each produces 1 cytoplasmic NADH	—
2(1,3-bisphosphoglycerate ⟶ 3-phosphoglycerate)	each produces 1 ATP	+2
2(phosphoenolpyruvate ⟶ pyruvate)	each produces 1 ATP	+2
Oxidation of Pyruvate		
2(pyruvate ⟶ acetyl CoA + CO_2)	each produces 1 NADH	—
Citric Acid Cycle		
2(isocitrate ⟶ α-ketoglutarate + CO_2)	each produces 1 NADH	—
2(α-ketoglutarate ⟶ succinyl CoA + CO_2)	each produces 1 NADH	—
2(succinyl CoA ⟶ succinate)	each produces 1 GTP	+2
2(succinate ⟶ fumarate)	each produces 1 $FADH_2$	—
2(malate ⟶ oxaloacetate)	each produces 1 NADH	—
Electron Transport Chain and Oxidative Phosphorylation		
2 cytoplasmic NADH formed in glycolysis	each yields 1.5 ATP	+3
2 NADH formed in the oxidation of pyruvate	each yields 2.5 ATP	+5
2 $FADH_2$ formed in the citric acid cycle	each yields 1.5 ATP	+3
6 NADH formed in the citric acid cycle	each yields 2.5 ATP	+15
	Net yield of ATP	+30

Reactions Involved in "Processing" a Glucose Molecule for Its Energy Content

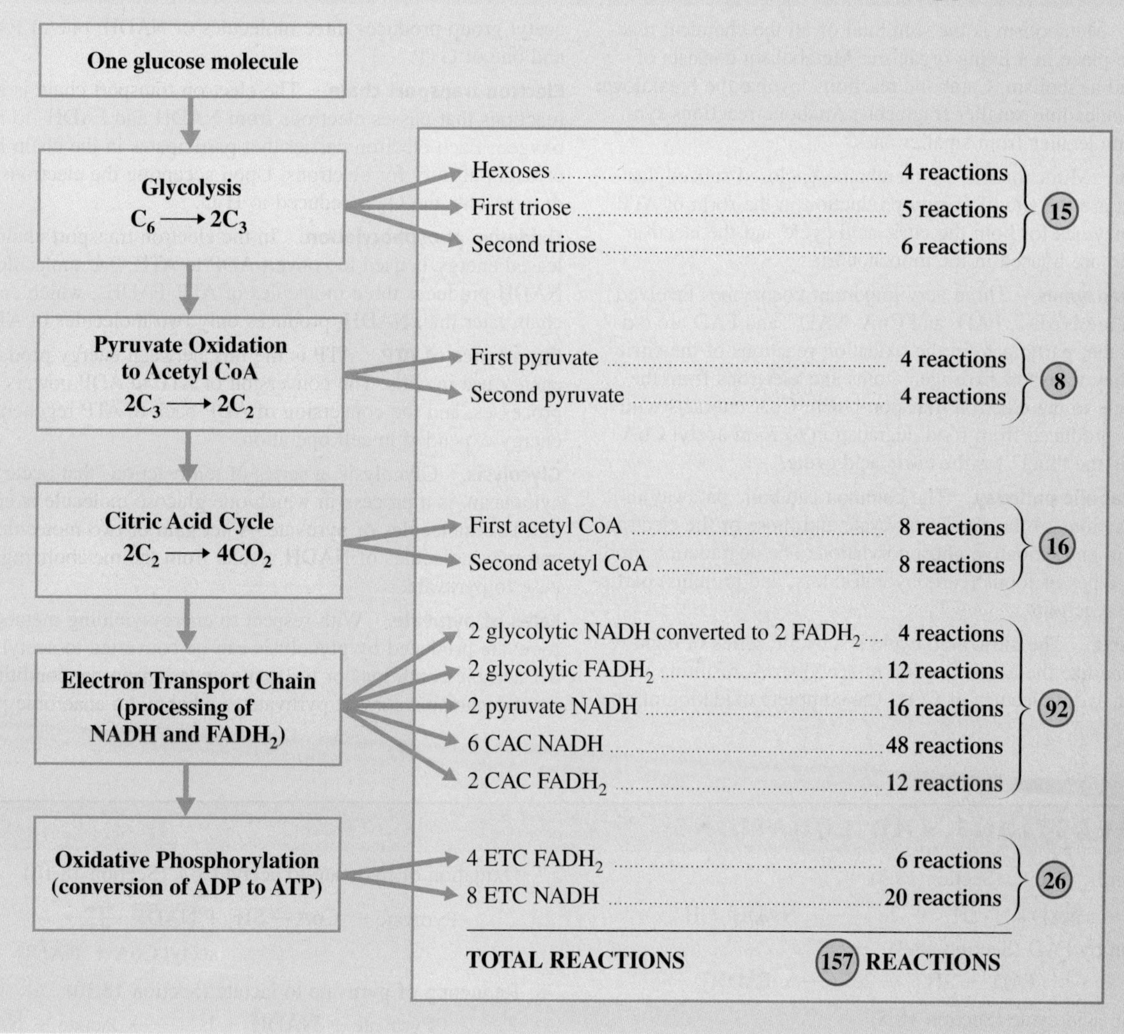

Practice Problems and Questions

18.73 Contrast, in terms of ATP production, the oxidation of glucose to CO_2 and H_2O with the oxidation of glucose to
a. pyruvate b. ethanol

18.74 Indicate how many of the 30 ATP molecules produced from the complete oxidation of 1 molecule of glucose are
a. produced during glycolysis
b. produced from oxidation of pyruvate to acetyl CoA

18.75 How many ATPs are ultimately produced when an acetyl CoA passes through the citric acid cycle?

18.76 Explain the statement "An NADH molecule is equivalent to 2.5 ATPs, and a $FADH_2$ molecule is equivalent to only 1.5 ATPs."

18.77 Indicate how many turns of the citric acid cycle would be required to process the acetyl CoA
a. that comes from 1 glucose molecule
b. generated from 1 pyruvate molecule

CONCEPTS TO REMEMBER

Metabolism. Metabolism is the sum total of all the chemical reactions that take place in a living organism. Metabolism consists of catabolism and anabolism. Catabolic reactions involve the breakdown of large molecules into smaller fragments. Anabolic reactions synthesize large molecules from smaller ones.

Mitochondria. Mitochondria are membrane-enclosed subcellular structures that are the site of energy production in the form of ATP molecules. Enzymes for both the citric acid cycle and the electron transport chain are housed in the mitochondria.

Important coenzymes. Three very important coenzymes involved in catabolism are NAD^+, FAD, and CoA. NAD^+ and FAD are oxidizing agents that participate in the oxidation reactions of the citric acid cycle. They transport hydrogen atoms and electrons from the citric acid cycle to the electron transport chain. CoA interacts with acetyl groups produced from food degradation to form acetyl CoA. Acetyl CoA is the "fuel" for the citric acid cycle.

Common catabolic pathway. The common catabolic pathway includes the reactions of the citric acid cycle and those of the electron transport chain and oxidative phosphorylation. The degradation products from all types of foods (carbohydrates, fats, and proteins) participate in these reactions.

Citric acid cycle. The citric acid cycle is a cyclic series of eight reactions that oxidize the acetyl portion of acetyl CoA, resulting in the production of two molecules of CO_2. The complete oxidation of one acetyl group produces three molecules of NADH, one of $FADH_2$, and one of GTP.

Electron transport chain. The electron transport chain is a series of reactions that passes electrons from NADH and $FADH_2$ to molecular oxygen. Each electron carrier that participates in the chain has an increasing affinity for electrons. Upon accepting the electrons and hydrogen ions, the O_2 is reduced to H_2O.

Oxidative phosphorylation. In the electron transport chain, released energy is used to convert ADP to ATP. One molecule of NADH produces three molecules of ATP. $FADH_2$, which enters the chain later than NADH, produces only two molecules of ATP.

Importance of ATP. ATP is the link between energy production and energy use in cells. The conversion of ATP to ADP powers life processes, and the conversion of ADP back to ATP regenerates the energy expended in cell operation.

Glycolysis. Glycolysis, a series of ten reactions that occur in the cytoplasm, is a process in which one glucose molecule is converted into two molecules of pyruvate. A net gain of two molecules of ATP and two molecules of NADH results from the metabolizing of glucose to pyruvate.

Fates of pyruvate. With respect to energy-yielding metabolism, the pyruvate produced by glycolysis can be converted to acetyl CoA under aerobic conditions or to lactate under anaerobic conditions. Some microorganisms convert pyruvate to ethanol, an anaerobic process.

KEY REACTIONS AND EQUATIONS

1. Oxidation by NAD^+ (Section 18.3)

$$NAD^+ + 2H^+ + 2e^- \longrightarrow NADH + H^+$$

2. Oxidation by FAD (Section 18.3)

$$FAD + 2H^+ + 2e^- \longrightarrow FADH_2$$

3. The citric acid cycle (Section 18.5)

$$\text{Acetyl CoA} + 3NAD^+ + FAD + GDP + P_i + 2H_2O \longrightarrow 2CO_2 + CoA + 3NADH + 2H^+ + FADH_2 + GTP$$

4. The electron transport chain (Section 18.6)

$$NADH + H^+ \longrightarrow NAD^+ + 2H^+ + 2e^-$$

$$FADH_2 \longrightarrow FAD + 2H^+ + 2e^-$$

$$O_2 + 4H^+ + 4e^- \longrightarrow 2H_2O$$

5. Oxidative phosphorylation (Section 18.7)

$$ADP + P_i \xrightarrow[\text{from ETC}]{\text{Energy}} ATP$$

6. Glycolysis (Section 18.9)

$$\text{Glucose} + 2P_i + 2ADP + 2NAD^+ \longrightarrow 2\text{ pyruvate} + 2ATP + 2NADH + 2H^+ + 2H_2O$$

7. Oxidation of pyruvate to acetyl CoA (Section 18.10)

$$\text{Pyruvate} + CoA{-}SH + NAD^+ \xrightarrow[\text{steps}]{\text{Four}} \text{acetyl CoA} + NADH + CO_2$$

8. Reduction of pyruvate to lactate (Section 18.10)

$$\text{Pyruvate} + NADH + H^+ \longrightarrow \text{lactate} + NAD^+$$

9. Reduction of pyruvate to ethanol (Section 18.10)

$$\text{Pyruvate} + 2H^+ + NADH \xrightarrow[\text{steps}]{\text{Two}} \text{ethanol} + NAD^+ + CO_2$$

10. Oxidation of glucose to CO_2 and H_2O (Section 18.11)

$$\text{Glucose} + 6O_2 + 30ADP + 30P_i \longrightarrow 6CO_2 + 6H_2O + 30ATP$$

KEY TERMS

Alcohol fermentation (18.10)
Anabolism (18.1)
Catabolism (18.1)
Citric acid cycle (18.5)
Common metabolic pathway (18.4)
Coupled reactions (18.7)
Cytochrome (18.6)
Electron transport chain (18.6)
Glycolysis (18.9)
Metabolic pathway (18.1)
Metabolism (18.1)
Mitochondrion (18.2)
Organelle (18.2)
Oxidative phosphorylation (18.7)
Substrate-level phosphorylation (18.9)

ADDITIONAL PROBLEMS

18.78 Classify each of the following substances as a reactant in the citric acid cycle, a reactant in the electron transport chain, or a reactant in both the CAC and the ETC.
a. NAD^+ b. NADH c. O_2 d. Fumarate

18.79 Which of these substances—ATP, CoA, FAD, or NAD^+—contains the following subunits of structure? More than one choice may apply in a given situation.
a. Contains two ribose subunits
b. Contains two phosphate subunits
c. Contains one adenine subunit
d. Contains one ribitol subunit

18.80 At which of these locations—first fixed enzyme site, second fixed enzyme site, third fixed enzyme site, or mobile enzyme site—does each of the following ETC reactions occur?
a. $FADH_2$ + CoQ b. NADH + FMN
c. cyt a_3 + O_2 d. FeSP + CoQ

18.81 Classify each of the following substances as a reactant in the citric acid cycle, a reactant in glycolysis, or a reactant in both the citric acid cycle and glycolysis.
a. Oxaloacetate b. Dihydroxyacetone phosphate
c. 2-Phosphoglycerate d. Succinate

18.82 Classify each of the substances in Problem 18.80 as a C_6 molecule, a C_5 molecule, a C_4 molecule, or a C_3 molecule.

18.83 What is the difference between oxidative phosphorylation and substrate-level phosphorylation?

18.84 In the processing of 1 glucose molecule, what is the yield of ATP molecules from each of the following?
a. NADH produced during glycolysis
b. NADH produced during the oxidation of pyruvate to acetyl CoA
c. NADH produced during the citric acid cycle
d. $FADH_2$ produced during the citric acid cycle

18.85 Indicate how many reaction steps there are in
a. glycolysis
b. the citric acid cycle
c. the electron transport chain
d. the oxidation of pyruvate to acetyl CoA

PRACTICE TEST ▶ True/False

18.86 In catabolic reactions, large molecules are broken down into smaller molecules.

18.87 In mitochondria, both the matrix and the intermembrane space interface with the inner membrane.

18.88 ADP, NAD^+, and $FADH_2$ all have structures that contain two phosphate groups.

18.89 Both ATP and coenzyme A have structures that contain one of the B vitamins.

18.90 Glycolysis and the citric acid cycle are both components of the common metabolic pathway.

18.91 The intermediates in the citric acid cycle are all either C_6 or C_5 molecules.

18.92 CO_2 is produced in the citric acid cycle and also in the oxidation of pyruvate to acetyl CoA.

18.93 The first product in the citric acid cycle is citrate, and the first product in glycolysis is glucose 6-phosphate.

18.94 The "fuel" for the citric acid cycle is acetyl CoA.

18.95 In the electron transport chain, all of the enzymes are located at fixed sites.

18.96 Cytochromes are enzymes that contain iron atoms.

18.97 One of the end products of the electron transport chain is molecular O_2.

18.98 Most of the intermediates in glycolysis are C_3 molecules.

18.99 The amount of ATP produced during glycolysis exceeds that produced during operation of the electron transport chain and oxidative phosphorylation.

18.100 In the human body, under oxygen-deficient conditions, pyruvate is reduced to ethanol rather than oxidized to acetyl CoA.

PRACTICE TEST ▶ Multiple Choice

18.101 The correct notation for the reduced form of nicotinamide adenine dinucleotide is
a. NAD^+ b. NAD c. NADH d. $NADH_2$

18.102 Which of the following is a correct skeletal equation for a hydrolysis reaction involving adenosine phosphates?
a. ATP + H_2O $\longrightarrow$ ADP + 2 P_i
b. ADP + H_2O $\longrightarrow$ AMP + 2 P_i
c. ADP + H_2O $\longrightarrow$ ATP + P_i
d. ADP + H_2O $\longrightarrow$ AMP + P_i

18.103 Which of the following is a correct general description of the reaction that occurs in the first step of the citric acid cycle?
a. $C_2 + C_4 \longrightarrow C_6$
b. $C_3 + C_3 \longrightarrow C_6$
c. $C_2 + C_2 + C_2 \longrightarrow C_6$
d. $C_4 + C_4 \longrightarrow C_6 + C_2$

18.104 The first two intermediates in the citric acid cycle are, respectively,
a. isocitrate and α-ketoglutarate
b. citrate and α-ketoglutarate
c. citrate and isocitrate
d. isocitrate and succinate

18.105 At which step in the electron transport chain does O_2 participate?
a. First step b. Second step
c. Next to last step d. Last step

18.106 How many fixed-enzyme sites are associated with the electron transport chain?
a. One b. Two c. Three d. Four

18.107 In which of the following listings of electron transport chain electron carriers are the electron carriers listed

in the order in which they are encountered in the reactions of the electron transport chain?
a. FeSP, CoQ, FMN b. FMN, CoQ, FeSP
c. FMN, FeSP, CoQ d. CoQ, FMN, FeSP

18.108 How many molecules of ATP result from the entry of 1 molecule of $FADH_2$ into the electron transport chain?
a. 1 b. 1.5 c. 2 d. 2.5

18.109 The first two intermediates in the process of glycolysis are, respectively,
a. glucose 6-phosphate and glucose 1-phosphate
b. glucose 1-phosphate and glucose 6-phosphate
c. glucose 6-phosphate and fructose 6-phosphate
d. glucose 1-phosphate and fructose 1-phosphate

18.110 The net yield of ATP molecules from the complete oxidation of 1 molecule of glucose is
a. 10 b. 26 c. 30 d. 36

Answers to Selected Exercises

Chapter 1 **1.1** (a) matter (b) matter (c) energy (d) energy (e) matter (f) matter **1.3** (a) shape (indefinite and definite) (b) volume (indefinite and definite) **1.5** (a) no (b) no (c) yes (d) yes **1.7** (a) physical (b) physical (c) chemical (d) chemical **1.9** (a) chemical (b) physical (c) chemical (d) physical **1.11** (a) heterogeneous mixture (b) homogeneous mixture (c) pure substance (d) heterogeneous mixture **1.13** (a) true (b) false (c) false (d) true **1.15** (a) A, cannot classify; B, cannot classify; C, compound (b) D, compound; E, cannot classify; F, cannot classify **1.17** (a) true (b) false (c) true (d) true **1.19** (a) yes (b) no (c) no (d) yes **1.21** (a) heteroatomic, diatomic, compound (b) heteroatomic, triatomic, compound (c) homoatomic, diatomic, element (d) heteroatomic, triatomic, compound **1.23** (a) 3, 6 (b) 4, 10 (c) 3, 6 (d) 2, 14 (e) 3, 5 (f) 3, 9 **1.25** (a) XQ (b) XQ_2 (c) Q_3 (d) X_2Q **1.27** (a) element (b) mixture (c) mixture (d) compound **1.29** (a) an element and a compound (b) an element and a compound (c) two elements (d) a single pure substance **1.31** (a) same, both 4 (b) more, 6 and 5 (c) same, both 5 (d) fewer, 13 and 15 **1.33** (a) 2 (N_2, NH_3) (b) 4 (N, H, C, Cl) (c) 110, 5(2 + 6 + 4 + 5 + 5) (d) 56, 4(4 + 3 + 4 + 3)

Chapter 2 **2.1** mass, volume, length, time, temperature, pressure, and concentration **2.3** (a) kilo- (b) milli- (c) micro- (d) deci- **2.5** (a) nanogram, milligram, centigram (b) kilometer, megameter, gigameter (c) picoliter, microliter, deciliter (d) microgram, milligram, kilogram **2.7** (a) inexact (b) exact (c) exact (d) inexact **2.9** Because only one estimated digit may be recorded as part of a measured value. **2.11** (a) the 1 (b) the 0 (c) the 5 (d) the 3 **2.13** (a) yes (4 and 4) (b) no (4 and 3) (c) yes (2 and 2) (d) yes (4 and 4) **2.15** (a) two (b) two (c) two (d) two **2.17** (a) 162 (b) 9.3 (c) 1261 (d) 20.0 **2.19** (a) 1.0×10^{-3} (b) 1.0×10^{3} (c) 6.3×10^{4} (d) 6.3×10^{-4} **2.21** (a) 1 kg/10^3 g and 10^3 g/1 kg (b) 1 nm/10^{-9} m and 10^{-9} m/1 nm (c) 1 mL/10^{-3} L and 10^{-3} L/1 mL (d) 1 cg/10^{-2} g and 10^{-2} g/1 cg **2.23** (a) 60 sec = 1 min (b) 12 in. = 1 ft (c) 2.54 cm = 1.00 in. (d) 454 g = 1.00 lb **2.25** (a) 6.4 g × (1.00 lb/454 g) = 0.014 lb (b) 6.4 lb × (454 g/1.00 lb) = 2900 g (c) 53 cm × (1.00 in./2.54 cm) = 21 in. (d) 3.5 qt × (0.946 L/1.00 qt) = 3.3 L **2.27** 2500 mL × (10^{-3} L/1 mL) = 2.5 L **2.29** 524.5 g/38.72 cm^3 = 13.55 g/cm^3 **2.31** 15 cm^3 × (8.90 g/1 cm^3) = 130 g **2.33** 95°F **2.35** (a) 2.0 calories (b) 1.0 kilocalorie (c) 100 Calories (d) 1000 kilocalories **2.37** (a) 3.00×10^{-3} (b) 9.4×10^{5} (c) 2.35×10^{1} (d) 4.50000×10^{8} **2.39** (a) four significant figures (b) four significant figures (c) three significant figures (d) exact **2.41** (a) two (b) two (c) three (d) three **2.43** −10°C **2.45** An exact number cannot have digits to the right of the decimal point.

Chapter 3 **3.1** (a) electron (b) neutron (c) proton (d) proton **3.3** Protons present in a nucleus give it its positive charge. **3.5** (a) true (b) false (c) false (d) false **3.7** (a) 8 protons, 8 neutrons, 8 electrons (b) 8 protons, 10 neutrons, 8 electrons (c) 20 protons, 24 neutrons, 20 electrons (d) 100 protons, 157 neutrons, 100 electrons **3.9** (a) number of protons or number of electrons (b) number of nucleons (c) number of neutrons (d) number of subatomic particles **3.11** atomic number = 12; mass number = 24 **3.13** (a) 40 protons, 40 electrons, 50 neutrons (b) 40 protons, 40 electrons, 52 neutrons (c) 40 protons, 40 electrons, 55 neutrons (d) 40 protons, 40 electrons, 56 neutrons **3.15** ${}^{12}_{6}C$, ${}^{13}_{6}C$, ${}^{14}_{6}C$ **3.17** 6.95 amu **3.19** (a) Ca (b) Mo (c) Li (d) Sn **3.21** (a) 6 (b) 28.09 amu (c) 39 (d) 9.01 amu **3.23** (a) no (b) no (c) yes (d) yes **3.25** (a) metallic (b) nonmetallic (c) metallic (d) nonmetallic **3.27** (a) 1 (b) 2 (c) 3 (d) 4 **3.29** (a) 2 (b) 2 (c) 2 (d) 2 **3.31** (a) true (b) true (c) false (d) true **3.33** (a) oxygen (b) neon (c) aluminum (d) calcium **3.35** (a) true (b) true (c) true (d) false **3.37** (a) *s* area (b) *d* area (c) *f* area (d) *p* area **3.39** (a) representative element (b) noble-gas element (c) transition element (d) inner-transition element **3.41** spontaneous emission of radiation **3.43** low atomic number, about 1; high atomic number, about 1.5 **3.45** (a) 1/4 (b) 1/64 (c) 1/8 (d) 1/64 **3.47** 3.75 g **3.49** (a) +2, 4 amu (b) −1, 0.00055 amu (c) 0, 0 amu **3.51** They are identical. **3.53** (a) ${}^{10}_{4}Be \longrightarrow {}^{0}_{-1}\beta + {}^{10}_{5}B$ (b) ${}^{77}_{32}Ge \longrightarrow {}^{0}_{-1}\beta + {}^{77}_{33}As$ (c) ${}^{60}_{26}Fe \longrightarrow {}^{0}_{-1}\beta + {}^{60}_{27}Co$ (d) ${}^{25}_{11}Na \longrightarrow {}^{0}_{-1}\beta + {}^{25}_{12}Mg$ **3.55** no change in mass number; increase of 1 unit in atomic number **3.57** (a) alpha decay (b) beta decay **3.59** Alpha does not penetrate; beta penetrates outer layers; gamma completely penetrates. **3.61** alpha, 6 cm with 40,000 collisions; beta, 1000 cm with 2000 collisions **3.63** so that they remain in the body only a short period of time **3.65** yttrium-90, implanted in the body; cobalt-60, external exposure **3.67** (a) ${}^{44}_{20}Ca$ (b) ${}^{9}_{4}Be$ (c) ${}^{110}_{47}Ag$ (d) ${}^{9}_{4}Be$ **3.69** (a) no (b) no (c) yes (d) yes **3.71** All fluorine atoms are identical (only one naturally occurring fluorine isotope exists); more than one isotope exists for iron (55.847 amu is the average mass for all isotopes). **3.73** (a) $1s^22s^22p^1$ (b) $1s^22s^22p^63s^23p^1$ (c) $1s^22s^1$ (d) $1s^22s^2$ **3.75** (a) boron (b) scandium (c) lithium (d) phosphorus **3.77** (a) 54 hr (b) 90 hr (c) 108 hr (d) 126 hr

Chapter 4 **4.1** electron **4.3** Ionic compounds have high melting points and are good conductors of electricity in solution and in the molten state; molecular compounds have lower melting points than ionic compounds and do not conduct electricity. **4.5** (a) IA, one (b) VIIIA, eight (c) IIA, two (d) VIIA, seven **4.7** (a) IIA, $\dot{Mg}\cdot$ (b) IA, K· (c) VA, $\cdot\ddot{P}\cdot$ (d) VIIIA, $:\ddot{Kr}:$ **4.9** (a) Be and Mg (b) N and P (c) O and S (d) Na and K **4.11** most unreactive of all elements **4.13** He has only two valence electrons. **4.15** (a) O^{2-} (b) Mg^{2+} (c) F^{-} (d) Al^{3+} **4.17** (a) 15p, 18e (b) 16p, 18e (c) 12p, 10e (d) 3p, 2e **4.19** (a) four (b) three (c) two (d) one **4.21** (a) +2 (b) −1 (c) −3 (d) +3 **4.23** (a) IIA (b) VIA (c) VA (d) IA **4.25** It becomes a negative ion through electron gain. **4.27** (a) Be: → :Ö: (b) Mg: → :S̈: (c) K·, K·, K· → :N: (d) Ca: → :F̈:, :F̈: **4.29** (a) MgF_2 (b) BeF_2 (c) LiF (d) AlF_3 **4.31** The positive ion always appears first in the chemical formula. **4.33** smallest whole-number repeating ratio of ions present in the ionic compound **4.35** (a) yes (b) yes (c) no (d) yes **4.37** (a) +1 (b) +2 (c) +4 (d) +2 **4.39** (a) gold(I) chloride (b) potassium chloride (c) silver chloride (d) copper(II) chloride **4.41** (a) CoO (b) Co_2O_3 (c) SnI_4 (c) Pb_3N_2 **4.43** ionic, electron transfer; covalent, electron sharing **4.45** (a) :B̈r:B̈r: (b) H:Ï: (c) :Ï:B̈r: (d) :B̈r:F̈: **4.47** (a) H_2O (b) CBr_4 (c) PI_3 (d) SiH_4 **4.49** three terms that designate the same thing **4.51** (a) 0, 0, 1 (b) 1, 0, 1 (c) 0, 2, 0 (d) 1, 0, 0 **4.53** (a) :C:::O: (b) H:N̈:N̈:H with H under each N (c) :Ö:S:Ö: with :Ö: under S (d) H:C:::C:H **4.55** (a) H:P̈:H with H under P (b) :C̈l:P̈:C̈l: with :C̈l: under P (c) :B̈r:Si:B̈r: with :B̈r: above and below Si (d) :F̈:Ö:F̈:

4.57 (a)
```
 :Cl:C:Cl:
     :O
```
(b) :F:N::N:F: (c)
```
H:C::C:H
  H   H
```
(d) H:C:::C:H **4.59** (a) linear (b) angular (c) angular (d) angular **4.61** The shape does not change; it remains tetrahedral. **4.63** 0.5 unit **4.65** (a) $\overset{\delta^+}{B}—\overset{\delta^-}{N}$ (b) $\overset{\delta^+}{Cl}—\overset{\delta^-}{F}$ (c) $\overset{\delta^-}{N}—\overset{\delta^+}{C}$ (d) $\overset{\delta^-}{F}—\overset{\delta^+}{O}$ **4.67** (a) zero difference (b) greater than zero, less than 2.0 (c) 2.0 or greater **4.69** (a) nonpolar (b) polar (c) polar (d) polar **4.71** (a) nonpolar (b) polar (c) nonpolar (d) polar **4.73** (a) dichlorine monoxide (b) carbon monoxide (c) phosphorus triiodide (d) hydrogen iodide **4.75** (a) H_2O_2 (b) NH_3 (c) CH_4 (d) N_2H_4 **4.77** (a) SO_4^{2-} and SO_3^{2-} (b) NO_3^- and NO_2^- (c) CO_3^{2-} and HCO_3^- (d) HPO_4^{2-} and $H_2PO_4^-$ **4.79** (a) $Fe(OH)_3$ (b) $Be(NO_3)_2$ (c) $(NH_4)_2S$ (d) $(NH_4)_3PO_4$ **4.81** (a) copper(I) phosphate (b) iron(III) nitrate (c) iron(II) sulfate (d) gold(I) cyanide **4.83** (a) C: $1s^22s^22p^2$ (b) F: $1s^22s^22p^5$ (c) Mg: $1s^22s^22p^63s^2$ (d) P: $1s^22s^22p^63s^23p^3$ **4.85** (a) XZ_2 (b) X_2Z (c) XZ (d) ZX **4.87** (a) ionic (b) molecular (c) ionic (d) molecular **4.89** (a) beryllium chloride (b) nitrogen trichloride (c) aluminum chloride (d) iron(III) chloride **4.91** (a)

```
   H          H             :F:          :F:           :F:
   |          |              |            |             |
H—C—H,    H—C—F:,     H—C—F:,     H—C—F:,     :F—C—F:
   |          |              |            |             |
   H          H              H           :F:           :F:
```

(b) All are tetrahedral. (c) nonpolar, polar, polar, polar, nonpolar **4.93** (a) polar covalent, ionic, ionic, polar covalent (b) BA, CA, DB, DA (c) BA, CA, DB, DA

Chapter 5 **5.1** (a) 16.05 amu (b) 80.07 amu (c) 32.06 amu (d) 101.96 amu **5.3** (a) 106.88 amu (b) 68.17 amu (c) 331.31 amu (d) 158.18 amu **5.5** (a) 6.02×10^{23} molecules (b) 12.04×10^{23} molecules (c) 3.01×10^{23} molecules (d) 3.74×10^{23} molecules **5.7** They are the same. **5.9** (a) 28.01 g (b) 44.01 g (c) 58.44 g (d) 342.34 g **5.11** (a) 0.293 mole NH_3 (b) 0.147 mole H_2O_2 (c) 0.0780 mole SO_2 (d) 0.0765 mole Zn **5.13** (a) 2.00 moles S, 4.00 moles O (b) 2.00 moles S, 6.00 moles O (c) 3.00 moles N, 9.00 moles H (d) 6.00 moles N, 12.0 moles H **5.15** (a) 8.02 g S (b) 16.0 g SO_2 (c) 20.0 g SO_3 (d) 24.5 g H_2SO_4 **5.17** (a) 8.22 g N (b) 2.59 g N (c) 8.74 g N (d) 1.97 g N **5.19** (a) balanced (b) balanced (c) not balanced (d) balanced **5.21** (a) (s) means solid, (g) means gas (b) (g) means gas, (l) means liquid, (aq) means aqueous **5.23** (a) $BaCl_2 + Na_2S \longrightarrow BaS + 2NaCl$ (b) $Mg + 2HBr \longrightarrow MgBr_2 + H_2$ (c) $2Co + 3HgCl_2 \longrightarrow 2CoCl_3 + 3Hg$ (d) $2Na + 2H_2O \longrightarrow 2NaOH + H_2$ **5.25** (a) 1 molecule N_2 reacts with 3 molecules H_2 to produce 2 molecules NH_3. 1 mole N_2 reacts with 3 moles H_2 to produce 2 moles NH_3. (b) 1 molecule CH_4 reacts with 2 molecules O_2 to produce 1 molecule CO_2 and 2 molecules H_2O. 1 mole CH_4 reacts with 2 moles O_2 to produce 1 mole CO_2 and 2 moles H_2O. **5.27** (a) not consistent (b) consistent (c) consistent (d) not consistent **5.29** (a) 39.9 g H_2O_2 (b) 37.6 g S (c) 0.294 mole H_2S (d) 6.80 moles H_2O **5.31** $y = 8$ **5.33** (a) 1.00 mole Au (b) 1.00 mole S (c) 1.00 mole Cl_2 (d) 6.02×10^{23} Ne atoms **5.35** C_4H_6

Chapter 6 **6.1** (a) kinetic energy (b) potential energy **6.3** (a) tend to cause disorder (b) tend to cause order **6.5** through inelastic collisions **6.7** (a) A liquid has a definite volume because of significant cohesive forces; a gas has an indefinite volume because of the lack of significant cohesive forces. (b) The particles in solids and liquids are already very close together; there is very little empty space in a solid or liquid. **6.9** (a) 0.967 atm (b) 403 mm Hg (c) 535 torr (d) 0.816 atm **6.11** As one variable increases, there is a proportional decrease in the other variable. **6.13** (a) 1.0 atm (b) 0.57 L (c) 20 atm (d) 2.0 L **6.15** 2.71 L **6.17** As one variable increases, there is a proportional increase in the other variable **6.19** (a) 927°C (b) 1.00 L (c) −198°C (d) 8.00 L **6.21** 144°C **6.23** (a) 17.1 L (b) 48.3 L (c) 18.6 L (d) 3.35 L **6.25** 1.12 L **6.27** −209°C **6.29** Each of the individual components of air (O_2, N_2, etc.) contributes to the total pressure exerted by the air. The individual contributions to the total pressure are the partial pressures of the components. **6.31** 98 mm Hg **6.33** (a) endothermic (b) endothermic (c) exothermic (d) exothermic **6.35** (a) evaporation (b) sublimation (c) melting (d) deposition **6.37** amount of liquid decreases, temperature of liquid decreases **6.39** Gaseous molecules of a substance at a temperature and pressure at which we ordinarily would think of the substance as a liquid or solid are referred to as vapor. **6.41** (a) true (b) true (c) true (d) true **6.43** (a) At a given temperature, liquids that have strong attractive forces between molecules have lower vapor pressures than liquids that have weak attractive forces between molecules. (b) At a higher temperature, more molecules have the minimum energy needed to overcome the attractive forces that prevent escape from the liquid. **6.45** (a) true (b) false (c) true (d) false **6.47** (a) occurs only between polar molecules (b) occurs between all molecules (polar and nonpolar) (c) occurs only between molecules that contain hydrogen bonded to a small, very electronegative element **6.49** Hydrogen bonds are much stronger than dipole–dipole interactions. **6.51** Intermolecular forces are only about one-tenth as strong as intramolecular forces. **6.53** (a) boils (b) does not boil (c) does not boil (d) does not boil **6.55** (a) 0.871 atm (b) 0.981 atm (c) 298°C (d) 526°C **6.57** (a) PBr_3 (b) PI_3 (c) PI_3 (d) PI_3 **6.59** (a) not possible (b) possible (c) possible (d) not possible

Chapter 7 **7.1** (a) true (b) true (c) true (d) false **7.3** Mixtures are of two types: homogeneous and heterogeneous. Homogeneous mixtures are solutions; heterogeneous mixtures are not solutions. **7.5** (a) saturated (b) unsaturated (c) unsaturated (d) saturated **7.7** (a) aqueous (b) nonaqueous (c) aqueous (d) nonaqueous **7.9** (a) decrease (b) increase (c) increase (d) increase **7.11** (a) soluble with exceptions (b) soluble (c) insoluble with exceptions (d) soluble **7.13** (a) 7.10%(m/m) (b) 6.19%(m/m) (c) 9.06%(m/m) (d) 0.27%(m/m) **7.15** (a) 4.21%(v/v) (b) 4.60%(v/v) **7.17** (a) 2.0%(m/v) (b) 15%(m/v) **7.19** (a) 6.0 M KNO_3 (b) 0.456 M $C_{12}H_{22}O_{11}$ (c) 0.342 M NaCl (d) 0.500 M $NaHCO_3$ **7.21** (a) 0.183 M NaCl (b) 0.0733 M NaCl (c) 0.0120 M NaCl (d) 0.00275 M NaCl **7.23** (a) 3.0 M NaCl (b) 3.0 M $AgNO_3$ (c) 4.5 M NaCl (d) 1.5 M $AgNO_3$ **7.25** The boiling point increases and the freezing point decreases. **7.27** It is lower. **7.29** (a) the same as (b) greater than (c) the same as (d) greater than **7.31** (a) hemolyze (b) remain unaffected (c) hemolyze (d) crenate **7.33** In osmosis, only solvent passes through the membrane. In dialysis, both solvent and small solute particles pass through the membrane. **7.35** (a) like solubility; both soluble (b) unlike solubility (c) unlike solubility (d) unlike solubility **7.37** 0.0700 qt H_2O **7.39** (a) 0.472 M K_2SO_4 (b) 0.708 M K_2SO_4 (c) 1.04 M K_2SO_4 (d) 1.60 M K_2SO_4 **7.41** (a) NaCl (b) $MgCl_2$

Chapter 8 **8.1** (a) $XY \longrightarrow X + Y$ (b) $X + Y \longrightarrow XY$ (c) $AX + BY \longrightarrow AY + BX$ (d) $X + YZ \longrightarrow Y + XZ$ **8.3** (a) decomposition (b) combination (c) double-replacement (d) combustion **8.5** (a) oxidation (b) reduction (c) oxidation (d) oxidation **8.7** (a) Li, oxidized; F_2, reduced (b) Fe, oxidized; $CuSO_4$, reduced (c) C_4H_8, oxidized; O_2, reduced (d) Ca, oxidized; S, reduced **8.9** (a) reducing (b) gains **8.11** (a) F_2, oxidizing agent; Li, reducing agent (b) Fe_2O_3, oxidizing agent; CO, reducing agent (c) S, oxidizing agent; Ca, reducing agent (d) Ag_2SO_4, oxidizing agent; Fe, reducing agent **8.13** (a) Reactant particles must collide with a certain minimum combined kinetic energy, which is the activation energy. (b) Orientation relative to one another at the moment of collision is a factor in determining whether a collision produces a reaction. **8.15** (a) exothermic

(b) energy released **8.17** (a) exothermic (b) endothermic (c) endothermic (d) exothermic **8.19** Lowering the temperature decreases the rate of the "spoiling" reaction. **8.21** (a) Molecules have more energy and collide more frequently. (b) A catalyst lowers the activation energy for the reaction. (c) Physical state affects the freedom of movement of molecules. (d) At higher concentrations, there are more collisions between reactants. **8.23** They are equal. **8.25** (a) always applies (b) does not apply (c) always applies (d) always applies **8.27** (a) shifts left (b) shifts left (c) shifts left (d) shifts left **8.29** (a) single-replacement redox (b) combination redox (c) decomposition redox (d) double-replacement nonredox **8.31** (a) H_2O (b) Cl_2 (c) Cl_2 (d) H_2O **8.33** Constant concentrations are not changing, but they do not have to be equal.

Chapter 9 **9.1** (a) Arrhenius acid (b) Arrhenius base (c) Arrhenius acid (d) Arrhenius base **9.3** (a) ionization (b) dissociation **9.5** (a) Brønsted–Lowry acid (b) Brønsted–Lowry base (c) Brønsted–Lowry acid (d) Brønsted–Lowry acid **9.7** (a) HSO_3^- (b) HCN (c) $C_2O_4^{2-}$ (d) $H_2PO_4^-$ **9.9** proton **9.11** (1) It is not restricted to aqueous solution, and (2) it can explain why some compounds that do not contain hydroxide ion can produce basic solutions. **9.13** (a) 1, 0 (b) 2, 4 (c) 1, 7 (d) 0, 4 **9.15** to show that only one of the six hydrogen atoms present is acidic **9.17** 100% or nearly 100% of the molecules of a strong acid ionize; less than 5% of the molecules of a weak acid ionize. **9.19** (a) strong acid (b) strong acid (c) weak acid (d) weak acid **9.21** (a) strong base (b) strong base (c) weak base (d) weak base **9.23** (a) base (b) salt (c) acid (d) salt **9.25** (a) not a neutralization (b) neutralization (c) neutralization (d) not a neutralization **9.27** (a) $HNO_3 + KOH \longrightarrow KNO_3 + H_2O$ (b) $HNO_3 + LiOH \longrightarrow LiNO_3 + H_2O$ (c) $2HNO_3 + Ba(OH)_2 \longrightarrow Ba(NO_3)_2 + 2H_2O$ (d) $H_2SO_4 + Ba(OH)_2 \longrightarrow BaSO_4 + 2H_2O$ **9.29** (a) H_3O^+ ion and OH^- ion (b) self-ionization reaction of H_2O (c) both are 1.0×10^{-7} M (d) because one cannot be produced without the other **9.31** (a) 2.5×10^{-10} M (b) 1.3×10^{-8} M (c) 1.2×10^{-5} M (d) 8.3×10^{-12} M **9.33** (a) acidic (b) acidic (c) acidic (d) basic **9.35** (a) 4.00 (b) 11.00 (c) 7.00 (d) 5.00 **9.37** (a) 1.0×10^{-2} M (b) 1.0×10^{-4} M (c) 1.0×10^{-7} M (d) 1.0×10^{-9} M **9.39** (a) acidic (b) basic (c) basic (d) acidic **9.41** (a) no (b) yes (c) no (d) yes **9.43** (a) $H_3O^+ + F^- \longrightarrow HF + H_2O$ (b) $H_2CO_3 + OH^- \longrightarrow HCO_3^- + H_2O$ (c) $H_3O^+ + CO_3^{2-} \longrightarrow HCO_3^- + H_2O$ (d) $H_3PO_4 + OH^- \longrightarrow H_2PO_4^- + H_2O$ **9.45** (a) 35.0 mL (b) 750 mL (c) 25.0 mL (d) 0.23 mL **9.47** (a) no (b) yes (c) no (d) yes **9.49** (a) neutral (b) acidic (c) basic (d) basic **9.51** (a) 4.00 (b) 6.00 (c) 11.00 (d) 3.00 **9.53** HCl, HCN, NaCl, NaOH

Chapter 10 **10.1** (a) true (b) false (c) false (d) false **10.3** study of hydrocarbons and their derivatives **10.5** (a) meets (b) does not meet (c) meets (d) does not meet **10.7** hydrocarbon: contains only carbon and hydrogen; hydrocarbon derivative: contains one or more additional elements besides carbon and hydrogen **10.9** (a) 2 (b) 1 (c) 3 **10.11** (a) 0, 4 (b) 1, 6 (c) 2, 8 **10.13** (a) 2 (b) 4 (c) 7 (d) 10 **10.15**

```
       H  H  H  H
       |  |  |  |
(a) H—C—C—C—C—H  (b) CH3—CH2—CH2—CH3  (c) C4H10
       |  |  |  |
       H  H  H  H
```

```
              H  H  H  H  H
              |  |  |  |  |
10.17 (a) H—C—C—C—C—C—H
              |  |  |  |  |
              H  H  H  H  H

       H  H  H  H  H
       |  |  |  |  |
(b) H—C—C—C—C—C—H
       |  |  |  |  |
       H  H  H  H  H
```

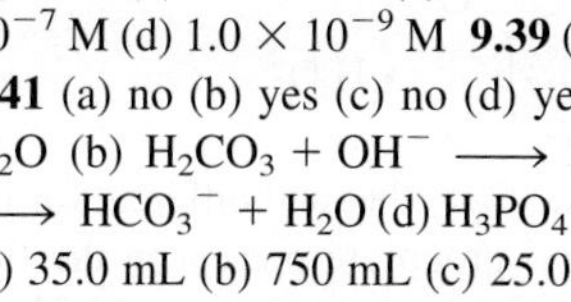

10.19 (a) C_4H_{10} (b) C_5H_{12} (c) C_7H_{16} (d) C_6H_{14} **10.21** They are different. **10.23** (a) 7 (b) 7 (c) 8 (d) 9 **10.25** butane and 2-methylpropane **10.27** (a) 2-methylpentane (b) 2,4,5-trimethylheptane (c) 3-ethyl-2,3-dimethylpentane (d) 3-ethyl-2,4-dimethylhexane
10.29

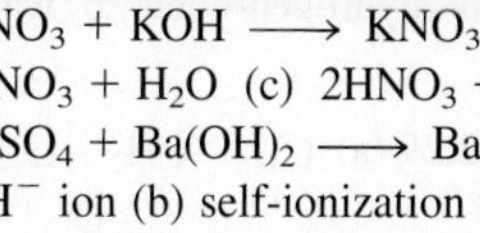

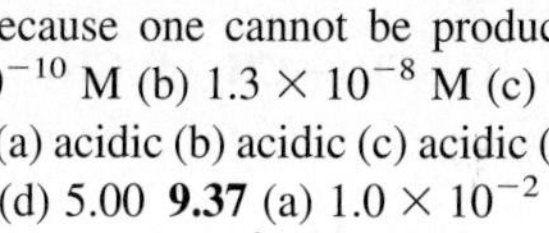

```
(a) CH3—CH2—CH2—CH—CH2—CH2—CH3
                 |
                 CH3

(b) CH3—CH2—CH2—CH—CH2—CH2—CH3
                 |
                 CH2
                 |
                 CH3

(c) CH3—CH2—CH2—CH—CH2—CH2—CH3
                 |
                 CH2
                 |
                 CH2
                 |
                 CH3

(d) CH3—CH2—CH2—CH—CH2—CH2—CH3
                 |
                 CH—CH3
                 |
                 CH3
```

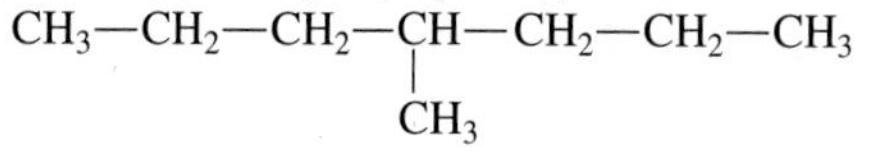

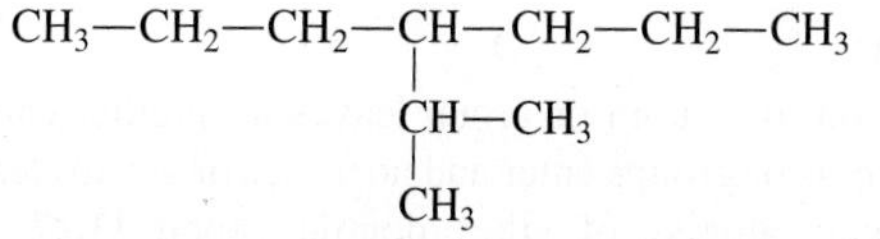

10.31 (a) 1 (b) 2 (c) 1 (d) 4 **10.33** (a) yes (b) no (c) yes (d) no **10.35** (a) C_3H_6 (b) C_8H_{16} (c) C_5H_{10} (d) C_7H_{14} **10.37** (a) cyclopentane (b) methylcyclobutane (c) 1,1-dimethylcyclopropane (d) 1,4-dimethylcyclohexane **10.39** (a) 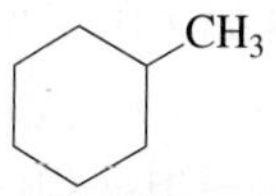 [cyclohexane ring]—CH_3 (b) 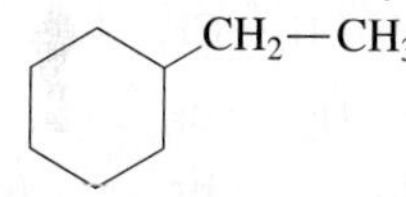 [cyclohexane ring]—$CH_2—CH_3$ (c) 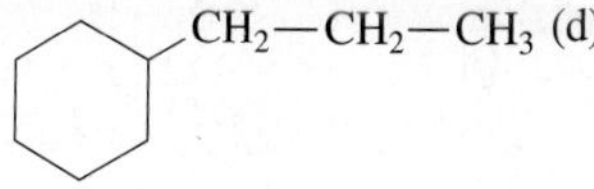 [cyclohexane ring]—$CH_2—CH_2—CH_3$ (d) 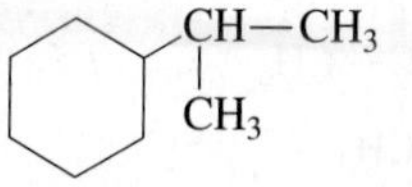 [cyclohexane ring]—$CH—CH_3$ with CH_3 on the CH **10.41** (a) no (b) no (c) yes, [cyclohexane rings: CH_3, H / CH_3, H at two carbons — *Cis* isomer; CH_3, H / H, CH_3 — *Trans* isomer] (d) no

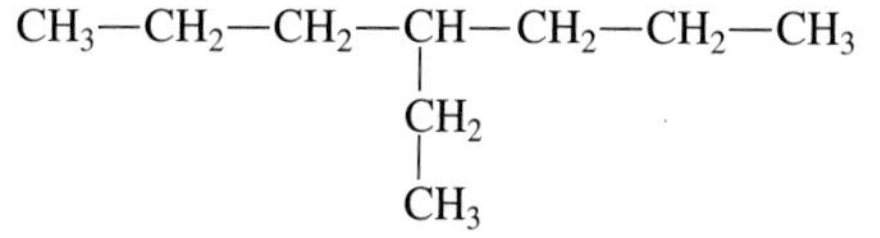

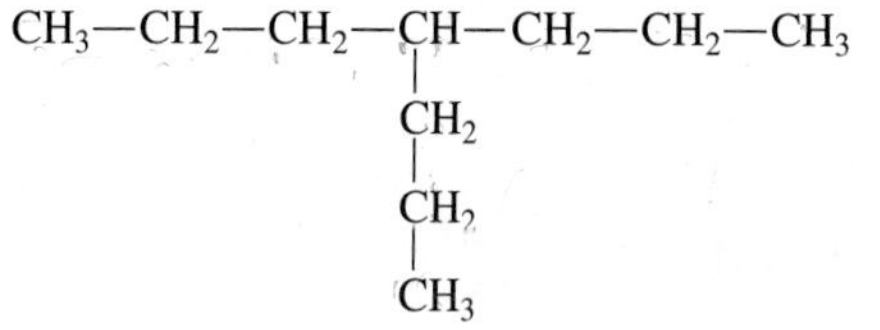

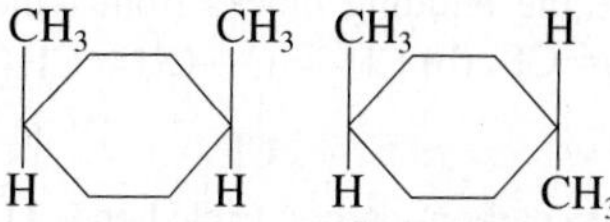

Cis isomer *Trans* isomer

10.43 dome-shaped rock formations **10.45** (a) octane (b) cyclopentane (c) pentane (d) cyclopentane **10.47** (a) less than (b) less than (c) less than (d) less than **10.49** CH_3Br, CH_2Br_2, $CHBr_3$, and CBr_4 **10.51** (a) 2 (b) 1 (c) 3 (d) 1 **10.53** (a) 16 (b) 5 (c) 14 (d) 21

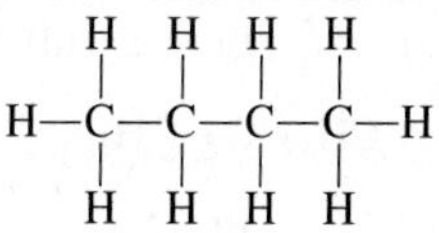

```
              H  H  H  H          H  H  H  H  H
              |  |  |  |          |  |  |  |  |
10.55 (a) H—C—C—C—C—H  (b) H—C—C—C—C—C—H
              |  |  |  |          |  |  |  |  |
              H  H  H  H          H  H  H  H  H
```

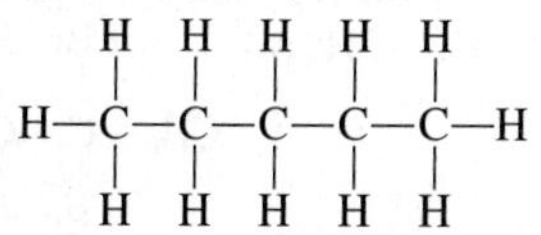

(c) $CH_3—(CH_2)_{10}—CH_3$ (d) C_8H_{18} **10.57** (a) Location of alkyl groups not given; should be 1,1-, 1,2-, 1,3- or 1,4-dimethylcyclohexane. (b) Ring not numbered correctly; should be 1,2-dimethylcyclobutane. (c) Number not needed to locate alkyl group; should be ethylcyclobutane. (d) Ring not numbered correctly; should be 1-ethyl-2-methylcyclopentane. **10.59** 1,1-, 1-2, 1,3-, and 1,4-dimethylcyclohexane **10.61** (a) 16 (b) 6 (c) 5 (d) 22 (e) liquid (f) less dense (g) insoluble (h) flammable

Chapter 11 **11.1** (a) similar (b) different **11.3** (a) carbon–carbon double bond (b) carbon–carbon triple bond (c) six-membered carbon ring with "delocalized" bonding (d) functional group not present **11.5** C_nH_{2n-2} **11.7** They are an older and a newer name for the same thing. **11.9** (a) cyclohexene (b) cyclobutene (c) 4-methylcyclopentene (d) 3-methylcyclohexene **11.11** (a) 1,4-cyclohexadiene (b) 1,3-cyclohexadiene (c) 3-methyl-1,4-cyclohexadiene (d) 2-methyl-1,3-cyclohexadiene **11.13** (a) $CH_2=CH—CH_2—CH_2—CH_3$

(b) $CH_3—CH{=}CH—CH_2—CH_3$ (c) $CH_2{=}CH—CH{=}CH—CH_3$ (d) $CH_2{=}CH—CH_2—CH{=}CH_2$ **11.15** in the location of the carbon–carbon double bond **11.17** orientation in space of the groups attached to the double-bonded carbon atoms **11.19** (a) *cis*-2-pentene (b) *trans*-3-methyl-2-pentene (c) *trans*-2-butene (d) 2-methylpropene **11.21** (a)

$CH_3—CH_2$ H
C=C
CH_3 $CH_2—CH_3$

(b) CH_3 $CH_2—CH_3$
C=C
H   H

(c) CH_3 H
C=C
H $CH_2—CH—CH_2—CH_3$
            CH_3

(d) CH_3
CH_3 $CH—CH_3$
C=C
H   H

11.23 In substitution, one atom or group leaves as another enters; in addition, two atoms or groups enter and no atoms or groups leave. **11.25** the carbon atoms of the double bond **11.27** (a) $CH_2—CH—CH_3$ (with H, H) (b) $CH_2—CH—CH_3$ (with Cl, Cl) (c) $CH_2—CH—CH_3$ (with H, OH)

(d) $CH_2—CH—CH_3$ (with H, Br) **11.29** (a) $CH_3—CH—CH—CH_3$ (with Cl, Cl)

(b) $CH_3—C(Br)(CH_3)—CH_3$ (c) $CH_3—CH_2—CH—CH_3$ (with Cl) (d) [cyclopentane ring]

11.31 Monomers are the building blocks from which polymers are made. **11.33** (a) $CF_2{=}CF_2$ (b) $CH_2{=}C—CH{=}CH_2$ (with Cl) (c) $CH_2{=}CH$ (with Cl)

(d) $CH_2{=}CH$ (with benzene ring) **11.35** carbon–carbon triple bond **11.37** (a) 1-hexyne

(b) 2-butyne (c) 4-methyl-2-pentyne (d) 2,2-dimethyl-3-heptyne **11.39** (a) $CH_3—CH_3$ (b) $CH_3—C(Br)_2—CH(Br)_2$ (c) $CH_3—C(Br)_2—CH_3$ (d) $CH_2{=}CH$ (with Cl)

11.41 a delocalized bond that involves all six carbon atoms **11.43** (a) 1-methyl-3-propylbenzene (b) 1-isopropyl-2-propylbenzene (c) 1,4-diethylbenzene (d) 1-ethyl-3-methylbenzene **11.45** (a) 1,3-diethyl-5-methylbenzene (b) 2-isopropyl-l-methyl-4-propylbenzene (c) 1,2-dimethyl-3-propylbenzene (d) 1,2,4,5-tetramethylbenzene **11.47** (a) no (b) yes (c) no (d) yes **11.49** (a) carbon–carbon double bond (b) carbon–carbon double bond (c) carbon–carbon triple bond (d) aromatic ring system **11.51** (a) (1) (b) (1) (c) (1) (d) (2) **11.53** (a) two (b) one (c) two (d) four **11.55** (a) (3) (b) (3) (c) (4) (d) (1) **11.57** (a) no (b) yes (c) no (d) yes **11.59** (a) ethyne (b) propene (c) ethene (d) methylbenzene

Chapter 12 **12.1** (a) oxygen (b) oxygen (c) nitrogen (d) sulfur **12.3** (a) halogenated hydrocarbon (b) alcohol (c) thiol (d) ether **12.5** halogen atom; carbon–carbon double bond **12.7** (a) 2-chloro-2-butene (b) 3-bromo-1,4-hexadiene (c) 1-chlorocyclohexene (d) 1,2-dibromo-4,5-dichlorobenzene **12.9** (a) $CH_2{=}CH—CH—CH_2—CH_3$ (with Cl)

(b) $CH{=}CH—CH{=}CH$ (with Br, Br) (c) [cyclohexene ring with Br, Br] (d) [benzene ring with Cl, Cl para]

12.11 (a) false (b) false (c) false (d) true **12.13** chlorofluorocarbon; CFCs can adversely affect the stratospheric ozone layer. **12.15** hydroxyl group; —OH **12.17** (a) alcohol (b) neither an alcohol nor a phenol (c) alcohol (d) neither an alcohol nor a phenol **12.19** (a) methanol (b) 1-propanol (c) 2-propanol (d) cyclobutanol **12.21** (a) $CH_3—CH_2—CH—CH_2—CH_3$ (with OH)

(b) $CH_3—CH_2—C(OH)(CH_2CH_3)—CH_2—CH_2—CH_3$ (c) $CH_2—CH—CH_3$ (with OH, CH_3)

(d) [cyclobutane ring with OH and CH_3] **12.23** (a) 3-ethylphenol (b) 2-chlorophenol (c) 2-methylphenol (d) 2-bromo-3-ethylphenol **12.25** Alcohols can hydrogen-bond to water; alkanes cannot hydrogen-bond to water.

12.27 (a) 1-heptanol (b) 1,2-ethanediol **12.29** (a) $CH_3—CH_2—C({=}O)—H$

(b) $CH_3—CH(CH_3)—C({=}O)—H$ (c) $CH_3—C({=}O)—CH_2—CH_3$ (d) $CH_3—C({=}O)—CH_3$

12.31 [phenol, OH] $+ H_2O \rightleftharpoons$ [phenoxide, O^-] $+ H_3O^+$

12.33 (a) R groups are identical for the first ether and different for the second ether. (b) An alkyl and an aryl group are present in the first ether, and two aryl groups are present in the second ether. **12.35** (a) phenoxybenzene (b) propoxycyclohexane (c) methoxybenzene (d) cyclohexoxycyclohexane **12.37** (a) diphenyl ether (b) cyclohexyl propyl ether (c) methyl phenyl ether (d) dicyclohexyl ether **12.39** Ethyl alcohol has a higher boiling point because of hydrogen bonding; ether molecules cannot hydrogen-bond to each other. **12.41** flammability and hyperperoxide/peroxide formation **12.43** (a) $CH_3—SH$ (b) $CH_3—CH—CH_3$ (with SH) (c) [cyclopentane ring with SH] (d) $CH_2—CH_2$ (with SH, SH) **12.45** their strong strong odors **12.47** (a) $CH_3—CH_2—S—S—CH_2—CH_3$ (b) $CH_3—CH_2—SH$ **12.49** (a) primary (b) tertiary (c) secondary (d) primary **12.51** (a) 3-pentanamine (b) 2-methyl-3-pentanamine (c) *N*-methyl-3-pentanamine (d) 2,3-butanediamine **12.53** (a) $CH_3—CH_2—NH_2$

(b) $CH_3—C(NH_2)(CH_3)—CH_2—CH_3$ (c) $CH_2—CH_2—CH_2$ (with NH_2, NH_2)

(d) [benzene ring]$—NH—CH_3$ **12.55** (a) $CH_3—CH_2—NH_2$; solubility decreases with increasing carbon chain length. (b) the diamine; more hydrogen bonding to water is possible when two amino groups are present. **12.57** (a) $CH_3-\overset{+}{N}H_3\ Cl^-$ (b) $CH_3-CH_2-CH_2-\overset{+}{N}H_3\ Cl^-$ (c) $CH_3-\overset{+}{N}H_2-CH_3\ Cl^-$ (d) $CH_3-CH_2-\overset{+}{N}H_2-CH_2-CH_3\ Cl^-$ **12.59** (a) propylammonium chloride (b) methylpropylammonium chloride (c) ethyldimethylammonium chloride (d) cyclohexyldimethylammonium chloride **12.61** (a) no (b) yes (c) no (d) no **12.63** (a) yes (b) no (c) yes (d) no **12.65** (a) alcohol (b) amine (c) thiol (d) ether **12.67** (a) two (b) one (c) one (d) one **12.69** (a) one (b) two (c) one (d) one **12.71** (a) OH on benzene ring (b) OH on cyclohexane ring (c) NH_2 on benzene ring (d) NH_2 on cyclohexane ring

12.73 (a) ethanethiol (b) ethanol (c) ethanamine (d) methoxyethane **12.75** (a) aldehyde (b) amine salt (c) alkene (d) thiol **12.77** (a) four (b) two (c) one (d) four

Chapter 13 **13.1** a carbon atom and an oxygen atom joined by a double bond **13.3** (a) C, H, O (b) C, H, O, N (c) C, H, O (d) C, H, O

13.5 (a) $H-\overset{\overset{O}{\|}}{C}-H$ (b) $H-\overset{\overset{O}{\|}}{C}-O-CH_3$ (c) $CH_3-\overset{\overset{O}{\|}}{C}-CH_3$

(d) $H-\overset{\overset{O}{\|}}{C}-OH$ **13.7** (a) no (b) yes (c) no (d) no **13.9**

(a) $CH_3-CH_2-\overset{\overset{CH_3}{|}}{CH}-CH_2-\overset{\overset{O}{\|}}{C}-H$

(b) $CH_3-CH_2-CH_2-\overset{\overset{CH_3}{|}}{CH}-\overset{\overset{CH_3}{|}}{CH}-CH_2-\overset{\overset{O}{\|}}{C}-H$

(c) $CH_3-CH_2-CH_2-CH_2-\underset{\underset{CH_2-CH_3}{|}}{CH}-\overset{\overset{O}{\|}}{C}-H$ (d) $CH_3-\underset{\underset{Cl}{|}}{\overset{\overset{Cl}{|}}{C}}-\overset{\overset{O}{\|}}{C}-H$

13.11 (a) propionaldehyde (b) 2-methylpropionaldehyde (c) 2, 2-dichlorocacetaldehyde (d) 2-chlorobenzaldehyde **13.13** Alcohols can hydrogen-bond; aldehydes cannot hydrogen-bond. **13.15** Octanal; higher-molecular-mass aldehydes tend to be more "fragrant." **13.17** (a) CH_3-CH_2-OH (b) $CH_3-CH_2-CH_2-CH_2-CH_2-OH$

(c) CH_3-OH (d) $CH_3-CH_2-\overset{\overset{Cl}{|}}{CH}-\overset{\overset{Cl}{|}}{CH}-CH_2-OH$ **13.19** (a) acetal (b) hemiacetal (c) neither a hemiacetal nor an acetal (d) neither a hemiacetal nor an acetal **13.21** (a) yes (b) yes (c) no (d) yes **13.23**

(a) $CH_3-CH_2-\overset{\overset{O}{\|}}{C}-CH_2-CH_2-CH_3$

(b) $CH_3-\overset{\overset{O}{\|}}{C}-\overset{\overset{CH_3}{|}}{CH}-CH_2-CH_3$ (c) cyclobutanone (d) $\overset{\overset{Cl}{|}}{CH_2}-\overset{\overset{O}{\|}}{C}-\overset{\overset{Cl}{|}}{CH_2}$

13.25 (a) dipropyl ketone (b) ethyl propyl ketone (c) isopropyl methyl ketone (d) ethyl phenyl ketone **13.27** Oxidation of an aldehyde involves breaking a C—H bond, and oxidation of a ketone involves breaking a C—C bond. The former is much easier to accomplish than the latter. **13.29** (a) neither a hemiketal nor a ketal (b) hemiketal (c) ketal (d) neither a hemiketal nor a ketal **13.31** (a) yes (b) no (c) no (d) yes **13.33** (a) $CH_3-CH_2-\underset{\underset{CH_3-CH_2}{|}}{CH}-\overset{\overset{O}{\|}}{C}-OH$ (b) $CH_3-\overset{\overset{CH_3}{|}}{CH}-\overset{\overset{CH_3}{|}}{CH}-\overset{\overset{O}{\|}}{C}-OH$

(c) $CH_3-\overset{\overset{CH_3}{|}}{CH}-\overset{\overset{O}{\|}}{C}-OH$ (d) $Cl-\overset{\overset{Cl}{|}}{CH}-\overset{\overset{O}{\|}}{C}-OH$

13.35 (a) formic acid (b) acetic acid (c) propionic acid (d) butyric acid **13.37** Hydrogen-bonding opportunities are more extensive for carboxylic acids. **13.39** the negative ion produced when a carboxylic acid loses its acidic hydrogen atom or atoms **13.41** (a) sodium acetate (b) sodium butanoate (c) sodium oxalate (d) sodium succinate **13.43** (a) $H-COOH + NaOH \longrightarrow H-COO^-\ Na^+ + H_2O$ (b) $H-COOH + NaOH \longrightarrow H-COO^-Na^+ + H_2O$ (c) $CH_3-COOH + NaOH \longrightarrow CH_3-COO^-Na^+ + H_2O$ (d) $CH_3-COOH + NaOH \longrightarrow CH_3-COO^-Na^+ + H_2O$ **13.45** (a) $CH_3-CH_2-\overset{\overset{O}{\|}}{C}-O-CH_3$

(b) $CH_3-\overset{\overset{O}{\|}}{C}-O-CH_2-CH_2-CH_3$

(c) $CH_3-CH_2-\overset{\overset{CH_3}{|}}{CH}-\overset{\overset{O}{\|}}{C}-O-\overset{\overset{CH_3}{|}}{CH}-CH_3$

(d) $CH_3-CH_2-CH_2-CH_2-\overset{\overset{O}{\|}}{C}-O-\overset{\overset{CH_3}{|}}{CH}-CH_2-CH_3$

13.47 (a) $CH_3-CH_2-\overset{\overset{O}{\|}}{C}-OH;\ CH_3-CH_2-OH$

(b) $CH_3-CH_2-CH_2-\overset{\overset{O}{\|}}{C}-OH;\ CH_3-OH$

(c) $CH_3-\overset{\overset{O}{\|}}{C}-OH$; phenol (OH on benzene ring) (d) benzene ring$-\overset{\overset{O}{\|}}{C}-OH;\ CH_3-OH$

13.49 (a) methyl propionate (b) methyl formate (c) methyl acetate (d) propyl acetate **13.51** (a) 8, 10 (b) 3, 2 (c) 4, 2 (d) 6, 7 **13.53** (a) orange (b) apple (c) pear (d) pineapple **13.55** (a) ethanol (b) ethyl alcohol (c) methanol (d) isopropyl alcohol **13.57** The alcohol is salicylic acid; the carboxylic acid is acetic acid. **13.59** (a) yes (b) yes (c) no (d) no **13.61** (a) *N*-ethylacetamide (b) *N,N*-dimethylpropionamide (c) butyramide (d) *N*-methylformamide **13.63** (a) monosubstituted (b) disubstituted (c) unsubstituted (d) monosubstituted **13.65** The polarity of the carbonyl group makes the lone pair of electrons on nitrogen unavailable for bonding to a H^+ ion. **13.67** Carboxylic acids and amines (or ammonia) undergo an acid–base reaction. **13.69** (a) ammonia (b) methyl amine (c) dimethyl amine (d) dimethyl amine **13.71** In addition polymerization, the polymer is the only product; in condensation polymerization, a small molecule is produced in addition to the product. **13.73** poly(ethylene terephthalate) **13.75** (a) amino acids (b) diacid and dialcohol (c) diacid and diamine (d) diacid and diamine **13.77** (a) $R-\overset{\overset{O}{\|}}{C}-OR$ (b) $R-\overset{\overset{O}{\|}}{C}-NH_2$ (c) $R-\overset{\overset{O}{\|}}{C}-OH$

(d) $R-\overset{\overset{O}{\|}}{C}-H$ **13.79** (a) CH_2O (b) CH_2O_2 (c) $C_2H_4O_2$ (d) CH_3ON

13.81 (a) $H-\overset{\overset{O}{\|}}{C}-H$ (b) $CH_3-\overset{\overset{O}{\|}}{C}-CH_3$ (c) $H-\overset{\overset{O}{\|}}{C}-O-CH_3$

(d) $H-\overset{\overset{O}{\|}}{C}-NH_2$ **13.83** (a) carboxylic acid, primary amine (b) carboxylic acid, alcohol (c) acetal, water (d) ester, water **13.85** (a) one (b) two (c) one (d) four

Chapter 14 **14.1** 75% bioinorganic substances and 25% bioorganic substances by mass **14.3** (a) proteins (b) proteins (c) lipids (d) lipids **14.5** photosynthesis; CO_2 from air and water from the soil are reactants, and sunlight is the energy source. **14.7** hydroxyl groups **14.9** (a) two (b) four (c) two to ten (d) many **14.11** Pentose has five carbon atoms; hexose has six carbon atoms. **14.13** ketone and alcohol functional groups; six carbon atoms **14.15** A chiral object has a nonsuperimposable mirror image; an achiral object has a superimposable mirror image. **14.17** (a) achiral (b) chiral (c) achiral (d) achiral **14.19** (a) left-handed (b) left-handed (c) right-handed (d) left-handed **14.21** (a) three (b) two (c) four (d) three **14.23**

(a) CHO; H—OH; HO—H; H—OH; H—OH; CH_2OH
(b) CHO; H—OH; HO—H; HO—H; H—OH; CH_2OH
(c) CH_2OH; C=O; HO—H; H—OH; H—OH; CH_2OH
(d) CHO; H—OH; H—OH; H—OH; CH_2OH

14.25 (a) glucose (b) glucose (c) galactose (d) fructose **14.27** (a) 1 and 5 (b) 1 and 5 (c) 2 and 5 (d) 1 and 4 **14.29** orientation of the —OH group on carbon 4 **14.31** (a) alpha (b) alpha (c) beta (d) alpha **14.33** CHO; H—OH; HO—H; HO—H; H—OH; CH_2OH **14.35** (a) yes (b) yes (c) yes (d) yes **14.37** (a) yes (b) no (c) yes (d) yes **14.39** (a) no (b) no (c) no (d) yes **14.41** (a) $\beta(1 \rightarrow 4)$ (b) α, $\beta(1 \rightarrow 2)$ (c) $\beta(1 \rightarrow 6)$ (d) $\alpha(1 \rightarrow 4)$ **14.43** (a) reducing sugar (b) nonreducing sugar (c) reducing sugar (d) reducing sugar **14.45** (a) galactose, glucose (b) glucose, fructose (c) glucose, glucose **14.47** (a) sucrose (b) lactose (c) maltose **14.49** Animal starch (glycogen) is a more highly branched glucose polymer than plant starch (amylose and amylopectin). **14.51** Bacteria in the intestinal tract produce enzymes that can break the $\beta(1 \rightarrow 4)$ glycosidic linkages. **14.53** Cellulose is a polymer of glucose; chitin is a polymer of a glucose derivative. **14.55** (a) one (b) two (c) many (d) many **14.57** (a) same (b) same (c) same (d) different **14.59** (a) yes (b) no (c) no (d) yes **14.61** α-D-Galactose is in equilibrium with open-chain D-galactose, which in turn is in equilibrium with the β-D-galactose.

Chapter 15 **15.1** relatively insoluble in water **15.3** nonpolar fatty-acid-containing lipids, polar fatty-acid-containing lipids, and non-fatty-acid-containing lipids **15.5** (a) unsaturated (b) polyunsaturated (c) monounsaturated (d) unsaturated **15.7** produces a 30° bend in the carbon chain **15.9** The greater the number of *cis* double bonds, the lower the melting point. **15.11** three ester functional groups

15.13

Glycerol — Fatty acid; — Fatty acid; — Fatty acid

15.15

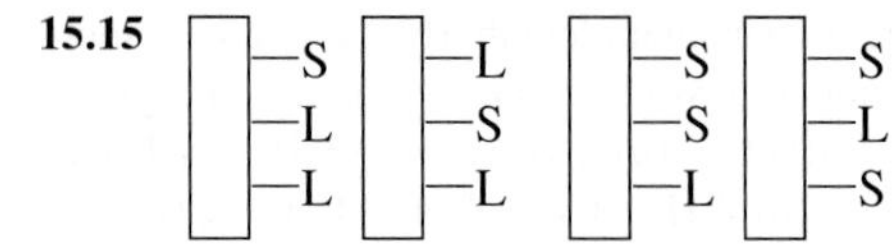

15.17 (a) no difference (b) Fat is a "solid" triacylglycerol. (c) Mixed triacylglycerol is a triacylglycerol in which at least two different fatty acids are present. (d) Fat is a "solid" triacylglycerol and oil is a "liquid" triacylglycerol. **15.19** In general, the fatty acids in oils have a higher degree of unsaturation than the fatty acids in fats. **15.21** (a) glycerol and three fatty acids (b) a triacylglycerol in which all three fatty acids are saturated **15.23** Antioxidants prevent rancidity; they are more easily oxidized than the fatty acid carbon–carbon double bonds and thus undergo oxidation.

15.25 $CH_2(OH)—CH(OH)—CH_2(OH)$; $CH_3—(CH_2)_{14}—COOH$; $CH_3—(CH_2)_{12}—COOH$; $CH_3—(CH_2)_7—CH{=}CH—(CH_2)_7—COOH$

15.27 In a triacylglycerol, all three of glycerol's —OH groups are esterified with fatty acids; in a phosphoacylglycerol, two of glycerol's —OH groups are esterified with fatty acids and the other is esterified with phosphoric acid, which in turn is esterified with an alcohol. **15.29** In a lecithin, choline is esterified to phosphoric acid; in cephalins, an ethanol amine or serine is esterified to phosphoric acid. **15.31** carbon, hydrogen, oxygen, nitrogen, and phosphorous. **15.33** The amide linkage involves the sphingosine $–NH_2$ group and the fatty acid; the ester or glycosidic linkage involves the terminal —OH group of sphingosine and the additional component. **15.35** No. Some sphingolipids have a monosaccharide as the additional component.

15.37

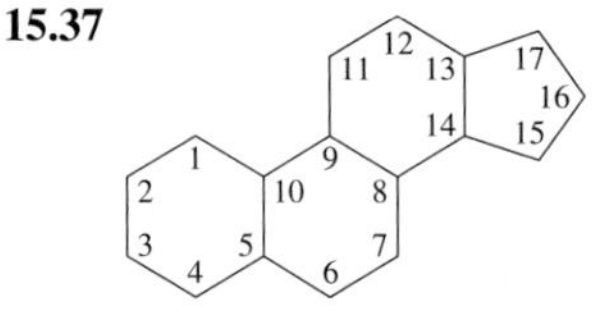

15.39 an —OH group on carbon 3, methyl groups on carbons 10 and 13, and a small branched hydrocarbon chain on carbon 17 **15.41** They solubilize lipids. **15.43** sex hormones and adrenocortical hormones **15.45** phosphoacylglycerols and sphingolipids **15.47** —Polar; >Nonpolar; —Polar

15.49 controls the rigidityof the bilayer **15.51** function as markers important in recognition processes **15.53** (a) glycerol-based (b) glycerol-based (c) neither glycerol-based nor sphingosine-based (d) sphingosine-based **15.55** (a) triacylglycerol (b) sphingomyelin (c) lecithin (d) cerebroside **15.57** (a) 3,0,0 (b) 2,1,0 (c) 4,0,0 (d) 4,0,0 **15.59** (a) both saturated and unsaturated (b) both saturated and unsaturated (c) both saturated and unsaturated (d) both saturated and unsaturated

Chapter 16 **16.1** nitrogen **16.3** amino acids **16.5** nonpolar, polar neutral, polar acidic, and polar basic **16.7** Its side chain is bonded to both the α-carbon atom and the amino nitrogen atom. **16.9** isoleucine (Ile), asparagine (Asn), glutamine (Gln), and tryptophan (Trp) **16.11** (a) no (b) yes (c) yes (d) yes **16.13** It does not possess handedness; no chiral center is present. **16.15** Both an acidic group (—COOH) and a basic group ($—NH_2$) are present.

16.17 (a) H_2N-CH_2-COOH, $H_3\overset{+}{N}-CH_2-COO^-$
(b) $H_2N-CH(CH_3)-COOH$, $H_3\overset{+}{N}-CH(CH_3)-COO^-$
(c) $H_2N-CH(CH(CH_3)-CH_3)-COOH$, $H_3\overset{+}{N}-CH(CH(CH_3)-CH_3)-COO^-$
(d) $H_2N-CH(CH(OH)-CH_3)-COOH$, $H_3\overset{+}{N}-CH(CH(OH)-CH_3)-COO^-$
16.19 (a) $H_3\overset{+}{N}-CH(CH_2-OH)-COOH$ (b) $H_3\overset{+}{N}-CH(CH_2-SH)-COOH$
(c) $H_3\overset{+}{N}-CH(CH_2-CH(CH_3)-CH_3)-COOH$ (d) $H_3\overset{+}{N}-CH(CH(CH_3)-CH_2-CH_3)-COOH$
16.21 Two —COOH groups are present, and they are deprotonated at different pH values. **16.23** the bond between the carboxyl group of one amino acid and the amino group of another amino acid **16.25** The N-terminal end is the peptide end with a free $H_3\overset{+}{N}$ group: the C-terminal end is the peptide end with a free COO^- group. **16.27** Ser is the N-terminal end in the first dipeptide, and Cys is the N-terminal end in the second dipeptide. **16.29** (a) Ser–Val–Cys (b) Asp–Thr–Asn **16.31** (a) six (b) six **16.33** primary, secondary, tertiary, quaternary **16.35** They differ in amino acid sequence. **16.37** six (Ala–Ala–Gly–Gly, Gly–Gly–Ala–Ala, Ala–Gly–Ala–Gly, Gly–Ala–Gly–Ala, Ala–Gly–Gly–Ala, and Gly–Ala–Ala–Gly) **16.39** carbonyl and amino groups **16.41** Yes, a section of the peptide chain can be α helix and another section β pleated sheet. **16.43** remains the same **16.45** (a) noncovalent interaction (b) noncovalent interaction (c) covalent bond (d) noncovalent interaction **16.47** (a) hydrophobic interaction (b) electrostatic interaction (c) disulfide bond (d) hydrogen bond **16.49** the associations among the separate chains in an oligomeric protein **16.51** a conjugated protein has a non-amino-acid component **16.53** (a) lipid (b) carbohydrate (c) phosphate group (d) metal ion **16.55** quaternary, tertiary, secondary, and primary (all levels) **16.57** Yes; both yield Ala and Val. **16.59** quaternary, tertiary, and secondary **16.61** They are the same in primary structure. **16.63** Each biochemical reaction requires a different catalyst. **16.65** Heat can denature the protein present. **16.67** (a) oxidation of cytochromes (b) dehydrogenation of alcohols (c) reduction of L-amino acids (d) hydrolysis of lactose **16.69** (a) pyruvate (b) galactose (c) succinate (d) L-amino acids **16.71** A holoenzyme contains an apoenzyme plus a cofactor. **16.73** Cofactors provide additional chemically reactive functional groups. **16.75** an intermediate reaction species that is formed when a substrate binds to the active site of the enzyme **16.77** The substrate is the only molecule with a geometry complementary to that of the active site. **16.79** electrostatic interactions, hydrogen bonds, and hydrophobic interactions **16.81** protein (enzyme) denaturation occurs **16.83** number of molecules processed per second by one molecule of enzyme functioning under optimum conditions **16.85** (a) tertiary (b) tertiary (c) secondary (d) primary **16.87** (a) −1 (b) −1 (c) −4 (d) −1 **16.89** enzyme + substrate $\rightleftharpoons$ enzyme-substrate complex $\longrightarrow$ enzyme + product

Chapter 17 **17.1** *d*eoxyribo*n*ucleic *a*cid (DNA) and *r*ibo*n*ucleic *a*cid (RNA) **17.3** DNA: storage and transfer of genetic information; RNA: synthesis of proteins **17.5** the carbon-2 —OH group in ribose is changed to a —H atom in deoxyribose **17.7** (a) pyrimidine (b) pyrimidine (c) purine (d) purine **17.9** (a) one (b) four (c) one (d) one **17.11** (a) ribose (b) 2′-deoxyribose (c) 2′-deoxyribose (d) ribose **17.13** the alternating chain of sugar and phosphate residues **17.15** the sugar residues **17.17** The 5′-end carries a free phosphate group; the 3′-end carries a free hydroxyl group on carbon 3 of the sugar. **17.19** phosphate and two sugar molecules **17.21** (a) Two polynucleotide strands are coiled around each other in a manner somewhat like a spiral staircase. (b) The outside involves the sugar–phosphate backbones; the inside involves base pairs. **17.23** thymine–thymine–adenine–guanine–cytosine **17.25** T is the 5′end in T–A; A is the 5′end in A–T. **17.27** One strand runs in the 5′-to-3′ direction, and the other strand runs in the 3′-to-5′ direction. **17.29** governs the unwinding of the DNA double helix **17.31** Each of the daughter molecules contains one of the parent strands. **17.33** The unwound strands are antiparallel (5′-to-3′ and 3′-to-5′); only the 5′-to-3′ strand can grow continuously. **17.35** transcription **17.37** process by which RNA molecules direct the synthesis of protein molecules **17.39** ribosomal RNA (rRNA), messenger RNA (mRNA), primary transcript RNA (ptRNA), and transfer RNA (tRNA) **17.41** causes the DNA double helix to unwind at a particular location **17.43** ptRNA **17.45** 3′ U–A–C–G–A–A–U 5′ **17.47** (a) a segment of a gene that carries genetic information (b) a segment of a gene that does not carry genetic information **17.49** a sequence of three nucleotides in an mRNA molecule that codes for a specific amino acid **17.51** There are only 16 two-base code possibilities, and there are 20 standard amino acids. **17.53** (a) CUC, CUA, CUG, UUA, UUG (b) AAC (c) AGC, UCU, UCC, UCA, UCG (d) GGU, GGC, GGA **17.55** Met–Lys–Glu–Asp–Leu **17.57** the hairpin loop opposite the open end of the cloverleaf structure **17.59** complementary base-pairing **17.61** (a) UCU (b) GCA (c) AAA (d) GUU **17.63** (a) Ser (b) Ala (c) Lys (d) Val **17.65** (a) An amino acid is activated and then forms a complex with a tRNA molecule. (b) mRNA attaches itself to the P site of a ribosome. (c) With amino acids in place at the P and A ribosomal sites, the P site amino acid is linked to the A site amino acid. (d) A stop codon terminates the protein synthesis process. **17.67** A site **17.69** Gly: GGU, GGC, GGA or GGG; Ala: GCU, GCC, GCA or GCG; Cys: UGU or UGC; Val: GUU, GUC, GUA or GUG; Tyr: UAU or UAC. **17.71** DNA is present as plasmids that replicate independently of the chromosomes; plasmids are relatively easily transferred from cell to cell. **17.73** DNA is cleaved in a special manner that produces "sticky ends." **17.75** Recombinant DNA is incorporated in a host cell. **17.77** (a) Thymine is a methyl uracil; the methyl group is on carbon 5′. (b) Adenine is the 6-amino derivative of purine, and guanine is the 2-amino-6-oxo derivative of purine. **17.79** (a) mRNA (b) tRNA (c) ptRNA (d) tRNA

17.81 A chromosome is a DNA–protein complex; a gene is a segment of a DNA strand. **17.83** (a) Single strands of DNA serve as templates; new strands are formed in the 5′-to-3′ direction; base-pairing occurs. (b) replication: entire DNA strand copied, transcription: a segment of a DNA strand copied; replication: new strand remains with the parent strand, transcription: new strand leaves the parent strand; replication: deoxynucleotides are used, transcription: ribonucleotides are used

Chapter 18 **18.1** (a) Catabolism: large molecules are broken down into smaller ones; anabolism: small molecules are put together to form larger ones. (b) Catabolism: energy is usually released; anabolism: energy is usually consumed. **18.3** linear pathway: a series of reactions that generates a final product; cyclic pathway: a series of reactions that regenerates the first reactant **18.5** a minute structure within the cell cytoplasm that carries out a specific cellular function **18.7** The outer membrane is permeable to small molecules; the inner membrane is highly impermeable to most substances. **18.9** (a) adenosine tri phosphate (b) adenosine di phosphate (c) adenosine mono phosphate (d) guanine tri phosphate **18.11** (a) $ATP + H_2O \longrightarrow ADP + P_i$ (b) $ADP + H_2O \longrightarrow AMP + P_i$ (c) $ATP + 2H_2O \longrightarrow AMP + 2P_i$ **18.13** FAD (oxidized form) and $FADH_2$ (reduced form) **18.15** flavin subunit

18.17

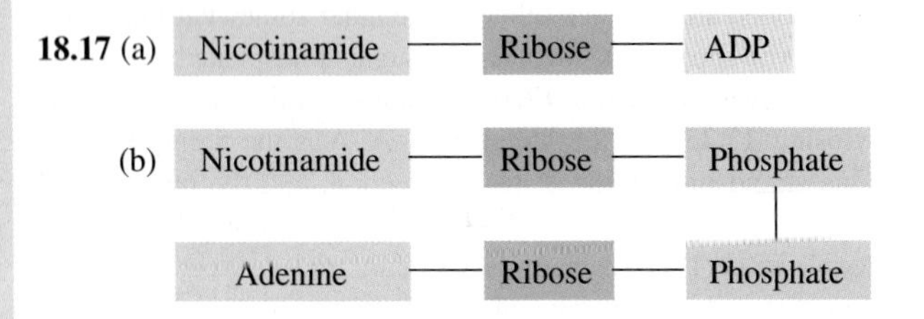

18.19 (a) reduced (b) oxidized (c) oxidized (d) reduced **18.21** sulfhydryl group of 2-aminoethanethiol **18.23** (a) stage 1 (b) stage 3 (c) stage 4 (d) stage 2 **18.25** (a) inside mitochondria (b) inside mitochondria (c) inside mitochondria (d) cytoplasm of cell and inside mitochondria **18.27** It is the first product of the cycle. **18.29** exhaled in the process of respiration **18.31** (a) 2 (b) 1 (c) 3 (d) 1 **18.33** (a) Steps 3 and 4 (b) Steps 6 and 8 (c) Step 5 (d) Step 1 **18.35** (a) fumarate, malate (b) malate, oxaloacetate (c) isocitrate, α-ketoglutarate (d) citrate, isocitrate **18.37** NADH and $FADH_2$ **18.39** mitochondrion inner membrane **18.41** FeSP, $FADH_2$, cyt c_1, cyt a_3 **18.43** (a) $FADH_2$, CoQ, $2Fe^{3+}$ (b) $FMNH_2$, 2Fe(II)SP, $CoQH_2$ **18.45** all three fixed enzyme sites **18.47** 10 **18.49** ATP synthase **18.51** proton flow through ATP synthase **18.53** 2.5 ATP molecules **18.55** They enter the electron transport chain at different stages. **18.57** (a) glucose (b) pyruvate **18.59** (a) dihydroxyacetone and glyceraldehyde 3-phosphate (b) It is converted into the one that can be degraded. **18.61** (a) 9 (b) 2 (c) 3 (d) 2 **18.63** (a) glucose 6-phosphate (b) 2-phosphoglycerate (c) phosphoglyceromutase (d) ADP **18.65** $\text{Glucose} + 2\ NAD^+ + 2\ ADP + 2\ P_i \longrightarrow 2\ \text{pyruvate} + 2\ NADH + 2\ ATP + 2H^+ + 2H_2O$ **18.67** (a) $\text{Pyruvate} + CoA + NAD^+ \longrightarrow \text{acetyl CoA} + NADH + CO_2$ (b) $\text{Pyruvate} + NADH + H^+ \longrightarrow \text{lactate} + NAD^+$ **18.69** NAD^+ is regenerated from NADH. **18.71** Pyruvate is the keto 3-carbon acid, and lactate is the hydroxy 3-carbon acid. **18.73** (a) 30 ATP versus 2 ATP (b) 30 ATP versus 2 ATP **18.75** 26 ATP **18.77** (a) two (b) one **18.79** (a) NAD^+ (b) CoA, FAD, and NAD^+ (c) ATP, CoA, FAD, and NAD^+ (d) FAD **18.81** (a) CAC reactant (b) glycolysis reactant (c) glycolysis reactant (d) CAC reactant **18.83** Oxidative phosphorylation produces ATP from ADP and P_i; substrate-level phosphorylation produces ATP from ADP and a phosphate-containing intermediate compound. **18.85** (a) ten steps (b) eight steps (c) eight steps (d) four steps

Answers to Practice Tests

Chapter 1 **1.34** false **1.35** true **1.36** false **1.37** false **1.38** false **1.39** false **1.40** true **1.41** true **1.42** false **1.43** false **1.44** false **1.45** true **1.46** false **1.47** false **1.48** true **1.49** b **1.50** b **1.51** c **1.52** c **1.53** d **1.54** b **1.55** c **1.56** b **1.57** d **1.58** d

Chapter 2 **2.46** false **2.47** true **2.48** false **2.49** false **2.50** false **2.51** false **2.52** true **2.53** false **2.54** true **2.55** true **2.56** false **2.57** true **2.58** false **2.59** true **2.60** false **2.61** c **2.62** c **2.63** c **2.64** d **2.65** d **2.66** c **2.67** b **2.68** b **2.69** c **2.70** d

Chapter 3 **3.78** true **3.79** false **3.80** false **3.81** true **3.82** false **3.83** true **3.84** false **3.85** false **3.86** true **3.87** true **3.88** true **3.89** false **3.90** false **3.91** true **3.92** false **3.93** c **3.94** d **3.95** c **3.96** a **3.97** b **3.98** d **3.99** b **3.100** a **3.101** b **3.102** d

Chapter 4 **4.94** true **4.95** true **4.96** true **4.97** false **4.98** false **4.99** true **4.100** false **4.101** true **4.102** false **4.103** true **4.104** false **4.105** true **4.106** true **4.107** true **4.108** false **4.109** c **4.110** c **4.111** a **4.112** b **4.113** c **4.114** a **4.115** d **4.116** c **4.117** d **4.118** c

Chapter 5 **5.37** false **5.38** false **5.39** true **5.40** false **5.41** true **5.42** true **5.43** true **5.44** true **5.45** false **5.46** true **5.47** true **5.48** true **5.49** true **5.50** true **5.51** false **5.52** b **5.53** b **5.54** c **5.55** c **5.56** a **5.57** c **5.58** d **5.59** a **5.60** b **5.61** c

Chapter 6 **6.60** false **6.61** true **6.62** true **6.63** false **6.64** true **6.65** true **6.66** true **6.67** false **6.68** true **6.69** false **6.70** false **6.71** false **6.72** false **6.73** false **6.74** true **6.75** d **6.76** b **6.77** b **6.78** b **6.79** a **6.80** c **6.81** c **6.82** c **6.83** b **6.84** c

Chapter 7 **7.42** false **7.43** false **7.44** true **7.45** true **7.46** true **7.47** false **7.48** false **7.49** true **7.50** false **7.51** true **7.52** true **7.53** true **7.54** true **7.55** true **7.56** false **7.57** c **7.58** c **7.59** c **7.60** b **7.61** a **7.62** b **7.63** c **7.64** d **7.65** c **7.66** d

Chapter 8 **8.35** true **8.36** true **8.37** true **8.38** false **8.39** true **8.40** false **8.41** true **8.42** false **8.43** true **8.44** false **8.45** true **8.46** true **8.47** true **8.48** true **8.49** true **8.50** c **8.51** a **8.52** b **8.53** b **8.54** c **8.55** c **8.56** c **8.57** c **8.58** d **8.59** b

Chapter 9 **9.54** true **9.55** false **9.56** true **9.57** false **9.58** false **9.59** false **9.60** false **9.61** false **9.62** true **9.63** false **9.64** true **9.65** true **9.66** true **9.67** true **9.68** true **9.69** a **9.70** a **9.71** b **9.72** c **9.73** c **9.74** a **9.75** c **9.76** b **9.77** d **9.78** c

Chapter 10 **10.62** true **10.63** true **10.64** false **10.65** true **10.66** false **10.67** true **10.68** false **10.69** false **10.70** false **10.71** false **10.72** true **10.73** true **10.74** false **10.75** false **10.76** true **10.77** d **10.78** c **10.79** b **10.80** a **10.81** b **10.82** c **10.83** d **10.84** c **10.85** d **10.86** a

Chapter 11 **11.60** false **11.61** false **11.62** true **11.63** false **11.64** true **11.65** false **11.66** false **11.67** true **11.68** true **11.69** true **11.70** true **11.71** true **11.72** false **11.73** true **11.74** false **11.75** c **11.76** b **11.77** a **11.78** c **11.79** b **11.80** a **11.81** c **11.82** a **11.83** c **11.84** b

Chapter 12 **12.78** true **12.79** false **12.80** false **12.81** true **12.82** true **12.83** false **12.84** false **12.85** true **12.86** true **12.87** true **12.88** true **12.89** false **12.90** true **12.91** true **12.92** false **12.93** a **12.94** c **12.95** c **12.96** c **12.97** d **12.98** a **12.99** b **12.100** b **12.101** c **12.102** d

Chapter 13 **13.87** true **13.88** false **13.89** true **13.90** false **13.91** true **13.92** true **13.93** true **13.94** false **13.95** true **13.96** true **13.97** true **13.98** false **13.99** true **13.100** true **13.101** false **13.102** true **13.103** a **13.104** b **13.105** a **13.106** a **13.107** a **13.108** b **13.109** a **13.110** b **13.111** b **13.112** b

Chapter 14 **14.62** true **14.63** false **14.64** false **14.65** true **14.66** false **14.67** false **14.68** true **14.69** true **14.70** true **14.71** true **14.72** false **14.73** true **14.74** false **14.75** false **14.76** true **14.77** c **14.78** c **14.79** b **14.80** a **14.81** b **14.82** c **14.83** d **14.84** c **14.85** a **14.86** c

Chapter 15 **15.60** true **15.61** true **15.62** false **15.63** false **15.64** true **15.65** true **15.66** false **15.67** true **15.68** true **15.69** true **15.70** false **15.71** true **15.72** false **15.73** false **15.74** true **15.75** b **15.76** c **15.77** a **15.78** a **15.79** a **15.80** b **15.81** b **15.82** d **15.83** c **15.84** a

Chapter 16 **16.91** true **16.92** false **16.93** false **16.94** false **16.95** true **16.96** false **16.97** false **16.98** true **16.99** true **16.100** true **16.101** true **16.102** true **16.103** true **16.104** false **16.105** true **16.106** d **16.107** b **16.108** c **16.109** a **16.110** d **16.111** c **16.112** b **16.113** a **16.114** a **16.115** d

Chapter 17 **17.85** false **17.86** true **17.87** false **17.88** true **17.89** false **17.90** false **17.91** true **17.92** false **17.93** true **17.94** false **17.95** true **17.96** true **17.97** true **17.98** true **17.99** false **17.100** d **17.101** a **17.102** a **17.103** b **17.104** b **17.105** a **17.106** c **17.107** b **17.108** d **17.109** d

Chapter 18 **18.86** true **18.87** true **18.88** true **18.89** false **18.90** false **18.91** false **18.92** true **18.93** true **18.94** true **18.95** false **18.96** true **18.97** false **18.98** true **18.99** false **18.100** false **18.101** c **18.102** d **18.103** a **18.104** c **18.105** d **18.106** c **18.107** c **18.108** b **18.109** c **18.110** c

▶ Glossary/Index

Mixed triacylglycerol(s) A triester formed from the reaction of glycerol with more than one kind of fatty acid molecule. 414
structural formulas for, 415
Mixture(s) A physical combination of two or more substances in which each substance retains its own chemical identity. 6
characteristics of, 6–9
comparison with compounds, 8
heterogeneous, 6–9
homogeneous, 6–9
separation of, by physical means, 6
types of, 6–9
Molar mass A mass in grams that is numerically equal to a substance's formula mass. 129
relationship to formula mass, 129
use of, in calculations, 129–130
Molarity A ratio giving the number of moles of solute per liter of solution. 184
calculations involving, 185–186
mathematical equation for, 184
Mole(s) 6.02×10^{23} objects; the amount of substance that contains as many elementary particles (atoms, molecules or formula units) as there are carbon-12 atoms in exactly 12 grams of carbon-12. 127, 131
as a counting unit, 127–128
Avogadro's number and, 127
chemical equations and, 140
chemical formulas and, 131–132
equation-based calculations involving, 141–144
general calculations involving, 132–134
important relationships involving, summary of, 141
mass of, 128–131
number of objects in, calculation of, 127–128
Molecular collisions, elastic, 148
inelastic, 148
binary, naming of, 116–117
general properties of, 82
Molecular compound(s), Lewis structures for, 99–106
polarity of, 114–116
Molecular geometry The three-dimensional arrangement of atoms in a molecule. 107
effects of nonbonding electron pairs on, 107–109
electron pair repulsions and, 107–109
prediction of, using VSEPR theory, 107–110
role of central atom in determining, 107–109
Molecular mass, calculation of, 126
relationship to formula mass, 126
Molecular polarity A measure of the degree of inequality in the attraction of bonding electrons to various locations within a molecule. 114
diatomic molecules and, 115
factors which determine, 115
tetraatomic molecules and, 115–116
triatomic molecules and, 115
Molecule(s) A group of two or more atoms that functions as a unit because the atoms are tightly bound together. 13
achiral, 385–386
characteristics of, 14–16
chiral, characteristics of, 385–386
chiral, interactions between, 386–387
classification of, by number of atoms, 15–16
classification of, by types of atoms, 15–16
diatomic, 15–16
geometry of, 107–110
handedness in, recognition of, 385–386
heteroatomic, 15–16
homoatomic, 15–16
limit of physical subdivision and, 16
nonpolar, 114–115
polar, 114–115
triatomic, 15–16
Monatomic ion(s) A single atom that has lost or gained electrons and thus acquired a charge. 117–118
charges on, 88–89
formulas for compounds containing, 91–92
naming compounds containing, 94–98
Monomer(s) The small molecule that is repeated many times in a polymer. 293
in addition polymerization, table of, 294
in condensation polymerization, 373–376
Monoprotic acid(s) An acid that transfers one H^+ ion (proton) per molecule during an acid-base reaction. 231
acidic and nonacidic hydrogen atoms in, 231
Monosaccharide(s) A carbohydrate that contains a single polyhydroxy aldehyde or polyhydroxy ketone unit. 382
biochemically important, 387–390
chemical reactions of, 392–395
classification of, by functional group, 383–384
classification of, by number of carbon atoms, 383–384
cyclic forms, alpha and beta, 391–392
cyclic forms, formation of, 390–392
cyclic forms, notation for, 391–392
disaccharide formation from, 396–398
glycoside formation and, 393–394
handedness in, 384–387
hemiacetal forms of, 390
oxidation of, 393
phosphate ester formation and, 394
reduction of, 393
Monounsaturated fatty acid(s) A fatty acid whose carbon chain contains one carbon-carbon double bond. 409
common, listing of, 410
structural formula notation for, 409–410
Morphine, structure of, 334
Myoglobin, primary structure of, 446
tertiary structure of, 451

NAD^+, as an oxidizing agent, 506
citric acid cycle and, 510–513
structure of, 505–506
NADH, electron transport chain and, 515–516
structure of, 506
Naphthalene, Chemical Portrait of, 301
Naproxen, structure of, 358
Natural gas, chemical composition of, 271
Neon, Chemical Portrait of, 66
Neutral solution(s) A solution in which the concentrations of H_3O^+ and OH^- ions are equal. 238
hydronium ion concentration and, 238
pH value and, 241
Neutralization The reaction between an acid and a hydroxide base to produce a salt and water. 234
acid-base, balancing equations for, 235
Neutron(s) A subatomic particle that has no charge associated with it. 47
location of, within atom, 47
properties of, 47
Nickel, as a strategic element, 11
Chemical Portrait of, 11
Nicotinamide adenine dinucleotide, general structural characteristics of, 505–506
oxidized form of, 505–506
reduced form of, 506
Nicotine, structure of, 334
Nitric acid, Chemical Portrait for, 233
Nitrogen, Chemical Portraits of, 15, 176
Nitrogen atom(s), bonding characteristics of, 326
Nitrogen-containing bases, as nucleotide subunits, 469–470
Nitrogen oxide (NO_x) pollution, Chemical Portrait of, 203
Noble gas(es) The elements found in the far right column of the periodic table. 65
electronic characteristics of, 65
periodic table location of, 53, 65–67
periodic table positions of, 65–66
stability of electron configurations of, 85–86
Nomenclature, alcohols, IUPAC rules for, 312–313
aldehydes, IUPAC rules for, 342–343
alkanes, IUPAC rules for, 260–264
alkenes and cycloalkenes, IUPAC rules for, 281–284
alkynes, IUPAC rules for, 297
amides, IUPAC rules for, 369–370
amine salts, 332
amines, IUPAC rules for, 328–329
binary ionic compounds, 95–98
binary molecular compounds, 116–117
carboxylate ions, 360–361
carboxylic acid salts, 361
carboxylic acids, IUPAC rules for, 356–358
common name use in, 117
cycloalkanes, IUPAC rules for, 266–267
enzymes, 458
esters, IUPAC rules for, 364–365
ethers, IUPAC rules for, 322–323
glycosides, 394
halogenated hydrocarbons, IUPAC rules for, 308
IUPAC naming system, 260
ketones, IUPAC rules for, 349–351
nucleotides, 471–472
ortho-meta-para, use of, 300
phenols, IUPAC rules for, 314
polyatomic-ion containing compounds, 120
substituted ammonium ions, 331
thiols, IUPAC rules for, 325
Non-oxidation-reduction reaction(s) A chemical reaction in which there is no transfer of electrons between reactants. 207
Nonaqueous solution(s) A solution in which water is not the solvent. 177
Nonbonding electrons Pairs of valence electrons that are not involved in electron sharing. 100
Nonmetal(s) An element characterized by the absence of the properties of luster, thermal conductivity, electrical conductivity, and malleability. 55
abundances for, 56
general properties of, 55, 67
periodic table locations for, 55–56
Nonpolar amino acid(s) An amino acid that contains one amino group, one carboxyl group, and a nonpolar side chain. 435
structures of, 436

Single-replacement reaction(s) A reaction in which an atom or molecule replaces an atom or group of atoms from a compound. 201
examples of, 201–202
general equation for, 201
Sodium chloride, Chemical Portrait of, 95
Solid(s) The physical state of matter characterized by a definite shape and a definite volume. 2
distinguishing characteristics of, 2–3, 148
kinetic molecular theory of matter applied to, 149
melting of, 160–162
sublimation of, 160–162
Solid state The physical state characterized by a dominance of potential energy (cohesive forces) over kinetic energy (disruptive forces). 149
Solubility(ies) The maximum amount of solute that will dissolve in a given amount of solvent. 176
pressure and, for gases, 177
rules for, 179–180
temperature and, for selected solutes, 176
terminology associated with, 176
values of, for selected solutes, 176
Solubility rules, for solutes, 179–180
Solute(s) Solution component that is present in a small amount relative to that of solvent. 174
ionic, dissolving process for, 178–179
solubility rules for, 179–180
Solution(s) A homogeneous combination of two or more substances in which each substance retains its own chemical identity. 174
acidic, 237–238, 241
aqueous, 177
basic, 237–238, 241
colligative properties of, 189–196
concentrated, 177
concentration units for, 181–187
dilute, 177
dilution of, 188–189
formation of, ionic solutes and, 178–179
general characteristics of, 175
hypertonic, 193–194
hypotonic, 193–194
isotonic, 194
neutral, 237–238, 241
nonaqueous, 177
pH of, 239–241
rate of formation of, factors affecting, 179
saturated, 177
solubility rules for solutes, 179–180
unsaturated, 177
Solvent(s) Component of a solution that is present in the greatest amount. 174
general characteristics of, 174–175
Sphingolipid(s) A polar fatty-acid-containing lipid derived from the amino dialcohol sphingosine. 422
cerebrosides, 422
general structure for, 422
head and two-tail structure of, 422–423
sphingomyelins, 422
structure of, 422
Sphingosine, component of sphingolipids, 422–423
structure of, 422
Stable nucleus A nucleus that does not easily undergo change. 68
Standard amino acid(s) One of the 20 alpha-amino acids normally found in proteins. 435
name abbreviations for, 436–437
structures of, 436
Starch, amylopectin form of, 402
amylose form of, 401
animal, 402
properties of, 401–402
structure of, 401–402
State(s), gaseous, 3–4
liquid, 3–4
solid, 3–4
States of matter, compressibility and, 147–148
kinetic molecular theory and, 147–151
property differences among, 147–148
thermal expansion and, 147–148
Steroid(s) A lipid with a structure that is based on a fused-ring system involving three 6-membered rings and one 5-membered ring. 423
adrenocortical hormones, 426–427
athletes and anabolic agents, 427
bile salts, 425
cholesterol, 433
general structural characteristics of, 423
sex hormones, 425–426
synthetic, 426
Strong acid(s) An acid that transfers 100% or very nearly 100% of its protons to water. 232
commonly encountered, 232
Strong base(s), commonly encountered, 233
Structural formula(s), generation of, from IUPAC names, 264
condensed, 255–256
expanded, 255
line-angle drawings for, 266
Structural isomers Compounds with the same molecular formula but different structural formulas. 256
alkanes, number possible, 258
alkenes, 285
Subatomic particle(s) Very small particles that are the building blocks from which atoms are made. 45
arrangement of, within atom, 46
properties, table of, 46
types of, 45–46
Sublimation, process of, 160–162
Substance, use of the term, 6
Substituent(s) An atom or group of atoms attached to a chain (or ring) of carbon atoms. 260
alkyl, 260–261
Substituted ammonium ion(s) An ammonium ion in which one or more alkyl, cycloalkyl, or aryl groups have been substituted for hydrogen atoms. 331
generalizations concerning, 331
nomenclature for, 331
Substitution reaction(s) A reaction in which part of a small reacting molecule replaces an atom or a group of atoms on a hydrocarbon or hydrocarbon derivative. 273
alkanes and cycloalkanes, 273–275
aromatic hydrocarbons, 303–304
halogenation, 273–275
Substrate(s) The reactant in an enzyme-catalyzed reaction. 458
enzyme nomenclature and, 458
Substrate-level phosphorylation The direct transfer of a high-energy phosphate group from an intermediate compound (substrate) to an ADP molecule to produce ATP. 525
glycolysis and, 525
Succinate, citric acid cycle and, 512–513
Succinyl CoA, citric acid cycle and, 512
Sucrose, Chemical Portrait of, 399
hydrolysis of, 398
structure of, 398
Sugar(s) General designation for either a monosaccharide or disaccharide. 383
pentose, as nucleotides subunit, 469
reducing, 393
Sugar alcohols, 393
Sulfhydryl group(s) The —SH functional group. 325
in thiols, 325
Sulfur dioxide, Chemical Portrait of, 150
Sulfur oxide (SO_x) pollution, Chemical Portrait of, 203
Sulfuric acid, Chemical Portrait for, 233
Symmetrical addition reaction(s) A reaction in which identical atoms (or groups of atoms) are added to each carbon of a carbon-carbon multiple bond. 288
halogenation of alkenes, 288–289
hydrogenation of alkenes, 288

Temperature, absolute zero, Charles's law and, 154
enzyme activity and, 462
enzyme denaturation and, 462
lowest possible value of, 40
optimum, for enzymes, 462
scales for measuring, 40–42
solubility of solutes and, 176–177
vapor pressure magnitude and, 164
Temperature scales, Celsius, characteristics of, 40–41
conversions between, 41–42
Fahrenheit, characteristics of, 41
Kelvin, characteristics of, 40–41
Tertiary alcohol(s) An alcohol in which the hydroxyl-bearing carbon atom is attached to three other carbon atoms. 319
Tertiary amide(s), structural characteristics of, 368
Tertiary amine(s), structural characteristics of, 327
Tertiary protein structure The overall three-dimensional shape of a protein that results from the attractive forces between amino acid side chains (R groups) that are widely separated from each other within the chain. 449
interactions responsible for, 449–451
Thermal expansion A measure of the volume change resulting from a temperature change. 147
states of matter and, 147–148
Thiol(s) A hydrocarbon derivative in which a sulfhydryl group (—SH) is present. 325
chemical properties of, 326
generalized formula for, 325
nomenclature of, 325
oxidation of, 326
physical properties of, 325–326
Thymine, as nucleotide subunit, 470
Titration(s), acid-base, 246–248
Tollens test, aldehydes and, 346
reducing sugars and, 393

Photo Credits

Photographs from the Edgar Fahs Smith Collection, Van Pelt-Dietrich Library, University of Pennsylvania: Pages 52, 85, 128 (bottom), 152, 154, 219, 290, 298

Computer-generated models by James P. Birk and Kara M. Birk: Pages 254 (six models), 257 (three models), 281 (four models), 286 (four models), 296 (two models), 311 (three models), 316 (two models), 327 (four models), 413 (four models), 416 (two models), 424, 470

Credits for all other photographs are listed below, with page numbers in boldface.

1 Lou Jacobs Jr./Grant Heilman Photography; **2 (top)** Phil Degginger/Color-Pic; **3 (top)** Stone/Ed Simpson; **4** Andy Levin/Photo Researchers; **5 (top)** Phil Degginger/Color-Pic; **(bottom)** Stone/Andy Sacks; **6 (two photos)** James Scherer; **7 (right)** Bruce Iverson; **9 (emerald)** National Museum of Natural History © Smithsonian Institution; **13** Digital Instruments; **21** Dan McCoy/Rainbow; **22 (left)** Stone/Don and Pat Valenti; **23 (left)** E.R. Degginger; **45** Novastock/PhotoEdit; **77** Simon Fraser/Science Photo Library/Custom Medical Stock Photo; **78** Yoav Levy/Phototake; **82** M.S. Davidson/Photo Researchers; **93** E.R. Degginger; **111** Archive Photos; **125** Stone/Ralph H. Wetmore II; **147** Stone/Bruce Forster; **148** Betty Weiser/Photo Researchers; **158** Courtesy Manchester Literary and Philosophical Society; **161 (two photos)** James Scherer; **166** Brian Bailey/Network Aspen; **174** Steve Allen/Peter Arnold, Inc.; **177 (bottom)** Coco McCoy/Rainbow; **180** Stone/David Woodfall; **193** John Mead/Science Photo Library/Photo Researchers; **194 (left and center)** David M. Phillips/Visuals Unlimited; **(right)** Stanley Flegler/Visuals Unlimited; **200** Science Photo Library/Photo Researchers; **201** James Scherer; **202** James Scherer; **213 (top left)** Stone/Vince Streano; **(top right)** Cecile Brunswick/Peter Arnold, Inc.; **(bottom left)** S.C. Fried/Photo Researchers; **(bottom right)** Myrleen Ferguson/Photo Edit; **217** Mark Gibson; **226** Dr. Paul A. Zahl/Photo Researchers; **229** Ken O'Donoghue; **231** Ken O'Donoghue; **245 (two photos)** Ken O'Donoghue; **251** Stone/Arnuls Husmo; **271** Richard Megna/Fundamental Photos; **272** Daryl Solomon/Envision; **273** Phil Degginger/Color-Pic; **279** Junebug Clark/Photo Researchers; **293** James Scherer; **295 (bottom)** Bill Stanton/Rainbow; **307** Ed Reschke/Peter Arnold, Inc.; **320** D. & I. MacDonald/Envision; **325** Jeff Lapore/Photo Researchers; **334** Scott Camazine/Photo Researchers; **337** © 2000 PhotoDisc Inc.; **344** Norbert Wu; **358** Hans Pfletschinger/Peter Arnold, Inc.; **360** AP/Wide World Photos; **362** Sherman Thomson/Visuals Unlimited; **366** L. Clarke/CORBIS; **375 (top)** Peter Skinner/Photo Researchers; **(bottom)** Dan McCoy/Rainbow; **380** Norris Blake/Visuals Unlimited; **382** Bobby Long/The Stock Shop; **388** Tom Raymond/Medichrome; **403 (top)** Dr. Dennis Kunkel/Phototake NYC; **(bottom)** Don Kreuter/Rainbow; **408** Dan Guravich/Photo Researchers; **417** Manfred Kage/Peter Arnold, Inc.; **418** BRI/Vision/Photo Researchers; **424** C. James Webb/Phototake NYC; **434** F.H. Kolwicz/Visuals Unlimited; **455** E.R. Degginger; **457** Steven Needham/Envision; **468** Dr. Nikas/Jason Burns/Phototake NYC; **481** Erika Stone/Peter Arnold, Inc.; **500** Luiz C. Marigo/Peter Arnold, Inc.; **502** R. Bhatnagar/Visuals Unlimited

Common Functional Groups

Name of class	Structural feature
Alkane	$-\overset{\vert}{\underset{\vert}{C}}-$
Alkene	$>C=C<$
Alkyne	$-C\equiv C-$
Aromatic hydrocarbon	(benzene ring) or (benzene ring with circle)
Alcohol	$-\overset{\vert}{\underset{\vert}{C}}-OH$
Phenol	(benzene ring)–OH
Ether	$-\overset{\vert}{\underset{\vert}{C}}-O-\overset{\vert}{\underset{\vert}{C}}-$
Thiol	$-\overset{\vert}{\underset{\vert}{C}}-SH$
Aldehyde	$-\overset{O}{\overset{\Vert}{C}}-H$
Ketone	$-\overset{\vert}{\underset{\vert}{C}}-\overset{O}{\overset{\Vert}{C}}-\overset{\vert}{\underset{\vert}{C}}-$
Carboxylic acid	$-\overset{O}{\overset{\Vert}{C}}-OH$
Ester	$-\overset{O}{\overset{\Vert}{C}}-O-\overset{\vert}{\underset{\vert}{C}}-$
Amine	$-\overset{\vert}{\underset{\vert}{C}}-NH_2$
Amide	$-\overset{O}{\overset{\Vert}{C}}-NH_2$